I. N. Bronshtein
A Guide Book t

Bronshtein · Semendyayev

A Guide Book to Mathematics

Fundamental Formulas · Tables · Graphs · Methods

Verlag Harri Deutsch · Zürich · Frankfurt/Main
Springer-Verlag New York Inc.

Translated from the Russian on basis of the enlarged and improved German edition by Jan Jaworowski, Professor at the University of Warsaw, and Michael N. Bleicher, Professor at the University of Wisconsin.

Published by
Verlag Harri Deutsch, D-6000 Frankfurt/M. 90, Gräfstraße 47
Distributed in the United States and Canada by
Springer Verlag, New York Inc., 175 Fifth Ave., New York, N. Y. 10010

ISBN-13: 978-1-4684-6290-6 e-ISBN-13: 978-1-4684-6288-3
DOI: 10.1007/ 978-1-4684-6288-3

 Library of Congress Catalog Card Number 60-16788

CONTENTS

IV. Trigonometry

Part three

ANALYTIC AND DIFFERENTIAL GEOMETRY

I. Analytic geometry

II. Differential geometry

Part four

FOUNDATIONS OF MATHEMATICAL ANALYSIS

I. Introduction to analysis

II. Differential calculus

III. Integral calculus

FROM THE AUTHORS' PREFACE TO THE FIRST RUSSIAN EDITION

It was a very difficult task to write a guide-book of a small size designed to contain the fundamental knowledge of mathematics which is most necessary to engineers and students of higher technical schools. In our tendency to the compactness and brevity of the exposition, we attempted, however, to produce a guide-book which would be easy to understand, convenient to use and as accurate as possible (as much as it is required in engineering).

It should be pointed out that this book is neither a handbook nor a compendium, but a guide-book. Therefore it is not written as systematically as a handbook should be written. Hence the reader should not be surprised to find, for example, l'Hôpital's rule in the section devoted to computation of limits which is a part of the chapter "Introduction to the analysis" placed before the concept of the derivative, or information about the Gamma function in the chapter "Algebra"—just after the concept of the factorial. There are many such "imperfections" in the book. Thus a reader who wants to acquire certain information is advised to use not only the table of contents but also the alphabetical index inserted at the end of the book. If a problem mentioned in the text is explained in detail in another place of the book, then the corresponding page is indicated in a footnote.

PREFACE TO THE ENGLISH EDITION

The English translation is based on the enlarged German edition and completed by two chapters on calculus of variations and integral equations. Both are translated from German.

MATHEMATICAL NOTATIONS[1]

I. Relations between quantities

$=$	equal
$\equiv$	identically equal
$\neq$	not equal
$\approx$	approximately equal
$<$	less
$>$	greater
$\leq$	less or equal
$\geq$	greater or equal

II. Algebra

$\lvert a \rvert$	absolute value of the number a
$+$	(plus)—addition
$-$	(minus)—subtraction
$\cdot$ or $\times$	multiplication, for example, $a \cdot b$ or $a \times b$; the multiplication sign is often omitted, for example, ab
$:$ or —, or $/$	division ($a : b$ or $\frac{a}{b}$, or a/b)
a^m	"a to the power m"
$\sqrt{\ }$	square root, for example, $\sqrt{a}$
$\sqrt[n]{\ }$	root of the nth degree, for example, $\sqrt[n]{a}$
$\log_b$	logarithm to the base b, for example, $5 = \log_2 32$ (p. 156)
log	common logarithm, for example, $2 = \log 100$ (p. 156)
ln	natural logarithm, for example, $1 = \ln e$ (p.156)
(), [], {}	parentheses of brackets (denote succession of operations)
!	factorial, for example, $a!$; $6! = 1 \cdot 2 \cdot 3 \cdot 4 \cdot 5 \cdot 6 = 720$ (p. 190)

---

(1) Numbers in brackets (p. ...) denote the pages on which the corresponding notions are explained.

III. Geometry

Symbol	Meaning
$\perp$	perpendicular
$\parallel$	parallel
$\#$	equal and parallel
$\sim$	similar, for example, $\triangle ABC \sim \triangle DEF$
$\triangle$	triangle
$\sphericalangle$	angle (sometimes $\angle$), for example, $\sphericalangle ABC$, $\angle ABC$
$\smile$	arc, for example, $\overset{\smile}{AB}$
$\circ$	degree
$'$	minute
$''$	second

(degree, minute, second: in the degree measure, for example, $32°14'11''.5$)

IV. Trigonometry, hyperbolic functions

Symbol	Meaning	
sin	the sine	(p. 213)
cos	the cosine	
tan	the tangent	
cot	the cotangent	
sec	the secant	
cosec	the cosecant	
Arc sin	the inverse sine	(p. 223)
Arc cos	the inverse cosine	
Arc tan	the inverse tangent	
Arc cot	the inverse cotangent	
arc sin	the principal branch of the inverse sine	(p. 223)
arc cos	the principal branch of the inverse cosine	
arc tan	the principal branch of the inverse tangent	
arc cot	the principal branch of the inverse cotangent	
sinh	the hyperbolic sine	(p. 230)
cosh	the hyperbolic cosine	
tanh	the hyperbolic tangent	
coth	the hyperbolic cotangent	
sech	the hyperbolic secant	(p. 230)
cosech	the hyperbolic cosecant	
ar sinh	the inverse hyperbolic sine	(p. 232)
ar cosh	the inverse hyperbolic cosine	
ar tanh	the inverse hyperbolic tangent	
ar coth	the inverse hyperbolic cotangent	

V. Notations for constants

const	a constant quantity
$\pi = 3.14159\ldots$	the ratio of the length of a circumference to its diameter (p. 199)
$e = 2.71828\ldots$	base of the natural logarithms (p. 331)
$C = 0.57722\ldots$	Euler's constant (p. 331)

VI. Mathematical analysis

lim	limit ⎫ for example, $\lim\limits_{N\to\infty}\left(1+\frac{1}{N}\right)^N = e$ (pp. 318, 328)
$\to$	tends to ... ⎬
∞	infinity ⎭
Σ	sum
$\sum\limits_{i=1}^{n}$	sum in which i varies from 1 to n
$f(), \varphi()$	notations for functions, for example, $y = f(x)$, $u = \varphi(x, y, z)$
Δ	increment, for example, Δx
d	differential, for example, dx (p. 363)
d_x, d_y etc.	partial differential, for example, $d_x u$ (p. 363)
$'$, $''$, $'''$, ^{IV} or $^{\cdot}$, $^{\cdot\cdot}$, $^{\cdot\cdot\cdot}$, $^{\cdot\cdot\cdot\cdot}$	notation for successive derivatives of functions of one variable; for example, if $y = f(x)$: $f'(x)$, $f''(x)$, $f'''(x)$, $f^{\mathrm{IV}}(x)$, $y', y'', y''', y^{\mathrm{IV}}$, $y^{\cdot}, y^{\cdot\cdot}, y^{\cdot\cdot\cdot}$ (pp. 360, 364, 365)
$\frac{d}{dx}, \frac{d^2}{dx^2}, \ldots$	first derivative, second derivative etc. — for example, $\frac{dy}{dx}, \frac{d^2y}{dx^2}$ etc. (pp. 360, 364)
D	symbol of derivative (differentiation operator), for example, $Dy = y'$, $D^2y = y''$ etc. (pp. 360, 364)
f'_x, f''_{xx}, f''_{xy} or $\frac{\partial}{\partial x}, \frac{\partial^2}{\partial x^2}, \frac{\partial^2}{\partial x \partial y}$	partial derivatives, for example, $f'_x(u), \frac{\partial f}{\partial x}, \frac{\partial^2 f}{\partial x^2}$ etc. (pp. 362, 365)
$\int$	integral (p. 394)
$\int\limits_a^b$	definite integral from the lower limit a to the upper limit b (p. 455)
$\int\limits_{(K)}$	line integral taken over the arc K or over a projection of K (pp. 486, 490)
$\int\limits_S, \int\limits_V$	integral over the surface S or over the volume V (pp. 495–497)
$\iint$	double integral ⎫ (pp. 495–497)
$\iiint$	triple integral ⎭

VII. Complex numbers

i (sometimes j)	imaginary unit ($i^2 = -1$) (p. 585)
re a	the real part of the number a (p. 585)
im a	the imaginary part of the number a (p. 585)
$\|a\|$	absolute value (modulus) of a (p. 586)
arg a	argument of a (p. 586)
$\bar{a}$	the conjugate of a, for example, $a = 2 + 3i$, $\bar{a} = 2 - 3i$ (p. 587)
Ln	(natural) logarithm of a complex number (p. 592)

VIII. Vector calculus

$\boldsymbol{a}, \boldsymbol{b}, \boldsymbol{c}$, or $\bar{a}, \bar{b}, \bar{c}$	symbols of vectors (p. 613)
$\boldsymbol{a}^0$	unit vector of the direction of the vector $\boldsymbol{a}$ (p. 614)
$\boldsymbol{i}, \boldsymbol{j}, \boldsymbol{k}$	unit vectors of the rectangular coordinate system (p. 615)
$\|\boldsymbol{a}\|$ or a	length (absolute value) of the vector $\boldsymbol{a}$ (p. 613)
$\boldsymbol{a} = \boldsymbol{b}$, $\boldsymbol{a} + \boldsymbol{b}$, $\boldsymbol{a} - \boldsymbol{b}$	equality, composition and subtraction of vectors (pp. 613, 614)
$\alpha\boldsymbol{a}$	multiplication of a vector by a scalar (p. 614)
$\boldsymbol{ab}$	scalar product of vectors (p. 616)
$\boldsymbol{a} \times \boldsymbol{b}$ or $[\boldsymbol{ab}]$	vector product of vectors (p. 616)
$\boldsymbol{abc} = \boldsymbol{a}(\boldsymbol{b} \times \boldsymbol{c})$	box product of three vectors (p. 618)
a_x, a_y, a_z	coordinates of the vector $\boldsymbol{a}$ in the Cartesian coordinate system (p. 615)
∇	Hamilton's differential operator (nabla) (p. 643)
Δ	Laplace's operator (p. 645)
grad	gradient of a scalar field (grad $\varphi = \nabla\varphi$) (p. 632)
div	divergence of a vector field (div $\boldsymbol{V} = \nabla\boldsymbol{V}$) (p. 640)
rot (or curl)	rotation (or curl) of a vector field (curl $\boldsymbol{V} = \nabla \times \boldsymbol{V}$) (p. 641)
$\frac{\partial U}{\partial \boldsymbol{c}}$	derivative of a scalar field in the direction $\boldsymbol{c}$ (p. 632)

PART ONE

TABLES AND GRAPHS

I. TABLES

Interpolation. Most of the tables inserted below give the values of functions to four significant figures for three significant figures of the argument. In the cases, when the argument is given with a greater accuracy, and the desired value of the function cannot be obtained directly from the tables, interpolation should be used. The simplest form is *linear interpolation* in which we assume that the increment of the function is proportional to the increment of the argument. If the desired value of the argument x lies between the values x_0 and $x_1 = x_0 + h$ in the tables and the corresponding values of the function are

$$y_0 = f(x_0) \quad \text{and} \quad y_1 = f(x_1) = y_0 + \Delta,$$

then we assume that

$$f(x) = f(x_0) + \frac{x - x_0}{h}\Delta.$$

The *interpolation correction* $\frac{x - x_0}{h}\Delta$ can be easily computed by using the tables of proportional parts on pp. 84, 85 and also by using the supplement to this guide which gives the products of the difference Δ (from 11 to 90) times 0.1, 0.2, ..., 0.9.

Examples. (1) Find 1.6754^2. We find in the table (p. 23) $1.67^2 = 2.789$, $1.68^2 = 2.822$, $\Delta = 33$ [1]. From the tables of proportional parts we have $0.5 \cdot 33 = 16.5$, $0.04 \cdot 33 = 1.3$, $\frac{x - x_0}{h}\Delta = 16.5 + 1.3 \approx 18$, hence $1.6754^2 = 2.807$.

(2) Find tan 79°24′. From the tables (pp. 60 and 85), tan 79°20′ = 5.309, tan 79°30′ = 5.396, $\Delta = 87$; $0.4 \cdot 87 \approx 35$, hence tan 79°24′ = 5.344.

The error in linear interpolation does not exceed one unit of the last significant figure, provided that the two consecutive

---

(1) The difference Δ is usually expressed in the units of the last order of the value of the function, without the first zeros or the decimal point.

differences Δ_0 and Δ_1 do not differ by more than 4 units of the last figure. If this condition is not satisfied (as, for example, in the table of $\tan x$ for $x > 80°$, p. 61), we have to use more complicated interpolation formulas. In most cases, *Bessel's quadratic interpolation* is sufficient:

$$f(x) = f(x_0) + k\Delta_0 - k_1(\Delta_1 - \Delta_{-1}),$$

where

$$k = \frac{x - x_0}{h}, \quad k_1 = \frac{k(1-k)}{4};$$

the value k_1 is given in the table on p. 86.

$x_{-1} = x_0 - h$	y_{-1}	
x_0	y_0	Δ_{-1}
$x_1 = x_0 + h$	y_1	Δ_0
$x_2 = x_0 + 2h$	y_2	Δ_1

Example. Find $\tan 85°33'$ (the table on p. 61). We find ($h = 10'$): $k = 0.3$, $k_1 = 0.052$; the correction is $0.3 \cdot 491 - 0.052 \cdot 75 \approx 143$, $\tan 85°33' = 12.849$.

x	$\tan x$	Δ
85°20′	12.251	
85°30′	12.706	455
85°40′	13.197	491
85°50′	13.727	530

A. TABLES OF ELEMENTARY FUNCTIONS

1. Some frequently occurring constants

Constant	n	$\log n$	Constant	n	$\log n$
π	3.141593	0.49715	$1:\pi$	0.318310	$\bar{1}.50285$
2π	6.283185	0.79818	$1:2\pi$	0.159155	$\bar{1}.20182$
3π	9.424778	0.97427	$1:3\pi$	0.106103	$\bar{1}.02573$
4π	12.566371	1.09921	$1:4\pi$	0.079577	$\bar{2}.90079$
$\pi:2$	1.570796	0.19612	$2:\pi$	0.636620	$\bar{1}.80388$
$\pi:3$	1.047198	0.02003	$3:\pi$	0.954930	$\bar{1}.97997$
$\pi:4$	0.785398	$\bar{1}.89509$	$4:\pi$	1.273240	0.10491
$\pi:6$	0.523599	$\bar{1}.71900$	$6:\pi$	1.909859	0.28100
$\pi:180\ (=1^\circ)$	0.017453	$\bar{2}.24188$	$180^\circ:\pi$	$57^\circ.295780$	1.75812
$\pi:10800\ (=1')$	0.000291	$\bar{4}.46373$	$10800':\pi$	$3437'.7468$	3.53627
$\pi:648000 (=1'')$	0.000005	$\bar{6}.68557$	$648000'':\pi$	$206264''.81$	5.31443
π^2	9.869604	0.99430	$1:\pi^2$	0.101321	$\bar{1}.00570$
$\sqrt{\pi}$	1.772454	0.24857	$\sqrt{1:\pi}$	0.564190	$\bar{1}.75143$
$\sqrt{2\pi}$	2.506628	0.39909	$\sqrt{1:2\pi}$	0.398942	$\bar{1}.60091$
$\sqrt{\pi:2}$	1.253314	0.09806	$\sqrt{2:\pi}$	0.797885	$\bar{1}.90194$
$\sqrt[3]{\pi}$	1.464592	0.16572	$\sqrt[3]{1:\pi}$	0.682784	$\bar{1}.83428$
$\sqrt[3]{4\pi:3}$	1.611992	0.20736	$\sqrt[3]{3:4\pi}$	0.620350	$\bar{1}.79264$
e	2.718282	0.43429	$1:e$	0.367879	$\bar{1}.56571$
e^2	7.389056	0.86859	$1:e^2$	0.135335	$\bar{1}.13141$
$\sqrt{e}$	1.648721	0.21715	$\sqrt{1:e}$	0.606531	$\bar{1}.78285$
$\sqrt[3]{e}$	1.395612	0.14476	$\sqrt[3]{1:e}$	0.716532	$\bar{1}.85524$
$e^{\pi/2}$	4.810477	0.68219	$e^{-\pi/2}$	0.207880	$\bar{1}.31781$
e^{π}	23.140693	1.36438	$e^{-\pi}$	0.043214	$\bar{2}.63562$
$e^{2\pi}$	535.491656	2.72875	$e^{-2\pi}$	0.001867	$\bar{3}.27125$
C (1)	0.577216	$\bar{1}.76134$	$\ln \pi$	1.144730	0.05870
$M = \log e$	0.434294	$\bar{1}.63778$	$1:M = \ln 10$	2.302585	0.36222
g (2)	9.81	0.99167	$1:g$	0.10194	$\bar{1}.00833$
g^2	96.2361	1.98334	$1:2g$	0.050968	$\bar{2}.70730$
$\sqrt{g}$	3.13209	0.49583	$\pi\sqrt{g}$	9.83976	0.99298
$\sqrt{2g}$	4.42945	0.64635	$\pi\sqrt{2g}$	13.91552	1.14350

(1) C denotes Euler's constant.

(2) g denotes the acceleration of gravity; the values of acceleration given here are for sea level at latitude 45-50°.

2. Squares, cubes, roots

The table on pp. 21–43 enables us to find squares, cubes, square and cube roots to four significant figures. For the arguments n contained between 1 and 10, the values of n^2 and n^3 can be found directly in the table, provided that the argument is given to three significant figures. For example, $1.79^2 = 3.204$ (p. 23). If, however, the argument is given to more than three significant figures, an interpolation is necessary (see p. 17). The error of linear interpolation, for this table, never exceeds one unit of the last significant figure.

When seeking n^2 and n^3, we observe that if n is multiplied by 10^k, then n^2 increases 10^{2k} times and n^3 increases 10^{3k} times, i.e., moving the decimal point in the number n by k places to the right involves moving by $2k$ places in n^2 and by $3k$ places in n^3; moreover, to the number taken from the table, we add zeros on the right or on the left side, if necessary. For example, $0.179^2 = 0.03204$, $179^3 = 5\,735\,000$ (1).

Square roots for n contained between 1 and 100 can be obtained directly from the table (by applying the linear interpolation, p. 17), and for arbitrary n, according to the following rules:

(1) We divide the number under the root sign into groups of two figures, on the left and on the right side of the decimal point. (2) According to whether the highest non-zero group contains one or two significant figures, we find the value of the square root in the $\sqrt{n}$ column or in the $\sqrt{10n}$ column. (3) The position of the decimal point in the value of the square root is determined by the fact that each of the two figure groups on the left side of the decimal point gives one figure in the value of the root, and, for numbers less than 1, each group composed of two zeros on the right side of the decimal point gives one zero in the value of the root on the right side of its decimal point.

Examples. (1) $\sqrt{23.9} = 4.889$; (2) $\sqrt{0.00|02|39} = 0.01546$; (3) $\sqrt{23|90|00} = 488.9$; (4) $\sqrt{0.00|3} = 0.05477$. (In the latter example, one more zero at the end under the root sign should be mentally added, to complete the last two figure group; hence the root should be sought in the $\sqrt{10n}$ column.)

Cube roots of the numbers n contained between 1 and 1000 can be found directly from the table (by applying the linear interpolation), and for arbitrary n, according to the following rules:

(1) It is better to write $179^3 = 5.735 \cdot 10^6$, avoiding unnecessary zeros put instead of unknown figures (exactly, $179^3 = 5\,735\,339$).

(1) We divide the number under the root sign into groups of three figures on the left and on the right side of the decimal point. (2) According to whether the highest non-zero group contains one, two or three significant figures, we find the value of the root in the $\sqrt[3]{n}$, $\sqrt[3]{10n}$ or $\sqrt[3]{100n}$ column, respectively. (3) The position of the decimal point in the calculated value of the root can be determined by a method similar to that for square roots.

Examples. (1) $\sqrt[3]{23.9} = 2.880$ [1]. (2) $\sqrt[3]{239|000} = 62.06$. (3) $\sqrt[3]{0.000|002|39} = 0.01337$. (4) $\sqrt[3]{0.000|3} = 0.06694$. (5) $\sqrt[3]{0.03} = 0.3107$. (In the last two examples, two or one zero, respectively, should be added mentally, to complete the last three figure group.)

n	n^2	n^3	$\sqrt{n}$	$\sqrt{10n}$	$\sqrt[3]{n}$	$\sqrt[3]{10n}$	$\sqrt[3]{100n}$
1.00	1.000	1.000	1.000	3.162	1.000	2.154	4.642
1.01	1.020	1.030	1.005	3.178	1.003	2.162	4.657
1.02	1.040	1.061	1.010	3.194	1.007	2.169	4.672
1.03	1.061	1.093	1.015	3.209	1.010	2.176	4.688
1.04	1.082	1.125	1.020	3.225	1.013	2.183	4.703
1.05	1.102	1.158	1.025	3.240	1.016	2.190	4.718
1.06	1.124	1.191	1.030	3.256	1.020	2.197	4.733
1.07	1.145	1.225	1.034	3.271	1.023	2.204	4.747
1.08	1.166	1.260	1.039	3.286	1.026	2.210	4.762
1.09	1.188	1.295	1.044	3.302	1.029	2.217	4.777
1.10	1.210	1.331	1.049	3.317	1.032	2.224	4.791
1.11	1.232	1.368	1.054	3.332	1.035	2.231	4.806
1.12	1.254	1.405	1.058	3.347	1.038	2.237	4.820
1.13	1.277	1.443	1.063	3.362	1.042	2.244	4.835
1.14	1.300	1.482	1.068	3.376	1.045	2.251	4.849
1.15	1.322	1.521	1.072	3.391	1.048	2.257	4.863
1.16	1.346	1.561	1.077	3.406	1.051	2.264	4.877
1.17	1.369	1.602	1.082	3.421	1.054	2.270	4.891
1.18	1.392	1.643	1.086	3.435	1.057	2.277	4.905
1.19	1.416	1.685	1.091	3.450	1.060	2.283	4.919
1.20	1.440	1.728	1.095	3.464	1.063	2.289	4.932

[1] Zero at the end should be preserved, for it is a significant figure and indicates the accuracy of the obtained value of the root.

n	n^2	n^3	$\sqrt{n}$	$\sqrt{10n}$	$\sqrt[3]{n}$	$\sqrt[3]{10n}$	$\sqrt[3]{100n}$
1.20	1.440	1.728	1.095	3.464	1.063	2.289	4.932
1.21	1.464	1.772	1.100	3.479	1.066	2.296	4.946
1.22	1.488	1.816	1.105	3.493	1.069	2.302	4.960
1.23	1.513	1.861	1.109	3.507	1.071	2.308	4.973
1.24	1.538	1.907	1.114	3.521	1.074	2.315	4.987
1.25	1.562	1.953	1.118	3.536	1.077	2.321	5.000
1.26	1.588	2.000	1.122	3.550	1.080	2.327	5.013
1.27	1.613	2.048	1.127	3.564	1.083	2.333	5.027
1.28	1.638	2.097	1.131	3.578	1.086	2.339	5.040
1.29	1.664	2.147	1.136	3.592	1.089	2.345	5.053
1.30	1.690	2.197	1.140	3.606	1.091	2.351	5.066
1.31	1.716	2.248	1.145	3.619	1.094	2.357	5.079
1.32	1.742	2.300	1.149	3.633	1.097	2.363	5.092
1.33	1.769	2.353	1.153	3.647	1.100	2.369	5.104
1.34	1.796	2.406	1.158	3.661	1.102	2.375	5.117
1.35	1.822	2.460	1.162	3.674	1.105	2.381	5.130
1.36	1.850	2.515	1.166	3.688	1.108	2.387	5.143
1.37	1.877	2.571	1.170	3.701	1.111	2.393	5.155
1.38	1.904	2.628	1.175	3.715	1.113	2.399	5.168
1.39	1.932	2.686	1.179	3.728	1.116	2.404	5.180
1.40	1.960	2.744	1.183	3.742	1.119	2.410	5.192
1.41	1.988	2.803	1.187	3.755	1.121	2.416	5.205
1.42	2.016	2.863	1.192	3.768	1.124	2.422	5.217
1.43	2.045	2.924	1.196	3.782	1.127	2.427	5.229
1.44	2.074	2.986	1.200	3.795	1.129	2.433	5.241
1.45	2.102	3.049	1.204	3.808	1.132	2.438	5.254
1.46	2.132	3.112	1.208	3.821	1.134	2.444	5.266
1.47	2.161	3.177	1.212	3.834	1.137	2.450	5.278
1.48	2.190	3.242	1.217	3.847	1.140	2.455	5.290
1.49	2.220	3.308	1.221	3.860	1.142	2.461	5.301
1.50	2.250	3.375	1.225	3.873	1.145	2.466	5.313
1.51	2.280	3.443	1.229	3.886	1.147	2.472	5.325
1.52	2.310	3.512	1.233	3.899	1.150	2.477	5.337
1.53	2.341	3.582	1.237	3.912	1.152	2.483	5.348
1.54	2.372	3.652	1.241	3.924	1.155	2.488	5.360
1.55	2.402	3.724	1.245	3.937	1.157	2.493	5.372
1.56	2.434	3.796	1.249	3.950	1.160	2.499	5.383
1.57	2.465	3.870	1.253	3.962	1.162	2.504	5.395
1.58	2.496	3.944	1.257	3.975	1.165	2.509	5.406
1.59	2.528	4.020	1.261	3.987	1.167	2.515	5.418
1.60	2.560	4.096	1.265	4.000	1.170	2.520	5.429

n	n^2	n^3	$\sqrt{n}$	$\sqrt{10n}$	$\sqrt[3]{n}$	$\sqrt[3]{10n}$	$\sqrt[3]{100n}$
1.60	2.560	4.096	1.265	4.000	1.170	2.520	5.429
1.61	2.592	4.173	1.269	4.012	1.172	2.525	5.440
1.62	2.624	4.252	1.273	4.025	1.174	2.530	5.451
1.63	2.657	4.331	1.277	4.037	1.177	2.535	5.463
1.64	2.690	4.411	1.281	4.050	1.179	2.541	5.474
1.65	2.722	4.492	1.285	4.062	1.182	2.546	5.485
1.66	2.756	4.574	1.288	4.074	1.184	2.551	5.496
1.67	2.789	4.657	1.292	4.087	1.186	2.556	5.507
1.68	2.822	4.742	1.296	4.099	1.189	2.561	5.518
1.69	2.856	4.827	1.300	4.111	1.191	2.566	5.529
1.70	2.890	4.913	1.304	4.123	1.193	2.571	5.540
1.71	2.924	5.000	1.308	4.135	1.196	2.576	5.550
1.72	2.958	5.088	1.311	4.147	1.198	2.581	5.561
1.73	2.993	5.178	1.315	4.159	1.200	2.586	5.572
1.74	3.028	5.268	1.319	4.171	1.203	2.591	5.583
1.75	3.062	5.359	1.323	4.183	1.205	2.596	5.593
1.76	3.098	5.452	1.327	4.195	1.207	2.601	5.604
1.77	3.133	5.545	1.330	4.207	1.210	2.606	5.615
1.78	3.168	5.640	1.334	4.219	1.212	2.611	5.625
1.79	3.204	5.735	1.338	4.231	1.214	2.616	5.636
1.80	3.240	5.832	1.342	4.243	1.216	2.621	5.646
1.81	3.276	5.930	1.345	4.254	1.219	2.626	5.657
1.82	3.312	6.029	1.349	4.266	1.221	2.630	5.667
1.83	3.349	6.128	1.353	4.278	1.223	2.635	5.677
1.84	3.386	6.230	1.356	4.290	1.225	2.640	5.688
1.85	3.422	6.332	1.360	4.301	1.228	2.645	5.698
1.86	3.460	6.435	1.364	4.313	1.230	2.650	5.708
1.87	3.497	6.539	1.367	4.324	1.232	2.654	5.718
1.88	3.534	6.645	1.371	4.336	1.234	2.659	5.729
1.89	3.572	6.751	1.375	4.347	1.236	2.664	5.739
1.90	3.610	6.859	1.378	4.359	1.239	2.668	5.749
1.91	3.648	6.968	1.382	4.370	1.241	2.673	5.759
1.92	3.686	7.078	1.386	4.382	1.243	2.678	5.769
1.93	3.725	7.189	1.389	4.393	1.245	2.682	5.779
1.94	3.764	7.301	1.393	4.405	1.247	2.687	5.789
1.95	3.802	7.415	1.396	4.416	1.249	2.692	5.799
1.96	3.842	7.530	1.400	4.427	1.251	2.696	5.809
1.97	3.881	7.645	1.404	4.438	1.254	2.701	5.819
1.98	3.920	7.762	1.407	4.450	1.256	2.705	5.828
1.99	3.960	7.881	1.411	4.461	1.258	2.710	5.838
2.00	4.000	8.000	1.414	4.472	1.260	2.714	5.848

n	n^2	n^3	$\sqrt{n}$	$\sqrt{10n}$	$\sqrt[3]{n}$	$\sqrt[3]{10n}$	$\sqrt[3]{100n}$
2.00	4.000	8.000	1.414	4.472	1.260	2.714	5.848
2.01	4.040	8.121	1.418	4.483	1.262	2.719	5.858
2.02	4.080	8.242	1.421	4.494	1.264	2.723	5.867
2.03	4.121	8.365	1.425	4.506	1.266	2.728	5.877
2.04	4.162	8.490	1.428	4.517	1.268	2.732	5.887
2.05	4.202	8.615	1.432	4.528	1.270	2.737	5.896
2.06	4.244	8.742	1.435	4.539	1.272	2.741	5.906
2.07	4.285	8.870	1.439	4.550	1.274	2.746	5.915
2.08	4.326	8.999	1.442	4.561	1.277	2.750	5.925
2.09	4.368	9.129	1.446	4.572	1.279	2.755	5.934
2.10	4.410	9.261	1.449	4.583	1.281	2.759	5.944
2.11	4.452	9.394	1.453	4.593	1.283	2.763	5.953
2.12	4.494	9.528	1.456	4.604	1.285	2.768	5.963
2.13	4.537	9.664	1.459	4.615	1.287	2.772	5.972
2.14	4.580	9.800	1.463	4.626	1.289	2.776	5.981
2.15	4.622	9.938	1.466	4.637	1.291	2.781	5.991
2.16	4.666	10.08	1.470	4.648	1.293	2.785	6.000
2.17	4.709	10.22	1.473	4.658	1.295	2.789	6.009
2.18	4.752	10.36	1.476	4.669	1.297	2.794	6.018
2.19	4.796	10.50	1.480	4.680	1.299	2.798	6.028
2.20	4.840	10.65	1.483	4.690	1.301	2.802	6.037
2.21	4.884	10.79	1.487	4.701	1.303	2.806	6.046
2.22	4.928	10.94	1.490	4.712	1.305	2.811	6.055
2.23	4.973	11.09	1.493	4.722	1.306	2.815	6.064
2.24	5.018	11.24	1.497	4.733	1.308	2.819	6.073
2.25	5.062	11.39	1.500	4.743	1.310	2.823	6.082
2.26	5.108	11.54	1.503	4.754	1.312	2.827	6.091
2.27	5.153	11.70	1.507	4.764	1.314	2.831	6.100
2.28	5.198	11.85	1.510	4.775	1.316	2.836	6.109
2.29	5.244	12.01	1.513	4.785	1.318	2.840	6.118
2.30	5.290	12.17	1.517	4.796	1.320	2.844	6.127
2.31	5.336	12.33	1.520	4.806	1.322	2.848	6.136
2.32	5.382	12.49	1.523	4.817	1.324	2.852	6.145
2.33	5.429	12.65	1.526	4.827	1.326	2.856	6.153
2.34	5.476	12.81	1.530	4.837	1.328	2.860	6.162
2.35	5.522	12.98	1.533	4.848	1.330	2.864	6.171
2.36	5.570	13.14	1.536	4.858	1.331	2.868	6.180
2.37	5.617	13.31	1.539	4.868	1.333	2.872	6.188
2.38	5.664	13.48	1.543	4.879	1.335	2.876	6.197
2.39	5.712	13.65	1.546	4.889	1.337	2.880	6.206
2.40	5.760	13.82	1.549	4.899	1.339	2.884	6.214

n	n^2	n^3	$\sqrt{n}$	$\sqrt{10n}$	$\sqrt[3]{n}$	$\sqrt[3]{10n}$	$\sqrt[3]{100n}$
2.40	5.760	13.82	1.549	4.899	1.339	2.884	6.214
2.41	5.808	14.00	1.552	4.909	1.341	2.888	6.223
2.42	5.856	14.17	1.556	4.919	1.343	2.892	6.232
2.43	5.905	14.35	1.559	4.930	1.344	2.896	6.240
2.44	5.954	14.53	1.562	4.940	1.346	2.900	6.249
2.45	6.002	14.71	1.565	4.950	1.348	2.904	6.257
2.46	6.052	14.89	1.568	4.960	1.350	2.908	6.266
2.47	6.101	15.07	1.572	4.970	1.352	2.912	6.274
2.48	6.150	15.25	1.575	4.980	1.354	2.916	6.283
2.49	6.200	15.44	1.578	4.990	1.355	2.920	6.291
2.50	6.250	15.62	1.581	5.000	1.357	2.924	6.300
2.51	6.300	15.81	1.584	5.010	1.359	2.928	6.308
2.52	6.350	16.00	1.587	5.020	1.361	2.932	6.316
2.53	6.401	16.19	1.591	5.030	1.363	2.936	6.325
2.54	6.452	16.39	1.594	5.040	1.364	2·940	6.333
2.55	6.502	16.58	1.597	5.050	1.366	2.943	6.341
2.56	6.554	16.78	1.600	5.060	1.368	2.947	6.350
2.57	6.605	16.97	1.603	5.070	1.370	2.951	6.358
2.58	6.656	17.17	1.606	5.079	1.372	2.955	6.366
2.59	6.708	17.37	1.609	5.089	1.373	2.959	6.374
2.60	6.760	17.58	1.612	5.099	1.375	2.962	6.383
2.61	6.812	17.78	1.616	5.109	1.377	2.966	6.391
2.62	6.864	17.98	1.619	5.119	1.379	2.970	6.399
2.63	6.917	18.19	1.622	5.128	1.380	2.974	6.407
2.64	6.970	18.40	1.625	5.138	1.382	2.978	6.415
2.65	7.022	18.61	1.628	5.148	1.384	2.981	6.432
2.66	7.076	18.82	1.631	5.158	1.386	2.985	6.431
2.67	7.129	19.03	1.634	5.167	1.387	2.989	6.439
2.68	7.182	19.25	1.637	5.177	1.389	2.993	6.447
2.69	7.236	19.47	1.640	5.187	1.391	2.996	6.455
2.70	7.290	19.68	1.643	5.196	1.392	3.000	6.463
2.71	7.344	19.90	1.646	5.206	1.394	3.004	6.471
2.72	7.398	20.12	1.649	5.215	1.396	3.007	6.479
2.73	7.453	20.35	1.652	5.225	1.398	3.011	6.487
2.74	7.508	20.57	1.655	5.235	1.399	3.015	6.495
2.75	7.562	20.80	1.658	5.244	1.401	3.018	6.503
2.76	7.618	21.02	1.661	5.254	1.403	3.022	6.511
2.77	7.673	21.25	1.664	5.263	1.404	3.026	6.519
2.78	7.728	21.48	1.667	5.273	1.406	3.029	6.527
2.79	7.784	21.72	1.670	5.282	1.408	3.033	6.534
2.80	7.840	21.95	1.673	5.292	1.409	3.037	6.542

n	n^2	n^3	$\sqrt{n}$	$\sqrt{10n}$	$\sqrt[3]{n}$	$\sqrt[3]{10n}$	$\sqrt[3]{100n}$
2.80	7.840	21.95	1.673	5.292	1.409	3.037	6.542
2.81	7.896	22.19	1.676	5.301	1.411	3.040	6.550
2.82	7.952	22.43	1.679	5.310	1.413	3.044	6.558
2.83	8.009	22.67	1.682	5.320	1.414	3.047	6.565
2.84	8.066	22.91	1.685	5.329	1.416	3.051	6.573
2.85	8.122	23.15	1.688	5.339	1.418	3.055	6.581
2.86	8.180	23.39	1.691	5.348	1.419	3.058	6.589
2.87	8.237	23.64	1.694	5.357	1.421	3.062	6.596
2.88	8.294	23.89	1.697	5.367	1.423	3.065	6.604
2.89	8.352	24.14	1.700	5.376	1.424	3.069	6.611
2.90	8.410	24.39	1.703	5.385	1.426	3.072	6.619
2.91	8.468	24.64	1.706	5.394	1.428	3.076	6.627
2.92	8.526	24.90	1.709	5.404	1.429	3.079	6.634
2.93	8.585	25.15	1.712	5.413	1.431	3.083	6.642
2.94	8.644	25.41	1.715	5.422	1.433	3.086	6.649
2.95	8.702	25.67	1.718	5.431	1.434	3.090	6.657
2.96	8.762	25.93	1.720	5.441	1.436	3.093	6.664
2.97	8.821	26.20	1.723	5.450	1.437	3.097	6.672
2.98	8.880	26.46	1.726	5.459	1.439	3.100	6.679
2.99	8.940	26.73	1.729	5.468	1.441	3.104	6.687
3.00	9.000	27.00	1.732	5.477	1.442	3.107	6.694
3.01	9.060	27.27	1.735	5.486	1.444	3.111	6.702
3.02	9.120	27.54	1.738	5.495	1.445	3.114	6.709
3.03	9.181	27.82	1.741	5.505	1.447	3.118	6.717
3.04	9.242	28.09	1.744	5.514	1.449	3.121	6.724
3.05	9.302	28.37	1.746	5.523	1.450	3.124	6.731
3.06	9.364	28.65	1.749	5.532	1.452	3.128	6.739
3.07	9.425	28.93	1.752	5.541	1.453	3.131	6.746
3.08	9.486	29.22	1.755	5.550	1.455	3.135	6.753
3.09	9.548	29.50	1.758	5.559	1.457	3.138	6.761
3.10	9.610	29.79	1.761	5.568	1.458	3.141	6.768
3.11	9.672	30.08	1.764	5.577	1.460	3.145	6.775
3.12	9.734	30.37	1.766	5.586	1.461	3.148	6.782
3.13	9.797	30.66	1.769	5.595	1.463	3.151	6.790
3.14	9.860	30.96	1.772	5.604	1.464	3.155	6.797
3.15	9.922	31.26	1.775	5.612	1.466	3.158	6.804
3.16	9.986	31.55	1.778	5.621	1.467	3.162	6.811
3.17	10.05	31.86	1.780	5.630	1.469	3.165	6.818
3.18	10.11	32.16	1.783	5.639	1.471	3.168	6.826
3.19	10.18	32.46	1.786	5.648	1.472	3.171	6.833
3.20	10.24	32.77	1.789	5.657	1.474	3.175	6.840

n	n^2	n^3	$\sqrt{n}$	$\sqrt{10n}$	$\sqrt[3]{n}$	$\sqrt[3]{10n}$	$\sqrt[3]{100n}$
3.20	10.24	32.77	1.789	5.657	1.474	3.175	6.840
3.21	10.30	33.08	1.792	5.666	1.475	3.178	6.847
3.22	10.37	33.39	1.794	5.675	1.477	3.181	6.854
3.23	10.43	33.70	1.797	5.683	1.478	3.185	6.861
3.24	10.50	34.01	1.800	5.692	1.480	3.188	6.868
3.25	10.56	34.33	1.803	5.701	1.481	3.191	6.875
3.26	10.63	34.65	1.806	5.710	1.483	3.195	6.882
3.27	10.69	34.97	1.808	5.718	1.484	3.198	6.889
3.28	10.76	35.29	1.811	5.727	1.486	3.201	6.896
3.29	10.82	35.61	1.814	5.736	1.487	3.204	6.903
3.30	10.89	35.94	1.817	5.745	1.489	3.208	6.910
3.31	10.96	36.26	1.819	5.753	1.490	3.211	6.917
3.32	11.02	36.59	1.822	5.762	1.492	3.214	6.924
3.33	11.09	36.93	1.825	5.771	1.493	3.217	6.931
3.34	11.16	37.26	1.828	5.779	1.495	3.220	6.938
3.35	11.22	37.60	1.830	5.788	1.496	3.224	6.945
3.36	11.29	37.93	1.833	5.797	1.498	3.227	6.952
3.37	11.36	38.27	1.836	5.805	1.499	3.230	6.959
3.38	11.42	38.61	1.838	5.814	1.501	3.233	6.966
3.39	11.49	38.96	1.841	5.822	1.502	3.236	6.973
3.40	11.56	39.30	1.844	5.831	1.504	3.240	6.980
3.41	11.63	39.65	1.847	5.840	1.505	3.243	6.986
3.42	11.70	40.00	1.849	5.848	1.507	3.246	6.993
3.43	11.76	40.35	1.852	5.857	1.508	3.249	7.000
3.44	11.83	40.71	1.855	5.865	1.510	3.252	7.007
3.45	11.90	41.06	1.857	5.874	1.511	3.255	7.014
3.46	11.97	41.42	1.860	5.882	1.512	3.259	7.020
3.47	12.04	41.78	1.863	5.891	1.514	3.262	7.027
3.48	12.11	42.14	1.865	5.899	1.515	3.265	7.034
3.49	12.18	42.51	1.868	5.908	1.517	3.268	7.041
3.50	12.25	42.88	1.871	5.916	1.518	3.271	7.047
3.51	12.32	43.24	1.873	5.925	1.520	3.274	7.054
3.52	12.39	43.61	1.876	5.933	1.521	3.277	7.061
3.53	12.46	43.99	1.879	5.941	1.523	3.280	7.067
3.54	12.53	44.36	1.881	5.950	1.524	3.283	7.074
3.55	12.60	44.74	1.884	5.958	1.525	3.287	7.081
3.56	12.67	45.12	1.887	5.967	1.527	3.290	7.087
3.57	12.74	45.50	1.889	5.975	1.528	3.293	7.094
3.58	12.82	45.88	1.892	5.983	1.530	3.296	7.101
3.59	12.89	46.27	1.895	5.992	1.531	3.299	7.107
3.60	12.96	46.66	1.897	6.000	1.533	3.302	7.114

n	n^2	n^3	$\sqrt{n}$	$\sqrt{10n}$	$\sqrt[3]{n}$	$\sqrt[3]{10n}$	$\sqrt[3]{100n}$
3.60	12.96	46.66	1.897	6.000	1.533	3.302	7.114
3.61	13.03	47.05	1.900	6.008	1.534	3.305	7.120
3.62	13.10	47.44	1.903	6.017	1.535	3.308	7.127
3.63	13.18	47.83	1.905	6.025	1.537	3.311	7.133
3.64	13.25	48.23	1.908	6.033	1.538	3.314	7.140
3.65	13.32	48.63	1.910	6.042	1.540	3.317	7.147
3.66	13.40	49.03	1.913	6.050	1.541	3.320	7.153
3.67	13.47	49.43	1.916	6.058	1.542	3.323	7.160
3.68	13.54	49.84	1.918	6.066	1.544	3.326	7.166
3.69	13.62	50.24	1.921	6.075	1.545	3.329	7.173
3.70	13.69	50.65	1.924	6.083	1.547	3.332	7.179
3.71	13.76	51.06	1.926	6.091	1.548	3.335	7.186
3.72	13.84	51.48	1.929	6.099	1.549	3.338	7.192
3.73	13.91	51.90	1.931	6.107	1.551	3.341	7.198
3.74	13.99	52.31	1.934	6.116	1.552	3.344	7.205
3.75	14.06	52.73	1.936	6.124	1.554	3.347	7.211
3.76	14.14	53.16	1.939	6.132	1.555	3.350	7.218
3.77	14.21	53.58	1.942	6.140	1.556	3.353	7.224
3.78	14.29	54.01	1.944	6.148	1.558	3.356	7.230
3.79	14.36	54.44	1.947	6.156	1.559	3.359	7.237
3.80	14.44	54.87	1.949	6.164	1.560	3.362	7.243
3.81	14.52	55.31	1.952	6.173	1.562	3.365	7.250
3.82	14.59	55.74	1.954	6.181	1.563	3.368	7.256
3.83	14.67	56.18	1.957	6.189	1.565	3.371	7.262
3.84	14.75	56.62	1.960	6.197	1.566	3.374	7.268
3.85	14.82	57.07	1.962	6.205	1.567	3.377	7.275
3.86	14.90	57.51	1.965	6.213	1.569	3.380	7.281
3.87	14.98	57.96	1.967	6.221	1.570	3.382	7.287
3.88	15.05	58.41	1.970	6.229	1.571	3.385	7.294
3.89	15.13	58.86	1.972	6.237	1.573	3.388	7.300
3.90	15.21	59.32	1.975	6.245	1.574	3.391	7.306
3.91	15.29	59.78	1.977	6.253	1.575	3.394	7.312
3.92	15.37	60.24	1.980	6.261	1.577	3.397	7.319
3.93	15.44	60.70	1.982	6.269	1.578	3.400	7.325
3.94	15.52	61.16	1.985	6.277	1.579	3.403	7.331
3.95	15.60	61.63	1.987	6.285	1.581	3.406	7.337
3.96	15.68	62.10	1.990	6.293	1.582	3.409	7.343
3.97	15.76	62.57	1.992	6.301	1.583	3.411	7.350
3.98	15.84	63.04	1.995	6.309	1.585	3.414	7.356
3.99	15.92	63.52	1.997	6.317	1.586	3.417	7.362
4.00	16.00	64.00	2.000	6.325	1.587	3.420	7.368

n	n^2	n^3	$\sqrt{n}$	$\sqrt{10n}$	$\sqrt[3]{n}$	$\sqrt[3]{10n}$	$\sqrt[3]{100n}$
4.00	16.00	64.00	2.000	6.325	1.587	3.420	7.368
4.01	16.08	64.48	2.002	6.332	1.589	3.423	7.374
4.02	16.16	64.96	2.005	6.340	1.590	3.426	7.380
4.03	16.24	65.45	2.007	6.348	1.591	3.428	7.386
4.04	16.32	65.94	2.010	6.356	1.593	3.431	7.393
4.05	16.40	66.43	2.012	6.364	1.594	3.434	7.399
4.06	16.48	66.92	2.015	6.372	1.595	3.437	7.405
4.07	16.56	67.42	2.017	6.380	1.597	3.440	7.411
4.08	16.65	67.92	2.020	6.387	1.598	3.443	7.417
4.09	16.73	68.42	2.022	6.395	1.599	3.445	7.423
4.10	16.81	68.92	2.025	6.403	1.601	3.448	7.429
4.11	16.89	69.43	2.027	6.411	1.602	3.451	7.435
4.12	16.97	69.93	2.030	6.419	1.603	3.454	7.441
4.13	17.06	70.44	2.032	6.427	1.604	3.457	7.447
4.14	17.14	70.96	2.035	6.434	1.606	3.459	7.453
4.15	17.22	71.47	2.037	6.442	1.607	3.462	7.459
4.16	17.31	71.99	2.040	6.450	1.608	3.465	7.465
4.17	17.39	72.51	2.042	6.458	1.610	3.468	7.471
4.18	17.47	73.03	2.045	6.465	1.611	3.471	7.477
4.19	17.56	73.56	2.047	6.473	1.612	3.473	7.483
4.20	17.64	74.09	2.049	6.481	1.613	3.476	7.489
4.21	17.72	74.62	2.052	6.488	1.615	3.479	7.495
4.22	17.81	75.15	2.054	6.496	1.616	3.482	7.501
4.23	17.89	75.69	2.057	6.504	1.617	3.484	7.507
4.24	17.98	76.23	2.059	6.512	1.619	3.487	7.513
4.25	18.06	76.77	2.062	6.519	1.620	3.490	7.518
4.26	18.15	77.31	2.064	6.527	1.621	3.493	7.524
4.27	18.23	77.85	2.066	6.535	1.622	3.495	7.530
4.28	18.32	78.40	2.069	6.542	1.624	3.498	7.536
4.29	18.40	78.95	2.071	6.550	1.625	3.501	7.542
4.30	18.49	79.51	2.074	6.557	1.626	3.503	7.548
4.31	18.58	80.06	2.076	6.565	1.627	3.506	7.554
4.32	18.66	80.62	2.078	6.573	1.629	3.509	7.560
4.33	18.75	81.18	2.081	6.580	1.630	3.512	7.565
4.34	18.84	81.75	2.083	6.588	1.631	3.514	7.571
4.35	18.92	82.31	2.086	6.595	1.632	3.517	7.577
4.36	19.01	82.88	2.088	6.603	1.634	3.520	7.583
4.37	19.10	83.45	2.090	6.611	1.635	3.522	7.589
4.38	19.18	84.03	2.093	6.618	1.636	3.525	7.594
4.39	19.27	84.60	2.095	6.626	1.637	3.528	7.600
4.40	19.36	85.18	2.098	6.633	1.639	3.530	7.606

n	n^2	n^3	$\sqrt{n}$	$\sqrt{10n}$	$\sqrt[3]{n}$	$\sqrt[3]{10n}$	$\sqrt[3]{100n}$
4.40	19.36	85.18	2.098	6.633	1.639	3.530	7.606
4.41	19.45	85.77	2.100	6.641	1.640	3.533	7.612
4.42	19.54	86.35	2.102	6.648	1.641	3.536	7.617
4.43	19.62	86.94	2.105	6.656	1.642	3.538	7.623
4.44	19.71	87.53	2.107	6.663	1.644	3.541	7.629
4.45	19.80	88.12	2.110	6.671	1.645	3.544	7.635
4.46	19.89	88.72	2.112	6.678	1.646	3.546	7.640
4.47	19.98	89.31	2.114	6.686	1.647	3.549	7.646
4.48	20.07	89.92	2.117	6.693	1.649	3.552	7.652
4.49	20.16	90.52	2.119	6.701	1.650	3.554	7.657
4.50	20.25	91.12	2.121	6.708	1.651	3.557	7.663
4.51	20.34	91.73	2.124	6.716	1.652	3.560	7.669
4.52	20.43	92.35	2.126	6.723	1.653	3.562	7.674
4.53	20.52	92.96	2.128	6.731	1.655	3.565	7.680
4.54	20.61	93.58	2.131	6.738	1.656	3.567	7.686
4.55	20.70	94.20	2.133	6.745	1.657	3.570	7.691
4.56	20.79	94.82	2.135	6.753	1.658	3.573	7.697
4.57	20.88	95.44	2.138	6.760	1.659	3.575	7.703
4.58	20.98	96.07	2.140	6.768	1.661	3.578	7.708
4.59	21.07	96.70	2.142	6.775	1.662	3.580	7.714
4.60	21.16	97.34	2.145	6.782	1.663	3.583	7.719
4.61	21.25	97.97	2.147	6.790	1.664	3.586	7.725
4.62	21.34	98.61	2.149	6.797	1.666	3.588	7.731
4.63	21.44	99.25	2.152	6.804	1.667	3.591	7.736
4.64	21.53	99.90	2.154	6.812	1.668	3.593	7.742
4.65	21.62	100.5	2.156	6.819	1.669	3.596	7.747
4.66	21.72	101.2	2.159	6.826	1.670	3.599	7.753
4.67	21.81	101.8	2.161	6.834	1.671	3.601	7.758
4.68	21.90	102.5	2.163	6.841	1.673	3.604	7.764
4.69	22.00	103.2	2.166	6.848	1.674	3.606	7.769
4.70	22.09	103.8	2.168	6.856	1.675	3.609	7.775
4.71	22.18	104.5	2.170	6.863	1.676	3.611	7.780
4.72	22.28	105.2	2.173	6.870	1.677	3.614	7.786
4.73	22.37	105.8	2.175	6.877	1.679	3.616	7.791
4.74	22.47	106.5	2.177	6.885	1.680	3.619	7.797
4.75	22.56	107.2	2.179	6.892	1.681	3.622	7.802
4.76	22.66	107.9	2.182	6.899	1.682	3.624	7.808
4.77	22.75	108.5	2.184	6.907	1.683	3.627	7.813
4.78	22.85	109.2	2.186	6.914	1.685	3.629	7.819
4.79	22.94	109.9	2.189	6.921	1.686	3.632	7.824
4.80	23.04	110.6	2.191	6.928	1.687	3.634	7.830

n	n^2	n^3	$\sqrt{n}$	$\sqrt{10n}$	$\sqrt[3]{n}$	$\sqrt[3]{10n}$	$\sqrt[3]{100n}$
4.80	23.04	110.6	2.191	6.928	1.687	3.634	7.830
4.81	23.14	111.3	2.193	6.935	1.688	3.637	7.835
4.82	23.23	112.0	2.195	6.943	1.689	3.639	7.841
4.83	23.33	112.7	2.198	6.950	1.690	3.642	7.846
4.84	23.43	113.4	2.200	6.957	1.692	3.644	7.851
4.85	23.52	114.1	2.202	6.964	1.693	3.647	7.857
4.86	23.62	114.8	2.205	6.971	1.694	3.649	7.862
4.87	23.72	115.5	2.207	6.979	1.695	3.652	7.868
4.88	23.81	116.2	2.209	6.986	1.696	3.654	7.873
4.89	23.91	116.9	2.211	6.993	1.697	3.657	7.878
4.90	24.01	117.6	2.214	7.000	1.698	3.659	7.884
4.91	24.11	118.4	2.216	7.007	1.700	3.662	7.889
4.92	24.21	119.1	2.218	7.014	1.701	3.664	7.894
4.93	24.30	119.8	2.220	7.021	1.702	3.667	7.900
4.94	24.40	120.6	2.223	7.029	1.703	3.669	7.905
4.95	24.50	121.3	2.225	7.036	1.704	3.672	7.910
4.96	24.60	122.0	2.227	7.043	1.705	3.674	7.916
4.97	24.70	122.8	2.229	7.050	1.707	3.677	7.921
4.98	24.80	123.5	2.232	7.057	1.708	3.679	7.926
4.99	24.90	124.3	2.234	7.064	1.709	3.682	7.932
5.00	25.00	125.0	2.236	7.071	1.710	3.684	7.937
5.01	25.10	125.8	2.238	7.078	1.711	3.686	7.942
5.02	25.20	126.5	2.241	7.085	1.712	3.689	7.948
5.03	25.30	127.3	2.243	7.092	1.713	3.691	7.953
5.04	25.40	128.0	2.245	7.099	1.715	3.694	7.958
5.05	25.50	128.8	2.247	7.106	1.716	3.696	7.963
5.06	25.60	129.6	2.249	7.113	1.717	3.699	7.969
5.07	25.70	130.3	2.252	7.120	1,718	3.701	7.974
5.08	25.81	131.1	2.254	7.127	1.719	3.704	7.979
5.09	25.91	131.9	2.256	7.134	1.720	3.706	7.984
5.10	26.01	132.7	2.258	7.141	1.721	3.708	7.990
5.11	26.11	133.4	2.261	7.148	1.722	3.711	7.995
5.12	26.21	134.2	2.263	7.155	1.724	3.713	8.000
5.13	26.32	135.0	2.265	7.162	1.725	3.716	8.005
5.14	26.42	135.8	2.267	7.169	1.726	3.718	8.010
5.15	26.52	136.6	2.269	7.176	1.727	3.721	8.016
5.16	26.63	137.4	2.272	7.183	1.728	3.723	8.021
5.17	26.73	138.2	2.274	7.190	1.729	3.725	8.026
5.18	26.83	139.0	2.276	7.197	1.730	3.728	8.031
5.19	26.94	139.8	2.278	7.204	1.731	3.730	8.036
5.20	27.04	140.6	2.280	7.211	1.732	3.733	8.041

n	n^2	n^3	$\sqrt{n}$	$\sqrt{10n}$	$\sqrt[3]{n}$	$\sqrt[3]{10n}$	$\sqrt[3]{100n}$
5.20	27.04	140.6	2.280	7.211	1.732	3.733	8.041
5.21	27.14	141.4	2.283	7.218	1.734	3.735	8.047
5.22	27.25	142.2	2.285	7.225	1.735	3.737	8.052
5.23	27.35	143.1	2.287	7.232	1.736	3.740	8.057
5.24	27.46	143.9	2.289	7.239	1.737	3.742	8.062
5.25	27.56	144.7	2.291	7.246	1.738	3.744	8.067
5.26	27.67	145.5	2.293	7.253	1.739	3.747	8.072
5.27	27.77	146.4	2.296	7.259	1.740	3.749	8.077
5.28	27.88	147.2	2.298	7.266	1.741	3.752	8.082
5.29	27.98	148.0	2.300	7.273	1.742	3.754	8.088
5.30	28.09	148.9	2.302	7.280	1.744	3.756	8.093
5.31	28.20	149.7	2.304	7.287	1.745	3.759	8.098
5.32	28.30	150.6	2.307	7.294	1.746	3.761	8.103
5.33	28.41	151.4	2.309	7.301	1.747	3.763	8.108
5.34	28.52	152.3	2.311	7.308	1.748	3.766	8.113
5.35	28.62	153.1	2.313	7.314	1.749	3.768	8.118
5.36	28.73	154.0	2.315	7.321	1.750	3.770	8.123
5.37	28.84	154.9	2.317	7.328	1.751	3.773	8.128
5.38	28.94	155.7	2.319	7.335	1.752	3.775	8.133
5.39	29.05	156.6	2.322	7.342	1.753	3.777	8.138
5.40	29.16	157.5	2.324	7.348	1.754	3.780	8.143
5.41	29.27	158.3	2.326	7.355	1.755	3.782	8.148
5.42	29.38	159.2	2.328	7.362	1.757	3.784	8.153
5.43	29.48	160.1	2.330	7.369	1.758	3.787	8.158
5.44	29.59	161.0	2.332	7.376	1.759	3.789	8.163
5.45	29.70	161.9	2.335	7.382	1.760	3.791	8.168
5.46	29.81	162.8	2.337	7.389	1.761	3.794	8.173
5.47	29.92	163.7	2.339	7.396	1.762	3.796	8.178
5.48	30.03	164.6	2.341	7.403	1.763	3.798	8.183
5.49	30.14	165.5	2.343	7.409	1.764	3.801	8.188
5.50	30.25	166.4	2.345	7.416	1.765	3.803	8.193
5.51	30.36	167.3	2.347	7.423	1.766	3.805	8.198
5.52	30.47	168.2	2.349	7.430	1.767	3.808	8.203
5.53	30.58	169.1	2.352	7.436	1.768	3.810	8.208
5.54	30.69	170.0	2.354	7.443	1.769	3.812	8.213
5.55	30.80	171.0	2.356	7.450	1.771	3.814	8.218
5.56	30.91	171.9	2.358	7.457	1.772	3.817	8.223
5.57	31.02	172.8	2.360	7.463	1.773	3.819	8.228
5.58	31.14	173.7	2.362	7.470	1.774	3.821	8.233
5.59	31.25	174.7	2.364	7.477	1.775	3.824	8.238
5.60	31.36	175.6	2.366	7.483	1.776	3.826	8.243

n	n^2	n^3	$\sqrt{n}$	$\sqrt{10n}$	$\sqrt[3]{n}$	$\sqrt[3]{10n}$	$\sqrt[3]{100n}$
5.60	31.36	175.6	2.366	7.483	1.776	3.826	8.243
5.61	31.47	176.6	2.369	7.490	1.777	3.828	8.247
5.62	31.58	177.5	2.371	7.497	1.778	3.830	8.252
5.63	31.70	178.5	2.373	7.503	1.779	3.833	8.257
5.64	31.81	179.4	2.375	7.510	1.780	3.835	8.262
5.65	31.92	180.4	2.377	7.517	1.781	3.837	8.267
5.66	32.04	181.3	2.379	7.523	1.782	3.839	8.272
5.67	32.15	182.3	2.381	7.530	1.783	3.842	8.277
5.68	32.26	183.3	2.383	7.537	1.784	3.844	8.282
5.69	32.38	184.2	2.385	7.543	1.785	3.846	8.286
5.70	32.49	185.2	2.387	7.550	1.786	3.849	8.291
5.71	32.60	186.2	2.390	7.556	1.787	3.851	8.296
5.72	32.72	187.1	2.392	7.563	1.788	3.853	8.301
5.73	32.83	188.1	2.394	7.570	1.789	3.855	8.306
5.74	32.95	189.1	2.396	7.576	1.790	3.857	8.311
5.75	33.06	190.1	2.398	7.583	1.792	3.860	8.316
5.76	33.18	191.1	2.400	7.589	1.793	3.862	8.320
5.77	33.29	192.1	2.402	7.596	1.794	3.864	8.325
5.78	33.41	193.1	2.404	7.603	1.795	3.866	8.330
5.79	33.52	194.1	2.406	7.609	1.796	3.869	8.335
5.80	33.64	195.1	2.408	7.616	1.797	3.871	8.340
5.81	33.76	196.1	2.410	7.622	1.798	3.873	8.344
5.82	33.87	197.1	2.412	7.629	1.799	3.875	8.349
5.83	33.99	198.2	2.415	7.635	1.800	3.878	8.354
5.84	34.11	199.2	2.417	7.642	1.801	3.880	8.359
5.85	34.22	200.2	2.419	7.649	1.802	3.882	8.363
5.86	34.34	201.2	2.421	7.655	1.803	3.884	8.368
5.87	34.46	202.3	2.423	7.662	1.804	3.886	8.373
5.88	34.57	203.3	2.425	7.668	1.805	3.889	8.378
5.89	34.69	204.3	2.427	7.675	1.806	3.891	8.382
5.90	34.81	205.4	2.429	7.681	1.807	3.893	8.387
5.91	34.93	206.4	2.431	7.688	1.808	3.895	8.392
5.92	35.05	207.5	2.433	7.694	1.809	3.897	8.397
5.93	35.16	208.5	2.435	7.701	1.810	3.900	8.401
5.94	35.28	209.6	2.437	7.707	1.811	3.902	8.406
5.95	35.40	210.6	2.439	7.714	1.812	3.904	8.411
5.96	35.52	211.7	2.441	7.720	1.813	3.906	8.416
5.97	35.64	212.8	2.443	7.727	1.814	3.908	8.420
5.98	35.76	213.8	2.445	7.733	1.815	3.911	8.425
5.99	35.88	214.9	2.447	7.740	1.816	3.913	8.430
6.00	36.00	216.0	2.449	7.746	1.817	3.915	8.434

n	n^2	n^3	$\sqrt{n}$	$\sqrt{10n}$	$\sqrt[3]{n}$	$\sqrt[3]{10n}$	$\sqrt[3]{100n}$
6.00	36.00	216.0	2.449	7.746	1.817	3.915	8.434
6.01	36.12	217.1	2.452	7.752	1.818	3.917	8.439
6.02	36.24	218.2	2.454	7.759	1.819	3.919	8.444
6.03	36.36	219.3	2.456	7.765	1.820	3.921	8.448
6.04	36.48	220.3	2.458	7.772	1.821	3.924	8.453
6.05	36.60	221.4	2.460	7.778	1.822	3.926	8.458
6.06	36.72	222.5	2.462	7.785	1.823	3.928	8.462
6.07	36.84	223.6	2.464	7.791	1.824	3.930	8.467
6.08	36.97	224.8	2.466	7.797	1.825	3.932	8.472
6.09	37.09	225.9	2.468	7.804	1.826	3.934	8.476
6.10	37.21	227.0	2.470	7.810	1.827	3.936	8.481
6.11	37.33	228.1	2.472	7.817	1.828	3.939	8.486
6.12	37.45	229.2	2.474	7.823	1.829	3.941	8.490
6.13	37.58	230.3	2.476	7.829	1.830	3.943	8.495
6.14	37.70	231.5	2.478	7.836	1.831	3.945	8.499
6.15	37.82	232.6	2.480	7.842	1.832	3.947	8.504
6.16	37.95	233.7	2.482	7.849	1.833	3.949	8.509
6.17	38.07	234.9	2.484	7.855	1.834	3.951	8.513
6.18	38.19	236.0	2.486	7.861	1.835	3.954	8.518
6.19	38.32	237.2	2.488	7.868	1.836	3.956	8.522
6.20	38.44	238.3	2.490	7.874	1.837	3.958	8.527
6.21	38.56	239.5	2.492	7.880	1.838	3.960	8.532
6.22	38.69	240.6	2.494	7.887	1.839	3.962	8.536
6.23	38.81	241.8	2.496	7.893	1.840	3.964	8.541
6.24	38.94	243.0	2.498	7.899	1.841	3.966	8.545
6.25	39.06	244.1	2.500	7.906	1.842	3.969	8.550
6.26	39.19	245.3	2.502	7.912	1.843	3.971	8.554
6.27	39.31	246.5	2.504	7.918	1.844	3.973	8.559
6.28	39.44	247.7	2.506	7.925	1.845	3.975	8.564
6.29	39.56	248.9	2.508	7.931	1.846	3.977	8.568
6.30	39.69	250.0	2.510	7.937	1.847	3.979	8.573
6.31	39.82	251.2	2.512	7.944	1.848	3.981	8.577
6.32	39.94	252.4	2.514	7.950	1.849	3.983	8.582
6.33	40.07	252.6	2.516	7.956	1.850	3.985	8.586
6.34	40.20	254.8	2.518	7.962	1.851	3.987	8.591
6.35	40.32	256.0	2.520	7.969	1.852	3.990	8.595
6.36	40.45	257.3	2.522	7.975	1.853	3.992	8.600
6.37	40.58	258.5	2.524	7.981	1.854	3.994	8.604
6.38	40.70	259.7	2.526	7.987	1.855	3.996	8.609
6.39	40.83	260.9	2.528	7.994	1.856	3.998	8.613
6.40	40.96	262.1	2.530	8.000	1.857	4.000	8.618

n	n^2	n^3	$\sqrt{n}$	$\sqrt{10n}$	$\sqrt[3]{n}$	$\sqrt[3]{10n}$	$\sqrt[3]{100n}$
6.40	40.96	262.1	2.530	8.000	1.857	4.000	8.618
6.41	41.09	263.4	2.532	8.006	1.858	4.002	8.622
6.42	41.22	264.6	2.534	8.012	1.859	4.004	8.627
6.43	41.34	265.8	2.536	8.019	1.860	4.006	8.631
6.44	41.47	267.1	2.538	8.025	1.860	4.008	8.636
6.45	41.60	268.3	2.540	8.031	1.861	4.010	8.640
6.46	41.73	269.6	2.542	8.037	1.862	4.012	8.645
6.47	41.86	270.8	2.544	8.044	1.863	4.015	8.649
6.48	41.99	272.1	2.546	8.050	1.864	4.017	8.653
6.49	42.12	273.4	2.548	8.056	1.865	4.019	8.658
6.50	42.25	274.6	2.550	8.062	1.866	4.021	8.662
6.51	42.38	275.9	2.551	8.068	1.867	4.023	8.667
6.52	42.51	277.2	2.553	8.075	1.868	4.025	8.671
6.53	42.64	278.4	2.555	8.081	1.869	4.027	8.676
6.54	42.77	279.7	2.557	8.087	1.870	4.029	8.680
6.55	42.90	281.0	2.559	8.093	1.871	4.031	8.685
6.56	43.03	282.3	2.561	8.099	1.872	4.033	8.689
6.57	43.16	283.6	2.563	8.106	1.873	4.035	8.693
6.58	43.30	284.9	2.565	8.112	1.874	4.037	8.698
6.59	43.43	286.2	2.567	8.118	1.875	4.039	8.702
6.60	43.56	287.5	2.569	8.124	1.876	4.041	8.707
6.61	43.69	288.8	2.571	8.130	1.877	4.043	8.711
6.62	43.82	290.1	2.573	8.136	1.878	4.045	8.715
6.63	43.96	291.4	2.575	8.142	1.879	4.047	8.720
6.64	44.09	292.8	2.577	8.149	1.880	4.049	8.724
6.65	44.22	294.1	2.579	8.155	1.881	4.051	8.729
6.66	44.36	295.4	2.581	8.161	1.881	4.053	8.733
6.67	44.49	296.7	2.583	8.167	1.882	4.055	8.737
6.68	44.62	298.1	2.585	8.173	1.883	4.058	8.742
6.69	44.76	299.4	2.587	8.179	1.884	4.060	8.746
6.70	44.89	300.8	2.588	8.185	1.885	4.062	8.750
6.71	45.02	302.1	2.590	8.191	1.886	4.064	8.755
6.72	45.16	303.5	2.592	8.198	1.887	4.066	8.759
6.73	45.29	304.8	2.594	8.204	1.888	4.068	8.763
6.74	45.43	306.2	2.596	8.210	1.889	4.070	8.768
6.75	45.56	307.5	2.598	8.216	1.890	4.072	8.772
6.76	45.70	308.9	2.600	8.222	1.891	4.074	8.776
6.77	45.83	310.3	2.602	8.228	1.892	4.076	8.781
6.78	45.97	311.7	2.604	8.234	1.893	4.078	8.785
6.79	46.10	313.0	2.606	8.240	1.894	4.080	8.789
6.80	46.24	314.4	2.608	8.246	1.895	4.082	8.794

n	n^2	n^3	$\sqrt{n}$	$\sqrt{10n}$	$\sqrt[3]{n}$	$\sqrt[3]{10n}$	$\sqrt[3]{100n}$
6.80	46.24	314.4	2.608	8.246	1.895	4.082	8.794
6.81	46.38	315.8	2.610	8.252	1.895	4.084	8.798
6.82	46.51	317.2	2.612	8.258	1.896	4.086	8.802
6.83	46.65	318.6	2.613	8.264	1.897	4.088	8.807
6.84	46.79	320.0	2.615	8.270	1.898	4.090	8.811
6.85	46.92	321.4	2.617	8.276	1.899	4.092	8.815
6.86	47.06	322.8	2.619	8.283	1.900	4.094	8.819
6.87	47.20	324.2	2.621	8.289	1.901	4.096	8.824
6.88	47.33	325.7	2.623	8.295	1.902	4.098	8.828
6.89	47.47	327.1	2.625	8.301	1.903	4.100	8.832
6.90	47.61	328.5	2.627	8.307	1.904	4.102	8.837
6.91	47.75	329.9	2.629	8.313	1.905	4.104	8.841
6.92	47.89	331.4	2.631	8.319	1.906	4.106	8.845
6.93	48.02	332.8	2.632	8.325	1.907	4.108	8.849
6.94	48.16	334.3	2.634	8.331	1.907	4.109	8.854
6.95	48.30	335.7	2.636	8.337	1.908	4.111	8.858
6.96	48.44	337.2	2.638	8.343	1.909	4.113	8.862
6.97	48.58	338.6	2.640	8.349	1.910	4.115	8.866
6.98	48.72	340.1	2.642	8.355	1.911	4.117	8.871
6.99	48.86	341.5	2.644	8.361	1.912	4.119	8.875
7.00	49.00	343.0	2.646	8.367	1.913	4.121	8.879
7.01	49.14	344.5	2.648	8.373	1.914	4.123	8.883
7.02	49.28	345.9	2.650	8.379	1.915	4.125	8.887
7.03	49.42	347.4	2.651	8.385	1.916	4.127	8.892
7.04	49.56	348.9	2.653	8.390	1.917	4.129	8.896
7.05	49.70	350.4	2.655	8.396	1.917	4.131	8.900
7.06	49.84	351.9	2.657	8.402	1.918	4.133	8.904
7.07	49.98	353.4	2.659	8.408	1.919	4.135	8.909
7.08	50.13	354.9	2.661	8.414	1.920	4.137	8.913
7.09	50.27	356.4	2.663	8.420	1.921	4.139	8.917
7.10	50.41	357.9	2.665	8.426	1.922	4.141	8.921
7.11	50.55	359.4	2.666	8.432	1.923	4.143	8.925
7.12	50.69	360.9	2.668	8.438	1.924	4.145	8.929
7.13	50.84	362.5	2.670	8.444	1.925	4.147	8.934
7.14	50.98	364.0	2.672	8.450	1.926	4.149	8.938
7.15	51.12	365.5	2.674	8.456	1.926	4.151	8.942
7.16	51.27	367.1	2.676	8.462	1.927	4.152	8.946
7.17	51.41	368.6	2.678	8.468	1.928	4.154	8.950
7.18	51.55	370.1	2.680	8.473	1.929	4.156	8.955
7.19	51.70	371.7	2.681	8.479	1.930	4.158	8.959
7.20	51.84	373.2	2.683	8.485	1.931	4.160	8.963

n	n^2	n^3	$\sqrt{n}$	$\sqrt{10n}$	$\sqrt[3]{n}$	$\sqrt[3]{10n}$	$\sqrt[3]{100n}$
7.20	51.84	373.2	2.683	8.485	1.931	4.160	8.963
7.21	51.98	374.8	2.685	8.491	1.932	4.162	8.967
7.22	52.13	376.4	2.687	8.497	1.933	4.164	8.971
7.23	52.27	377.9	2.689	8.503	1.934	4.166	8.975
7.24	52.42	379.5	2.691	8.509	1.935	4.168	8.979
7.25	52.56	381.1	2.693	8.515	1.935	4.170	8.984
7.26	52.71	382.7	2.694	8.521	1.936	4.172	8.988
7.27	52.85	384.2	2.696	8.526	1.937	4.174	8.992
7.28	53.00	385.8	2.698	8.532	1.938	4.176	8.996
7.29	53.14	387.4	2.700	8.538	1.939	4.177	9.000
7.30	53.29	389.0	2.702	8.544	1.940	4.179	9.004
7.31	53.44	390.6	2.704	8.550	1.941	4.181	9.008
7.32	53.58	392.2	2.706	8.556	1.942	4.183	9.012
7.33	53.73	393.8	2.707	8.562	1.943	4.185	9.016
7.34	53.88	395.4	2.709	8.567	1.943	4.187	9.021
7.35	54.02	397.1	2.711	8.573	1.944	4.189	9.025
7.36	54.17	398.7	2.713	8.579	1.945	4.191	9.029
7.37	54.32	400.3	2.715	8.585	1.946	4.193	9.033
7.38	54.46	401.9	2.717	8.591	1.947	4.195	9.037
7.39	54.61	403.6	2.718	8.597	1.948	4.196	9.041
7.40	54.76	405.2	2.720	8.602	1.949	4.198	9.045
7.41	54.91	406.9	2.722	8.608	1.950	4.200	9.049
7.42	55.06	408.5	2.724	8.614	1.950	4.202	9.053
7.43	55.20	410.2	2.726	8.620	1.951	4.204	9.057
7.44	55.35	411.8	2.728	8.626	1.952	4.206	9.061
7.45	55.50	413.5	2.729	8.631	1.953	4.208	9.065
7.46	55.65	415.2	2.731	8.637	1.954	4.210	9.069
7.47	55.80	416.8	2.733	8.643	1.955	4.212	9.073
7.48	55.95	418.5	2.735	8.649	1.956	4.213	9.078
7.49	56.10	420.2	2.737	8.654	1.957	4.215	9.082
7.50	56.25	421.9	2.739	8.660	1.957	4.217	9.086
7.51	56.40	423.6	2.740	8.666	1.958	4.219	9.090
7.52	56.55	425.3	2.742	8.672	1.959	4.221	9.094
7.53	56.70	427.0	2.744	8.678	1.960	4.223	9.098
7.54	56.85	428.7	2.746	8.683	1.961	4.225	9.102
7.55	57.00	430.4	2.748	8.689	1.962	4.227	9.106
7.56	57.15	432.1	2.750	8.695	1.963	4.228	9.110
7.57	57.30	433.8	2.751	8.701	1.964	4.230	9.114
7.58	57.46	435.5	2.753	8.706	1.964	4.232	9.118
7.59	57.61	437.2	2.755	8.712	1.965	4.234	9.122
7.60	57.76	439.0	2.757	8.718	1.966	4.236	9.126

n	n^2	n^3	$\sqrt{n}$	$\sqrt{10n}$	$\sqrt[3]{n}$	$\sqrt[3]{10n}$	$\sqrt[3]{100n}$
7.60	57.76	439.0	2.757	8.718	1.966	4.236	9.126
7.61	57.91	440.7	2.759	8.724	1.967	4.238	9.130
7.62	58.06	442.5	2.760	8.729	1.968	4.240	9.134
7.63	58.22	444.2	2.762	8.735	1.969	4.241	9.138
7.64	58.37	445.9	2.764	8.741	1.970	4.243	9.142
7.65	58.52	447.7	2.766	8.746	1.970	4.245	9.146
7.66	58.68	449.5	2.768	8.752	1.971	4.247	9.150
7.67	58.83	451.2	2.769	8.758	1.972	4.249	9.154
7.68	58.98	453.0	2.771	8.764	1.973	4.251	9.158
7.69	59.14	454.8	2.773	8.769	1.974	4.252	9.162
7.70	59.29	456.5	2.775	8.775	1.975	4.254	9.166
7.71	59.44	458.3	2.777	8.781	1.976	4.256	9.170
7.72	59.60	460.1	2.778	8.786	1.976	4.258	9.174
7.73	59.75	461.9	2.780	8.792	1.977	4.260	9.178
7.74	59.91	463.7	2.782	8.798	1.978	4.262	9.182
7.75	60.06	465.5	2.784	8.803	1.979	4.264	9.185
7.76	60.22	467.3	2.786	8.809	1.980	4.265	9.189
7.77	60.37	469.1	2.787	8.815	1.981	4.267	9.193
7.78	60.53	470.9	2.789	8.820	1.981	4.269	9.197
7.79	60.68	472.7	2.791	8.826	1.982	4.271	9.201
7.80	60.84	474.6	2.793	8.832	1.983	4.273	9.205
7.81	61.00	476.4	2.795	8.837	1.984	4.274	9.209
7.82	61.15	478.2	2.796	8.843	1.985	4.276	9.213
7.83	61.31	480.0	2.798	8.849	1.986	4.278	9.217
7.84	61.47	481.9	2.800	8.854	1.987	4.280	9.221
7.85	61.62	483.7	2.802	8.860	1.987	4.282	9.225
7.86	61.78	485.6	2.804	8.866	1.988	4.284	9.229
7.87	61.94	487.4	2.805	8.871	1.989	4.285	9.233
7.88	62.09	489.3	2.807	8.877	1.990	4.287	9.237
7.89	62.25	491.2	2.809	8.883	1.991	4.289	9.240
7.90	62.41	493.0	2.811	8.888	1.992	4.291	9.244
7.91	62.57	494.9	2.812	8.894	1.992	4.293	9.248
7.92	62.73	496.8	2.814	8.899	1.993	4.294	9.252
7.93	62.88	498.7	2.816	8.905	1.994	4.296	9.256
7.94	63.04	500.6	2.818	8.911	1.995	4.298	9.260
7.95	63.20	502.5	2.820	8.916	1.996	4.300	9.264
7.96	63.36	504.4	2.821	8.922	1.997	4.302	9.268
7.97	63.52	506.3	2.823	8.927	1.997	4.303	9.272
7.98	63.68	508.2	2.825	8.933	1.998	4.305	9.275
7.99	63.84	510.1	2.827	8.939	1.999	4.307	9.279
8.00	64.00	512.0	2.828	8.944	2.000	4.309	9.283

n	n^2	n^3	$\sqrt{n}$	$\sqrt{10n}$	$\sqrt[3]{n}$	$\sqrt[3]{10n}$	$\sqrt[3]{100n}$
8.00	64.00	512.0	2.828	8.944	2.000	4.309	9.283
8.01	64.16	513.9	2.830	8.950	2.001	4.311	9.287
8.02	64.32	515.8	2.832	8.955	2.002	4.312	9.291
8.03	64.48	517.8	2.834	8.961	2.002	4.314	9.295
8.04	64.64	519.7	2.835	8.967	2.003	4.316	9.299
8.05	64.80	521.7	2.837	8.972	2.004	4.318	9.302
8.06	64.96	523.6	2.839	8.978	2.005	4.320	9.306
8.07	65.12	525.6	2.841	8.983	2.006	4.321	9.310
8.08	65.29	527.5	2.843	8.989	2.007	4.323	9.314
8.09	65.45	529.5	2.844	8.994	2.007	4.325	9.318
8.10	65.61	531.4	2.846	9.000	2.008	4.327	9.322
8.11	65.77	533.4	2.848	9.006	2.009	4.329	9.326
8.12	65.93	535.4	2.850	9.011	2.010	4.330	9.329
8.13	66.10	537.4	2.851	9.017	2.011	4.332	9.333
8.14	66.26	539.4	2.853	9.022	2.012	4.334	9.337
8.15	66.42	541.3	2.855	9.028	2.012	4.336	9.341
8.16	66.59	543.3	2.857	9.033	2.013	4.337	9.345
8.17	66.75	545.3	2.858	9.039	2.014	4.339	9.348
8.18	66.91	547.3	2.860	9.044	2.015	4.341	9.352
8.19	67.08	549.4	2.862	9.050	2.016	4.343	9.356
8.20	67.24	551.4	2.864	9.055	2.017	4.344	9.360
8.21	67.40	553.4	2.865	9.061	2.017	4.346	9.364
8.22	67.57	555.4	2.867	9.066	2.018	4.348	9.368
8.23	67.73	557.4	2.869	9.072	2.019	4.350	9.371
8.24	67.90	559.5	2.871	9.077	2.020	4.352	9.375
8.25	68.06	561.5	2.872	9.083	2.021	4.353	9.379
8.26	68.23	563.6	2.874	9.088	2.021	4.355	9.383
8.27	68.39	565.6	2.876	9.094	2.022	4.357	9.386
8.28	68.56	567.7	2.877	9.099	2.023	4.359	9.390
8.29	68.72	569.7	2.879	9.105	2.024	4.360	9.394
8.30	68.89	571.8	2.881	9.110	2.025	4.362	9.398
8.31	69.06	573.9	2.883	9.116	2.026	4.364	9.402
8.32	69.22	575.9	2.884	9.121	2.026	4.366	9.405
8.33	69.39	578.0	2.886	9.127	2.027	4.367	9.409
8.34	69.56	580.1	2.888	9.132	2.028	4.369	9.413
8.35	69.72	582.2	2.890	9.138	2.029	4.371	9.417
8.36	69.89	584.3	2.891	9.143	2.030	4.373	9.420
8.37	70.06	586.4	2.893	9.149	2.030	4.374	9.424
8.38	70.22	588.5	2.895	9.154	2.031	4.376	9.428
8.39	70.39	590.6	2.897	9.160	2.032	4.378	9.432
8.40	70.56	592.7	2.898	9.165	2.033	4.380	9.435

n	n^2	n^3	$\sqrt{n}$	$\sqrt{10n}$	$\sqrt[3]{n}$	$\sqrt[3]{10n}$	$\sqrt[3]{100n}$
8.40	70.56	592.7	2.898	9.165	2.033	4.380	9.435
8.41	70.73	594.8	2.900	9.171	2.034	4.381	9.439
8.42	70.90	596.9	2.902	9.176	2.034	4.383	9.443
8.43	71.06	599.1	2.903	9.182	2.035	4.385	9.447
8.44	71.23	601.2	2.905	9.187	2.036	4.386	9.450
8.45	71.40	603.4	2.907	9.192	2.037	4.388	9.454
8.46	71.57	605.5	2.909	9.198	2.038	4.390	9.458
8.47	71.74	607.6	2.910	9.203	2.038	4.392	9.462
8.48	71.91	609.8	2.912	9.209	2.039	4.393	9.465
8.49	72.08	612.0	2.914	9.214	2.040	4.395	9.469
8.50	72.25	614.1	2.915	9.220	2.041	4.397	9.473
8.51	72.42	616.3	2.917	9.225	2.042	4.399	9.476
8.52	72.59	618.5	2.919	9.230	2.042	4.400	9.480
8.53	72.76	620.7	2.921	9.236	2.043	4.402	9.484
8.54	72.93	622.8	2.922	9.241	2.044	4.404	9.488
8.55	73.10	625.0	2.924	9.247	2.045	4.405	9.491
8.56	73.27	627.2	2.926	9.252	2.046	4.407	9.495
8.57	73.44	629.4	2.927	9.257	2.046	4.409	9.499
8.58	73.62	631.6	2.929	9.263	2.047	4.411	9.502
8.59	73.79	633.8	2.931	9.268	2.048	4.412	9.506
8.60	73.96	636.1	2.933	9.274	2.049	4.414	9.510
8.61	74.13	638.3	2.934	9.279	2.050	4.416	9.513
8.62	74.30	640.5	2.936	9.284	2.050	4.417	9.517
8.63	74.48	642.7	2.938	9.290	2.051	4.419	9.521
8.64	74.65	645.0	2.939	9.295	2.052	4.421	9.524
8.65	74.82	647.2	2.941	9.301	2.053	4.423	9.528
8.66	75.00	649.5	2.943	9.306	2.054	4.424	9.532
8.67	75.17	651.7	2.944	9.311	2.054	4.426	9.535
8.68	75.34	654.0	2.946	9.317	2.055	4.428	9.539
8.69	75.52	656.2	2.948	9.322	2.056	4.429	9.543
8.70	75.69	658.5	2.950	9.327	2.057	4.431	9.546
8.71	75.86	660.8	2.951	9.333	2.057	4.433	9.550
8.72	76.04	663.1	2.953	9.338	2.058	4.434	9.554
8.73	76.21	665.3	2.955	9.343	2.059	4.436	9.557
8.74	76.39	667.6	2.956	9.349	2.060	4.438	9.561
8.75	76.56	669.9	2.958	9.354	2.061	4.440	9.565
8.76	76.74	672.2	2.960	9.359	2.061	4.441	9.568
8.77	76.91	674.5	2.961	9.365	2.062	4.443	9.572
8.78	77.09	676.8	2.963	9.370	2.063	4.445	9.576
8.79	77.26	679.2	2.965	9.375	2.064	4.446	9.579
8.80	77.44	681.5	2.966	9.381	2.065	4.448	9.583

n	n^2	n^3	$\sqrt{n}$	$\sqrt{10n}$	$\sqrt[3]{n}$	$\sqrt[3]{10n}$	$\sqrt[3]{100n}$
8.80	77.44	681.5	2.966	9.381	2.065	4.448	9.583
8.81	77.62	683.8	2.968	9.386	2.065	4.450	9.586
8.82	77.79	686.1	2.970	9.391	2.066	4.451	9.590
8.83	77.97	688.5	2.972	9.397	2.067	4.453	9.594
8.84	78.15	690.8	2.973	9.402	2.068	4.455	9.597
8.85	78.32	693.2	2.975	9.407	2.068	4.456	9.601
8.86	78.50	695.5	2.977	9.413	2.069	4.458	9.605
8.87	78.68	697.9	2.978	9.418	2.070	4.460	9.608
8.88	78.85	700.2	2.980	9.423	2.071	4.461	9.612
8.89	79.03	702.6	2.982	9.429	2.072	4.463	9.615
8.90	79.21	705.0	2.983	9.434	2.072	4.465	9.619
8.91	79.39	707.3	2.985	9.439	2.073	4.466	9.623
8.92	79.57	709.7	2.987	9.445	2.074	4.468	9.626
8.93	79.74	712.1	2.988	9.450	2.075	4.470	9.630
8.94	79.92	714.5	2.990	9.455	2.075	4.471	9.633
8.95	80.10	716.9	2.992	9.460	2.076	4.473	9.637
8.96	80.28	719.3	2.993	9.466	2.077	4.475	9.641
8.97	80.46	721.7	2.995	9.471	2.078	4.476	9.644
8.98	80.64	724.2	2.997	9.476	2.079	4.478	9.648
8.99	80.82	726.6	2.998	9.482	2.079	4.480	9.651
9.00	81.00	729.0	3.000	9.487	2.080	4.481	9.655
9.01	81.18	731.4	3.002	9.492	2.081	4.483	9.658
9.02	81.36	733.9	3.003	9.497	2.082	4.485	9.662
9.03	81.54	736.3	3.005	9.503	2.082	4.486	9.666
9.04	81.72	738.8	3.007	9.508	2.083	4.488	9.669
9.05	81.90	741.2	3.008	9.513	2.084	4.490	9.673
9.06	82.08	743.7	3.010	9.518	2.085	4.491	9.676
9.07	82.26	746.1	3.012	9.524	2.085	4.493	9.680
9.08	82.45	748.6	3.013	9.529	2.086	4.495	9.683
9.09	82.63	751.1	3.015	9.534	2.087	4.496	9.687
9.10	82.81	753.6	3.017	9.539	2.088	4.498	9.691
9.11	82.99	756.1	3.018	9.545	2.089	4.500	9.694
9.12	83.17	758.6	3.020	9.550	2.089	4.501	9.698
9.13	83.36	761.0	3.022	9.555	2.090	4.503	9.701
9.14	83.54	763.6	3.023	9.560	2.091	4.505	9.705
9.15	83.72	766.1	3.025	9.566	2.092	4.506	9.708
9.16	83.91	768.6	3.027	9.571	2.092	4.508	9.712
9.17	84.09	771.1	3.028	9.576	2.093	4.509	9.715
9.18	84.27	773.6	3.030	9.581	2.094	4.511	9.719
9.19	84.46	776.2	3.032	9.586	2.095	4.513	9.722
9.20	84.64	778.7	3.033	9.592	2.095	4.514	9.726

n	n^2	n^3	$\sqrt{n}$	$\sqrt{10n}$	$\sqrt[3]{n}$	$\sqrt[3]{10n}$	$\sqrt[3]{100n}$
9.20	84.64	778.7	3.033	9.592	2.095	4.514	9.726
9.21	84.82	781.2	3.035	9.597	2.096	4.516	9.729
9.22	85.01	783.8	3.036	9.602	2.097	4.518	9.733
9.23	85.19	786.3	3.038	9.607	2.098	4.519	9.736
9.24	85.38	788.9	3.040	9.612	2.098	4.521	9.740
9.25	85.56	791.5	3.041	9.618	2.099	4.523	9.743
9.26	85.75	794.0	3.043	9.623	2.100	4.524	9.747
9.27	85.93	796.6	3.045	9.628	2.101	4.526	9.750
9.28	86.12	799.2	3.046	9.633	2.101	4.527	9.754
9.29	86.30	801.8	3.048	9.638	2.102	4.529	9.758
9.30	86.49	804.4	3.050	9.644	2.103	4.531	9.761
9.31	86.68	807.0	3.051	9.649	2.104	4.532	9.764
9.32	86.86	809.6	3.053	9.654	2.104	4.534	9.768
9.33	87.05	812.2	3.055	9.659	2.105	4.536	9.771
9.34	87.24	814.8	3.056	9.664	2.106	4.537	9.775
9.35	87.42	817.4	3.058	9.670	2.107	4.539	9.778
9.36	87.61	820.0	3.059	9.675	2.107	4.540	9.782
9.37	87.80	822.7	3.061	9.680	2.108	4.542	9.785
9.38	87.98	825.3	3.063	9.685	2.109	4.544	9.789
9.39	88.17	827.9	3.064	9.690	2.110	4.545	9.792
9.40	88.36	830.6	3.066	9.695	2.110	4.547	9.796
9.41	88.55	833.2	3.068	9.701	2.111	4.548	9.799
9.42	88.74	835.9	3.069	9.706	2.112	4.550	9.803
9.43	88.92	838.6	3.071	9.711	2.113	4.552	9.806
9.44	89.11	841.2	3.072	9.716	2.113	4.553	9.810
9.45	89.30	843.9	3.074	9.721	2.114	4.555	9.813
9.46	89.49	846.6	3.076	9.726	2.115	4.556	9.817
9.47	89.68	849.3	3.077	9.731	2.116	4.558	9.820
9.48	89.87	852.0	3.079	9.737	2.116	4.560	9.824
9.49	90.06	854.7	3.081	9.742	2.117	4.561	9.827
9.50	90.25	857.4	3.082	9.747	2.118	4.563	9.830
9.51	90.44	860.1	3.084	9.752	2.119	4.565	9.834
9.52	90.63	862.8	3.085	9.757	2.119	4.566	9.837
9.53	90.82	865.5	3.087	9.762	2.120	4.568	9.841
9.54	91.01	868.3	3.089	9.767	2.121	4.569	9.844
9.55	91.20	871.0	3.090	9.772	2.122	4.571	9.848
9.56	91.39	873.7	3.092	9.778	2.122	4.572	9.851
9.57	91.58	876.5	3.094	9.783	2.123	4.574	9.855
9.58	91.78	879.2	3.095	9.788	2.124	4.576	9.858
9.59	91.97	882.0	3.097	9.793	2.125	4.577	9.861
9.60	92.16	884.7	3.098	9.798	2.125	4.579	9.865

n	n^2	n^3	$\sqrt{n}$	$\sqrt{10n}$	$\sqrt[3]{n}$	$\sqrt[3]{10n}$	$\sqrt[3]{100n}$
9.60	92.16	884.7	3.098	9.798	2.125	4.579	9.865
9.61	92.35	887.5	3.100	9.803	2.126	4.580	9.868
9.62	92.54	890.3	3.102	9.808	2.127	4.582	9.872
9.63	92.74	893.1	3.103	9.813	2.128	4.584	9.875
9.64	92.93	895.8	3.105	9.818	2.128	4.585	9.879
9.65	93.12	898.6	3.106	9.823	2.129	4.587	9.882
9.66	93.32	901.4	3.108	9.829	2.130	4.588	9.885
9.67	93.51	904.2	3.110	9.834	2.130	4.590	9.889
9.68	93.70	907.0	3.111	9.839	2.131	4.592	9.892
9.69	93.90	909.9	3.113	9.844	2.132	4.593	9.896
9.70	94.09	912.7	3.114	9.849	2.133	4.595	9.899
9.71	94.28	915.5	3.116	9.854	2.133	4.596	9.902
9.72	94.48	918.3	3.118	9.859	2.134	4.598	9.906
9.73	94.67	921.2	3.119	9.864	2.135	4.599	9.909
9.74	94.87	924.0	3.121	9.869	2.136	4.601	9.913
9.75	95.06	926.9	3.122	9.874	2.136	4.603	9.916
9.76	95.26	929.7	3.124	9.879	2.137	4.604	9.919
9.77	95.45	932.6	3.126	9.884	2.138	4.606	9.923
9.78	95.65	935.4	3.127	9.889	2.139	4.607	9.926
9.79	95.84	938.3	3.129	9.894	2.139	4.609	9.930
9.80	96.04	941.2	3.130	9.899	2.140	4.610	9.933
9.81	96.24	944.1	3.132	9.905	2.141	4.612	9.936
9.82	96.43	947.0	3.134	9.910	2.141	4.614	9.940
9.83	96.63	949.9	3.135	9.915	2.142	4.615	9.943
9.84	96.83	952.8	3.137	9.920	2.143	4.617	9.946
9.85	97.02	955.7	3.138	9.925	2.144	4.618	9.950
9.86	97.22	958.6	3.140	9.930	2.144	4.620	9.953
9.87	97.42	961.5	3.142	9.935	2.145	4.621	9.956
9.88	97.61	964.4	3.143	9.940	2.146	4.623	9.960
9.89	97.81	967.4	3.145	9.945	2.147	4.625	9.963
9.90	98.01	970.3	3.146	9.950	2.147	4.626	9.967
9.91	98.21	973.2	3.148	9.955	2.148	4.628	9.970
9.92	98.41	976.2	3.150	9.960	2.149	4.629	9.973
9.93	98.60	979.1	3.151	9.965	2.149	4.631	9.977
9.94	98.80	982.1	3.153	9.970	2.150	4.632	9.980
9.95	99.00	985.1	3.154	9.975	2.151	4.634	9.983
9.96	99.20	988.0	3.156	9.980	2.152	4.635	9.987
9.97	99.40	991.0	3.158	9.985	2.152	4.637	9.990
9.98	99.60	994.0	3.159	9.990	2.153	4.638	9.993
9.99	99.80	997.0	3.161	9.995	2.154	4.640	9.997
10.00	100.00	1000.0	3.162	10.000	2.154	4.642	10.000

3. Powers of integers from $n=1$ to $n=100$

n	n^2	n^3	n^4	n^5
1	1	1	1	1
2	4	8	16	32
3	9	27	81	243
4	16	64	256	1 024
5	25	125	625	3 125
6	36	216	1 296	7 776
7	49	343	2 401	16 807
8	64	512	4 096	32 768
9	81	729	6 561	59 049
10	100	1 000	10 000	100 000
11	121	1 331	14 641	161 051
12	144	1 729	20 736	248 832
13	169	2 197	28 561	371 293
14	196	2 744	38 416	537 824
15	225	3 375	50 625	759 375
16	256	4 096	65 536	1 048 576
17	289	4 913	83 521	1 419 857
18	324	5 832	104 976	1 889 568
19	361	6 859	130 321	2 476 099
20	400	8 000	160 000	3 200 000
21	441	9 261	194 481	4 084 101
22	484	10 648	234 256	5 153 632
23	529	12 167	279 841	6 436 343
24	576	13 824	331 776	7 962 624
25	625	15 625	390 625	9 765 625
26	676	17 576	456 976	11 881 376
27	729	19 683	531 441	14 348 907
28	784	21 952	614 656	17 210 368
29	841	24 389	707 281	20 511 149
30	900	27 000	810 000	24 300 000
31	961	29 791	923 521	28 629 151
32	1 024	32 768	1 048 576	33 554 432
33	1 089	35 937	1 185 921	39 135 393
34	1 156	39 304	1 336 336	45 435 424
35	1 225	42 875	1 500 625	52 521 875
36	1 296	46 656	1 679 616	60 466 176
37	1 369	50 653	1 874 161	69 343 957
38	1 444	54 872	2 085 136	79 235 168
39	1 521	59 319	2 313 441	90 224 199

n	n^2	n^3	n^4	n^5
40	1 600	64 000	2 560 000	102 400 000
41	1 681	68 921	2 825 761	115 856 201
42	1 764	74 088	3 111 696	130 691 232
43	1 849	79 507	3 418 801	147 008 443
44	1 936	85 184	3 748 096	164 916 224
45	2 025	91 125	4 100 625	184 528 125
46	2 116	97 336	4 477 456	205 962 976
47	2 209	103 823	4 879 681	229 345 007
48	2 304	110 592	5 308 416	254 803 968
49	2 401	117 649	5 764 801	282 475 249
50	2 500	125 000	6 250 000	312 500 000
51	2 601	132 651	6 765 201	345 025 251
52	2 704	140 608	7 311 616	380 204 032
53	2 809	148 877	7 890 481	418 195 493
54	2 916	157 464	8 503 056	459 165 024
55	3 025	166 375	9 150 625	503 284 375
56	3 136	175 616	9 834 496	550 731 776
57	3 249	185 193	10 556 001	601 692 057
58	3 364	195 112	11 316 496	656 356 768
59	3 481	205 379	12 117 361	714 924 299
60	3 600	216 000	12 960 000	777 600 000
61	3 721	226 981	13 845 841	844 596 301
62	3 844	238 328	14 776 336	916 132 832
63	3 969	250 047	15 752 961	992 436 543
64	4 096	262 144	16 777 216	1 073 741 824
65	4 225	274 625	17 850 635	1 160 290 625
66	4 356	287 496	18 974 736	1 252 332 576
67	4 489	300 763	20 151 121	1 350 125 107
68	4 624	314 432	21 381 376	1 453 933 568
69	4 761	328 509	22 667 121	1 564 031 349
70	4,900	343 000	24 010 000	1 680 700 000
71	5 041	357 911	25 411 681	1 804 229 351
72	5 184	373 248	26 873 856	1 934 917 632
73	5 329	389 017	28 398 241	2 073 071 593
74	5 476	405 224	29 986 576	2 219 006 624
75	5 625	421 875	31 640 625	2 373 046 875
76	5 776	438 976	33 362 176	2 535 525 376
77	5 929	456 533	35 153 041	2 706 784 157
78	6 084	474 552	37 015 056	2 887 174 368
79	6 241	493 039	38 950 081	3 077 056 399

n	n^2	n^3	n^4	n^5
80	6 400	512 000	40 960 000	3 276 800 000
81	6 561	531 441	43 046 721	3 486 784 401
82	6 724	551 368	45 212 176	3 707 398 432
83	6 889	571 787	47 458 321	3 939 040 643
84	7 056	592 704	49 787 136	4 182 119 424
85	7 225	614 125	52 200 625	4 437 053 125
86	7 396	636 056	54 700 816	4 704 270 176
87	7 569	658 503	57 289 761	4 984 209 207
88	7 744	681 472	59 969 536	5 277 319 168
89	7 921	704 969	62 742 241	5 584 059 449
90	8 100	729 000	65 610 000	5 904 900 000
91	8 281	753 571	68 574 961	6 240 321 451
92	8 464	778 688	71 639 296	6 590 815 232
93	8 649	804 357	74 805 201	6 956 883 693
94	8 836	830 584	78 074 896	7 339 040 224
95	9 025	857 375	81 450 625	7 737 809 375
96	9 216	884 736	84 934 656	8 153 726 976
97	9 409	912 673	88 529 281	8 587 340 257
98	9 604	941 192	92 236 816	9 039 207 968
99	9 801	970 299	96 059 601	9 509 900 499
100	10 000	1 000 000	100 000 000	10 000 000 000

4. Reciprocals of numbers

This table give the values of $10000 : n$, correct to four significant figures, for values of n from 1 to 10, given to three significant figures. Each number in the table lies in the row marked by the first two figures of the argument (the column of n) and in the column corresponding to the third figure of the argument. For example, $10000 : 2.26 = 4425$. If the argument is given to four significant figures, linear interpolation should be applied (see p. 17). It should be remembered that the interpolation corrections are s u b t r a c t e d, not added.

The numbers in the table can be regarded as the right-hand figures in the decimal fraction of $1 : n$; for example, $1 : 2.26 = 0.4425$. To find the value of $1 : n$ for $n > 10$ or $n < 1$, it should be noted that, by multiplying n by 10^k, the reciprocal $1 : n$ is multiplied by 10^{-k}, i.e., moving the decimal point k places in either direction in the number n requires moving it the same number of places in the opposite direction in the reciprocal $1 : n$. For example, $1 : 22.6 = 0.04425$, and $1 : 0.0226 = 44.25$.

n	0	1	2	3	4	5	6	7	8	9
1.0	10000	9901	9804	9709	9615	9524	9434	9346	9259	9174
1.1	9091	9009	8929	8850	8772	8696	8621	8547	8475	8403
1.2	8333	8264	8197	8130	8065	8000	7937	7874	7812	7752
1.3	7692	7634	7576	7519	7463	7407	7353	7299	7246	7194
1.4	7143	7092	7042	6993	6944	6897	6849	6803	6757	6711
1.5	6667	6623	6579	6536	6494	6452	6410	6369	6329	6289
1.6	6250	6211	6173	6135	6098	6061	6024	5988	5952	5917
1.7	5882	5848	5814	5780	5747	5714	5682	5650	5618	5587
1.8	5556	5525	5495	5464	5435	5405	5376	5348	5319	5291
1.9	5263	5236	5208	5181	5155	5128	5102	5076	5051	5025
2.0	5000	4975	4950	4926	4902	4878	4854	4831	4808	4785
2.1	4762	4739	4717	4695	4673	4651	4630	4608	4587	4566
2.2	4545	4525	4505	4484	4464	4444	4425	4405	4386	4367
2.3	4348	4329	4310	4292	4274	4255	4237	4219	4202	4184
2.4	4167	4149	4132	4115	4098	4082	4065	4049	4032	4016
2.5	4000	3984	3968	3953	3937	3922	3906	3891	3876	3861
2.6	3846	3831	3817	3802	3788	3774	3759	3745	3731	3717
2.7	3704	3690	3676	3663	3650	3636	3623	3610	3597	3584
2.8	3571	3559	3546	3534	3521	3509	3497	3484	3472	3460
2.9	3448	3436	3425	3413	3401	3390	3378	3367	3356	3344
3.0	3333	3322	3311	3300	3289	3279	3268	3257	3247	3236
3.1	3226	3215	3205	3195	3185	3175	3165	3155	3145	3135
3.2	3125	3115	3106	3096	3086	3077	3067	3058	3049	3040
3.3	3030	3021	3012	3003	2994	2985	2976	2967	2959	2950
3.4	2941	2933	2924	2915	2907	2899	2890	2882	2874	2865
3.5	2857	2849	2841	2833	2825	2817	2809	2801	2793	2786
3.6	2778	2770	2762	2755	2747	2740	2732	2725	2717	2710
3.7	2703	2695	2688	2681	2674	2667	2660	2653	2646	2639
3.8	2632	2625	2618	2611	2604	2597	2591	2584	2577	2571
3.9	2564	2558	2551	2545	2538	2532	2525	2519	2513	2506
4.0	2500	2494	2488	2481	2475	2469	2463	2457	2451	2445
4.1	2439	2433	2427	2421	2415	2410	2404	2398	2392	2387
4.2	2381	2375	2370	2364	2358	2353	2347	2342	2336	2331
4.3	2326	2320	2315	2309	2304	2299	2294	2288	2283	2278
4.4	2273	2268	2262	2257	2252	2247	2242	2237	2232	2227
4.5	2222	2217	2212	2208	2203	2198	2193	2188	2183	2179
4.6	2174	2169	2165	2160	2155	2151	2146	2141	2137	2132
4.7	2128	2123	2119	2114	2110	2105	2101	2096	2092	2088
4.8	2083	2079	2075	2070	2066	2062	2058	2053	2049	2045
4.9	2041	2037	2033	2028	2024	2020	2016	2012	2008	2004

n	0	1	2	3	4	5	6	7	8	9
5.0	2000	1996	1992	1988	1984	1980	1976	1972	1969	1965
5.1	1961	1957	1953	1949	1946	1942	1938	1934	1931	1927
5.2	1923	1919	1916	1912	1908	1905	1901	1898	1894	1890
5.3	1887	1883	1880	1876	1873	1869	1866	1862	1859	1855
5.4	1852	1848	1845	1842	1838	1835	1832	1828	1825	1821
5.5	1818	1815	1812	1808	1805	1802	1799	1795	1792	1789
5.6	1786	1783	1779	1776	1773	1770	1767	1764	1761	1757
5.7	1754	1751	1748	1745	1742	1739	1736	1733	1730	1727
5.8	1724	1721	1718	1715	1712	1709	1706	1704	1701	1698
5.9	1695	1692	1689	1686	1684	1681	1678	1675	1672	1669
6.0	1667	1664	1661	1658	1656	1653	1650	1647	1645	1642
6.1	1639	1637	1634	1631	1629	1626	1623	1621	1618	1616
6.2	1613	1610	1608	1605	1603	1600	1597	1595	1592	1590
6.3	1587	1585	1582	1580	1577	1575	1572	1570	1567	1565
6.4	1562	1560	1558	1555	1553	1550	1548	1546	1543	1541
6.5	1538	1536	1534	1531	1529	1527	1524	1522	1520	1517
6.6	1515	1513	1511	1508	1506	1504	1502	1499	1497	1495
6.7	1493	1490	1488	1486	1484	1481	1479	1477	1475	1473
6.8	1471	1468	1466	1464	1462	1460	1458	1456	1453	1451
6.9	1449	1447	1445	1443	1441	1439	1437	1435	1433	1431
7.0	1429	1427	1425	1422	1420	1418	1416	1414	1412	1410
7.1	1408	1406	1404	1403	1401	1399	1397	1395	1393	1391
7.2	1389	1387	1385	1383	1381	1379	1377	1376	1374	1372
7.3	1370	1368	1366	1364	1362	1361	1359	1357	1355	1353
7.4	1351	1350	1348	1346	1344	1342	1340	1339	1337	1335
7.5	1333	1332	1330	1328	1326	1325	1323	1321	1319	1318
7.6	1316	1314	1312	1311	1309	1307	1305	1304	1302	1300
7.7	1299	1297	1295	1294	1292	1290	1289	1287	1285	1284
7.8	1282	1280	1279	1277	1276	1274	1272	1271	1269	1267
7.9	1266	1264	1253	1261	1259	1258	1256	1255	1253	1252
8.0	1250	1248	1247	1245	1244	1242	1241	1239	1238	1236
8.1	1235	1233	1232	1230	1229	1227	1225	1224	1222	1221
8.2	1220	1218	1217	1215	1214	1212	1211	1209	1208	1206
8.3	1205	1203	1202	1200	1199	1198	1196	1195	1193	1192
8.4	1190	1189	1188	1186	1185	1183	1182	1181	1179	1178
8.5	1176	1175	1174	1172	1171	1170	1168	1167	1166	1164
8.6	1163	1161	1160	1159	1157	1156	1155	1153	1152	1151
8.7	1149	1148	1147	1145	1144	1143	1142	1140	1139	1138
8.8	1136	1135	1134	1133	1131	1130	1129	1127	1126	1125
8.9	1124	1122	1121	1120	1119	1117	1116	1115	1114	1112

n	0	1	2	3	4	5	6	7	8	9
9.0	1111	1110	1109	1107	1106	1105	1104	1103	1101	1100
9.1	1099	1098	1096	1095	1094	1093	1092	1091	1089	1088
9.2	1087	1086	1085	1083	1082	1081	1080	1079	1078	1076
9.3	1075	1074	1073	1072	1071	1070	1068	1067	1066	1065
9.4	1064	1063	1062	1060	1059	1058	1057	1056	1055	1054
9.5	1053	1052	1050	1049	1048	1047	1046	1045	1044	1043
9.6	1042	1041	1040	1038	1037	1036	1035	1034	1033	1032
9.7	1031	1030	1029	1028	1027	1026	1025	1024	1022	1021
9.8	1020	1019	1018	1017	1016	1015	1014	1013	1012	1011
9.9	1010	1009	1008	1007	1006	1005	1004	1003	1002	1001

5. Factorials and their reciprocals

Factorials

n	$n!$	n	$n!$
1	1	11	39 916 800
2	2	12	479 001 600
3	6	13	6 227 020 800
4	24	14	87 178 291 200
5	120	15	1 307 674 368 000
6	720	16	20 922 789 888 000
7	5 040	17	355 687 428 096 000
8	40 320	18	6 402 373 705 728 000
9	362 880	19	121 645 100 408 832 000
10	3 628 800	20	2 432 902 008 176 640 000

Reciprocals of factorials (1)

n	$\frac{1}{n!}$	n	$\frac{1}{n!}$	n	$\frac{1}{n!}$
1	1.000000	11	$0.0^{7}25052$	21	$0.0^{19}19573$
2	0.500000	12	$0.0^{8}20877$	22	$0.0^{21}88968$
3	0.166667	13	$0.0^{9}16059$	23	$0.0^{22}38682$
4	0.041667	14	$0.0^{10}11471$	24	$0.0^{23}16117$
5	$0.0^{2}83333$	15	$0.0^{12}76472$	25	$0.0^{25}64470$
6	$0.0^{2}13889$	16	$0.0^{13}47795$	26	$0.0^{26}24796$
7	$0.0^{3}19841$	17	$0.0^{14}28115$	27	$0.0^{28}91837$
8	$0.0^{4}24802$	18	$0.0^{15}15619$	28	$0.0^{29}32799$
9	$0.0^{5}27557$	19	$0.0^{17}82206$	29	$0.0^{30}11310$
10	$0.0^{6}27557$	20	$0.0^{18}41103$	30	$0.0^{32}37700$

(1) For $1:n!$, an abbreviation for right-hand zeros is used; thus $1:8! = 0.000024802 = 0.0^{4}24802$.

6. Some powers of the numbers 2, 3 and 5

n	2^n	3^n	5^n
1	2	3	5
2	4	9	25
3	8	27	125
4	16	81	625
5	32	243	3 125
6	64	729	15 625
7	128	2 187	78 125
8	256	6 561	390 625
9	512	19 683	1 953 125
10	1 024	59 049	9 765 625
11	2 048	177 147	48 828 125
12	4 096	531 441	244 140 625
13	8 192	1 594 323	1 220 703 125
14	16 384	4 782 969	6 103 515 625
15	32 768	14 348 907	30 517 578 125
16	65 536	43 046 721	152 587 890 625
17	131 072	129 140 163	762 939 453 125
18	262 144	387 420 489	3 814 697 265 625
19	524 288	1 162 261 467	19 073 486 328 125
20	1 048 576	3 486 784 401	95 367 431 640 625

7. Common logarithms

This table is used to find the common logarithms of numbers. For a given number, we first determine the characteristic of its logarithm according to the rules on p. 157 and then we find the mantissa from the tables. The mantissa of a number with three significant figures should be sought in the row marked by its first two significant figures (the column N) and in the column which corresponds to the third figure. If the given number has more than three significant figures, interpolation should be applied (see p. 17). The interpolation correction should be determined only for the fourth significant figure; determining a correction for the fifth significant figure has a sense only when the first figure is 1 or 2.

Example: $\log 254.3 = 2.4053$ ($0.3 \cdot 17 = 5.1$ is added to 4048).

N	0	1	2	3	4	5	6	7	8	9
10	0000	0043	0086	0128	0170	0212	0253	0294	0334	0374
11	0414	0453	0492	0531	0569	0607	0645	0682	0719	0755
12	0792	0828	0864	0899	0934	0969	1004	1038	1072	1106
13	1139	1173	1206	1239	1271	1303	1335	1367	1399	1430
14	1461	1492	1523	1553	1584	1614	1644	1673	1703	1732
15	1761	1790	1818	1847	1875	1903	1931	1959	1987	2014
16	2041	2068	2095	2122	2148	2175	2201	2227	2253	2279
17	2304	2330	2355	2380	2405	2430	2455	2480	2504	2529
18	2553	2577	2601	2625	2648	2672	2695	2718	2742	2765
19	2788	2810	2833	2856	2878	2900	2923	2945	2967	2989
20	3010	3032	3054	3075	3096	3118	3139	3160	3181	3201
21	3222	3243	3263	3284	3304	3324	3345	3365	3385	3404
22	3424	3444	3464	3483	3502	3522	3541	3560	3579	3598
23	3617	3636	3655	3674	3692	3711	3729	3747	3766	3784
24	3802	3820	3838	3856	3874	3892	3909	3927	3945	3962
25	3979	3997	4014	4031	4048	4065	4082	4099	4116	4133
26	4150	4166	4183	4200	4216	4232	4249	4265	4281	4298
27	4314	4330	4346	4362	4378	4393	4409	4425	4440	4456
28	4472	4487	4502	4518	4533	4548	4564	4579	4594	4609
29	4624	4639	4654	4669	4683	4698	4713	4728	4742	4757
30	4771	4786	4800	4814	4829	4843	4857	4871	4886	4900
31	4914	4928	4942	4955	4969	4983	4997	5011	5024	5038
32	5051	5065	5079	5092	5105	5119	5132	5145	5159	5172
33	5185	5198	5211	5224	5237	5250	5263	5276	5289	5302
34	5315	5328	5340	5353	5366	5378	5391	5403	5416	5428
35	5441	5453	5465	5478	5490	5502	5514	5527	5539	5551
36	5563	5575	5587	5599	5611	5623	5635	5647	5658	5670
37	5682	5694	5705	5717	5729	5740	5752	5763	5775	5786
38	5798	5809	5821	5832	5843	5855	5866	5877	5888	5899
39	5911	5922	5933	5944	5955	5966	5977	5988	5999	6010
40	6021	6031	6042	6053	6064	6075	6085	6096	6107	6117
41	6128	6138	6149	6160	6170	6180	6191	6201	6212	6222
42	6232	6243	6253	6263	6274	6284	6294	6304	6314	6325
43	6335	6345	6355	6365	6375	6385	6395	6405	6415	6425
44	6435	6444	6454	6464	6474	6484	6493	6503	6513	6522
45	6532	6542	6551	6561	6571	6580	6590	6599	6609	6618
46	6628	6637	6646	6656	6665	6675	6684	6693	6702	6712
47	6721	6730	6739	6749	6758	6767	6776	6785	6794	6803
48	6812	6821	6830	6839	6848	6857	6866	6875	6884	6893
49	6902	6911	6920	6928	6937	6946	6955	6964	6972	6981
50	6990	6998	7007	7016	7024	7033	7042	7050	7059	7067
51	7076	7084	7093	7101	7110	7118	7126	7135	7143	7152
52	7160	7168	7177	7185	7193	7202	7210	7218	7226	7235
53	7243	7251	7259	7267	7275	7284	7292	7300	7308	7316
54	7324	7332	7340	7348	7356	7364	7372	7380	7388	7396

N	0	1	2	3	4	5	6	7	8	9
55	7404	7412	7419	7427	7435	7443	7451	7459	7466	7474
56	7482	7490	7497	7505	7513	7520	7528	7536	7543	7551
57	7559	7566	7574	7582	7589	7597	7604	7612	7619	7627
58	7634	7642	7649	7657	7664	7672	7679	7686	7694	7701
59	7709	7716	7723	7731	7738	7745	7752	7760	7767	7774
60	7782	7789	7796	7803	7810	7818	7825	7832	7839	7846
61	7853	7860	7868	7875	7882	7889	7896	7903	7910	7917
62	7924	7931	7938	7945	7952	7959	7966	7973	7980	7987
63	7993	8000	8007	8014	8021	8028	8035	8041	8048	8055
64	8062	8069	8075	8082	8089	8096	8102	8109	8116	8122
65	8129	8136	8142	8149	8156	8162	8169	8176	8182	8189
66	8195	8202	8209	8215	8222	8228	8235	8241	8248	8254
67	8261	8267	8274	8280	8287	8293	8299	8306	8312	8319
68	8325	8331	8338	8344	8351	8357	8363	8370	8376	8382
69	8388	8395	8401	8407	8414	8420	8426	8432	8439	8445
70	8451	8457	8463	8470	8476	8482	8488	8494	8500	8506
71	8513	8519	8525	8531	8537	8543	8549	8555	8561	8567
72	8573	8579	8585	8591	8597	8603	8609	8615	8621	8627
73	8633	8639	8645	8651	8657	8663	8669	8675	8681	8686
74	8692	8698	8704	8710	8716	8722	8727	8733	8739	8745
75	8751	8756	8762	8768	8774	8779	8785	8791	8797	8802
76	8808	8814	8820	8825	8831	8837	8842	8848	8854	8859
77	8865	8871	8876	8882	8887	8893	8899	8904	8910	8915
78	8921	8927	8932	8938	8943	8949	8954	8960	8965	8971
79	8976	8982	8987	8993	8998	9004	9009	9015	9020	9025
80	9031	9036	9042	9047	9053	9058	9063	9069	9074	9079
81	9085	9090	9096	9101	9106	9112	9117	9122	9128	9133
82	9138	9143	9149	9154	9159	9156	9170	9175	9180	9186
83	9191	9196	9201	9206	9212	9217	9222	9227	9232	9238
84	9243	9248	9253	9258	9263	9269	9274	9279	9284	9289
85	9294	9299	9304	9309	9315	9320	9325	9330	9335	9340
86	9345	9350	9355	9360	9365	9370	9375	9380	9385	9390
87	9395	9400	9405	9410	9415	9420	9425	9430	9435	9440
88	9445	9450	9455	9460	9465	9469	9474	9479	9484	9489
89	9494	9499	9504	9509	9513	9518	9523	9528	9533	9538
90	9542	9547	9552	9557	9562	9566	9571	9576	9581	9586
91	9590	9595	9600	9605	9609	9614	9619	9624	9628	9633
92	9638	9643	9647	9652	9657	9661	9666	9671	9675	9680
93	9685	9689	9694	9699	9703	9708	9713	9717	9722	9727
94	9731	9736	9741	9745	9750	9754	9759	9763	9768	9773
95	9777	9782	9786	9791	9795	9800	9805	9808	9814	9818
96	9823	9827	9832	9836	9841	9845	9850	9854	9859	9863
97	9868	9872	9877	9881	9886	9890	9894	9899	9903	9908
98	9912	9917	9921	9926	9930	9934	9939	9943	9948	9952
99	9956	9961	9965	9969	9974	9978	9983	9987	9991	9996

8. Antilogarithms

The table of antilogarithms ([1]) is used to find a number corresponding to its common logarithm. The antilogarithm, or a system of significant figures corresponding to the mantissa of the given logarithm should be sought in the row marked by the first two figures of the mantissa (the column m) and in the row corresponding to the third figure of the mantissa; then a correction for the fourth figure of the mantissa should be computed. The position of the decimal point in the number thus found is determined by the characteristic of the logarithm according to the rules given on p. 157.

Examples. $\log x = 1.2763$; $x = 18.89$ (we add $0.3 \cdot 4 = 1.2$ to the number 1888 obtained from the table). If $\log x = \bar{2}.2763$, then $x = 0.01889$. These results can also be written as follows: $10^{1.2763} = 18.89$; $10^{-1.7237} = 0.01889$ (since $\bar{2}.2763 = -1.7237$).

m	0	1	2	3	4	5	6	7	8	9
00	1000	1002	1005	1007	1009	1012	1014	1016	1019	1021
01	1023	1026	1028	1030	1033	1035	1038	1040	1042	1045
02	1047	1050	1052	1054	1057	1059	1062	1064	1067	1069
03	1072	1074	1076	1079	1081	1084	1086	1089	1091	1094
04	1096	1099	1102	1104	1107	1109	1112	1114	1117	1119
05	1122	1125	1127	1130	1132	1135	1138	1140	1143	1146
06	1148	1151	1153	1156	1159	1161	1164	1167	1169	1172
07	1175	1178	1180	1183	1186	1189	1191	1194	1197	1199
08	1202	1205	1208	1211	1213	1216	1219	1222	1225	1227
09	1230	1233	1236	1239	1242	1245	1247	1250	1253	1256
10	1259	1262	1265	1268	1271	1274	1276	1279	1282	1285
11	1288	1291	1294	1297	1300	1303	1306	1309	1312	1315
12	1318	1321	1324	1327	1330	1334	1337	1340	1343	1346
13	1349	1352	1355	1358	1361	1365	1368	1371	1374	1377
14	1380	1384	1387	1390	1393	1396	1400	1403	1406	1409
15	1413	1416	1419	1422	1426	1429	1432	1435	1439	1442
16	1445	1449	1452	1455	1459	1462	1466	1469	1472	1476
17	1479	1483	1486	1489	1493	1496	1500	1503	1507	1510
18	1514	1517	1521	1524	1528	1531	1535	1538	1542	1545
19	1549	1552	1556	1560	1563	1567	1570	1574	1578	1581

([1]) The number y whose common logarithm is equal to x is called the *antilogarithm* of x. By definition of the logarithm (see p. 156), this function coincides with the exponential function $y = 10^x$.

m	0	1	2	3	4	5	6	7	8	9
20	1585	1589	1592	1596	1600	1603	1607	1611	1614	1618
21	1622	1626	1629	1633	1637	1641	1644	1648	1652	1656
22	1660	1663	1667	1671	1675	1679	1683	1687	1690	1694
23	1698	1702	1706	1710	1714	1718	1722	1726	1730	1734
24	1738	1742	1746	1750	1754	1758	1762	1766	1770	1774
25	1778	1782	1786	1791	1795	1799	1803	1807	1811	1816
26	1820	1824	1828	1832	1837	1841	1845	1849	1854	1858
27	1862	1866	1871	1875	1879	1884	1888	1892	1897	1901
28	1905	1910	1914	1919	1923	1928	1932	1936	1941	1945
29	1950	1954	1959	1963	1968	1972	1977	1982	1986	1991
30	1995	2000	2004	2009	2014	2018	2023	2028	2032	2037
31	2042	2046	2051	2056	2061	2065	2070	2075	2080	2084
32	2089	2094	2099	2104	2109	2113	2118	2123	2128	2133
33	2138	2143	2148	2153	2158	2163	2168	2173	2178	2183
34	2188	2193	2198	2203	2208	2213	2218	2223	2228	2234
35	2239	2244	2249	2254	2259	2265	2270	2275	2280	2286
36	2291	2296	2301	2307	2312	2317	2323	2328	2333	2339
37	2344	2350	2355	2360	2366	2371	2377	2382	2388	2393
38	2399	2404	2410	2415	2421	2427	2432	2438	2443	2449
39	2455	2460	2466	2472	2477	2483	2489	2495	2500	2506
40	2512	2518	2523	2529	2535	2541	2547	2553	2559	2564
41	2570	2576	2582	2588	2594	2600	2606	2612	2618	2624
42	2630	2636	2642	2649	2655	2661	2667	2673	2679	2685
43	2692	2698	2704	2710	2716	2723	2729	2735	2742	2748
44	2754	2761	2767	2773	2780	2786	2793	2799	2805	2812
45	2818	2825	2831	2838	2844	2851	2858	2864	2871	2877
46	2884	2891	2897	2904	2911	2917	2924	2931	2938	2944
47	2951	2958	2965	2972	2979	2985	2992	2999	3006	3013
48	3020	3027	3034	3041	3048	3055	3062	3069	3076	3083
49	3090	3097	3105	3112	3119	3126	3133	3141	3148	3155
50	3162	3170	3177	3184	3192	3199	3206	3214	3221	3228
51	3236	3243	3251	3258	3266	3273	3281	3289	3296	3304
52	3311	3319	3327	3334	3342	3350	3357	3365	3373	3381
53	3388	3396	3404	3412	3420	3428	3436	3443	3451	3459
54	3467	3475	3483	3491	3499	3508	3516	3524	3532	3540
55	3548	3556	3565	3573	3581	3589	3597	3606	3614	3622
56	3631	3639	3648	3656	3664	3673	3681	3690	3698	3707
57	3715	3724	3733	3741	3750	3758	3767	3776	3784	3793
58	3802	3811	3819	3828	3837	3846	3855	3864	3873	3882
59	3890	3899	3908	3917	3926	3936	3945	3954	3963	3972

m	0	1	2	3	4	5	6	7	8	9
60	3981	3990	3999	4009	4018	4027	4036	4046	4055	4064
61	4074	4083	4093	4102	4111	4121	4130	4140	4150	4159
62	4169	4178	4188	4198	4207	4217	4227	4236	4246	4256
63	4266	4276	4285	4295	4305	4315	4325	4335	4345	4355
64	4365	4375	4385	4395	4406	4416	4426	4436	4446	4457
65	4467	4477	4487	4498	4508	4519	4529	4539	4550	4560
66	4571	4581	4592	4603	4613	4624	4634	4645	4656	4667
67	4677	4688	4699	4710	4721	4732	4742	4753	4764	4775
68	4786	4797	4808	4819	4831	4842	4853	4864	4875	4887
69	4898	4909	4920	4932	4943	4955	4966	4977	4989	5000
70	5012	5023	5035	5047	5058	5070	5082	5093	5105	5117
71	5129	5140	5152	5164	5176	5188	5200	5212	5224	5236
72	5248	5260	5272	5284	5297	5309	5321	5333	5346	5358
73	5370	5383	5395	5408	5420	5433	5445	5458	5470	5483
74	5495	5508	5521	5534	5546	5559	5572	5585	5598	5610
75	5623	5636	5649	5662	5675	5689	5702	5715	5728	5741
76	5754	5768	5781	5794	5808	5821	5834	5848	5861	5875
77	5888	5902	5916	5929	5943	5957	5970	5984	5998	6012
78	6026	6039	6053	6067	6081	6095	6109	6124	6138	6152
79	6166	6180	6194	6209	6223	6237	6252	6266	6281	6295
80	6310	6324	6339	6353	6368	6383	6397	6412	6427	6442
81	6457	6471	6486	6501	6516	6531	6546	6561	6577	6592
82	6607	6622	6637	6653	6668	6683	6699	6714	6730	6745
83	6761	6776	6792	6808	6823	6839	6855	6871	6887	6902
84	6918	6934	6950	6966	6982	6998	7015	7031	7047	7063
85	7079	7096	7112	7129	7145	7161	7178	7194	7211	7228
86	7244	7261	7278	7295	7311	7328	7345	7362	7379	7396
87	7413	7430	7447	7464	7482	7499	7516	7534	7551	7568
88	7586	7603	7621	7638	7656	7674	7691	7709	7727	7745
89	7762	7780	7798	7816	7834	7852	7870	7889	7907	7925
90	7943	7962	7980	7998	8017	8035	8054	8072	8091	8110
91	8128	8147	8106	8185	8204	8222	8241	8260	8279	8299
92	8318	8337	8356	8375	8395	8414	8433	8453	8472	8492
93	8511	8531	8551	8570	8590	8610	8630	8650	8670	8690
94	8710	8730	8750	8770	8790	8810	8831	8851	8872	8892
95	8913	8933	8954	8974	8995	9016	9036	9057	9078	9099
96	9120	9141	9162	9183	9204	9226	9247	9268	9290	9311
97	9333	9354	9376	9397	9419	9441	9462	9484	9506	9528
98	9550	9572	9594	9616	9638	9661	9683	9705	9727	9750
99	9772	9795	9817	9840	9863	9886	9908	9931	9954	9977

9. Natural values of trigonometric functions

Sines

Degrees	0′	10′	20′	30′	40′	50′	60′	→
0 ↓	0.0000	0.0029	0.0058	0.0087	0.0116	0.0145	0.0175	89
1	0.0175	0.0204	0.0233	0.0262	0.0291	0.0320	0.0349	88
2	0.0349	0.0378	0.0407	0.0436	0.0465	0.0494	0.0523	87
3	0.0523	0.0552	0.0581	0.0610	0.0640	0.0669	0.0698	86
4	0.0698	0.0727	0.0756	0.0785	0.0814	0.0843	0.0872	85
5	0.0872	0.0901	0.0929	0.0958	0.0987	0.1016	0.1045	84
6	0.1045	0.1074	0.1103	0.1132	0.1161	0.1190	0.1219	83
7	0.1219	0.1248	0.1276	0.1305	0.1334	0.1363	0.1392	82
8	0.1392	0.1421	0.1449	0.1478	0.1507	0.1536	0.1564	81
9	0.1564	0.1593	0.1622	0.1650	0.1679	0.1708	0.1736	80
10	0.1736	0.1765	0.1794	0.1822	0.1851	0.1880	0.1908	79
11	0.1908	0.1937	0.1965	0.1994	0.2022	0.2051	0.2079	78
12	0.2079	0.2108	0.2136	0.2164	0.2193	0.2221	0.2250	77
13	0.2250	0.2278	0.2306	0.2334	0.2363	0.2391	0.2419	76
14	0.2419	0.2447	0.2476	0.2504	0.2532	0.2560	0.2588	75
15	0.2588	0.2616	0.2644	0.2672	0.2700	0.2728	0.2756	74
16	0.2756	0.2784	0.2812	0.2840	0.2868	0.2896	0.2924	73
17	0.2924	0.2952	0.2979	0.3007	0.3035	0.3062	0.3090	72
18	0.3090	0.3118	0.3145	0.3173	0.3201	0.3228	0.3256	71
19	0.3256	0.3283	0.3311	0.3338	0.3365	0.3393	0.3420	70
20	0.3420	0.3448	0.3475	0.3502	0.3529	0.3557	0.3584	69
21	0.3584	0.3611	0.3638	0.3665	0.3692	0.3719	0.3746	68
22	0.3746	0.3773	0.3800	0.3827	0.3854	0.3881	0.3907	67
23	0.3907	0.3934	0.3961	0.3987	0.4014	0.4041	0.4067	66
24	0.4067	0.4094	0.4120	0.4147	0.4173	0.4200	0.4226	65
25	0.4226	0.4253	0.4279	0.4305	0.4331	0.4358	0.4384	64
26	0.4384	0.4410	0.4436	0.4462	0.4488	0.4514	0.4540	63
27	0.4540	0.4566	0.4592	0.4617	0.4643	0.4669	0.4695	62
28	0.4695	0.4720	0.4746	0.4772	0.4797	0.4823	0.4848	61
29	0.4848	0.4874	0.4899	0.4924	0.4950	0.4975	0.5000	60
30	0.5000	0.5025	0.5050	0.5075	0.5100	0.5125	0.5150	59
31	0.5150	0.5175	0.5200	0.5225	0.5250	0.5275	0.5299	58
32	0.5299	0.5324	0.5348	0.5373	0.5398	0.5422	0.5446	57
33	0.5446	0.5471	0.5495	0.5519	0.5544	0.5568	0.5592	56
34	0.5592	0.5616	0.5640	0.5664	0.5688	0.5712	0.5736	55
35	0.5736	0.5760	0.5783	0.5807	0.5831	0.5854	0.5878	↑ 54
←	60′	50′	40′	30′	20′	10′	0′	Degrees

Cosines

Sines

Degrees	0′	10′	20′	30′	40′	50′	60′	→
35 ↓	0.5736	0.5760	0.5783	0.5807	0.5831	0.5854	0.5878	54
36	0.5878	0.5901	0.5925	0.5948	0.5972	0.5995	0.6018	53
37	0.6018	0.6041	0.6065	0.6088	0.6111	0.6134	0.6157	52
38	0.6157	0.6180	0.6202	0.6225	0.6248	0.6271	0.6293	51
39	0.6293	0.6316	0.6338	0.6361	0.6383	0.6406	0.6428	50
40	0.6428	0.6450	0.6472	0.6494	0.6517	0.6539	0.6561	49
41	0.6561	0.6583	0.6604	0.6626	0.6648	0.6670	0.6691	48
42	0.6691	0.6713	0.6734	0.6756	0.6777	0.6799	0.6820	47
43	0.6820	0.6841	0.6862	0.6884	0.6905	0.6926	0.6947	46
44	0.6947	0.6967	0.6988	0.7009	0.7030	0.7050	0.7071	45
45	0.7071	0.7092	0.7112	0.7133	0.7153	0.7173	0.7193	44
46	0.7193	0.7214	0.7234	0.7254	0.7274	0.7294	0.7314	43
47	0.7314	0.7333	0.7353	0.7373	0.7392	0.7412	0.7431	42
48	0.7431	0.7451	0.7470	0.7490	0.7509	0.7528	0.7547	41
49	0.7547	0.7566	0.7585	0.7604	0.7623	0.7642	0.7660	40
50	0.7660	0.7679	0.7698	0.7716	0.7735	0.7753	0.7771	39
51	0.7771	0.7790	0.7808	0.7826	0.7844	0.7862	0.7880	38
52	0.7880	0.7898	0.7916	0.7934	0.7951	0.7969	0.7986	37
53	0.7986	0.8004	0.8021	0.8039	0.8056	0.8073	0.8090	36
54	0.8090	0.8107	0.8124	0.8141	0.8158	0.8175	0.8192	35
55	0.8192	0.8208	0.8225	0.8241	0.8258	0.8274	0.8290	34
56	0.8290	0.8307	0.8323	0.8339	0.8355	0.8371	0.8387	33
57	0.8387	0.8403	0.8418	0.8434	0.8450	0.8465	0.8480	32
58	0.8480	0.8496	0.8511	0.8526	0.8542	0.8557	0.8572	31
59	0.8572	0.8587	0.8601	0.8616	0.8631	0.8646	0.8660	30
60	0.8660	0.8675	0.8689	0.8704	0.8718	0.8732	0.8746	29
61	0.8746	0.8760	0.8774	0.8788	0.8802	0.8816	0.8829	28
62	0.8829	0.8843	0.8857	0.8870	0.8884	0.8897	0.8910	27
63	0.8910	0.8923	0.8936	0.8949	0.8962	0.8975	0.8988	26
64	0.8988	0.9001	0.9013	0.9026	0.9038	0.9051	0.9063	25
65	0.9063	0.9075	0.9088	0.9100	0.9112	0.9124	0.9135	24
66	0.9135	0.9147	0.9159	0.9171	0.9182	0.9194	0.9205	23
67	0.9205	0.9216	0.9228	0.9239	0.9250	0.9261	0.9272	22
68	0.9272	0.9283	0.9293	0.9304	0.9315	0.9325	0.9336	21
69	0.9336	0.9346	0.9356	0.9367	0.9377	0.9387	0.9397	20
70	0.9397	0.9407	0.9417	0.9426	0.9436	0.9446	0.9455	↑ 19
←	60′	50′	40′	30′	20′	10′	0′	Degrees

Cosines

Sines

Degrees	0′	10′	20′	30′	40′	50′	60′	→
70 ↓	0.9397	0.9407	0.9417	0.9426	0.9436	0.9446	0.9455	**19**
71	0.9455	0.9465	0.9474	0.9483	0.9492	0.9502	0.9511	18
72	0.9511	0.9520	0.9528	0.9537	0.9546	0.9555	0.9563	17
73	0.9563	0.9572	0.9580	0.9588	0.9596	0.9605	0.9613	16
74	0.9613	0.9621	0.9628	0.9636	0.9644	0.9652	0.9659	**15**
75	0.9659	0.9667	0.9674	0.9681	0.9689	0.9696	0.9703	14
76	0.9703	0.9710	0.9717	0.9724	0.9730	0.9737	0.9744	13
77	0.9744	0.9750	0.9757	0.9763	0.9769	0.9775	0.9781	12
78	0.9781	0.9787	0.9793	0.9799	0.9805	0.9811	0.9816	11
79	0.9816	0.9822	0.9827	0.9833	0.9838	0.9843	0.9848	**10**
80	0.9848	0.9853	0.9858	0.9863	0.9868	0.9872	0.9877	9
81	0.9877	0.9881	0.9886	0.9890	0.9894	0.9899	0.9903	8
82	0.9903	0.9907	0.9911	0.9914	0.9918	0.9922	0.9925	7
83	0.9925	0.9929	0.9932	0.9936	0.9939	0.9942	0.9945	6
84	0.9945	0.9948	0.9951	0.9954	0.9957	0.9959	0.9962	**5**
85	0.9962	0.9964	0.9967	0.9969	0.9971	0.9974	0.9976	4
86	0.9976	0.9978	0.9980	0.9981	0.9983	0.9985	0.9986	3
87	0.9986	0.9988	0.9989	0.9990	0.9992	0.9993	0.9994	2
88	0.9994	0.9995	0.9996	0.9997	0.9997	0.9998	0.9998	1
89	0.9998	0.9999	0.9999	1.0000	1.0000	1.0000	1.0000	↑ **0**

Cosines

Tangents

Degrees	0′	10′	20′	30′	40′	50′	60′	→
0 ↓	0.0000	0.0029	0.0058	0.0087	0 0116	0.0145	0.0175	89
1	0.0175	0.0204	0.0233	0.0262	0.0291	0.0320	0.0349	88
2	0.0349	0.0378	0.0407	0.0437	0.0466	0.0459	0.0524	87
3	0.0524	0.0553	0.0582	0.0612	0.0641	0.0670	0.0699	86
4	0.0699	0.0729	0.0758	0.0787	0.0816	0.0846	0.0875	**85**
5	0.0875	0.0904	0.0934	0.0963	0.0992	0.1022	0.1051	84
6	0.1051	0.1080	0.1110	0.1139	0.1169	0.1198	0.1228	83
7	0.1228	0.1257	0.1287	0.1317	0.1346	0.1376	0.1405	82
8	0.1405	0.1435	0.1465	0.1495	0.1524	0.1554	0.1584	81
9	0.1584	0.1614	0.1644	0.1673	0.1703	0.1733	0.1763	**80**
10	0.1763	0.1793	0.1823	0.1853	0.1883	0.1914	0.1944	↑ 79
←	60′	50′	40′	30′	20′	10′	0′	Degrees

Cotangents

Tangents

Degrees	0′	10′	20′	30′	40′	50′	60′	→
10 ↓	0.1763	0.1793	0.1823	0.1853	0.1883	0.1914	0.1944	79
11	0.1944	0.1974	0.2004	0.2035	0.2065	0.2095	0.2126	78
12	0.2126	0.2156	0.2186	0.2217	0.2247	0.2278	0.2309	77
13	0.2309	0.2339	0.2370	0.2401	0.2432	0.2462	0.2493	76
14	0.2493	0.2524	0.2555	0.2586	0.2617	0.2648	0.2679	75
15	0.2679	0.2711	0.2742	0.2773	0.2805	0.2836	0.2867	74
16	0.2867	0.2899	0.2931	0.2962	0.2994	0.3026	0.3057	73
17	0.3057	0.3089	0.3121	0.3153	0.3185	0.3217	0.3249	72
18	0.3249	0.3281	0.3314	0.3346	0.3378	0.3411	0.3443	71
19	0.3443	0.3476	0.3508	0.3541	0.3574	0.3607	0.3640	70
20	0.3640	0.3673	0.3706	0.3739	0.3772	0.3805	0.3839	69
21	0.3839	0.3872	0.3906	0.3939	0.3973	0.4006	0.4040	68
22	0.4040	0.4074	0.4108	0.4142	0.4176	0.4210	0.4245	67
23	0.4245	0.4279	0.4314	0.4348	0.4383	0.4417	0.4452	66
24	0.4452	0.4487	0.4522	0.4557	0.4592	0.4628	0.4663	65
25	0.4663	0.4699	0.4734	0.4770	0.4806	0.4841	0.4877	64
26	0.4877	0.4913	0.4950	0.4986	0.5022	0.5059	0.5095	63
27	0.5095	0.5132	0.5169	0.5206	0.5243	0.5280	0.5317	62
28	0.5317	0.5354	0.5392	0.5430	0.5467	0.5505	0.5543	61
29	0.5543	0.5581	0.5619	0.5658	0.5696	0.5735	0.5774	60
30	0.5774	0.5812	0.5851	0.5890	0.5930	0.5969	0.6009	59
31	0.6009	0.6048	0.6088	0.6128	0.6168	0.6208	0.6249	58
32	0.6249	0.6289	0.6330	0.6371	0.6412	0.6453	0.6494	57
33	0.6494	0.6536	0.6577	0.6619	0.6661	0.6703	0.6745	56
34	0.6745	0.6787	0.6830	0.6873	0.6916	0.6959	0.7002	55
35	0.7002	0.7046	0.7089	0.7133	0.7177	0.7221	0.7265	54
36	0.7265	0.7310	0.7355	0.7400	0.7445	0.7490	0.7536	53
37	0.7536	0.7581	0.7627	0.7673	0.7720	0.7766	0.7813	52
38	0.7813	0.7860	0.7907	0.7954	0.8002	0.8050	0.8098	51
39	0.8098	0.8146	0.8195	0.8243	0.8292	0.8342	0.8391	50
40	0.8391	0.8441	0.8491	0.8541	0.8591	0.8642	0.8693	49
41	0.8693	0.8744	0.8796	0.8847	0.8899	0.8952	0.9004	48
42	0.9004	0.9057	0.9110	0.9163	0.9217	0.9271	0.9325	47
43	0.9325	0.9380	0.9435	0.9490	0.9545	0.9601	0.9657	46
44	0.9657	0.9713	0.9770	0.9827	0.9884	0.9942	1.0000	45
45	1.0000	1.0058	1.0117	1.0176	1.0235	1.0295	1.0355	↑ 44
←	60′	50′	40′	30′	20′	10′	0′	Degrees

Cotangents

Tangents

Degrees	0′	10′	20′	30′	40′	50′	60′	→
45 ↓	1.000	1.006	1.012	1.018	1.024	1.030	1.036	44
46	1.036	1.042	1.048	1.054	1.060	1.066	1.072	43
47	1.072	1.079	1.085	1.091	1.098	1.104	1.111	42
48	1.111	1.117	1.124	1.130	1.137	1.144	1.150	41
49	1.150	1.157	1.164	1.171	1.178	1.185	1.192	**40**
50	1.192	1.199	1.206	1.213	1.220	1.228	1.235	39
51	1.235	1.242	1.250	1.257	1.265	1.272	1.280	38
52	1.280	1.288	1.295	1.303	1.311	1.319	1.327	37
53	1.327	1.335	1.343	1.351	1.360	1.368	1.376	36
54	1.376	1.385	1.393	1.402	1.411	1.419	1.428	**35**
55	1.428	1.437	1.446	1.455	1.464	1.473	1.483	34
56	1.483	1.492	1.501	1.511	1.520	1.530	1.540	33
57	1.540	1.550	1.560	1.570	1.580	1.590	1.600	32
58	1.600	1.611	1.621	1.632	1.643	1.653	1.664	31
59	1.664	1.675	1.686	1.698	1.709	1.720	1.732	**30**
60	1.732	1.744	1.756	1.767	1.780	1.792	1.804	29
61	1.804	1.816	1.829	1.842	1.855	1.868	1.881	28
62	1.881	1.894	1.907	1.921	1.935	1.949	1.963	27
63	1.963	1.977	1.991	2.006	2.020	2.035	2.050	26
64	2.050	2.066	2.081	2.097	2.112	2.128	2.145	**25**
65	2.145	2.161	2.177	2.194	2.211	2.229	2.246	24
66	2.246	2.264	2.282	2.300	2.318	2.337	2.356	23
67	2.356	2.375	2.394	2.414	2.434	2.455	2.475	22
68	2.475	2.496	2.517	2.539	2.560	2.583	2.605	21
69	2.605	2.628	2.651	2.675	2.699	2.723	2.747	**20**
70	2.747	2.773	2.798	2.824	2.850	2.877	2.904	19
71	2.904	2.932	2.960	2.989	3.018	3.047	3.078	18
72	3.078	3.108	3.140	3.172	3.204	3.237	3.271	17
73	3.271	3.305	3.340	3.376	3.412	3.450	3.487	16
74	3.487	3.526	3.566	3.606	3.647	3.689	3.732	**15**
75	3.732	3.776	3.821	3.867	3.914	3.962	4.011	14
76	4.011	4.061	4.113	4.165	4.219	4.275	4.331	13
77	4.331	4.390	4.449	4.511	4.574	4.638	4.705	12
78	4.705	4.773	4.843	4.915	4.989	5.066	5.145	11
79	5.145	5.226	5.309	5.396	5.485	5.576	5.671	↑ **10**
←	60′	50′	40′	30′	20′	10′	0′	Degrees

Cotangents

Tangents

Degrees	0′	10′	20′	30′	40′	50′	60′	→
80 ↓	5.671	5.769	5.871	5.976	6.084	6.197	6.314	9
81	6.314	6.435	6.561	6.691	6.827	6.968	7.115	8
82	7.115	7.269	7.429	7.596	7.770	7.953	8.144	7
83	8.144	8.345	8.556	8.777	9.010	9.255	9.514	6
84	9.514	9.788	10.078	10.385	10.712	11.059	11.430	**5**
85	11.430	11.826	12.251	12.706	13.197	13.727	14.301	4
86	14.301	14.924	15.605	16.350	17.169	18.075	19.081	3
87	19.081	20.206	21.470	22.904	24.542	26.432	28.636	2
88	28.636	31.242	34.368	38.188	42.964	49.104	57.290	1
89	57.290	68.750	85.940	114.59	171.89	343.77	∞	↑ **0**
←	60′	50′	40′	30′	20′	10′	0′	Degrees

Cotangents

10. Exponential, hyperbolic and trigonometric functions

(for x from 0 to 1.6)

This table contains the values of the exponential (e^x, e^{-x}), hyperbolic ($\sinh x$, $\cosh x$, $\tanh x$), and trigonometric ($\sin x$, $\cos x$, $\tan x$) functions, for the values of x from 0 to 1.6 (the argument is given in radian measure). The values of the exponential and hyperbolic functions for $x > 1.6$ should be determined from table 11 and the values of the trigonometric functions for $x > 1.6$—by using the table of multiples of $\frac{1}{2}\pi$ and π.

Examples. (1) $\sin 7.5 = \sin(5 \cdot \frac{1}{2}\pi - 0.35398) = \cos 0.35398 = 0.9380$ (linear interpolation). (2) $\sin 29 = \sin(9\pi + 0.72567) = -\sin 0.72567 = -0.6637$ (linear interpolation).

x	e^x	e^{-x}	$\sinh x$	$\cosh x$	$\tanh x$	$\sin x$	$\cos x$	$\tan x$
0.00	1.0000	1.0000	0.0000	1.0000	0.0000	0.0000	1.0000	0.0000
01	1.0101	0.9900	0.0100	1.0001	0.0100	0.0100	1.0000	0.0100
02	1.0202	0.9802	0.0200	1.0002	0.0200	0.0200	0.9998	0.0200
03	1.0305	0.9704	0.0300	1.0005	0.0300	0.0300	0.9996	0.0300
04	1.0408	0.9608	0.0400	1.0008	0.0400	0.0400	0.9992	0.0400
0.05	1.0513	0.9512	0.0500	1.0013	0.0500	0.0500	0.9988	0.0500
06	1.0618	0.9418	0.0600	1.0018	0.0599	0.0600	0.9982	0.0601
07	1.0725	0.9324	0.0701	1.0025	0.0699	0.0699	0.9976	0.0701
08	1.0833	0.9231	0.0801	1.0032	0.0798	0.0799	0.9968	0.0802
09	1.0942	0.9139	0.0901	1.0041	0.0898	0.0899	0.9960	0.0902
0.10	1.1052	0.9048	0.1002	1.0050	0.0997	0.0998	0.9950	0.1003

x	e^x	e^{-x}	$\sinh x$	$\cosh x$	$\tanh x$	$\sin x$	$\cos x$	$\tan x$
0.10	1.1052	0.9048	0.1002	1.0050	0.0997	0.0998	0.9950	0.1003
11	1.1163	0.8958	0.1102	1.0061	0.1096	0.1098	0.9940	0.1104
12	1.1275	0.8869	0.1203	1.0072	0.1194	0.1197	0.9928	0.1206
13	1.1388	0.8781	0.1304	1.0085	0.1293	0.1296	0.9916	0.1307
14	1.1503	0.8694	0.1405	1.0098	0.1391	0.1395	0.9902	0.1409
0.15	1.1618	0.8607	0.1506	1.0113	0.1489	0.1494	0.9888	0.1511
16	1.1735	0.8521	0.1607	1.0128	0.1586	0.1593	0.9872	0.1614
17	1.1853	0.8437	0.1708	1.0145	0.1684	0.1692	0.9856	0.1717
18	1.1972	0.8353	0.1810	1.0162	0.1781	0.1790	0.9838	0.1820
19	1.2092	0.8270	0.1911	1.0181	0.1877	0.1889	0.9820	0.1923
0.20	1.2214	0.8187	0.2013	1.0201	0.1974	0.1987	0.9801	0.2027
21	1.2337	0.8106	0.2115	1.0221	0.2070	0.2085	0.9780	0.2131
22	1.2461	0.8025	0.2218	1.0243	0.2165	0.2182	0.9759	0.2236
23	1.2586	0.7945	0.2320	1.0266	0.2260	0.2280	0.9737	0.2341
24	1.2712	0.7866	0.2423	1.0289	0.2355	0.2377	0.9713	0.2447
0.25	1.2840	0.7788	0.2526	1.0314	0.2449	0.2474	0.9689	0.2553
26	1.2969	0.7711	0.2629	1.0340	0.2543	0.2571	0.9664	0.2660
27	1.3100	0.7634	0.2733	1.0367	0.2636	0.2667	0.9638	0.2768
28	1.3231	0.7558	0.2837	1.0395	0.2729	0.2764	0.9611	0.2876
29	1.3364	0.7483	0.2941	1.0423	0.2821	0.2860	0.9582	0.2984
0.30	1.3499	0.7408	0.3045	1.0453	0.2913	0.2955	0.9553	0.3093
31	1.3634	0.7334	0.3150	1.0484	0.3004	0.3051	0.9523	0.3203
32	1.3771	0.7261	0.3255	1.0516	0.3095	0.3146	0.9492	0.3314
33	1.3910	0.7189	0.3360	1.0549	0.3185	0.3240	0.9460	0.3425
34	1.4049	0.7118	0.3466	1.0584	0.3275	0.3335	0.9428	0.3537
0.35	1.4191	0.7047	0.3572	1.0619	0.3364	0.3429	0.9394	0.3650
36	1.4333	0.6977	0.3678	1.0655	0.3452	0.3523	0.9359	0.3764
37	1.4477	0.6907	0.3785	1.0692	0.3540	0.3616	0.9323	0.3879
38	1.4623	0.6839	0.3892	1.0731	0.3627	0.3709	0.9287	0.3994
39	1.4770	0.6771	0.4000	1.0770	0.3714	0.3802	0.9249	0.4111
0.40	1.4918	0.6703	0.4108	1.0811	0.3799	0.3894	0.9211	0.4228
41	1.5068	0.6637	0.4216	1.0852	0.3885	0.3986	0.9171	0.4346
42	1.5220	0.6570	0.4325	1.0895	0.3969	0.4078	0.9131	0.4466
43	1.5373	0.6505	0.4434	1.0939	0.4053	0.4169	0.9090	0.4586
44	1.5527	0.6440	0.4543	1.0984	0.4136	0.4259	0.9048	0.4708
0.45	1.5683	0.6376	0.4653	1.1030	0.4219	0.4350	0.9004	0.4831
46	1.5841	0.6313	0.4764	1.1077	0.4301	0.4439	0.8961	0.4954
47	1.6000	0.6250	0.4875	1.1125	0.4382	0.4529	0.8916	0.5080
48	1.6161	0.6188	0.4986	1.1174	0.4462	0.4618	0.8870	0.5206
49	1.6323	0.6126	0.5098	1.1225	0.4542	0.4706	0.8823	0.5334
0.50	1.6487	0.6065	0.5211	1.1276	0.4621	0.4794	0.8776	0.5463

x	e^x	e^{-x}	sinh x	cosh x	tanh x	sin x	cos x	tan x
0.50	1.6487	0.6065	0.5211	1.1276	0.4621	0.4794	0.8776	0.5463
51	1.6653	0.6005	0.5324	1.1329	0.4699	0.4882	0.8727	0.5594
52	1.6820	0.5945	0.5438	1.1383	0.4777	0.4969	0.8678	0.5726
53	1.6989	0.5886	0.5552	1.1438	0.4854	0.5055	0.8628	0.5859
54	1.7160	0.5827	0.5666	1.1494	0.4930	0.5141	0.8577	0.5994
0.55	1.7333	0.5769	0.5782	1.1551	0.5005	0.5227	0.8525	0.6131
56	1.7507	0.5712	0.5897	1.1609	0.5080	0.5312	0.8473	0.6269
57	1.7683	0.5655	0.6014	1.1669	0.5154	0.5396	0.8419	0.6410
58	1.7860	0.5599	0.6131	1.1730	0.5227	0.5480	0.8365	0.6552
59	1.8040	0.5543	0.6248	1.1792	0.5299	0.5564	0.8309	0.6696
0.60	1.8221	0.5488	0.6367	1.1855	0.5370	0.5646	0.8253	0.6841
61	1.8404	0.5434	0.6485	1.1919	0.5441	0.5729	0.8196	0.6989
62	1.8589	0.5379	0.6605	1.1984	0.5511	0.5810	0.8139	0.7139
63	1.8776	0.5326	0.6725	1.2051	0.5581	0.5891	0.8080	0.7291
64	1.8965	0.5273	0.6846	1.2119	0.5649	0.5972	0.8021	0.7445
0.65	1.9155	0.5220	0.6967	1.2188	0.5717	0.6052	0.7961	0.7602
66	1.9348	0.5169	0.7090	1.2258	0.5784	0.6131	0.7900	0.7761
67	1.9542	0.5117	0.7213	1.2330	0.5850	0.6210	0.7838	0.7923
68	1.9739	0.5066	0.7336	1.2402	0.5915	0.6288	0.7776	0.8087
69	1.9937	0.5016	0.7461	1.2476	0.5980	0.6365	0.7712	0.8253
0.70	2.0138	0.4966	0.7586	1.2552	0.6044	0.6442	0.7648	0.8423
71	2.0340	0.4916	0.7712	1.2628	0.6107	0.6518	0.7584	0.8595
72	2.0544	0.4868	0.7838	1.2706	0.6169	0.6594	0.7518	0.8771
73	2.0751	0.4819	0.7966	1.2785	0.6231	0.6669	0.7452	0.8949
74	2.0959	0.4771	0.8094	1.2865	0.6291	0.6743	0.7385	0.9181
0.75	2.1170	0.4724	0.8223	1.2947	0.6351	0.6816	0.7317	0.9316
76	2.1383	0.4677	0.8353	1.3030	0.6411	0.6889	0.7248	0.9505
77	2.1598	0.4630	0.8484	1.3114	0.6469	0.6961	0.7179	0.9697
78	2.1815	0.4584	0.8615	1.3199	0.6527	0.7033	0.7109	0.9893
79	2.2034	0.4538	0.8748	1.3286	0.6584	0.7104	0.7038	1.0092
0.80	2.2255	0.4493	0.8881	1.3374	0.6640	0.7174	0.6967	1.0296
81	2.2479	0.4449	0.9015	1.3464	0.6696	0.7243	0.6895	1.0505
82	2.2705	0.4404	0.9150	1.3555	0.6751	0.7311	0.6822	1.0717
83	2.2933	0.4360	0.9286	1.3647	0.6805	0.7379	0.6749	1.0934
84	2.3164	0.4317	0.9423	1.3740	0.6858	0.7446	0.6675	1.1156
0.85	2.3396	0.4274	0.9561	1.3835	0.6911	0.7513	0.6600	1.1383
86	2.3632	0.4232	0.9700	1.3932	0.6963	0.7578	0.6524	1.1616
87	2.3869	0.4190	0.9840	1.4029	0.7014	0.7643	0.6448	1.1853
88	2.4109	0.4148	0.9981	1.4128	0.7064	0.7707	0.6372	1.2097
89	2.4351	0.4107	1.0122	1.4229	0.7114	0.7771	0.6294	1.2346
0.90	2.4596	0.4066	1.0265	1.4331	0.7163	0.7833	0.6216	1.2602

x	e^x	e^{-x}	$\sinh x$	$\cosh x$	$\tanh x$	$\sin x$	$\cos x$	$\tan x$
0.90	2.4596	0.4066	1.0265	1.4331	0.7163	0.7833	0.6216	1.2602
91	2.4843	0.4025	1.0409	1.4434	0.7211	0.7895	0.6137	1.2864
92	2.5093	0.3985	1.0554	1.4539	0.7259	0.7956	0.6058	1.3133
93	2.5345	0.3946	1.0700	1.4645	0.7306	0.8016	0.5978	1.3409
94	2.5600	0.3906	1.0847	1.4753	0.7352	0.8076	0.5898	1.3692
0.95	2.5857	0.3867	1.0995	1.4862	0.7398	0.8134	0.5817	1.3984
96	2.6117	0.3829	1.1144	1.4973	0.7443	0.8192	0.5735	1.4284
97	2.6379	0.3791	1.1294	1.5085	0.7487	0.8249	0.5653	1.4592
98	2.6645	0.3753	1.1446	1.5199	0.7531	0.8305	0.5570	1.4910
99	2.6912	0.3716	1.1598	1.5314	0.7574	0.8360	0.5487	1.5237
1.00	2.7183	0.3679	1.1752	1.5431	0.7616	0.8415	0.5403	1.5574
01	2.7456	0.3642	1.1907	1.5549	0.7658	0.8468	0.5319	1.5922
02	2.7732	0.3606	1.2063	1.5669	0.7699	0.8521	0.5234	1.6281
03	2.8011	0.3570	1.2220	1.5790	0.7739	0.8573	0.5148	1.6652
04	2.8292	0.3535	1.2379	1.5913	0.7779	0.8624	0.5062	1.7036
1.05	2.8577	0.3499	1.2539	1.6038	0.7818	0.8674	0.4976	1.7433
06	2.8864	0.3465	1.2700	1.6164	0.7857	0.8724	0.4889.	1.7844
07	2.9154	0.3430	1.2862	1.6292	0.7895	0.8772	0.4801	1.8270
08	2.9447	0.3396	1.3025	1.6421	0.7932	0.8820	0.4713	1.8712
09	2.9743	0.3362	1.3190	1.6552	0.7969	0.8866	0.4625	1.9171
1.10	3.0042	0.3329	1.3356	1.6685	0.8005	0.8912	0.4536	1.9648
11	3.0344	0.3296	1.3524	1.6820	0.8041	0.8957	0.4447	2.0143
12	3.0649	0.3263	1.3693	1.6956	0.8076	0.9001	0.4357	2.0660
13	3.0957	0.3230	1.3863	1.7093	0.8110	0.9044	0.4267	2.1198
14	3.1268	0.3198	1.4035	1.7233	0.8144	0.9086	0.4176	2.1759
1.15	3.1582	0.3166	1.4208	1.7374	0.8178	0.9128	0.4085	2.2345
16	3.1899	0.3135	1.4382	1.7517	0.8210	0.9168	0.3993	2.2958
17	3.2220	0.3104	1.4558	1.7662	0.8243	0.9208	0.3902	2.3600
18	3.2544	0.3073	1.4735	1.7808	0.8275	0.9246	0.3809	2.4273
19	3.2871	0.3042	1.4914	1.7957	0.8306	0.9284	0.3717	2.4979
1.20	3.3201	0.3012	1.5095	1.8107	0.8337	0.9320	0.3624	2.5722
21	3.3535	0.2982	1.5276	1.8258	0.8367	0.9356	0.3530	2.6503
22	3.3872	0.2952	1.5460	1.8412	0.8397	0.9391	0.3436	2.7328
23	3.4212	0.2923	1.5645	1.8568	0.8426	0.9425	0.3342	2.8198
24	3.4556	0.2894	1.5831	1.8725	0.8455	0.9458	0.3248	2.9119
1.25	3.4903	0.2865	1.6019	1.8884	0.8483	0.9490	0.3153	3.0096
26	3.5254	0.2837	1.6209	1.9045	0.8511	0.9521	0.3058	3.1133
27	3.5609	0.2808	1.6400	1.9208	0.8538	0.9551	0.2963	3.2236
28	3.5966	0.2780	1.6593	1.9373	0.8565	0.9580	0.2867	3.3413
29	3.6328	0.2753	1.6788	1.9540	0.8591	0.9608	0.2771	3.4672
1.30	3.6693	0.2725	1.6984	1.9709	0.8617	0.9636	0.2675	3.6021

x	e^x	e^{-x}	sinh x	cosh x	tanh x	sin x	cos x	tan x
1.30	3.6693	0.2725	1.6984	1.9709	0.8617	0.9636	0.2675	3.6021
31	3.7062	0.2698	1.7182	1.9880	0.8643	0.9662	0.2579	3.7471
32	3.7434	0.2671	1.7381	2.0053	0.8668	0.9687	0.2482	3.9033
33	3.7810	0.2645	1.7583	2.0228	0.8692	0.9711	0.2385	4.0723
34	3.8190	0.2618	1.7786	2.0404	0.8717	0.9735	0.2288	4.2556
1.35	3.8574	0.2592	1.7991	2.0583	0.8741	0.9757	0.2190	4.4552
36	3.8962	0.2567	1.8198	2.0764	0.8764	0.9779	0.2092	4.6734
37	3.9354	0.2541	1.8406	2.0947	0.8787	0.9799	0.1994	4.9131
38	3.9749	0.2516	1.8617	2.1132	0.8810	0.9819	0.1896	5.1774
39	4.0149	0.2491	1.8829	2.1320	0.8832	0.9837	0.1798	5.4707
1.40	4.0552	0.2466	1.9043	2.1509	0.8854	0.9854	0.1700	5.7979
41	4.0960	0.2441	1.9259	2.1700	0.8875	0.9871	0.1601	6.1654
42	4.1371	0.2417	1.9477	2.1894	0.8896	0.9887	0.1502	6.5811
43	4.1787	0.2393	1.9697	2.2090	0.8917	0.9901	0.1403	7.0555
44	4.2207	0.2369	1.9919	2.2288	0.8937	0.9915	0.1304	7.6018
1.45	4.2631	0.2346	2.0143	2.2488	0.8957	0.9927	0.1205	8.2381
46	4.3060	0.2322	2.0369	2.2691	0.8977	0.9939	0.1106	8.9886
47	4.3492	0.2299	2.0597	2.2896	0.8996	0.9949	0.1006	9.8874
48	4.3929	0.2276	2.0827	2.3103	0.9015	0.9959	0.0907	10.983
49	4.4371	0.2254	2.1059	2.3312	0.9033	0.9967	0.0807	12.350
1.50	4.4817	0.2231	2.1293	2.3524	0.9051	0.9975	0.0707	14.101
51	4.5267	0.2209	2.1529	2.3738	0.9069	0.9982	0.0608	16.428
52	4.5722	0.2187	2.1768	2.3955	0.9087	0.9987	0.0508	19.670
53	4.6182	0.2165	2.2008	2.4174	0.9104	0.9992	0.0408	24.498
54	4.6646	0.2144	2.2251	2.4395	0.9121	0.9995	0.0308	32.461
1.55	4.7115	0.2122	2.2496	2.4619	0.9138	0.9998	0.0208	48.078
56	4.7588	0.2101	2.2743	2.4845	0.9154	0.9999	0.0108	92.620
57	4.8066	0.2080	2.2993	2.5073	0.9170	1.0000	+0.0008	1255.8
58	4.8550	0.2060	2.3245	2.5305	0.9186	1.0000	−0.0092	−108.65
59	4.9037	0.2039	2.3499	2.5538	0.9201	0.9998	−0.0192	−52.067
1.60	4.9530	0.2019	2.3756	2.5775	0.9217	0.9996	−0.0292	−34.233

Multiples of $\frac{1}{2}\pi$ and π for evaluation of trigonometric functions for $x > 1.6$

n	$n \cdot \frac{1}{2}\pi$	$n \cdot \pi$	n	$n \cdot \frac{1}{2}\pi$	$n \cdot \pi$
1	1.57080	3.14159	6	9.42478	18.84956
2	3.14159	6.28319	7	10.99557	21.99115
3	4.71239	9.42478	8	12.56637	25.13274
4	6.28319	12.56637	9	14.13717	28.27433
5	7.85398	15.70796	10	15.70796	31.41593

11. Exponential functions (continued)

(for x from 1.6 to 10)

This table contains values of the exponential functions e^x and e^{-x} for x from 1.6 to 10. It can also be used to compute values of the hyperbolic functions by applying the formulas (p. 230):

$$\sinh x = \frac{e^x - e^{-x}}{2}, \quad \cosh x = \frac{e^x + e^{-x}}{2}, \quad \tanh x = \frac{\sinh x}{\cosh x} = \frac{1 - e^{-2x}}{1 + e^{-2x}}.$$

x	e^x	e^{-x}	x	e^x	e^{-x}	x	e^x	e^{-x}
1.60	4.9530	0.2019	**1.90**	6.6859	0.1496	**2.20**	9.0250	0.1108
1.61	5.0028	0.1999	1.91	6.7531	0.1481	2.21	9.1157	0.1097
1.62	5.0531	0.1979	1.92	6.8210	0.1466	2.22	9.2073	0.1086
1.63	5.1039	0.1959	1.93	6.8895	0.1451	2.23	9.2999	0.1075
1.64	5.1552	0.1940	1.94	6.9588	0.1437	2.24	9.3933	0.1065
1.65	5.2070	0.1920	**1.95**	7.0287	0.1423	**2.25**	9.4877	0.1054
1.66	5.2593	0.1901	1.96	7.0993	0.1409	2.26	9.5831	0.1044
1.67	5.3122	0.1882	1.97	7.1707	0.1395	2.27	9.6794	0.1033
1.68	5.3656	0.1864	1.98	7.2427	0.1381	2.28	9.7767	0.1023
1.69	5.4195	0.1845	1.99	7.3155	0.1367	2.29	9.8749	0.1013
1.70	5.4739	0.1827	**2.00**	7.3891	0.1353	**2.30**	9.9742	0.10026
1.71	5.5290	0.1809	2.01	7.4633	0.1340	2.31	10.074	0.09926
1.72	5.5845	0.1791	2.02	7.5383	0.1327	2.32	10.176	0.09827
1.73	5.6407	0.1773	2.03	7.6141	0.1313	2.33	10.278	0.09730
1.74	5.6973	0.1755	2.04	7.6906	0.1300	2.34	10.381	0.09633
1.75	5.7546	0.1738	**2.05**	7.7679	0.1287	**2.35**	10.486	0.09537
1.76	5.8124	0.1720	2.06	7.8466	0.1275	2.36	10.591	0.09442
1.77	5.8709	0.1703	2.07	7.9248	0.1262	2.37	10.697	0.09348
1.78	5.9299	0.1686	2.08	8.0045	0.1249	2.38	10.805	0.09255
1.79	5.9895	0.1670	2.09	8.0849	0.1237	2.39	10.913	0.09163
1.80	6.0496	0.1653	**2.10**	8.1662	0.1225	**2.40**	11.023	0.09072
1.81	6.1104	0.1637	2.11	8.2482	0.1212	2.41	11.134	0.08982
1.82	6.1719	0.1620	2.12	8.3311	0.1200	2.42	11.246	0.08892
1.83	6.2339	0.1604	2.13	8.4149	0.1188	2.43	11.359	0.08804
1.84	6.2965	0.1588	2.14	8.4994	0.1177	2.44	11.473	0.08716
1.85	6.3598	0.1572	**2.15**	8.5849	0.1165	**2.45**	11.588	0.08629
1.86	6.4237	0.1557	2.16	8.6711	0.1153	2.46	11.705	0.08543
1.87	6.4883	0.1541	2.17	8.7583	0.1142	2.47	11.822	0.08458
1.88	6.5535	0.1526	2.18	8.8463	0.1130	2.48	11.941	0.08374
1.89	6.6194	0.1511	2.19	8.9352	0.1119	2.49	12.061	0.08291
1.90	6.6859	0.1496	**2.20**	9.0250	0.1108	**2.50**	12.182	0.08208

x	e^x	e^{-x}	x	e^x	e^{-x}	x	e^x	e^{-x}
2.50	12.182	0.08208	**2.90**	18.174	0.05502	**3.30**	27.113	0.03688
2.51	12.305	0.08127	2.91	18.357	0.05448	3.31	27.385	0.03652
2.52	12.429	0.08046	2.92	18.541	0.05393	3.32	27.660	0.03615
2.53	12.554	0.07966	2.93	18.728	0.05340	3.33	27.938	0.03579
2.54	12.680	0.07887	2.94	18.916	0.05287	3.34	28.219	0.03544
2.55	12.807	0.07808	**2.95**	19.106	0.05234	**3.35**	28.503	0.03508
2.56	12.936	0.07730	2.96	19.298	0.05182	3.36	28.789	0.03474
2.57	13.066	0.07654	2.97	19.492	0.05130	3.37	29.079	0.03439
2.58	13.197	0.07577	2.98	19.688	0.05079	3.38	29.371	0.03405
2.59	13.330	0.07502	2.99	19.886	0.05029	3.39	29.666	0.03371
2.60	13.464	0.07427	**3.00**	20.086	0.04979	**3.40**	29.964	0.03337
2.61	13.599	0.07353	3.01	20.287	0.04929	3.41	30.265	0.03304
2.62	13.736	0.07280	3.02	20.491	0.04880	3.42	30.569	0.03271
2.63	13.874	0.07208	3.03	20.697	0.04832	3.43	30.877	0.03239
2.64	14.013	0.07136	3.04	20.905	0.04783	3.44	31.187	0.03206
2.65	14.154	0.07065	**3.05**	21.115	0.04736	**3.45**	31.500	0.03175
2.66	14.296	0.06995	3.06	21.328	0.04689	3.46	31.817	0.03143
2.67	14.440	0.06925	3.07	21.542	0.04642	3.47	32.137	0.03112
2.68	14.585	0.06856	3.08	21.758	0.04596	3.48	32.460	0.03081
2.69	14.732	0.06788	3.09	21.977	0.04550	3.49	32.786	0.03050
2.70	14.880	0.06721	**3.10**	22.198	0.04505	**3.50**	33.115	0.03020
2.71	15.029	0.06654	3.11	22.421	0.04460	3.51	33.448	0.02990
2.72	15.180	0.06587	3.12	22.646	0.04416	3.52	33.784	0.02960
2.73	15.333	0.06522	3.13	22.874	0.04372	3.53	34.124	0.02930
2.74	15.487	0.06457	3.14	23.104	0.04328	3.54	34.467	0.02901
2.75	15.643	0.06393	**3.15**	23.336	0.04285	**3.55**	34.813	0.02872
2.76	15.800	0.06329	3.16	23.571	0.04243	3.56	35.163	0.02844
2.77	15.959	0.06266	3.17	23.807	0.04200	3.57	35.517	0.02816
2.78	16.119	0.06204	3.18	24.047	0.04159	3.58	35.874	0.02788
2.79	16.281	0.06142	3.19	24.288	0.04117	3.59	36.234	0.02760
2.80	16.445	0.06081	**3.20**	24.533	0.04076	**3.60**	36.598	0.02732
2.81	16.610	0.06020	3.21	24.779	0.04036	3.61	36.966	0.02705
2.82	16.777	0.05961	3.22	25.028	0.03996	3.62	37.338	0.02678
2.83	16.945	0.05901	3.23	25.280	0.03956	3.63	37.713	0.02652
2.84	17.116	0.05843	3.24	25.534	0.03916	3.64	38.092	0.02625
2.85	17.288	0.05784	**3.25**	25.790	0.03877	**3.65**	38.475	0.02599
2.86	17.462	0.05727	3.26	26.050	0.03839	3.66	38.861	0.02573
2.87	17.637	0.05670	3.27	26.311	0.03801	3.67	39.252	0.02548
2.88	17.814	0.05613	3.28	26.576	0.03763	3.68	39.646	0.02522
2.89	17.993	0.05558	3.29	26.843	0.03725	3.69	40.045	0.02497
2.90	18.174	0.05502	**3.30**	27.113	0.03688	**3.70**	40.447	0.02472

x	e^x	e^{-x}	x	e^x	e^{-x}	x	e^x	e^{-x}
3.70	40.447	0.02472	**4.0**	54.598	0.01832	**7.0**	1096.6	0.000912
3.71	40.854	0.02448	4.1	60.340	0.01657	7.1	1212.0	0.000825
3.72	41.264	0.02423	4.2	66.686	0.01500	7.2	1339.4	0.000747
3.73	41.679	0.02399	4.3	73.700	0.01357	7.3	1480.3	0.000676
3.74	42.098	0.02375	4.4	81.451	0.01228	7.4	1636.0	0.000611
3.75	42.521	0.02352	**4.5**	90.017	0.01111	**7.5**	1808.0	0.000553
3.76	42.948	0.02328	4.6	99.484	0.01005	7.6	1998.2	0.000500
3.77	43.380	0.02305	4.7	109.95	0.00910	7.7	2208.3	0.000453
3.78	43.816	0.02282	4.8	121.51	0.00823	7.8	2440.6	0.000410
3.79	44.256	0.02260	4.9	134.29	0.00745	7.9	2697.3	0.000371
3.80	44.701	0.02237	**5.0**	148.41	0.00674	**8.0**	2981.0	0.000335
3.81	45.150	0.02215	5.1	164.02	0.00610	8.1	3294.5	0.000304
3.82	45.604	0.02193	5.2	181.27	0.00552	8.2	3641.0	0.000275
3.83	46.063	0.02171	5.3	200.34	0.00499	8.3	4023.9	0.000249
3.84	46.525	0.02149	5.4	221.41	0.00452	8.4	4447.1	0.000225
3.85	46.993	0.02128	**5.5**	244.69	0.00409	**8.5**	4914.8	0.000203
3.86	47.465	0.02107	5.6	270.43	0.00370	8.6	5431.7	0.000184
3.87	47.942	0.02086	5.7	298.87	0.00335	8.7	6002.9	0.000167
3.88	48.424	0.02065	5.8	330.30	0.00303	8.8	6634.2	0.000151
3.89	48.911	0.02045	5.9	365.04	0.00274	8.9	7332.0	0.000136
3.90	49.402	0.02024	**6.0**	403.43	0.002479	**9.0**	8103.1	0.000123
3.91	49.899	0.02004	6.1	445.86	0.002243	9.1	8955.3	0.000112
3.92	50.400	0.01984	6.2	492.75	0.002029	9.2	9897.1	0.000101
3.93	50.907	0.01964	6.3	544.57	0.001836	9.3	10938	0.000091
3.94	51.419	0.01945	6.4	601.85	0.001662	9.4	12088	0.000083
3.95	51.935	0.01925	**6.5**	665.14	0.001503	**9.5**	13360	0.000075
3.96	52.457	0.01906	6.6	735.10	0.001360	9.6	14765	0.000068
3.97	52.985	0.01887	6.7	812.41	0.001231	9.7	16318	0.000061
3.98	53.517	0.01869	6.8	897.85	0.001114	9.8	18034	0.000055
3.99	54.055	0.01850	6.9	992.27	0.001008	9.9	19930	0.000050
4.0	54.598	0.01832	**7.0**	1096.6	0.000912	**10.0**	22026	0.000045

12. Natural logarithms

This table, unlike that of common logarithms, contains the characteristics as well as the mantissas. Logarithms of numbers from 1 to 9.99 can be obtained directly from the table (the interpolation correction should be introduced for the third and fourth figure on the right side of the decimal point, see p. 17).

To compute the natural logarithm of a number $M < 1$ or $M \geqslant 10$, this number should be written in the form $M = N/10^m$ or

$M = N \cdot 10^m$, where N lies between 1.00 and 9.99 and then the table of values of $\ln 10^m$ inserted at the end should be used.

Examples.

(1) $\ln 862 = \ln 8.62 + \ln 10^2 = 2.1541 + 4.6052 = 6.7593.$

(2) $\ln 0.0862 = \ln 8.62 - \ln 10^2 = 2.1541 - 4.6052 = -2.4511.$

N	0	1	2	3	4	5	6	7	8	9
1.0	0.0000	0.0100	0.0198	0.0296	0.0392	0.0488	0.0583	0.0677	0.0770	0.0862
1.1	0.0953	0.1044	0.1133	0.1222	0.1310	0.1398	0.1484	0.1570	0.1655	0.1740
1.2	0.1823	0.1906	0.1989	0.2070	0.2151	0.2231	0.2311	0.2390	0.2469	0.2546
1.3	0.2624	0.2700	0.2776	0.2852	0.2927	0.3001	0.3075	0.3148	0.3221	0.3293
1.4	0.3365	0.3436	0.3507	0.3577	0.3646	0.3716	0.3784	0.3853	0.3920	0.3988
1.5	0.4055	0.4121	0.4187	0.4253	0.4318	0.4383	0.4447	0.4511	0.4574	0.4637
1.6	0.4700	0.4762	0.4824	0.4886	0.4947	0.5008	0.5068	0.5128	0.5188	0.5247
1.7	0.5306	0.5365	0.5423	0.5481	0.5539	0.5596	0.5653	0.5710	0.5766	0.5822
1.8	0.5878	0.5933	0.5988	0.6043	0.6098	0.6152	0.6206	0.6259	0.6313	0.6366
1.9	0.6419	0.6471	0.6523	0.6575	0.6627	0.6678	0.6729	0.6780	0.6831	0.6881
2.0	0.6931	0.6981	0.7031	0.7080	0.7129	0.7178	0.7227	0.7275	0.7324	0.7372
2.1	0.7419	0.7467	0.7514	0.7561	0.7608	0.7655	0.7701	0.7747	0.7793	0.7839
2.2	0.7885	0.7930	0.7975	0.8020	0.8065	0.8109	0.8154	0.8198	0.8242	0.8286
2.3	0.8329	0.8372	0.8416	0.8459	0.8502	0.8544	0.8587	0.8629	0.8671	0.8713
2.4	0.8755	0.8796	0.8838	0.8879	0.8920	0.8961	0.9002	0.9042	0.9083	0.9123
2.5	0.9163	0.9203	0.9243	0.9282	0.9322	0.9361	0.9400	0.9439	0.9478	0.9517
2.6	0.9555	0.9594	0.9632	0.9670	0.9708	0.9746	0.9783	0.9821	0.9858	0.9895
2.7	0.9933	0.9969	1.0006	1.0043	1.0080	1.0116	1.0152	1.0188	1.0225	1.0260
2.8	1.0296	1.0332	1.0367	1.0403	1.0438	1.0473	1.0508	1.0543	1.0578	1.0613
2.9	1.0647	1.0682	1.0716	1.0750	1.0784	1.0818	1.0852	1.0886	1.0919	1.0953
3.0	1.0986	1.1019	1.1053	1.1086	1.1119	1.1151	1.1184	1.1217	1.1249	1.1282
3.1	1.1314	1.1346	1.1378	1.1410	1.1442	1.1474	1.1506	1.1537	1.1569	1.1600
3.2	1.1632	1.1663	1.1694	1.1725	1.1756	1.1787	1.1817	1.1848	1.1878	1.1909
3.3	1.1939	1.1969	1.2000	1.2030	1.2060	1.2090	1.2119	1.2149	1.2179	1.2208
3.4	1.2238	1.2267	1.2296	1.2326	1.2355	1.2384	1.2413	1.2442	1.2470	1.2499
3.5	1.2528	1.2556	1.2585	1.2613	1.2641	1.2669	1.2698	1.2726	1.2754	1.2782
3.6	1.2809	1.2837	1.2865	1.2892	1.2920	1.2947	1.2975	1.3002	1.3029	1.3056
3.7	1.3083	1.3110	1.3137	1.3164	1.3191	1.3218	1.3244	1.3271	1.3297	1.3324
3.8	1.3350	1.3376	1.3403	1.3429	1.3455	1.3481	1.3507	1.3533	1.3558	1.3584
3.9	1.3610	1.3635	1.3661	1.3686	1.3712	1.3737	1.3762	1.3788	1.3813	1.3838
4.0	1.3863	1.3888	1.3913	1.3938	1.3962	1.3987	1.4012	1.4036	1.4061	1.4085
4.1	1.4110	1.4134	1.4159	1.4183	1.4207	1.4231	1.4255	1.4279	1.4303	1.4327
4.2	1.4351	1.4375	1.4398	1.4422	1.4446	1.4469	1.4493	1.4516	1.4540	1.4563
4.3	1.4586	1.4609	1.4633	1.4656	1.4679	1.4702	1.4725	1.4748	1.4770	1.4793
4.4	1.4816	1.4839	1.4861	1.4884	1.4907	1.4929	1.4951	1.4974	1.4996	1.5019

N	0	1	2	3	4	5	6	7	8	9
4.5	1.5041	1.5063	1.5085	1.5107	1.5129	1.5151	1.5173	1.5195	1.5217	1.5239
4.6	1.5261	1.5282	1.5304	1.5326	1.5347	1.5369	1.5390	1.5412	1.5433	1.5454
4.7	1.5476	1.5497	1.5518	1.5539	1.5560	1.5581	1.5602	1.5623	1.5644	1.5665
4.8	1.5686	1.5707	1.5728	1.5748	1.5769	1.5790	1.5810	1.5831	1.5851	1.5872
4.9	1.5892	1.5913	1.5933	1.5953	1.5974	1.5994	1.6014	1.6034	1.6054	1.6074
5.0	1.6094	1.6114	1.6134	1.6154	1.6174	1.6194	1.6214	1.6233	1.6253	1.6273
5.1	1.6292	1.6312	1.6332	1.6351	1.6371	1.6390	1.6409	1.6429	1.6448	1.6467
5.2	1.6487	1.6506	1.6525	1.6544	1.6563	1.6582	1.6601	1.6620	1.6639	1.6658
5.3	1.6677	1.6696	1.6715	1.6734	1.6752	1.6771	1.6790	1.6808	1.6827	1.6845
5.4	1.6864	1.6882	1.6901	1.6919	1.6938	1.6956	1.6974	1.6993	1.7011	1.7029
5.5	1.7047	1.7066	1.7084	1.7102	1.7120	1.7138	1.7156	1.7174	1.7192	1.7210
5.6	1.7228	1.7246	1.7263	1.7281	1.7299	1.7317	1.7334	1.7352	1.7370	1.7387
5.7	1.7405	1.7422	1.7440	1.7457	1.7475	1.7492	1.7509	1.7527	1.7544	1.7561
5.8	1.7579	1.7596	1.7613	1.7630	1.7647	1.7664	1.7681	1.7699	1.7716	1.7733
5.9	1.7750	1.7766	1.7783	1.7800	1.7817	1.7834	1.7851	1.7867	1.7884	1.7901
6.0	1.7918	1.7934	1.7951	1.7967	1.7984	1.8001	1.8017	1.8034	1.8050	1.8066
6.1	1.8083	1.8099	1.8116	1.8132	1.8148	1.8165	1.8181	1.8197	1.8213	1.8229
6.2	1.8245	1.8262	1.8278	1.8294	1.8310	1.8326	1.8342	1.8358	1.8374	1.8390
6.3	1.8405	1.8421	1.8437	1.8453	1.8469	1.8485	1.8500	1.8516	1.8532	1.8547
6.4	1.8563	1.8579	1.8594	1.8610	1.8625	1.8641	1.8656	1.8672	1.8687	1.8703
6.5	1.8718	1.8733	1.8749	1.8764	1.8779	1.8795	1.8810	1.8825	1.8840	1.8856
6.6	1.8871	1.8886	1.8901	1.8916	1.8931	1.8946	1.8961	1.8976	1.8991	1.9006
6.7	1.9021	1.9036	1.9051	1.9066	1.9081	1.9095	1.9110	1.9125	1.9140	1.9155
6.8	1.9169	1.9184	1.9199	1.9213	1.9228	1.9242	1.9257	1.9272	1.9286	1.9301
6.9	1.9315	1.9330	1.9344	1.9359	1.9373	1.9387	1.9402	1.9416	1.9430	1.9445
7.0	1.9459	1.9473	1.9488	1.9502	1.9516	1.9530	1.9544	1.9559	1.9573	1.9587
7.1	1.9601	1.9615	1.9629	1.9643	1.9657	1.9671	1.9685	1.9699	1.9713	1.9727
7.2	1.9741	1.9755	1.9769	1.9782	1.9796	1.9810	1.9824	1.9838	1.9851	1.9865
7.3	1.9879	1.9892	1.9906	1.9920	1.9933	1.9947	1.9961	1.9974	1.9988	2.0001
7.4	2.0015	2.0028	2.0042	2.0055	2.0069	2.0082	2.0096	2.0109	2.0122	2.0136
7.5	2.0149	2.0162	2.0176	2.0189	2.0202	2.0215	2.0229	2.0242	2.0255	2.0268
7.6	2.0281	2.0295	2.0308	2.0321	2.0334	2.0347	2.0360	2.0373	2.0386	2.0399
7.7	2.0412	2.0425	2.0438	2.0451	2.0464	2.0477	2.0490	2.0503	2.0516	2.0528
7.8	2.0541	2.0554	2.0567	2.0580	2.0592	2.0605	2.0618	2.0631	2.0643	2.0656
7.9	2.0669	2.0681	2.0694	2.0707	2.0719	2.0732	2.0744	2.0757	2.0769	2.0782
8.0	2.0794	2.0807	2.0819	2.0832	2.0844	2.0857	2.0869	2.0882	2.0894	2.0906
8.1	2.0919	2.0931	2.0943	2.0956	2.0968	2.0980	2.0992	2.1005	2.1017	2.1029
8.2	2.1041	2.1054	2.1066	2.1078	2.1090	2.1102	2.1114	2.1126	2.1138	2.1150
8.3	2.1163	2.1175	2.1187	2.1199	2.1211	2.1223	2.1235	2.1247	2.1258	2.1270
8.4	2.1282	2.1294	2.1306	2.1318	2.1330	2.1342	2.1353	2.1365	2.1377	2.1389
8.5	2.1401	2.1412	2.1424	2.1436	2.1448	2.1459	2.1471	2.1483	2.1494	2.1506
8.6	2.1518	2.1529	2.1541	2.1552	2.1564	2.1576	2.1587	2.1599	2.1610	2.1622
8.7	2.1633	2.1645	2.1656	2.1668	2.1679	2.1691	2.1702	2.1713	2.1725	2.1736
8.8	2.1748	2.1759	2.1770	2.1782	2.1793	2.1804	2.1815	2.1827	2.1838	2.1849
8.9	2.1861	2.1872	2.1883	2.1894	2.1905	2.1917	2.1928	2.1939	2.1950	2.1961

N	0	1	2	3	4	5	6	7	8	9
9.0	2.1972	2.1983	2.1994	2.2006	2.2017	2.2028	2.2039	2.2050	2.2061	2.2072
9.1	2.2083	2.2094	2.2105	2.2116	2.2127	2.2138	2.2148	2.2159	2.2170	2.2181
9.2	2.2192	2.2203	2.2214	2.2225	2.2235	2.2246	2.2257	2.2268	2.2279	2.2289
9.3	2.2300	2.2311	2.2322	2.2332	2.2343	2.2354	2.2364	2.2375	2.2386	2.2396
9.4	2.2407	2.2418	2.2428	2.2439	2.2450	2.2460	2.2471	2.2481	2.2492	2.2502
9.5	2.2513	2.2523	2.2534	2.2544	2.2555	2.2565	2.2576	2.2586	2.2597	2.2607
9.6	2.2618	2.2628	2.2638	2.2649	2.2659	2.2670	2.2680	2.2690	2.2701	2.2711
9.7	2.2721	2.2732	2.2742	2.2752	2.2762	2.2773	2.2783	2.2793	2.2803	2.2814
9.8	2.2824	2.2834	2.2844	2.2854	2.2865	2.2875	2.2885	2.2895	2.2905	2.2915
9.9	2.2925	2.2935	2.2946	2.2956	2.2966	2.2976	2.2986	2.2996	2.3006	2.3016

m	1	2	3	4	5
$\ln 10^m$	2.3026	4.6052	6.9078	9.2103	11.5129

13. Length of circumference of a circle with diameter d

This table gives the lengths of circumferences of circles with diameter d to four significant figures, where d is contained in the interval from 1.00 to 9.99. An interpolation correction should be introduced for the third and fourth figure on the right side of the decimal point of d (see p. 17).

If a diameter $D < 1$ or $D > 10$, then we write it in the form $D = d/10^k$ or $D = d \cdot 10^k$, where d lies in the interval from 1.00 to 9.99, and then the value found from the table should be divided or multiplied by 10^k.

Examples. (1) For $d = 69.3$, the length of the circumference is 217.7. (2) For $d = 0.693$, the length is 2.177.

d	0	1	2	3	4	5	6	7	8	9
1.0	3.142	3.173	3.204	3.236	3.267	3.299	3.330	3.362	3.393	3.424
1.1	3.456	3.487	3.519	3.550	3.581	3.613	3.644	3.676	3.707	3.738
1.2	3.770	3.801	3.833	3.864	3.896	3.927	3.958	3.990	4.021	4.053
1.3	4.084	4.115	4.147	4.178	4.210	4.241	4.273	4.304	4.335	4.367
1.4	4.398	4.430	4.461	4.492	4.524	4.555	4.587	4.618	4.650	4.681
1.5	4.712	4.744	4.775	4.807	4.838	4.869	4.901	4.932	4.964	4.995
1.6	5.027	5.058	5.089	5.121	5.152	5.184	5.215	5.246	5.278	5.309
1.7	5.341	5.372	5.404	5.435	5.466	5.498	5.529	5.561	5.592	5.623
1.8	5.655	5.686	5.718	5.749	5.781	5.812	5.843	5.875	5.906	5.938
1.9	5.969	6.000	6.032	6.063	6.095	6.126	6.158	6.189	6.220	6.252

d	0	1	2	3	4	5	6	7	8	9
2.0	6.283	6.315	6.346	6.377	6.409	6.440	6.472	6.503	6.535	6.566
2.1	6.597	6.629	6.660	6.692	6.723	6.754	6.786	6.817	6.849	6.880
2.2	6.912	6.943	6.974	7.006	7.037	7.069	7.100	7.131	7.163	7.194
2.3	7.226	7.257	7.288	7.320	7.351	7.383	7.414	7.446	7.477	7.508
2.4	7.540	7.571	7.603	7.634	7.665	7.697	7.728	7.760	7.791	7.823
2.5	7.854	7.885	7.917	7.948	7.980	8.011	8.042	8.074	8.105	8.137
2.6	8.168	8.200	8.231	8.262	8.294	8.325	8.357	8.388	8.419	8.451
2.7	8.482	8.514	8.545	8.577	8.608	8.639	8.671	8.702	8.734	8.765
2.8	8.796	8.828	8.859	8.891	8.922	8.954	8.985	9.016	9.048	9.079
2.9	9.111	9.142	9.173	9.205	9.236	9.268	9.299	9.331	9.362	9.393
3.0	9.425	9.456	9.488	9.519	9.550	9.582	9.613	9.645	9.676	9.708
3.1	9.739	9.770	9.802	9.833	9.865	9.896	9.927	9.959	9.990	10.02
3.2	10.05	10.08	10.12	10.15	10.18	10.21	10.24	10.27	10.30	10.34
3.3	10.37	10.40	10.43	10.46	10.49	10.52	10.56	10.59	10.62	10.65
3.4	10.68	10.71	10.74	10.78	10.81	10.84	10.87	10.90	10.93	10.96
3.5	11.00	11.03	11.06	11.09	11.12	11.15	11.18	11.22	11.25	11.28
3.6	11.31	11.34	11.37	11.40	11.44	11.47	11.50	11.53	11.56	11.59
3.7	11.62	11.66	11.69	11.72	11.75	11.78	11.81	11.84	11.88	11.91
3.8	11.94	11.97	12.00	12.03	12.06	12.10	12.13	12.16	12.19	12.22
3.9	12.25	12.28	12.32	12.35	12.38	12.41	12.44	12.47	12.50	12.53
4.0	12.57	12.60	12.63	12.66	12.69	12.72	12.75	12.79	12.82	12.85
4.1	12.88	12.91	12.94	12.97	13.01	13.04	13.07	13.10	13.13	13.16
4.2	13.19	13.23	13.26	13.29	13.32	13.35	13.38	13.41	13.45	13.48
4.3	13.51	13.54	13.57	13.60	13.63	13.67	13.70	13.73	13.76	13.79
4.4	13.82	13.85	13.89	13.92	13.95	13.98	14.01	14.04	14.07	14.11
4.5	14.14	14.17	14.20	14.23	14.26	14.29	14.33	14.36	14.39	14.42
4.6	14.45	14.48	14.51	14.55	14.58	14.61	14.64	14.67	14.70	14.73
4.7	14.77	14.80	14.83	14.86	14.89	14.92	14.95	14.99	15.02	15.05
4.8	15.08	15.11	15.14	15.17	15.21	15.24	15.27	15.30	15.33	15.36
4.9	15.39	15.43	15.46	15.49	15.52	15.55	15.58	15.61	15.65	15.68
5.0	15.71	15.74	15.77	15.80	15.83	15.87	15.90	15.93	15.96	15.99
5.1	16.02	16.05	16.08	16.12	16.15	16.18	16.21	16.24	16.27	16.30
5.2	16.34	16.37	16.40	16.43	16.46	16.49	16.52	16.56	16.59	16.62
5.3	16.65	16.68	16.71	16.74	16.78	16.81	16.84	16.87	16.90	16.93
5.4	16.96	17.00	17.03	17.06	17.09	17.12	17.15	17.18	17.22	17.25
5.5	17.28	17.31	17.34	17.37	17.40	17.44	17.47	17.50	17.53	17.56
5.6	17.59	17.62	17.66	17.69	17.72	17.75	17.78	17.81	17.84	17.88
5.7	17.91	17.94	17.97	18.00	18.03	18.06	18.10	18.13	18.16	18.19
5.8	18.22	18.25	18.28	18.32	18.35	18.38	18.41	18.44	18.47	18.50
5.9	18.54	18.57	18.60	18.63	18.66	18.69	18.72	18.76	18.79	18.82

d	0	1	2	3	4	5	6	7	8	9
6.0	18.85	18.88	18.91	18.94	18.98	19.01	19.04	19.07	19.10	19.13
6.1	19.16	19.20	19.23	19.26	19.29	19.32	19.35	19.38	19.42	19.45
6.2	19.48	19.51	19.54	19.57	19.60	19.63	19.67	19.70	19.73	19.76
6.3	19.79	19.82	19.85	19.89	19.92	19.95	19.98	20.01	20.04	20.07
6.4	20.11	20.14	20.17	20.20	20.23	20.26	20.29	20.33	20.36	20.39
6.5	20.42	20.45	20.48	20.51	20.55	20.58	20.61	20.64	20.67	20.70
6.6	20.73	20.77	20.80	20.83	20.86	20.89	20.92	20.95	20.99	21.02
6.7	21.05	21.08	21.11	21.14	21.17	21.21	21.24	21.27	21.30	21.33
6.8	21.36	21.39	21.43	21.46	21.49	21.52	21.55	21.58	21.61	21.65
6.9	21.68	21.71	21.74	21.77	21.80	21.83	21.87	21.90	21.93	21.96
7.0	21.99	22.02	22.05	22.09	22.12	22.15	22.18	22.21	22.24	22.27
7.1	22.31	22.34	22.37	22.40	22.43	22.46	22.49	22.53	22.56	22.59
7.2	22.62	22.65	22.68	22.71	22.75	22.78	22.81	22.84	22.87	22.90
7.3	22.93	22.97	23.00	23.03	23.06	23.09	23.12	23.15	23.19	23.22
7.4	23.25	23.28	23.31	23.34	23.37	23.40	23.44	23.47	23.50	23.53
7.5	23.56	23.59	23.62	23.66	23.69	23.72	23.75	23.78	23.81	23.84
7.6	23.88	23.91	23.94	23.97	24.00	24.03	24.06	24.10	24.13	24.16
7.7	24.19	24.22	24.25	24.28	24.32	24.35	24.38	24.41	24.44	24.47
7.8	24.50	24.54	24.57	24.60	24.63	24.66	24.69	24.72	24.76	24.79
7.9	24.82	24.85	24.88	24.91	24.94	24.98	25.01	25.04	25.07	25.10
8.0	25.13	25.16	25.20	25.23	25.26	25.29	25.32	25.35	25.38	25.42
8.1	25.45	25.48	25.51	25.54	25.57	25.60	25.64	25.67	25.70	25.73
8.2	25.76	25.79	25.82	25.86	25.89	25.92	25.95	25.98	26.01	26.04
8.3	26.08	26.11	26.14	26.17	26.20	26.23	26.26	26.30	26.33	26.36
8.4	26.39	26.42	26.45	26.48	26.52	26.55	26.58	26.61	26.64	26.67
8.5	26.70	26.73	26.77	26.80	26.83	26.86	26.89	26.92	26.95	26.99
8.6	27.02	27.05	27.08	27.11	27.14	27.17	27.21	27.24	27.27	27.30
8.7	27.33	27.36	27.39	27.43	27.46	27.49	27.52	27.55	27.58	27.61
8.8	27.65	27.68	27.71	27.74	27.77	27.80	27.83	27.87	27.90	27.93
8.9	27.96	27.99	28.02	28.05	28.09	28.12	28.15	28.18	28.21	28.24
9.0	28.27	28.31	28.34	28.37	28.40	28.43	28.46	28.49	28.53	28.56
9.1	28.59	28.62	28.65	28.68	28.71	28.75	28.78	28.81	28.84	28.87
9.2	28.90	28.93	28.97	29.00	29.03	29.06	29.09	29.12	29.15	29.19
9.3	29.22	29.25	29.28	29.31	29.34	29.37	29.41	29.44	29.47	29.50
9.4	29.53	29.56	29.59	29.63	29.66	29.69	29.72	29.75	29.78	29.81
9.5	29.85	29.88	29.91	29.94	29.97	30.00	30.03	30.07	30.10	30.13
9.6	30.16	30.19	30.22	30.25	30.28	30.32	30.35	30.38	30.41	30.44
9.7	30.47	30.50	30.54	30.57	30.60	30.63	30.66	30.69	30.72	30.76
9.8	30.79	30.82	30.85	30.88	30.91	30.94	30.98	31.01	31.04	31.07
9.9	31.10	31.13	31.16	31.20	31.23	31.26	31.29	31.32	31.35	31.38
10.0	31.42									

14. Area of a circle with diameter d

This table gives the areas of circles with diameter d to four significant figures, where d lies in the interval from 1.00 to 9.99. As in table 13, a interpolation correction should be introduced for the third and fourth figure on the right side of the decimal point of d (see p. 17).

If a diameter $D < 1$ or $D > 10$, we write it in the form $D = d/10^k$ or $D = d \cdot 10^k$, where $1.00 < d < 9.99$, and then the value of the area taken from the table we divide or multiply by 10^{2k}.

Examples. (1) For $d = 69.3$, the area is 3772. (2) For $d = 0.693$, the area is 0.3772.

d	0	1	2	3	4	5	6	7	8	9
1.0	0.7854	0.8012	0.8171	0.8332	0.8495	0.8659	0.8825	0.8992	0.9161	0.9331
1.1	0.9503	0.9677	0.9852	1.003	1.021	1.039	1.057	1.075	1.094	1.112
1.2	1.131	1.150	1.169	1.188	1.208	1.227	1.247	1.267	1.287	1.307
1.3	1.327	1.348	1.368	1.389	1.410	1.431	1.453	1.474	1.496	1.517
1.4	1.539	1.561	1.584	1.606	1.629	1.651	1.674	1.697	1.720	1.744
1.5	1.767	1.791	1.815	1.839	1.863	1.887	1.911	1.936	1.961	1.986
1.6	2.011	2.036	2.061	2.087	2.112	2.138	2.164	2.190	2.217	2.243
1.7	2.270	2.297	2.324	2.351	2.378	2.405	2.433	2.461	2.488	2.516
1.8	2.545	2.573	2.602	2.630	2.659	2.688	2.717	2.746	2.776	2.806
1.9	2.835	2.865	2.895	2.926	2.956	2.986	3.017	3.048	3.079	3.110
2.0	3.142	3.173	3.205	3.237	3.269	3.301	3.333	3.365	3.398	3.431
2.1	3.464	3.497	3.530	3.563	3.597	3.631	3.664	3.698	3.733	3.767
2.2	3.801	3.836	3.871	3.906	3.941	3.976	4.011	4.047	4.083	4.119
2.3	4.155	4.191	4.227	4.264	4.301	4.337	4.374	4.412	4.449	4.486
2.4	4.524	4.562	4.600	4.638	4.676	4.714	4.753	4.792	4.831	4.870
2.5	4.909	4.948	4.988	5.027	5.067	5.107	5.147	5.187	5.228	5.269
2.6	5.309	5.350	5.391	5.433	5.474	5.515	5.557	5.599	5.641	5.683
2.7	5.726	5.768	5.811	5.853	5.896	5.940	5.983	6.026	6.070	6.114
2.8	6.158	6.202	6.246	6.290	6.335	6.379	6.424	6.469	6.514	6.560
2.9	6.605	6.651	6.697	6.743	6.789	6.835	6.881	6.928	6.975	7.022
3.0	7.069	7.116	7.163	7.211	7.258	7.306	7.354	7.402	7.451	7.499
3.1	7.548	7.596	7.645	7.694	7.744	7.793	7.843	7.892	7.942	7.992
3.2	8.042	8.093	8.143	8.194	8.245	8.296	8.347	8.398	8.450	8.501
3.3	8.553	8.605	8.657	8.709	8.762	8.814	8.867	8.920	8.973	9.026
3.4	9.079	9.133	9.186	9.240	9.294	9.348	9.402	9.457	9.511	9.566

d	0	1	2	3	4	5	6	7	8	9
3.5	9.621	9.676	9.731	9.787	9.842	9.898	9.954	10.01	10.07	10.12
3.6	10.18	10.24	10.29	10.35	10.41	10.46	10.52	10.58	10.64	10.69
3.7	10.75	10.81	10.87	10.93	10.99	11.04	11.10	11.16	11.22	11.28
3.8	11.34	11.40	11.46	11.52	11.58	11.64	11.70	11.76	11.82	11.88
3.9	11.95	12.01	12.07	12.13	12.19	12.25	12.32	12.38	12.44	12.50
4.0	12.57	12.63	12.69	12.76	12.82	12.88	12.95	13.01	13.07	13.14
4.1	13.20	13.27	13.33	13.40	13.46	13.53	13.59	13.66	13.72	13.79
4.2	13.85	13.92	13.99	14.05	14.12	14.19	14.25	14.32	14.39	14.45
4.3	14.52	14.59	14.66	14.73	14.79	14.86	14.93	15.00	15.07	15.14
4.4	15.21	15.27	15.34	15.41	15.48	15.55	15.62	15.69	15.76	15.83
4.5	15.90	15.98	16.05	16.12	16.19	16.26	16.33	16.40	16.47	16.55
4.6	16.62	16.69	16.76	16.84	16.91	16.98	17.06	17.13	17.20	17.28
4.7	17.35	17.42	17.50	17.57	17.65	17.72	17.80	17.87	17.95	18.02
4.8	18.10	18.17	18.25	18.32	18.40	18.47	18.55	18.63	18.70	18.78
4.9	18.86	18.93	19.01	19.09	19.17	19.24	19.32	19.40	19.48	19.56
5.0	19.63	19.71	19.79	19.87	19.95	20.03	20.11	20.19	20.27	20.35
5.1	20.43	20.51	20.59	20.67	20.75	20.83	20.91	20.99	21.07	21.16
5.2	21.24	21.32	21.40	21.48	21.57	21.65	21.73	21.81	21.90	21.98
5.3	22.06	22.15	22.23	22.31	22.40	22.48	22.56	22.65	22.73	22.82
5.4	22.90	22.99	23.07	23.16	23.24	23.33	23.41	23.50	23.59	23.67
5.5	23.76	23.84	23.93	24.02	24.11	24.19	24.28	24.37	24.45	24.54
5.6	24.63	24.72	24.81	24.89	24.98	25.07	25.16	25.25	25.34	25.43
5.7	25.52	25.61	25.70	25.79	25.88	25.97	26.06	26.15	26.24	26.33
5.8	26.42	26.51	26.60	26.69	26.79	26.88	26.97	27.06	27.15	27.25
5.9	27.34	27.43	27.53	27.62	27.71	27.81	27.90	27.99	28.09	28.18
6.0	28.27	28.37	28.46	28.56	28.65	28.75	28.84	28.94	29.03	29.13
6.1	29.22	29.32	29.42	29.51	29.61	29.71	29.80	29.90	30.00	30.09
6.2	30.19	30.29	30.39	30.48	30.58	30.68	30.78	30.88	30.97	31.07
6.3	31.17	31.27	31.37	31.47	31.57	31.67	31.77	31.87	31.97	32.07
6.4	32.17	32.27	32.37	32.47	32.57	32.67	32.78	32.88	32.98	33.08
6.5	33.18	33.29	33.39	33.49	33.59	33.70	33.80	33.90	34.00	34.11
6.6	34.21	34.32	34.42	34.52	34.63	34.73	34.84	34.94	35.05	35.15
6.7	35.26	35.36	35.47	35.57	35.68	35.78	35.89	36.00	36.10	36.21
6.8	36.32	36.42	36.53	36.64	36.75	36.85	36.96	37.07	37.18	37.28
6.9	37.39	37.50	37.61	37.72	37.83	37.94	38.05	38.16	38.26	38.37
7.0	38.48	38.59	38.70	38.82	38.93	39.04	39.15	39.26	39.37	39.48
7.1	39.59	39.70	39.82	39.93	40.04	40.15	40.26	40.38	40.49	40.60
7.2	40.72	40.83	40.94	41.06	41.17	41.28	41.40	41.51	41.62	41.74
7.3	41.85	41.97	42.08	42.20	42.31	42.43	42.54	42.66	42.78	42.89
7.4	43.01	43.12	43.24	43.36	43.47	43.59	43.71	43.83	43.94	44.06
7.5	44.18	44.30	44.41	44.53	44.65	44.77	44.89	45.01	45.13	45.25
7.6	45.36	45.48	45.60	45.72	45.84	45.96	46.08	46.20	46.32	46.45
7.7	46.57	46.69	46.81	46.93	47.05	47.17	47.29	47.42	47.54	47.66
7.8	47.78	47.91	48.03	48.15	48.27	48.40	48.52	48.65	48.77	48.89
7.9	49.02	49.14	49.27	49.39	49.51	49.64	49.76	49.89	50.01	50.14

d	0	1	2	3	4	5	6	7	8	9
8.0	50.27	50.39	50.52	50.64	50.77	50.90	51.02	51.15	51.28	51.40
8.1	51.53	51.66	51.78	51.91	52.04	52.17	52.30	52.42	52.55	52.68
8.2	52.81	52.94	53.07	53.20	53.33	53.46	53.59	53.72	53.85	53.98
8.3	54.11	54.24	54.37	54.50	54.63	54.76	54.89	55.02	55.15	55.29
8.4	55.42	55.55	55.68	55.81	55.95	56.08	56.21	56.35	56.48	56.61
8.5	56.75	56.88	57.01	57.15	57.28	57.41	57.55	57.68	57.82	57.95
8.6	58.09	58.22	58.36	58.49	58.63	58.77	58.90	59.04	59.17	59.31
8.7	59.45	59.58	59.72	59.86	59.99	60.13	60.27	60.41	60.55	60.68
8.8	60.82	60.96	61.10	61.24	61.38	61.51	61.65	61.79	61.93	62.07
8.9	62.21	62.35	62.49	62.63	62.77	62.91	63.05	63.19	63.33	63.48
9.0	63.62	63.76	63.90	64.04	64.18	64.33	64.47	64.61	64.75	64.90
9.1	65.04	65.18	65.33	65.47	65.61	65.76	65.90	66.04	66.19	66.33
9.2	66.48	66.62	66.77	66.91	67.06	67.20	67.35	67.49	67.64	67.78
9.3	67.93	68.08	68.22	68.37	68.51	68.66	68.81	68.96	69.10	69.25
9.4	69.40	69.55	69.69	69.84	69.99	70.14	70.29	70.44	70.58	70.73
9.5	70.88	71.03	71.18	71.33	71.48	71.63	71.78	71.93	72.08	72.23
9.6	72.38	72.53	72.68	72.84	72.99	73.14	73.29	73.44	73.59	73.75
9.7	73.90	74.05	74.20	74.36	74.51	74.66	74.82	74.97	75.12	75.28
9.8	75.43	75.58	75.74	75.89	76.05	76.20	76.36	76.51	76.67	76.82
9.9	76.98	77.13	77.29	77.44	77.60	77.76	77.91	78.07	78.23	78.38
10.0	78.54									

15. Elements of the segment of a circle ([1])

(a) Length l of arc and area S of a segment with a chord equal to 1. This table gives the length l of arc and the area S of the segment with a constant chord $a = 1$, when its height h varies from 0.00 to 0.50. When the chord $a \neq 1$, the argument h should be divided by a while the value of l should be multiplied by a and the value of S by a^2. An interpolation correction should be introduced for the third and fourth figure beyond the decimal point of h.

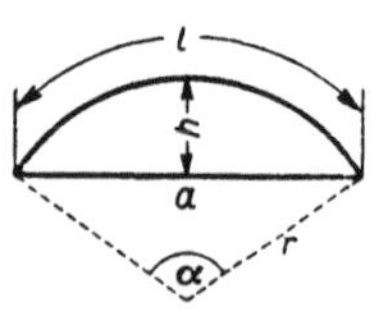

FIG. 1

Example. If the chord of a segment is $a = 40$ cm and the height $h = 6$ cm, then, to find the length of arc l_1, we use table 15a. From the proportion $h_1/a_1 = h/1$ we obtain $h = 6/40 = 0.15$ and, from the table, the corresponding value for $l = 1.0590$. Then, from the proportion $l_1/l = a_1/1$, we obtain $l_1 = la_1 = 1.0590 \cdot 40 = 42.36$ cm.

([1]) **For the corresponding formulas see pp. 200, 201.**

h	l	S	h	l	S
—	—	—	**0.25**	1.1591	0.1747
0.01	1.0003	0.0067	0.26	1.1715	0.1824
0.02	1.0011	0.0133	0.27	1.1843	0.1901
0.03	1.0024	0.0200	0.28	1.1975	0.1979
0.04	1.0043	0.0267	0.29	1.2110	0.2058
0.05	1.0067	0.0334	**0.30**	1.2250	0.2137
0.06	1.0096	0.0401	0.31	1.2393	0.2218
0.07	1.0130	0.0468	0.32	1.2539	0.2299
0.08	1.0170	0.0536	0.33	1.2689	0.2381
0.09	1.0215	0.0604	0.34	1.2843	0.2464
0.10	1.0265	0.0672	**0.35**	1.3000	0.2548
0.11	1.0320	0.0740	0.36	1.3160	0.2633
0.12	1.0380	0.0809	0.37	1.3323	0.2719
0.13	1.0445	0.0878	0.38	1.3490	0.2806
0.14	1.0515	0.0948	0.39	1.3660	0.2893
0.15	1.0590	0.1018	**0.40**	1.3832	0.2982
0.16	1.0669	0.1088	0.41	1.4008	0.3072
0.17	1.0754	0.1159	0.42	1.4186	0.3162
0.18	1.0843	0.1231	0.43	1.4367	0.3254
0.19	1.0936	0.1303	0.44	1.4551	0.3347
0.20	1.1035	0.1375	**0.45**	1.4738	0.3441
0.21	1.1137	0.1448	0.46	1.4927	0.3536
0.22	1.1244	0.1522	0.47	1.5118	0.3632
0.23	1.1356	0.1596	0.48	1.5313	0.3729
0.24	1.1471	0.1671	0.49	1.5509	0.3828
0.25	1.1591	0.1747	**0.50**	1.5708	0.3927

(b) Length l of arc, height h, chord a and area S of a segment of a circle with radius $r = 1$. This table gives elements of different segments of a circle whose radius is equal to 1 (Fig. 1). The argument is the angle α expressed in degrees. For every angle α, the length l of arc (i.e., the radian measure of α), the height h, the ratio l/a, the chord a, the ratio a/h and the area S of the segment are given. If the radius of the circle is $r \neq 1$, then the tabulated values of l, h and a should be multiplied by r, the area S by r^2 and the ratios l/a and a/h remain unchanged. To obtain the radius r, for a given length l_1 of arc and height h_1, we compute the ratio $l_1/h_1 = s$ (which determines the shape of the segment) and, taking s from the column l/h, we obtain the corresponding tabulated value of l; then, from the proportion $l_1/l = r/1$, we obtain the radius $r = l_1/l$. The procedure of finding r, when the chord a_1 and the height h_1 are given, is analogous.

Example. If the chord of a segment is $a = 40$ cm, and the height $h = 6$ cm, then we obtain the angle and the radius r from table 15b; we find $a_1/h_1 = a/h = 6.67$ and, in the corresponding row (using interpolation), we have $\alpha = 66.8°$, $a_1 = 1.1010$ and $l = 1.1661$; from the proportion $a_1/a = r/1$, we obtain $r = 40/1.1010 = 36.33$ cm. Similarly, from the proportion $l_1/l = r/1$, we obtain $l_1 = 1.1661 \cdot 36.33 = 42.36$ cm.

$\alpha°$	l	h	$\frac{l}{h}$	a	$\frac{a}{h}$	S
1	0.0175	0.0000	458.37	0.0175	458.36	0.00000
2	0.0349	0.0002	229.19	0.0349	229.18	0.00000
3	0.0524	0.0003	152.80	0.0524	152.78	0.00001
4	0.0698	0.0006	114.60	0.0698	114.58	0.00003
5	0.0873	0.0010	91.69	0.0872	91.66	0.00006
6	0.1047	0.0014	76.41	0.1047	76.38	0.00010
7	0.1222	0.0019	65.50	0.1221	65.46	0.00015
8	0.1396	0.0024	57.32	0.1395	57.27	0.00023
9	0.1571	0.0031	50.96	0.1569	50.90	0.00032
10	0.1745	0.0038	45.87	0.1743	45.81	0.00044
11	0.1920	0.0046	41.70	0.1917	41.64	0.00059
12	0.2094	0.0055	38.23	0.2091	38.16	0.00076
13	0.2269	0.0064	35.30	0.2264	35.22	0.00097
14	0.2443	0.0075	32.78	0.2437	32.70	0.00121
15	0.2618	0.0086	30.60	0.2611	30.51	0.00149
16	0.2793	0.0097	28.69	0.2783	28.60	0.00181
17	0.2967	0.0110	27.01	0.2956	26.91	0.00217
18	0.3142	0.0123	25.52	0.3129	25.41	0.00257
19	0.3316	0.0137	24.18	0.3301	24.07	0.00302
20	0.3491	0.0152	22.98	0.3473	22.86	0.00352
21	0.3665	0.0167	21.89	0.3645	21.77	0.00408
22	0.3840	0.0184	20.90	0.3816	20.77	0.00468
23	0.4014	0.0201	20.00	0.3987	19.86	0.00535
24	0.4189	0.0219	19.17	0.4158	19.03	0.00607
25	0.4363	0.0237	18.41	0.4329	18.26	0.00686
26	0.4538	0.0256	17.71	0.4499	17.55	0.00771
27	0.4712	0.0276	17.06	0.4669	16.90	0.00862
28	0.4887	0.0297	16.45	0.4838	16.29	0.00961
29	0.5061	0.0319	15.89	0.5008	15.72	0.01067
30	0.5236	0.0341	15.37	0.5176	15.19	0.01180
31	0.5411	0.0364	14.88	0.5345	14.70	0.01301
32	0.5585	0.0387	14.42	0.5513	14.23	0.01429
33	0.5760	0.0412	13.99	0.5680	13.79	0.01566
34	0.5934	0.0437	13.58	0.5847	13.38	0.01711
35	0.6109	0.0463	13.20	0.6014	12.99	0.01864

$\alpha°$	l	h	$\frac{l}{h}$	a	$\frac{a}{h}$	S
35	0.6109	0.0463	13.20	0.6014	12.99	0.01864
36	0.6283	0.0489	12.84	0.6180	12.63	0.02027
37	0.6458	0.0517	12.50	0.6346	12.28	0.02198
38	0.6632	0.0545	12.17	0.6511	11.95	0.02378
39	0.6807	0.0574	11.87	0.6676	11.64	0.02568
40	0.6981	0.0603	11.58	0.6840	11.34	0.02767
41	0.7156	0.0633	11.30	0.7004	11.06	0.02976
42	0.7330	0.0664	11.04	0.7167	10.79	0.03195
43	0.7505	0.0696	10.79	0.7330	10.53	0.03425
44	0.7679	0.0728	10.55	0.7492	10.29	0.03664
45	0.7854	0.0761	10.32	0.7654	10.05	0.03915
46	0.8029	0.0795	10.10	0.7815	9.83	0.04176
47	0.8203	0.0829	9.89	0.7975	9.62	0.04448
48	0.8378	0.0865	9.69	0.8135	9.41	0.04731
49	0.8552	0.0900	9.50	0.8294	9.21	0.05025
50	0.8727	0.0937	9.31	0.8452	9.02	0.05331
51	0.8901	0.0974	9.14	0.8610	8.84	0.05649
52	0.9076	0.1012	8.97	0.8767	8.66	0.05978
53	0.9250	0.1051	8.80	0.8924	8.49	0.06319
54	0.9425	0.1090	8.65	0.9080	8.33	0.06673
55	0.9599	0.1130	8.50	0.9235	8.17	0.07039
56	0.9774	0.1171	8.35	0.9389	8.02	0.07417
57	0.9948	0.1212	8.21	0.9543	7.88	0.07808
58	1.0123	0.1254	8.07	0.9696	7.73	0.08212
59	1.0297	0.1296	7.94	0.9848	7.60	0.08629
60	1.0472	0.1340	7.82	1.0000	7.46	0.09059
61	1.0647	0.1384	7.69	1.0151	7.34	0.09502
62	1.0821	0.1428	7.58	1.0301	7.21	0.09958
63	1.0996	0.1474	7.46	1.0450	7.09	0.10428
64	1.1170	0.1520	7.35	1.0598	6.97	0.10911
65	1.1345	0.1566	7.24	1.0746	6.86	0.11408
66	1.1519	0.1613	7.14	1.0893	6.75	0.11919
67	1.1694	0.1661	7.04	1.1039	6.65	0.12443
68	1.1868	0.1710	6.94	1.1184	6.54	0.12982
69	1.2043	0.1759	6.85	1.1328	6.44	0.13535
70	1.2217	0.1808	6.76	1.1472	6.34	0.14102
71	1.2392	0.1859	6.67	1.1614	6.25	0.14683
72	1.2566	0.1910	6.58	1.1756	6.16	0.15279
73	1.2741	0.1961	6.50	1.1896	6.07	0.15889
74	1.2915	0.2014	6.41	1.2036	5.98	0.16514
75	1.3090	0.2066	6.33	1.2175	5.89	0.17154

$\alpha°$	l	h	$\frac{l}{h}$	a	$\frac{a}{h}$	S
75	1.3090	0.2066	6.33	1.2175	5.89	0.17154
76	1.3265	0.2120	6.26	1.2313	5.81	0.17808
77	1.3439	0.2174	6.18	1.2450	5.73	0.18477
78	1.3614	0.2229	6.11	1.2586	5.65	0.19160
79	1.3788	0.2284	6.04	1.2722	5.57	0.19859
80	1.3963	0.2340	5.97	1.2856	5.49	0.20573
81	1.4137	0.2396	5.90	1.2989	5.42	0.21301
82	1.4312	0.2453	5.83	1.3121	5.35	0.22045
83	1.4486	0.2510	5.77	1.3252	5.28	0.22804
84	1.4661	0.2569	5.71	1.3383	5.21	0.23578
85	1.4835	0.2627	5.65	1.3512	5.14	0.24367
86	1.5010	0.2686	5.59	1.3640	5.08	0.25171
87	1.5184	0.2746	5.53	1.3767	5.01	0.25990
88	1.5359	0.2807	5.47	1.3893	4.95	0.26825
89	1.5533	0.2867	5.42	1.4018	4.89	0.27675
90	1.5708	0.2929	5.36	1.4142	4.83	0.28540
91	1.5882	0.2991	5.31	1.4265	4.77	0.29420
92	1.6057	0.3053	5.26	1.4387	4.71	0.30316
93	1.6232	0.3116	5.21	1.4507	4.66	0.31226
94	1.6406	0.3180	5.16	1.4627	4.60	0.32152
95	1.6581	0.3244	5.11	1.4746	4.55	0.33093
96	1.6755	0.3309	5.06	1.4863	4.49	0.34050
97	1.6930	0.3374	5.02	1.4979	4.44	0.35021
98	1.7104	0.3439	4.97	1.5094	4.39	0.36008
99	1.7279	0.3506	4.93	1.5208	4.34	0.37009
100	1.7453	0.3572	4.89	1.5321	4.29	0.38026
101	1.7628	0.3639	4.84	1.5432	4.24	0.39058
102	1.7802	0.3707	4.80	1.5543	4.19	0.40104
103	1.7977	0.3775	4.76	1.5652	4.15	0.41166
104	1.8151	0.3843	4.72	1.5760	4.10	0.42242
105	1.8326	0.3912	4.68	1.5867	4.06	0.43333
106	1.8500	0.3982	4.65	1.5973	4.01	0.44439
107	1.8675	0.4052	4.61	1.6077	3.97	0.45560
108	1.8850	0.4122	4.57	1.6180	3.93	0.46695
109	1.9024	0.4193	4.54	1.6282	3.88	0.47845
110	1.9199	0.4264	4.50	1.6383	3.84	0.49008
111	1.9373	0.4336	4.47	1.6483	3.80	0.50187
112	1.9548	0.4408	4.43	1.6581	3.76	0.51379
113	1.9722	0.4481	4.40	1.6678	3.72	0.52586
114	1.9897	0.4554	4.37	1.6773	3.68	0.53806
115	2.0071	0.4627	4.34	1.6868	3.65	0.55041

$\alpha°$	l	h	$\frac{l}{h}$	a	$\frac{a}{h}$	S
115	2.0071	0.4627	4.34	1.6868	3.65	0.55041
116	2.0246	0.4701	4.31	1.6961	3.61	0.56289
117	2.0420	0.4775	4.28	1.7053	3.57	0.57551
118	2.0595	0.4850	4.25	1.7143	3.53	0.58827
119	2.0769	0.4925	4.22	1.7233	3.50	0.60116
120	2.0944	0.5000	4.19	1.7321	3.46	0.61418
121	2.1118	0.5076	4.16	1.7407	3.43	0.62734
122	2.1293	0.5152	4.13	1.7492	3.40	0.64063
123	2.1468	0.5228	4.11	1.7576	3.36	0,65404
124	2.1642	0.5305	4.08	1.7659	3.33	0.66759
125	2.1817	0.5383	4.05	1.7740	3.30	0.68125
126	2.1991	0.5460	4.03	1.7820	3.26	0.69505
127	2.2166	0.5538	4.00	1.7899	3.23	0.70897
128	2.2340	0.5616	3.98	1.7976	3.20	0.72301
129	2.2515	0.5695	3.95	1.8052	3.17	0.73716
130	2.2689	0.5774	3.93	1.8126	3.14	0.75144
131	2.2864	0.5853	3.91	1.8199	3.11	0.76584
132	2.3038	0.5933	3.88	1.8271	3.08	0.78034
133	2.3213	0.6013	3.86	1.8341	3.05	0.79497
134	2.3387	0.6093	3.84	1.8410	3.02	0.80970
135	2.3562	0.6173	3.82	1.8478	2.99	0.82454
136	2.3736	0.6254	3.80	1.8544	2.97	0.83949
137	2.3911	0.6335	3.77	1.8608	2.94	0.85455
138	2.4086	0.6416	3.75	1.8672	2.91	0.86971
139	2.4260	0.6498	3.73	1.8733	2.88	0.88497
140	2.4435	0.6580	3.71	1.8794	2.86	0.90034
141	2.4609	0.6662	3.69	1.8853	2.83	0.91580
142	2.4784	0.6744	3.67	1.8910	2.80	0.93135
143	2.4958	0.6827	3.66	1.8966	2.78	0.94700
144	2.5133	0.6910	3.64	1.9021	2.75	0.96274
145	2.5307	0.6993	3.62	1.9074	2.73	0.97858
146	2.5482	0.7076	3.60	1.9126	2.70	0.99449
147	2.5656	0.7160	3.58	1.9176	2.68	1.01050
148	2.5831	0.7244	3.57	1.9225	2.65	1.02658
149	2.6005	0.7328	3.55	1.9273	2.63	1.04275
150	2.6180	0.7412	3.53	1.9319	2.61	1.05900
151	2.6354	0.7496	3.52	1.9363	2.58	1.07532
152	2.6529	0.7581	3.50	1.9406	2.56	1.09171
153	2.6704	0.7666	3.48	1.9447	2.54	1.10818
154	2.6878	0.7750	3.47	1.9487	2.51	1.12472
155	2.7053	0.7836	3.45	1.9526	2.49	1.14132

$\alpha°$	l	h	$\frac{l}{h}$	a	$\frac{a}{h}$	S
155	2.7053	0.7836	3.45	1.9526	2.49	1.14132
156	2.7227	0.7921	3.44	1.9563	2.47	1.15799
157	2.7402	0.8006	3.42	1.9598	2.45	1.17472
158	2.7576	0.8092	3.41	1.9633	2.43	1.19151
159	2.7751	0.8178	3.39	1.9665	2.40	1.20835
160	2.7925	0.8264	3.38	1.9696	2.38	1.22525
161	2.8100	0.8350	3.37	1.9726	2.36	1.24221
162	2.8274	0.8436	3.35	1.9754	2.34	1.25921
163	2.8449	0.8522	3.34	1.9780	2.32	1.27626
164	2.8623	0.8608	3.33	1.9805	2.30	1.29335
165	2.8798	0.8695	3.31	1.9829	2.28	1.31049
166	2.8972	0.8781	3.30	1.9851	2.26	1.32766
167	2.9147	0.8868	3.29	1.9871	2.24	1.34487
168	2.9322	0.8955	3.27	1.9890	2.22	1.36212
169	2.9496	0.9042	3.26	1.9908	2.20	1.37940
170	2.9671	0.9128	3.25	1.9924	2.18	1.39671
171	2.9845	0.9215	3.24	1.9938	2.16	1.41404
172	3.0020	0.9302	3.23	1.9951	2.14	1.43140
173	3.0194	0.9390	3.22	1.9963	2.13	1.44878
174	3.0369	0.9477	3.20	1.9973	2.11	1.46617
175	3.0543	0.9564	3.19	1.9981	2.09	1.48359
176	3.0718	0.9651	3.18	1.9988	2.07	1.50101
177	3.0892	0.9738	3.17	1.9993	2.05	1.51845
178	3.1067	0.9825	3.16	1.9997	2.04	1.53589
179	3.1241	0.9913	3.15	1.9999	2.02	1.55334
180	3.1416	1.0000	3.14	2.0000	2.00	1.57080

16. Sexagesimal measure of angles expressed in radians

The following examples explain the method of using table 16.

(1) 52°37′23″

50°	= 0.872665
2°	= 0.034907
30′	= 0.008727
7′	= 0.002036
20″	= 0.000097
3″	= 0.000015
	0.918447
52°37′23″	= 0.91845 rad

(2) 5.645 rad

5.235988	= 300°
0.409012	
0.401426	= 23°
0.007586	
0.005818	= 20′
0.001768	
0.001745	= 6′
0.000023	= 5″
5.645 rad	= 323°26′5″

It is convenient to perform the computations on an abacus.

Length of arc of a circle with radius equal to 1

Angle	Arc	Angle	Arc	Angle	Arc
1″	0.000005	1°	0.017453	31°	0.541052
2	0.000010	2	0.034907	32	0.558505
3	0.000015	3	0.052360	33	0.575959
4	0.000019	4	0.069813	34	0.593412
5	0.000024	5	0.087266	35	0.610865
6	0.000029	6	0.104720	36	0.628319
7	0.000034	7	0.122173	37	0.645772
8	0.000039	8	0.139626	38	0.663225
9	0.000044	9	0.157080	39	0.680678
10	0.000048	10	0.174533	40	0.698132
20	0.000097	11	0.191986	45	0.785398
30	0.000145	12	0.209440	50	0.872665
40	0.000194	13	0.226893	55	0.959931
50	0.000242	14	0.244346	60	1.047198
		15	0.261799	65	1.134464
1′	0.000291	16	0.279253	70	1.221730
2	0.000582	17	0.296706	75	1.308997
3	0.000873	18	0.314159	80	1.396263
4	0.001164	19	0.331613	85	1.483530
5	0.001454	20	0.349066	90	1.570796
6	0.001745	21	0.366519	100	1.745329
7	0.002036	22	0.383972	120	2.094395
8	0.002327	23	0.401426	150	2.617994
9	0.002618	24	0.418879	180	3.141593
10	0.002909	25	0.436332	200	3.490659
20	0.005818	26	0.453786	250	4.363323
30	0.008727	27	0.471239	270	4.712389
40	0.011636	28	0.488692	300	5.235988
50	0.014544	29	0.506145	360	6.283185
		30	0.523599	400	6.981317

The arc equal to the radius subtends an angle 57°17′44″.8 (it is equal to 1 radian).

17. Proportional parts

	11	12	13	14	15	16	17	18	19	20	
1	1.1	1.2	1.3	1.4	1.5	1.6	1.7	1.8	1.9	2.0	1
2	2.2	2.4	2.6	2.8	3.0	3.2	3.4	3.6	3.8	4.0	2
3	3.3	3.6	3.9	4.2	4.5	4.8	5.1	5.4	5.7	6.0	3
4	4.4	4.8	5.2	5.6	6.0	6.4	6.8	7.2	7.6	8.0	4
5	5.5	6.0	6.5	7.0	7.5	8.0	8.5	9.0	9.5	10.0	5
6	6.6	7.2	7.8	8.4	9.0	9.6	10.2	10.8	11.4	12.0	6
7	7.7	8.4	9.1	9.8	10.5	11.2	11.9	12.6	13.3	14.0	7
8	8.8	9.6	10.4	11.2	12.0	12.8	13.6	14.4	15.2	16.0	8
9	9.9	10.8	11.7	12.6	13.5	14.4	15.3	16.2	17.1	18.0	9

	21	22	23	24	25	26	27	28	29	30	
1	2.1	2.2	2.3	2.4	2.5	2.6	2.7	2.8	2.9	3.0	1
2	4.2	4.4	4.6	4.8	5.0	5.2	5.4	5.6	5.8	6.0	2
3	6.3	6.6	6.9	7.2	7.5	7.8	8.1	8.4	8.7	9.0	3
4	8.4	8.8	9.2	9.6	10.0	10.4	10.8	11.2	11.6	12.0	4
5	10.5	11.0	11.5	12.0	12.5	13.0	13.5	14.0	14.5	15.0	5
6	12.6	13.2	13.8	14.4	15.0	15.6	16.2	16.8	17.4	18.0	6
7	14.7	15.4	16.1	16.8	17.5	18.2	18.9	19.6	20.3	21.0	7
8	16.8	17.6	18.4	19.2	20.0	20.8	21.6	22.4	23.2	24.0	8
9	18.9	19.8	20.7	21.6	22.5	23.4	24.3	25.2	26.1	27.0	9

	31	32	33	34	35	36	37	38	39	40	
1	3.1	3.2	3.3	3.4	3.5	3.6	3.7	3.8	3.9	4.0	1
2	6.2	6.4	6.6	6.8	7.0	7.2	7.4	7.6	7.8	8.0	2
3	9.3	9.6	9.9	10.2	10.5	10.8	11.1	11.4	11.7	12.0	3
4	12.4	12.8	13.2	13.6	14.0	14.4	14.8	15.2	15.6	16.0	4
5	15.5	16.0	16.5	17.0	17.5	18.0	18.5	19.0	19.5	20.0	5
6	18.6	19.2	19.8	20.4	21.0	21.6	22.2	22.8	23.4	24.0	6
7	21.7	22.4	23.1	23.8	24.5	25.2	25.9	26.6	27.3	28.0	7
8	24.8	25.6	26.4	27.2	28.0	28.8	29.6	30.4	31.2	32.0	8
9	27.9	28.8	29.7	30.6	31.5	32.4	33.3	34.2	35.1	36.0	9

	41	42	43	44	45	46	47	48	49	50	
1	4.1	4.2	4.3	4.4	4.5	4.6	4.7	4.8	4.9	5.0	1
2	8.2	8.4	8.6	8.8	9.0	9.2	9.4	9.6	9.8	10.0	2
3	12.3	12.6	12.9	13.2	13.5	13.8	14.1	14.4	14.7	15.0	3
4	16.4	16.8	17.2	17.6	18.0	18.4	18.8	19.2	19.6	20.0	4
5	20.5	21.0	21.5	22.0	22.5	23.0	23.5	24.0	24.5	25.0	5
6	24.6	25.2	25.8	26.4	27.0	27.6	28.2	28.8	29.4	30.0	6
7	28.7	29.4	30.1	30.8	31.5	32.2	32.9	33.6	34.3	35.0	7
8	32.8	33.6	34.4	35.2	36.0	36.8	37.6	38.4	39.2	40.0	8
9	36.9	37.8	38.7	39.6	40.5	41.4	42.3	43.2	44.1	45.0	9

	51	52	53	54	55	56	57	58	59	60	
1	5.1	5.2	5.3	5.4	5.5	5.6	5.7	5.8	5.9	6.0	1
2	10.2	10.4	10.6	10.8	11.0	11.2	11.4	11.6	11.8	12.0	2
3	15.3	15.6	15.9	16.2	16.5	16.8	17.1	17.4	17.7	18.0	3
4	20.4	20.8	21.2	21.6	22.0	22.4	22.8	23.2	23.6	24.0	4
5	25.5	26.0	26.5	27.0	27.5	28.0	28.5	29.0	29.5	30.0	5
6	30.6	31.2	31.8	32.4	33.0	33.6	34.2	34.8	35.4	36.0	6
7	35.7	36.4	37.1	37.8	38.5	39.2	39.9	40.6	41.3	42.0	7
8	40.8	41.6	42.4	43.2	44.0	44.8	45.6	46.4	47.2	48.0	8
9	45.9	46.8	47.7	48.6	49.5	50.4	51.3	52.2	53.1	54.0	9

	61	62	63	64	65	66	67	68	69	70	
1	6.1	6.2	6.3	6.4	6.5	6.6	6.7	6.8	6.9	7.0	1
2	12.2	12.4	12.6	12.8	13.0	13.2	13.4	13.6	13.8	14.0	2
3	18.3	18.6	18.9	19.2	19.5	19.8	20.1	20.4	20.7	21.0	3
4	24.4	24.8	25.2	25.6	26.0	26.4	26.8	27.2	27.6	28.0	4
5	30.5	31.0	31.5	32.0	32.5	33.0	33.5	34.0	34.5	35.0	5
6	36.6	37.2	37.8	38.4	39.0	39.6	40.2	40.8	41.4	42.0	6
7	42.7	43.4	44.1	44.8	45.5	46.2	46.9	47.6	48.3	49.0	7
8	48.8	49.6	50.4	51.2	52.0	52.8	53.6	54.4	55.2	56.0	8
9	54.9	55.8	56.7	57.6	58.5	59.4	60.3	61.2	62.1	63.0	9

	71	72	73	74	75	76	77	78	79	80	
1	7.1	7.2	7.3	7.4	7.5	7.6	7.7	7.8	7.9	8.0	1
2	14.2	14.4	14.6	14.8	15.0	15.2	15.4	15.6	15.8	16.0	2
3	21.3	21.6	21.9	22.2	22.5	22.8	23.1	23.4	23.7	24.0	3
4	28.4	28.8	29.2	29.6	30.0	30.4	30.8	31.2	31.6	32.0	4
5	35.5	36.0	36.5	37.0	37.5	38.0	38.5	39.0	39.5	40.0	5
6	42.6	43.2	43.8	44.4	45.0	45.6	46.2	46.8	47.4	48.0	6
7	49.7	50.4	51.1	51.8	52.5	53.2	53.9	54.6	55.3	56.0	7
8	56.8	57.6	58.4	59.2	60.0	60.8	61.6	62.4	63.2	64.0	8
9	63.9	64.8	65.7	66.6	67.5	68.4	69.3	70.2	71.1	72.0	9

	81	82	83	84	85	86	87	88	89	90	
1	8.1	8.2	8.3	8.4	8.5	8.6	8.7	8.8	8.9	9.0	1
2	16.2	16.4	16.6	16.8	17.0	17.2	17.4	17.6	17.8	18.0	2
3	24.3	24.6	24.9	25.2	25.5	25.8	26.1	26.4	26.7	27.0	3
4	32.4	32.8	33.2	33.6	34.0	34.4	34.8	35.2	35.6	36.0	4
5	40.5	41.0	41.5	42.0	42.5	43.0	43.5	44.0	44.5	45.0	5
6	48.6	49.2	49.8	50.4	51.0	51.6	52.2	52.8	53.4	54.0	6
7	56.7	57.4	58.1	58.8	59.5	60.2	60.9	61.6	62.3	63.0	7
8	64.8	65.6	66.4	67.2	68.0	68.8	69.6	70.4	71.2	72.0	8
9	72.9	73.8	74.7	75.6	76.5	77.4	78.3	79.2	80.1	81.0	9

18. Table of quadratic interpolation

This table gives values of the coefficients k_1 used in the quadratic interpolation of Bessel (p. 18). To every value of k lying between two successive numbers of the column k (the left-hand one as well as the right one) there corresponds the same value of k_1 placed between them. The value of k_1 lying above the value of k should always be taken.

Examples: (1) For $k = 0.8$ (and for every value of k lying between 0.797 and 0.804 or between 0.196 and 0.203), we have $k_1 = 0.040$. (2) For $k = 0.3$ (or $k = 0.7$), we have $k_1 = 0.052$.

k	k_1	k	k	k_1	k	k	k_1	k	k	k_1	k
0.000		1.000	0.066		0.934	0.147		0.853	0.255		0.745
	0.000			0.016			0.032			0.048	
0.002		0.998	0.071		0.929	0.153		0.847	0.263		0.737
	0.001			0.017			0.033			0.049	
0.006		0.994	0.075		0.925	0.159		0.841	0.271		0.729
	0.002			0.018			0.034			0.050	
0.010		0.990	0.080		0.920	0.165		0.835	0.280		0.720
	0.003			0.019			0.035			0.051	
0.014		0.986	0.085		0.915	0.171		0.829	0.290		0.710
	0.004			0.020			0.036			0.052	
0.018		0.982	0.090		0.910	0.177		0.823	0.300		0.700
	0.005			0.021			0.037			0.053	
0.022		0.978	0.095		0.905	0.183		0.817	0.310		0.690
	0.006			0.022			0.038			0.054	
0.026		0.974	0.100		0.900	0.190		0.810	0.321		0.679
	0.007			0.023			0.039			0.055	
0.030		0.970	0.105		0.895	0.196		0.804	0.332		0.668
	0.008			0.024			0.040			0.056	
0.035		0.965	0.110		0.890	0.203		0.797	0.345		0.655
	0.009			0.025			0.041			0.057	
0.039		0.961	0.115		0.885	0.210		0.790	0.358		0.642
	0.010			0.026			0.042			0.058	
0.043		0.957	0.120		0.880	0.217		0.783	0.373		0.627
	0.011			0.027			0.043			0.059	
0.048		0.952	0.125		0.875	0.224		0.776	0.390		0.610
	0.012			0.028			0.044			0.060	
0.052		0.948	0.131		0.869	0.231		0.769	0.410		0.590
	0.013			0.029			0.045			0.061	
0.057		0.943	0.136		0.864	0.239		0.761	0.436		0.564
	0.014			0.030			0.046			0.062	
0.061		0.939	0.142		0.858	0.247		0.753	0.500		0.500
	0.015			0.031			0.047				
0.066		0.934	0.147		0.853	0.255		0.745			

B. TABLES OF SPECIAL FUNCTIONS

19. The Gamma function

This table gives the values of the function $\Gamma(x)$ (see p. 191) for x from 1 to 2. The values of $\Gamma(x)$ for $x < 1$ and $x > 2$ can be computed from the formulas

$$\Gamma(x) = \frac{\Gamma(x+1)}{x}, \qquad \Gamma(x) = (x-1)\,\Gamma(x-1).$$

Examples. (1) $\Gamma(0.7) = \frac{\Gamma(1.7)}{0.7} = \frac{0.90864}{0.7} = 1.2981$. (2) $\Gamma(3.5) = 2.5 \cdot \Gamma(2.5) = 2.5 \cdot 1.5 \cdot \Gamma(1.5) = 2.5 \cdot 1.5 \cdot 0.88623 = 3.32336$.

x	$\Gamma(x)$	x	$\Gamma(x)$	x	$\Gamma(x)$	x	$\Gamma(x)$
1.00	1.00000	**1.25**	0.90640	**1.50**	0.88623	**1.75**	0.91906
01	0.99433	26	0.90440	51	0.88659	76	0.92137
02	0.98884	27	0.90250	52	0.88704	77	0.92376
03	0.98355	28	0.90072	53	0.88757	78	0.92623
04	0.97844	29	0.89904	54	0.88818	79	0.92877
1.05	0.97350	**1.30**	0.89747	**1.55**	0.88887	**1.80**	0.93138
06	0.96874	31	0.89600	56	0.88964	81	0.93408
07	0.96415	32	0.89464	57	0.89049	82	0.93685
08	0.95973	33	0.89338	58	0.89142	83	0.93969
09	0.95546	34	0.89222	59	0.89243	84	0.94261
1.10	0.95135	**1.35**	0.89115	**1.60**	0.89352	**1.85**	0.94561
11	0.94740	36	0.89018	61	0.89468	86	0.94869
12	0.94359	37	0.88931	62	0.89592	87	0.95184
13	0.93993	38	0.88854	63	0.89724	88	0.95507
14	0.93642	39	0.88785	64	0.89864	89	0.95838
1.15	0.93304	**1.40**	0.88726	**1.65**	0.90012	**1.90**	0.96177
16	0.92980	41	0.88676	66	0.90167	91	0.96523
17	0.92670	42	0.88636	67	0.90330	92	0.96877
18	0.92373	43	0.88604	68	0.90500	93	0.97240
19	0.92089	44	0.88581	69	0.90678	94	0.97610
1.20	0.91817	**1.45**	0.88566	**1.70**	0.90864	**1.95**	0.97988
21	0.91558	46	0.88560	71	0.91057	96	0.98374
22	0.91311	47	0.88563	72	0.91258	97	0.98768
23	0.91075	48	0.88575	73	0.91467	98	0.99171
24	0.90852	49	0.88595	74	0.91683	99	0.99581
1.25	0.90640	**1.50**	0.88623	**1.75**	0.91906	**2.00**	1.00000

20. Bessel's cylindrical functions (¹)

x	$J_0(x)$	$J_1(x)$	$Y_0(x)$	$Y_1(x)$	$I_0(x)$	$I_1(x)$	$K_0(x)$	$K_1(x)$
0.0	+1.0000	+0.0000	$-\infty$	$-\infty$	1.000	0.0000	∞	∞
0.1	0.9975	0.0499	−1.5342	−6.4590	1.003	0.0501	2.4271	9.8538
0.2	0.9900	0.0995	1.0811	3.3238	1.010	0.1005	1.7527	4.7760
0.3	0.9776	0.1483	0.8073	2.2931	1.023	0.1517	1.3725	3.0560
0.4	0.9604	0.1960	0.6060	1.7809	1.040	0.2040	1.1145	2.1844
0.5	+0.9385	0.2423	−0.4445	−1.4715	1.063	0.2579	0.9244	1.6564
0.6	0.9120	0.2867	0.3085	1.2604	1.092	0.3137	0.7775	1.3028
0.7	0.8812	0.3290	0.1907	1.1032	1.126	0.3719	0.6605	1.0503
0.8	0.8463	0.3688	−0.0868	0.9781	1.167	0.4329	0.5658	0.8618
0.9	0.8075	0.4059	+0.0056	0.8731	1.213	0.4971	0.4867	0.7165
1.0	+0.7652	+0.4401	+0.0883	−0.7812	1.266	0.5652	0.4210	0.6019
1.1	0.7196	0.4709	0.1622	0.6981	1.326	0.6375	0.3656	0.5098
1.2	0.6711	0.4983	0.2281	0.6211	1.394	0.7147	0.3185	0.4346
1.3	0.6201	0.5220	0.2865	0.5485	1.469	0.7973	0.2782	0.3725
1.4	0.5669	0.5419	0.3379	0.4791	1.553	0.8861	0.2437	0.3208
1.5	+0.5118	+0.5579	+0.3824	−0.4123	1.647	0.9817	0.2138	0.2774
1.6	0.4554	0.5699	0.4204	0.3476	1.750	1.085	0.1880	0.2406
1.7	0.3980	0.5778	0.4520	0.2847	1.864	1.196	0.1655	0.2094
1.8	0.3400	0.5815	0.4774	0.2237	1.990	1.317	0.1459	0.1826
1.9	0.2818	0.5812	0.4968	0.1644	2.128	1.448	0.1288	0.1597
2.0	+0.2239	+0.5767	+0.5104	−0.1070	2.280	1.591	0.1139	0.1399
2.1	0.1666	0.5683	0.5183	−0.0517	2.446	1.745	0.1008	0.1227
2.2	0.1104	0.5560	0.5208	+0.0015	2.629	1.914	0.08927	0.1079
2.3	0.0555	0.5399	0.5181	0.0523	2.830	2.098	0.07914	0.09498
2.4	0.0025	0.5202	0.5104	0.1005	3.049	2.298	0.07022	0.08372
2.5	−0.0484	+0.4971	+0.4981	+0.1459	3.290	2.517	0.06235	0.07389
2.6	0.0968	0.4708	0.4813	0.1884	3.553	2.755	0.05540	0.06528
2.7	0.1424	0.4416	0.4605	0.2276	3.842	3.016	0.04926	0.05774
2.8	0.1850	0.4097	0.4359	0.2635	4.157	3.301	0.04382	0.05111
2.9	0.2243	0.3754	0.4079	0.2959	4.503	3.613	0.03901	0.04529
3.0	−0.2601	+0.3391	+0.3769	+0.3247	4.881	3.953	0.03474	0.04016
3.1	0.2921	0.3009	0.3431	0.3496	5.294	4.326	0.03095	0.03563
3.2	0.3202	0.2613	0.3070	0.3707	5.747	4.734	0.02759	0.03164
3.3	0.3443	0.2207	0.2691	0.3879	6.243	5.181	0.02461	0.02812
3.4	0.3643	0.1792	0.2296	0.4010	6.785	5.670	0.02196	0.02500
3.5	−0.3801	+0.1374	+0.1890	+0.4102	7.378	6.206	0.01960	0.02224
3.6	0.3918	0.0955	0.1477	0.4154	8.028	6.793	0.01750	0.01979
3.7	0.3992	0.0538	0.1061	0.4167	8.739	7.436	0.01563	0.01763
3.8	0.4026	+0.0128	0.0645	0.4141	9.517	8.140	0.01397	0.01571
3.9	0.4018	−0.0272	+0.0234	0.4078	10.37	8.913	0.01248	0.01400
4.0	−0.3971	−0.0660	−0.0169	+0.3979	11.30	9.759	0.01116	0.01248

(¹) For definitions, formulas and graphs see pp. 549–551.

x	$J_0(x)$	$J_1(x)$	$Y_0(x)$	$Y_1(x)$	$I_0(x)$	$I_1(x)$	$K_0(x)$ [1]	$K_1(x)$
4.0	−0.3971	−0.0660	−0.0169	+0.3979	11.30	9.759	0.01116	0.01248
4.1	0.3887	0.1033	0.0561	0.3846	12.32	10.69	$0.0^{2}9980$	0.01114
4.2	0.3766	0.1386	0.0938	0.3680	13.44	11.71	$0.0^{2}8927$	$0.0^{2}9938$
4.3	0.3610	0.1719	0.1296	0.3484	14.67	12.82	$0.0^{2}7988$	$0.0^{2}8872$
4.4	0.3423	0.2028	0.1633	0.3260	16.01	14.05	$0.0^{2}7149$	$0.0^{2}7923$
4.5	−0.3205	−0.2311	−0.1947	+0.3010	17.48	15.39	$0.0^{2}6400$	$0.0^{2}7078$
4.6	0.2961	0.2566	0.2235	0.2737	19.09	16.86	$0.0^{2}5730$	$0.0^{2}6325$
4.7	0.2693	0.2791	0.2494	0.2445	20.86	18.48	$0.0^{2}5132$	$0.0^{2}5654$
4.8	0.2404	0.2985	0.2723	0.2136	22.79	20.25	$0.0^{2}4597$	$0.0^{2}5055$
4.9	0.2097	0.3147	0.2921	0.1812	24.91	22.20	$0.0^{2}4119$	$0.0^{2}4521$
5.0	−0.1776	−0.3276	−0.3085	+0.1479	27.24	24.34	$0.0^{2}3691$	$0.0^{2}4045$
5.1	0.1443	0.3371	0.3216	0.1137	29.79	26.68	$0.0^{2}3308$	$0.0^{2}3619$
5.2	0.1103	0.3432	0.3313	0.0792	32.58	29.25	$0.0^{2}2966$	$0.0^{2}3239$
5.3	0.0758	0.3460	0.3374	0.0445	35.65	32.08	$0.0^{2}2659$	$0.0^{2}2900$
5.4	0.0412	0.3453	0.3402	+0.0101	39.01	35.18	$0.0^{2}2385$	$0.0^{2}2597$
5.5	−0.0068	−0.3414	−0.3395	−0.0238	42.69	38.59	$0.0^{2}2139$	$0.0^{2}2326$
5.6	+0.0270	0.3343	0.3354	0.0568	46.74	42.33	$0.0^{2}1918$	$0.0^{2}2083$
5.7	0.0599	0.3241	0.3282	0.0887	51.17	46.44	$0.0^{2}1721$	$0.0^{2}1866$
5.8	0.0917	0.3110	0.3177	0.1192	56.04	50.95	$0.0^{2}1544$	$0.0^{2}1673$
5.9	0.1220	0.2951	0.3044	0.1481	61.38	55.90	$0.0^{2}1386$	$0.0^{2}1499$
6.0	+0.1506	−0.2767	−0.2882	−0.1750	67.23	61.34	$0.0^{2}1244$	$0.0^{2}1344$
6.1	0.1773	0.2559	0.2694	0.1998	73.66	67.32	$0.0^{2}1117$	$0.0^{2}1205$
6.2	0.2017	0.2329	0.2483	0.2223	80.72	73.89	$0.0^{2}1003$	$0.0^{2}1081$
6.3	0.2238	0.2081	0.2251	0.2422	88.46	81.10	$0.0^{3}9001$	$0.0^{3}9691$
6.4	0.2433	0.1816	0.1999	0.2596	96.96	89.03	$0.0^{3}8083$	$0.0^{3}8693$
6.5	+0.2601	−0.1538	−0.1732	−0.2741	106.3	97.74	$0.0^{3}7259$	$0.0^{3}7799$
6.6	0.2740	0.1250	0.1452	0.2857	116.5	107.3	$0.0^{3}6520$	$0.0^{3}6998$
6.7	0.2851	0.0953	0.1162	0.2945	127.8	117.8	$0.0^{3}5857$	$0.0^{3}6280$
6.8	0.2931	0.0652	0.0864	0.3002	140.1	129.4	$0.0^{3}5262$	$0.0^{3}5636$
6.9	0.2981	0.0349	0.0563	0.3029	153.7	142.1	$0.0^{3}4728$	$0.0^{3}5059$
7.0	+0.3001	−0.0047	−0.0259	−0.3027	168.6	156.0	$0.0^{3}4248$	$0.0^{3}4542$
7.1	0.2991	+0.0252	+0.0042	0.2995	185.0	171.4	$0.0^{3}3817$	$0.0^{3}4078$
7.2	0.2951	0.0543	0.0339	0.2934	202.9	188.3	$0.0^{3}3431$	$0.0^{3}3662$
7.3	0.2882	0.0826	0.0628	0.2846	222.7	206.8	$0.0^{3}3084$	$0.0^{3}3288$
7.4	0.2786	0.1096	0.0907	0.2731	244.3	227.2	$0.0^{3}2772$	$0.0^{3}2953$
7.5	+0.2663	+0.1352	+0.1173	−0.2591	268.2	249.6	$0.0^{3}2492$	$0.0^{3}2653$
7.6	0.2516	0.1592	0.1424	0.2428	294.3	274.2	$0.0^{3}2240$	$0.0^{3}2383$
7.7	0.2346	0.1813	0.1658	0.2243	323.1	301.3	$0.0^{3}2014$	$0.0^{3}2141$
7.8	0.2154	0.2014	0.1872	0.2039	354.7	331.1	$0.0^{3}1811$	$0.0^{3}1924$
7.9	0.1944	0.2192	0.2065	0.1817	389.4	363.9	$0.0^{3}1629$	$0.0^{3}1729$
8.0	+0.1717	+0.2346	+0.2235	−0.1581	427.6	399.9	$0.0^{3}1465$	$0.0^{3}1554$

[1] We write $0.0^{2}9980$ for 0.009980.

x	$J_0(x)$	$J_1(x)$	$Y_0(x)$	$Y_1(x)$	$I_0(x)$	$I_1(x)$	$K_0(x)$	$K_1(x)$
8.0	+0.1717	+0.2346	+0.2235	−0.1581	427.6	399.9	$0.0^{3}1465$	$0.0^{3}1554$
8.1	0.1475	0.2476	0.2381	0.1331	469.5	439.5	$0.0^{3}1317$	$0.0^{3}1396$
8.2	0.1222	0.2580	0.2501	0.1072	515.6	483.0	$0.0^{3}1185$	$0.0^{3}1255$
8.3	0.0960	0.2657	0.2595	0.0806	566.3	531.0	$0.0^{3}1066$	$0.0^{3}1128$
8.4	0.0692	0.2708	0.2662	0.0535	621.9	583.7	$0.0^{4}9588$	$0.0^{3}1014$
8.5	+0.0419	+0.2731	+0.2702	−0.0262	683.2	641.6	$0.0^{4}8626$	$0.0^{4}9120$
8.6	+0.0146	0.2728	0.2715	+0.0011	750.5	705.4	$0.0^{4}7761$	$0.0^{4}8200$
8.7	−0.0125	0.2697	0.2700	0.0280	824.4	775.5	$0.0^{4}6983$	$0.0^{4}7374$
8.8	0.0392	0.2641	0.2659	0.0544	905.8	852.7	$0.0^{4}6283$	$0.0^{4}6631$
8.9	0.0653	0.2559	0.2592	0.0799	995.2	937.5	$0.0^{4}5654$	$0.0^{4}5964$
9.0	−0.0903	+0.2453	+0.2499	+0.1043	1094	1031	$0.0^{4}5088$	$0.0^{4}5364$
9.1	0.1142	0.2324	0.2383	0.1275	1202	1134	$0.0^{4}4579$	$0.0^{4}4825$
9.2	0.1367	0.2174	0.2245	0.1491	1321	1247	$0.0^{4}4121$	$0.0^{4}4340$
9.3	0.1577	0.2004	0.2086	0.1691	1451	1371	$0.0^{4}3710$	$0.0^{4}3904$
9.4	0.1768	0.1816	0.1907	0.1871	1595	1508	$0.0^{4}3339$	$0.0^{4}3512$
9.5	−0.1939	+0.1613	+0.1712	+0.2032	1753	1658	$0.0^{4}3006$	$0.0^{4}3160$
9.6	0.2090	0.1395	0.1502	0.2171	1927	1824	$0.0^{4}2706$	$0.0^{4}2843$
9.7	0.2218	0.1166	0.1279	0.2287	2119	2006	$0.0^{4}2436$	$0.0^{4}2559$
9.8	0.2323	0.0928	0.1045	0.2379	2329	2207	$0.0^{4}2193$	$0.0^{4}2302$
9.9	0.2403	0.0684	0.0804	0.2447	2561	2428	$0.0^{4}1975$	$0.0^{4}2072$
10.0	−0.2459	+0.0435	+0.0557	+0.2490	2816	2671	$0.0^{4}1778$	$0.0^{4}1865$

21. Legendre's polynomials [1]

$$P_0(x) = 1,$$

$$P_1(x) = x,$$

$$P_2(x) = \tfrac{1}{2}(3x^2 - 1),$$

$$P_3(x) = \tfrac{1}{2}(5x^3 - 3x),$$

$$P_4(x) = \tfrac{1}{8}(35x^4 - 30x^2 + 3),$$

$$P_5(x) = \tfrac{1}{8}(63x^5 - 70x^3 + 15x),$$

$$P_6(x) = \tfrac{1}{16}(231x^6 - 315x^4 + 105x^2 - 5),$$

$$P_7(x) = \tfrac{1}{16}(429x^7 - 693x^5 + 315x^3 - 35x).$$

---

(1) For definitions and graphs see pp. 551, 552.

$x = P_1(x)$	$P_2(x)$	$P_3(x)$	$P_4(x)$	$P_5(x)$	$P_6(x)$	$P_7(x)$
0.00	−0.5000	0.0000	0.3750	0.0000	−0.3125	0.0000
0.05	−0.4962	−0.0747	0.3657	0.0927	−0.2962	−0.1069
0.10	−0.4850	−0.1475	0.3379	0.1788	−0.2488	−0.1995
0.15	−0.4662	−0.2166	0.2928	0.2523	−0.1746	−0.2649
0.20	−0.4400	−0.2800	0.2320	0.3075	−0.0806	−0.2935
0.25	−0.4062	−0.3359	0.1577	0.3397	+0.0243	−0.2799
0.30	−0.3650	−0.3825	+0.0729	0.3454	0.1292	−0.2241
0.35	−0.3162	−0.4178	−0.0187	0.3225	0.2225	−0.1318
0.40	−0.2600	−0.4400	−0.1130	0.2706	0.2926	−0.0146
0.45	−0.1962	−0.4472	−0.2050	0.1917	0.3290	+0.1106
0.50	−0.1250	−0.4375	−0.2891	+0.0898	0.3232	0.2231
0.55	−0.0462	−0.4091	−0.3590	−0.0282	0.2708	0.3007
0.60	+0.0400	−0.3600	−0.4080	−0.1526	0.1721	0.3226
0.65	0.1338	−0.2884	−0.4284	−0.2705	+0.0347	0.2737
0.70	0.2350	−0.1925	−0.4121	−0.3652	−0.1253	+0.1502
0.75	0.3438	−0.0703	−0.3501	−0.4164	−0.2808	−0.0342
0.80	0.4600	+0.0800	−0.2330	−0.3995	−0.3918	−0.2397
0.85	0.5838	0.2603	−0.0506	−0.2857	−0.4030	−0.3913
0.90	0.7150	0.4725	+0.2079	−0.0411	−0.2412	−0.3678
0.95	0.8538	0.7184	0.5541	+0.3727	+0.1875	+0.0112
1.00	1.0000	1.0000	1.0000	1.0000	1.0000	1.0000

22. Elliptic integrals [1]

(a) Elliptic integrals of the first kind

$$F(k,\varphi)=\int_0^{\varphi}\frac{d\psi}{\sqrt{1-k^2\sin^2\psi}}=\int_0^{\sin\varphi}\frac{dt}{\sqrt{1-t^2}\sqrt{1-k^2t^2}},\qquad k=\sin\alpha$$

$\varphi°$ \ $\alpha°$	0	10	20	30	40	50	60	70	80	90
0	0.0000	0.0000	0.0000	0.0000	0.0000	0.0000	0.0000	0.0000	0.0000	0.0000
10	0.1745	0.1746	0.1746	0.1748	0.1749	0.1751	0.1752	0.1753	0.1754	0.1754
20	0.3491	0.3493	0.3499	0.3508	0.3520	0.3533	0.3545	0.3555	0.3561	0.3564
30	0.5236	0.5243	0.5263	0.5294	0.5334	0.5379	0.5422	0.5459	0.5484	0.5493
40	0.6981	0.6997	0.7043	0.7116	0.7213	0.7323	0.7436	0.7535	0.7604	0.7629
50	0.8727	0.8756	0.8842	0.8982	0.9173	0.9401	0.9647	0.9876	1.0044	1.0107
60	1.0472	1.0519	1.0660	1.0896	1.1226	1.1643	1.2126	1.2619	1.3014	1.3170
70	1.2217	1.2286	1.2495	1.2853	1.3372	1.4068	1.4944	1.5959	1.6918	1.7354
80	1.3963	1.4056	1.4344	1.4846	1.5597	1.6660	1.8125	2.0119	2.2653	2.4362
90	1.5708	1.5828	1.6200	1.6858	1.7868	1.9356	2.1565	2.5046	3.1534	∞

[1] Definitions are given on pp. 407, 408

(b) Elliptic integrals of the second kind

$$E(k,\varphi)=\int_0^{\varphi}\sqrt{1-k^2\sin^2\psi}\,d\psi=\int_0^{\sin\varphi}\sqrt{\frac{1-k^2t^2}{1-t^2}}\,dt,\qquad k=\sin\alpha$$

$\varphi°$ \ $\alpha°$	0	10	20	30	40	50	60	70	80	90
0	0.0000	0.0000	0.0000	0.0000	0.0000	0.0000	0.0000	0.0000	0.0000	0.0000
10	0.1745	0.1745	0.1744	0.1743	0.1742	0.1740	0.1739	0.1738	0.1737	0.1736
20	0.3491	0.3489	0.3483	0.3473	0.3462	0.3450	0.3438	0.3429	0.3422	0.3420
30	0.5236	0.5229	0.5209	0.5179	0.5141	0.5100	0.5061	0.5029	0.5007	0.5000
40	0.6981	0.6966	0.6921	0.6851	0.6763	0.6667	0.6575	0.6497	0.6446	0.6428
50	0.8727	0.8698	0.8614	0.8483	0.8317	0.8134	0.7954	0.7801	0.7697	0.7660
60	1.0472	1.0426	1.0290	1.0076	0.9801	0.9493	0.9184	0.8914	0.8728	0.8660
70	1.2217	1.2149	1.1949	1.1632	1.1221	1.0750	1.0266	0.9830	0.9514	0.9397
80	1.3963	1.3870	1.3597	1.3161	1.2590	1.1926	1.1225	1.0565	1.0054	0.9848
90	1.5708	1.5589	1.5238	1.4675	1.3931	1.3055	1.2111	1.1184	1.0401	1.0000

(c) Complete elliptic integrals

$$\mathrm{K}=F\left(k,\frac{\pi}{2}\right)=\int_0^{\pi/2}\frac{d\psi}{\sqrt{1-k^2\sin^2\psi}}=\int_0^1\frac{dt}{\sqrt{1-t^2}\sqrt{1-k^2t^2}},\quad k=\sin\alpha,$$

$$\mathrm{E}=E\left(k,\frac{\pi}{2}\right)=\int_0^{\pi/2}\sqrt{1-k^2\sin^2\psi}\,d\psi=\int_0^1\sqrt{\frac{1-k^2t^2}{1-t^2}}\,dt,\qquad k=\sin\alpha$$

$\alpha°$	K	E	$\alpha°$	K	E	$\alpha°$	K	E
0	1.5708	1.5708	**15**	1.5981	1.5442	**30**	1.6858	1.4675
1	1.5709	1.5707	16	1.6020	1.5405	31	1.6941	1.4608
2	1.5713	1.5703	17	1.6061	1.5367	32	1.7028	1.4539
3	1.5719	1.5697	18	1.6105	1.5326	33	1.7119	1.4469
4	1.5727	1.5689	19	1.6151	1.5283	34	1.7214	1.4397
5	1.5738	1.5678	**20**	1.6200	1.5238	**35**	1.7312	1.4323
6	1.5751	1.5665	21	1.6252	1.5191	36	1.7415	1.4248
7	1.5767	1.5649	22	1.6307	1.5141	37	1.7522	1.4171
8	1.5785	1.5632	23	1.6365	1.5090	38	1.7633	1.4092
9	1.5805	1.5611	24	1.6426	1.5037	39	1.7748	1.4013
10	1.5828	1.5589	**25**	1.6490	1.4981	**40**	1.7868	1.3931
11	1.5854	1.5564	26	1.6557	1.4924	41	1.7992	1.3849
12	1.5882	1.5537	27	1.6627	1.4864	42	1.8122	1.3765
13	1.5913	1.5507	28	1.6701	1.4803	43	1.8256	1.3680
14	1.5946	1.5476	29	1.6777	1.4740	44	1.8396	1.3594
15	1.5981	1.5442	**30**	1.6858	1.4675	**45**	1.8541	1.3506

α°	K	E	α°	K	E	α°	K	E
45	1.8541	1.3506	**60**	2.1565	1.2111	**75**	2.7681	1.0764
46	1.8691	1.3418	61	2.1842	1.2015	76	2.8327	1.0686
47	1.8848	1.3329	62	2.2132	1.1920	77	2.9026	1.0611
48	1.9011	1.3238	63	2.2435	1.1826	78	2.9786	1.0538
49	1.9180	1.3147	64	2.2754	1.1732	79	3.0617	1.0468
50	1.9356	1.3055	**65**	2.3088	1.1638	**80**	3.1534	1.0401
51	1.9539	1.2963	66	2.3439	1.1545	81	3.2553	1.0338
52	1.9729	1.2870	67	2.3809	1.1453	82	3.3699	1.0278
53	1.9927	1.2776	68	2.4198	1.1362	83	3.5004	1.0223
54	2.0133	1.2681	69	2.4610	1.1272	84	3.6519	1.0172
55	2.0347	1.2587	**70**	2.5046	1.1184	**85**	3.8317	1.0127
56	2.0571	1.2492	71	2.5507	1.1096	86	4.0528	1.0086
57	2.0804	1.2397	72	2.5998	1.1011	87	4.3387	1.0053
58	2.1047	1.2301	73	2.6521	1.0927	88	4.7427	1.0026
59	2.1300	1.2206	74	2.7081	1.0844	89	5.4349	1.0008
60	2.1565	1.2111	**75**	2.7681	1.0764	**90**	∞	1.0000

23. Probability integral

$$\Phi(x) = \frac{2}{\sqrt{2\pi}} \int_0^x e^{-t^2/2}\, dt \ (^1)$$

x	$\Phi(x)$	x	$\Phi(x)$	x	$\Phi(x)$	x	$\Phi(x)$
0.00	0.0000	**0.10**	0.0797	**0.20**	0.1585	**0.30**	0.2358
01	0.0080	11	0.0876	21	0.1663	31	0.2434
02	0.0160	12	0.0955	22	0.1741	32	0.2510
03	0.0239	13	0.1034	23	0.1819	33	0.2586
04	0.0319	14	0.1113	24	0.1897	34	0.2661
0.05	0.0399	**0.15**	0.1192	**0.25**	0.1974	**0.35**	0.2737
06	0.0478	16	0.1271	26	0.2051	36	0.2812
07	0.0558	17	0.1350	27	0.2128	37	0.2886
08	0.0638	18	0.1428	28	0.2205	38	0.2961
09	0.0717	19	0.1507	29	0.2282	39	0.3035
0.10	0.0797	**0.20**	0.1585	**0.30**	0.2358	**0.40**	0.3108

(1) The graph of $\Phi(x)$ and some applications of it are given on pp. 747, 748. Sometimes the function

$$\operatorname{Erf} x = \frac{2}{\sqrt{\pi}} \int_0^x e^{-t}dt^2 = \Phi\left(x\sqrt{2}\right)$$

is called the probability integral.

x	$\Phi(x)$	x	$\Phi(x)$	x	$\Phi(x)$	x	$\Phi(x)$
0.40	0.3108	**0.80**	0.5763	**1.20**	0.7699	**1.60**	0.8904
41	0.3182	81	0.5821	21	0.7737	61	0.8926
42	0.3255	82	0.5878	22	0.7775	62	0.8948
43	0.3328	83	0.5935	23	0.7813	63	0.8969
44	0.3401	84	0.5991	24	0.7850	64	0.8990
0.45	0.3473	**0.85**	0.6047	**1.25**	0.7887	**1.65**	0.9011
46	0.3545	86	0.6102	26	0.7923	66	0.9031
47	0.3616	87	0.6157	27	0.7959	67	0.9051
48	0.3688	88	0.6211	28	0.7995	68	0.9070
49	0.3759	89	0.6265	29	0.8029	69	0.9090
0.50	0.3829	**0.90**	0.6319	**1.30**	0.8064	**1.70**	0.9109
51	0.3899	91	0.6372	31	0.8098	71	0.9127
52	0.3969	92	0.6424	32	0.8132	72	0.9146
53	0.4039	93	0.6476	33	0.8165	73	0.9164
54	0.4108	94	0.6528	34	0.8198	74	0.9181
0.55	0.4177	**0.95**	0.6579	**1.35**	0.8230	**1.75**	0.9199
56	0.4245	96	0.6629	36	0.8262	76	0.9216
57	0.4313	97	0.6680	37	0.8293	77	0.9233
58	0.4381	98	0.6729	38	0.8324	78	0.9249
59	0.4448	99	0.6778	39	0.8355	79	0.9265
0.60	0.4515	**1.00**	0.6827	**1.40**	0.8385	**1.80**	0.9281
61	0.4581	01	0.6875	41	0.8415	81	0.9297
62	0.4647	02	0.6923	42	0.8444	82	0.9312
63	0.4713	03	0.6970	43	0.8473	83	0.9328
64	0.4778	04	0.7017	44	0.8501	84	0.9342
0.65	0.4843	**1.05**	0.7063	**1.45**	0.8529	**1.85**	0.9357
66	0.4907	06	0.7109	46	0.8557	86	0.9371
67	0.4971	07	0.7154	47	0.8584	87	0.9385
68	0.5035	08	0.7199	48	0.8611	88	0.9399
69	0.5098	09	0.7243	49	0.8638	89	0.9412
0.70	0.5161	**1.10**	0.7287	**1.50**	0.8664	**1.90**	0.9426
71	0.5223	11	0.7330	51	0.8690	91	0.9439
72	0.5285	12	0.7373	52	0.8715	92	0.9451
73	0.5346	13	0.7415	53	0.8740	93	0.9464
74	0.5407	14	0.7457	54	0.8764	94	0.9476
0.75	0.5467	**1.15**	0.7499	**1.55**	0.8789	**1.95**	0.9488
76	0.5527	16	0.7540	56	0.8812	96	0.9500
77	0.5587	17	0.7580	57	0.8836	97	0.9512
78	0.5646	18	0.7620	58	0.8859	98	0.9523
79	0.5705	19	0.7660	59	0.8882	99	0.9534
0.80	0.5763	**1.20**	0.7699	**1.60**	0.8904	**2.00**	0.9545

x	$\Phi(x)$	x	$\Phi(x)$	x	$\Phi(x)$	x	$\Phi(x)$
2.00	0.9545	**2.50**	0.9876	**3.00**	0.99730	**4.00**	0.99994
05	0.9596	55	0.9892	10	0.99806		
10	0.9643	60	0.9907	20	0.99863	**4.417**	$1-10^{-5}$
15	0.9684	65	0.9920	30	0.99903		
20	0.9722	70	0.9931	40	0.99933	**4.892**	$1-10^{-6}$
2.25	0.9756	**2.75**	0.9940	**3.50**	0.99953	**5.327**	$1-10^{-7}$
30	0.9786	80	0.9949	60	0.99968		
35	0.9812	85	0.9956	70	0.99978		
40	0.9836	90	0.9963	80	0.99986		
45	0.9857	95	0.9968	90	0.99990		
2.50	0.9876	**3.00**	0.99730	**4.00**	0.99994		

II. GRAPHS

A. ELEMENTARY FUNCTIONS

1. Polynomials

Linear function: $y = ax + b$ (Fig. 2a). The graph—*a straight line*. The function increases monotonically for $a > 0$ and decreases monotonically for $a < 0$; it is constant for $a = 0$. Intersections with

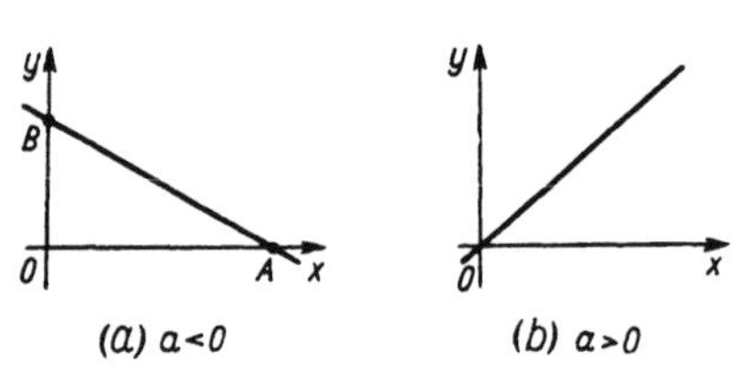

FIG. 2

the axes: $A\left(-\frac{a}{b}, 0\right)$, $B(0, b)$. For details see p. 239. For $b = 0$, it represents a *direct proportionality*: $y = ax$; the graph is a straight line passing through the origin (Fig. 2b).

Quadratic function: $y = ax^2 + bx + c$ (Fig. 3). The graph—a *parabola* with a vertical axis of symmetry $x = -\frac{b}{2a}$. For $a > 0$, the function first decreases, reaches a minimum and then increases; for $a < 0$, it increases, passes through a maximum and then decreases. Intersections with the x axis: $A_1, A_2\left(\frac{-b \pm \sqrt{b^2 - 4ac}}{2a}, 0\right)$; with the y axis: $B(0, c)$. Extreme: $C\left(-\frac{b}{2a}, \frac{4ac - b^2}{4a}\right)$. For the parabola see pp. 251–253.

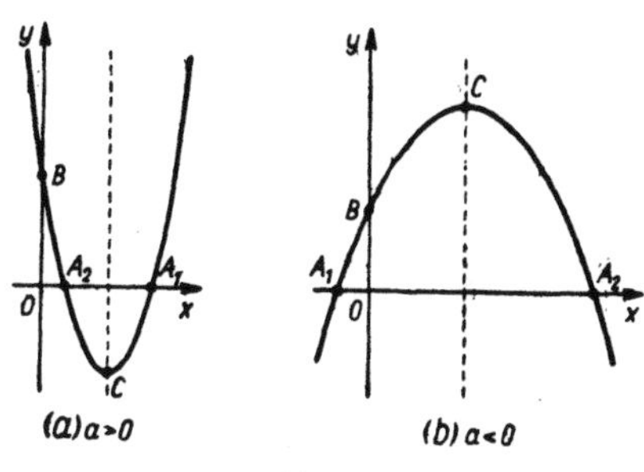

FIG. 3

Polynomial of the third degree: $y = ax^3 + bx^2 + cx + d$ (Fig 4). The graph—a *cubical parabola*. Behaviour of the function depends on the signs of a and $\Delta = 3ac - b^2$. If $\Delta > 0$ (Fig. 4a, b), the function is monotone increasing if $a > 0$ and monotone decreasing if $a < 0$.

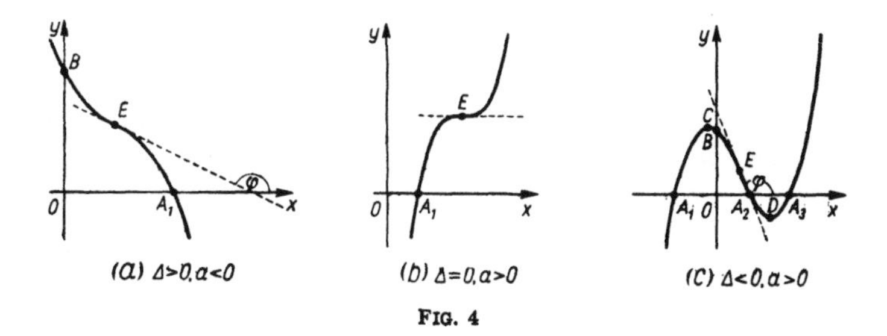

FIG. 4

If $\Delta < 0$, the function has one maximum and one minimum (Fig. 4c): for $a > 0$, it first increases from $-\infty$ to a maximum, then decreases to a minimum and again increases to $+\infty$; for $a < 0$, it decreases from $+\infty$ to a minimum, increases to a maximum and then decreases to $-\infty$. Intersections with the x axis are the roots of the equation $y = 0$ [1]; there can be one, two or three roots A_1, A_2, A_3 (in the second case, the x axis is tangent at one of the roots). Intersection with the y axis: $B(0, d)$. Turning values $C, D\left(-\dfrac{b \pm \sqrt{-\Delta}}{3a},\right.$ $\left. d + \dfrac{2b^3 - 9abc \mp (6ac - 2b^2)\sqrt{-\Delta}}{27a^2}\right)$. The symmetry centre of the curve

---

(1) For the solution of the cubic equation see pp. 161–163.

is a point of inflection: $E\left(-\frac{b}{3a}, \frac{2b^3-9abc}{27a^2}+d\right)$; the slope of the tangent at E is $\tan\varphi = \left(\frac{dy}{dx}\right)_E = \frac{\Delta}{3a}$.

Polynomial of the n**-th degree** (Fig. 5): $y = a_0x^n + a_1x^{n-1} + \ldots + a_{n-1}x + a_n$. The graph is a *curve of the n-th degree* [1] *of the parabolic type.*

(a) n is odd. y varies continuously from $-\infty$ to $+\infty$ for $a_0 > 0$ and from $+\infty$ to $-\infty$ for $a_0 < 0$; the curve can intersect the x axis (or be tangent to it) from 1 to n times [2]. There can be no

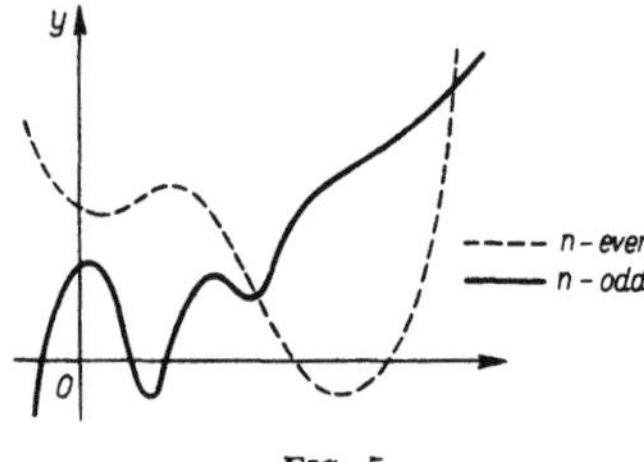

FIG. 5

turning values or an even number of them (from 2 to $n-1$) maxima and minima alternate. There can be an odd number of points of inflection (from 1 to $n-2$).

(b) n is even. y varies continuously from $+\infty$ to $+\infty$ for $a_0 > 0$ and from $-\infty$ to $-\infty$, for $a_0 < 0$. The curve either does not intersect the x axis at all or intersects it (or is tangent to it) from 1 to n times. There is an odd number of turning values (from 1 to $n-1$), maxima and minima alternate. There is an even number of points of inflection (from 0 to $n-2$). Asymptotes and singular points do not occur in either case.

To plot the graph, we first determine turning values and points of inflection (with the values of the derivatives), plot these points and the tangent lines at them and then draw the continuous curve.

When a portion of the graph of a polynomial of the fourth degree $y = ax^4 + bx^3 + cx^2 + dx + e$ with a fixed a is used, it is convenient to plot the graph of $y = ax^4$ (see below) and then, to reduce the function to the form $Y = aX^4 + a'X^2 + b'X + c'$ translating the coordinate origin to the point $(-b/4a, 0)$; finally, the graph of

---

[1] For degree of a curve see p. 239.

[2] For solution of the algebraic equation of the n-th degree see pp. 164–167 and pp. 169–171.

the given function can be found by the geometric composition of the ordinates of the curve $Y = aX^4$ with those of the parabola $Y = a'X^2 + b'X + c'$.

Power function: $y = ax^n$ (n an integer > 1) (Fig. 6). The graph—a *parabola of the n-th degree.* (1) $a=1$; the graph passes through the points $O(0, 0)$ and $A(1, 1)$ and is tangent to the x axis at the origin. If n is even (Fig. 6a), the curve is symmetric with respect

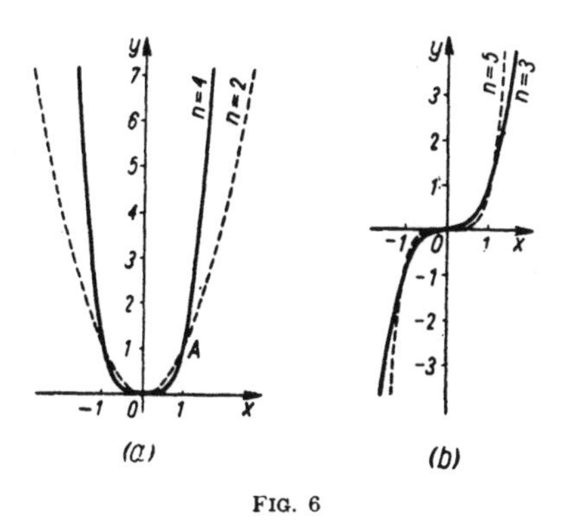

FIG. 6

to the y axis and has a minimum at the origin; if n is odd (Fig 6b), the curve is symmetric with respect to the origin and has the origin as a point of inflection. No asymptotes. (2) General case. The curve $y = ax^n$ is obtained from $y = x^n$ by stretching it in the direction of the y axis in the ratio $|a| : 1$; if $a < 0$, the curve is a reflection of the curve $y = |a|x^n$ in the x axis.

2. Rational functions

Inverse proportionality: $y = \frac{a}{x}$ (Fig. 7). The graph—a *rectangular hyperbola* with the coordinate axes as the asymptotes. Discontinuity at $x=0$ ($y = \pm\infty$). If $a > 0$, the function decreases monotonically from 0 to $-\infty$ and from $+\infty$ to 0 (continuously drawn curve in the first and third quadrants); if $a < 0$, the function increases from 0 to $+\infty$ and from $-\infty$ to 0

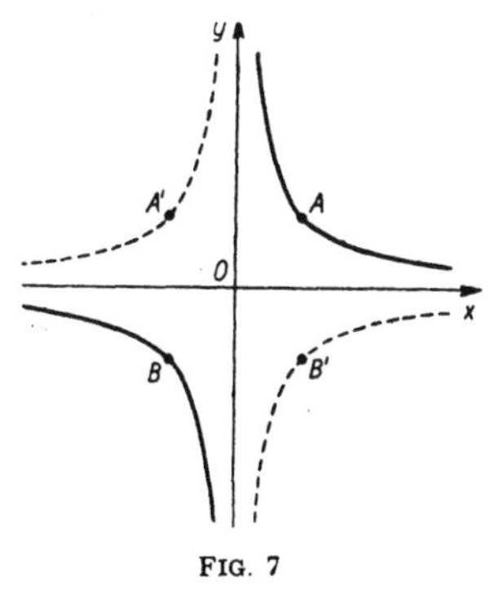

FIG. 7

(the dotted curve in the second and fourth quadrants). The vertices of hyperbola: A, B $(\pm\sqrt{|a|}, \pm\sqrt{|a|})$. No turning values. For details see pp. 247-250.

Homographic (rational-linear) function: $y = \dfrac{a_1x + b_1}{a_2x + b_2}$ (Fig 8). The graph—a *rectangular hyperbola* with the asymptotes parallel to the coordinate axes; the centre $C\left(-\dfrac{b_2}{a_2}, \dfrac{a_1}{a_2}\right)$. The parameter corresponding to a in the previous function: $a = -D/a_2$, where $D = \begin{vmatrix} a_1 & b_1 \\ a_2 & b_2 \end{vmatrix}$; the vertices of the hyperbola: $A, B\left(-\dfrac{b_2 \pm \sqrt{|D|}}{a_2}, \dfrac{a_1 \pm \sqrt{|D|}}{a_2}\right)$. Discontinuity at $x = -b_2/a_2$. If $D < 0$, the function decreases from a_1/a_2 to $-\infty$ and from $+\infty$ to a_1/a_2; if $D > 0$, the function increases from a_1/a_2 to $+\infty$ and from $-\infty$ to a_1/a_2.

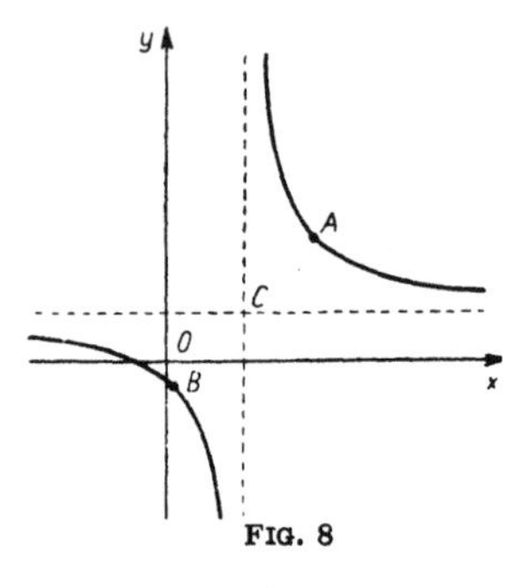

FIG. 8

The function $y = a + \dfrac{b}{x} + \dfrac{c}{x^2}\left(= \dfrac{ax^2 + bx + c}{x^2}\right)$ (Fig. 9) $(b \neq 0$, $c \neq 0$. For $b = 0$ see p. 103, for $c = 0$ see above). The graph —a *curve*

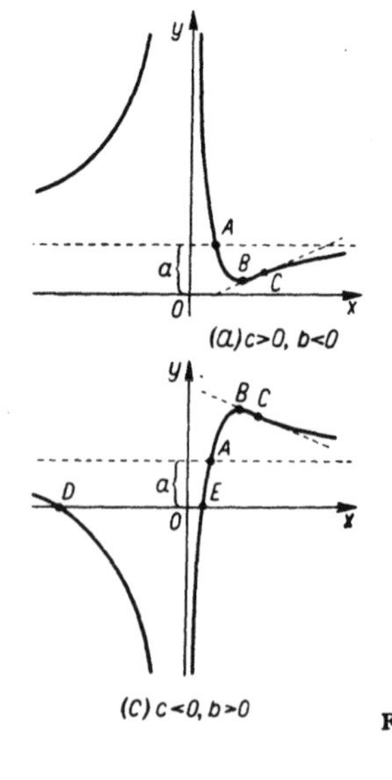

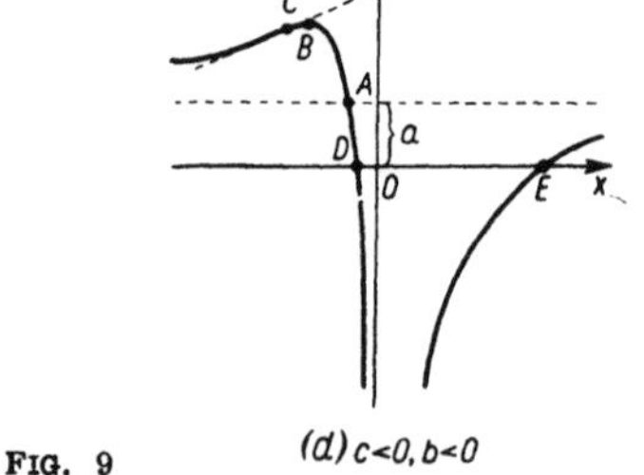

FIG. 9

of the third degree with two asymptotes: $x = 0$ and $y = a$; it consists of two branches: the one which corresponds to the monotone variation of y from a to $+\infty$ (or to $-\infty$) and the second passing through the 3 characteristic points: the intersection with the asymptote $A(-c/b, a)$, the turning point $B(-2c/b, a - b^2/4c)$ and the point of inflection $C(-3c/b, a - 2b^2/9c)$. Four possibilities of dislocation of these branches depend on the signs of b and c (Fig. 9). There can be two points of intersection with the x axis, one (the curve is tangent to it) or none: $D, E\left(\frac{-b \pm \sqrt{b^2 - 4ac}}{2a}, 0\right)$, according to the sign of $b^2 - 4ac$.

The function $y = \frac{1}{ax^2 + bx + c}$ (Fig. 10). The graph is a *curve of the third degree* symmetric with respect to the vertical line $x = -b/2a$ and has the x axis as an asymptote. The form of the function depends on the signs of a and $\Delta = 4ac - b^2$. Only the case when

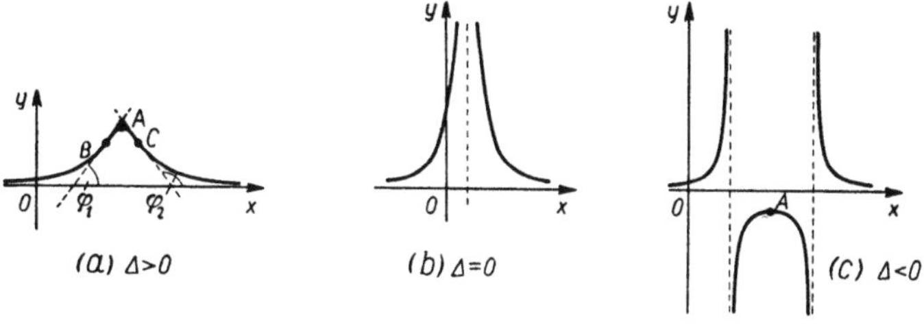

FIG. 10

$a > 0$ will be considered; the case when $a < 0$ can be reduced to the former one by considering the curve $y = \frac{1}{(-a)x^2 - bx - c}$ and then reflecting it in the x axis.

(a) $\Delta > 0$ (Fig. 10a). The function is continuous and positive for every x. It increases from 0 to a maximum and then decreases to 0. Themaximum $A\left(-\frac{b}{2a}, \frac{4a}{\Delta}\right)$, points of inflection $B, C\left(-\frac{b}{2a} \pm \frac{\sqrt{\Delta}}{2a\sqrt{3}}, \frac{3a}{\Delta}\right)$; the slopes of the tangents at these points are $\tan \varphi = \mp a^2 \left(\frac{3}{\Delta}\right)^{3/2}$

(b) $\Delta = 0$. The function is positive for every x. It increases from 0 to $+\infty$, has an infinite discontinuity at $x = -b/2a$ and decreases from $+\infty$ to 0 (Fig. 10b).

(c) $\Delta < 0$. The function increases from 0 to $+\infty$, has an infinite discontinuity, passes from $-\infty$ to $-\infty$ through a maximum, admits a second discontinuity and then decreases from $+\infty$ to 0. Maxima $A\left(-\frac{b}{2a}, \frac{4a}{\Delta}\right)$, points of discontinuity $x = \frac{-b \pm \sqrt{-\Delta}}{2a}$ (Fig. 10c).

The function $y = \frac{x}{ax^2 + bx + c}$ (Fig. 11). The graph is a *curve of the third degree,* passes through the origin and has the x axis as an asymptote. The form depends on the signs of a and $\Delta = 4ac - b^2$

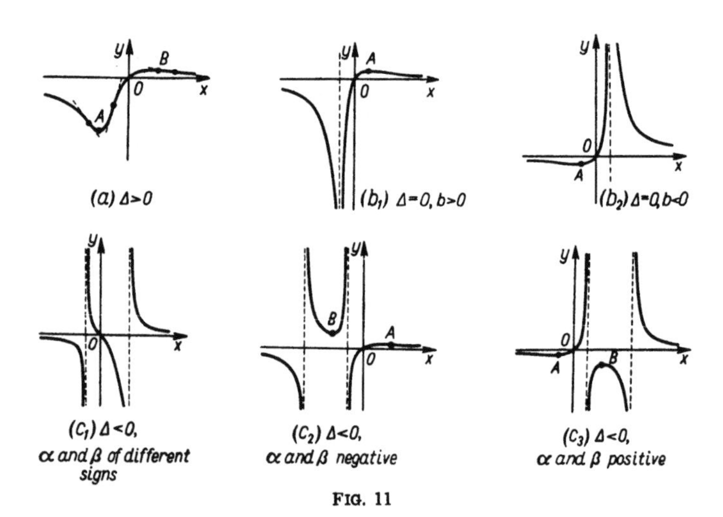

FIG. 11

as well as on the signs of the roots α and β of the equation $ax^2 + bx + c = 0$ (if $\Delta < 0$) and on the sign of b, if $\Delta = 0$. Only the case $a > 0$ will be considered; the case $a < 0$ is reduced to the former one by considering the curve $y = \frac{x}{(-a)x^2 - bx - c}$; this curve should be then reflected in the x axis.

(a) $\Delta > 0$. The function is continuous, decreases from 0 to a minimum, increases to a maximum and again decreases to 0. The minimum and the maximum $A, B\left(\mp\sqrt{\frac{c}{a}}, \frac{-b \mp 2\sqrt{ac}}{\Delta}\right)$, three points of inflection (Fig. 11a).

(b) $\Delta = 0$. The form of the function depends on the sign of b:

(1) $b > 0$; the function decreases from 0 to $-\infty$, has a discontinuity, increases from $-\infty$ to a maximum and decreases to 0 (Fig. $11b_1$); maximum $A\left(+\sqrt{\frac{c}{a}}, \frac{1}{2\sqrt{ac}+b}\right)$;

(2) $b < 0$; the function decreases from 0 to a minimum, passes through 0 and increases to $+\infty$, has an infinite discontinuity and decreases from $+\infty$ to 0 (Fig. $11b_2$), minimum $A\left(-\sqrt{\frac{c}{a}}, -\frac{1}{2\sqrt{ac}-b}\right)$.

In both cases the function has one point of inflection and a discontinuity at $x = -\frac{b}{2a}$.

(c) $\Delta < 0$. Two points of inflection: $x = \alpha$ and $x = \beta$; the form of the function depends on the signs of α and β. (1) α and β of different signs; the function decreases from 0 to $-\infty$, from $+\infty$ to $-\infty$ and from $+\infty$ to 0; no extremes (Fig. $11c_1$). (2) α and β negative; the function decreases from 0 to $-\infty$, then passes from $+\infty$ to $+\infty$ through a minimum and finally increases from $-\infty$ to a maximum and decreases to 0; the points of maximum and of minimum are the same as in (a) (Fig. $11c_2$). (3) α and β positive; the function decreases from 0 to a minimum and increases to $+\infty$, then passes from $-\infty$ to $-\infty$ through a maximum and finally decreases from $+\infty$ to 0; the points of maximum and of minimum are the same as in (a) (Fig. $11c_3$).

Power function: $y = \frac{a}{x^n} = ax^{-n}$ (n a positive integer) (Fig. 12). The graph is a *curve of the hyperbolic type* with the coordinate axes as the asymptotes. Discontinuity at $x = 0$. If $a > 0$, the function decreases from 0 to $-\infty$ and from $+\infty$ to 0, when n is odd, and increases from 0 to $+\infty$ and decreases from $+\infty$ to 0 being always positive, when n is even. If $a < 0$, the function increases from 0 to $+\infty$ and from $-\infty$ to 0, when n is odd, and decreases from 0 to $-\infty$ and increases from $-\infty$ to 0 being always negative, when n is even. No turning points. The greater n is, the quicker the function approaches asymptotically the x axis and the slower—the y axis. If n is even, the curve is symmetric with respect to the y axis, and if n is odd, it is symmetric with respect to the origin. Fig. 12 represents two cases: $n = 2$ and $n = 3$.

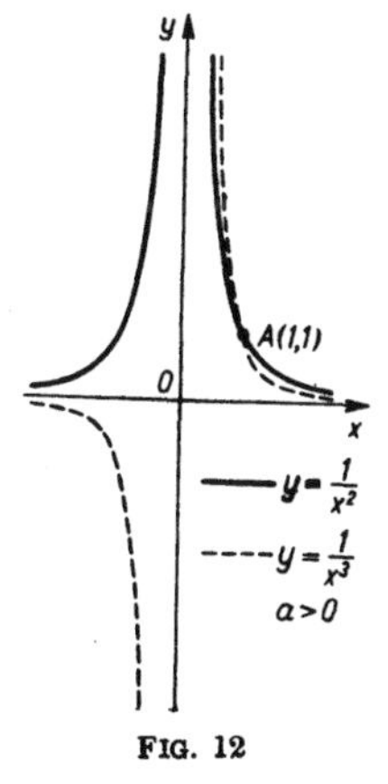

FIG. 12

3. Irrational functions

The square root of a linear function: $y = \pm\sqrt{ax+b}$ (Fig. 13). The graph is a *parabola* symmetric with respect to the x axis, with the vertex $A(-b/a, 0)$ and with the parameter $p = a/2$. The

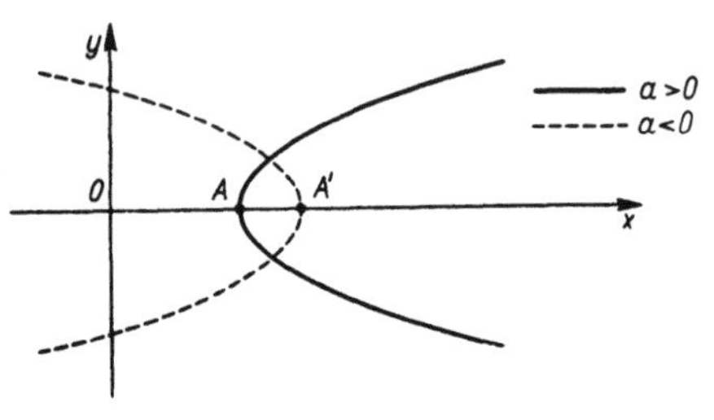

FIG. 13

range of existence and the form of the function depend on the sign of a (Fig. 13). The function is two-valued, there are no turning values. For details about the parabola see pp. 251–253.

The square root of a quadratic function: $y = \pm\sqrt{ax^2+bx+c}$ (Fig. 14). The graph is an *ellipse* for $a < 0$ and a *hyperbola* for $a > 0$; one of the axes is the x axis and the second is the straight line

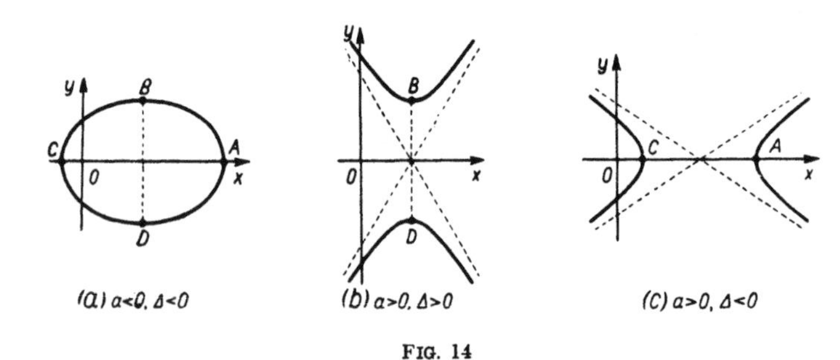

FIG. 14

$x = -b/2a$; the vertices $A, C\left(-\frac{b \pm \sqrt{-\Delta}}{2a}, 0\right)$, $B, D\left(-\frac{b}{2a}, \pm\sqrt{\frac{\Delta}{4a}}\right)$, where $\Delta = 4ac - b^2$. The range of existence and the form of the function depend on the signs of a and Δ (see Fig. 14). The function is two-valued; it has two turning values, if Δ and a are of the same sign (the points B and D). If $a < 0$ and $\Delta > 0$, the function admits only imaginary values and the curve does not exist. For details about the ellipse and the hyperbola see pp. 244–250.

Power function: $y = ax^k = ax^{\pm m/n}$ (m, n integers without a common factor). The case $a = 1$ will be considered (for $a \neq 1$, the curve is stretched in comparison with $y = x^k$ in the direction of the y axis in the ratio $|a|:1$ and, if a is negative, it is reflected in the x axis).

(1) $k > 0$, $y = x^{m/n}$. The graph (Fig. 15) passes through the points (0, 0) and (1,1). If $k > 1$, it is tangent (in the origin) to the x

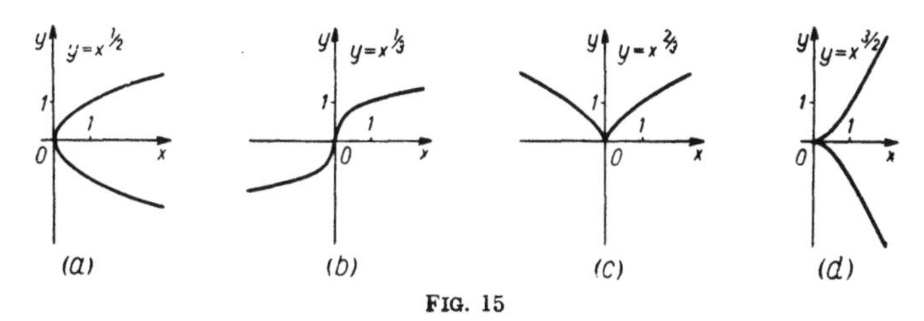

FIG. 15

axis; if $k < 1$, it is tangent (in the origin) to the y axis. If n is even, the curve is symmetric with respect to the x axis (the function is two-valued); if m is even, it is symmetric with respect to the y axis; if both m and n are odd, it is symmetric with respect to the origin. According to that, the function can have a vertex, a point of inflection or a cusp at the origin (see Fig. 15). No asymptotes.

(2) $k < 0$, $y = x^{-m/n}$. The graph is a curve of the hyperbolic type with the coordinate axes as the asymptotes (Fig. 16). Discontinuity at $x = 0$. The greater $|k|$ is, the quicker the function approaches asymptotically the x axis and the slower the y axis. The symmetry with respect to the coordinate axes or with respect to the origin depends on whether m and n are even or odd, as in the case $k > 0$ (see above); this determines the form of the function (see Fig. 16). No turning values.

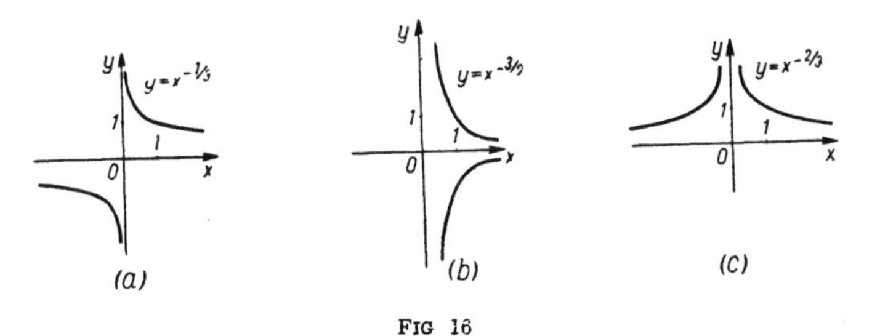

FIG 16

4. Exponential and logarithmic functions

Exponential function: $y = a^x = e^{bx}$ $(a > 0, b = \ln a)$ (Fig. 17). The graph is an *exponential curve* (for $a = e$—the natural exponential curve $y = e^x$). The function is everywhere positive. For $a > 1$ (i.e., $b > 0$) it increases monotonically from 0 to ∞, for $a < 1$—decreases from ∞ to 0, the quicker, the greater is $|b|$. The curve passes through $A(0,1)$ and approaches asymptotically the x axis (for $b > 0$—from the left, for $b < 0$—from the right) the quicker, the greater $|b|$ is. The function $y = a^{-x} = (1/a)^x$ increases for $a < 1$ and decreases for $a > 1$.

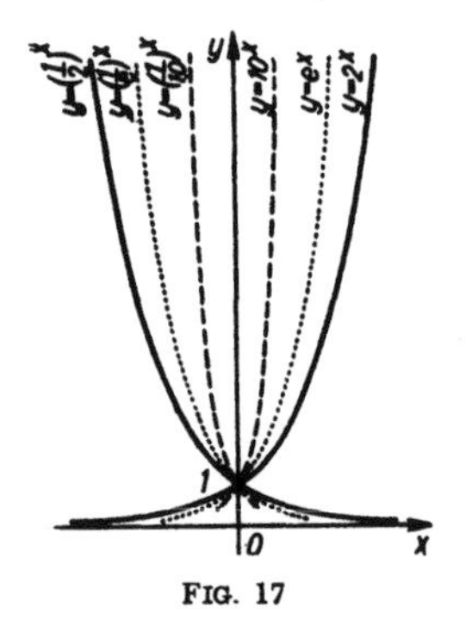

FIG. 17

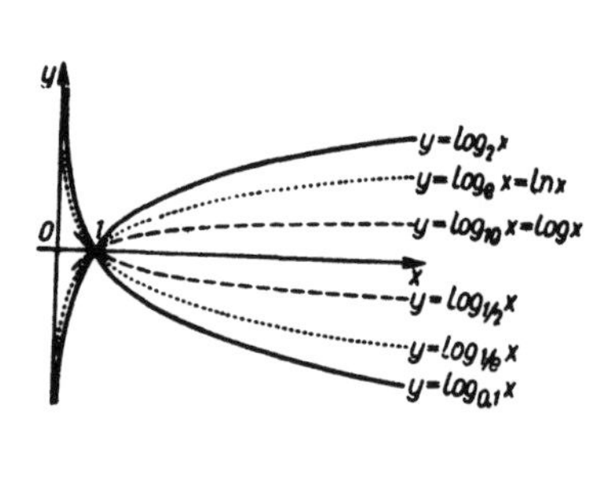

FIG. 18

Logarithmic function: $y = \log_a x$ $(a > 0)$ (Fig. 18). The graph is a *logarithmic curve* (it may be obtained from the exponential curve by the reflection in the bisector $y = x$); for $a = e$—the *natural logarithmic curve* $y = \ln x$. The function exists only for $x > 0$. If $a > 1$, it increases monotonically from $-\infty$ to $+\infty$, and if $a < 1$, it decreases monotonically from $+\infty$ to $-\infty$, the slower, the greater $|\ln a|$ is. The curve passes through $A(1,0)$ and approaches asymptotically the y axis (downwards, if $a > 0$ and upwards, if $a < 0$) the quicker, the greater $|\ln a|$ is.

The function $y = e^{-(ax)^2}$ (Fig. 19). The function increases from 0 to 1 and decreases from 1 to 0. It is symmetric with respect to the y axis and approaches asymptotically the x axis the quicker, the greater a is. A maximum at $A(0,1)$; points of inflection

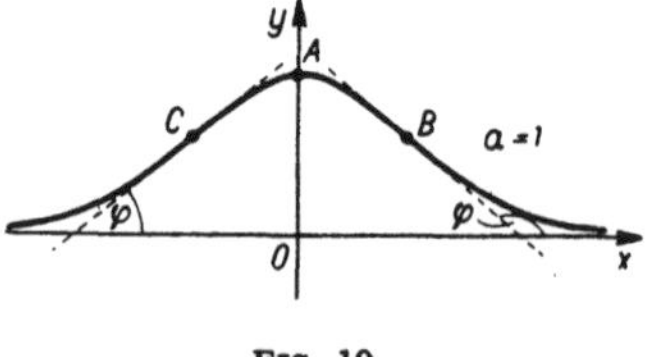

FIG. 19

$B, C\left(\pm\frac{1}{a\sqrt{2}}, \frac{1}{\sqrt{e}}\right)$ with the slopes of the tangent lines $\tan\varphi = \mp a\sqrt{2/e}$. An important case: the *curve of the normal distribution of errors* (the so-called *Gauss' curve*) $y = \varphi(x) = \frac{1}{\sigma\sqrt{2\pi}} e^{-x^2/2\sigma^2}$ (for its graph and application in theory of probability see p. 748).

The function $y = ae^{bx} + ce^{dx}$ (Fig. 20). It is convenient to plot the curve by composing the ordinates of the curves $y_1 = ae^{bx}$ and $y_2 = ce^{dx}$ (see p. 106) drawn by thin lines (the continuous line and the dotted line). The function is continuous. In the case, when each of the numbers a, b, c and d is different from zero, the curve has one of the following four forms (according to the signs of the parameters, the graph in Fig. 20 should be reflected in the coordinate axes):

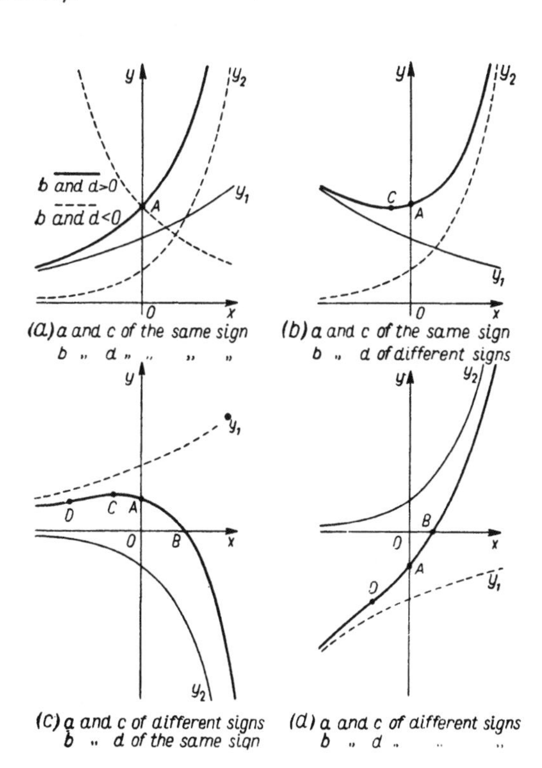

FIG. 20

(a) a and c are of the same sign, b and d are of the same sign. The function varies monotonically either from 0 to $+\infty(-\infty)$ or from $+\infty(-\infty)$ to 0 without changing sign. No points of inflection, the x axis is an asymptote (Fig. 20a).

(b) a and c are of the same sign, and b and d are of different signs. The function varies from $+\infty$ to $+\infty$ or from $-\infty$ to $-\infty$ and passes through an extreme without changing sign. No points of inflection (Fig. 20b).

(c) a and c are of different signs and b and d are of the same sign. The function varies from 0 to $+\infty(-\infty)$ or from $+\infty(-\infty)$ to 0, changes the sign once, has one extreme C and one point of inflection D. The x axis is an asymptote (Fig. 20c).

(d) a and c are of different signs and b and d are of different signs. The function varies monotonically from $-\infty$ to $+\infty$ or from $+\infty$ to $-\infty$ through a point of inflection. No turning points (Fig. 20d). Intersection with the y axis: $A(0, a+c)$. Intersection with the x axis: $B\left[x=\frac{1}{d-b}\ln\left(-\frac{a}{c}\right)\right]$. Turning value: $C\left[x=\frac{1}{d-b}\ln\left(-\frac{ab}{cd}\right)\right]$. Point of inflection: $D\left[x=\frac{1}{d-b}\ln\left(-\frac{ab^2}{cd^2}\right)\right]$.

The function $y=ae^{bx+cx^2}$ (Fig. 21). The curve is symmetric with respect to the vertical line $x=-b/2c$, does not intersect the x axis and intersects the y axis at the point $D(0, a)$. The form of

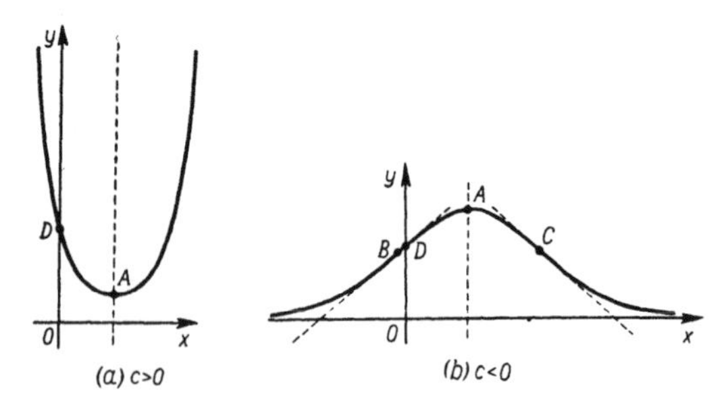

FIG. 21

the function depends on the sign of a and c. Only the case $a>0$ will be considered; if $a<0$, the curve should be reflected in the x axis.

(a) $c>0$. The function decreases from $+\infty$ to a minimum and then increases to $+\infty$ being always positive. The minimum $A(-b/2c, ae^{-b^2/4c})$. No points of inflection or asymptotes (Fig. 21a).

(b) $c < 0$. The function increases from 0 to a maximum and then decreases to 0. The maximum $A\left(-\frac{b}{2c}, ae^{-b^2/4c}\right)$, two points of inflection $B, C\left(\frac{-b \pm \sqrt{-2c}}{2c}, ae^{-(b^2+2c)/4c}\right)$. The x axis is an asymptote (Fig. 21b).

The function $y = ax^b e^{cx}$ (Fig. 22). Only the case $a > 0$ will be considered (if $a < 0$, the curve should be reflected in the x axis) and only for positive x. If $b > 0$, the curve passes through the origin; as a tangent line at the origin it has: the x axis, for $b > 1$,

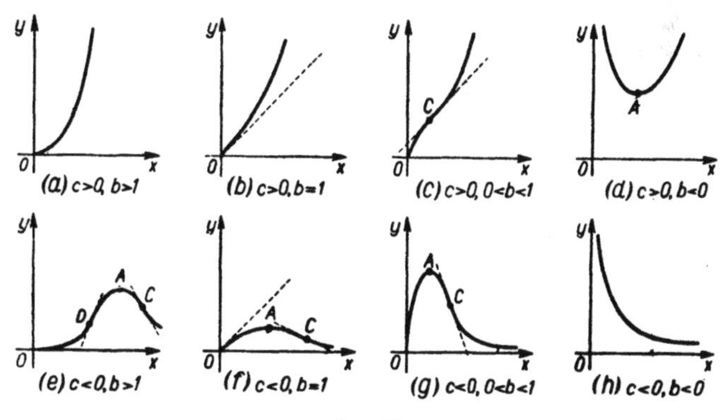

FIG. 22

the bisector $y = x$ for $b = 1$ and the y axis for $b < 1$. If $b < 0$, the y axis is an asymptote. If $c > 0$, the function increases infinitely; if $c < 0$, it decreases to 0, approaching the x axis asymptotically. If b and c are of different signs, the function has a turning value $A(x = -b/c)$. There can be 0, 1 or 2 points of inflection: $C, D\left(x = -\frac{b \pm \sqrt{b}}{c};\text{ if } b = 1, \text{ then } x = -2/c\right)$.

The function $y = Ae^{-ax}\sin(\omega x + \varphi_0)$ (Fig. 23). This is a *damped vibration curve*. It oscillates about the x axis approaching it asymptotically. It is contained between two curves $y = \pm Ae^{-ax}$ and is tangent to them at the points with the coordinates

$$\left(\frac{(k + \frac{1}{2})\pi - \varphi_0}{\omega}, \ (-1)^k Ae^{-ax}\right).$$

Intersections with the axes: $B(0, A \sin \varphi_0)$, $C_1, C_2, \ldots \left(\frac{k\pi - \varphi_0}{\omega}, 0\right)$. Turning values: $D_1, D_2, \ldots$ for $x = \frac{k\pi - \varphi_0 + \alpha}{\omega}$. Points of inflection: $E_1, E_2, \ldots$ for $x = \frac{k\pi - \varphi_0 + 2\alpha}{\omega}$, where $\tan \alpha = \omega / a$.

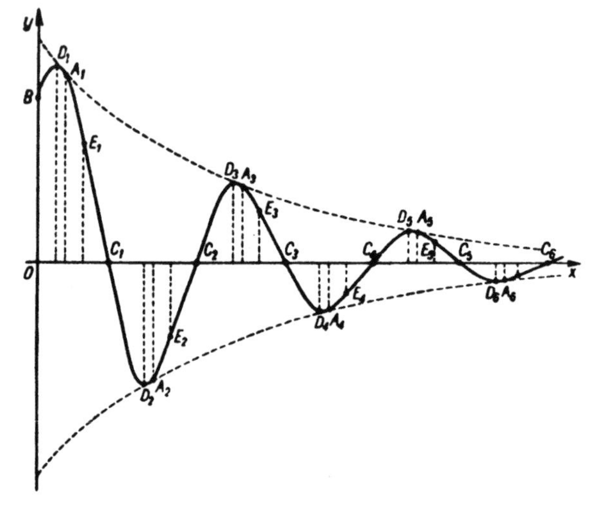

FIG. 23

The number $\delta = \ln \left| \frac{y_i}{y_{i+1}} \right| = a \frac{\pi}{\omega}$ (where y_i and y_{i+1} are two consecutive extreme ordinates) is called the *logarithmic decrement of damping.*

5. Trigonometric functions

The sine: $y = A \sin(\omega x + \varphi_0)$ (Fig. 24). The graph is a *sine curve*. If $A = \omega = 1$ and $\varphi_0 = 0$, it is the usual sine curve $y = \sin x$ (Fig. 24a)—a continuous curve with the period $T = 2\pi$. Intersections with the x axis: $B_1, B_2, \ldots (k\pi, 0)$; these are also points of inflection with the slopes $\tan \varphi = \pm 1$. Turning points $C_1, C_2, \ldots \left((k + \frac{1}{2})\pi, (-1)^k\right)$. In the general case, (Fig. 24b), the curve is stretched in the ratio $|A| : 1$ ($|A|$—the *amplitude*) in the direction of the y axis, is

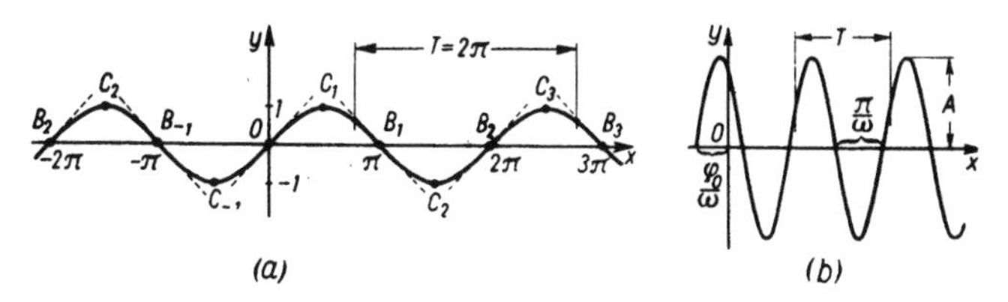

FIG. 24

contracted in the ratio ω:1 in the direction of the x axis (ω—the *frequency*) and is translated to the left by φ_0/ω (φ_0—the *original phase*). The period $T = 2\pi/\omega$; intersections with the x axis B_1, B_2, ... $\left(\frac{k\pi - \varphi_0}{\omega}, 0\right)$; extremes $C_1, C_2, \ldots \left(\frac{(k+\frac{1}{2})\pi - \varphi_0}{\omega}, (-1)^k A\right)$ (see also p. 219).

The cosine: $y = A\cos(\omega x + \varphi_0)$ (Fig. 25). This can be written in the form $y = A\sin(\omega x + \varphi_0 + \frac{1}{2}\pi)$; the graph is a *sine curve* (see above). The usual cosine curve $y = \cos x = \sin(x + \frac{1}{2}\pi)$. Inter-

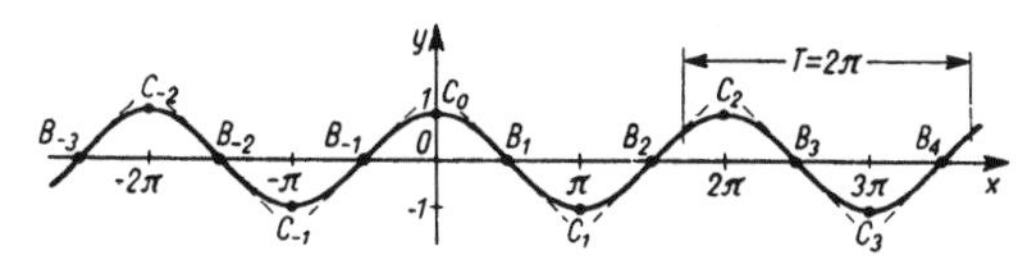

FIG. 25

sections with the y axis: $B_1, B_2, \ldots ((k+\frac{1}{2})\pi, 0)$; these are also the points of inflection with the slopes $\tan\varphi = \pm 1$. Extremes C_1, C_2, ... $(k\pi, (-1)^k)$.

The tangent: $y = \tan x$ (Fig. 26). The graph is a *tangent curve*. It is periodic with the period $T = \pi$ and has points of discontinuity at $x = (k+\frac{1}{2})\pi$. The function increases monotonically from $-\frac{1}{2}\pi$ to $+\frac{1}{2}\pi$ in the interval from $-\infty$ to $+\infty$ and then the values repeat over each interval of length π. Intersections with the x axis: 0, A_1, A_{-1}, A_2, A_{-2}, ... $(k\pi, 0)$ and these are also points of inflection with the slopes $\tan\varphi = +1$.

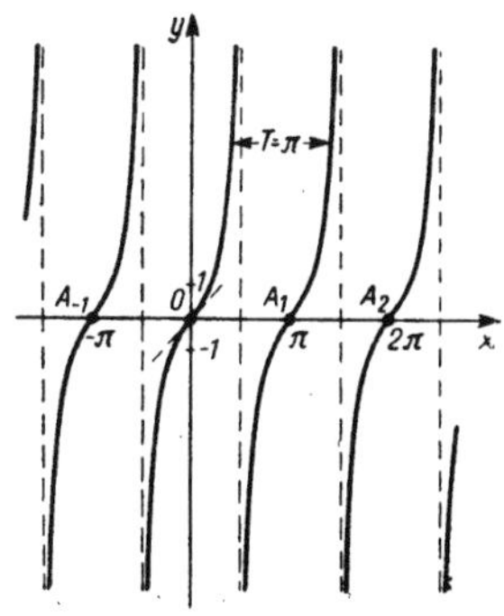

FIG. 26

The cotangent: $y = \cot x$ (Fig. 27) or $y = -\tan(\frac{1}{2}\pi + x)$. The graph is a reflection of the tangent curve translated to the left by $\frac{1}{2}\pi$. The function decreases monotonically from $+\infty$ to $-\infty$ in the interval from 0 to π and then the values repeat. Intersection with the x axis: $A_1, A_{-1}, A_2, A_{-2}, \ldots \left((k + \frac{1}{2})\pi, 0\right)$; these are also points of inflection with the slopes $\tan \varphi = -1$.

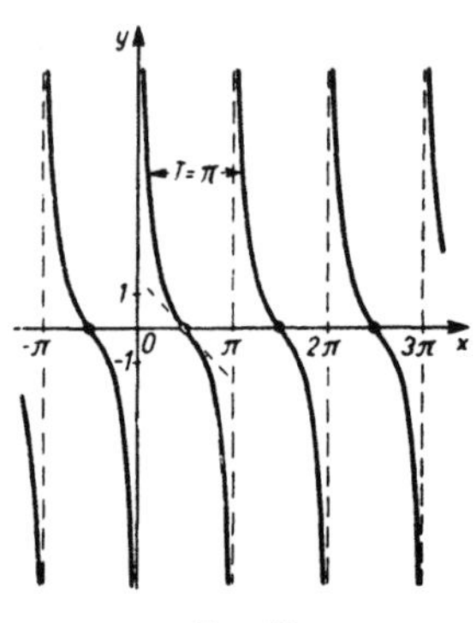

FIG. 27

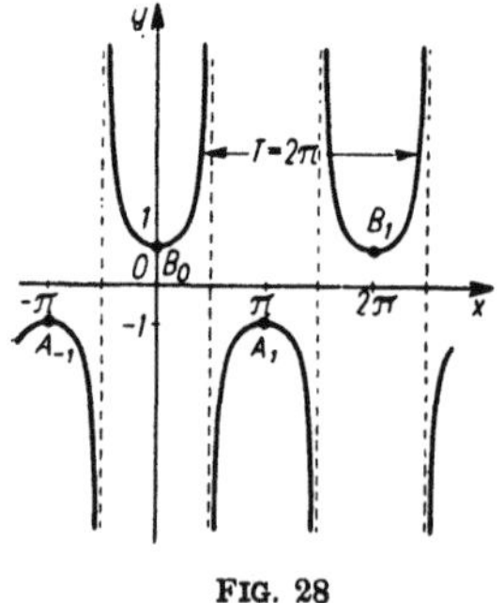

FIG. 28

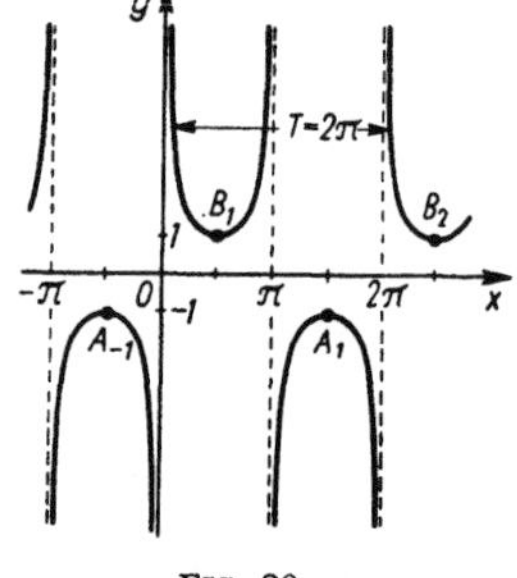

FIG. 29

The secant: $y = \sec x = \dfrac{1}{\cos x}$ (Fig. 28). The graph is a periodic curve with the period $T = 2\pi$ and with discontinuities at $x = (k + \frac{1}{2})\pi$; $|y| > 1$. Maxima $A_1, A_2, \ldots \left((2k + 1)\pi, -1\right)$, minima $B_1, B_2, \ldots (2k\pi, +1)$.

The cosecant: $y = \operatorname{cosec} x = \dfrac{1}{\sin x}$ (Fig. 29) or $y = \sec(x - \frac{1}{2}\pi)$. The graph is a secant curve translated to the left by $x = \frac{1}{2}\pi$. Discontinuities at $x = k\pi$. Maxima $A_1, A_2, \ldots \left(\frac{4k+3}{2}\pi, -1\right)$, minima $B_1, B_2, \ldots \left(\frac{4k+1}{2}\pi, 1\right)$.

6. Inverse trigonometric functions ([1])

The graphs of these functions may be obtained from those of the trigonometric functions by reflecting them in the bisector $y = x$.

The inverse sine: $y = \text{Arc} \sin x$ (Fig. 30). The function exists only for $|x| \leqslant 1$ and is multiple-valued. The principal value $y = \text{arc} \sin x$ (marked by the continuous line) increases monotonically from $A(-1, -\frac{1}{2}\pi)$ to $B(+1, +\frac{1}{2}\pi)$. The origin is a point of inflection with the slope $\tan \varphi = \frac{1}{2}\pi$ and a centre of symmetry the curve.

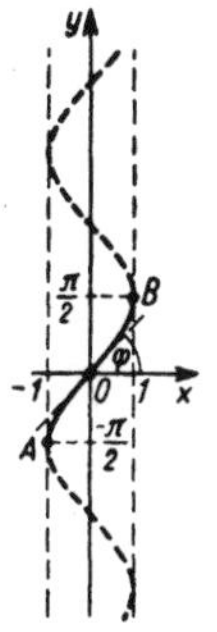

FIG. 30 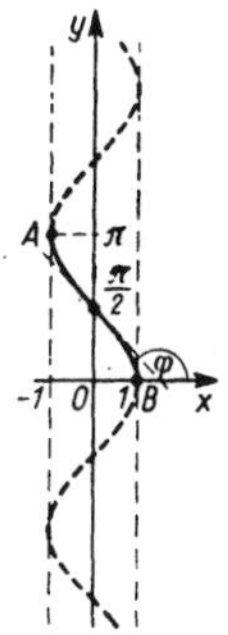 FIG. 31

The inverse cosine: $y = \text{Arc} \cos x$ (Fig. 31). The same curve as for Arc sin x, but lowered by $\frac{1}{2}\pi$. The function exists only for $|x| < 1$ and is multiple-valued. The principal value, $y = \text{arc} \cos x$, decreases monotonically from $A(-1, +\pi)$ to $B(+1, 0)$. The point $(0, \frac{1}{2}\pi)$ is a centre of symmetry and a point of inflection with the slope $\tan \varphi = -\frac{1}{2}\pi$.

The inverse tangent: $y = \text{Arc} \tan x$ (Fig. 32). The function is multiple-valued. The principal value $y = \text{arc} \tan x$ increases monotonically from $(-\infty, -\frac{1}{2}\pi)$ to $(+\infty, +\frac{1}{2}\pi)$. The origin is a centre

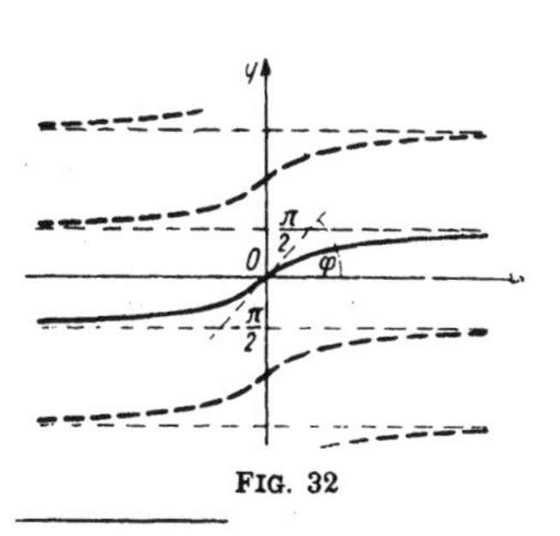

FIG. 32

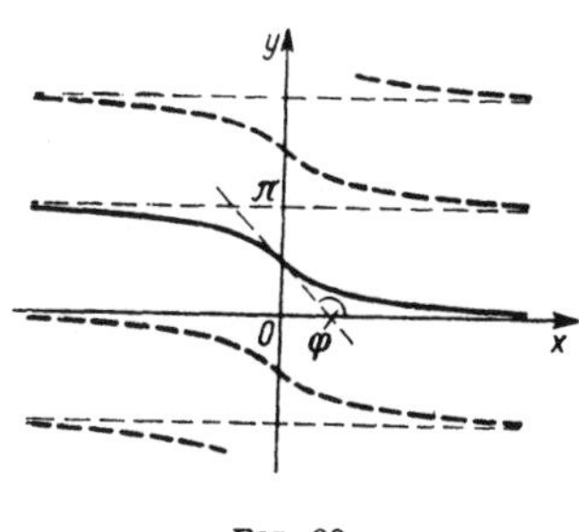

FIG. 33

([1]) For the definitions and formulas see pp. 223–225.

of symmetry and a point of inflection with the slope $\tan\varphi = \frac{1}{2}\pi$. Other values of the function are obtained from the principal one by adding $\pm k\pi$. Asymptotes $x = \pm k \cdot \frac{1}{2}\pi$.

The inverse cotangent: $y = \text{Arc}\cot x$ (Fig. 33). The function is multiple-valued. The principal value $y = \text{arc}\cot x$ decreases monotonically from $(-\infty, \pi)$ to $(+\infty, 0)$. The point $A(0, \frac{1}{2}\pi)$ is a centre of symmetry and a point of inflection with slope $\tan\varphi = -\frac{1}{2}\pi$. Other values of the function are obtained from the principal one by adding $\pm k\pi$. Asymptotes $y = \pm k\pi$.

7. Hyperbolic functions [1]

The hyperbolic sine: $y = \sinh x$ (Fig. 34). The function is odd and increases monotonically from $-\infty$ to $+\infty$. The origin is a centre of symmetry and a point of inflection $(\varphi = \frac{1}{4}\pi)$. No asymptotes.

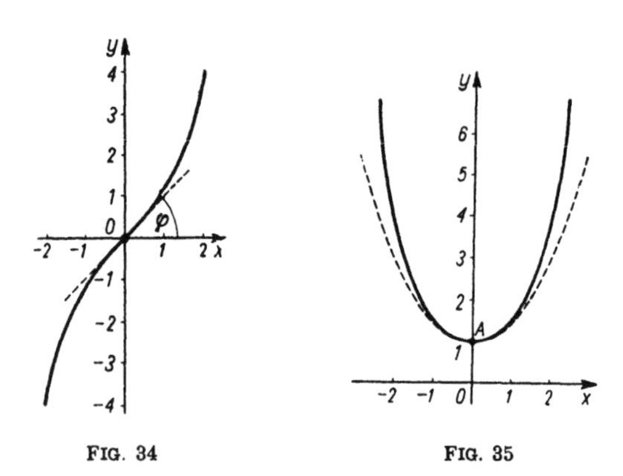

FIG. 34 FIG. 35

The hyperbolic cosine: $y = \cosh x$ (Fig. 35). The graph is like a hanging chain (a *catenary*, see p. 130). The function is even. It decreases from $+\infty$ to 1, for $x < 0$ and increases from 1 to $+\infty$, for $x > 0$. A minimum is at $A(0, 1)$ and there are no asymptotes. The curve is symmetric with respect to the y axis and lies above the parabola $y = 1 + \frac{1}{2}x^2$ dotted in Fig. 35.

---

[1] For theoretical discussion of hyperbolic functions see pp. 229–230; for the tables see pp. 61–65.

The hyperbolic tangent: $y = \tanh x$ (Fig. 36). The function is odd and increases monotonically from -1 to $+1$. It has a point of inflection ($\varphi = \frac{1}{4}\pi$) and a centre of symmetry at the origin. Two asymptotes: $y = \pm 1$.

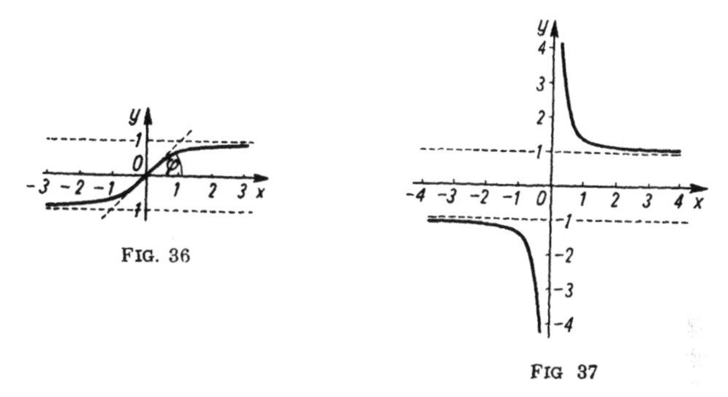

FIG. 36

FIG 37

The hyperbolic cotangent: $y = \coth x$ (Fig. 37). The function is odd and has a discontinuity at $x = 0$. It decreases from -1 to $-\infty$ for $x < 0$ and decreases from $+\infty$ to $+1$ for $x > 0$. No turning points. Three asymptotes: $x = 0$, $y = \pm 1$.

8. Inverse hyperbolic functions [1]

The graphs are obtained from those of hyperbolic functions by the reflection in the bisector $y = x$.

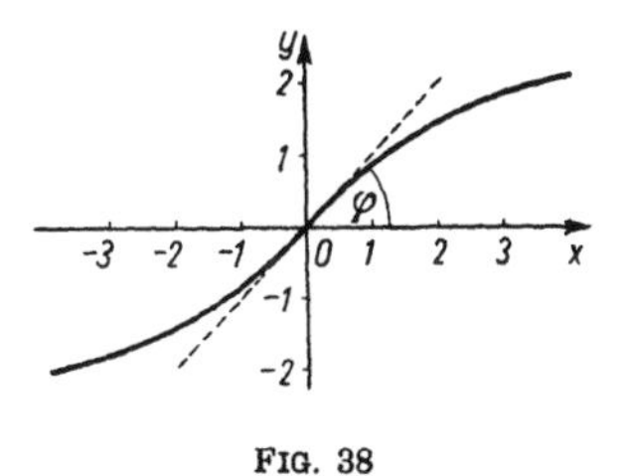

FIG. 38

The inverse hyperbolic sine: $y = \text{ar sinh}\, x = \ln(x + \sqrt{x^2 + 1})$ (Fig. 38). The function increases monotonically from $-\infty$ to $+\infty$. There is a point of inflection ($\varphi = \frac{1}{4}\pi$) and a centre of symmetry at the origin. No asymptotes.

---

[1] For theoretical discussion see pp. 232–233.

The inverse hyperbolic cosine: $y = \operatorname{ar}\cosh x = \ln(x \pm \sqrt{x^2 - 1})$ (Fig. 39). The function is two-valued and exists only for $x > 1$. The curve is symmetric with respect to the x axis. There is a vertical tangent line $x = 1$ at $A(1, 0)$, hence the absolute value of y increases.

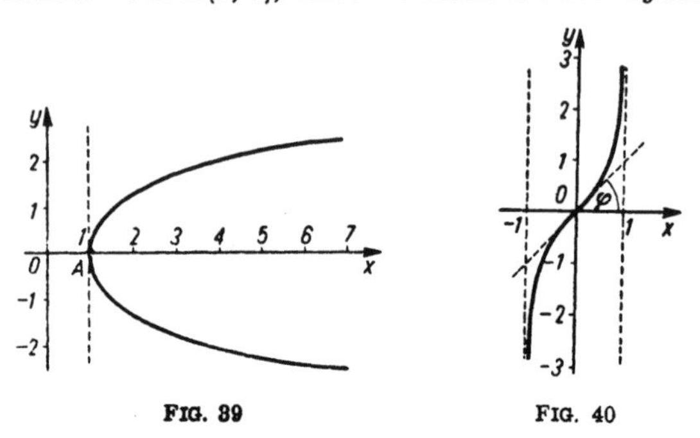

FIG. 39 FIG. 40

The inverse hyperbolic tangent: $y = \operatorname{ar}\tanh x = \frac{1}{2}\ln\frac{1+x}{1-x}$ (Fig. 40). The function is odd and exists only for $|x| < 1$. It increases monotonically from $-\infty$ to $+\infty$. There is point of inflection $(\varphi = \frac{1}{4}\pi)$ and a centre of symmetry at the origin. Two asymptotes: $x = \pm 1$.

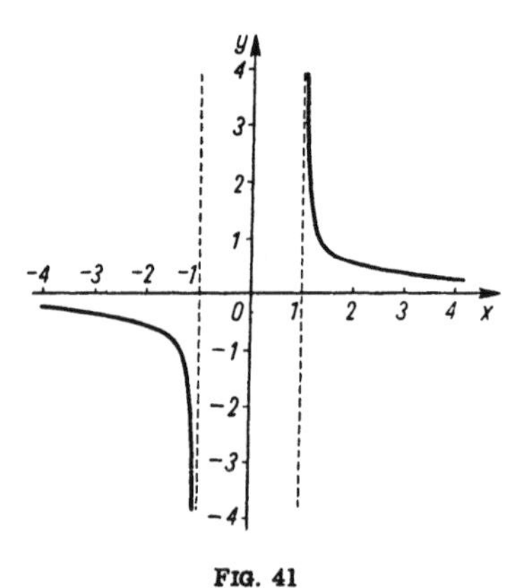

FIG. 41

The inverse hyperbolic cotangent: $y = \operatorname{ar}\coth x = \frac{1}{2}\ln\frac{x+1}{x-1}$ (Fig. 41). The function is odd and exists only for $|x| > 1$. It decreases from 0 to $-\infty$ in the interval $-\infty < x < -1$ and decreases from $+\infty$ to 0 in the interval $+1 < x < +\infty$. No extremes or inflection points. Three asymptotes: $y = 0$, $x = \pm 1$.

B. IMPORTANT CURVES

In this section we consider some properties of important curves which occur in practice. These properties include: definitions of the curve as the locus of a point, coordinates of characteristic points, length of the curve or of a part of it, area bounded by the curve or by a part of it, radius of curvature at the characteristic points.

For general information about plotting a curve from its equation see pp. 292–293.

For the curves of the second degree (ellipse, hyperbola, parabola) see pp. 244–256.

9. Curves of the third degree

Neil's parabola or the semicubic parabola (Fig. 42). Equation: $y = ax^{3/2}$; in the parametric form: $x = at^2$, $y = at^3$. There is a cusp at the origin. No asymptotes. The curvature $K = \frac{6a}{\sqrt{x}(4+9a^2x)^{3/2}}$ admits all the values from ∞ to 0. The length of the curve from the origin to an arbitrary point $M(x, y)$ (¹): $L = \frac{1}{27a^2}\left((4+9a^2x)^{3/2} - 8\right)$.

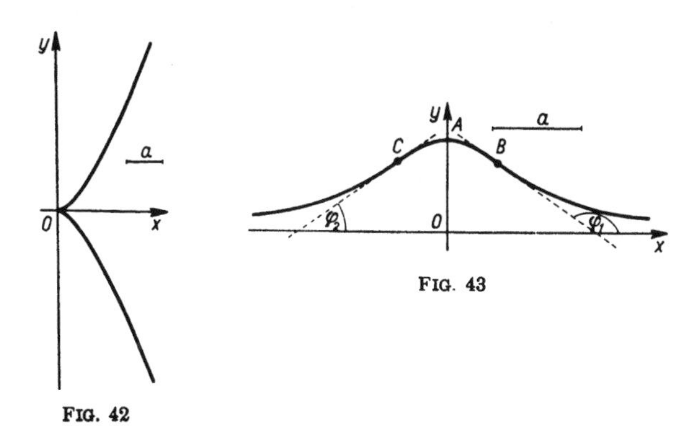

FIG. 42

FIG. 43

The witch of Agnesi (Fig. 43). Equation: $y = \frac{a^3}{a^2 + x^2}$. Asymptote: $y = 0$. Maximum $A(0, a)$; radius of curvature at $A: r = \frac{1}{2}a$. Points of inflection $B, C(\pm\frac{1}{3}a\sqrt{3}, \frac{3}{4}a)$, with slopes $\tan\varphi = \mp\frac{3}{8}\sqrt{3}$. The area between the curve and its asymptote $S = \pi a^2$.

(¹) In the following M will denote an arbitrary point of the curve with the coordinates x, y.

The folium of Descartes (Fig. 44). Equation: $x^3 + y^3 = 3axy$; parametric form: $x = \frac{3at}{1+t^3}$, $y = \frac{3at^2}{1+t^3}$ ($t = \tan MOx$). The origin is a double point with the coordinate axes as tangent lines; the radius of curvature of the two branches of the curve at the

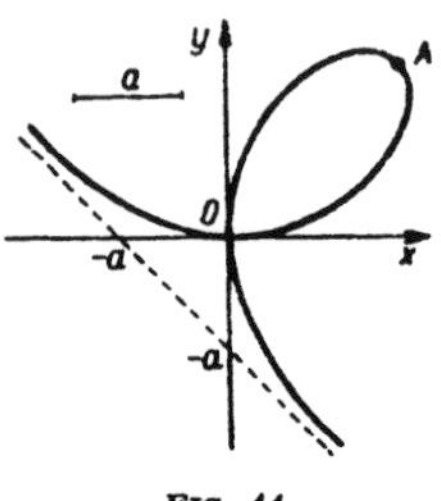

FIG. 44

origin is $r = 3a/2$. Asymptote $x + y + a = 0$. Vertex $A(\frac{3}{2}a, \frac{3}{2}a)$. Area of the loop $S_1 = \frac{3}{2}a^2$; area between the curve and its asymptote $S_2 = \frac{3}{2}a^2$.

Cissoid (Fig. 45). It is the locus of a point M for which $OM = PQ$ (P is a point of the generating circle with the diameter a).

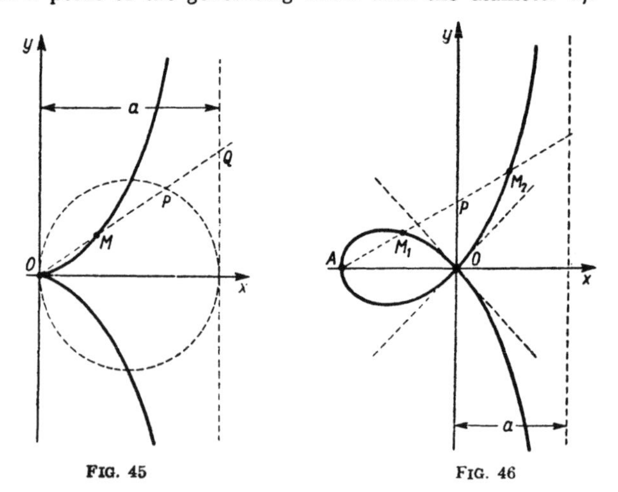

FIG. 45 FIG. 46

Equation: $y^2 = \frac{x^3}{a - x}$; the parametric form: $x = \frac{at^2}{1+t^2}$, $y = \frac{at^3}{1+t^2}$ ($t = \tan MOx$); in the polar coordinates: $\varrho = \frac{a \sin^2 \varphi}{\cos \varphi}$

The origin is a cusp. Asymptote: $x = a$. Area between the curve and the asymptote $S = \frac{3}{4}\pi a^2$.

Strofoid (Fig. 46). The locus of points M_1 and M_2 for which $PM_1 = PM_2 = OP$.

Equation: $y^2 = x^2 \dfrac{a+x}{a-x}$; the parametric form $x = a\dfrac{t^2-1}{t^2+1}$, $y = at\dfrac{t^2-1}{t^2+1}$ ($t = \tan MOx$); in the polar coordinates: $\varrho = -a\dfrac{\cos 2\varphi}{\cos\varphi}$. The origin is a double point with tangents $y = \pm x$. Asymptote $x = a$. Vertex $A(-a, 0)$. Area of the loop $S_1 = a^2 - \frac{1}{4}\pi a^2$, the area between the curve and the asymptote $S_2 = a^2 + \frac{1}{4}\pi a^2$.

10. Curves of the fourth degree

Conchoid of Nicomedes (Fig. 47). The locus of a point M, for which $OM = OP \pm l$ (the sign "+" corresponds to the exterior branch, the sign "−" to the interior one) [1].

Equation: $(x-a)^2(x^2+y^2) - l^2x^2 = 0$; in parametric form $x = a + l\cos\varphi$, $y = a\tan\varphi + l\sin\varphi$; in polar coordinates $\varrho = \dfrac{a}{\cos\varphi} \pm l$.

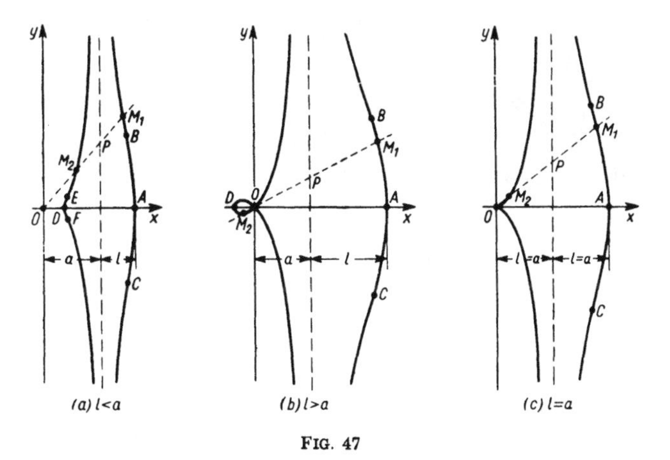

FIG. 47

[1] In general, the *conchoid* of a curve is the curve obtained from the given one by increasing or decreasing the radius vector of the curve by a constant segment l. If the equation of the given curve in the polar coordinates is $\varrho = f(\varphi)$, then the *equation of its conchoid* is $\varrho = f(\varphi) \pm l$. The conchoid of Nicomedes is the conchoid of the straight line.

Exterior branch: Asymptote $x = a$. Vertex $A(a + l, 0)$. Two points of inflections B, C (x is equal to the greatest root of the equation $x^3 - 3a^2x + 2a(a^2 - l^2) = 0$) [1]. Area between the branch and its asymptote $S = \infty$.

Interior branch: Asymptote $x = a$. Vertex $D(a - l, 0)$. There is a double point at the origin; its type depends on a and l.

(a) If $l < a$—an isolated point (Fig. 47a). The curve has two points of inflection E, F (x is equal to the second positive root of the equation $x^3 - 3a^2x + 2a(a^2 - l^2) = 0$).

(b) If $l > a$—a branch point (Fig. 47b). The curve has a maximum and a minimum at $x = a - \sqrt[3]{al^2}$. The slopes of the tangent lines at the origin: $\tan\alpha = \pm\sqrt{(l^2 - a^2)/a}$; the radius of curvature $r = \frac{1}{2}l\sqrt{l^2 - a^2}$.

(c) If $l = a$—a cusp (Fig. 47c).

Pascal's snail (Fig. 48). The conchoid of the circle [2]: $OM = OP \pm l$ (the pole is the end-point of one diameter).

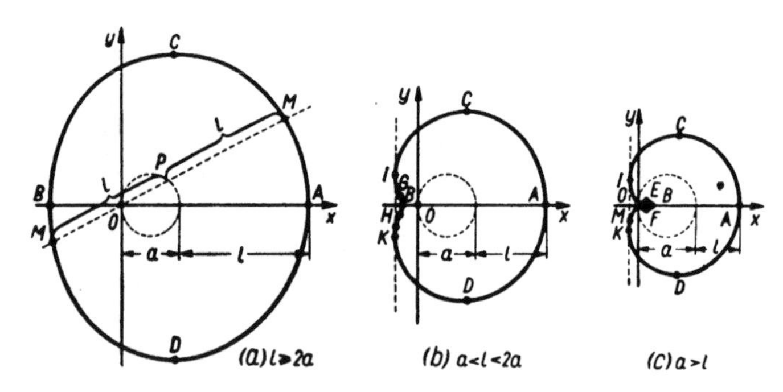

FIG. 48

Equation: $(x^2 + y^2 - ax)^2 = l^2(x^2 + y^2)$; parametric form $x = a\cos^2\varphi + l\cos\varphi$, $y = a\cos\varphi\sin\varphi + l\sin\varphi$; in polar coordinates $\varrho = a\cos\varphi + l$ (a is diameter of the circle). Vertices $A, B(a \pm l, 0)$. The form of the curve depends on a and l, as is shown in Figs. 48 and 49. There are four turning points, if $a > l$ and two, if $a < l$:

---

[1] For solution of such equations see pp. 161–162.

[2] See footnote on p. 119.

$C, D, E, F\left(\cos\varphi = \frac{-l \pm \sqrt{l^2 + 8a^2}}{4a}\right)$. Points of inflection $G, H\left(\cos\varphi = -\frac{2a^2 + l^2}{3al}\right)$ exist, if $a < l < 2a$. A double tangent line at the points $I, K\left(-\frac{l^2}{4a}, \pm\frac{l\sqrt{4a^2 - l^2}}{4a}\right)$ exists, if $l < 2a$. There is a double point at the origin: an isolated point, if $a < l$, a branch point if $a > l$ (with the slopes of the tangent lines $\tan\alpha = \pm\frac{\sqrt{a^2 - l^2}}{l}$, and the radius of curvature $r = \frac{1}{2}\sqrt{a^2 - l^2}$) and a cusp if $a = l$ (see below).

Area of the snail $S = \frac{1}{2}\pi a^2 + \pi l^2$ (in the case (c) (Fig. 48c) the area of the interior loop is counted twice).

Cardioid (Fig. 49). This curve can be regarded as: (1) a particular case of the Pascal snail: $OM = OP \pm a$ (a is a diameter of the circle) or as (2) an epicycloid (see p. 124) in which the radii of both rolling and fixed circles are equal ($=a$).

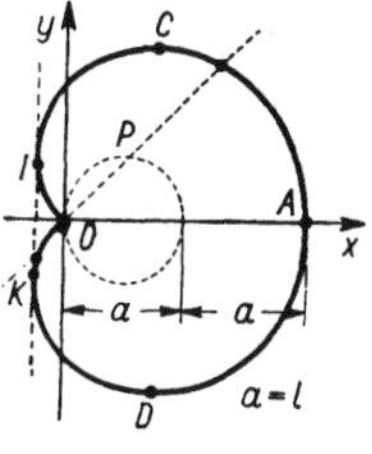

FIG. 49

Equation: $(x^2 + y^2)^2 - 2ax(x^2 + y^2) = a^2y^2$; in parametric form $x = a\cos\varphi(1 + \cos\varphi)$, $y = a\sin\varphi(1 + \cos\varphi)$; in polar coordinates $\varrho = a(1 + \cos\varphi)$.

A cusp at the origin. Vertex $A(2a, 0)$. Maximum and minimum ($\cos\varphi = \frac{1}{2}$): $C, D\left(\frac{3}{4}a, \pm\frac{3}{4}\sqrt{3}\,a\right)$. Area $S = \frac{3}{2}\pi a^2$ (6 times the area of a circle of diameter a). Length of the curve $L = 8a$.

Cassini's ovals (Fig. 50). The locus of a point M for which the product $F_1M \cdot F_2M = a^2$ (F_1, F_2 are two fixed foci, a is a constant).

Equation: $(x^2 + y^2)^2 - 2c^2(x^2 - y^2) = a^4 - c^4$, where $F_1, F_2(\pm c, 0)$; in polar coordinates $\varrho^2 = c^2\cos 2\varphi \pm \sqrt{c^4\cos^2 2\varphi + (a^4 - c^4)}$. The form depends on a and c as follows:

(a) $a > c\sqrt{2}$; an ellipse-like oval (Fig. 50a). Intersections with the x axis: $A, C(\pm\sqrt{a^2 + c^2}, 0)$; intersections with the y axis: $B, D(0, \pm\sqrt{a^2 - c^2})$. If $a = c\sqrt{2}$, the oval has the same type; in this case, $A, C(\pm c\sqrt{3}, 0)$, $B, D(0, \pm c)$; the curvature at B and D is equal to 0 (the straight lines $y = \pm c$ are closely tangent to the curve).

(b) $c < a < c\sqrt{2}$; an oval with two swellings (Fig. 50b). Intersection with the axes are the same as in the case (a); maxima and minima B, D (with the coordinates given above) and E, G, K, I

$\left(\pm\frac{\sqrt{4c^4-a^4}}{2c}, \pm\frac{a^2}{2c}\right)$; four points of inflection: P, L, M, N $\left(\pm\sqrt{\frac{1}{2}(m-n)}, \pm\sqrt{\frac{1}{2}(m+n)}\right)$, where $n=\frac{a^4-c^4}{3c^2}, m=\sqrt{\frac{a^4-c^4}{3}}$.

(c) $a=c$; Bernoulli's lemniscate (see below).

(d) $a<c$; two ovals (Fig. 50c). Intersections with the x axis: $A, C\,(\pm\sqrt{a^2+c^2}, 0)$ and $P, Q\,(\pm\sqrt{c^2-a^2}, 0)$; maxima and minima: $E, G, K, I\left(\pm\frac{\sqrt{4c^4-a^4}}{2c}, \pm\frac{a^2}{2c}\right)$. Radius of curvature $r=\frac{2a^2\varrho^3}{c^4-a^4+3\varrho^4}$ (ϱ—the radius-vector).

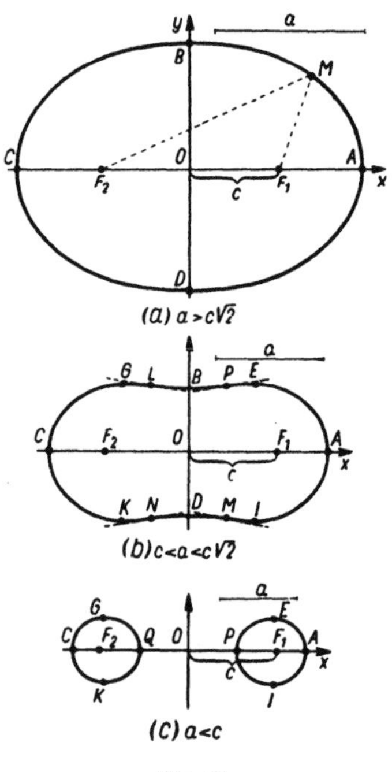

FIG. 50

Bernoulli's lemniscate (Fig. 51). A particular case of Cassini's oval ($a=c$): $F_1M\cdot F_2M=(\frac{1}{2}F_1F_2)^2$, where $F_1, F_2\,(\pm a, 0)$.

Equation: $(x^2+y^2)^2-2a^2(x^2-y^2)=0$; in polar coordinates $\varrho=a\sqrt{2\cos 2\varphi}$. There is a branch point at the origin with the tangent lines $y=\pm x$; this is also a point of inflection. Intersections with the x axis: $A, C(\pm a\sqrt{2}, 0)$; maxima and minima $E, G, K, I(\pm \frac{1}{2}a\sqrt{3}, \pm\frac{1}{2}a)$, i.e., $\varphi = \pm\frac{1}{6}\pi$. Radius of curvature $r = 2a^2/3\varrho$. Area of each loop $S = a^2$.

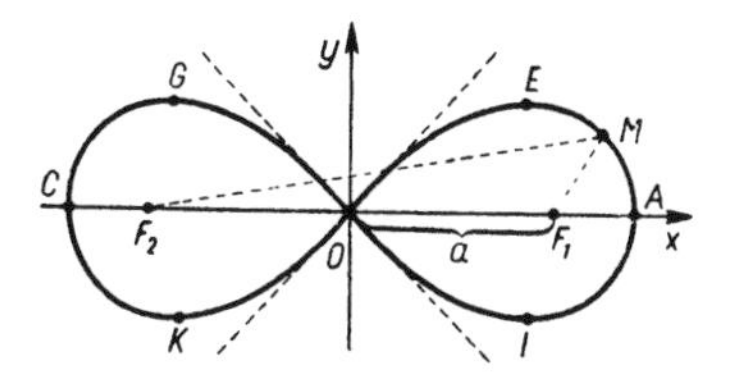

FIG. 51

11. Cycloids

Usual cycloid (Fig. 52). The curve traced out by a fixed point of a circle which rolls without slipping on a straight line.

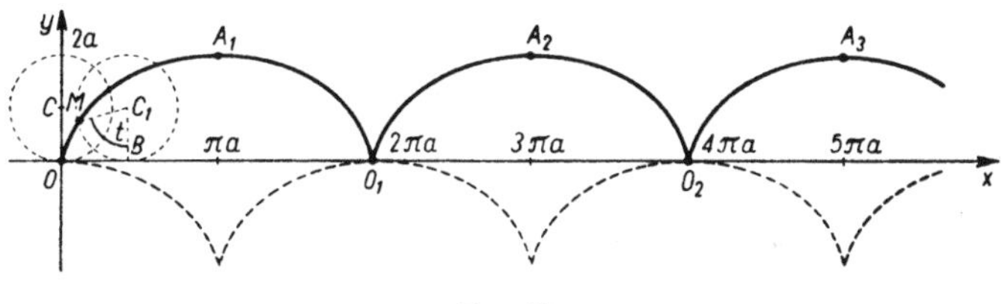

FIG. 52

Equation in parametric form: $x = a(t - \sin t)$, $y = a(1 - \cos t)$ (a—radius of the circle, $t = \sphericalangle MC_1B$); in the Cartesian coordinates: $x + \sqrt{y(2a - y)} = a \arccos \dfrac{a-y}{a}$. The curve is periodic with the period $OO_1 = 2\pi a$ (the base of the cycloid). Cusps $O, O_1, O_2, \ldots$ $(2k\pi a, 0)$; vertices $A_1, A_2, \ldots$ $((2k+1)\pi a, 2a)$. Length of OM: $L = 8a \sin^2 \frac{1}{4}t$; length of one period $L_{OO_1A_1} = 8a$; area of OA_1O_1O: $S = 3\pi a^2$. Radius of curvature $r = 4a \sin\frac{1}{2}t$. Evolute of the cycloid is also a cycloid (marked by the dotted line).

The lengthened (Fig. 53a) **and the shortened** (Fig. 53b) **cycloid.** The curve formed by a point lying outside or inside on a radius of a circle which rolls without slipping on a straight line.

Equation in parametric form: $x = a(t - \lambda \sin t)$, $y = a(1 - \lambda \cos t)$, where a is the radius of the circle, $t = \sphericalangle MC_1P$, $\lambda a = C_1M$ ($\lambda > 1$ for the lengthened cycloid and $\lambda < 1$ for the shortened one). The curves are periodic with the period (the base) $OO_1 = 2\pi a$; maxima $A_1, A_2, \ldots \big((2k+1)\pi a, (1+\lambda)a\big)$, minima $B_1, B_2, \ldots \big(2k\pi a, (1-\lambda)a\big)$.

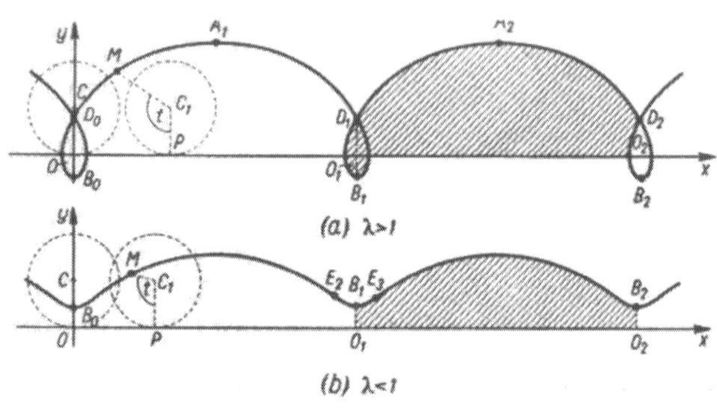

(a) $\lambda > 1$

(b) $\lambda < 1$

FIG. 53

Branch points of the lengthened cycloid $D_0, D_1, D_2, \ldots$ $\big(2k\pi a, a(1-\sqrt{\lambda^2 - t_0^2}\big)$, where t_0 is the least positive root of the equation $t = \lambda \sin t$ [1].

Points of inflection of the shortened cycloid $E_1, E_2, \ldots$ $\big(a(\arccos\lambda - \lambda\sqrt{1-\lambda^2}), a(1-\lambda^2)\big)$. Length of one period $L = a\int_0^{2\pi}\sqrt{1+\lambda^2 - 2\lambda\cos t}\,dt$; area marked on Fig. 53: $S = \pi a^2(2+\lambda^2)$. Radius of curvature $r = a\dfrac{(1+\lambda^2-2\lambda\cos t)^{3/2}}{\lambda(\cos t - \lambda)}$ in the points of maximum $r_A = -a\dfrac{(1+\lambda)^2}{\lambda}$, in the points of minimum $r_B = a\dfrac{(1-\lambda)^2}{\lambda}$.

Epicycloid (Fig. 54). The curve traced out by a point of a circle rolling without slipping on the outside of a fixed circle.

Equation in parametric form: $x = (A+a)\cos\varphi - a\cos\dfrac{A+a}{a}\varphi$, $y = (A+a)\sin\varphi - a\sin\dfrac{A+a}{a}\varphi$ (a—radius of the rolling circle, A—radius of the fixed circle, $\varphi = \sphericalangle COx$). The form of the curve depends on the ratio $A/a = m$. For $m = 1$ the curve is a cardioid (see above, p. 121).

[1] For solution of such equations see pp. 168–169.

(a) If m is an integer, the curve is composed of m equal branches (Fig. 54a) surrounding the fixed circle; cusps $A_1, A_2, \ldots, A_m \left(\varrho = A,\ \varphi = \frac{2k\pi}{m}\ (k = 0, 1, \ldots, m-1)\right)$; vertices $B_1, B_2, \ldots, B_m \left(\varrho = A + 2a,\ \varphi = \frac{2\pi}{m}\left(k + \frac{1}{2}\right)\right)$.

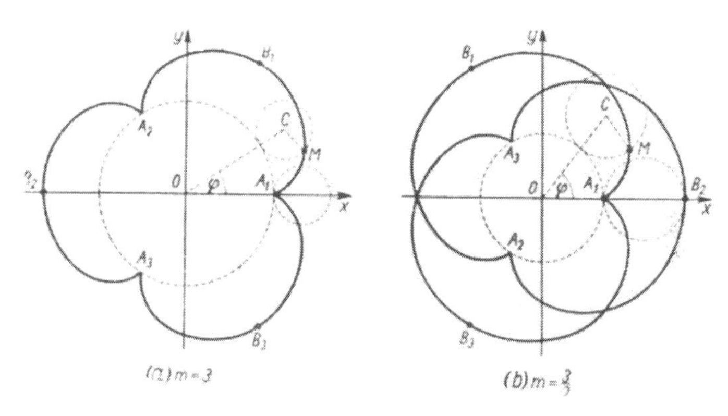

FIG. 54

(b) If m is a fraction, the branches cross one another (Fig. 54b), but the moving point M, after passing a finite number of branches, returns to the starting position. If m is irrational, the number of branches is infinite and the point M never returns to the starting position.

The length of one branch $L_{A_1B_1A_2} = \frac{8(A+a)}{m}$; when m is an integer, the length of the whole curve is $L = 8(A + a)$. The area of one sector (without the sector of the fixed circle): $S = \pi a^2 \left(\frac{3A + 2a}{A}\right)$.

Radius of curvature $r = \frac{4a(A+a)}{2a+A} \sin \frac{A\varphi}{2a}$; in the vertices $r_B = \frac{4a(A+a)}{2a+A}$.

Hypocycloid (Fig. 55). The curve formed by a point of a circle rolling without slipping on the inside of a fixed circle.

Equation of the hypocycloid, coordinates of its vertices and points of return, formulas for the length, area and radius of curvature can be obtained from those of the epicycloid by changing "$+a$" into "$-a$"; the number of cusps, when m is integral, fractional or irrational (m is always > 1) is the same as for the epicycloid. If $m = 2$, the curve degenerates to one diameter of the fixed circle. If $m = 3$, a hypocycloid with three branches (Fig. 55a): $x = a(2\cos\varphi + \cos 2\varphi)$, $y = a(2\sin\varphi - \sin 2\varphi)$; $L = 16a$,

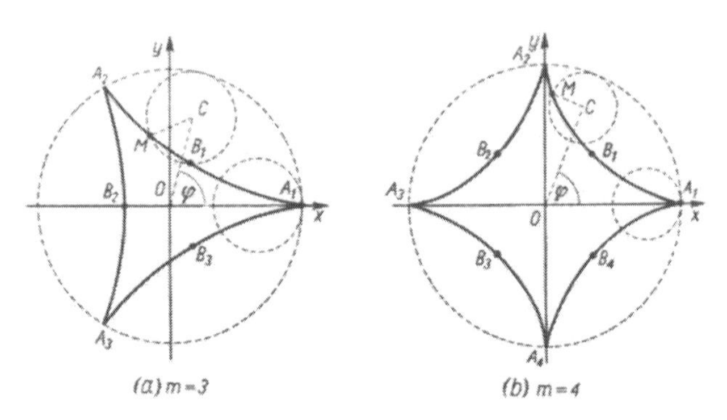

FIG. 55

$S_{\text{total}} = 2\pi a^2$. If $m = 4$ (Fig. 55b), a hypocycloid with four branches (*astroid*): $x = A\cos^3\varphi$, $y = A\sin^3\varphi$; in the Cartesian coordinates $x^{2/3} + y^{2/3} = A^{2/3}$; $L = 24a = 6A$; $S = \frac{3}{8}\pi A^2$.

The lengthened and the shortened epicycloid and hypocycloid (Figs. 56 and 57). The curve formed by a point lying outside or inside of a circle which rolls without slipping on the outside (epicycloid, Fig. 56) or on the inside (hypocycloid, Fig. 57) of fixed circle.

Equation in parametric form: $x = (A + a)\cos\varphi - \lambda a \cos\left(\frac{A+a}{a}\varphi\right)$,

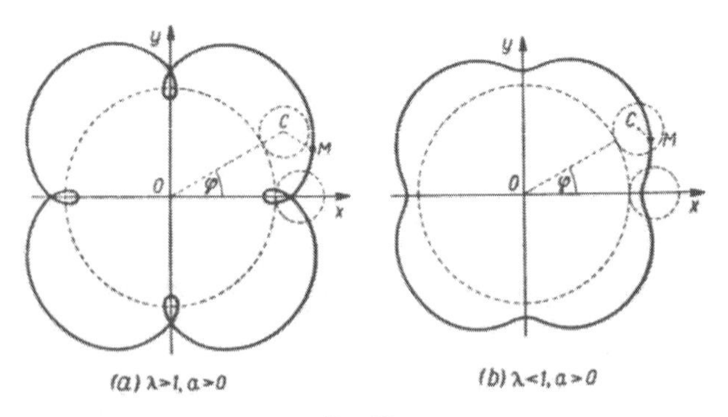

FIG. 56

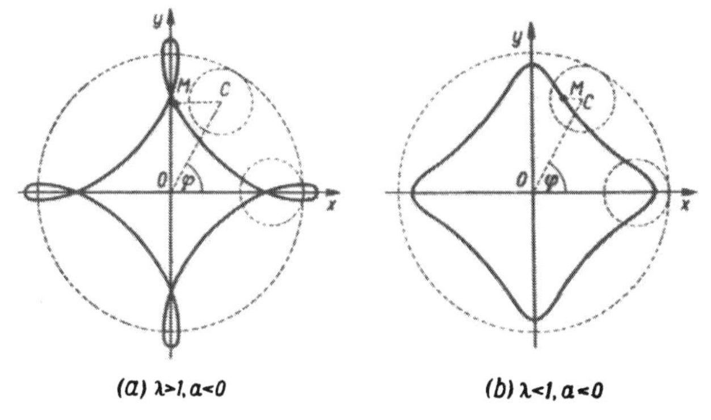

FIG. 57

$y = (A+a)\sin\varphi - \lambda a \sin\left(\frac{A+a}{a}\varphi\right)$. A—radius of the fixed circle, a—radius of the rolling circle (in the case of hypocycloid "$+a$" should be changed into "$-a$"), $\lambda a = CM$ (for the lengthened hypocycloid $\lambda > 1$, for the shortened one $\lambda < 1$). For $A = 2a$ (λ—arbitrary) the hypocycloid $x = a(1+\lambda)\cos\varphi$, $y = a(1-\lambda)\sin\varphi$ changes into an ellipse with half-axes $a(1+\lambda)$ and $a(1-\lambda)$. For $a = A$, we get the Pascal snail (see p. 120) (¹):

$$x = a(2\cos\varphi - \lambda\cos 2\varphi), \quad y = a(2\sin\varphi - \lambda\sin 2\varphi).$$

12. Spirals

Spiral of Archimedes (Fig. 58). The curve formed by a point moving with a constant velocity v on a straight line when this line revolves with a constant angular velocity ω about a pole.

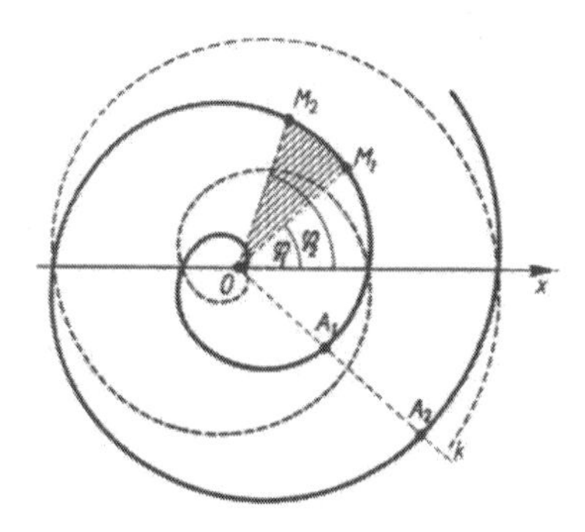

FIG. 58

Equation in polar coordinates: $\varrho = a\varphi$; $a = v/\omega$. The curve is composed of two branches situated symmetrically with respect to the x axis. Each half straight line OK beginning at

(¹) The constant denoted on p. 120 by a is here denoted by $2\lambda a$ and l denoted the diameter $2a$. The coordinate system has been changed.

the origin intersects the curve in the points $O, A_1, A_2, \ldots, A_n$, lying on mutual distances $A_iA_{i+1} = 2\pi a$. The length of the arc OM: $L = \frac{1}{2}a(\varphi\sqrt{\varphi^2+1} + \operatorname{Ar}\sinh\varphi)$, for large φ: $L \approx \frac{1}{2}a\varphi^2$. The area of the sector M_1OM_2: $S = \frac{1}{6}a^2(\varphi_2^3 - \varphi_1^3)$. Radius of curvature $r = a\dfrac{(\varphi^2+1)^{3/2}}{\varphi^2+2}$, at the origin $r = \frac{1}{2}a$.

Hyperbolic spiral (Fig. 59). Equation in the polar coordinates: $\varrho = a/\varphi$. The curve is composed of two branches situated symmetrically with respect to the y axis; each of the branches has the line

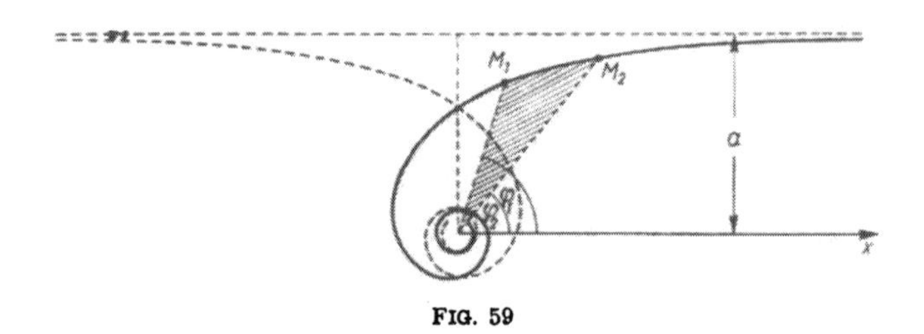

FIG. 59

$y = a$ as an asymptote and the origin as an asymptotic point. The area of the sector M_1OM_2: $S = \dfrac{a^2}{2}\left(\dfrac{1}{\varphi_1} - \dfrac{1}{\varphi_2}\right)$; $S \to \dfrac{a^2}{2\varphi}$ for $\varphi_2 \to \infty$. Radius of curvature $r = \dfrac{a}{\varphi}\left(\dfrac{\sqrt{1+\varphi^2}}{\varphi}\right)^3$.

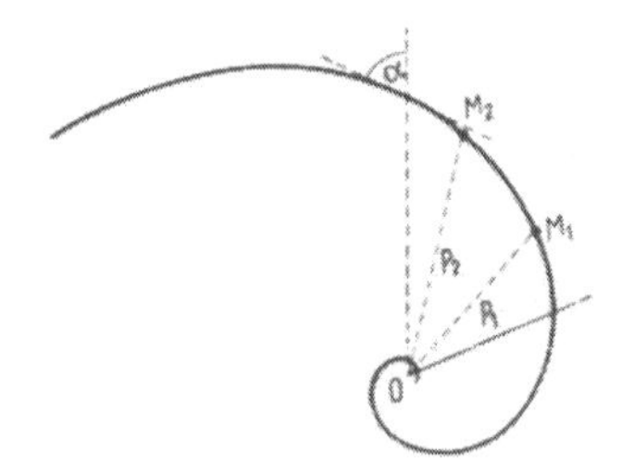

FIG. 60

Logarithmic spiral (Fig. 60). The curve intersecting each radius vector at the same angle α.

Equation in polar coordinates: $\varrho = ae^{k\varphi}$ ($k = \cot\alpha$; if $\alpha = \frac{1}{2}\pi$ then $k = 0$ and the curve is a circle). The origin is an asymptotic point. The length of the arc M_1M_2: $L = \frac{a\sqrt{1+k^2}}{k}(\varrho_2 - \varrho_1)$, limit of the length of the arc OM from the origin: $L_0 = \frac{a\sqrt{1+k^2}}{k}\varrho$. Radius of curvature $r = a\sqrt{1+k^2}\,\varrho = L_0 k$.

Involute [1] **of the circle** (Fig. 61). The curve formed by an end-point of a thread winding off a circle ($AB = BM$).

Equation in parametric form: $x = a\cos\varphi + a\varphi\sin\varphi$, $y = a\sin\varphi - a\varphi\cos\varphi$ (a—radius of the circle, $\varphi = \sphericalangle MOx$). The curve

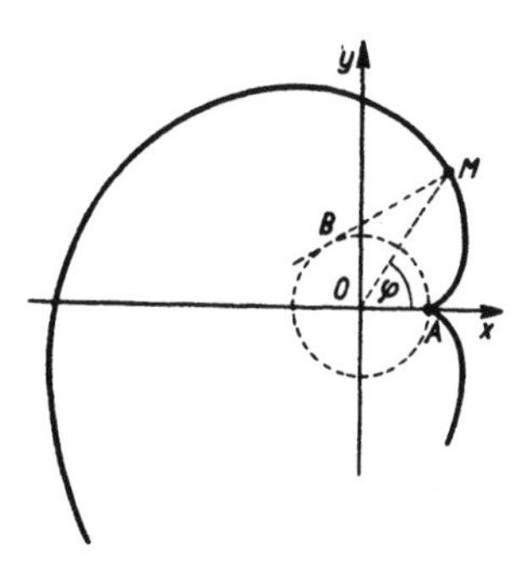

FIG. 61

has two branches situated symmetrically with respect to the x axis; cusp $A(a, 0)$, intersection with Ox: $x = \frac{a}{\cos\varphi_i}$, where φ_i are the roots of the equation $\tan\varphi = \varphi$ [2]. Length of the arc AM: $L = \frac{1}{2}a\varphi^2$. Radius of curvature: $r = a\varphi = \sqrt{2aL}$; centre of curvature lies on the circle.

Clotoid (Fig. 62). The curve for which the radius of curvature is inversely proportional to the length of arc: $r = a^2 : s$.

Equation in parametric form:

$$x = a\sqrt{\pi}\int_0^t \cos\tfrac{1}{2}\pi t^2\,dt, \qquad y = a\sqrt{\pi}\int_0^t \sin\tfrac{1}{2}\pi t^2\,dt$$

(this cannot be expressed in terms of elementary functions), where $t = s/a\sqrt{\pi}$, $s = \overset{\frown}{OM}$.

---

[1] For the involute see p. 294.

[2] For solution of such equations see pp. 168–169.

The curve is symmetric with respect to the origin which is a point of inflection (the x axis is the tangent line at it); two asymptotic points: $A(+\frac{1}{2}a\sqrt{\pi}, +\frac{1}{2}a\sqrt{\pi})$ and $B(-\frac{1}{2}a\sqrt{\pi}, -\frac{1}{2}a\sqrt{\pi})$.

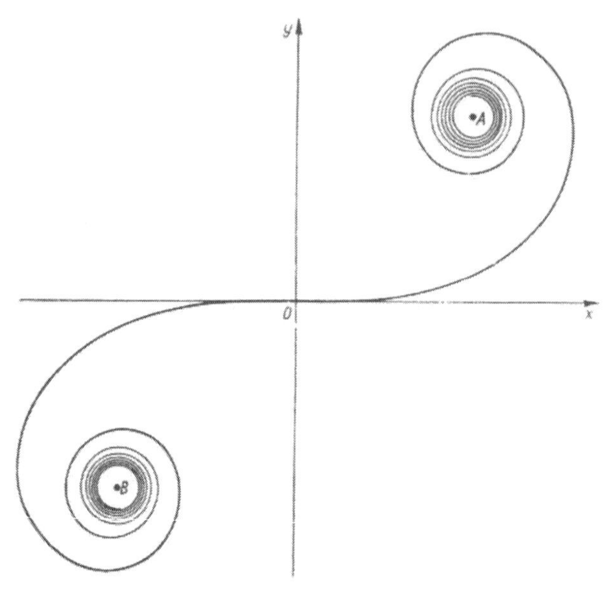

FIG. 62

13. Some other curves

The catenary (Fig. 63). A heavy flexible chain hanging between two points has the form of the catenary.

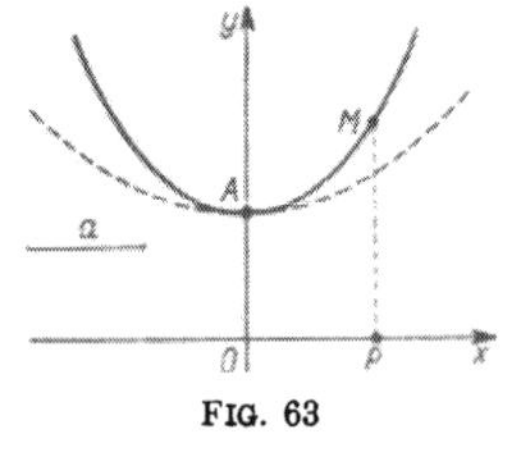

FIG. 63

Equation: $y = a \cosh \frac{x}{a} = a \frac{e^{x/a} + e^{-x/a}}{2}$.
The curve is symmetric with respect to the y axis and lies above the parabola $y = a + x^2/2a$ (dotted on the figure). Vertex $A(0, a)$. Length of the arc AM: $L = a \sinh \frac{x}{a} = a \frac{e^{x/a} - e^{-x/a}}{2}$; area $OAMP$: $S = aL = a^2 \sinh \frac{x}{a}$. Radius of curvature $r = \frac{y^2}{a} = a \cosh^2 \frac{x}{a}$.

Tractrix (Fig. 64). A curve such that the length of the segment MP of the tangent between the point of contact M and the intersection P with a given straight line (the x axis in Fig. 64) is constant [1].

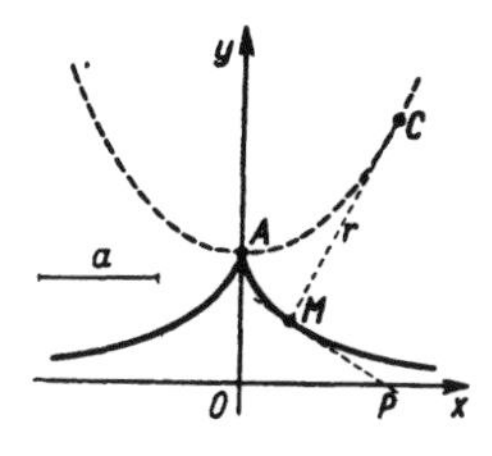

FIG. 64

Tractrix is an involute (p. 294) of the catenary, when winding off begins at the vertex A.

Equation: $x = a \operatorname{ar}\cosh \frac{a}{y} \pm \sqrt{a^2-y^2} = a \ln \frac{a \pm \sqrt{a^2-y^2}}{y} \mp \sqrt{a^2-y^2}$.
The x axis is the asymptote and $A(0, a)$ is a cusp with vertical tangent line. The curve is symmetric with respect to the y axis.
Length of the arc AM: $L = a \ln \frac{a}{y}$; when the length L is large, the difference $L - x$ (where x is the abscissa of the point M) $\approx a(1 - \ln 2) \approx 0.307\,a$. Radius of curvature $r = a \cot \frac{x}{y}$.

---

[1] In other words: If one end point of a non-expansible thread of a given length a is fastened to a material point M and the other end point P is dragged along a straight line (Ox), then the point M draws the tractrix (whence comes the name of the curve; *tractus*, in Latin, means dragged).

PART TWO

ELEMENTARY MATHEMATICS

I. APPROXIMATE COMPUTATIONS

1. Rules of approximate computations

Approximate computations. In any computation, we should always bear in mind the accuracy which is desirable and which it is possible to acquire. It is absolutely impossible to do the computation with a greater accuracy than that allowed by the data; it is aimless to do the computation with a greater accuracy than is desirable (for example, one should not use seven figure logarithms, when the data are to 5 significant figures only). Every one who does computations should be well acquainted with the rules of approximate computation.

Errors. The difference between the correct value of x and its approximate value a is called the *error* of this approximation. If it is known that $|x-a|<\Delta_a$, then Δ_a is called the *absolute limit error* of the approximate value a; the ratio $\Delta_a : a = \delta_a$ is called the *relative limit error* of a. The relative error is often expressed as a percentage.

Example. 3.14 is an approximate value of π. Its error is equal to 0.00159...; the absolute limit error can be assumed to be 0.0016 and the relative limit error—to be $\frac{0.0016}{3.14} = 0.00051$ $= 0.051\%$.

The word "limit" is usually omitted, for brevity. For observational errors see p. 747.

Significant figures. If the absolute error of a does not exceed one unit of the last figure of a, then all the figures of a are said to be correct [1]. In the approximation of a number, only the correct figures should be retained. If, for example, the absolute error of 52400 is 100, then this number should be written in the

---

[1] It is often required in this definition that the error should not exceed one half of one unit of the last figure of the approximate value. In connection with this, see p. 134 ("rounding").

form $524 \cdot 10^2$ or $5.24 \cdot 10^4$. The error can be estimated by giving the *number of correct figures* (the zeros from the left side are not taken into account).

Examples. (1) 1 cube foot $= 0.0283\ m^3$—correct to three significant figures. (2) 1 inch $= 2.5400$ cm—correct to five significant figures.

If the number a has n correct significant figures, then its relative error $\delta_a < \frac{1}{z \cdot 10^{n-1}}$, where z is the first significant figure of a. The number a with the relative error δ_a has n significant figures, where n is the greatest integral number satisfying $(1 + z)\,\delta_a < 10^{1-n}$ [1].

Example. If the number $a = 47.542$ is obtained as a result of operations with approximate numbers (see below) and it is known that $\delta_a = 0.1\%$, then a has three correct figures, for $(1 + 4) \cdot 0.001 < 10^{-2}$.

Rounding. If an approximate number has superfluous (or non-correct) figures, then it should be *rounded off*. In rounding off, only correct figures should be retained. The superfluous figures are discarded, but, if the first discarded figure is greater than 4, the preceding figure should be increased by 1. If the discarded part consists of the single figure 5, then the last figure of the rounding off should be made even. An additional error arises in rounding off; it does not exceed one half of one unit of the last significant figure. Therefore, in order that all figures of the rounding off are correct, the error before rounding off should not exceed one half of one unit of the last correct figure.

Operations with approximate numbers. The results of operations with approximate numbers are also approximate. The following theorems express the error of the result by means of the errors of data:

(1) The absolute limit error of a sum is equal to the sum of absolute errors of the summands.

(2) The relative error of a sum lies between the greatest and the least of absolute errors of the summands.

(3) The relative error of a product or a quotient is equal to the sum of relative errors of the involved approximate numbers.

(4) The relative error of the n-th power of an approximate number is n times greater than the relative error of the base (for an integral or fractional n).

Using these theorems, we can estimate the error of an arbitrary combination of arithmetic operations with approximate numbers.

---

[1] If a possible error of rounding is taken into account, then it should be put $(1 + z)\,\delta_a \leqslant 0.5 \cdot 10^{1-n}$.

Examples. (1) $V = r^2 h$; $\Delta_V = V\delta_V = V(2\delta_r + \delta_h)$. (2) $z = \sqrt{\frac{x}{1+y}}$; $\delta_z = \frac{1}{2}(\delta_x + \delta_{1+y}) = \frac{1}{2}\left(\frac{\Delta_x}{x} + \frac{\Delta_y}{1+y}\right)$.

Error of a function. The error in approximate evaluation of values of a function whose arguments are given approximately can be estimated by using, besides the above rules, the differential of this function. The error of a function is the same as its increment corresponding to the increment of the argument equal to its error. Since the errors are usually sufficiently small, the increments can be practically replaced by differentials (see p. 363). If only the absolute limit errors are known, their absolute values for all derivatives should be necessarily taken.

Examples. (1) $\tan\varphi = \frac{a}{b}$; $d\varphi = \frac{b\,da - a\,db}{a^2 + b^2}$; $\Delta_\varphi = \frac{b\Delta_a + a\Delta_b}{a^2 + b^2}$.
(2) $z = \sqrt{x^2 + y^2}$; $\frac{dz}{z} = \frac{x\,dx + y\,dy}{x^2 + y^2}$; $\delta_z = \frac{\Delta_z}{z} = \frac{x\Delta_x + y\Delta_y}{x^2 + y^2}$.

For functions whose values are obtained from tables, estimation of error can be made very simply. If the argument is given with the error Δ_x, then the error of the function $f(x)$ should be estimated by applying the linear interpolation (see p. 17) for the increment of the function corresponding to $\pm\Delta_x$.

Absolute value of this increment gives the absolute limit error of $f(x)$.

Examples. (1) If the diameter of a circle is $D = 5.92$ cm and has an error $\Delta_D = 0.005$, then the corresponding errors in the circumference of the circle and in the area of the circle are, respectively (see pp. 70 and 74) 0.015 cm and 0.05 cm^2. (2) If $\tan\alpha = 0.818 \pm \pm 0.002$, then (see p. 59) $\alpha = 39°17' \pm 0°4'$.

The inverse problem. If we want to obtain the result with a desired accuracy, we first find a formula for evaluation of the error of the result and then, using one of the methods given above, we compute what the admissible errors of the data can be. The solution of this problem is not unique and requires additional assumptions.

Example. One adjacent side of a right triangle is about three times greater than the other. With what accuracy should they be measured so that the error of the angle determined by means of the tangent does not exceed 1′? It follows from $\tan\varphi = a/b$ that (see below) $\Delta_\varphi = \frac{b\Delta_a + a\Delta_b}{a^2 + b^2}$, whence $1' = 0.00029 = 0.4\frac{\Delta_a}{a}$, or $\delta_a = 0.0007$. Thus, assuming that the errors of measurement of both adjacent sides are the same, we have obtained 0.07% as the relative error for the less one.

Approximate evaluations without calculation of errors. By the method given above the absolute limit error can be estimated; it certainly exceeds the absolute value of the true error. It is assumed all the time that particular errors accumulate, although this rarely happens in practice. In doing a great mass of calculations, when the errors are not estimated for each result separately, the following rules of counting the significant figures are used. When observing these rules, the results obtained can be reckoned to have correct figures, although, in particular cases, errors up to several units in the last significant figure are possible.

1. In addition and subtraction, the number of decimal figures retained in the result should be as much as that in the approximation with the least number of decimal figures.

2. In multiplication and division, the number of significant figures retained should be as great as in the approximation with the least number of significant figures.

3. In the square or in the cube of an approximate value, the number of significant figures in the result should be as great as that in the base. (The last figure of the square and, especially, of the cube is less certain than the last figure of the base.)

4. In the square or in the cube root, the number of the retained significant figures in the result should be as great as that in the approximate value of the number under the root sign. (The last figure of the square root and, especially, of the cube root is here more certain than the last figure of the number under the root sign.)

5. In all intermediate results, the number of figures retained should be greater by one than that allowed by the rules given above. This extra figure should be discarded in the final result.

6. If certain data have more decimal figures (in addition and subtraction) or more significant figures (in multiplication, division, powers and roots), than other approximate values, then they should be rounded off before calculation, with one extra figure retained.

7. If the data can be taken with an arbitrary accuracy, then, in order to obtain a result with k figures, the number of figures taken in the data should be that which, according to the rules 1-4, provides $k+1$ figures in the result.

8. In calculating the values of quantities by means of logarithms, the number of decimal figures of the logarithmic tables

used in computation should be greater by one than the number of significant figures of that datum which has the least number of significant figures. In the final result, the last figure should be discarded.

Division and multiplication of approximate numbers. In order to avoid superfluous figures, multiplication and division of approximate numbers should be performed as follows:

In multiplication, the number given with the less accuracy should be taken as the multiplier. We perform the multiplication starting from the highest order and, in each partial product, we cancel the last figure of the multiplicand; the last but one figure should then, if necessary, be increased by 1.

In division, according to rule 6, we retain in the dividend one more significant figure than in the divisor (if this is possible). Instead of annexing a zero in the successive steps of division, we should cancel the last figure of the divisor, introducing a correction to the last but one figure, if this should be necessary.

Example

Multiply 4.128 by 2.953.

$$\begin{array}{r} 4.128 \\ \times\ 2.953 \\ \hline 8.256 \\ 3.715 \\ +\ 206 \\ 12 \\ \hline 12.189 \approx 12.19. \end{array}$$

Divide 12.189 by 4.128.

$$\begin{array}{r|l} 12.189 & 4.128 \\ -\ 8.256 & \overline{2.953} \\ \hline 3.933 & \\ -\ 3.715 & \\ \hline 218 & \\ -\ 206 & \\ \hline 12 & \\ -\ 12 & \\ \hline - & \end{array}$$

2. Approximate formulas

In many cases, rather complicated functions can be replaced by simpler ones giving the results with an admissible error. To do this, we can take the first terms of the expansion of the function into Taylor's series (see p. 385) or use the least squares method (see p. 755). In the latter case, the formula will depend essentially on the interval in which it is used. In the table, several commonly-used formulas are given together with the accuracy which they provide.

Formula	Relative error does not exceed		
	0.1 %	1 %	10 %
	when x varies between		
$\sin x = x$	$\mp 0.077 = \mp 4°.4$	$\mp 0.245 = \mp 14°.0$	$\mp 0.786 = \mp 45°.0$
$\sin x = x - \frac{x^3}{6}$	$\mp 0.580 = \mp 33°.2$	$\mp 1.005 = \mp 57°.6$	$\mp 1.632 = \mp 93°.5$
$\cos x = 1$	$\mp 0.045 = \mp 2°.6$	$\mp 0.141 = \mp 8°.1$	$\mp 0.451 = \mp 25°.8$
$\cos x = 1 - \frac{x^2}{2}$	$\mp 0.386 = \mp 22°.1$	$\mp 0.662 = \mp 37°.9$	$\mp 1.036 = \mp 59°.3$
$\tan x = x$	$\mp 0.054 = \mp 3°.1$	$\mp 0.172 = \mp 9°.8$	$\mp 0.517 = \mp 29°.6$
$\tan x = x + \frac{x^3}{3}$	$\mp 0.293 = \mp 16°.8$	$\mp 0.519 = \mp 29°.7$	$\mp 0.895 = \mp 51°.3$
$\sqrt{a^2 + x} = a + \frac{x}{2a}$ [1]	$-0.085a^2$ $0.093a^2$	$-0.247a^2$ $0.328a^2$	$-0.607a^2$ $1.545a^2$
$\frac{1}{\sqrt{a^2+x}} = \frac{1}{a} - \frac{x}{2a^3}$	$-0.051a^2$ $0.052a^2$	$-0.157a^2$ $0.166a^2$	$-0.448a^2$ $0.530a^2$
$\frac{1}{a+x} = \frac{1}{a} - \frac{x}{a^2}$	$\mp 0.031a$	$\mp 0.099a$	$\mp 0.301a$
$e^x = 1 + x$	∓ 0.045	-0.134 0.148	-0.375 0.502
$\ln(1 + x) = x$	∓ 0.002	∓ 0.020	-0.176 0.230

3. Slide rule

Use of the slide rule. Elementary computations involving multiplication, division, raising to a square or to a cube, extracting square or cube roots, taking logarithms of given numbers and operations with trigonometric functions can be approximately carried out with a slide rule. The accuracy of computation is different in particular cases. However, the results obtained with a 25 cm slide rule correspond, on the average, to computations with three significant figures, i.e., the relative error is contained between 0.1 and 1 %. In cases where this accuracy is sufficient, we can therefore use a slide rule.

---

[1] This formula can be written in the form $\sqrt{a^2+x} = \frac{1}{2}\left(a + \frac{a^2+x}{a}\right)$, used in practice. Since a is an approximate value of the root ("the first approximation"), this formula implies, that to obtain the value of the root, we should take the arithmetic mean of the first approximation and of the quotient of the number by the first approximation; the number of correct figures of the result can here be assumed to be twice the number of correct figures of the first approximation.

It should be pointed out that the formula $\sqrt{a^2 + b^2} = 0.960a + 0.398b$ (where $a > b > 0$) obtained by the principle of uniform approximation (see p. 745) gives the error not exceeding 4 %.

Logarithmic scale. The slide rule is based on a logarithmic scale, constructed in the following way: Selecting a segment as a unit of measure on a number scale and starting from an initial point, we lay off the segments equal to the common logarithms of a sequence of numbers (Fig. 65). We mark the end point

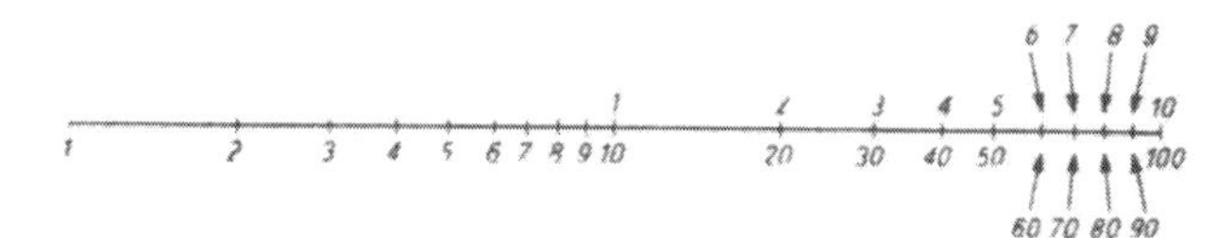

FIG. 65

of the segment $\log a$ by a (Fig. 66). The initial point should be marked by 1 ($\log 1 = 0$). Thus the distance from the point 1 to the point a on the logarithmic scale is equal to $\log a$ in the chosen scale. Since $\log 10a = 1 + \log a$, hence to each number of the interval from 10 to 100, there corresponds the number 10 times less on the logarithmic scale. The same reasoning also holds for the next intervals of the scale. It follows that the segment equal to the chosen unit of length and corresponding to the interval of numbers from 1 to 10 can represent the whole infinite logarithmic scale. Numbers with identical systems of ciphers, i.e., differing only by a factor 10^n (for example, 7.15, 0.0715, 71500), are represented by the same point of the scale.

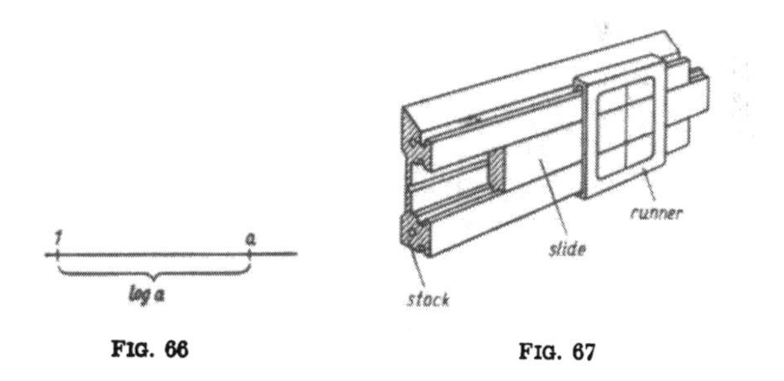

FIG. 66 FIG. 67

Scales of the slide rule. The slide rule consists of the stock, of the slide which is free to move in grooves of the stock and of the runner which is a glass in frames with one or three hair lines (Fig. 67). On the stock and on two sides of the slide, there are several scales; we denote them by A, B, C, D, I, K (Fig. 68). Some forms of slide rules have no scales I or K and scale L is on the opposite side of the slide. Before calculating with the slide rule, we should

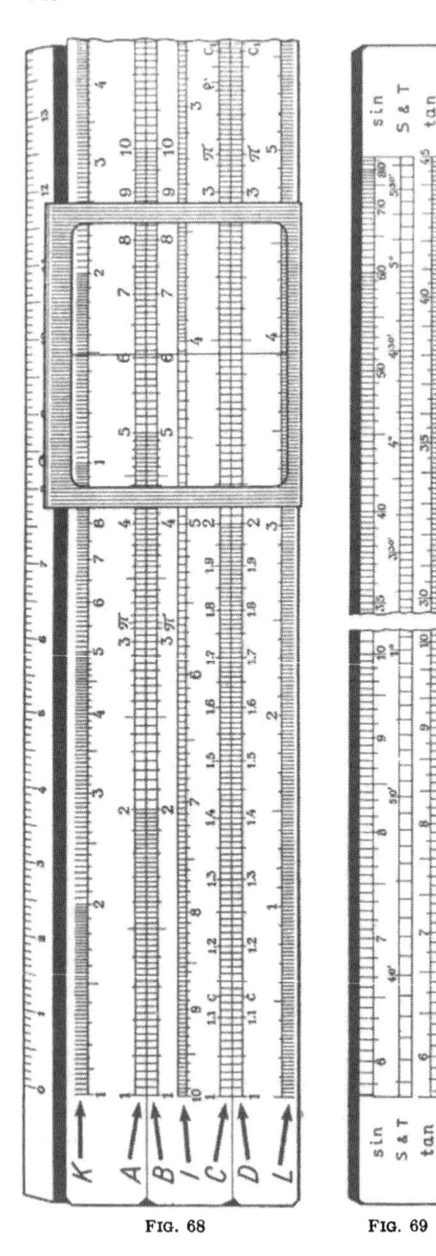

FIG. 68

FIG. 69

first become familiar with the scales. The scales A, B, C, D, I, K are logarithmic. The unit of measure chosen for the scales C, D, I is 25 cm and, for the scale I (in contrast to the other scales) the left-hand direction is taken as positive. The unit of measure for the scales A and B is equal to 12.5 cm and, for the scale K—$8\frac{1}{3}$ cm; the scales A and B are composed of two, and the scale K — of three equal segments. The partition of logarithmic scales is not uniform and its condensation varies from one place to another. To determine the position of a number which is not marked on the logarithmic scale, we assume that the scale is uniform between every two consecutive marked points; thus, for example, we assume that the number 235 lies in the middle of the numbers 234 and 236.

The scale L is uniform; it is divided into 0.002 of the unit of measure (equal to 25 cm).

On the reverse side of the slide there is a logarithmic scale for trigonometric functions (Fig. 69): T or tan (the tangent), S or sin (the sine) and S & T (the sine and the tangent). The initial point of the scale T at the right-hand end; the corresponding angle is 45°, for $\log \tan 45° = 0$. The angles $T°$ corresponding to the points of the scale are less than 45°, hence $\log \tan T° < 0$. The distance between the point $T°$ and the initial point is equal to $\log \tan T°$, in the given unit of measure (Fig. 70). For the left-hand end point $T_1°$ of the scale, $\log \tan T_1° = -1$, whence $\tan T_1° = 0.1$, $T_1° \approx 5°43'$.

The angle corresponding to the initial point of the scale S is $S_1° = 90°$, for $\log \sin 90° = 0$. The distance from the initial

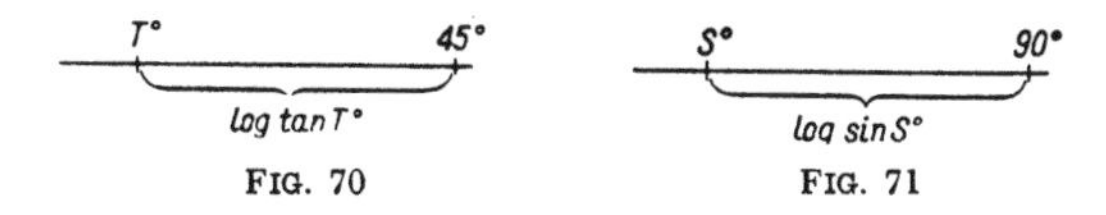

FIG. 70 FIG. 71

point to $S_1°$ is $\log \sin S_1°$, in the given unit of measure (Fig. 71). The left-hand end point is $S_1°$, for which $\log \sin S_1° = -1$, whence $\sin S_1° = 0.1$, $S_1° \approx 5°44'$.

For angles less than 5°44', the values of the tangent and the sine coincide (within the accuracy of the slide rule); thus a common scale S & T is added with the right-hand end point $T_1° \approx S_1°$, such that $\log \sin S_1° = -1$, whence $\sin S_1° = 0.1$, $S_1° \approx 5°43'$; its left-hand end point is $T_2° \approx S_2°$, such that $\log \sin S_2° = -2$, whence $\sin S_2° = 0.01$, $S_2° \approx 0°35'$. (Certain types of slide rules have 12.5 cm as the unit of measure for the scale S and then the whole scale contains the angles from 0°35' to 90°; in this case, the principles of calculation given below should be correspondingly changed.)

Principles of calculation with the slide rule. The operation of a slide rule is based on setting two numbers on two different scales one over another and on reading the result at a definite point of one scale lying opposite a definite point of the second scale. This is done by using the runner. The calculations are performed according to the following schemes.

General principles: (1) The slide rule gives only a system of significant figures so that the factor 10^n (n is an integer), i.e., the position of the decimal point, should be, in each case, suitably chosen. The best way of doing this is estimating the result mentally, in order to estimate the order of magnitude of the result.

(2) In complicated calculation, the medial results should not be read off every time, but only the middle line of the runner

should be set on them. Therefore, the work should be arranged so that the results of successive operations or groups of operations could be read off on the fixed scales and not on the slide.

(3) When a result is to be read off on a fixed scale opposite a point a of the slide and a falls beyond the fixed scale, then we

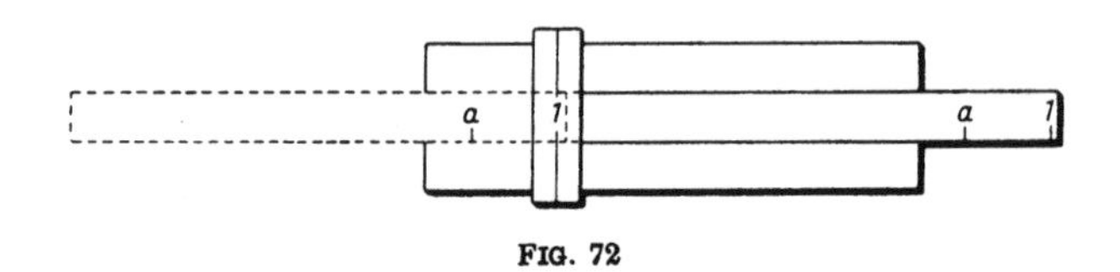

FIG. 72

should set the line of the runner on one end point of the slide and shift the slide so that its other end point falls under the line (Fig. 72). The desired result will then lie inside the fixed scale opposite a.

Schemes. In the schemes, only the setting of two points one opposite another plays a role, not the relative position of two different pairs. The slide can be moved to the right as well as to the left.

Multiplication, division, proportions

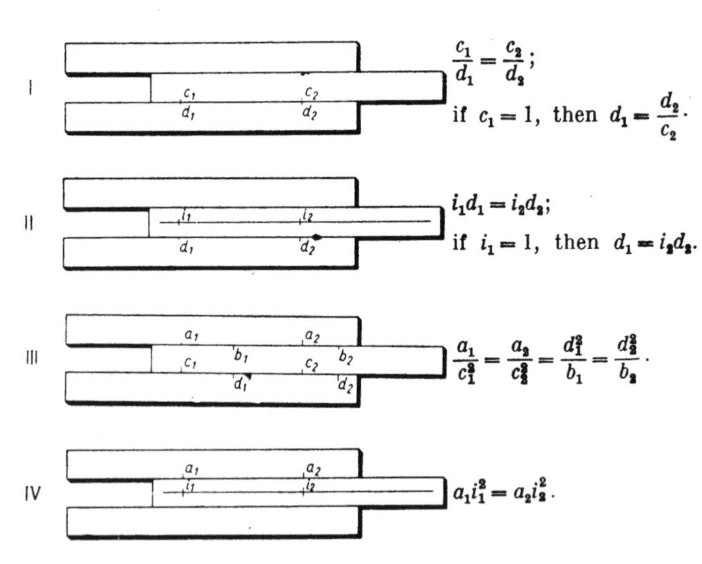

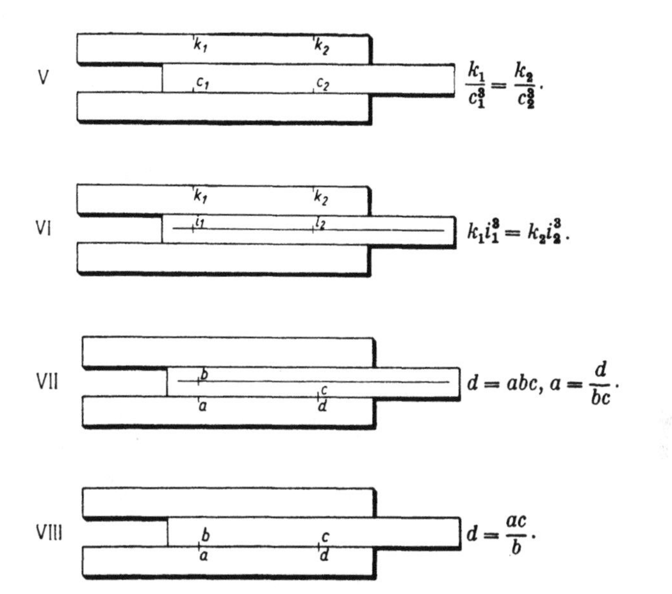

In multiplication and division, we generally use schemes I and II. Schemes III-VI enable us to do multiplication and division by means of squares and cubes of given numbers. In computation involving successive multiplications and divisions, schemes VII and VIII should be applied several times. For example, the calculation of $\frac{a \cdot b \cdot c}{d \cdot e \cdot f \cdot g}$ is done in three steps: twice, according to scheme VIII and once according to VII: $\frac{a \cdot b}{d} \cdot \frac{c}{e} \cdot \frac{1}{f \cdot g}$.

Powers and roots

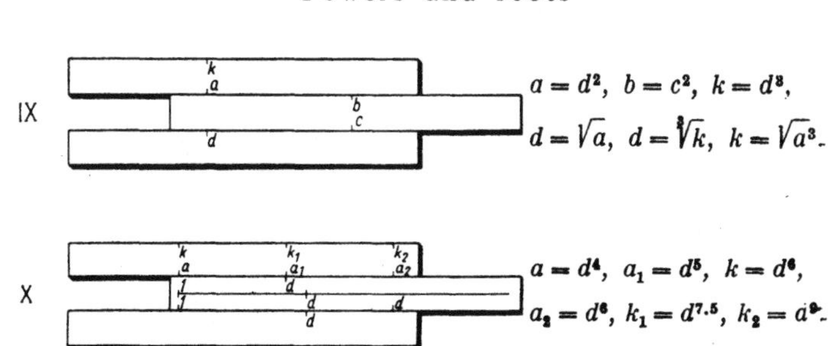

Calculations of scheme IX are done by aid of the runner, without using the slide. The same scheme IX can be applied to calculate square and cube roots; the number under the root sign should be first divided into groups of two or three figures on the left and on the right side of the decimal point (see p. 20). Then, according to the number of significant figures in the highest non-zero group, we determine the part of the scale A or K, where the number under the root sign is to be sought (Fig. 73). For example, in computing the square root of the number 37|50 or 0.00|37|5, the highest non-zero group 37 contains two significant figures, hence the number 3.75 should be sought in the second part of the scale A; for the numbers 3|75 or 0.03|75, the highest group 03 has one significant figure, hence the number 3.75 should be sought in the first part of the scale A.

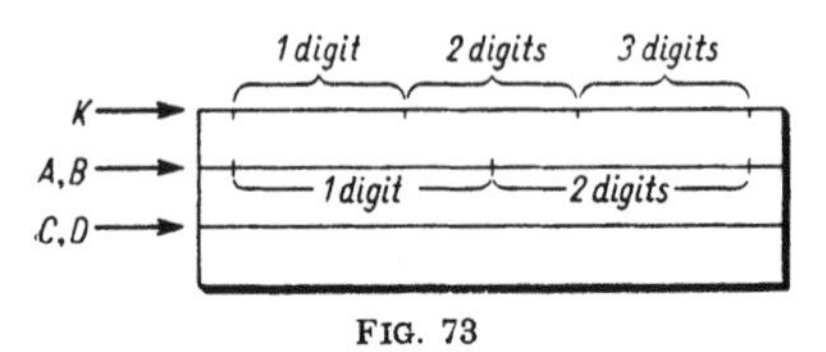

Fig. 73

Similarly, in extracting the cube root from the numbers 375 or 0.375, the number 3.75 should be sought in the third part of the scale K; for the cube roots of 37.5 or 0.037|5, the number 3.75 is sought in the second part of K and for 3|750 or 0.003|75, it is sought in the first part of K.

Logarithms

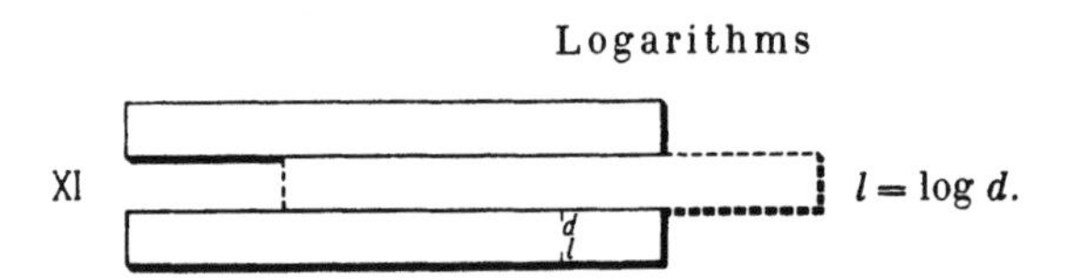

$$l = \log d.$$

The computations of the scheme XI are performed by aid of the runner alone, without using the slide. The scale L gives only the mantissa of a logarithm. The characteristic is determined by the known rules (see p. 157).

Trigonometric computations

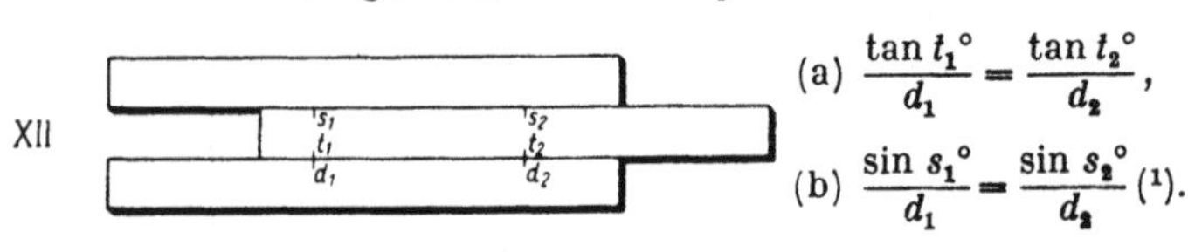

$$\text{(a)} \quad \frac{\tan t_1^\circ}{d_1} = \frac{\tan t_2^\circ}{d_2},$$

$$\text{(b)} \quad \frac{\sin s_1^\circ}{d_1} = \frac{\sin s_2^\circ}{d_2} \; [1].$$

[1] This scheme should be changed for the slide rules with the scale S of the length 12.5 cm (see p. 141).

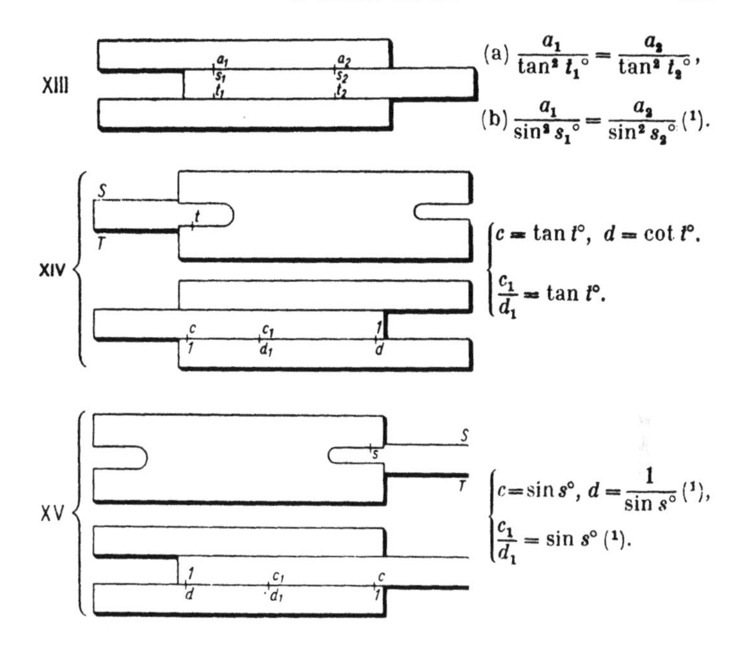

XIII (a) $\dfrac{a_1}{\tan^2 t_1{}^\circ} = \dfrac{a_2}{\tan^2 t_2{}^\circ}$, (b) $\dfrac{a_1}{\sin^2 s_1{}^\circ} = \dfrac{a_2}{\sin^2 s_2{}^\circ}$ [1].

XIV $c = \tan t^\circ,\ d = \cot t^\circ$. $\dfrac{c_1}{d_1} = \tan t^\circ$.

XV $c = \sin s^\circ,\ d = \dfrac{1}{\sin s^\circ}$ [1], $\dfrac{c_1}{d_1} = \sin s^\circ$ [1].

Computation of schemes XII and XIII are performed by using the reverse side of the slide. Certain operations with trigonometric functions can be done without reversing the slide; in this case, we determine the points on the scales S and T by looking through the cuttings in the reverse side of the slide rule (schemes XIV and XV).

Special signs

Changing of degrees into radians and conversely

XVI

$\varrho^\circ = 57.30,$
$\varrho' = 3438,$
$\varrho'' = 206265.$

$$d^\circ = i_0 \,\mathrm{rad}\left(= \frac{d^\circ}{\varrho^\circ}\,\mathrm{rad}\right),\ d' = i_1 \,\mathrm{rad}\left(= \frac{d'}{\varrho'}\,\mathrm{rad}\right),\ d'' = i_2 \,\mathrm{rad}\left(= \frac{d''}{\varrho''}\,\mathrm{rad}\right)$$

(for example, $15^\circ = 0.262$ rad, $15' = 0.00436$ rad, $15'' = 0.0000727$ rad).

Scheme XVI may also be used instead of scheme I.

[1] See footnote on p. 140.

Area of a circle

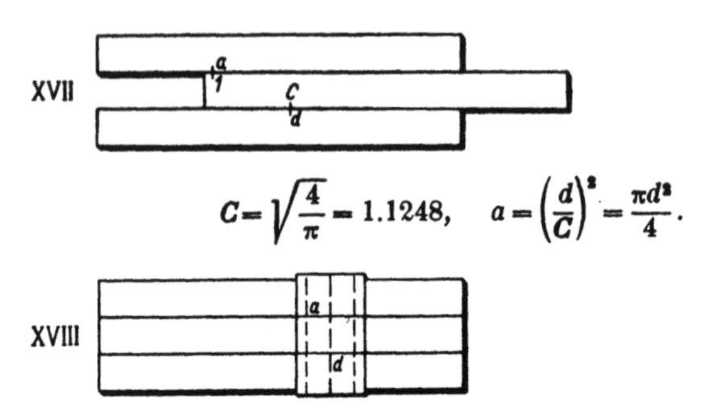

$$C = \sqrt{\frac{4}{\pi}} = 1.1248, \qquad a = \left(\frac{d}{C}\right)^2 = \frac{\pi d^2}{4}.$$

If the glass of the runner has three hair-lines, then the area of a circle can be found without using the slide, according to scheme XVIII.

II. ALGEBRA

A. IDENTITY TRANSFORMATIONS

1. Fundamental notions

Definitions. An *algebraic expression* is a collection of one or more algebraic quantities (numbers or letters) connected by the signs of operations ($+$, $-$, $:$, $\sqrt{\ }$ etc.) with brackets of different kinds indicating the succession of operations.

An equality of two algebraic expressions is called an ***identity***, if it is true for all substitutions of arbitrary numerical values for the letters occuring in it. An ***equation*** is an equality which is valid only for certain definite values [1].

An *identity transformation* is the process of obtaining one algebraic expression from another identically equal to the first one. It can be done in many ways, according to the aim of the transformation which should always be kept in mind. By a transformation, for example, an expression can be given a more compact form suitable for substitution of numerical values for letters, or a form suitable for solving equations, logarithmic calculation, differentiation, integration etc.

Classification of algebraic expressions. In every particular algebraic expression, we distinguish certain *fundamental* quantities according to which the expression is classified; the ***auxiliary*** quantities (the remaining letters) are called the ***parameters*** of the expression. The expression belongs to one or other class according to what operations are performed with the fundamental quantities of the expression. The ***integral rational*** expressions involve only addition, subtraction and multiplication of fundamental quantities (powers with natural exponents are also included here). The *fractional rational* expressions involve, moreover, division by an integral rational expression (or powers with

---

[1] For equations see p. 158.

negative integral exponents). The *irrational* expressions include roots of rational expressions (or powers with fractional exponents). The *exponential* expressions involve raising to a power whose exponent is a rational or irrational expression and the ***logarithmic*** expressions involve logarithms of (rational or irrational) expressions containing fundamental quantities.

In all the examples given below, the last letters of alphabet x, y, z denote the fundamental quantities, the initial $a, b, c, \ldots$ or middle $m, n, p, \ldots$ letters denote parameters; the middle letters assume only positive integral values.

2. Integral rational expressions

Representation in the form of a polynomial. Every integral rational expression can be represented in the form of a polynomial by means of elementary transformations: cancellation of similar terms, addition, subtraction and multiplication of monomials and polynomials.

Example.

$$(-a^3+2a^2x-x^3)(4a^2+8ax)+(a^3x^2+2a^2x^3-4ax^4)-(a^5+$$

$$+4a^3x^2-4ax^4)=-\underline{4a^5}+\underline{\underline{8a^4x}}-4a^2x^3-\underline{\underline{8a^4x}}+\underline{\underline{\underline{16a^3x^2}}}-8ax^4+$$

$$+\underline{\underline{\underline{a^3x^2}}}+2a^2x^3-4ax^4-\underline{a^5}-\underline{\underline{\underline{4a^3x^2}}}+4ax^4$$

$$=-5a^5+13a^3x^2-2a^2x^3-8ax^4.$$

Factorization of a polynomial. In many cases, a polynomial can be expressed in the form of a product of factors by taking a common factor out of the brackets, grouping, by aid of formulas for shortened multiplication and division or by using other properties of equations.

Examples. (1) Factoring out of the brackets:

$$8ax^2y-6bx^3y^2+4cx^5=2x^2(4ay-3bxy^2+2cx^3).$$

(2) Grouping:

$$6x^2+xy-y^2-10xz-5yz=\underline{6x^2+3xy}-\underline{2xy-y^2}-\underline{\underline{10xz-5yz}}$$

$$=3x(2x+y)-y(2x+y)-5z(2x+y)=(2x+y)(3x-y-5z).$$

(3) Using properties of algebraic equations [1]:

$$P(x) = x^6 - 2x^5 + 4x^4 + 2x^3 - 5x^2.$$

(a) We take x^2 outside of the bracket

(b) By testing, we find that $a_1 = 1$ and $a_2 = -1$ are roots of the equation $P(x) = 0$. Dividing $P(x)$ by $x^2(x-1)(x+1)$, i.e., by $x^4 - x^2$, we obtain as the quotient $x^2 - 2x + 5$. In this expression, $p = -2$, $q = 5$, $(\frac{1}{2}p)^2 - q < 0$, hence it cannot be decomposed into real factors. Consequently,

$$x^6 - 2x^5 + 4x^4 + 2x^3 - 5x^2 = x^2(x-1)(x+1)(x^2 - 2x + 5).$$

Formulas for shortened multiplication and division.

$$(x \pm y)^2 = x^2 \pm 2xy + y^2,$$

$$(x + y + z)^2 = x^2 + y^2 + z^2 + 2xy + 2xz + 2yz,$$

$$(x + y + z + \ldots + t + u)^2 = x^2 + y^2 + z^2 + \ldots + t^2 + u^2 + $$
$$+ 2xy + 2xz + \ldots + 2xu + 2yz + \ldots + 2yu + \ldots + 2tu,$$

$$(x \pm y)^3 = x^3 \pm 3x^2y + 3xy^2 \pm y^3,$$

$(x \pm y)^n$ is computed according to Newton's formula (see p. 193).

$$(x + y)(x - y) = x^2 - y^2,$$

$$(x^n - y^n) : (x - y) = x^{n-1} + x^{n-2}y + x^{n-3}y^2 + \ldots + xy^{n-2} + y^{n-1},$$

$$(x^n + y^n) : (x + y) = x^{n-1} - x^{n-2}y + x^{n-3}y^2 - \ldots - xy^{n-2} + y^{n-1}$$

(only for an odd n),

$$(x^n - y^n) : (x + y) = x^{n-1} - x^{n-2}y + x^{n-3}y^2 - \ldots + xy^{n-2} - y^{n-1}$$

(only for an even n).

Determining the highest common factor of two polynomials. Two polynomials $P(x)$ (of degree n) and $Q(x)$ (of degree m), $(n > m)$, can have common factors containing x; the product of all such factors is called the *highest common divisor* of the polynomials $P(x)$ and $Q(x)$. If $P(x)$ and $Q(x)$ have no common factors, then they are said to be *relative prime* (their highest common divisor is a constant).

The highest common divisor of polynomials $P(x)$ and $Q(x)$ can be determined without factorizing them in the following way (*Euclid's algorithm*):

(1) Divide $P(x)$ by $Q(x)$; denote the quotient by $T_1(x)$ and the remainder by $R_1(x)$:

$$P(x) = Q(x) \cdot T_1(x) + R_1(x).$$

---

[1] See p. 164.

(2) Divide $Q(x)$ by $R_1(x)$; denote the quotient by $T_2(x)$ and the remainder by $R_2(x)$:

$$Q(x) = R_1(x) \cdot T_2(x) + R_2(x)$$

etc. The last remainder $R_n(x)$ different from zero is just the highest common divisor of $P(x)$ and $Q(x)$.

Determining the highest common divisor is important in solving equations (separation of multiple roots, see p. 164, Sturm method—see p. 165, interpolation method of Ostrogradsky—see pp. 402–404) and in many other problems.

3. Rational fractional expressions

Reduction to the simplest form. Every fractional expression can be transformed into a product of two polynomials without common factors by means of elementary transformations (addition, subtraction, multiplication and division of polynomial and fractions and cancelling of fractions).

Example. Reduce to the simplest form:

$$\frac{3x + \dfrac{2x+y}{z}}{x\left(x^2 + \dfrac{1}{z^2}\right)} - y^2 + \frac{x+z}{z} = \frac{(3xz + 2x + y)z^2}{(x^2z^2 + x)z} + \frac{-y^2z + x + z}{z}$$

$$= \frac{3xz^3 + 2xz^2 + yz^2 + (x^2z^2 + x)(-y^2z + x + z)}{x^2z^3 + xz}$$

$$= \frac{3xz^3 + 2xz^2 + yz^2 - x^2y^2z^3 - xy^2z + x^3z^2 + x^2 + x^2z^3 + xz}{x^2z^3 + xz}$$

Excluding the integral part. A quotient of two polynomials with a common fundamental quantity x is called a *proper algebraic fraction,* if the degree m of the highest term [1] of the numerator is less than the degree n of the highest term of the denominator; if $m > n$, the quotient is called *improper.* Every improper fraction can be transformed into a sum of a polynomial and a proper fraction by excluding the integral part (division of a polynomial by a polynomial).

Example. Exclude the integral part from

$$R(x) = \frac{3x^4 - 10ax^3 + 22a^2x^2 - 24a^3x + 10a^4}{x^2 - 2ax + 3a^2}.$$

---

[1] I.e., the term containing x in the highest power.

$$\begin{array}{l}
3x^2 - 4ax + 5a^2 \\
\hline
(3x^4 - 10ax^3 + 22a^2x^2 - 24a^3x + 10a^4) : (x^2 - 2ax + 3a^2) \\
-3x^4 + 6ax^3 - 9a^2x^2 \\
\hline
\quad - 4ax^3 + 13a^2x^2 - 24a^3x \\
\quad + 4ax^3 - 8a^2x^2 + 12a^3x \\
\hline
\qquad + 5a^2x^2 - 12a^3x + 10a^4 \\
\qquad - 5a^2x^2 + 10a^3x - 15a^4 \\
\hline
\qquad\qquad - 2a^3x - 5a^4
\end{array}$$

$$R(x) = 3x^2 - 4ax + 5a^2 + \frac{-2a^3x - 5a^4}{x^2 - 2ax + 3a^2}.$$

Resolving into partial fractions. Every proper fraction

$$R(x) = \frac{Q(x)}{P(x)} = \frac{b_0x^m + b_1x^{m-1} + \ldots + b_m}{x^n + a_1x^{n-1} + \ldots + a_n},$$

reduced to simplest form (1), where the coefficients $b_0, b_1, \ldots, b_m$ and $a_1, a_2, \ldots, a_n$ are real numbers (the coefficient at x^n in the denominator can be made equal to 1 by dividing the numerator and denominator by this coefficient) can be uniquely transformed into a sum of *partial fractions* of the form $\dfrac{A}{(x-\alpha)^k}$ or $\dfrac{Dx+E}{(x^2+px+q)^l}$, where $(\frac{1}{2}p)^2 - q < 0$. The following four cases can occur here (2):

(1) The denominator $P(x)$ is such that the equation $P(x) = 0$ has only simple roots (3) $\alpha_1, \ldots, \alpha_n$. Then $R(x)$ is resolved according to the formula

$$\frac{Q(x)}{P(x)} = \frac{b_0x^m + \ldots + b_m}{(x-\alpha_1)(x-\alpha_2)\ldots(x-\alpha_n)} = \frac{A}{x-\alpha_1} + \frac{B}{x-\alpha_2} + \ldots + \frac{C}{x-\alpha_n},$$

where the coefficients $A, B, \ldots, C$ are determined by the formulas

$$A = \frac{Q(\alpha_1)}{P'(\alpha_1)}, \quad B = \frac{Q(\alpha_2)}{P'(\alpha_2)}, \quad \ldots, \quad C = \frac{Q(\alpha_n)}{P'(\alpha_n)}$$

(in the denominators, there are values of the derivative dP/dx for $x = \alpha_1, x = \alpha_2, \ldots, x = \alpha_n$).

---

(1) I.e., denominator and numerator have no common factor containing x.

(2) If we do not confine ourselves to the real numbers, then the case (3) will not differ from (1) and the case (4) from (2). From this point of view, each fraction $R(x)$ can be transformed into a sum of partial fractions of the form $\dfrac{A}{(x-\alpha)^k}$, where A and α are complex numbers. This is applied in solving linear differential equations (see p. 543).

(3) For simple and multiple roots see p. 164.

Example.

$$\frac{6x^2 - x + 1}{x^3 - x} = \frac{A}{x} + \frac{B}{x-1} + \frac{C}{x+1}.$$

$$\alpha_1 = 0,\ \alpha_2 = 1,\ \alpha_3 = -1,\ Q(x) = 6x^2 - x + 1,\ P'(x) = 3x^2 - 1,$$

whence

$$A = \frac{Q(0)}{P'(0)} = -1, \quad B = \frac{Q(1)}{P'(1)} = 3, \quad C = \frac{Q(-1)}{P'(-1)} = 4;$$

$$\frac{Q(x)}{P(x)} = -\frac{1}{x} + \frac{3}{x-1} + \frac{4}{x+1}.$$

Another way of determining the coefficients $A, B, \dots, C$ is the *method of undeterminate coefficients* applicable in each of the four cases.

Example.

$$\frac{6x^2 - x + 1}{x^3 - x} = \frac{A}{x} + \frac{B}{x-1} + \frac{C}{x+1} = \frac{A(x^2-1) + Bx(x+1) + Cx(x-1)}{x(x^2-1)}.$$

Comparing the coefficients of equal powers of x in the numerators of both sides of the equality, we obtain: $6 = A + B + C$, $-1 = B - C$, $1 = -A$; solving this system of equations, we find the same values for A, B, C as previously.

(2) The roots of the denominator are real, but some of them are multiple. The fraction is resolved according to the formula:

$$\frac{Q(x)}{P(x)} = \frac{b_0x^m + b_1x^{m-1} + \dots + b_m}{(x-\alpha_1)^{k_1}(x-\alpha_2)^{k_2}\dots(x-\alpha_i)^{k_i}} = \frac{A_1}{x-\alpha_1} + \frac{A_2}{(x-\alpha_1)^2} + \dots +$$

$$+ \frac{A_{k_1}}{(x-\alpha_1)^{k_1}} + \frac{B_1}{x-\alpha_2} + \frac{B_2}{(x-\alpha_2)^2} + \dots + \frac{B_{k_2}}{(x-\alpha_2)^{k_2}} + \dots + \frac{L_{k_i}}{(x-\alpha_i)^{k_i}}.$$

Example.

$$\frac{x+1}{x(x-1)^3} = \frac{A_1}{x} + \frac{B_1}{x-1} + \frac{B_2}{(x-1)^2} + \frac{B_3}{(x-1)^3}.$$

We find the coefficients A_1, B_1, B_2, B_3 by using the method of undetermined coefficients.

(3) There are simple complex roots among the roots of the denominator. The fraction is resolved as follows:

$$\frac{Q(x)}{P(x)} = \frac{b_0x^m + b_1x^{m-1} + \dots + b_m}{(x-\alpha_1)^{k_1}(x-\alpha_2)^{k_2}\dots(x^2+p_1x+q_1)(x^2+p_2x+q_2)\dots}$$

$$= \frac{A_1}{x-\alpha_1} + \frac{A_2}{(x-\alpha_1)^2} + \dots + \frac{Dx+E}{x^2+p_1x+q_1} + \frac{Fx+G}{x^2+p_2x+q_2} + \dots$$

Example.

$$\frac{3x^2-2}{(x^2+x+1)(x+1)}=\frac{A}{x+1}+\frac{Dx+E}{x^2+x+1}.$$

We find A, D, E by using the method of undetermined coefficients.

(4) Some complex roots of the denominator are multiple. The fraction is resolved according to the formula

$$\frac{Q(x)}{P(x)}=\frac{b_0x^m+b_1x^{m-1}+\ldots+b_m}{(x-\alpha_1)^{k_1}(x-\alpha_2)^{k_2}\ldots(x^2+p_1x+q_1)^{l_1}(x^2+p_2x+q_2)^{l_2}\ldots}$$

$$=\frac{A_1}{x-\alpha_1}+\frac{A_2}{(x-\alpha_1)^2}+\ldots+\frac{D_1x+E_1}{x^2+p_1x+q_1}+\frac{D_2x+E_2}{(x^2+p_1x+q_1)^2}+\ldots+$$

$$+\frac{D_{l_1}x+E_{l_1}}{(x^2+p_1x+q_1)^{l_1}}+\frac{F_1x+G_1}{x^2+p_2x+q_2}+\ldots+\frac{F_{l_2}x+G_{l_2}}{(x^2+p_2x+q_2)^{l_2}}+\ldots$$

Example.

$$\frac{5x^2-4x+16}{(x-3)(x^2-x+1)^2}=\frac{A}{x-3}+\frac{D_1x+E_1}{x^2-x+1}+\frac{D_2x+E_2}{(x^2-x+1)^2}.$$

We find A, D_1, E_1, D_2, E_2 by using the method of undetermined coefficients.

Transformation of a proportion. The following equalities follow from the proportion $\frac{a}{b}=\frac{c}{d}$:

$$ad=bc,\qquad \frac{a}{c}=\frac{b}{d},\qquad \frac{d}{b}=\frac{c}{a},\qquad \frac{b}{a}=\frac{d}{c},$$

and also so-called derivative proportions:

$$\frac{a\pm b}{b}=\frac{c\pm d}{d},\qquad \frac{a\pm b}{a}=\frac{c\pm d}{c},\qquad \frac{a\pm c}{c}=\frac{b\pm d}{d},\qquad \frac{a+b}{a-b}=\frac{c+d}{c-d}.$$

From an equality of several ratios $\frac{a_1}{b_1}=\frac{a_2}{b_2}=\cdots=\frac{a_n}{b_n}$, it follows that

$$\frac{a_1+a_2+\ldots+a_n}{b_1+b_2+\ldots+b_n}=\frac{a_1}{b_1}.$$

4. Irrational expressions; transformations of exponents and radicals

Reduction to the simplest form. An expression of the form $\sqrt[n]{A}$ is called a *radical* of the n-th degree. Then n is called the *index* and the A the *radicand*. Any irrational expression can be simplified by means of (1) cancelling the index, (2) factorizing the radicand, (3) rationalizing the denominator.

(1) Cancellation of the index. The index can bé cancelled by the highest common divisor of the index and of exponent of all factors of the radicand (which is assumed to be factored).

Example. $\sqrt[6]{16(x^{12}-2x^{11}+x^{10})}=\sqrt[6]{4^2x^{5\cdot 2}(x-1)^2}=\sqrt[3]{4x^5(x-1)}$.

(2) Factorizing the radicand. A factor X of the radicand with the exponent m equal or greater than the index n can be factored out of the radicand and placed before the radical. Then m is divided by n, X becomes a factor before the radical with the exponent equal to the quotient of the division and remains under the radical with the exponent equal to the remainder of the division.

Example. $\sqrt[3]{32x^4y^6z^{10}u^2}=2xy^2z^3\sqrt[3]{4xzu^2}$.

(3) Rationalizing of the denominator. The denominator of an irrational fractional expression can be rationalized in several ways.

Examples.

$$(1)\ \sqrt{\frac{x}{2y}}=\sqrt{\frac{2xy}{4y^2}}=\frac{\sqrt{2xy}}{2y};$$

$$(2)\ \sqrt[3]{\frac{x}{4yz^2}}=\sqrt[3]{\frac{2xy^2z}{8y^3z^3}}=\frac{\sqrt[3]{2xy^2z}}{2yz};$$

$$(3)\ \frac{1}{x+\sqrt{y}}=\frac{x-\sqrt{y}}{(x+\sqrt{y})(x-\sqrt{y})}=\frac{x-\sqrt{y}}{x^2-y};$$

$$(4)\ \frac{1}{x+\sqrt[3]{y}}=\frac{x^2-x\sqrt[3]{y}+\sqrt[3]{y^2}}{(x+\sqrt[3]{y})(x^2-x\sqrt[3]{y}+\sqrt[3]{y^2})}=\frac{x^2-x\sqrt[3]{y}+\sqrt[3]{y^2}}{x^3+y}.$$

Example. Reduce to the simplest form

$$\sqrt[4]{\frac{81x^6}{(\sqrt{2}-\sqrt{x})^4}}=\sqrt{\frac{9x^3}{(\sqrt{2}-\sqrt{x})^2}}=\frac{3x\sqrt{x}}{\sqrt{2}-\sqrt{x}}$$

$$=\frac{3x\sqrt{x}(\sqrt{2}+\sqrt{x})}{2-x}=\frac{3x\sqrt{2x}+3x^2}{2-x}.$$

Transformation of exponents into radicals.

$$(\bullet)\ x^m x^n = x^{m+n},\quad \frac{x^m}{x^n} = x^{m-n},\quad (xy)^n = x^n y^n,\quad \left(\frac{x}{y}\right)^n = \frac{x^n}{y^n},\quad (x^m)^n = x^{mn},$$

$$\sqrt[n]{xy} = \sqrt[n]{x}\,\sqrt[n]{y},\quad \sqrt[n]{\frac{x}{y}} = \frac{\sqrt[n]{x}}{\sqrt[n]{y}},$$

$$\sqrt[n]{x^m} = (\sqrt[n]{x})^m,\quad \sqrt[m]{\sqrt[n]{x}} = \sqrt[nm]{x}.$$

Generalization of the notion of a power. We assume the following definitions:

$$x^0 = 1,\quad x^{-n} = \frac{1}{x^n},\quad x^{1/n} = \sqrt[n]{x},\quad x^{m/n} = \sqrt[n]{x^m},\quad x^{-m/n} = \frac{1}{\sqrt[n]{x^m}}.$$

Powers with zero, negative or fractional exponents are transformed by the same rules as the powers with positive exponents; this often enables us to simplify the calculations.

Example.

$$(\sqrt{x} + \sqrt[3]{x^2} + \sqrt[4]{x^3} + \sqrt[12]{x^7})(\sqrt{x} - \sqrt[3]{x} + \sqrt[4]{x} - \sqrt[12]{x^5}) =$$

$$= (x^{1/2} + x^{2/3} + x^{3/4} + x^{7/12})(x^{1/2} - x^{1/3} + x^{1/4} - x^{5/12})$$

$$= x + x^{7/6} + x^{5/4} + x^{13/12} - x^{5/6} - x - x^{13/12} - x^{11/12} +$$

$$+ x^{3/4} + x^{11/12} + x + x^{5/6} - x^{11/12} - x^{13/12} - x^{7/6} - x$$

$$= x^{5/4} - x^{13/12} - x^{11/12} + x^{3/4} = \sqrt[4]{x^5} - \sqrt[12]{x^{13}} - \sqrt[12]{x^{11}} + \sqrt[4]{x^3}.$$

5. Exponential and logarithmic expressions

Exponential expressions of the form a^x are transformed according to formulas analogous to $(\bullet)$:

$$a^x a^y = a^{x+y};\quad \frac{a^x}{a^y} = a^{x-y};\quad (a^x)^y = a^{xy};\quad \sqrt[y]{a^x} = a^{x/y};$$

for arbitrary real numbers x and y. Exponential expressions with various bases $(a^x, b^y, c^z, \ldots)$ can be reduced to a common base by using the formula $b = a^{\log_a b}$.

Example. Express $(a^x b^y) : c^z$ as a power of a:

$$\frac{a^x b^y}{c^z} = \frac{a^x a^{y \log_a b}}{a^{z \log_a c}} = a^{x + y \log_a b - z \log_a c}.$$

Any expression which does not contain addition or subtraction can be reduced to a similar form.

The expression e^x, where e is the base of natural logarithms, (see below) is sometimes denoted by $\exp x$.

Logarithms. The *logarithm*, A, of a number, N, to the base a is the exponent of the power of a that is equal to N; we write $A = \log_a N$. Consequently, the equations $a^A = N$ and $A = \log_a N$ are equivalent. Any positive number has a logarithm to an arbitrary positive base (different from 1). Logarithms of various numbers to a given base a form the system of logarithms to the base a. Logarithms to any other base b can be computed from logarithms to the base a according to the formula $\log_b N = M \cdot \log_a N$, where $M = \frac{1}{\log_a b}$ (*modulus of the transformation*) [1].

Fundamental properties of logarithms to a common base a $(a \neq 1)$:

$$\log_a 1 = 0; \quad \log_a a = 1; \quad \log_a 0 = \begin{cases} -\infty \text{ for } a > 1, \\ +\infty \text{ for } a < 1; \end{cases}$$

$$(**)\begin{cases} \log(N_1 \cdot N_2) = \log N_1 + \log N_2; \quad \log\frac{N_1}{N_2} = \log N_1 - \log N_2; \\ \log(N^n) = n\log N; \quad \log\sqrt[n]{N} = \frac{1}{n}\log N. \end{cases}$$

Logarithms to the base 10 are called *common logarithms*. They are most frequently used in computation (notation: $\log_{10} N = \log N$). Logarithms to the base $e = 2.71828\ldots$ [2] (notation: $\log_e N = \ln N$) are called *natural, hyperbolic* or *Naperian logarithms*.

Modulus of transformation of natural logarithms into common ones:

$$M = \log e = \frac{1}{\ln 10} \approx 0.43429; \quad \log N = 0.43429 \ln N.$$

Modulus of transformation of common logarithms into natural ones:

$$M_1 = \frac{1}{M} \ln 10 = \frac{1}{\log e} \approx 2.30259; \quad \ln N = 2.30259 \log N.$$

Properties of common logarithms. Common logarithms are written as decimal fractions with accuracy to a definite signi-

---

[1] The following formula is easy to remember $\log_a N = \frac{\log N}{\log a}$, where the logarithms on the right hand side are taken to an arbitrary, but to the same base.

[2] For the number e see p. 381.

ficant figure. The integral part of it is called the *characteristic* and the decimal part is called the *mantissa*. For example, log 324 = 2.5105; the characteristic is 2 and the mantissa is 0.5105. When a number is multiplied or divided by 10^n (for example, 3240, 324000, 3.24, 0.0324), the mantissa remains unchanged (0.5105, the same as for 324). The mantissa can be found from the logarithmic tables (¹).

The characteristic is determined according to the rule: (1) If the number is greater than 1, then the characteristic is one less than the number of digits in the integral part of the number; (2) if the number is less than 1, then the characteristic is negative with the absolute value one greater than the number of zeros immediately following the decimal point. For example: $\log 3240 = 3.5105$; $\log 324\,000 = 5.5105$; $\log 3.24 = 0.5105$; $\log 0.0324 = \bar{2}.5105$. The sign "–" is usually written above the characteristic, since the mantissa remains positive; thus $\log 0.0324 = \bar{2}.5105$ means that $\log 0.0324 = -2 + 0.5105 = -1.4895$; $\log 0.0005272 = \bar{4}.7220$ means that $\log 0.0005272 = 0.7220 - 4 = -3.2780$.

Sometimes (in particular in tables), the number 10 is added to a given negative logarithm, to avoid the sign "–" above the characteristic. For example, $\bar{1}.324$ is written as 9.324.

Transformation of logarithmic expression is performed according to the formulas (**) (²).

Example. Find $\log \frac{3x^2 \sqrt[3]{y}}{2zu^3}$.

$$\log \frac{3x^2 \sqrt[3]{y}}{2zu^3} = \log (3x^2 \sqrt[3]{y}) - \log (2zu^3)$$

$$= \log 3 + 2 \log x + \tfrac{1}{3} \log y - \log 2 - \log z - 3 \log u.$$

Frequently the inverse transformation is used to represent an expression containing logarithms of several quantities in the form of a logarithm of one quantity. For example,

$$\log 3 + 2 \log x + \tfrac{1}{3} \log y - \log 2 - \log z - 3 \log u = \log \frac{3x^2 \sqrt[3]{y}}{2zu^3}$$

For the slide rule see pp. 138.

(¹) For tables of common logarithms see pp. 50–52; tables of antilogarithms pp. 53–55; tables of natural logarithms pp. 68–71.

(²) To find a logarithm of an expression involving addition and subtraction, this expression should be first transformed into a suitable form (i.e., into a form containing multiplication and division).

B. EQUATIONS

6. Transformation of algebraic equations into canonical form

Definition. An *equation with one unknown* is an equality of two functions of the same variable

$$F(x) = f(x)$$

which is valid only for certain values of the variable x [1]. The variable involved in the equation is called the *unknown* and the values $x_1, x_2, \ldots, x_n$ for which the equation is valid are called the *roots* or *solutions* of the equation. Two equations are called *equivalent* if all their roots coincide.

An equation is called *algebraic*, if the involved functions $F(x)$ and $f(x)$ are algebraic (rational or irrational). One of the functions $F(x)$ or $f(x)$ can be a constant value.

Any algebraic equation can be reduced, by an algebraic transformation, to the *canonical form*:

$$P(x) = a_0 x^n + a_1 x^{n-1} + \ldots + a_n = 0 \ [2]$$

(a_0 can be made equal to 1) which has the same roots as the given equation (certain superfluous roots can also occur, see below).

The exponent n is called the *degree* of the equation.

Example. Reduce the following equation to the canonical form

$$\frac{x - 1 + \sqrt{x^2 - 6}}{3(x-2)} = 1 + \frac{x-3}{x}.$$

Successive transformations give

$$x\left(x - 1 + \sqrt{x^2 - 6}\right) = 3x(x-2) + 3(x-2)(x-3),$$

$$x^2 - x + x\sqrt{x^2 - 6} = 3x^2 - 6x + 3x^2 - 15x + 18,$$

$$x\sqrt{x^2 - 6} = 5x^2 - 20x + 18,$$

$$x^2(x^2 - 6) = 25x^4 - 200x^3 + 580x^2 - 720x + 324,$$

$$24x^4 - 200x^3 + 586x^2 - 720x + 324 = 0.$$

The given equation is of the fourth degree.

---

[1] An equality that is true for arbitrary values of the variable x is called an *identity*.

[2] Here and in the sequel, the coefficients $a_0, a_1, \ldots, a_n$ are assumed to be real, with exception of certain cases which are discussed separately.

A system of n algebraic equations is a set of n equations which are valid only for certain definite sets $(x_1, y_1, \ldots, z_1, x_2, y_2, \ldots, z_2, \ldots)$ of values of the variables $x, y, \ldots, z$; such a set is called a *solution of the system*. Any system of algebraic equations can be reduced to the canonical form:

$$P_1(x, y, \ldots) = 0, \quad P_2(x, y, \ldots) = 0, \quad \ldots, \quad P_n(x, y, \ldots) = 0,$$

where P_i are polynomials with respect to $x, y, z, \ldots$

Example. Reduce the following system of equations to canonical form:

$$(1)\ \frac{x}{\sqrt{y}} = \frac{1}{z}, \quad (2)\ \frac{x-1}{y-1} = \sqrt{z}, \quad (3)\ xy = z.$$

We obtain

$$(1)\ x^2z^2 - y = 0, \quad (2)\ x^2 - 2x + 1 - y^2z + 2yz - z = 0, \quad (3)\ xy - z = 0.$$

Superfluous roots. In reducing an algebraic equation to the canonical form $P(x) = 0$, it can happen that the equation $P(x) = 0$ has certain roots which do not satisfy the original equation. The following cases can here occur:

I. Vanishing of the denominator. If the equation has the form

$$(1) \qquad \frac{P(x)}{Q(x)} = 0,$$

where $P(x)$ and $Q(x)$ are polynomials, then, multiplying by the denominator, we obtain an equation in the canonical form

$$(2) \qquad P(x) = 0.$$

This equation has all the roots of (1) but it may also have superfluous roots. This happens when the root $x = \alpha$ of (2) is also a root of the equation $Q(x) = 0$. Then the fraction should be cancelled by $x - \alpha$ (or by $(x-\alpha)^k$, if this is possible); otherwise $P(x) = 0$ will have the root $x = \alpha$ which does not occur among the roots of (1) or which occur with a less multiplicity([1]).

Example (1).

$$(1') \qquad \frac{x^3}{x-1} = \frac{1}{x-1} \quad \text{or} \quad \frac{x^3 - 1}{x-1} = 0.$$

([1]) For multiple roots see p. 164.

If we do not cancel by $x - 1$ and discard the denominator, then the root $x_1 = 1$ of the equation $x^3 - 1 = 0$ does not satisfy the original equation (1′) since the denominator vanishes for $x = 1$.
Example (2).

$$(2') \qquad \frac{x^3 - 3x^2 + 3x - 1}{x^2 - 2x + 1} = 0.$$

If we do not cancel by $(x - 1)^2$ and discard the denominator, then the equation $(x - 1)^3 = 0$ has a triple root $x_1 = 1$, while (2′) has a simple root $x = 1$.

II. Irrational equations. If an equation contains the unknown under the radical sign, then, after the reduction to canonical form, some superfluous roots may appear. Therefore, after solving the canonical equation, it should be verified which of the obtained roots are in fact roots of the given equation.

Example.

$$(1'') \qquad \sqrt{x + 7} + 1 = 2x \quad \text{or} \quad \sqrt{x + 7} = 2x - 1,$$

$$(2'') \qquad x + 7 = (2x - 1)^2 \quad \text{or} \quad 4x^2 - 5x - 6 = 0.$$

The roots of (2″) are $x_1 = 2$, $x_2 = -\frac{3}{4}$; x_1 satisfies (1″) and x_2 does not (the radical in the equation (1″) is understood in the arithmetical sense).

7. Equations of the first, second, third and fourth degree

Equation of the first degree (linear equation). Canonical form:

$$ax + b = 0.$$

Number of solutions: there is always one real solution $x = -b/a$.

Equation of the second degree (quadratic equation). Canonical form:

$$ax^2 + bx + c = 0 \quad \text{or (after dividing by } a), \quad x^2 + px + q = 0.$$

The number of real solutions depends on the sign of the *discriminant* D equal to $4ac - b^2$ or $q - \frac{1}{4}p^2$:

if $D < 0$, there are 2 solutions (2 real roots),
if $D = 0$, there is 1 solution (2 coincident roots),
if $D > 0$, there is no solution (2 imaginary roots).

Solution of quadratic equations. (1) Solution by factorizing the left member (if possible):

$ax^2 + bx + c = a(x-\alpha)(x-\beta)$ or $x^2 + px + q = (x-\alpha)(x-\beta)$.

Then $x_1 = \alpha$ and $x_2 = \beta$ are the roots of the equation.

Example. $x^2 + x - 6 = 0$, $x^2 + x - 6 = (x+3)(x-2)$, $x_1 = -3$, $x_2 = 2$.

(2) Solution by quadratic formula.

(a) For the form $ax^2 + bx + c = 0$,

$$x_{1,2} = \frac{-b \pm \sqrt{b^2 - 4ac}}{2a} \quad \text{or} \quad x_{1,2} = \frac{-\frac{1}{2}b \pm \sqrt{(\frac{1}{2}b)^2 - ac}}{a}.$$

(b) For the form $x^2 + px + q = 0$,

$$x_{1,2} = -\tfrac{1}{2}p \pm \sqrt{\tfrac{1}{4}p^2 - q}.$$

Properties of roots of the quadratic equation:

$$x_1 + x_2 = -\frac{b}{a} = -p, \quad x_1 \cdot x_2 = \frac{c}{a} = q.$$

Equation of the third degree (cubic equation). Canonical form:

$$ax^3 + bx^2 + cx + d = 0$$

or (by dividing by a and introducing a new variable $y = x + b/3a$ instead of x):

$$(*) \qquad y^3 + 3py + 2q = 0,$$

where

$$2q = \frac{2b^3}{27a^3} - \frac{bc}{3a^2} + \frac{d}{a} \quad \text{and} \quad 3p = \frac{3ac - b^2}{3a^2}.$$

The number of real solutions of the equation (*) depends on the sign of the discriminant $D = q^2 + p^3$:

if $D > 0$, then the equation has 1 solution (one real and two imaginary roots),

if $D < 0$, then the equation has 3 solutions (three different real roots),

if $D = 0$, then the equation has 1 solution for $p = q = 0$ (three coincident zero roots), and 2 solutions for $p^3 = -q^2 \neq 0$ (two of three real roots coincide).

Solution of cubic equations.

1st way: Solution by factorizing the left member (if possible):

$$ax^3 + bx^2 + cx + d = a(x-\alpha)(x-\beta)(x-\gamma).$$

Then the roots are $x_1 = \alpha$, $x_2 = \beta$, $x_3 = \gamma$.

Example: $x^3 + x^2 - 6x = 0$, $x^3 + x^2 - 6x = x(x + 3)(x - 2)$; $x_1 = 0$, $x_2 = -3$, $x_3 = 2$.

2nd way: By Cardan's formula (for the form (*)):

$$y_1 = u + v, \qquad y_2 = \varepsilon_1 u + \varepsilon_2 v, \qquad y_3 = \varepsilon_2 u + \varepsilon_1 v,$$

where $u = \sqrt[3]{-q + \sqrt{q^2 + p^3}}$, $v = \sqrt[3]{-q - \sqrt{q^2 + p^3}}$, and ε_1, ε_2 are roots of the equation $x^2 + x + 1 = 0$, i.e., $\varepsilon_{1,2} = -\frac{1}{2} \pm i\,\frac{1}{2}\sqrt{3}$.

In the case $D = q^2 + p^3 < 0$, three real roots can be expressed through complex numbers and the third way may be used.

Example. $y^3 + 6y + 2 = 0$. Here $p = 2, q = 1$, $q^2 + p^3 = 9$,

$$u = \sqrt[3]{-1 + 3} = \sqrt[3]{2} = 1.2599, \qquad v = \sqrt[3]{-1 - 3} = \sqrt[3]{-4} = -1.5874.$$

Real root: $y_1 = u + v = -0.3275$; complex roots:

$$y_{2,3} = -\tfrac{1}{2}(u + v) \pm i\tfrac{1}{2}\sqrt{3}(u - v) = 0.1638 \pm i \cdot 2.4659.$$

3rd way: By auxiliary quantities obtained from tables. In the equation (*), $r = \pm\sqrt{|p|}$; the signs of r and q should coincide. Then the auxiliary quantity φ and, by its aid, the roots y_1, y_2, y_3 are determined from the following table according to the signs of p and of the discriminant $D = q^2 + p^3$:

$p < 0$		$p > 0$
$q^2 + p^3 \leqslant 0$	$q^2 + p^3 > 0$	
$\cos\varphi = \frac{q}{r^3}$	$\cosh\varphi = \frac{q}{r^3}$	$\sinh\varphi = \frac{q}{r^3}$
$y_1 = -2r\cos\frac{1}{3}\varphi$	$y_1 = -2r\cosh\frac{1}{3}\varphi$	$y_1 = -2r\sinh\frac{1}{3}\varphi$
$y_2 = +2r\cos(60° - \frac{1}{3}\varphi)$	$y_2 = r\cosh\frac{1}{3}\varphi + i\sqrt{3}\,r\sinh\frac{1}{3}\varphi$	$y_2 = r\sinh\frac{1}{3}\varphi + i\sqrt{3}\,r\cosh\frac{1}{3}\varphi$
$y_3 = +2r\cos(60° + \frac{1}{3}\varphi)$	$y_3 = r\cosh\frac{1}{3}\varphi - i\sqrt{3}\,r\sinh\frac{1}{3}\varphi$	$y_3 = r\sin\frac{1}{3}\varphi - i\sqrt{3}\,r\cosh\frac{1}{3}\varphi$

Example.

$$y^3 - 9y + 4 = 0;$$

$$p = -3,\ q = 2,\ q^2 + p^3 < 0,$$

$$r = \sqrt{3} = 1.7321,\ \cos\varphi = 2 : 3\sqrt{3} = 0.3849,\ \varphi = 67°22',$$

$$y_1 = -2\sqrt{3}\cos 22°27' = -3.4641 \cdot 0.9242 = -3.201,$$

$$y_2 = 2\sqrt{3}\cos(60° - 22°27') = 3.4641 \cdot 0.7929 = 2.747,$$

$$y_3 = 2\sqrt{3}\cos(60° + 22°27') = 3.4641 \cdot 0.1314 = 0.455.$$

Verification (see properties of the roots):

$$y_1 + y_2 + y_3 = 0.001 \text{ instead of } 0.$$

4th way: approximate solution—see below, pp. 169–171.

Properties of the roots of the cubical equation:

$$x_1 + x_2 + x_3 = -\frac{b}{a}, \quad \frac{1}{x_1} + \frac{1}{x_2} + \frac{1}{x_3} = -\frac{c}{d}, \quad x_1x_2x_3 = -\frac{d}{a}.$$

Equation of the fourth degree. Canonical form:

$$ax^4 + bx^3 + cx^2 + dx + e = 0.$$

Number of (different) real solutions: from 0 to 4.

If $b = d = 0$, then the roots of the equation $ax^4 + cx^2 + e = 0$ are determined by the formulas

$$x_{1,2,3,4} = \pm\sqrt{y}, \quad y = \frac{-c \pm \sqrt{c^2 - 4ae}}{2a}.$$

If $a = e$, $b = d$, then the roots of the equation

$$ax^4 + bx^3 + cx^2 + bx + a = 0$$

(called *reflexive*) are determined by the formulas

$$x_{1,2,3,4} = \frac{y \pm \sqrt{y^2 - 4}}{2}, \quad y = \frac{-b \pm \sqrt{b^2 - 4ac + 8a^2}}{2a}.$$

Solution of the equation of the fourth degree in the general form:

1st way: Factorizing the left member (if possible)

$$ax^4 + bx^3 + cx^2 + dx + e = a(x - \alpha)(x - \beta)(x - \gamma)(x - \delta).$$

Then the roots of the equation are $x_1 = \alpha$, $x_2 = \beta$, $x_3 = \gamma$, $x_4 = \delta$.

Example.

$$x^4 - 2x^3 - x^2 + 2x = 0; \quad x(x^2 - 1)(x - 2) = x(x - 1)(x + 1)(x - 2);$$

$$x_1 = 0, \quad x_2 = 1, \quad x_3 = -1, \quad x_4 = 2.$$

2nd way: The roots of the equation $x^4 + bx^3 + cx^2 + dx + e = 0$ $(a = 1)$ coincide with the roots of two quadratic equations

$$x^2 + (b + A)\frac{x}{2} + \left(y + \frac{by - d}{A}\right) = 0,$$

where $A = \pm\sqrt{8y + b^2 - 4c}$ and y is an arbitrary real root of the cubical equation

$$8y^3 - 4cy^2 + (2bd - 8e)y + e(4c - b^2) - d^2 = 0.$$

3 rd way—approximate solution (see pp. 169–171).

Equations of the fifth and of higher degrees, in the general case, cannot be solved by means of radicals.

8. Equations of the n-th degree

General properties of algebraic equations. Denote the left member of the equation

(1) $$x^n + a_1x^{n-1} + \ldots + a_n = 0 \text{ [1]}$$

by $P(x)$; a root of the equation $P(x) = 0$ is also called a *zero* of the polynomial $P(x)$. If α is a zero of $P(x)$, then $P(x)$ is divisible by $(x - \alpha)$ without a remainder. If $P(x)$ is divisible by $(x - \alpha)^k$ but is not divisible by $(x - \alpha)^{k+1}$, then α is called a *multiple* (*k-fold*) *zero* of the polynomial $P(x)$; in this case, α is a common zero of the polynomial $P(x)$ and of its first $k - 1$ derivatives, inclusively. Other roots are called simple.

Fundamental Theorem of Algebra states that any equation of the n-th degree with real or complex coefficients has n, real or complex, roots, if k-fold roots are counted k times. If the roots of $P(x)$ are $\alpha, \beta, \gamma, \ldots$ with the multiplicities, respectively, $k, l, m, \ldots$, then

(*) $$P(x) = (x - \alpha)^k (x - \beta)^l (x - \gamma)^m \ldots$$

The process of finding roots of the equation $P(x) = 0$ can be simplified by abandoning multiple roots. This can be done as follows: we find the first derivative $P'(x)$ of $P(x)$, determine the highest common divisor $Q(x)$ of the polynomials $P(x)$ and $P'(x)$ and divide $P(x)$ by $Q(x)$. The remainder of this division is zero and the quotient is a polynomial $T(x)$ which has the same roots as $P(x)$, but only simple ones.

---

[1] The coefficient a_0 at the highest power of x has been made equal to 1 (by dividing the equation by this coefficient).

Relations between roots of an equation and the coefficients. If $x_1, x_2, \ldots, x_n$ are all the roots of the equation (1), then

$$x_1 + x_2 + \ldots + x_n = \sum_{i=1}^{n} x_i = -a_1,$$

$$x_1 x_2 + x_1 x_3 + \ldots + x_{n-1} x_n = \sum_{\substack{i,j=1 \\ (i<j)}}^{n} x_i x_j = a_2,$$

$$x_1 x_2 x_3 + x_1 x_2 x_4 + \ldots + x_{n-2} x_{n-1} x_n = \sum_{\substack{i,j,k=1 \\ (i<j<k)}}^{n} x_i x_j x_k = -a_3,$$

. . . .

$$x_1 x_2 \ldots x_n = (-1)^n a_n.$$

Equation with real coefficients. Complex roots of an equation with real coefficient must be pairwise conjugate ([1]), i.e., if $\alpha = a + bi$ is a root of the equation, then $\beta = a - bi$ is also a root of the equation and with the same multiplicity. Then we have

$$(x-\alpha)(x-\beta) = x^2 + px + q, \tag{2}$$

where $p = -(\alpha + \beta) = -2a$, $q = \alpha\beta = a^2 + b^2$, whence $(\frac{1}{2}p)^2 - q < 0$. Replacing in (*) the product of every pair of such factors by the formula (2), we obtain a factorization of a polynomial with real coefficients into real factors:

$$P(x) = (x-\alpha_1)^{k_1}(x-\alpha_2)^{k_2}\ldots(x^2+p_1x+q_1)^{l_1}(x^2+p_2x+q_2)^{l_2}\ldots, \tag{**}$$

where all the numbers α_i, p_i, q_i are real and $(\frac{1}{2}p_i)^2 - q_i < 0$.

Number of roots of an equation with real coefficients. It follows from the preceding remarks that any equation of an odd degree with real coefficients has at least one real root.

The number of real roots of the equation $P(x) = 0$ contained between any two numbers a and b $(a < b)$ which are not roots of the given equation, can be determined in the following way:

(1) Abandon multiple roots of the equation $P(x) = 0$, i.e., find an equation which has the same roots, but only simple ones ([2]).

([1]) For conjugate complex numbers see p. 587.

([2]) A way of obtaining such an equation has been given above. Practically, one can leave the multiple roots and at once form Sturm sequence of functions; if the last remainder P_m is not a constant, then $P(x)$ has multiple roots and they should be abandoned.

In the following, the equation $P(x) = 0$ is assumed to have only simple roots.

(2) Form *Sturm sequence of functions*

$$P(x),\ P'(x),\ P_1(x),\ P_2(x),\ \ldots,\ P_m = \text{const},$$

where $P(x)$ is the left member of the given equation, $P'(x)$ is the derivative, $P_1(x)$ is the remainder of the division of $P(x)$ by $P'(x)$, taken with the opposite sign, $P_2(x)$ is the remainder of the division of $P'(x)$ by $P_1(x)$, taken with the opposite sign and so on. Finally, P_m is the last remainder of this sequence of divisions equal to a certain constant [1].

(3) Count the number A of changes of the sign (i.e., how many times "+" changes into "−" and conversely) in the following sequence of numbers

$$P(a),\ P'(a),\ P_1(a),\ P_2(a),\ \ldots,\ P_m$$

and the number B of changes of the sign in the sequence

$$P(b),\ P'(b),\ P_1(b),\ P_2(b),\ \ldots,\ P_m \text{ [2]}.$$

The difference $A - B$ is equal to the desired number of real roots of the equation $P(x) = 0$ in the interval (a, b) (*Sturm's theorem*).

Example. Determine the number of roots of the equation $x^4 - 5x^2 + 8x - 8 = 0$ in the interval from 0 to 2.

Evaluation of Sturm's functions gives

$$P(x) = x^4 - 5x^2 + 8x - 8 = 0, \qquad P'(x) = 4x^3 - 10x + 8,$$

$$P_1(x) = 5x^2 - 12x + 16, \qquad P_2(x) = -3x + 284, \qquad P_3 = -1.$$

Substitution $x = 0$ gives the sequence of numbers: $-8, +8, +16, +284, -1$ (2 changes). Substitution $x = 2$ gives: $+4, +20, +12, +278, -1$ (1 change). $A - B = 2 - 1 = 1$, i.e., there is one root between 0 and 2.

The number of positive roots of the equation $P(x) = 0$ is not greater than the number of changes of the sign in the sequence of coefficients of $P(x)$ and can differ from this number by a certain even number (*Descartes' rule*).

Example. The equation $x^4 + 2x^3 - x^2 + 5x - 1 = 0$. The sequence of its coefficients has the signs: $+, +, -, +, -$, i.e., the sign changes three times. According to Descartes' rule, this

---

[1] To simplify the computation, the obtained remainders can be multiplied by constant positive factors; this will not change the result.

[2] If some of these number are zero, then they should be omitted in counting the changes of sign.

equation has three or one positive root. Since the roots of the equation change the signs after replacing x by $-x$ and they all decrease by h, after replacing x by $x+h$, then Descartes' rule can be also used to estimate the number of negative roots and also of the roots greater than h. In our example, replacing x by $-x$ gives $x^4 - 2x^3 - x^2 - 5x - 1 = 0$, i.e., the equation has one negative root; replacing x by $x+1$ gives $x^4 + 6x^3 + 11x^2 + 13x + 6 = 0$, hence the given equation has no roots greater than 1.

Solution of an equation of the n-th degree (when $n > 4$), can only be done approximately. In practice, approximate methods are also applied to solutions of equation of the third and fourth degree. For simultaneously approximating to all roots (including complex ones) of an algebraic equation of the n-th degree, the method of Lobatschewski can be applied. For determining separate roots of an algebraic equation, we can apply the methods of approximate solution of transcendental equations (see pp. 169–171).

9. Transcendental equations

Definition. The equation $F(x) = f(x)$ is called *transcendental* if at least one of the functions $F(x)$ or $f(x)$ is not algebraic.

Examples.

$$(1)\ 3^x = 4^{x-2} \cdot 2^x, \quad (2)\ 2^{x-1} = 8^{x-2} - 4^{x-2},$$

$$(3)\ 2\log_5(3x-1) - \log_5(12x+1) = 0, \quad (4)\ \sin x = \cos^2 x - \tfrac{1}{4},$$

$$(5)\ 3\cosh x = \sinh x + 9, \quad (6)\ x\cos x = \sin x.$$

In some cases, solution of transcendental equations reduce to algebraic equations and then tables can be applied. In general, transcendental equation can be solved only approximately.

Some cases of transcendental equations reducible to algebraic equations.

Exponential equations. An *exponential equation* is one which involves the unknown x or a polynomial $P(x)$ in exponents of powers of given bases $a, b, c, \ldots$ Such equations reduce to the algebraic ones in the following cases:

(1) If the powers $a^{P_1(x)}$, $b^{P_2(x)}$ are not submitted to addition or subtraction, then logarithms to an arbitrary base of both sides of the equation should be taken.

Example. $3^x = 4^{x-2}2^x$; $x \log 3 = (x-2)\log 4 + x \log 2$;

$$x = \frac{2\log 4}{\log 4 - \log 3 + \log 2}.$$

(2) If $a, b, c, \ldots$ are powers of the same number k with integral or fractional exponents (e.g., $a = k^\alpha$, $b = k^\beta$, $c = k^\gamma, \ldots$), then, substituting $k^x = y$, we obtain, in some cases, an algebraic equation with respect to y; after solving it, we find $x = \dfrac{\log y}{\log k}$.

Example. $2^{x-1} = 8^{x-2} - 4^{x-2}$; $\dfrac{2^x}{2} = \dfrac{2^{3x}}{64} - \dfrac{2^{2x}}{16}$. Putting $2^x = y$, we obtain $y^3 - 4y^2 - 32y = 0$ and $y_1 = 8$, $y_2 = -4$, $y_3 = 0$; $2^{x_1} = 8$, $2^{x_2} = -4$, $2^{x_3} = 0$, whence $x_1 = 3$. Other real solutions do not exist.

Logarithmic equations. A *logarithmic equation* is one which involves the unknown x or a polynomial $P(x)$ only under the logarithm sign. Such equations reduce to algebraic ones in the following cases:

(1) If the equation involves logarithms of one expression to the same base, then this logarithm can be taken as a new unknown. We obtain then an algebraic equation, solve it and find the result by raising the base to the obtained power.

Example. $m(\log_a P(x))^2 + n = a\sqrt{(\log_a P(x))^2 + b}$. Substitution $y = \log_a P(x)$ gives the equation $my^2 + n = a\sqrt{y^2 + b}$. We find y and then determine x from the equation $P(x) = a^y$.

(2) If the equation has the form

$$m_1 \log_a P_1(x) + m_2 \log_a P_2(x) + \ldots = 0,$$

where $m_1, m_2, \ldots$ are integers and $P_1(x), P_2(x), \ldots$ are polynomials of the variable x, then the left side of the equation reduces to a logarithm of one expression. Raising the base a to powers with exponents equal to both sides of the equation, we obtain an algebraic equation

$$(P_1(x))^{m_1}(P_2(x))^{m_2}\ldots = 1.$$

Example. $2\log_5(3x-1) - \log_5(12x+1) = 0$,

$$\log_5 \frac{(3x-1)^2}{12x+1} = \log_5 1; \quad \frac{(3x-1)^2}{12x+1} = 1, \quad x_1 = 0, \quad x_2 = 2.$$

We check that the solution $x_1 = 0$ leads to a logarithm of a negative number; hence we discard it.

Trigonometric equation (reducible to algebraic ones) is one in which the unknown x or the binomial $nx + a$ (n is an integer) is involved only under signs of trigonometric functions. Then, using trigonometric formulas, we reduce the expression to a one arbitrary function of the unknown x and, denoting this function by y, we obtain an algebraic equation. After solving it, we determine x from tables. It should be remembered that the solution can have a form of one or several periodic sequences.

Example. $\sin x = \cos^2 x - \frac{1}{4}$, or $\sin x = 1 - \sin^2 x - \frac{1}{4}$. Putting $\sin x = y$, we obtain $y^2 + y - \frac{3}{4} = 0$, whence $y_1 = \frac{1}{2}$, $y_2 = -\frac{3}{2}$. The solution y_2 does not yield real solutions of the given equation ($|\sin x| < 1$); $y_1 = \frac{1}{2}$ yields $x = \frac{1}{6}\pi + 2k\pi$ and $x = \frac{5}{6}\pi + 2k\pi$ (k are integral numbers).

Equations involving hyperbolic functions. Such equations are solved by expressing the hyperbolic functions of x by e^x and e^{-x} and then by substitutions $e^x = y$ and $e^{-x} = 1/y$. Then $x = \ln y$ is determined from tables.

Example. $3 \cosh x = \sinh x + 9$. $\frac{3}{2}(e^x + e^{-x}) = \frac{1}{2}(e^x - e^{-x}) + 9$; $e^x + 2e^{-x} - 9 = 0$. Putting $e^x = y$ and $e^{-x} = 1/y$, we obtain $y^2 - 9y + 2 = 0$, $y_{1,2} = \frac{1}{2}(9 \pm \sqrt{73})$, $x_1 = \ln \frac{1}{2}(9 + \sqrt{73}) \approx 2.1716$, $x_2 = \ln \frac{1}{2}(9 - \sqrt{73}) \approx -1.4784$.

Approximate solution of equations. The methods of approximate solution presented here are applicable to algebraic equations as well as to the transcendental ones. The process of determining the roots is composed of two steps: (1) rough estimation of approximate values of the roots, (2) sharpening the obtained approximations.

Rough estimation of roots. If $f(x)$ is a continuous [1] function and $f(a)$ and $f(b)$ are of different signs, then there exists at least one root of the equation $f(x) = 0$ between a and b. Taking various values for a and b, one can always obtain an interval containing only one root of the equation. The graphical method is applied when the equation can be written in the form $\varphi_1(x) = \varphi_2(x)$ such that the graphs of the functions $\varphi_1(x)$ and $\varphi_2(x)$ can be easily plotted. Then the roots are equal to the abscissae of the intersection points of the curves $y = \varphi_1(x)$ and $y = \varphi_2(x)$.

Example. Roots of the equation $x \cos x = \sin x$ (with exception of the evident root $x = 0$) are near to the value $\frac{1}{2}(2k + 1)\pi$ (where $k = \pm 1, \pm 2, \ldots$) since this equation can be written in the form $x = \tan x$ and the roots of the latter correspond to the intersection points of the line $y = x$ and the curve $y = \tan x$ (Fig. 74).

Methods of sharpening the approximations.

(1) *Newton's method.* If x_0 is an approximate value of a root α of the equation $f(x) = 0$, then we assume

---

[1] For continuity of a function see p. 335.

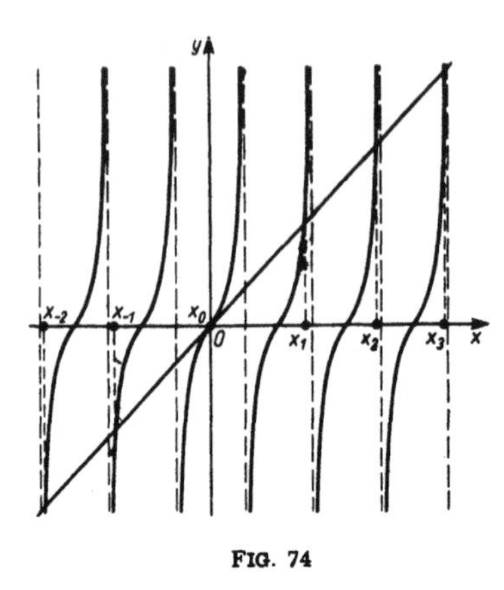

FIG. 74

$$x_1 = x_0 - \frac{f(x_0)}{f'(x_0)}$$

as a better approximation to α. Replacing x_0 by x_1 we obtain the next approximation x_2 and so on. These approximations always converge, provided that the root α is simple, i.e., that $f'(\alpha) \neq 0$ and that the first approximation is sufficiently accurate. Thus the root can be computed with an arbitrary accuracy.

Example—see below.

(2) *Linear interpolation* (*regula falsi*, i.e., the ***rule of a false assumption***). If a root α of the equation $f(x) = 0$ lies in the interval $a < x < b$, then

$$\tilde{x}_1 = a - f(a)\frac{b-a}{f(b)-f(a)}$$

can be taken as its approximate value. If $f''(x)$ does not change the sign in the interval (a, b) then the approximate values obtained by this method and by Newton's method will lie on different sides of the root. (In Newton's method, x_0 should be chosen from the numbers a and b so that $f(x_0)f''(x_0) > 0$.) Therefore, parallel application of both methods enables us to estimate the accuracy obtained.

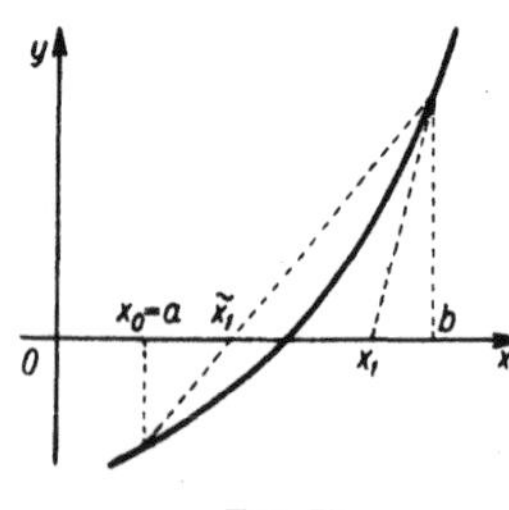

FIG. 75

Geometrically, in Newton's method, we replace the graph of the function $f(x)$ by the tangent at x_0 and in the linear interpolation—by the chord passing through the points $(a, f(a))$ and $(b, f(b))$ (Fig. 75).

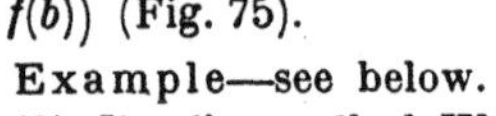

Example—see below.

(3) ***Iteration method.*** We reduce the given equation to the form $x = \varphi(x)$ and, having the first approximation x_0 of the root, we compute its best possible approximation using the formula $x_1 = \varphi(x_0)$; then we find a still better approximation using the formula $x_2 = \varphi(x_1)$ and so on. Iterating this procedure, we can obtain the value of the root with an arbitrary accuracy, provided that, in the interval between the root and its first approximation, the inequality $|\varphi'(x)| < 1$ holds. If this condition is not satisfied,

the equation should be transformed (e.g., the inverse function can be taken). For example, the iteration method is not applicable to the equation $x = \tan x$, but it can be applied to the equation $x = \arc\tan x$.

If $\varphi'(x) < 0$, then the successive approximations of the root obtained by the iteration method will, alternately, be less or greater than the root; this enables us to estimate the obtained degree of accuracy.

Example. Find the least positive root of the equation $\sin x - x \cos x = 0$.

We change this equation into an equivalent one $x = \tan x$ and find graphically (see p. 170) that the desired root is near to the number $\frac{3}{2}\pi = 4.71...$ We find a more accurate value by means of successive approximations:

(a) By Newton's method and linear approximation: for the function $f(x) = \sin x - x \cos x$, we have $f'(x) = x \sin x$. Substituting $x_0 = \frac{3}{2}\pi$, we obtain $f(x_0) = -1$, $f'(x_0) = -4.71$, $x_1 = 4.71 - \dfrac{1}{4.71} = 4.50$.

Since $f(x_1) = -0.029$ and $f(x_0)$ are of the same sign, the linear interpolation can not be applied. We calculate that $f(4.45) = 0.189$, hence the desired root lies between 4.45 and 4.50. Applying the linear interpolation, we obtain the following approximation:

$$\tilde{x}_1 = 4.50 - \frac{-0.029}{-0.029 - 0.189}(4.50 - 4.45) = 4.4930.$$

The next approximation computed by Newton's formula (the signs of $f''(x_1)$ and $f(x_1)$ are the same) gives

$$x_2 = 4.50 - \frac{-0.029}{-4.399} = 4.4934.$$

Since the approximations obtained from Newton's method and from the linear interpolation lie on different sides of the root, the error of the approximation x_2 does not exceed 0.0004.

(b) Iteration method: the equation $x = \tan x$ is not suitable to the linear interpolation since $(\tan x)' = 1 + \tan^2 x > 1$; replacing the function $\tan x$ by its inverse, we obtain the equation $x = \arc\tan x$ which can be iterated. Taking $x_0 = 4.7$, we find successively

$$x_1 = \arc\tan x_0 = \arc\tan 4.7 = 258° = 4.503,$$

$$x_2 = \arc\tan x_1 = 257°29' = 4.4942,$$

$$x_3 = \arc\tan x_2 = 257°27.3' = 4.4934,$$

$$x_4 = \arc\tan x_3 = 257°27.2' = 4.4934.$$

Evidently, all figures of x_4 can be considered as correct

10. Determinants

Definitions. A *determinant of order n* is a number D formed of n^2 numbers a_{ij} (*elements*) set into a square table composed of n *rows* and n *columns* as follows ([1]):

$$D = \begin{vmatrix} a_{11} & a_{12} & \cdots & a_{1n} \\ a_{21} & a_{22} & \cdots & a_{2n} \\ \cdots & \cdots & \cdots & \cdots \\ a_{n1} & a_{n2} & \cdots & a_{nn} \end{vmatrix} = \sum (-1)^k a_{1\alpha} a_{2\beta} \ldots a_{n\nu},$$

where the sum is extended over all possible $n!$ permutations $\alpha, \beta, \ldots, \nu$ ([2]) of the numbers $1, 2, \ldots, n$; the sign "+" or "−" before each summand is the same as that of $(-1)^k$, where k is the number of inversions in the corresponding permutation. For example, the summand $a_{13}a_{21}a_{34}a_{42}$ of the determinant of the fourth degree has the sign "−", for the permutation 3, 1, 4, 2 of second indices is composed of three inversions: (3,1), (3,2) and (4,2).

Minor of the element a_{ij} is the determinant of the $(n-1)$-st order obtained from the given determinant by cancelling its i-th row and j-th column. The *cofactor* A_{ij} of the element a_{ij} is its minor with the sign "+" or "−" according to the formula

$$A_{ij} = (-1)^{i+j} \begin{vmatrix} a_{11} & \cdots & a_{1j} & \cdots & a_{1n} \\ \cdots & \cdots & | & \cdots & \cdots \\ a_{i1} & — & a_{ij} & — & a_{in} \\ \cdots & \cdots & | & \cdots & \cdots \\ a_{n1} & \cdots & a_{nj} & \cdots & a_{nn} \end{vmatrix}$$

$$= (-1)^{i+j} \begin{vmatrix} a_{11} & \cdots & a_{1,j-1} & a_{1,j+1} & \cdots & a_{1n} \\ \cdots & \cdots & \cdots & \cdots & \cdots & \cdots \\ a_{i-1,1} & \cdots & a_{i-1,j-1} & a_{i-1,j+1} & \cdots & a_{i-1,n} \\ a_{i+1,1} & \cdots & a_{i+1,j-1} & a_{i+1,j+1} & \cdots & a_{i+1,n} \\ \cdots & \cdots & \cdots & \cdots & \cdots & \cdots \\ a_{n1} & \cdots & a_{n,j-1} & a_{n,j+1} & \cdots & a_{nn} \end{vmatrix}$$

A *linear combination* of some rows (columns) of the determinant is the row (column) $\bar{a}_1, \bar{a}_2, \ldots, \bar{a}_n$ whose elements are linear combinations of the corresponding elements of the given rows

---

([1]) The first index, i, of the element a_{ij} indicates that the element is taken from the i-th row of the determinant; the second index, j, indicates that the element is taken from the j-th column (i—number of the row counted from the top; j—number of the column counted from the left).

([2]) For permutations see p. 192.

(columns). For example, the linear combination $\bar{a}_1, \bar{a}_2, \ldots, \bar{a}_n$ of elements of the i-th, j-th and k-th rows is formed as follows

$$\bar{a}_1 = \alpha a_{i1} + \beta a_{j1} + \gamma a_{k1},$$
$$\bar{a}_2 = \alpha a_{i2} + \beta a_{j2} + \gamma a_{k2},$$
$$\cdots\cdots\cdots\cdots\cdots\cdots$$
$$\bar{a}_n = \alpha a_{in} + \beta a_{jn} + \gamma a_{kn};$$

α, β, γ are coefficients of the linear combination.

Properties of determinants:

(1) A determinant does not change its value when the rows are replaced by the columns and conversely (therefore all properties given below concerning the rows hold also for the columns).

(2) If elements of two rows of a determinant are respectively equal or proportional, or if one row is a linear combination of some other rows, then the determinant is equal to zero.

(3) A common factor of all elements of a row can be taken outside the determinant.

(4) If two determinants differ only by elements of the i-th row then its sum is the determinant whose i-th row is composed of sums of the corresponding elements of both determinants and other elements are the same as in either of the given determinants.

(5) A determinant does not change its value when elements of some row or linear combinations of some rows are added to (or subtracted from) the elements of another row.

(6) A determinant can be expanded in elements of an arbitrary (i-th) row according to the formula:

$$D = a_{i1} A_{i1} + a_{i2} A_{i2} + \ldots + a_{in} A_{in}$$

where A_{ij} is the cofactor of the element a_{ij}.

(7) The sum of products of elements a_{ik} of an arbitrary (i-th) row by the cofactor of the corresponding elements of another (j-th) row is zero:

$$a_{i1} A_{j1} + a_{i2} A_{j2} + \ldots + a_{in} A_{jn} = 0 \qquad (i \neq j).$$

Evaluation of determinants. Of the second order—according to the formula

$$\begin{vmatrix} a_{11} & a_{12} \\ a_{21} & a_{22} \end{vmatrix} = a_{11} a_{22} - a_{12} a_{21}.$$

Of the third order—according to Sarrus' formula (two first columns are annexed to the right):

$$\begin{vmatrix} a_{11} & a_{12} & a_{13} \\ a_{21} & a_{22} & a_{23} \\ a_{31} & a_{32} & a_{33} \end{vmatrix} \begin{matrix} a_{11} & a_{12} \\ a_{21} & a_{22} \\ a_{31} & a_{32} \end{matrix} = a_{11}a_{22}a_{33} + a_{12}a_{23}a_{31} + a_{13}a_{21}a_{32} -$$

$$- a_{13}a_{22}a_{31} - a_{11}a_{23}a_{32} - a_{12}a_{21}a_{33}.$$

A determinant of the n-th order can be reduced to a determinant of the $(n-1)$-th order by applying rule (6); but first one should try to transform the determinant by using other properties so as to obtain as many zeros as possible.

Example.

$$D = \begin{vmatrix} 2 & 9 & 9 & 4 \\ 2 & -3 & 12 & 8 \\ 4 & 8 & 3 & -5 \\ 1 & 2 & 6 & 4 \end{vmatrix} = \underset{\text{(property 5)}}{\begin{vmatrix} 2 & 5 & 9 & 4 \\ 2 & -7 & 12 & 8 \\ 4 & 0 & 3 & -5 \\ 1 & 0 & 6 & 4 \end{vmatrix}} = 3 \underset{\text{(property 3)}}{\begin{vmatrix} 2 & 5 & 3 & 4 \\ 2 & -7 & 4 & 8 \\ 4 & 0 & 1 & -5 \\ 1 & 0 & 2 & 4 \end{vmatrix}}$$

$$= 3\left\{-5 \begin{vmatrix} 2 & 4 & 8 \\ 4 & 1 & -5 \\ 1 & 2 & 4 \end{vmatrix} - 7 \begin{vmatrix} 2 & 3 & 4 \\ 4 & 1 & -5 \\ 1 & 2 & 4 \end{vmatrix} + 0\right\}$$

(property 6)

$$= 0 - 21 \underset{\text{(property 2)}}{\begin{vmatrix} 2 & 3 & 4 \\ 4 & 1 & -5 \\ 1 & 2 & 4 \end{vmatrix}} = -21 \underset{\text{(property 5)}}{\begin{vmatrix} 1 & 1 & 0 \\ 4 & 1 & -5 \\ 1 & 2 & 4 \end{vmatrix}}$$

$$= -21\left\{\begin{vmatrix} 1 & -5 \\ 2 & 4 \end{vmatrix} - \begin{vmatrix} 4 & -5 \\ 1 & 4 \end{vmatrix}\right\} = -21\{(4+10)-(16+5)\} = +147.$$

(property 6)

11. Solution of a system of linear equations

The case, when the number of unknowns is equal to the number of equations. Canonical form:

$$(*) \qquad \begin{aligned} a_{11}x_1 + a_{12}x_2 + \ldots + a_{1n}x_n &= b_1, \\ a_{21}x_1 + a_{22}x_2 + \ldots + a_{2n}x_n &= b_2, \\ \ldots\ldots\ldots\ldots\ldots\ldots\ldots\ldots \\ a_{n1}x_1 + a_{n2}x_2 + \ldots + a_{nn}x_n &= b_n. \end{aligned}$$

The determinant of the system is

$$D = \begin{vmatrix} a_{11} & \ldots & a_{1n} \\ \ldots & \ldots & \ldots \\ a_{n1} & \ldots & a_{nn} \end{vmatrix}.$$

Denote by D_j the determinant obtained from D by replacing its j-th column (containing the coefficients $a_{1j}, a_{2j}, \ldots, a_{nj}$) by the column of the right members $b_1, b_2, \ldots, b_n$. For example,

$$D_2 = \begin{vmatrix} a_{11} & b_1 & \ldots & a_{1n} \\ a_{21} & b_2 & \ldots & a_{2n} \\ \ldots & \ldots & \ldots & \ldots \\ a_{n1} & b_n & \ldots & a_{nn} \end{vmatrix}.$$

The system (•) is called *homogeneous* if all $b_k = 0$ (i.e., all $D_j = 0$) and *non-homogeneous*, if at least one of the numbers b_k is different from zero.

Solution of the system (•). If the determinant D of the system is $\neq 0$, then the system (•) is determinate, i.e., it has a unique solution. The roots x_j can be found from *Cramer's formulas*:

$$x_1 = \frac{D_1}{D}, \quad x_2 = \frac{D_2}{D}, \quad \ldots, \quad x_n = \frac{D_n}{D}.$$

If $D = 0$, but not all D_j are zero, then the system (•) is *inconsistent*, i.e., it has no solution (for a homogeneous system this case can not occur).

The case when $D = 0$ and all $D_j = 0$ is considered below (see the general case, example 4 on p. 179 and example 2 on p. 181).

Examples.

(1) $2x + y + 3z = 9, \quad x - 2y + z = -2, \quad 3x + 2y + 2z = 7.$

$$D = \begin{vmatrix} 2 & 1 & 3 \\ 1 & -2 & 1 \\ 3 & 2 & 2 \end{vmatrix} = 13,$$

$$D_x = \begin{vmatrix} 9 & 1 & 3 \\ -2 & -2 & 1 \\ 7 & 2 & 2 \end{vmatrix} = -13, \quad D_y = \begin{vmatrix} 2 & 9 & 3 \\ 1 & -2 & 1 \\ 3 & 7 & 2 \end{vmatrix} = 26, \quad D_z = \begin{vmatrix} 2 & 1 & 9 \\ 1 & -2 & -2 \\ 3 & 2 & 7 \end{vmatrix} = 39,$$

$$x = \frac{D_x}{D} = -\frac{13}{13} = -1, \quad y = \frac{D_y}{D} = \frac{26}{13} = 2, \quad z = \frac{D_z}{D} = \frac{39}{13} = 3$$

(the system is non-homogeneous and determinate).

(2) $2x + 3y - z = 1, \quad x - y + z = 2, \quad 3x + 2y = 5,$

$$D = \begin{vmatrix} 2 & 3 & -1 \\ 1 & -1 & 1 \\ 3 & 2 & 0 \end{vmatrix} = 0, \quad D_x = \begin{vmatrix} 1 & 3 & -1 \\ 2 & -1 & 1 \\ 5 & 2 & 0 \end{vmatrix} = 4$$

(the system is inconsistent).

Matrix and its rank. A system of mn numbers set into a rectangular table with m rows and n columns is called *matrix*. It is denoted by

$$[A] = \begin{bmatrix} a_{11} & a_{12} & \dots & a_{1n} \\ a_{21} & a_{22} & \dots & a_{2n} \\ \dots & \dots & \dots & \dots \\ a_{m1} & a_{m2} & \dots & a_{mn} \end{bmatrix}.$$

A *minor of order* k of the matrix $[A]$ $(k < m,\ k < n)$ is a determinant D composed of k^2 elements of the matrix (in the same ordering) lying in the intersection of certain k columns and k rows of the matrix (see the scheme):

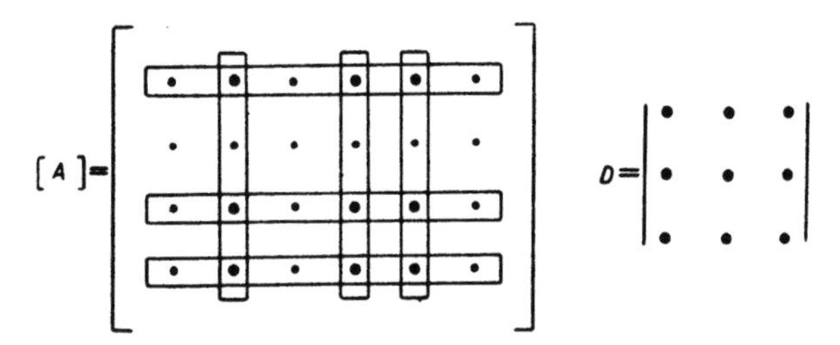

(a minor of the third degree)

The *rank* of the matrix $[A]$ is the maximal order of minors of $[A]$ which are different from zero. To determine the rank of a matrix, we should examine all its minors of order l, where l is the lesser one of the numbers m and n (or $l = m = n$, if the numbers m and n are equal); if one of these minors is different from zero, then the rank is equal to l; if all these minors are zero, then one should examine all minors of degree $l-1$ and so on. In practice, the inverse procedure is more suitable: to pass from minors of lower orders to those of higher orders using the following rule. If we find a minor D_k of order k different from zero, it remains to examine only those minors of order $k+1$ which are "borders" of D_k, for example,

$$\left|\underline{D_k}\,\vdots\right|, \quad \left|\vdots\,\underline{D_k}\right|, \quad \left|\overline{D_k}\,\vdots\right|, \quad \left|\vdots\,\overline{D_k}\right|.$$

If all such minors of order $k+1$ are equal to zero, then the rank of the matrix is k.

Example. Determine the rank of the matrix

$$[A]=\begin{bmatrix} 2 & -4 & 3 & 1 & 0 \\ 1 & -2 & 1 & -4 & 2 \\ 0 & 1 & -1 & 3 & 1 \\ 4 & -7 & 4 & -4 & 5 \end{bmatrix}.$$

The minor of the second order in the left top corner, $D_2 = \begin{vmatrix} 2 & -4 \\ 1 & -2 \end{vmatrix} = 0$. But there is in $[A]$ a minor of order 2 different from zero: $D_2' = \begin{vmatrix} -4 & 3 \\ -2 & 1 \end{vmatrix} \neq 0$. We border it from the left and from the bottom:

$$D_3 = \begin{vmatrix} 2 & -4 & 3 \\ 1 & -2 & 1 \\ 0 & 1 & -1 \end{vmatrix} = 1 \neq 0.$$

Then, bordering D_3 (this can be done only in two ways), we have

$$D_4 = \begin{vmatrix} 2 & -4 & 3 & 1 \\ 1 & -2 & 1 & -4 \\ 0 & 1 & -1 & 3 \\ 4 & -7 & 4 & -4 \end{vmatrix} = 0, \qquad D_4' = \begin{vmatrix} 2 & -4 & 3 & 0 \\ 1 & -2 & 1 & 2 \\ 0 & 1 & -1 & 1 \\ 4 & -7 & 4 & 5 \end{vmatrix} = 0.$$

Hence the rank of $[A]$ is 3.

General case of a system of linear equations.

Non-homogeneous equations. The system of non-homogeneous equations

$$(**) \qquad \begin{aligned} a_{11}x_1 + a_{12}x_2 + \ldots + a_{1n}x_n &= b_1, \\ a_{21}x_1 + a_{22}x_2 + \ldots + a_{2n}x_n &= b_2, \\ \ldots\ldots\ldots\ldots\ldots\ldots\ldots\ldots \\ a_{m1}x_1 + a_{m2}x_2 + \ldots + a_{mn}x_n &= b_m \end{aligned}$$

is called *consistent* if there exists at least one solution $\{\alpha_1, \alpha_2, \ldots, \alpha_n\}$ [1] satisfying all the equations (**). The system (**) is called *inconsistent*, if it has no solution.

A *test for consistency*: the system (**) is consistent if and only if the rank of the matrix

$$[A]=\begin{bmatrix} a_{11} & a_{12} & \ldots & a_{1n} \\ a_{21} & a_{22} & \ldots & a_{2n} \\ \ldots & \ldots & \ldots & \ldots \\ a_{m1} & a_{m2} & \ldots & a_{mn} \end{bmatrix}$$

[1] A system of solutions of the form $x_1 = \alpha_1, x_2 = \alpha_2, \ldots, x_n = \alpha_n$ will be denoted by $\{\alpha_1, \alpha_2, \ldots, \alpha_n\}$.

is equal to the rank of the augmented matrix

$$[B] = \begin{bmatrix} a_{11} & a_{12} & \dots & a_{1n} & b_1 \\ a_{21} & a_{22} & \dots & a_{2n} & b_2 \\ \dots & \dots & \dots & \dots & \dots \\ a_{m1} & a_{m2} & \dots & a_{mn} & b_m \end{bmatrix}.$$

A consistent system of equations is called *determinate*, if it has a unique solution; it is called *indeterminate*, if it has infinitely many solutions.

A consistent system of equations is solved as follows: We compute the rank r of the matrix $[A]$ and rearrange the equations (**) and the unknowns $x_1, x_2, \dots, x_n$ in the equations so that the minor different from zero of order r of the matrix $[A]$ is situated in the left top corner of the matrix. Then the following two cases can occur:

(1) $r = n, r < m$. We solve the system of n first equations with n unknowns and obtain a unique solution $\{\alpha_1, \alpha_2, \dots, \alpha_n\}$ since the determinant of this system is not equal to zero (see p. 175). If $n < m$, then the same solution also satisfies the $m - n$ remaining equations which follow from the first ones. The system (**) is determinate.

(2) $r < n, r < m$. We solve the system of the first equations with respect to the first r unknowns $x_1, x_2, \dots, x_r$, expressing them by the $n - r$ remaining unknowns $x_{r+1}, x_{r+2}, \dots, x_n$. We obtain the solution in the form of a system of linear functions:

$$(1) \qquad \begin{aligned} x_1 &= x_1(x_{r+1}, x_{r+2}, \dots, x_n), \\ x_2 &= x_2(x_{r+1}, x_{r+2}, \dots, x_n), \\ &\dots\dots\dots\dots\dots\dots \\ x_r &= x_r(x_{r+1}, x_{r+2}, \dots, x_n), \end{aligned}$$

since the determinant of the system of equations is not equal to zero. The unknowns $x_{r+1}, x_{r+2}, \dots, x_n$ can be given arbitrary values; then the unknowns $x_1, x_2, \dots, x_r$ are determined from formulas (1). The same solutions also satisfy the remaining $m - r$ equations (if $r < m$) which follow from the first ones. The system (**) is indeterminate.

Examples.

(1)

$$x - 2y + 3z - u + 2v = 2, \qquad 3x - y + 5z - 3u - v = 6,$$
$$2x + y + 2z - 2u - 3v = 8.$$

The rank of the matrix $[A]$ is 2, the rank of the matrix $[B]$ is 3. The system is inconsistent, there are no solutions.

(2)

$$x - y + 2z = 1, \quad x - 2y - z = 2, \quad 3x - y + 5z = 3,$$
$$-2x + 2y + 3z = -4.$$

The ranks of the matrix $[A]$ and of the matrix $[B]$ are both equal to 3; the system is consistent. Determinant of the third order in the left top corner, $D = \begin{vmatrix} 1 & -1 & 2 \\ 1 & -2 & -1 \\ 3 & -1 & 5 \end{vmatrix} \neq 0$. Therefore rearranging the equations and the unknowns is not necessary. $r = n$, hence the system of equations is determinate. We solve the system of first three equations: $x = \frac{10}{7}$, $y = -\frac{1}{7}$, $z = -\frac{2}{7}$; the same solution satisfies also the fourth equation.

(3)

$$x - y + z - u = 1, \qquad x - y - z + u = 0, \qquad x - y - 2z + 2u = -\tfrac{1}{2}.$$

The ranks of the matrices $[A]$ and $[B]$ are both equal to 2; the system is consistent. $r < n$, hence the system is indefinite. Determinant in the left top corner $D_2 = 0$; we put the column of x in the fourth place:

$$-y + z - u + x = 1, \quad -y - z + u + x = 0, \quad -y - 2z + 2u + x = -\tfrac{1}{2},$$

and solve the system of two first equations with respect to y and z:

$$-y + z = 1 - x + u, \qquad -y - z = -x - u.$$

The solutions $y = x - \frac{1}{2}$, $z = u + \frac{1}{2}$ satisfy all the equations, for arbitrary values of x and u.

(4)

$$\begin{aligned} x + 2y - z + u &= 1, \\ 2x - y + 2z + 2u &= 2, \\ 3x + y + z + 3u &= 3, \\ x - 3y + 3z + u &= 0. \end{aligned}$$

In this case the number of equations is equal to the number of unknowns. $D = 0$ and $D_x = D_y = D_z = D_u = 0$. The rank of the matrix $[A]$ is 2, the rank of $[B]$ is 3. The system is inconsistent, there are no solutions.

Homogeneous equations. A system of homogeneous equations

$$\begin{aligned} a_{11}x_1 + a_{12}x_2 + \dots + a_{1n}x_n &= 0, \\ a_{21}x_1 + a_{22}x_2 + \dots + a_{2n}x_n &= 0, \\ &\dots\dots\dots \\ a_{m1}x_1 + a_{m2}x_2 + \dots + a_{mn}x_n &= 0 \end{aligned} \tag{$\overset{**}{*}$}$$

has always a zero solution: $x_1 = x_2 = \ldots = x_n = 0$. If, moreover, the system $\binom{**}{*}$ has a non-zero solution $\{\alpha_1, \alpha_2, \ldots, \alpha_n\}$ [1] then it has also infinitely many proportional solutions $\{k\alpha_1, k\alpha_2, \ldots, k\alpha_n\}$, where k is an arbitrary number. If the system $\binom{**}{*}$ has p different, i.e., unproportional, non-zero solutions

$$(1)\qquad \{\alpha_1, \alpha_2, \ldots, \alpha_n\}, \quad \{\beta_1, \beta_2, \ldots, \beta_n\}, \quad \ldots, \quad \{\mu_1, \mu_2, \ldots, \mu_n\},$$

then it has also infinitely many solutions of the form

$$(2)\qquad \{k_1\alpha_1 + k_2\beta_1 + \ldots + k_p\mu_1, k_1\alpha_2 + k_2\beta_2 + \ldots + k_p\mu_2, \ldots, k_1\alpha_n + + k_2\beta_n + \ldots + k_p\mu_n\},$$

where $k_1, k_2, \ldots, k_p$ are arbitrary numbers which are not simultaneously equal to zero. Solution (2) is said to be a *linear combination* of solutions (1).

The solutions (1) of the system $\binom{**}{*}$ of equations are called *linearly independent* if none of them is a linear combination of the remaining ones; p linearly independent solutions form a *basic system of solutions*, if an arbitrary solution of the system $\binom{**}{*}$ is a linear combinations of these p solutions [2].

If the rank r of the matrix $[A]$ of coefficients of equations $\binom{**}{*}$ is less than the number n of unknowns, then the equations $\binom{**}{*}$ have a basic system of solutions; if $r = n$, then a basic system does not exist and the equations have only the zero solutions. If $r < n$, then a basic system of solutions consists of $n - r$ linearly independent solutions. To find a basic system of solutions, we rearrange the equations $\binom{**}{*}$ and the unknowns in these equations so that a minor of order r of the matrix $[A]$ is translated to the left top corner of the matrix. Then we solve the equations with respect to the first r unknowns $x_1, x_2, \ldots, x_r$ and express them by the remaining ones:

$$(3)\qquad \begin{aligned} x_1 &= x_1(x_{r+1}, x_{r+2}, \ldots, x_n), \\ x_2 &= x_2(x_{r+1}, x_{r+2}, \ldots, x_n), \\ &\ldots\ldots\ldots\ldots\ldots\ldots \\ x_r &= x_r(x_{r+1}, x_{r+2}, \ldots, x_n). \end{aligned}$$

The unknowns $x_{r+1}, x_{r+2}, \ldots, x_n$ can be given arbitrary values; thus, together with the corresponding values for $x_1, x_2, \ldots, x_r$ obtained from (3), we get one solution of the equations $\binom{**}{*}$. Choosing these solutions $n - r$ times

---

[1] See footnote on p. 177.

[2] There can be infinitely many basic systems of solutions (see below).

	x_{r+1}	x_{r+2}	...	x_n
1	b_{11}	b_{12}	...	$b_{1,n-r}$
2	b_{21}	b_{22}	...	$b_{2,n-r}$
...	...	...	...	...
$n-r$	$b_{n-r,1}$	$b_{n-r,2}$	...	$b_{n-r,n-r}$

so that the determinant $B=|b_{ik}|$ is different from zero, we obtain one of basic systems of solutions for the equations $(\ast\ast)$. In particular, we can substitute $b_{ik}=1$ for $i=k$ and $b_{ik}=0$, for $i\neq k$; then $B=1$ and the solutions

	x_{r+1}	x_{r+2}	...	x_n
1	1	0	...	0
2	0	1	...	0
...	...	...	...	...
$n-r$	0	0	...	1

together with formulas (3) determine a basic system of solutions of the equations $(\ast\ast)$ in the simplest way.

Examples. (1) Find a basic system of solutions of the equations

$$x-y+5z-u=0,$$

$$x+y-2z+3u=0,$$

$$3x-y+8z+u=0,$$

$$x+3y-9z+7u=0.$$

The rank of the matrix $[A]$ is 2; the determinant $\begin{vmatrix} 1 & -1 \\ 1 & 1 \end{vmatrix}\neq 0$, hence rearranging is not necessary. We solve the first two equations with respect to the unknowns x and y. Substituting $z=1$, $u=0$, we obtain first basic solution

$$x=-\tfrac{3}{2},\quad y=\tfrac{7}{2},\quad z=1,\quad u=0\quad \text{or}\quad \{-\tfrac{3}{2},\tfrac{7}{2},1,0\}.$$

Substituting $z=0$ and $u=1$, we obtain second basic solution

$$x=-1,\quad y=-2,\quad z=0,\quad u=1\quad \text{or}\quad \{-1,-2,0,1\}.$$

Therefore an arbitrary solution of the given system of equations is of the form $\{-\tfrac{3}{2}k_1-k_2,\tfrac{7}{2}k_1-2k_2,k_1,k_2\}$; where k_1, and k_2 are arbitrary numbers.

(2) $\quad 2x+3y-z=0,\quad x-y+z=0,\quad 3x+2y=0.$

The number of equations is equal to the number of unknowns, $D = D_x = D_y = D_z = 0$. The rank of the matrix $[A]$ is 2, the determinant $\begin{vmatrix} 2 & 3 \\ 1 & -1 \end{vmatrix} \neq 0$, rearranging of order is not necessary. We solve the system with respect to x and y: $x = -\frac{2}{5}z$, $y = \frac{3}{5}z$. Substituting $z = 1$, we get a single linearly independent basic solution: $x = -\frac{2}{5}$, $y = \frac{3}{5}$, $z = 1$. Hence an arbitrary solution is of the form

$$x = -\tfrac{2}{5}k, \quad y = \tfrac{3}{5}k, \quad z = k$$

or

$$x = -2k, \quad y = 3k, \quad z = 5k,$$

where k is an arbitrary number.

12. System of equations of higher degrees

Condition for independence of equations. Two equations

$$f(x, y) = 0 \quad \text{and} \quad \varphi(x, y) = 0$$

are *independent*, if their Jacobian (see p. 345)

$$\frac{D(f, \varphi)}{D(x, y)} = \begin{vmatrix} \dfrac{\partial f}{\partial x} & \dfrac{\partial f}{\partial y} \\[2ex] \dfrac{\partial \varphi}{\partial x} & \dfrac{\partial \varphi}{\partial y} \end{vmatrix}$$

is not equal identically to zero; otherwise one solution follows from another one and the system has infinitely many solutions.

The condition of independence for three equations is analogous:

$$\frac{D(f, \varphi, \psi)}{D(x, y, z)} \neq 0$$

and so on.

These conditions concern the algebraic equations as well as the transcendental ones.

Number of solutions of a system of two algebraic equations $P_1(x, y) = 0$ and $P_2(x, y) = 0$. If P_1 is a polynomial of degree m and P_2 is a polynomial of degree n with respect to x and y [1] then the system has mn solutions $x = \alpha$, $y = \beta$, where α and β are real or

---

[1] *Degree of a polynomial* of two variables x and y is the highest sum of exponents of these variables in members of the polynomial. For example, the polynomial $x^3 + x^2y^2 + y^3$ is of the fourth degree.

complex numbers. A system of three algebraic equations of degrees m, n, p has mnp solutions $x=\alpha$, $y=\beta$, $z=\gamma$.

Solution of a system of two algebraic equations is usually reduced to solution of one equation of degree mn with one unknown (a resolvent), by elimination of the second unknown. Having found the roots of the resolvent, we substitute them in one of the equations to determine the second unknown. The most simple to solve is a system of two equations such that one of them is linear. If the second one is of degree n, then solving the linear equation with respect to one unknown and substituting it in the second equation, we get a resolvent of degree n for the second unknown. In the case of a system of two equations each of which is of the second degree, we obtain a resolvent of the fourth degree. Sometimes such a system can be solved by certain artificial tricks.

Example. $x^2+y^2=a$, $xy=b$.

We obtain $(x+y)^2=a+2b$, $(x-y)^2=a-2b$; whence $x+y=+\sqrt{a+2b}$ or $x+y=-\sqrt{a+2b}$, $x-y=+\sqrt{a-2b}$, $x-y=-\sqrt{a-2b}$. Thus we find four pairs of solutions of the given system of equations: $(5, 3)$, $(3, 5)$, $(-5, -3)$, $(-3, -5)$.

Graphical method of solution of a system of two equations reduces to determining the intersection points of curves given by the equations.

C. SUPPLEMENTARY SECTIONS OF ALGEBRA

13. Inequalities

Definitions. An *inequality* is a statement composed of two literal or numerical expressions connected by one of the following signs:

(1) $>$ ("greater"),
(2) $<$ ("less"),
(3) $\neq$ ("not equal"),
(3a) $\gtrless$ ("greater or less"),
(4) $\geqslant$ ("greater or equal"),
(4a) $\not<$ ("not less"),
(5) $\leqslant$ ("less or equal"),
(5a) $\not>$ ("not greater").

The symbols (3) and (3a), (4) and (4a), (5) and (5a) are equivalent so that they can be replaced one by another [1].

Inequalities (1), (2), (3) are *sharp* and inequalities (4) and (5) are *non-sharp*.

An inequality is called *identical*, if it is valid for all values of the involved letters. A true inequality, containing numbers only is also called *identical*.

Inequalities, like equations, may involve unknown values (usually denoted by last letters of the alphabet). To *solve an inequality* or a *system of inequalities* means to determine in what intervals the values of unknowns should be contained in order that the inequality (or a system of inequalities) is true. We can seek a solution of an inequality of types (1)–(5), but the most frequently occurring cases are (1) and (2). Two inequalities of type (1) or two inequalities of type (2) are said to be *likewise directed*; if one inequality is of type (1) and the other one is of type (2), then they are said to be *oppositely directed*. Two inequalities are called *equivalent* if they are true for the same values of the unknowns

Main properties of inequalities of types (1) and (2).

(1) Asymmetry of inequality: if $a > b$, then $b < a$; if $a < b$ then $b > a$.

(2) Transitiveness of inequalities: if $a > b$ and $b > c$, then $a > c$.

(3) Monotony of inequalities: if $a > b$, then $a \pm c > b \pm c$; if $a < b$, then $a \pm c < b \pm c$. I.e., if the same number is added to both sides of an inequality, then its direction remains unchanged.

(4) Addition of inequalities: if $a > b$ and $c > d$, then $a + c > b + d$; if $a < b$ and $c < d$, then $a + c < b + d$. I.e., two likewise directed inequalities can be added by sides.

(5) Subtraction of inequalities: if $a > b$ and $c < d$, then $a - c > b - d$; if $a < b$ and $c > d$, then $a - c < b - d$. I.e., one can subtract by sides one inequality from another one with opposite direction; in the result, the sign of the first inequality is preserved (two likewise directed inequalities can not be subtracted).

(6) Multiplication and division of inequalities:

if $a > b$ and $c > 0$, then $ac > bc$, $\dfrac{a}{c} > \dfrac{b}{c}$,

---

[1] If the symbol (3) concerns the values for which the relations of "greater" or "less" are not defined, e.g., complex numbers (p. 585), vectors (p. 613), then it can not be replaced by the symbol (3a). In this section, we are concerned only with real numbers.

if $a < b$ and $c > 0$, then $ac < bc$, $\frac{a}{c} < \frac{b}{c}$,

if $a > b$ and $c < 0$, then $ac < bc$, $\frac{a}{c} < \frac{b}{c}$,

if $a < b$ and $c < 0$, then $ac > bc$, $\frac{a}{c} > \frac{b}{c}$.

I.e., both sides of an inequality can be multiplied or divided by the same positive number and the direction of inequality is preserved. Both sides of an inequality can be multiplied or divided by the same negative number but the direction of inequality should be changed into the opposite one.

Certain important inequalities.

(1) $|a+b| < |a| + |b|, \quad |a+b+\ldots+k| < |a| + |b| + \ldots + |k|$

(absolute value of a sum of two or several numbers is less or equal than the sum of their absolute values).

The equality occurs only when all numbers have the same sign.

(2) $|a| + |b| > |a-b| > |a| - |b|$

(absolute value of a difference of two numbers is less or equal than the sum and greater or equal than the difference of absolute values of these numbers).

(3) $\frac{a_1 + a_2 + \ldots + a_n}{n} > \sqrt[n]{a_1 a_2 \ldots a_n} \quad$ for $a_i > 0$

(*Cauchy's inequality*: arithmetical mean of n positive numbers is greater or equal than the n-th root of the product of these numbers) [1].

Equality occurs only when all the n numbers are equal.

(4) $\left| \frac{a_1 + a_2 + \ldots + a_n}{n} \right| < \sqrt{\frac{a_1^2 + a_2^2 + \ldots + a_n^2}{n}}$

(absolute value of the arithmetical mean of n numbers is less than or equal to the quadratic mean of these numbers, see p. 190).

---

[1] For a particular case ($n = 2$) of the inequality see p. 190.

(5) $a_1b_1 + a_2b_2 + \ldots + a_nb_n \leqslant \sqrt{a_1^2 + a_2^2 + \ldots + a_n^2}\ \sqrt{b_1^2 + b_2^2 + \ldots + b_n^2}$,

or

$$(a_1b_1 + a_2b_2 + \ldots + a_nb_n)^2 \leqslant (a_1^2 + a_2^2 + \ldots + a_n^2)(b_1^2 + b_2^2 + \ldots + b_n^2),$$

$$\left(\sum_{i=1}^{n} a_ib_i\right)^2 < \left(\sum_{i=1}^{n} a_i^2\right)\left(\sum_{i=1}^{n} b_i^2\right)$$

("*Buniakowsky-Cauchy*" *inequality*: if two finite sequences of n numbers are given, then the sum of products of the corresponding terms of these sequences does not exceed the product of the square roots of the sum of squares of these numbers [1]. Equality can occur only when $a_1 : b_1 = a_2 : b_2 = \ldots = a_n : b_n$.

(6) If $a_1, a_2, \ldots, a_n, b_1, b_2, \ldots, b_n$ are positive numbers then

$$(6_1)\quad \frac{a_1 + a_2 + \ldots + a_n}{n} \cdot \frac{b_1 + b_2 + \ldots + b_n}{n} < \frac{a_1b_1 + a_2b_2 + \ldots + a_nb_n}{n},$$

if $a_1 < a_2 < \ldots < a_n$ and $b_1 < b_2 < \ldots < b_n$ or if $a_1 > a_2 > \ldots > a_n$ and $b_1 > b_2 > \ldots > b_n$,

and

$$(6_2)\quad \frac{a_1 + a_2 + \ldots + a_n}{n} \cdot \frac{b_1 + b_2 + \ldots + b_n}{n} > \frac{a_1b_1 + a_2b_2 + \ldots + a_nb_n}{n},$$

if $a_1 < a_2 < \ldots < a_n$ and $b_1 > b_2 > \ldots > b_n$—*Tschebyscheff inequality*: if two finite sequences of n positive numbers are given, then the product of arithmetical means of these sequences is less than or equal (resp. greater or equal) to the arithmetical mean of the products, when both sequences are increasing or both are decreasing (resp., when one of the series is increasing and the other one is decreasing).

---

[1] If $n = 3$, then $\{a_1, a_2, a_3\}$ and $\{b_1, b_2, b_3\}$ may be regarded as rectangular Cartesian coordinates of a vector; then Buniakowsky-Cauchy inequality reads that the scalar product of vectors does not exceed the product of their moduli (see p. 617). For $n > 3$, this formulation is extended to vectors in the n-dimensional space.

An analogy of Buniakowsky-Cauchy inequality for infinite convergent series:

$$\left(\sum_{i=1}^{\infty} a_ib_i\right)^2 \leqslant \left(\sum_{i=1}^{\infty} a_i^2\right)\left(\sum_{i=1}^{\infty} b_i^2\right).$$

An analogy of the same inequality for definite integrals:

$$\left(\int_a^b f(x)\,\varphi(x)\,dx\right)^2 \leqslant \left(\int_a^b [f(x)]^2\,dx\right)\left(\int_a^b [\varphi(x)]^2\,dx\right).$$

(7) If $a_1, a_2, \ldots, a_n, b_1, b_2, \ldots, b_n$ are positive numbers, then

$$(7_1)\quad \sqrt[k]{\frac{a_1^k + a_2^k + \ldots + a_n^k}{n}}\sqrt[k]{\frac{b_1^k + b_2^k + \ldots + b_n^k}{n}} \leqslant \sqrt[k]{\frac{(a_1b_1)^k + (a_2b_2)^k + \ldots + (a_nb_n)^k}{n}}.$$

if $a_1 \leqslant a_2 \leqslant \ldots \leqslant a_n$ and $b_1 \leqslant b_2 \leqslant \ldots \leqslant b_n$ or if $a_1 \geqslant a_2 \geqslant \ldots \geqslant a_n$ and $b_1 \geqslant b_2 \geqslant \ldots \geqslant b_n$

and

$$(7_2)\quad \sqrt[k]{\frac{a_1^k + a_2^k + \ldots + a_n^k}{n}}\sqrt[k]{\frac{b_1^k + b_2^k + \ldots + b_n^k}{n}} \geqslant \sqrt[k]{\frac{(a_1b_1)^k + (a_2b_2)^k + \ldots + (a_nb_n)^k}{n}},$$

if $a_1 \leqslant a_2 \leqslant \ldots \leqslant a_n$ and $b_1 \geqslant b_2 \geqslant \ldots \geqslant b_n$ *(generalization of Tschebyscheff inequality)*.

Solution of inequalities of the first and second degree. In solving an inequality, we reduce it to other equivalent inequalities. As in solving equations, terms of an inequality can be transposed from one side to the other with the opposite sign. Both sides of an inequality can be multiplied or divided by the same number, different from zero; if this number is positive, then the direction of the inequality remains the same, and if it is negative, then the direction of the inequality should be changed. By aid of such transformations, an inequality of the first degree can always be reduced to the form $ax > b$, and an inequality of the second degree, in the simplest case, to the form $x^2 < m$ or $x^2 > m$, and, in the general case, to the form $ax^2 + bx + c < 0$ or $ax^2 + bx + c > 0$.

Inequality of the first degree $ax > b$ has the solution:

$$x > \frac{b}{a} \text{ if } a > 0 \quad \text{and} \quad x < \frac{b}{a} \text{ if } a < 0.$$

Example. $5x + 3 < 8x + 1$; $5x - 8x < 1 - 3$; $-3x < -2$, $x > \frac{2}{3}$.

The simplest inequalities of the second degree: $x^2 < m$ and $x^2 > m$ have the solutions:

(a) $x^2 < m$.

For $m > 0$, the solution $-\sqrt{m} < x < +\sqrt{m}$, or $|x| < \sqrt{m}$.

For $m \leqslant 0$, there is no solution.

(b) $x^2 > m$.

For $m > 0$, the solution $x > \sqrt{m}$ and $x < -\sqrt{m}$, or $|x| > \sqrt{m}$.

For $m = 0$, the solution $x > 0$ and $x < 0$, or $x \neq 0$.

For $m < 0$, the inequality is identically true.

General case of an inequality of the second degree $ax^2 + bx + c < 0$. We divide the inequality by a (changing the direction of inequality, if $a < 0$) and reduce it to the form $x^2 + px + q < 0$ or $x^2 + px + q > 0$. Then we transform these to one of the forms

$$\left(x + \frac{p}{2}\right)^2 < \left(\frac{p}{2}\right)^2 - q \quad \text{or} \quad \left(x + \frac{p}{2}\right)^2 > \left(\frac{p}{2}\right)^2 - q.$$

We denote $x + \frac{1}{2}p$ by z and $(\frac{1}{2}p)^2 - q$ by m and obtain the inequality $z^2 < m$ or $z^2 > m$. Solving it, we find x.

Examples. (1) $-2x^2 + 14x - 20 > 0$; $x^2 - 7x + 10 < 0$; $(x - \frac{7}{2})^2 < \frac{9}{4}$; $-\frac{3}{2} < x - \frac{7}{2} < \frac{3}{2}$; $-\frac{3}{2} + \frac{7}{2} < x < \frac{3}{2} + \frac{7}{2}$; solution: $2 < x < 5$.

(2) $x^2 + 6x + 15 > 0$; $(x + 3)^2 > -6$; inequality is identically true.

(3) $-2x^2 + 14x - 20 < 0$; $(x - \frac{7}{2})^2 > \frac{9}{4}$, $x - \frac{7}{2} > \frac{3}{2}$ and $x - \frac{7}{2} < -\frac{3}{2}$; solution: $x > 5$ and $x < 2$.

14. Progressions, finite series and mean values

Progressions. *Arithmetic progression* is a sequence of numbers $a_1, a_2, \ldots, a_n$ (called the *terms*) such that each term is obtained from the preceding one by adding a definite number r (called the *difference* of progression). If $r > 0$, then the progression is *increasing*; if $r < 0$ then it is *decreasing*.

Formulas for arithmetic progression:

$$a_n = a_1 + (n - 1)r, \qquad s_n = \frac{n(a_1 + a_n)}{2},$$

s_n is the sum of n terms.

Geometric progression is a sequence of numbers $a_1, a_2, \ldots, a_n$ (the *terms*) such that each term is obtained from the preceding one by dividing it by a definite number q (called the *ratio* of progression). If $q > 1$, then the progression is *increasing* and if $|q| < 1$, then the progression is *decreasing*.

Formulas for geometric progression:

$$a_n = a_1 q^{n-1} \quad \text{and} \quad s_n = \frac{a_1(q^n - 1)}{q - 1}.$$

For the sum of a decreasing geometric progression, the formula $s_n = \dfrac{a_1(1 - q^n)}{1 - q}$ is more convenient. If the number of terms of a decreasing geometric progression infinitely increases, then $q^n \to 0$ and s_n tends to the limit

$$\lim_{n\to\infty} s_n = s = \frac{a_1}{1-q}$$

(the sum of an infinitely decreasing geometric progression).

Example.

$$1+\frac{1}{2}+\frac{1}{2^2}+\dots+\frac{1}{2^n}+\dots=\frac{1}{1-\frac{1}{2}}=2.$$

Certain finite numerical series [1]:

1. $1+2+3+\dots+(n-1)+n=\dfrac{n(n+1)}{2}$.

2. $p+(p+1)+(p+2)+\dots+(q-1)+q=\dfrac{(q+p)(q-p+1)}{2}$.

3. $1+3+5+\dots+(2n-3)+(2n-1)=n^2$.

4. $2+4+6+\dots+(2n-2)+2n=n(n+1)$.

5. $1^2+2^2+3^2+\dots+(n-1)^2+n^2=\dfrac{n(n+1)(2n+1)}{6}$.

6. $1^3+2^3+3^3+\dots+(n-1)^3+n^3=\dfrac{n^2(n+1)^2}{4}$
$$=[1+2+3+\dots+(n-1)+n]^2.$$

7. $1^2+3^2+5^2+\dots+(2n-3)^2+(2n-1)^2=\dfrac{n(4n^2-1)}{3}$.

8. $1^3+3^3+5^3+\dots+(2n-3)^3+(2n-1)^3=n^2(2n^2-1)$.

9. $1^4+2^4+\dots+(n-1)^4+n^4=\dfrac{n(n+1)(2n+1)(3n^2+3n-1)}{30}$.

Mean values. The *arithmetic mean* of two numbers a and b is one half of its sum: $x=\frac{1}{2}(a+b)$; the numbers a, x, b form an arithmetic progression.

Arithmetic mean of n numbers $a_1, a_2, \dots, a_n$ is

$$x=\frac{a_1+a_2+\dots+a_n}{n}.$$

---

[1] For a table of infinite series see pp. 353–354.

The *quadratic mean* of n numbers $a_1, a_2, \ldots a_n$ (positive or negative ones) is

$$+\sqrt{\frac{1}{n}(a_1^2 + a_2^2 + \ldots + a_n^2)}\,;$$

it is important in the theory of errors (see p. 747).

The *geometric mean* (also called the *mean proportional*) of two numbers a and b is $x = \sqrt{ab}$; the numbers a, x, b form a geometric progression. The geometric mean of two different positive numbers is always less than their arithmetic mean. If a and b are lengths of two segments, then a segment of the length $x = \sqrt{ab}$ can be constructed as it is shown on Fig. 76a or b.

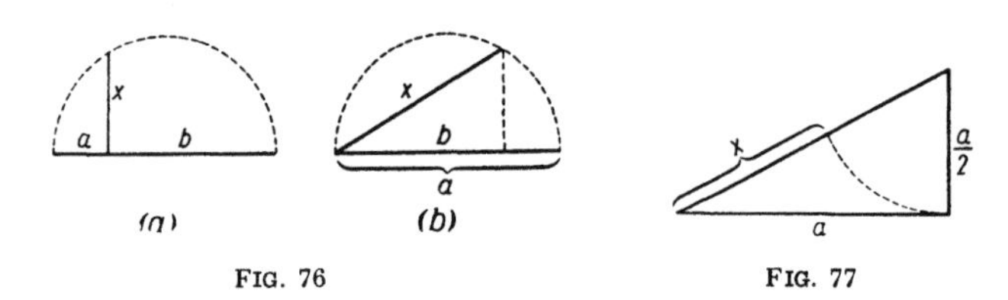

FIG. 76 FIG. 77

The *golden section* of a segment a (or the *division of a in extreme and mean ratio*) is the division of a into two parts x and $a-x$ such that x is the geometric mean between a and $a-x$:

$$x = \frac{\sqrt{5}-1}{2}a \approx 0.618a.$$

If a is the length of a segment, then a segment of length x can be determined as it is shown on Fig. 77. The segment x is equal to a side of a decagon inscribed in a circle of radius a.

15. Factorial and gamma function

Factorial. The *factorial* $n!$ of a natural number n is the product $1 \cdot 2 \cdot 3 \cdot \ldots \cdot n$.

Main property of the factorial: $n! = n\,(n-1)!$

Factorials of initial natural numbers and their reciprocals are given on p. 49.

Factorials of large numbers can be approximately expressed by Stirling's formula:

$$n! \approx \left(\frac{n}{e}\right)^n \sqrt{2\pi n}\left(1 + \frac{1}{12n} + \frac{1}{288n^2} + \ldots\right),$$

$$\ln(n!) \approx \left(n+\frac{1}{2}\right)\ln n - n + \ln\sqrt{2\pi}.$$

These formulas can also be applied for non-integral values of n (see below, the gamma function).

The gamma function. The notion of the factorial can be extended to arbitrary numbers x [1] by means of the *gamma function*, $\Gamma(x)$, defined as follows:

$$\Gamma(x) = \begin{cases} \int_0^\infty e^{-t} t^{x-1} dt & (\textit{Euler's integral} \text{ for } x > 0)\ [2], \\ \lim_{n\to\infty} \dfrac{n!\, n^{x-1}}{x(x+1)(x+2)\dots(x+n-1)} & \text{for arbitrary } x. \end{cases}$$

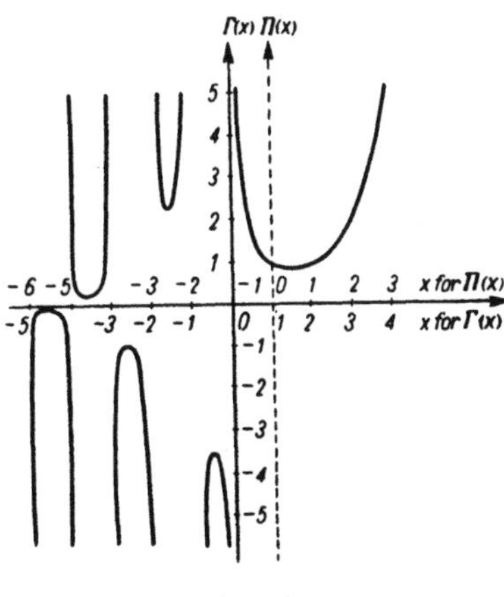

FIG. 78

Main properties of the gamma function:

$$\Gamma(x+1) = x\Gamma(x),$$

$\Gamma(n) = (n-1)!$, for integral positive n,

$$\Gamma(x)\,\Gamma(1-x) = \frac{\pi}{\sin \pi x},$$

$$\Gamma(x)\,\Gamma\left(x+\frac{1}{2}\right) = \frac{\sqrt{\pi}}{2^{2x-1}}\Gamma(2x).$$

---

[1] And also to the complex numbers.

[2] For complex x, if re $(x) > 0$.

Generalized notion of the factorial. The notion of the factorial $n!$ which has been first defined for a positive integral n can be generalized for arbitrary real x by means of the formula $\Pi(x) = \Gamma(x+1)$.

When x is a positive integer, then $\Pi(x) = x! = 1 \cdot 2 \cdot 3 \cdot \ldots \cdot x$.

When $x = 0$: $\Pi(0) = \Gamma(1) = 1$.

When x is a negative integer, then $\Pi(x) = \pm\infty$.

When $x = \frac{1}{2}$: $\Pi(\frac{1}{2}) = \Gamma(\frac{3}{2}) = \frac{1}{2}\sqrt{\pi}$.

When $x = -\frac{1}{2}$: $\Pi(-\frac{1}{2}) = \Gamma(\frac{1}{2}) = \sqrt{\pi}$.

When $x = -\frac{3}{2}$: $\Pi(-\frac{3}{2}) = \Gamma(-\frac{1}{2}) = -2\sqrt{\pi}$.

The graph of the functions $\Gamma(x)$ and $\Pi(x)$ are given on Fig. 78. For a table of $\Gamma(x)$ see p. 87.

16. Variations, permutations, combinations

Variations. An arrangement of n elements into a sequence consisting of k terms is called a *variation*. Two variations can differ either by their elements or by their order. For example, the variations of three elements a, b, c in two are: ab, ac, bc, ba, ca, cb. The number of all variations of n different elements in k is expressed by the formula

$$V_n^k = \underbrace{n(n-1)(n-2)\ldots(n-k+1)}_{k \text{ factors}} = \frac{n!}{(n-k)!} \ ^{(1)}$$

For example, $V_3^2 = 3 \cdot 2 = 6$.

Permutations. A *permutation* is an arrangement of n different elements (a sequence of n elements). Two permutations differ only by the order of their elements. For example, the permutations of three elements a, b, c are abc, bca, cab, cba, bac, acb. The number of all permutations of all different elements is

$$P_n = 1 \cdot 2 \cdot 3 \cdot \ldots \cdot n = n! = V_n^n.$$

If, among n elements $a, b, c, \ldots$, some are equal: a occurs α times, b occurs β times, c occurs γ times, and so on, then

$$P_n = \frac{n!}{\alpha!\,\beta!\,\gamma!\ldots}.$$

---

(1) For the symbol "$n!$" (the factorial) see p. 190.

Combinations. A *combination* is a set (group) of k elements taken from given n elements (without repetitions). Two combinations can differ only by their elements. For example, the combinations of three elements a, b, c into groups of two are ab, ac, bc. The number of all combinations of n different elements into groups of k elements is

$$C_n^k = \binom{n}{k} = \frac{n(n-1)(n-2)\ldots(n-k+1)}{1\cdot 2\cdot 3\cdot \ldots \cdot k} = \frac{V_n^k}{P_k} = \frac{n!}{k!\,(n-k)!}.$$

In particular,

$$C_n^1 = \binom{n}{1} = n, \quad C_n^n = \binom{n}{n} = 1.$$

Main property of combinations:

$$\binom{n}{k} = \binom{n}{n-k}, \quad \binom{n}{n} = \binom{n}{0} = 1.$$

17. Newton's binomial theorem

Newton's formula:

(*)

$$(a+b)^n = a^n + na^{n-1}b + \frac{n(n-1)}{2!}a^{n-2}b^2 + \frac{n(n-1)(n-2)}{3!}a^{n-3}b^3 + \\ + \ldots + \frac{n(n-1)\ldots(n-m+1)}{m!}a^{n-m}b^m + \ldots + nab^{n-1} + b^n,$$

or

$$(a+b)^n = \binom{n}{0}a^n + \binom{n}{1}a^{n-1}b + \binom{n}{2}a^{n-2}b^2 + \ldots + \\ + \binom{n}{k}a^{n-k}b^k + \ldots + \binom{n}{n-1}ab^{n-1} + \binom{n}{n}b^n.$$

Binomial coefficients $\binom{n}{k}$ can be determined from *Pascal's triangle*:

n	coefficients														
0								1							
1							1		1						
2						1		2		1					
3					1		3		3		1				
4				1		4		6		4		1			
5			1		5		10		10		5		1		
6		1		6		15		20		15		6		1	
7	1		7		21		35		35		21		7		1
...	....														

Each coefficient is the sum of two other ones lying above it (on the left and on the right).

Properties of the binomial coefficients.

(1) The coefficients in Newton's formula increase up to the middle of the formula and then decrease.

(2) The coefficients lying at equal distances from the begining and from the end are equal.

(3) The sum of coefficients in a binomial of degree n is 2^n.

(4) The sum of coefficients in the odd places is equal to the sum of coefficients in the even places.

Power of a difference:

$$(a-b)^n = a^n - na^{n-1}b + \frac{n(n-1)}{2!}a^{n-2}b^2 - \frac{n(n-1)(n-2)}{3!}a^{n-3}b^3 + \\ + \dots + (-1)^k \frac{n(n-1)\dots(n-k+1)}{k!}a^{n-k}b^k + \dots + (-1)^n b^n.$$

Generalization of Newton's formula to arbitrary powers. The formula (*) can be extended to negative and fractional exponents n. The power $(a+b)^n$, where $|b| < a$, can be then expressed in the form of an infinite series (see pp. 387, 388):

$$(a+b)^n = a^n + na^{n-1}b + \frac{n(n-1)}{2!}a^{n-2}b^2 + \frac{n(n-1)(n-2)}{3!}a^{n-3}b^3 + \\ + \dots + \frac{n(n-1)(n-2)\dots(n-k+1)}{k!}a^{n-k}b^k + \dots$$

III. GEOMETRY

A. PLANE GEOMETRY

1. Plane figures

Triangle. The sum of two sides of a triangle (Fig. 79) is always greater than the third side: $b + c > a$. The sum of angles of a triangle: $\alpha + \beta + \gamma = 180°$. A triangle is completely determined, when there are given: (1) three sides or (2) two sides and the angle included between them, or (3) one side and two angles adjacent to it. Two

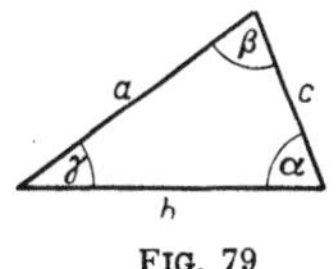

FIG. 79

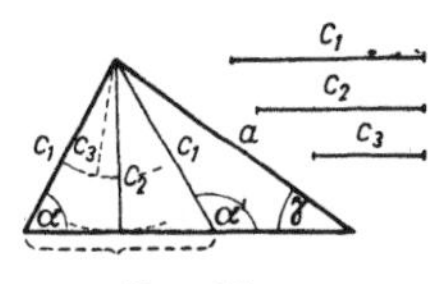

FIG. 80

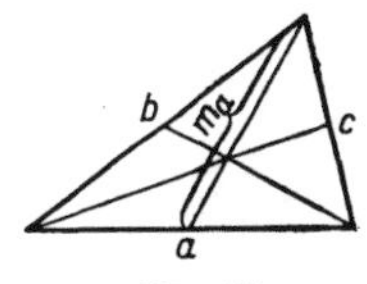

FIG. 81

sides and an angle opposite to one of them determine either two triangles or one, or no triangle, in various cases (see Fig. 80, and, for details, pp. 221, 222).

A *median* of a triangle is a segment joining a vertex with the middle point of the opposite side. The medians of a triangle intersect in one point which is the centre of gravity of the triangle (Fig. 81); each median is divided by this point in the ratio 2 : 1, from the vertex of the triangle. Length of the median of the side a: $m_a = \frac{1}{2}\sqrt{2(b^2 + c^2) - a^2}$ see p. 222).

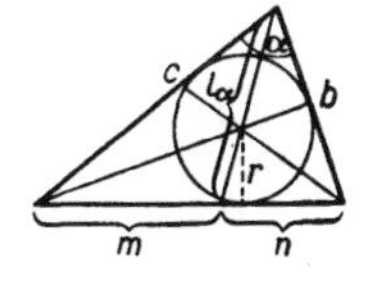

FIG. 82

A *bisector* of a triangle is a segment halving its interior angle. The bisectors of a triangle intersect in one point which is the centre of the inscribed circle (Fig. 82). Radius of the inscribed circle r—see p. 222. Length of the bisector of the angle α (see also p. 222) is $l_\alpha = \frac{\sqrt{bc(b + c)^2 - a^2}}{b + c}$. If a bisector divides a side a into segments m and n, then $m : n = c : b$.

The centre of the circumscribed circle lies in the intersection point of perpendiculars to the sides erected at their midpoints (Fig. 83). Radius of the circumscribed circle R—see p. 222.

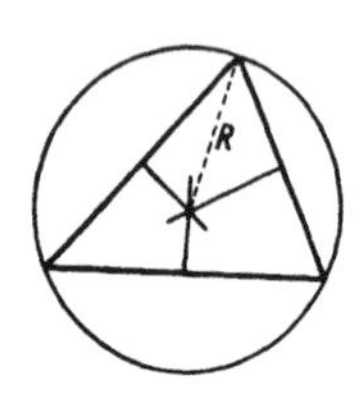

FIG. 83

An *altitude* of a triangle is a perpendicular drawn from a vertex to the opposite side. The altitudes of a triangle intersect in one point called the *orthocentre*. Length of the altitude—see p. 222.

Altitude, median and bisector corresponding to one side of a triangle coincide, if two other sides are equal (an *isosceles* triangle). Coinciding of two of these lines is sufficient for a triangle to be isosceles.

In an *equilateral* triangle ($a = b = c$) the centres of inscribed and circumscribed circle, centre of gravity and the orthocentre coincide.

A *middle line* is a segment joining the midpoints of two sides of a triangle; it is parallel to the third side and equal to one half of it.

Area of a triangle: $S = \frac{1}{2}bh_b$ [1] $= \frac{1}{2}ab\sin\gamma = \frac{1}{2}r(a+b+c) = \frac{abc}{4R} = \sqrt{p(p-a)(p-b)(p-c)}$, where $p = \frac{1}{2}(a+b+c)$.

Right triangle (Fig. 84). c—hypotenuse, a and b—cathetii. $a^2 + b^2 = c^2$ (the *Pythagorean theorem*). $h^2 = mn$, $a^2 = mc$, $b^2 = nc$. Area $S = \frac{1}{2}ab = \frac{1}{2}a^2\tan\beta = \frac{1}{4}c^2\sin 2\beta$.

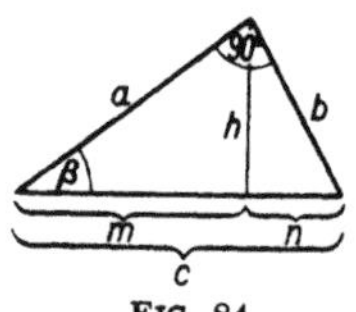

FIG. 84

For trigonometric formulas of triangles see pp. 220–222.

Two triangles (and also two polygons with the same number of sides) are called *similar*, if their corresponding angles are equal and the corresponding sides are proportional. Each of the following conditions is sufficient for two triangles to be similar: (1) Three sides of one triangle are proportional to the three sides of the other. (2) Two angles of one triangle are equal to the corresponding angles of the other. (3) Two sides of one triangle are proportional to the corresponding two sides of the other and the angles included between them are equal.

The areas of similar figures are proportional to the squares of their corresponding linear elements (such as sides, altitudes, diagonals etc.).

---

(1) h_b denote the altitude to the side b.

Parallelogram (Fig. 85). Main properties: (1) The opposite sides are pairwise equal. (2) The opposite sides are pairwise parallel. (3) Two sides are parallel and equal. (4) The diagonals bisect each other. (5) The opposite angles are equal. Each of these properties implies the remaining ones.

Relation between the diagonals and sides: $d_1^2 + d_2^2 = 2(a^2 + b^2)$. *Area* $S = ah$.

Rectangle and square. A parallelogram is a *rectangle* (Fig. 86) if (1) all its angles are right, (2) the diagonals are equal (either of these properties follows from the other). The *area* $S = ab$.

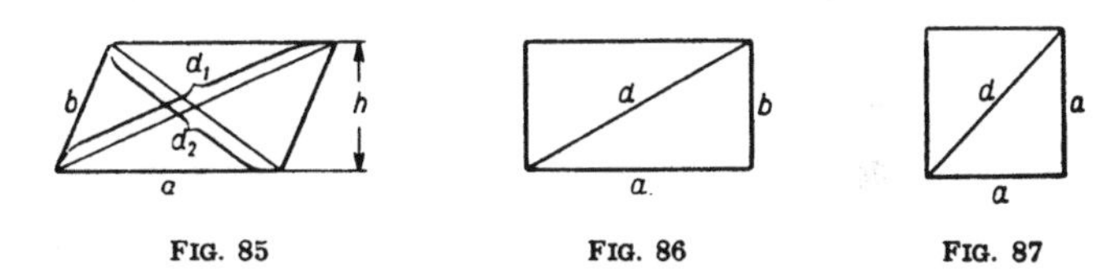

FIG. 85 FIG. 86 FIG. 87

A rectangle is a *square* (Fig. 87), if $a = b$; $d = \sqrt{2}\,a \approx 1.414a$; $a = \frac{1}{2}\sqrt{2}\,d \approx 0.707d$. The *area* $S = a^2 = \frac{1}{2}d^2$.

Rhombus. A parallelogram is a *rhombus* (Fig. 88), if (1) all sides are equal, (2) the diagonals are perpendicular, (3) the diagonals bisect the angles of the parallelogram. Each of these properties implies two remaining ones. $d_1 = 2a \sin \frac{1}{2}\alpha$; $d_2 = 2a \cos \frac{1}{2}\alpha$; $d_1^2 + d_2^2 = 4a^2$. The *area* $S = ah = a^2 \sin \alpha = \frac{1}{2} d_1 d_2$.

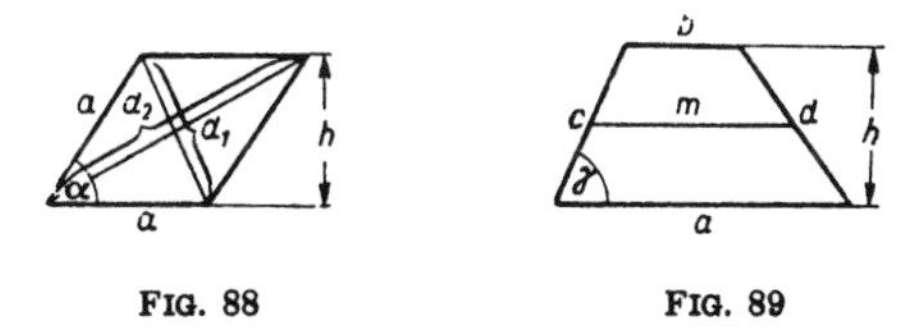

FIG. 88 FIG. 89

Trapezoid. A *trapezoid* is a quadrilateral with two parallel sides (Fig. 89). a and b are bases of the trapezoid, h—the altitude, m—the middle line (the segment joining the midpoints of two parallel sides). $m = \frac{1}{2}(a + b)$. The *area* $S = \frac{1}{2}(a + b)h = mh$.

A trapezoid is *isosceles*, if $d = c$. In this case, $S = (a - c \cos \gamma) \times$ $\times c \sin \gamma = (b + c \cos \gamma)\, c \sin \gamma$.

Quadrilateral (Fig. 90). The sum of angles of any convex quadrilateral is 360°. $a^2 + b^2 + c^2 + d^2 = d_1^2 + d_2^2 + 4m^2$, where m is the segment joining the midpoints of the diagonals. The *area* $S = \frac{1}{2} d_1 d_2 \sin \alpha$.

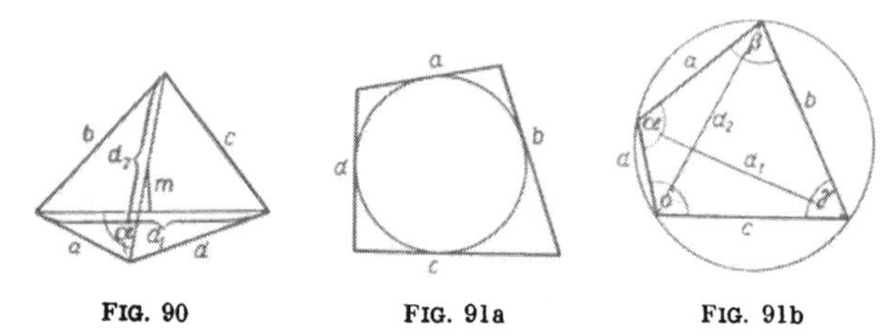

FIG. 90 FIG. 91a FIG. 91b

A circle can be inscribed in a quadrilateral (Fig. 91a), if and only if $a + c = b + d$. A circle can be circumscribed about a quadrilateral (Fig. 91b), if and only if $\alpha + \gamma = \beta + \delta = 180°$. For the inscribed quadrilateral $ac + bd = d_1 d_2$. The *area of an inscribed quadrilateral*, $S = \sqrt{(p-a)(p-b)(p-c)(p-d)}$, where $p = \frac{1}{2}(a + b + c + d)$.

Polygon (Fig. 92). If the number of sides is n, then the sum of interior angles is $180°(n-2)$. The sum of exterior angles is 360°. The area can be found by decomposition into triangles.

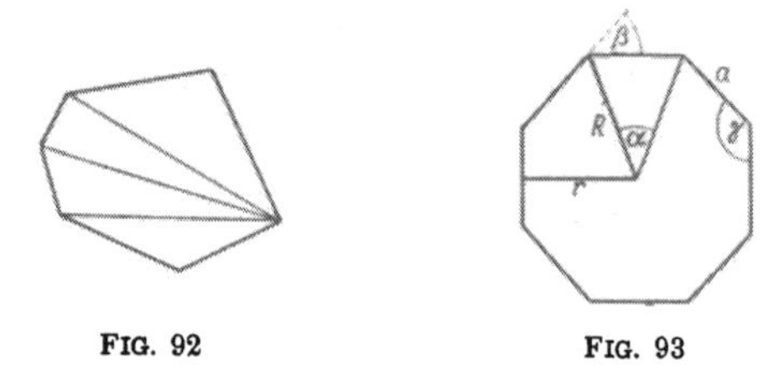

FIG. 92 FIG. 93

A polygon is *regular* if all its sides are equal and all its angles are equal. For regular polygons with n sides (Fig. 93): the central angle $\alpha = 360° : n$, exterior angle $\beta = 360° : n$, interior angle $\gamma = 180° - \beta$. If R is the radius of the circumscribed circle and r is that of the inscribed one (the *apothem*), then the side is $a = 2\sqrt{R^2 - r^2} = 2R \sin \frac{1}{2}\alpha = 2r \tan \frac{1}{2}\alpha$. The *area* $S = \frac{1}{2}\, nar = nr^2 \times \tan \frac{1}{2}\alpha = \frac{1}{2} nR^2 \sin \alpha = \frac{1}{4} na^2 \cot \frac{1}{2}\alpha$. For details about particular regular polygons see the table on p. 199.

Elements of the regular polygons. Notations: n—number of sides, a—one side, R—radius of the circumscribed circle, r—radius of the inscribed circle (the apothem).

n	$\frac{S}{a^2}$	$\frac{S}{R^2}$	$\frac{S}{r^2}$	$\frac{R}{a}$	$\frac{R}{r}$	$\frac{a}{R}$	$\frac{a}{r}$	$\frac{r}{R}$	$\frac{r}{a}$
3	0.4330	1.2990	5.1962	0.5774	2.0000	1.7321	3.4641	0.5000	0.2887
4	1.0000	2.0000	4.0000	0.7071	1.4142	1.4142	2.0000	0.7071	0.5000
5	1.7205	2.3776	3.6327	0.8507	1.2361	1.1756	1.4531	0.8090	0.6882
6	2.5981	2.5981	3.4641	1.0000	1.1547	1.0000	1.1547	0.8660	0.8660
7	3.6339	2.7364	3.3710	1.1524	1.1099	0.8678	0.9631	0.9010	1.0383
8	4.8284	2.8284	3.3137	1.3066	1.0824	0.7654	0.8284	0.9239	1.2071
9	6.1818	2.8925	3.2757	1.4619	1.0642	0.6840	0.7279	0.9397	1.3737
10	7.6942	2.9389	3.2492	1.6180	1.0515	0.6180	0.6498	0.9511	1.5388
12	11.196	3.0000	3.2154	1.9319	1.0353	0.5176	0.5359	0.9659	1.8660
15	17.642	3.0505	3.1883	2.4049	1.0223	0.4158	0.4251	0.9781	2.3523
16	20.109	3.0615	3.1826	2.5629	1.0196	0.3902	0.3978	0.9808	2.5137
20	31.569	3.0902	3.1677	3.1962	1.0125	0.3129	0.3168	0.9877	3.1569
24	45.575	3.1058	3.1597	3.8306	1.0086	0.2611	0.2633	0.9914	3.7979
32	81.225	3.1214	3.1517	5.1012	1.0048	0.1960	0.1970	0.9952	5.0766
48	183.08	3.1326	3.1461	7.6449	1.0021	0.1308	0.1311	0.9979	7.6285
64	325.69	3.1366	3.1441	10.190	1.0012	0.0981	0.0983	0.9988	10.178

Circumference. Notations: r—radius, $d=2r$—diameter. Angles related to a circle [1] (Fig. 94): inscribed angle $\alpha=\frac{1}{2}\breve{BC}$, angle between the chord and the tangent $\beta=\frac{1}{2}\breve{AC}$, angle between the

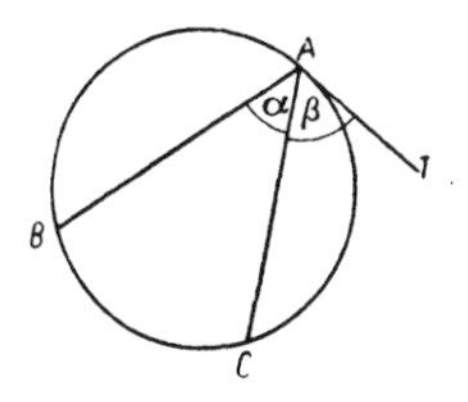

FIG. 94

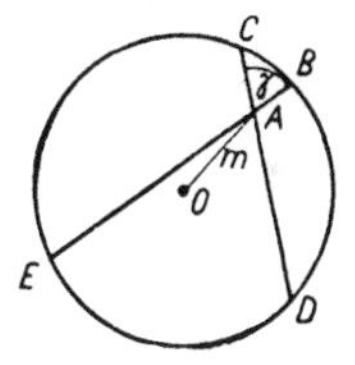

FIG. 95

chords (Fig. 95) $\gamma=\frac{1}{2}(\breve{CB}+\breve{ED})$, between the secants (Fig. 96) $\alpha=\frac{1}{2}(\breve{DE}-\breve{BC})$, angle between the tangent and the secant $\beta=$ $=\frac{1}{2}(\breve{TE}-\breve{TB})$, between the tangents (Fig. 97) $a=\frac{1}{2}(\breve{BDC}-\breve{BEC})$.

Intersecting chords (Fig. 95): $AC\cdot AD=AB\cdot AE=r^2-m^2$.

Secants (Fig. 96): $AB\cdot AE=AC\cdot AD=AT^2=m^2-r^2$.

Length C of circumference and *area S of circle*

$$\pi=\frac{C}{d}=3.141\,592\,653\,589\,793\ldots$$

[1] In these formulas, $\breve{AB}$ denotes the degree measure of the arc AB equal to the degree measure of the central angle AOB.

$$C = 2\pi r \approx 6.283r, \quad C = \pi d \approx 3.142d, \quad C = 2\sqrt{\pi S} \approx 3.545\sqrt{S},$$

$$S = \pi r^2 \approx 3.142r^2, \quad S = \frac{\pi d^2}{4} \approx 0.785d^2, \quad S = \frac{Cd}{4} = 0.25Cd,$$

$$r = \frac{C}{2\pi} \approx 0.159C, \quad d = 2\sqrt{\frac{S}{\pi}} \approx 1.128\sqrt{S}.$$

Also see the table on pp. 71–73.

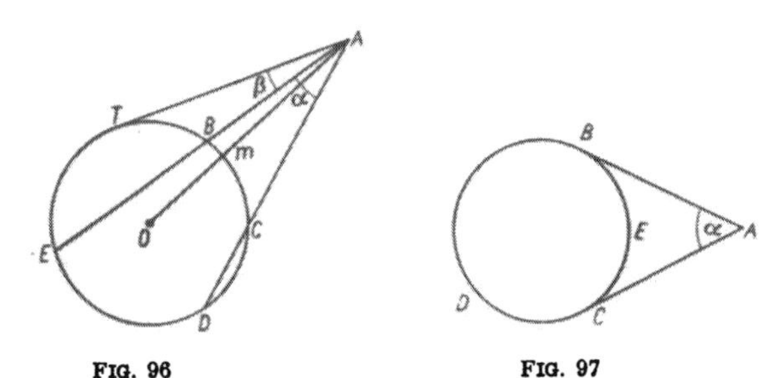

FIG. 96 FIG. 97

Segment and sector (Fig. 98). Notations: r—radius, l—length of arc, a—chord, α—central angle (in degrees), h—height.

$$a = 2\sqrt{2hr - h^2} = 2r \sin \tfrac{1}{2}\alpha;$$

$$h = r - \sqrt{r^2 - \tfrac{1}{4}a^2} = r(1 - \cos \tfrac{1}{2}\alpha) = \tfrac{1}{2}a \tan \tfrac{1}{4}\alpha; \quad l = \frac{2\pi r\alpha}{360} \approx 0.01745r\alpha.$$

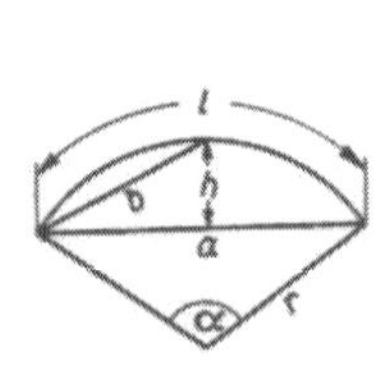

FIG. 98

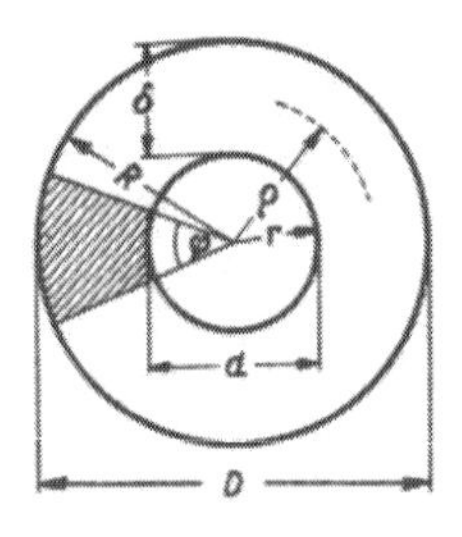

FIG. 99

Approximately,

$$l = \tfrac{1}{3}(8b - a) \quad \text{or} \quad l = \sqrt{a^2 + \tfrac{16}{3}h^2}.$$

Area of the sector $S = \dfrac{r^2\pi\alpha}{360} \approx 0.00873\, r^2\alpha.$

Area of the segment $S_1 = \frac{1}{2}r^2\left(\frac{\pi\alpha}{180} - \sin\alpha\right) = \frac{1}{2}(lr - a(r-h))$.
Approximately, $S_1 = \frac{1}{15}h(6a + 8b)$. For tables of $S_1, l, h,$ and a see pp. 76–82.

Circular annulus (Fig. 99). $D = 2R$—exterior diameter, $d = 2r$—interior diameter, $\varrho = \frac{1}{2}(R + r)$—mean radius, $\delta = R - r$—width of the annulus.

Area of the annulus

$$S = \pi(R^2 - r^2) = \frac{1}{4}\pi(D^2 - d^2) = 2\pi\varrho\delta.$$

Area of a sector of the annulus (marked on the Fig. 99) with the central angle φ (in degrees):

$$S = \frac{\varphi\pi}{360}(R^2 - r^2) = \frac{\varphi\pi}{90}(D^2 - d^2) = \frac{\varphi\pi}{180}\varrho\delta.$$

B. SOLID GEOMETRY

2. Straight lines and planes in space

Two straight lines lying on a plane have either one or no common point. In the latter case they are *parallel*. If two straight lines do not lie in one plane, then they are called *skew*.

An angle between two skew lines is measured by means of the angle between two lines parallel to them passing through one point (Fig. 100). The distance between two skew lines is determined by the segment of the line perpendicular to both of them.

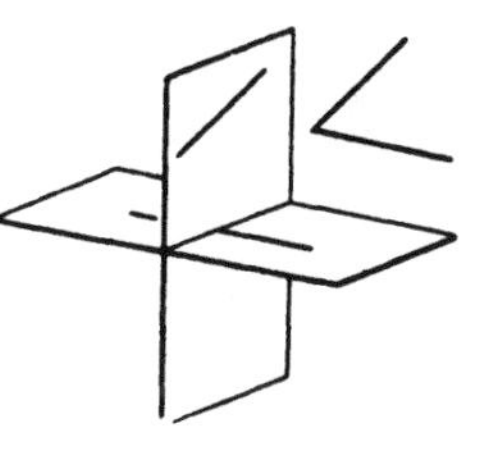
FIG. 100

Two planes either intersect in a straight line or have no common points, and then they are called *parallel*. If two planes are perpendicular to one straight line or if one plane contains two intersecting straight lines respectively parallel to two lines on the other plane, then the planes are parallel.

A straight line and a plane. A straight line can either lie on a plane or have one common point or have no common points with it. In the latter case, the line is parallel to the plane. The

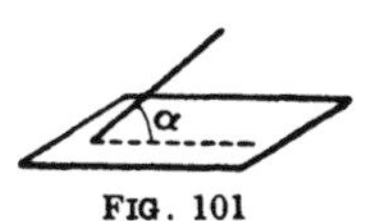

FIG. 101

angle between a straight line and a plane is the angle between the line and its projection on the plane (Fig. 101). If a straight line is perpendicular to two intersecting straight lines lying on a plane, then it is perpendicular to any straight line in the given plane (it is perpendicular to the plane).

3. Angles in space

A dihedral angle is a figure formed by two half-planes having a common edge. It is measured by means of its linear angle ABC (Fig. 102), i.e., the angle between two perpendiculars to the edge DE drawn from a point B of the edge on both faces of the dihedral angle.

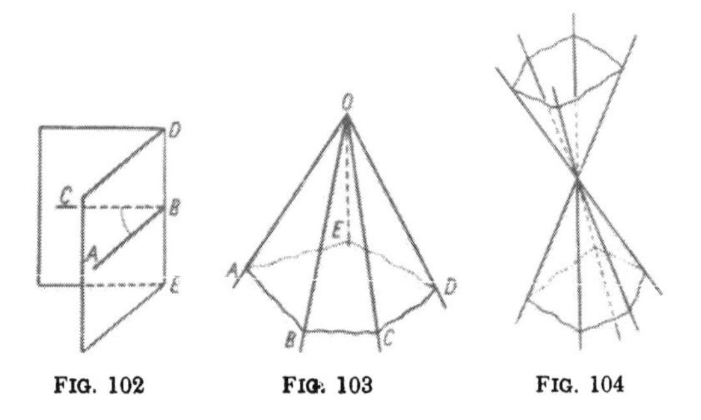

FIG. 102 FIG. 103 FIG. 104

Polyhedral angle $OABCDE$ (Fig. 103) is formed by several plane angles with a common vertex O and having pairwise a side in common. The plane angles with the common vertex are called the *faces* and their sides are called the *edges* of the polyhedral angle. Two consecutive edges of a polyhedral angle form a *plane angle* (a face) and two adjacent faces form a *dihedral angle* of the polyhedral angle. Two polyhedral angles are equal, if one of them can be made to coincide with the other by a rigid motion; a sufficient condition for that is that the corresponding plane and dihedral angles should be equal. If the corresponding elements of two polyhedral angles are equal in the reverse order then the angles are symmetrical, i.e., they can be brought in the position as in Fig. 104.

A *convex* polyhedral angle lies wholly on one side of each of its faces. The sum of plane angles of any convex polyhedral angle is less than 360°.

Two trihedral angles are equal, if: (1) two plane angles and the included dihedral angle of one are equal and equally placed respectively to two plane angles and the included dihedral angle of the other, or (2) two dihedral angles and the adjacent plane angle of one are equal and equally placed respectively to two dihedral angles and the adjacent plane angle of the other, or (3) three plane angles of one are equal and equally placed respectively to three plane angles of the other, or (4) three dihedral angles of one are equal and equally placed respectively to three dihedral angles of the other.

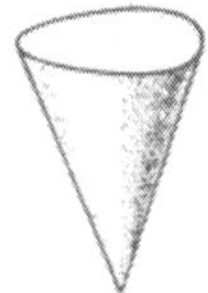

FIG. 105

A solid angle is a part of the space bounded by straight lines issuing from one point (the *vertex*) to all points of an arbitrary closed curve (Fig. 105). It characterizes the angle of seeing in which this curve is seen. The *measure* of a solid angle is the area cut out by the solid angle from the unit sphere having its centre in the vertex. For example, for a cone with the angle 120° at the vertex, the solid angle is equal to π (see formulas on p. 208).

4. Polyhedrons

Notations: V—volume, S—total area, M—lateral area, h—altitude, F—area of a base.

A polyhedron is a solid bounded by planes.

Prism (Fig. 106). The *bases* are equal polygons; the *lateral faces* are parallelograms. A prism is *right*, if its edge is perpendicular to the base. A prism is *regular*, if it is right and has regular polygons as bases.

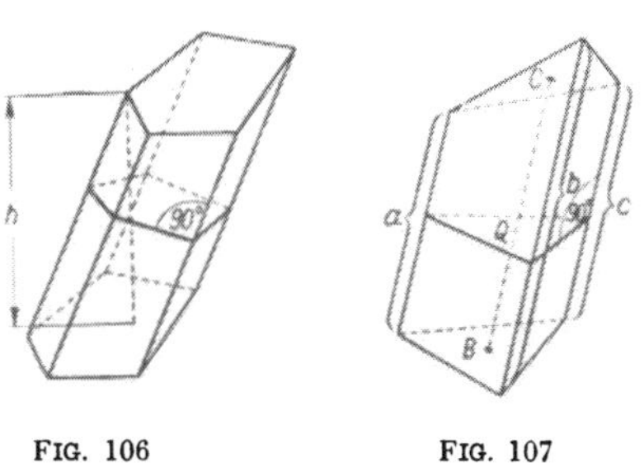

FIG. 106 FIG. 107

$M = pl$, where $l =$ an edge and $p =$ the perimeter of a section by a plane perpendicular to an edge. $S = M + 2F$, $V = Fh$.

For a truncated frustum of a triangular prism, $V = \frac{1}{3}(a + b + c)Q$ (Fig. 107), where a, b, c are lengths of the lateral edges and Q is the area of a perpendicular section. If a truncated frustum of a prism has n lateral faces, then $V = lQ$, where l is the length of the segment BC joining the centres of gravity of both bases and Q is the area of a section perpendicular to this segment.

A parallelepiped (Fig. 108) is a prism whose bases are parallelograms. All four diagonals of a parallelepiped intersect at one point and bisect at it.

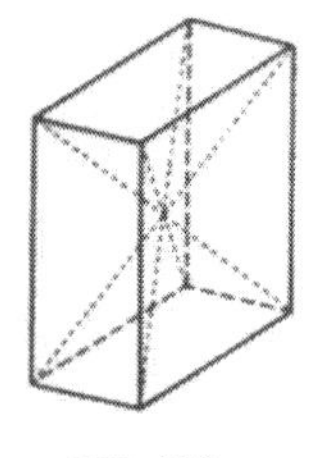

FIG. 108

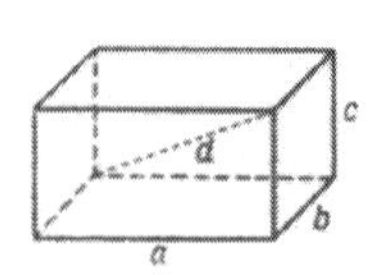

FIG. 109

A rectangular parallelepiped is a right prism whose bases (and sides) are rectangles. All diagonals of a rectangular parallelepiped (Fig. 109) are equal. If a, b, c are the edges of a rectangular parallele-

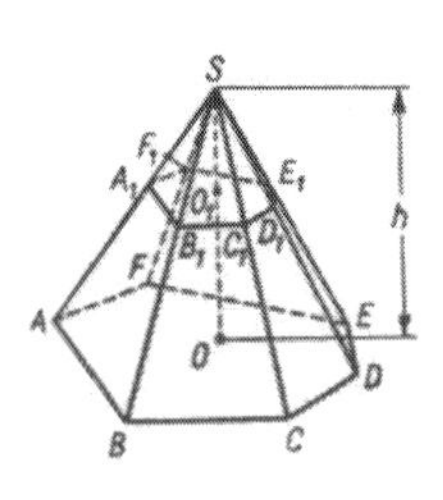

FIG. 110

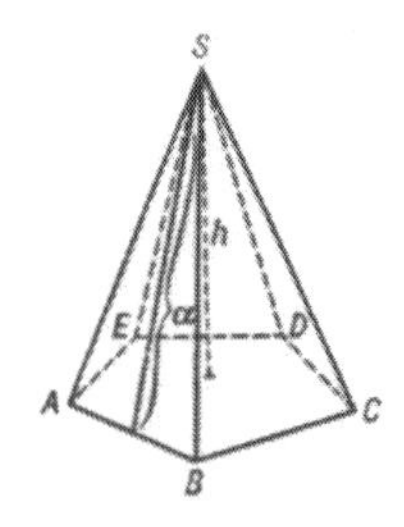

FIG. 111

piped and d is the diagonal, then $d^2 = a^2 + b^2 + c^2$, $V = abc$, $S = 2(ab + bc + ca)$.

A cube is a rectangular parallelepiped with equal edges: $a = b = c$, $V = a^3$, $S = 6a^2$, $d^2 = 3a^2$.

Pyramid (Fig. 110). Its base is an arbitrary polygon and its *lateral faces* are triangles meeting at a common *vertex*. A pyramid is called *n-angular*, if it has n lateral faces (it has $n+1$ faces together with the base). $V = \frac{1}{3}Fh$.

If a pyramid is cut by a plane parallel to the base, then

$$\frac{SA_1}{A_1A} = \frac{SB_1}{B_1B} = \frac{SC_1}{C_1C} = \dots = \frac{SO_1}{O_1O};$$

$$\frac{\text{Area of } ABCDEF}{\text{Area of } A_1B_1C_1D_1E_1F_1} = \left(\frac{SO}{SO_1}\right)^2,$$

where SO is the altitude of the pyramid (the perpendicular drawn from the vertex to the base).

A regular pyramid (Fig. 111) is one whose base is a regular polygon and whose altitude passes through the centre of the base. For a regular pyramid, $M = \frac{1}{2}p\alpha$, where p is the perimeter of the base, and α is the apothem of the regular pyramid (the altitude of any of its lateral faces).

A tetrahedron is a triangular pyramid (Fig. 112). If $OA = a$, $OB = b$, $OC = c$, $BC = p$, $CA = q$, $AB = r$, then (¹)

$$V^2 = \frac{1}{288}\begin{vmatrix} 0 & r^2 & q^2 & a^2 & 1 \\ r^2 & 0 & p^2 & b^2 & 1 \\ q^2 & p^2 & 0 & c^2 & 1 \\ a^2 & b^2 & c^2 & 0 & 1 \\ 1 & 1 & 1 & 1 & 1 \end{vmatrix}.$$

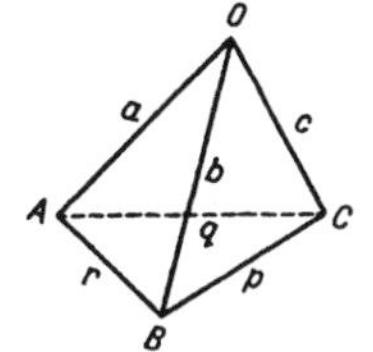

FIG. 112

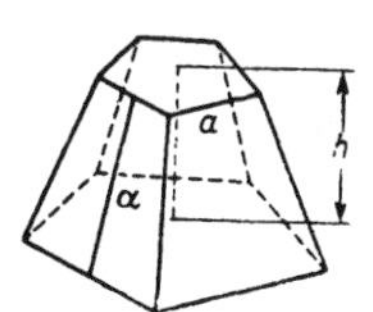

FIG. 113

Frustum of a pyramid (the cutting plane is parallel to the base, Fig. 113). If F and f are the areas of the bases, h is the altitude (distance between the bases), and a and A are two corresponding sides of the bases, then

$$V = \tfrac{1}{3}h\left[F + f + \sqrt{Ff}\right] = \tfrac{1}{3}hF\left[1 + \frac{a}{A} + \left(\frac{a}{A}\right)^2\right].$$

For a frustum of a regular pyramid, $M = \frac{1}{2}(P + p)\alpha$, where P and p are the perimeters of the bases and α is the apothem.

(¹) For determinants see p. 172.

Obelisk. An obelisk is a frustum of a pyramid whose bases are parallel rectangles and whose opposite lateral sides are equally sloping to the base, but do not intersect in one point (Fig. 114). If a, b, and a_1, b_1, are the sides of the bases and h is the altitude, then

$$V = \tfrac{1}{6}h\left[(2a + a_1)\,b + (2a_1 + a)\,b_1\right] = \tfrac{1}{6}h\left[ab + (a + a_1)(b + b_1) + a_1b_1\right].$$

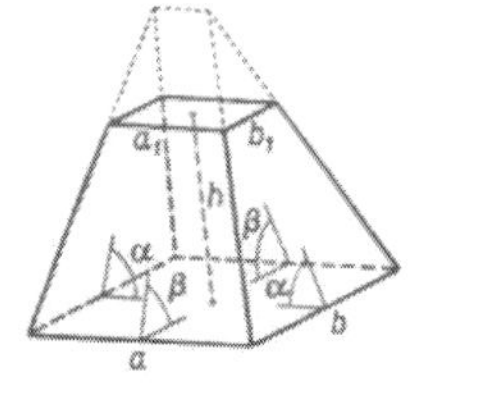

FIG. 114

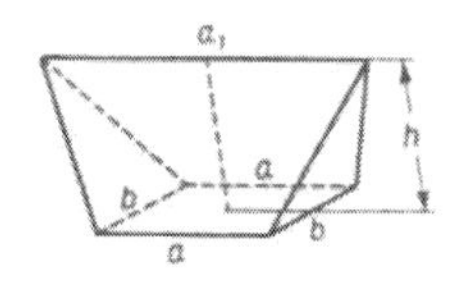

FIG. 115

Wedge. The base is a rectangle and the lateral faces are composed of two equilateral triangles and two equilateral trapezoids (Fig. 115).

$$V = \tfrac{1}{6}(2a + a_1)\,bh.$$

A regular polyhedron is a polyhedron all of whose faces are equal regular polygons and all of whose polyhedral angles are equal.

There exist five regular polyhedrons (Fig. 116). For details see the table.

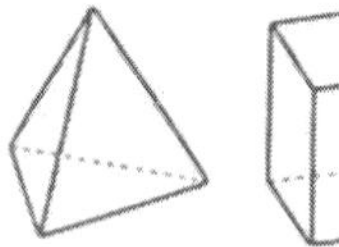 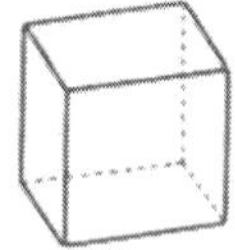 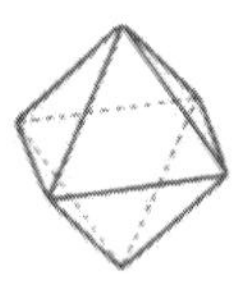 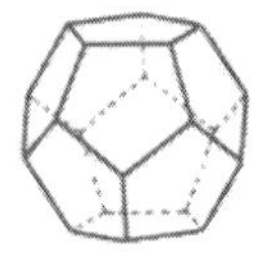 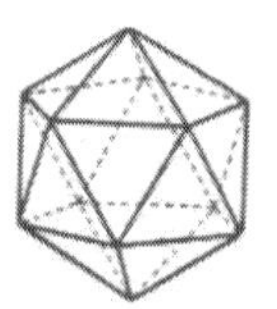

FIG. 116

Elements of regular polyhedrons (a = length of an edge)

Name	Number of faces and their shape	Number of edges	Number of vertices	Total area	Volume
Tetrahedron	4 triangles	6	4	$1.7321a^2$	$0.1179a^3$
Cube	6 squares	12	8	$6a^2$	a^3
Octahedron	8 triangles	12	6	$3.4641a^2$	$0.4714a^3$
Dodecahedron	12 pentagons	30	20	$20.6457a^2$	$7.6631a^3$
Icosahedron	20 triangles	30	12	$8.6603a^2$	$2.1817a^3$

Euler's theorem. If e is the number of vertices of a polyhedron, f—the number of faces and k—the number of edges, then $e - k + f = 2$ (provided that the polyhedron is convex or that it can be made convex by means of a continuous deformation). Examples—see in the table of regular polyhedrons.

5. Curvilinear solids

Notations: V—volume, S—total area, M—lateral area, h—altitude, F—area of a base.

Cylindrical surface (Fig. 117) is formed by a straight line (the *generator*) which moves parallel to a fixed direction while intersecting a fixed curve (the *directing curve*).

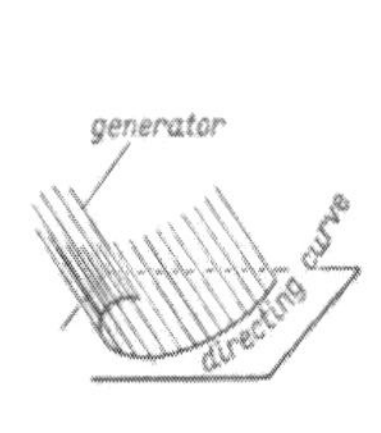

FIG. 117

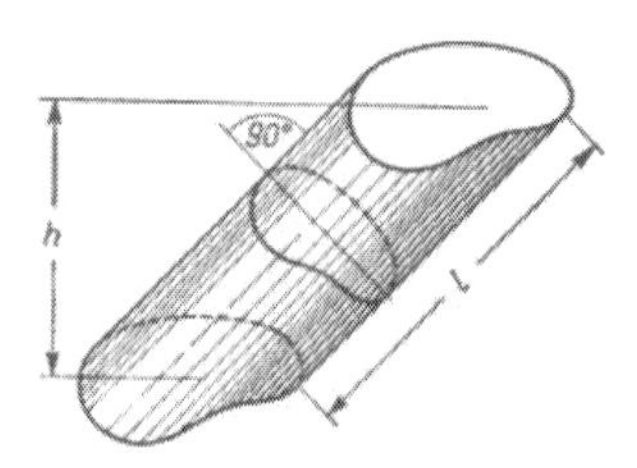

FIG. 118

A cylinder is a solid bounded by a cylindrical surface with a closed directing line and by two parallel planes which form the bases of the cylinder. For an arbitrary cylinder (Fig. 118),

$$M = ph = sl, \qquad V = Fh = Ql,$$

where p is the perimeter of a base, s is the perimeter of a section parallel to a generator, Q is the area of this section and l is the length of a generator.

Circular right cylinder has a circle as a base and its generators are perpendicular to the bases (Fig. 119). If R is the radius of a base, then

$$M = 2\pi Rh, \qquad S = 2\pi R(R + h), \qquad V = \pi R^2 h.$$

Truncated frustum of a cylinder (Fig. 120):

$$M = \pi R(h_1 + h_2), \qquad S = \pi R\left[h_1 + h_2 + R + \sqrt{R^2 + \tfrac{1}{4}(h_2 - h_1)^2}\right],$$

$$V = \pi R^2 \tfrac{1}{2}(h_1 + h_2).$$

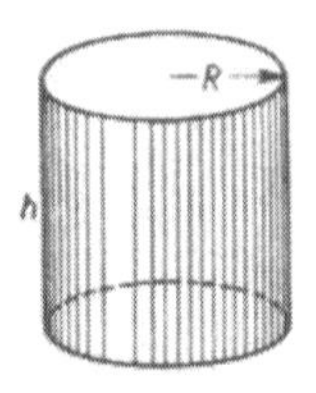

FIG. 119

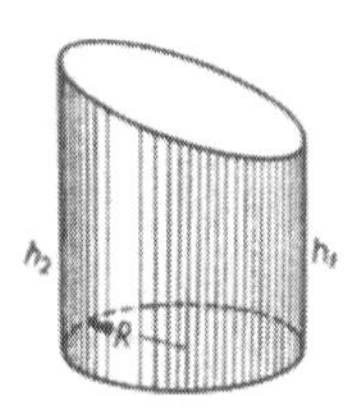

FIG. 120

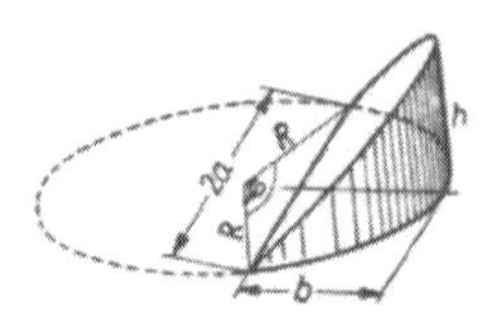

FIG. 121

Ungula of a cylinder (see Fig. 121; $\alpha = \frac{1}{2}\varphi$ in radians):

$$V = \frac{h}{3b}\left(a(3R^2 - a^2) + 3R^2(b - R)\alpha\right) = \frac{hR^3}{b}\left(\sin\alpha - \frac{\sin^3\alpha}{3} - \alpha\cos\alpha\right),$$

$$M = \frac{2Rh}{b}\left((b - R)\alpha + a\right).$$

Hollow cylinder (Fig. 122). R and r are the exterior and the interior radii, $\delta = R - r$, $\varrho = \frac{1}{2}(R + r)$ (mean radius):

$$V = \pi h(R^2 - r^2) = \pi h\delta(2R - \delta) = \pi h\delta(2r + \delta) = 2\pi h\delta\varrho.$$

A conical surface (Fig. 123) is a surface generated by a straight line (the *generator*) which passes through a fixed point (the *vertex*) and moves along a fixed curve (the *directing curve*).

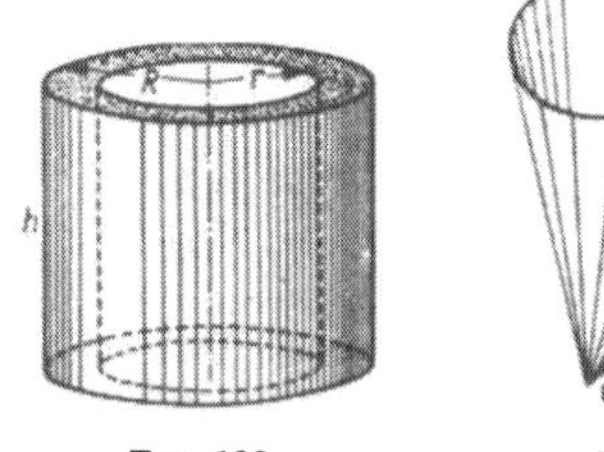

FIG. 122

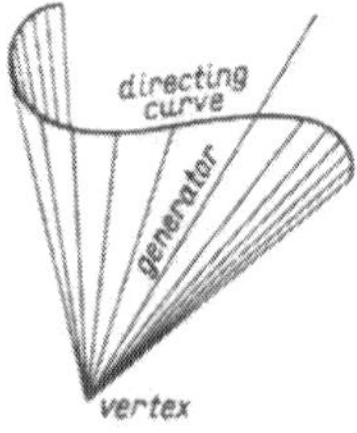

FIG. 123

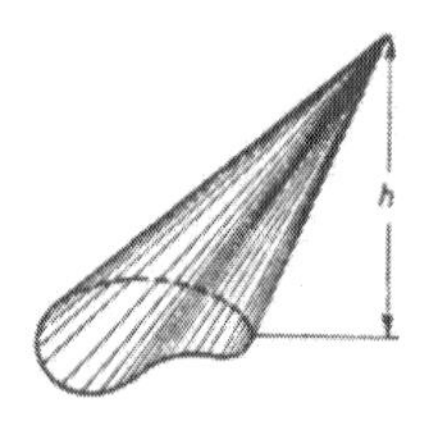

FIG. 124

A cone (Fig. 124) is bounded by a conical surface with a closed generating curve and by a plane which forms the base of the cone. For an arbitrary cone, $V = \frac{1}{3}hF$.

A right circular cone (Fig. 125) has a circle as its base and its altitude passes through the centre of the base circle. If l is the length of a generator and R is the radius of the base, then

$$M = \pi Rl = \pi R\sqrt{R^2 + h^2}, \qquad S = \pi R(R + l), \qquad V = \frac{1}{3}\pi R^2 h.$$

For a frustum of a right cone (Fig. 126):

$$l = \sqrt{h^2 + (R - r)^2}, \quad M = \pi l(R + r), \quad V = \frac{\pi h}{3}(R^2 + r^2 + Rr),$$

$$H = h + \frac{hr}{R - r} = \frac{hR}{R - r}.$$

Conic sections—see p. 253.

A sphere. R—radius, $D = 2R$—diameter (Fig. 127). Every plane section of a sphere is a circle. A *great circle* is a circle of radius R obtained as a section of the sphere by a plane passing

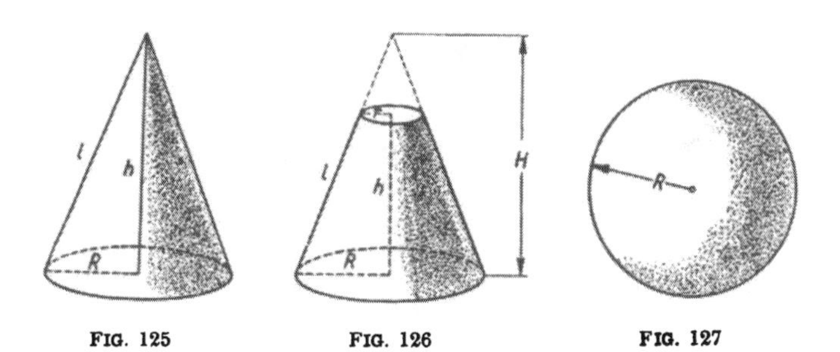

FIG. 125 FIG. 126 FIG. 127

through its centre. Any two points of a sphere (which are not opposite points of a diameter) determine an unique great circle passing through them. The less of two arcs of a great circle is the shortest distance on the sphere between two given points. For geometry on a sphere see pp. 226–227.

Area and volume of a sphere:

$$S = 4\pi R^2 \approx 12.57R^2, \quad S = \pi D^2 \approx 3.142D^2$$

$$S = \sqrt[3]{36\pi V^2} \approx 4.836\sqrt[3]{V^2},$$

$$V = \frac{4}{3}\pi R^3 \approx 4.189R^3, \quad V = \frac{\pi D^3}{6} \approx 0.5236D^3,$$

$$V = \frac{1}{6}\sqrt{\frac{S^3}{\pi}} \approx 0.09403\sqrt{S^3},$$

$$R = \frac{1}{2}\sqrt{\frac{S}{\pi}} \approx 0.2821\sqrt{S}, \quad R = \sqrt[3]{\frac{3V}{4\pi}} \approx 0.6204\sqrt[3]{V}.$$

Spherical sector (Fig. 128).

$$S = \pi R(2h + a), \quad V = \frac{2\pi R^2 h}{3}.$$

Spherical segment of one base (spherical cap) (Fig. 129).

$$a^2 = h(2R - h), \qquad M = 2\pi Rh = \pi(a^2 + h^2) = \pi l^2,$$

$$S = \pi(2Rh + a^2) = \pi(h^2 + 2a^2), \qquad V = \tfrac{1}{6}\pi h(3a^2 + h^2) = \tfrac{1}{3}\pi h^2(3R - h).$$

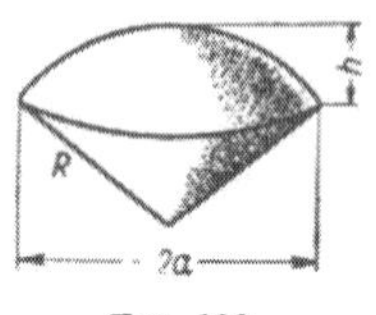

FIG. 128

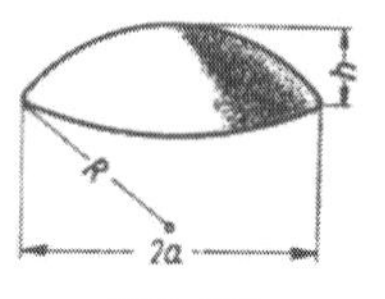

FIG. 129

Spherical segment of two bases (spherical zone) (Figs. 130 and 131):

$$R^2 = a^2 + \left(\frac{a^2 - b^2 - h^2}{2h}\right)^2, \qquad M = 2\pi Rh, \qquad S = \pi(2Rh + a^2 + b^2),$$

$$V = \tfrac{1}{6}\pi h(3a^2 + 3b^2 + h^2).$$

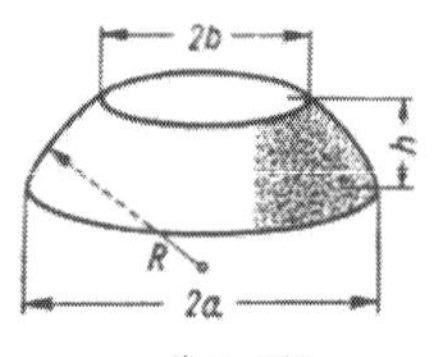

FIG. 130

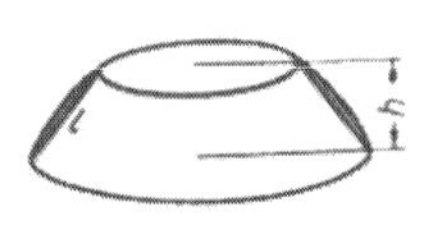

FIG. 131

If V_1 denotes the volume of the frustum of a cone inscribed in a spherical segment with two bases, then $V - V_1 = \frac{1}{6}\pi h l^2$, where l is a generator of the cone.

Torus (Fig. 132). A torus is a surface generated by a circle revolving about an axis lying in its plane and not intersecting the circle.

$$S = 4\pi^2 Rr \approx 39.48Rr, \qquad S = \pi^2 Dd \approx 9.870Dd,$$

$$V = 2\pi^2 Rr^2 \approx 19.74Rr^2, \qquad V = \tfrac{1}{4}\pi^2 Dd^2 \approx 2.467Dd^2.$$

Barrel or cask (Fig. 133). A barrel is generated by an arc of a circle revolving about an axis in its plane. Approximate formulas for its volume are

$$V = 0.262h(2D^2 + d^2) \qquad \text{or} \qquad V = 0.0873h(2D + d)^2.$$

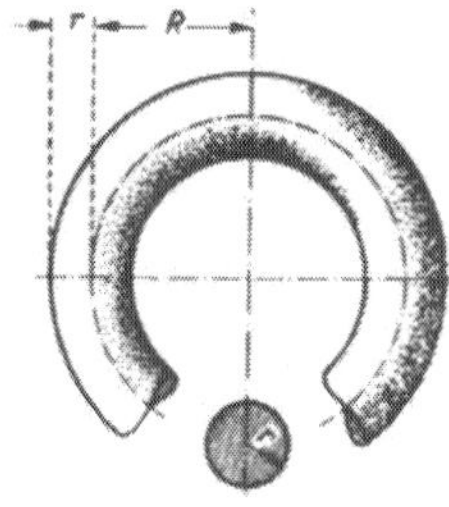

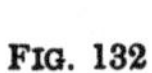
FIG. 132

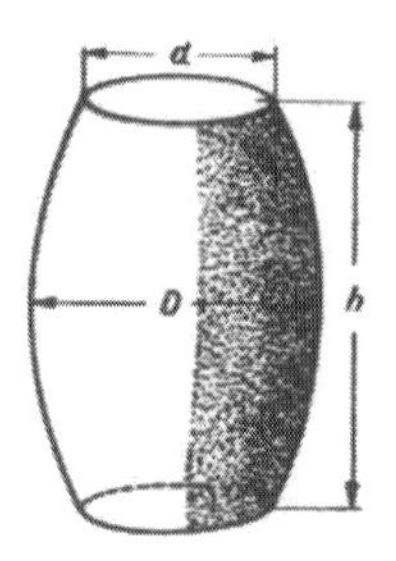

FIG. 133

For a barrel generated by an arc of a parabola

$$V = \frac{\pi h}{15}\left(2D^2 + Dd + \frac{3}{4}d^2\right) = 0.05236h\,(8D^2 + 4Dd + 3d^2).$$

IV. TRIGONOMETRY

A. PLANE TRIGONOMETRY

1. Trigonometric functions

Radian measure of angles. Besides the practical degree measure of angles, the *radian* or *circular measure* is also in use in theoretical considerations. A central angle α of an arbitrary circle subtended by an arc l is measured by the ratio of l to the radius r of the circle:

$$\alpha = \frac{l}{r}.$$

The unit angle of this measure is the angle subtended at the centre of a circle by an arc whose length is equal to the radius:

$$1 \text{ radian} = \frac{180^\circ}{\pi} = 57^\circ 17' 44.8''$$

$$\text{or} \quad = 57.2958^\circ = 3437.75' = 206264.8''.$$

The transition from one measure to the other is done according to the formulas

$$\alpha^\circ = \frac{180^\circ}{\pi}\alpha \text{ (radians)}\,(^1), \qquad \alpha \text{ (radians)} = \frac{\pi}{180}\alpha^\circ.$$

In particular, $360^\circ = 2\pi$, $180^\circ = \pi$, $90^\circ = \frac{1}{2}\pi$, $270^\circ = \frac{3}{2}\pi$ and so on. For a table of changing degrees into radians see pp. 82–83.

Definitions. *Trigonometric functions* are defined by aid of a trigonometric circle of radius 1 [2] or, for acute angles, by aid of a right triangle (Fig. 134a and b).

---

[1] There is no special symbol for a radian; the angle equal to α radians is denoted simply by α.

[2] The angle α is measured from the fixed radius OA to the moving radius OC in the counter-clockwise direction (the positive direction).

The *sine* $\sin\alpha = BC = \frac{a}{c}$,

The *cosine* $\cos\alpha = OB = \frac{b}{c}$,

The *tangent* $\tan\alpha = AD = \frac{a}{b}$,

The *cotangent* $\cot\alpha = EF = \frac{b}{a}$,

The *secant* $\sec\alpha = OD = \frac{c}{b}$,

The *cosecant* $\operatorname{cosec}\alpha = OF = \frac{c}{a}$.

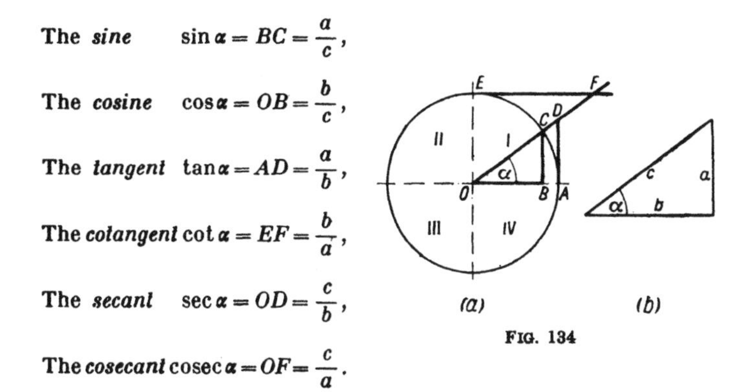

FIG. 134

Signs. The trigonometric functions are attributed definite signs depending on that in what quadrant the moving radius OC lies (Fig. 134a), as follows:

Quadrant	Angle	sin	cos	tan	cot	sec	cosec
I	from 0° to 90°	+	+	+	+	+	+
II	from 90° to 180°	+	−	−	−	−	+
III	from 180° to 270°	−	−	+	+	−	−
IV	from 270° to 360°	−	+	−	−	+	−

The range of values of trigonometric functions.

The sine and the cosine: from -1 to $+1$,

the tangent and the cotangent: from $-\infty$ to $+\infty$,

the secant and the cosecant: from $-\infty$ to -1 and from $+1$ to $+\infty$.

The values of trigonometric function for multiples of 30° and 45° are given in the table on p. 215.

The variation of the trigonometric functions for angles increasing from 0° to 360° is illustrated by the graphs in Fig. 135 [1].

The values of trigonometric functions for an arbitrary angle can be found from the following rules:

(1) If the angle is greater than 360° then the function is reduced to a function of an angle between 0° and 360° (and the tangent and the cotangent to an angle between 0° and 180°) according to the formulas (n denotes an integral number):

[1]) The graph of the sine is the usual sine curve (see p. 111).

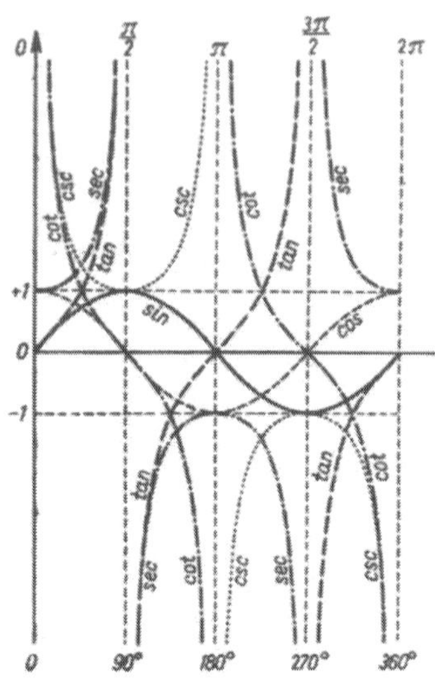

FIG. 135

$$\sin(360°n + \alpha) = \sin\alpha,$$
$$\cos(360°n + \alpha) = \cos\alpha,$$
$$\tan(180°n + \alpha) = \tan\alpha,$$
$$\cot(180°n + \alpha) = \cot\alpha.$$

(2) If the angle is negative, then the function is reduced to a function of a positive angle by the formulas:

$$\sin(-\alpha) = -\sin\alpha,$$
$$\cos(-\alpha) = \cos\alpha,$$
$$\tan(-\alpha) = -\tan\alpha,$$
$$\cot(-\alpha) = -\cot\alpha.$$

(3) If $90° < \alpha < 360°$, then the function is reduced to a function of an acute angle by the following *reducing formulas*:

The function	$\beta = 90° \pm \alpha$	$\beta = 180° \pm \alpha$	$\beta = 270° \pm \alpha$	$\beta = 360° - \alpha$
$\sin\beta$	$+\cos\alpha$	$\mp\sin\alpha$	$-\cos\alpha$	$-\sin\alpha$
$\cos\beta$	$\mp\sin\alpha$	$-\cos\alpha$	$\pm\sin\alpha$	$+\cos\alpha$
$\tan\beta$	$\mp\cot\alpha$	$\pm\tan\alpha$	$\mp\cot\alpha$	$-\tan\alpha$
$\cot\beta$	$\mp\tan\alpha$	$\pm\cot\alpha$	$\mp\tan\alpha$	$-\cot\alpha$

(4) If the angle is acute: $0° < \alpha < 90°$, then the value of the function can be found from tables (pp. 56–61).

For example,

$$\sin(-1000°) = -\sin 1000° = -\sin(2 \cdot 360° + 280°)$$
$$= -\sin 280° = +\cos 10° = +0.9848 \text{ (}^1\text{)}.$$

(1) Values of the functions of angles expressed in radians can be obtain from the tables on pp. 61–65; they give the values of the functions for arguments contained between 0.00 and 1.60. If an angle falls beyond the limits of the table, we use the same rules and reducing formulas as for the angles measured in degrees, e.g.,

$$\sin(2\pi + x) = \sin x, \quad \sin(2\pi - x) = -\sin x.$$

Values of trigonometric functions

for multiples of the angles 30° and 45° ($\frac{1}{6}\pi$ and $\frac{1}{4}\pi$)

Function	Angles of the first quadrant					Angles of the second quadrant			
	0°	30°	45°	60°	90°	120°	135°	150°	180°
	0	$\frac{1}{6}\pi$	$\frac{1}{4}\pi$	$\frac{1}{3}\pi$	$\frac{1}{2}\pi$	$\frac{2}{3}\pi$	$\frac{3}{4}\pi$	$\frac{5}{6}\pi$	π
sin	0	$\frac{1}{2}$	$\frac{\sqrt{2}}{2}$	$\frac{\sqrt{3}}{2}$	1	$\frac{\sqrt{3}}{2}$	$\frac{\sqrt{2}}{2}$	$\frac{1}{2}$	0
cos	1	$\frac{\sqrt{3}}{2}$	$\frac{\sqrt{2}}{2}$	$\frac{1}{2}$	0	$-\frac{1}{2}$	$-\frac{\sqrt{2}}{2}$	$-\frac{\sqrt{3}}{2}$	-1
tan	0	$\frac{\sqrt{3}}{3}$	1	$\sqrt{3}$	$\pm\infty$	$-\sqrt{3}$	-1	$-\frac{\sqrt{3}}{3}$	0
cot	$\mp\infty$	$\sqrt{3}$	1	$\frac{\sqrt{3}}{3}$	0	$-\frac{\sqrt{3}}{3}$	-1	$-\sqrt{3}$	$\mp\infty$
sec	1	$\frac{2\sqrt{3}}{3}$	$\sqrt{2}$	2	$\pm\infty$	-2	$-\sqrt{2}$	$-\frac{2\sqrt{3}}{3}$	-1
cosec	$\mp\infty$	2	$\sqrt{2}$	$\frac{2\sqrt{3}}{3}$	1	$\frac{2\sqrt{3}}{3}$	$\sqrt{2}$	2	$\pm\infty$

Function	Angles of the third quadrant					Angles of the fourth quadrant			
	180°	210°	225°	240°	270°	300°	315°	330°	360°
	π	$1\frac{1}{6}\pi$	$1\frac{1}{4}\pi$	$1\frac{1}{3}\pi$	$1\frac{1}{2}\pi$	$1\frac{2}{3}\pi$	$1\frac{3}{4}\pi$	$1\frac{5}{6}\pi$	2π
sin	0	$-\frac{1}{2}$	$-\frac{\sqrt{2}}{2}$	$-\frac{\sqrt{3}}{2}$	-1	$-\frac{\sqrt{3}}{2}$	$-\frac{\sqrt{2}}{2}$	$-\frac{1}{2}$	0
cos	-1	$-\frac{\sqrt{3}}{2}$	$-\frac{\sqrt{2}}{2}$	$-\frac{1}{2}$	0	$\frac{1}{2}$	$\frac{\sqrt{2}}{2}$	$\frac{\sqrt{3}}{2}$	1
tan	0	$\frac{\sqrt{3}}{3}$	1	$\sqrt{3}$	$\pm\infty$	$-\sqrt{3}$	-1	$-\frac{\sqrt{3}}{3}$	0
cot	$\mp\infty$	$\sqrt{3}$	1	$\frac{\sqrt{3}}{3}$	0	$-\frac{\sqrt{3}}{3}$	-1	$-\sqrt{3}$	$\mp\infty$
sec	-1	$-\frac{2\sqrt{3}}{3}$	$-\sqrt{2}$	-2	$\mp\infty$	2	$\sqrt{2}$	$\frac{2\sqrt{3}}{3}$	1
cosec	$\pm\infty$	-2	$-\sqrt{2}$	$-\frac{2\sqrt{3}}{3}$	-1	$-\frac{2\sqrt{3}}{3}$	$-\sqrt{2}$	-2	$\mp\infty$

2. Fundamental formulas of trigonometry

Functions of one angle:

$$\sin^2\alpha+\cos^2\alpha=1,\quad \frac{\sin\alpha}{\cos\alpha}=\tan\alpha,\quad \sin\alpha\cdot\operatorname{cosec}\alpha=1,$$

$$\sec^2\alpha-\tan^2\alpha=1,\qquad \cos\alpha\cdot\sec\alpha=1,$$

$$\operatorname{cosec}^2\alpha-\cot^2\alpha=1,\quad \frac{\cos\alpha}{\sin\alpha}=\cot\alpha,\quad \tan\alpha\cdot\cot\alpha=1.$$

One function expressed by another (of the same angle) ([1]):

$$\sin\alpha=\sqrt{1-\cos^2\alpha}=\frac{\tan\alpha}{\sqrt{1+\tan^2\alpha}}=\frac{1}{\sqrt{1+\cot^2\alpha}}=\frac{\sqrt{\sec^2\alpha-1}}{\sec\alpha}=\frac{1}{\operatorname{cosec}\alpha},$$

$$\cos\alpha=\sqrt{1-\sin^2\alpha}=\frac{1}{\sqrt{1+\tan^2\alpha}}=\frac{\cot\alpha}{\sqrt{1+\cot^2\alpha}}=\frac{1}{\sec\alpha}=\frac{\sqrt{\operatorname{cosec}^2\alpha-1}}{\operatorname{cosec}\alpha},$$

$$\tan\alpha=\frac{\sin\alpha}{\sqrt{1-\sin^2\alpha}}=\frac{\sqrt{1-\cos^2\alpha}}{\cos\alpha}=\frac{1}{\cot\alpha}=\sqrt{\sec^2\alpha-1}=\frac{1}{\sqrt{\operatorname{cosec}^2\alpha-1}},$$

$$\cot\alpha=\frac{\sqrt{1-\sin^2\alpha}}{\sin\alpha}=\frac{\cos\alpha}{\sqrt{1-\cos^2\alpha}}=\frac{1}{\tan\alpha}=\frac{1}{\sqrt{\sec^2\alpha-1}}=\sqrt{\operatorname{cosec}^2\alpha-1}.$$

Functions of sum and difference of angles:

$$\sin(\alpha\pm\beta)=\sin\alpha\cos\beta\pm\cos\alpha\sin\beta,$$

$$\cos(\alpha\pm\beta)=\cos\alpha\cos\beta\mp\sin\alpha\sin\beta,$$

$$\tan(\alpha\pm\beta)=\frac{\tan\alpha\pm\tan\beta}{1\mp\tan\alpha\tan\beta},\qquad \cot(\alpha\pm\beta)=\frac{\cot\alpha\cot\beta\mp1}{\cot\beta\pm\cot\alpha},$$

$$\sin(\alpha+\beta+\gamma)=\sin\alpha\cos\beta\cos\gamma+\cos\alpha\sin\beta\cos\gamma+\cos\alpha\cos\beta\sin\gamma-\\ -\sin\alpha\sin\beta\sin\gamma,$$

$$\cos(\alpha+\beta+\gamma)=\cos\alpha\cos\beta\cos\gamma-\sin\alpha\sin\beta\cos\gamma-\sin\alpha\cos\beta\sin\gamma-\\ -\cos\alpha\sin\beta\sin\gamma.$$

Functions of multiples of an angle:

$$\sin2\alpha=2\sin\alpha\cos\alpha,\qquad \sin3\alpha=3\sin\alpha-4\sin^3\alpha,$$

$$\cos2\alpha=\cos^2\alpha-\sin^2\alpha,\qquad \cos3\alpha=4\cos^3\alpha-3\cos\alpha,$$

---

[1] The sign "+" or "−" before the radicals should be chosen according to the quadrant in which the angle lies.

$$\sin 4\alpha = 8\cos^3\alpha\sin\alpha - 4\cos\alpha\sin\alpha,$$

$$\cos 4\alpha = 8\cos^4\alpha - 8\cos^2\alpha + 1,$$

$$\tan 2\alpha = \frac{2\tan\alpha}{1-\tan^2\alpha}, \quad \tan 3\alpha = \frac{3\tan\alpha - \tan^3\alpha}{1-3\tan^2\alpha},$$

$$\tan 4\alpha = \frac{4\tan\alpha - 4\tan^3\alpha}{1-6\tan^2\alpha+\tan^4\alpha},$$

$$\cot 2\alpha = \frac{\cot^2\alpha - 1}{2\cot\alpha}, \quad \cot 3\alpha = \frac{\cot^3\alpha - 3\cot\alpha}{3\cot^2\alpha - 1},$$

$$\cot 4\alpha = \frac{\cot^4\alpha - 6\cot^2\alpha + 1}{4\cot^3\alpha - 4\cot\alpha}.$$

It is more convenient to compute $\sin n\alpha$ and $\cos n\alpha$ for greater n from de Moivre's formula for complex numbers (p. 589)[1]:

$$\cos n\alpha + i\sin n\alpha = (\cos\alpha + i\sin\alpha)^n = \cos^n\alpha + in\cos^{n-1}\alpha\sin\alpha -$$

$$-\binom{n}{2}\cos^{n-2}\alpha\sin^2\alpha - i\binom{n}{3}\cos^{n-3}\alpha\sin^3\alpha + \binom{n}{4}\cos^{n-4}\alpha\sin^4\alpha + \ldots,$$

whence

$$\cos n\alpha = \cos^n\alpha - \binom{n}{2}\cos^{n-2}\alpha\sin^2\alpha + \binom{n}{4}\cos^{n-4}\alpha\sin^4\alpha -$$

$$-\binom{n}{6}\cos^{n-6}\alpha\sin^6\alpha + \ldots,$$

$$\sin n\alpha = n\cos^{n-1}\alpha\sin\alpha - \binom{n}{3}\cos^{n-3}\alpha\sin^3\alpha + \binom{n}{5}\cos^{n-5}\alpha\sin^5\alpha - \ldots$$

Functions of half angles:

$$\sin\frac{\alpha}{2} = \sqrt{\frac{1-\cos\alpha}{2}}, \quad \tan\frac{\alpha}{2} = \sqrt{\frac{1-\cos\alpha}{1+\cos\alpha}} = \frac{1-\cos\alpha}{\sin\alpha} = \frac{\sin\alpha}{1+\cos\alpha},$$

$$\cos\frac{\alpha}{2} = \sqrt{\frac{1+\cos\alpha}{2}}, \quad \cot\frac{\alpha}{2} = \sqrt{\frac{1+\cos\alpha}{1-\cos\alpha}} = \frac{1+\cos\alpha}{\sin\alpha} = \frac{\sin\alpha}{1-\cos\alpha}.$$

Sum and difference of functions:

$$\sin\alpha + \sin\beta = 2\sin\frac{\alpha+\beta}{2}\cos\frac{\alpha-\beta}{2},$$

$$\sin\alpha - \sin\beta = 2\cos\frac{\alpha+\beta}{2}\sin\frac{\alpha-\beta}{2},$$

[1] $\binom{n}{k}$ are Newton's binomial coefficients (see p. 193).

$$\cos\alpha + \cos\beta = 2\cos\frac{\alpha+\beta}{2}\cos\frac{\alpha-\beta}{2},$$

$$\cos\alpha - \cos\beta = -2\sin\frac{\alpha+\beta}{2}\sin\frac{\alpha-\beta}{2},$$

$$1+\cos\alpha = 2\cos^2\frac{\alpha}{2}, \qquad 1+\sin\alpha = 2\cos^2\left(\frac{\pi}{4}-\frac{\alpha}{2}\right) = 2\sin^2\left(\frac{\pi}{4}+\frac{\alpha}{2}\right),$$

$$1-\cos\alpha = 2\sin^2\frac{\alpha}{2}, \qquad 1-\sin\alpha = 2\sin^2\left(\frac{\pi}{4}-\frac{\alpha}{2}\right) = 2\cos^2\left(\frac{\pi}{4}+\frac{\alpha}{2}\right),$$

$$\tan\alpha \pm \tan\beta = \frac{\sin(\alpha\pm\beta)}{\cos\alpha\cos\beta}, \qquad \tan\alpha + \cot\beta = \frac{\cos(\alpha-\beta)}{\cos\alpha\sin\beta},$$

$$\cot\alpha \pm \cot\beta = \pm\frac{\sin(\alpha\pm\beta)}{\sin\alpha\sin\beta}, \qquad \cot\alpha - \tan\beta = \frac{\cos(\alpha+\beta)}{\sin\alpha\cos\beta}.$$

Product of functions:

$$\sin\alpha\sin\beta = \tfrac{1}{2}[\cos(\alpha-\beta) - \cos(\alpha+\beta)],$$

$$\cos\alpha\cos\beta = \tfrac{1}{2}[\cos(\alpha-\beta) + \cos(\alpha+\beta)],$$

$$\sin\alpha\cos\beta = \tfrac{1}{2}[\sin(\alpha-\beta) + \sin(\alpha+\beta)],$$

$$\sin\alpha\sin\beta\sin\gamma = \tfrac{1}{4}[\sin(\alpha+\beta-\gamma) + \sin(\beta+\gamma-\alpha) + \sin(\gamma+\alpha-\beta) - \sin(\alpha+\beta+\gamma)],$$

$$\sin\alpha\cos\beta\cos\gamma = \tfrac{1}{4}[\sin(\alpha+\beta-\gamma) - \sin(\beta+\gamma-\alpha) + \sin(\gamma+\alpha-\beta) - \sin(\alpha+\beta+\gamma)],$$

$$\sin\alpha\sin\beta\cos\gamma = \tfrac{1}{4}[-\cos(\alpha+\beta-\gamma) + \cos(\beta+\gamma-\alpha) + \cos(\gamma+\alpha-\beta) - \cos(\alpha+\beta+\gamma)],$$

$$\cos\alpha\cos\beta\cos\gamma = \tfrac{1}{4}[\cos(\alpha+\beta-\gamma) + \cos(\beta+\gamma-\alpha) + \cos(\gamma+\alpha-\beta) + \cos(\alpha+\beta+\gamma)].$$

Powers of functions:

$$\sin^2\alpha = \tfrac{1}{2}(1-\cos 2\alpha), \qquad \sin^3\alpha = \tfrac{1}{4}(3\sin\alpha - \sin 3\alpha),$$

$$\cos^2\alpha = \tfrac{1}{2}(1+\cos 2\alpha), \qquad \cos^3\alpha = \tfrac{1}{4}(\cos 3\alpha + 3\cos\alpha),$$

$$\sin^4\alpha = \tfrac{1}{8}(\cos 4\alpha - 4\cos 2\alpha + 3),$$

$$\cos^4\alpha = \tfrac{1}{8}(\cos 4\alpha + 4\cos 2\alpha + 3).$$

In computing $\sin^n\alpha$ and $\cos^n\alpha$ for greater n, the formulas of p. 217 for $\cos n\alpha$ and $\sin n\alpha$ can be successively applied.

3. Harmonic quantities

Definitions. In many problems of mechanics and physics one considers quantities depending on the time t according to the formula

$$(*) \qquad u = A \sin (\omega t + \varphi).$$

Such quantities are called *harmonic* and their variation in dependance on the time is called a *harmonic vibration*. The graph of the function $(*)$ is a general sine curve (Fig. 136) which differs from the usual sine curve ($y = \sin x$) as follows: (1) its *amplitude*, i.e., the maximal deviation from the t axis is equal to A, (2) its *period* T ("the *length of wave*") is equal to $2\pi/\omega$ (ω is called the *frequency of vibration*)(1), (3) its "*initial phase*" is the angle φ.

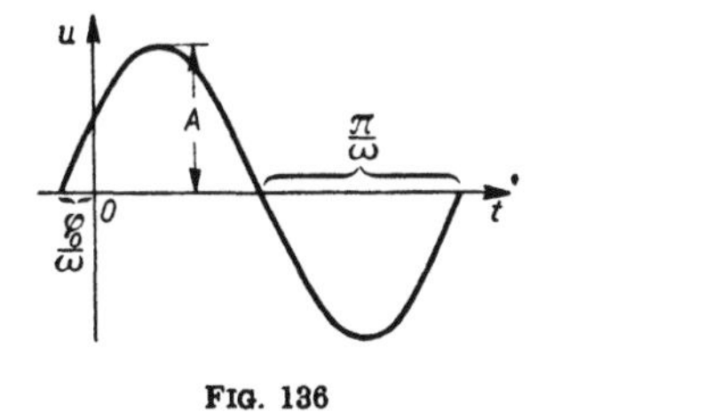

FIG. 136

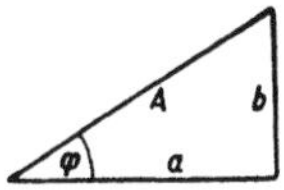

FIG. 137

The formula $(*)$ can be written in the form

$$(**) \qquad u = a \sin \omega t + b \cos \omega t,$$

where $A = \sqrt{a^2 + b^2}$, $\tan \varphi = b/a$. Relations between a, b, A and φ can be expressed as relations between elements of a right triangle (Fig. 137).

Operations with harmonic quantities. A sum of two harmonic quantities of the same frequency ω is also a harmonic quantity of the same frequency:

$$A_1 \sin (\omega t + \varphi_1) + A_2 \sin (\omega t + \varphi_2) = A \sin (\omega t + \varphi),$$

where

$$A = \sqrt{A_1^2 + A_2^2 + 2A_1 A_2 \cos (\varphi_2 - \varphi_1)},$$

$$\tan \varphi = \frac{A_1 \sin \varphi_1 + A_2 \sin \varphi_2}{A_1 \cos \varphi_1 + A_2 \cos \varphi_2};$$

(1) In the vibration theory this quantity is called the *cyclic* or *circular frequency*.

Linear combination of several harmonic quantities with a common frequency is also a harmonic quantity of the same frequency:

$$\sum c_i A_i \sin(\omega t + \varphi_i) = A \sin(\omega t + \varphi);$$

A and φ can be found graphically from a vector diagram.

Vector diagram of harmonic quantities. Harmonic quantities (*) or (**) can be represented on the plane as a radius vector $\mathbf{u}$ with polar coordinates $\varrho = A$ and φ or with Cartesian coordinates $x = a$ and $y = b$ (see pp. 235–236). A sum of two harmonic quantities is represented as the sum of vectors representing the sumands (Fig. 138) and a linear combination of harmonic quantities as the

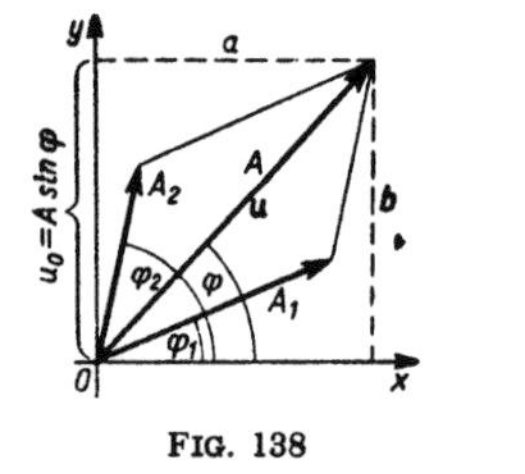

FIG. 138

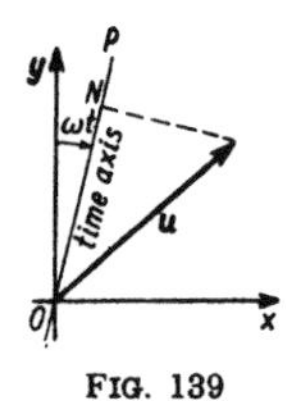

FIG. 139

linear combination of the corresponding vectors. Such representation of harmonic quantities is called a *vector diagram.*

The quantity u corresponding to a given time t can be found in the vector diagram as follows:

We draw the "*time axis*" OP (Fig. 139) which passes through the origin O and revolves about O with a constant angular velocity ω in the clockwise direction. At the initial time $t = 0$, the time axis coincides with the y axis. Then the projection ON of the vector $\mathbf{u}$ on the time axis gives, for a given time t, the value of the harmonic quantity $u = A \sin(\omega t + \varphi)$. At $t = 0$, $u_0 = A \sin \varphi$ is the projection of $\mathbf{u}$ on the y axis (Fig. 138).

4. Solution of triangles

Right triangle. Notation: a, b—catheti, c—hypotenuse, A, B—the angles opposite the sides a, b.

Fundamental relations:

$$a = c \sin A = c \cos B, \quad a = b \tan A = b \cot B.$$

Given	Formulas for the remaining elements
c, A	$B=90^\circ-A, \quad a=c\sin A, \quad b=c\cos A,$
a, A	$B=90^\circ-A, \quad b=a\cot A, \quad c=\frac{a}{\sin A},$
a, c	$\sin A=\frac{a}{c}, \quad b=c\cos A, \quad B=90^\circ-A,$
a, b	$\tan A=\frac{a}{b}, \quad c=\frac{a}{\sin A}, \quad B=90^\circ-A.$

Oblique triangle. Notation: a, b, c—sides, A, B, C—the corresponding opposite angles, S—area, R—radius of the circumscribed circle, r—radius of the inscribed circle, p—one half of the perimeter $\left(p=\frac{1}{2}(a+b+c)\right)$.

Fundamental relations:

1. $\dfrac{a}{\sin A}=\dfrac{b}{\sin B}=\dfrac{c}{\sin C}=2R$ (*"law of sines"*),

2. $a^2=b^2+c^2-2bc\cos A$ (*"law of cosines"*),

3. $\dfrac{a+b}{a-b}=\dfrac{\tan\frac{1}{2}(A+B)}{\tan\frac{1}{2}(A-B)}=\cot\frac{1}{2}C\cdot\cot\frac{1}{2}(A-B)$ (*"law of tangents"*),

4. $S=\frac{1}{2}ab\sin C=2R^2\sin A\sin B\sin C=pr$

$$=\sqrt{p(p-a)(p-b)(p-c)}.$$

Further relations:

$$\tan A=\frac{a\sin B}{c-a\cos B},$$

$$\sin\frac{A}{2}=\sqrt{\frac{(p-b)(p-c)}{bc}},$$

$$\cos\frac{A}{2}=\sqrt{\frac{p(p-a)}{bc}},$$

$$\tan\frac{A}{2}=\sqrt{\frac{(p-b)(p-c)}{p(p-a)}},$$

$$\frac{a+b}{c}=\frac{\cos\left[\frac{1}{2}(A-B)\right]}{\cos\left[\frac{1}{2}(A+B)\right]}=\frac{\cos\left[\frac{1}{2}(A-B)\right]}{\sin\frac{1}{2}C},$$

$$\frac{a-b}{c}=\frac{\sin\left[\frac{1}{2}(A-B)\right]}{\sin\left[\frac{1}{2}(A+B)\right]}=\frac{\sin\left[\frac{1}{2}(A-B)\right]}{\cos\frac{1}{2}C}.$$

Given	Formulas for the remaining elements
(1) One side and 2 angles (a, A, B)	$C = 180° - A - B, \quad b = \frac{a \sin B}{\sin A},$ $c = \frac{a \sin C}{\sin A}, \quad S = \frac{1}{2} ab \sin C$
(2) 2 sides and the included angle (a, b, C)	$\tan \frac{A-B}{2} = \frac{a-b}{a+b} \cot \frac{C}{2}, \quad \frac{A+B}{2} = 90° - \frac{1}{2} C.$ Having $A + B$ and $A - B$, we find A and B $c = \frac{a \sin C}{\sin A}, \quad S = \frac{1}{2} ab \sin C$
(3) 2 sides and the angle opposite one of them (a, b, A)	$\sin B = \frac{b \sin A}{a},$ if $a \geqslant b$, then $B < 90°$ and has only one value; if $a < b$, then (1) B has two values for $b \sin A < a$ $(B_2 = 180° - B_1)$ (2) B has only one value (90°) for $b \sin A = a$, (3) the triangle does not exist for $b \sin A > a$. $C = 180° - (A + B), \quad c = \frac{a \sin C}{\sin A}, \quad S = \frac{1}{2} ab \sin C$
(4) 3 sides (a, b, c)	$r = \sqrt{\frac{(p-a)(p-b)(p-c)}{p}},$ $\tan \frac{1}{2} A = \frac{r}{p-a}, \quad \tan \frac{1}{2} B = \frac{r}{p-b}, \quad \tan \frac{1}{2} C = \frac{r}{p-c},$ $S = rp = \sqrt{p(p-a)(p-b)(p-c)}$

Lengths of other segments of the triangle:

The altitude to the side a: $h_a = b \sin C = c \sin B$.

The median to the side a: $m_a = \frac{1}{2}\sqrt{b^2 + c^2 + 2bc \cos A}$.

The bisector of A: $l_A = \frac{2bc \cos \frac{1}{2} A}{b + c}$.

Radius of the circumscribed circle:

$$R = \frac{a}{2 \sin A} = \frac{b}{2 \sin B} = \frac{c}{2 \sin C}.$$

Radius of the inscribed circle:

$$r = \sqrt{\frac{(p-a)(p-b)(p-c)}{p}} = p \tan \tfrac{1}{2} A \tan \tfrac{1}{2} B \tan \tfrac{1}{2} C$$

$$= 4R \sin \tfrac{1}{2} A \sin \tfrac{1}{2} B \sin \tfrac{1}{2} C.$$

5. Inverse trigonometric functions

Definitions. *Inverse trigonometric functions* of x are defined by the following equalities:

$y = \text{Arc}\sin x$ (the *inverse sine*), if $x = \sin y$,
$y = \text{Arc}\cos x$ (the *inverse cosine*), if $x = \cos y$,
$y = \text{Arc}\tan x$ (the *inverse tangent*), if $x = \tan y$,
$y = \text{Arc}\cot x$ (the *inverse cotangent*), if $x = \cot y$.
(y is measured in radians).

Examples.

$$\text{Arc}\sin 0 = 0 \text{ or } \pi \text{ or } 2\pi; \text{ in general, } \text{Arc}\sin 0 = k\pi,$$

$$\text{Arc}\cos \tfrac{1}{2} = \tfrac{1}{3}\pi \text{ or } -\tfrac{1}{3}\pi \text{ or } \tfrac{1}{3}\pi + 2\pi; \text{ in general, } \text{Arc}\cos\tfrac{1}{2} = \pm\tfrac{1}{3}\pi + 2k\pi.$$

$$\text{Arc}\tan 1 = \tfrac{1}{4}\pi \text{ or } \tfrac{5}{4}\pi; \text{ in general, } \text{Arc}\tan 1 = \tfrac{1}{4}\pi + k\pi.$$

Principal values. The inverse trigonometric functions are *multiple-valued*. Their *principal values* (denoted: arc sin x, arc cos x, arc tan x, arc cot x) lie in the intervals:

$$-\tfrac{1}{2}\pi < \text{arc}\sin x < \tfrac{1}{2}\pi, \qquad 0 < \text{arc}\cos x < +\pi,$$

$$-\tfrac{1}{2}\pi < \text{arc}\tan x < +\tfrac{1}{2}\pi, \qquad 0 < \text{arc}\cot x < \pi.$$

For the graphs of the inverse trigonometric functions see p. 113; for the graphs of their principal values see Fig. 140.

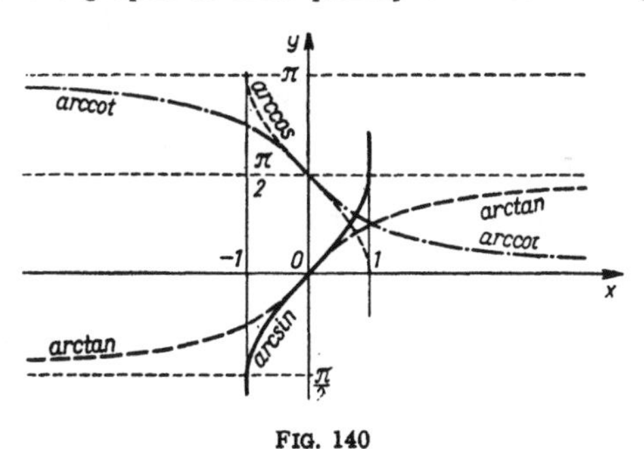

FIG. 140

Expression of one inverse trigonometric function by the other[1]:

$$\text{arc}\sin x = -\,\text{arc}\sin(-x) = \tfrac{1}{2}\pi - \text{arc}\cos x = \left[\text{arc}\cos\sqrt{1-x^2}\right]$$

$$= \text{arc}\tan\frac{x}{\sqrt{1-x^2}} = \left[\text{arc}\cot\frac{\sqrt{1-x^2}}{x}\right],$$

[1] These formulas are true only for the principal values of the inverse trigonometric functions and the formulas in rectangular brackets are true only for positive values of x (since the principal values lie in various intervals).

$$\arccos x = \pi - \arccos(-x) = \tfrac{1}{2}\pi - \arcsin x = \left[\arcsin\sqrt{1-x^2}\right]$$

$$= \left[\arctan\frac{\sqrt{1-x^2}}{x}\right] = \operatorname{arc\,cot}\frac{x}{\sqrt{1-x^2}},$$

$$\arctan x = -\arctan(-x) = \tfrac{1}{2}\pi - \operatorname{arc\,cot} x = \arcsin\frac{x}{\sqrt{1+x^2}}$$

$$= \left[\arccos\frac{1}{\sqrt{1+x^2}}\right] = \left[\operatorname{arc\,cot}\frac{1}{x}\right],$$

$$\operatorname{arc\,cot} x = \pi - \operatorname{arc\,cot}(-x) = \tfrac{1}{2}\pi - \arctan x$$

$$= \left[\arcsin\frac{1}{\sqrt{1+x^2}}\right] = \arccos\frac{x}{\sqrt{1+x^2}} = \left[\arctan\frac{1}{x}\right].$$

Fundamental relations between the inverse trigonometric functions:

$$\arcsin x + \arcsin y = \arcsin\left(x\sqrt{1-y^2} + y\sqrt{1-x^2}\right) \qquad (xy < 0 \text{ or } x^2 + y^2 < 1)$$

$$= \pi - \arcsin\left(x\sqrt{1-y^2} + y\sqrt{1-x^2}\right) \qquad (x > 0,\ y > 0 \text{ and } x^2 + y^2 > 1)$$

$$= -\pi - \arcsin\left(x\sqrt{1-y^2} + y\sqrt{1-x^2}\right) \qquad (x < 0,\ y < 0 \text{ and } x^2 + y^2 > 1),$$

$$\arcsin x - \arcsin y = \arcsin\left(x\sqrt{1-y^2} - y\sqrt{1-x^2}\right) \qquad (xy > 0 \text{ or } x^2 + y^2 < 1)$$

$$= \pi - \arcsin\left(x\sqrt{1-y^2} - y\sqrt{1-x^2}\right) \qquad (x > 0,\ y < 0 \text{ and } x^2 + y^2 > 1)$$

$$= -\pi - \arcsin\left(x\sqrt{1-y^2} - y\sqrt{1-x^2}\right) \qquad (x < 0,\ y > 0 \text{ and } x^2 + y^2 > 1),$$

$$\arccos x + \arccos y = \arccos\left(xy - \sqrt{1-x^2}\sqrt{1-y^2}\right) \qquad (x + y > 0)$$

$$= 2\pi - \arccos\left(xy - \sqrt{1-x^2}\sqrt{1-y^2}\right) \qquad (x + y < 0),$$

$$\arccos x - \arccos y = -\arccos\left(xy + \sqrt{1-x^2}\sqrt{1-y^2}\right) \qquad (x > y)$$

$$= \arccos\left(xy + \sqrt{1-x^2}\sqrt{1-y^2}\right) \qquad (x < y),$$

$$\arctan x + \arctan y = \arctan \frac{x+y}{1-xy} \qquad (xy < 1)$$

$$= \pi + \arctan \frac{x+y}{1-xy} \qquad (x > 0,\ xy > 1)$$

$$= -\pi + \arctan \frac{x+y}{1-xy} \qquad (x < 0,\ xy > 1),$$

$$\arctan x - \arctan y = \arctan \frac{x-y}{1+xy} \qquad (xy > -1)$$

$$= \pi + \arctan \frac{x-y}{1+xy} \qquad (x > 0,\ xy < -1)$$

$$= -\pi + \arctan \frac{x-y}{1+xy} \qquad (x < 0,\ xy < -1),$$

$$2 \arcsin x = \arcsin \left(2x\sqrt{1-x^2}\right) \qquad \left(|x| < \frac{1}{\sqrt{2}}\right)$$

$$= \pi - \arcsin \left(2x\sqrt{1-x^2}\right) \qquad \left(\frac{1}{\sqrt{2}} < x < 1\right)$$

$$= -\pi - \arcsin \left(2x\sqrt{1-x^2}\right) \qquad \left(-1 < x < -\frac{1}{\sqrt{2}}\right),$$

$$2 \arccos x = \arccos (2x^2 - 1) \qquad (0 < x < 1)$$

$$= 2\pi - \arccos (2x^2 - 1) \qquad (-1 < x < 0),$$

$$2 \arctan x = \arctan \frac{2x}{1-x^2} \qquad (|x| < 1)$$

$$= \pi + \arctan \frac{2x}{1-x^2} \qquad (x > 1)$$

$$= -\pi + \arctan \frac{2x}{1-x^2} \qquad (x < -1),$$

$$\cos(n \arccos x) = 2^{n-1} T_n(x) \qquad (n > 1)\ (^1),$$

where

$$T_n(x) = \frac{\left(x + \sqrt{x^2-1}\right)^n + \left(x - \sqrt{x^2-1}\right)^n}{2^n}.$$

When n is an integer, then $T_n(x)$ is a polynomial of x (*Tschebyscheff polynomial*).

---

(1) This formula holds also for non-integral values of n.

B. SPHERICAL TRIGONOMETRY

6. Geometry on a sphere

Geodesic lines on a sphere. A section of a sphere by a plane passing through its centre is a so-called *great circle* whose radius is equal to the radius of the sphere. Every two points A and B of the sphere determine (except for two opposite end points of a diameter of the sphere) a unique great circle passing through A and B. The lesser arc, AaB, of this great circle (Fig. 141) is the shortest of all lines (for example, AbB) on the sphere joining the points A and B, i.e., AaB is a *geodesic line* [1] on the sphere and plays on the surface of the sphere the same role as a straight line on the plane.

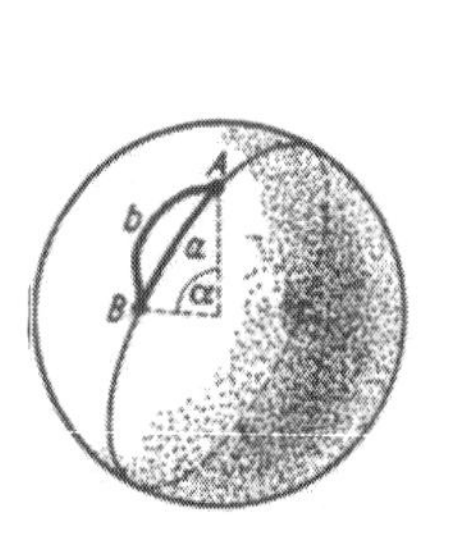

FIG. 141

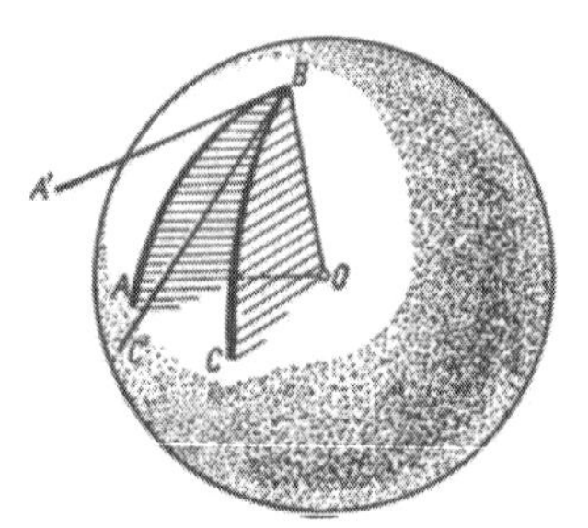

FIG. 142

Measurement of arcs and angles on a sphere. The length of an arc, $\smile a$, of a great circle with the central angle α (given in radians) is equal to $R\alpha$, where R is the radius of the sphere. For a given sphere, it is convenient to take R as a unit of measure of arcs; then $\smile a = \alpha$. In the following formulas, we assume this unit of measure.

A spherical angle ABC formed by two arcs of great circles (Fig. 142) is measured by the linear angle $A'BC'$ between the tangent to the corresponding arcs at the point B, that is to say, by the dihedral angle formed by the planes OBA and OBC.

Spherical triangles. Three great circles form several spherical triangles on the sphere. We consider this in the case of those triangles all of whose sides and angles are less than 180°. The

---

[1] See p. 312.

sides a, b, c of the triangle are measured by plane angles of the trihedral angle $OABC$ (Fig. 143), where O is the centre of the sphere, and the angles A, B, C of the triangle—by dihedral angles of the trihedral angle.

Fundamental property of a spherical triangle is that the sum of its angles $A + B + C$ is always greater than 180°. The difference $(A + B + C) - \pi = \delta$ expressed in radians is called the ***spherical excess*** of the given spherical triangle.

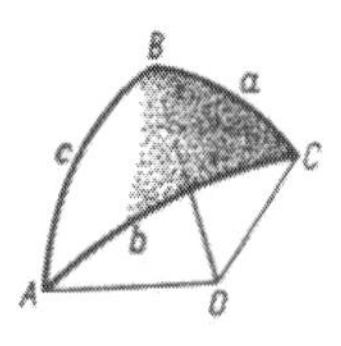

FIG. 143

FIG. 144

Area of a spherical triangle $S = R^2\delta$, where R is the radius of the sphere and δ is the spherical excess. *Area of a triangle* formed by two arcs of great circles (Fig. 144) is $S = 2R^2 A$ ($\sphericalangle A$ is expressed in radians).

7. Solution of spherical triangles

Right triangles. Notation: a, b—catheti, c—hypotenuse, A, B—the angles opposite a, b (Fig. 145).

Fundamental relations:

$$
\begin{array}{ll}
(1)\ \sin a = \sin c \sin A, & (6)\ \tan a = \tan c \cos B, \\
(2)\ \sin b = \sin c \sin B, & (7)\ \tan b = \tan c \cos A, \\
(3)\ \tan a = \sin b \tan A, & (8)\ \cos B = \cos b \sin A, \\
(4)\ \tan b = \sin a \tan B, & (9)\ \cos A = \cos a \sin B, \\
(5)\ \cos c = \cos a \cos b, & (10)\ \cos c = \cot A \cot B.
\end{array}
$$

Given	Numbers of formulas for determining the remaining elements		
A hypotenuse and an angle: c, A	a (1),	b (7),	B (10)
A cathetus and the opposite angle: a, A	b (3),	c (1),	B (9)
A cathetus and the adjacent angle: a, B	b (4),	c (6),	A (9)
Two catheti: a, b	c (5),	A (3),	B (4)
Two angles: A, B	a (9),	b (8),	c (10)

Formulas (1)–(10) can be obtained from the following *Neper's rule*: If we arrange the five elements of a spherical right triangle (except for the right angle) on a circle in the same ordering in

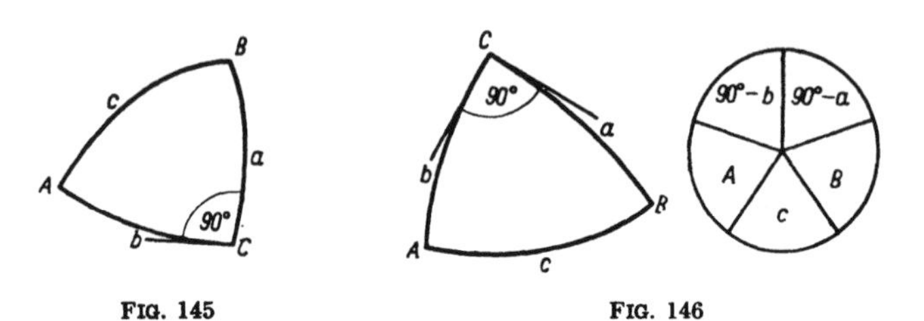

FIG. 145 FIG. 146

which they occur in the triangle, and replace there the sides a, b by their complements to 90° (Fig. 146), then

(1) The cosine of each element is equal to the product of the cotangents of two adjacent elements,

(2) The cosine of each element is equal to the product of the sines of two non-adjacent elements.

For example, $\cos A = \cot(90° - b)\cot c$, $\cos(90°-a) = \sin c \sin A$.

Oblique triangles. Notation: A, B, C—the angles of the triangle, a, b, c—the opposite sides (Fig. 147).

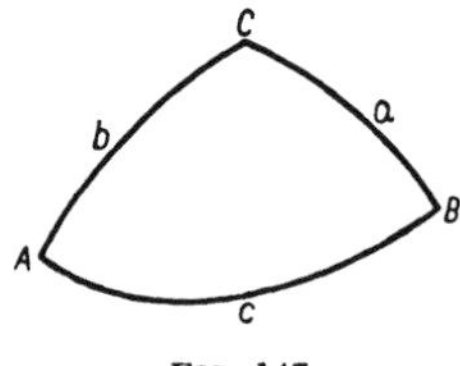

FIG. 147

Fundamental relations:

(1) $\dfrac{\sin a}{\sin A} = \dfrac{\sin b}{\sin B} = \dfrac{\sin c}{\sin C}$ (*"law of sines"*),

$$\left.\begin{aligned}&(2)\ \cos a = \cos b \cos c + \sin b \sin c \cos A\\&(3)\ \cos A = -\cos B \cos C + \sin B \sin C \cos a\end{aligned}\right\}\text{("laws of cosines")},$$

(4) $\sin a \cot b = \cot B \sin C + \cos a \cos C$,

(5) $\sin A \cot B = \cot b \sin c - \cos A \cos c$.

Given	Numbers of formulas for the remaining elements
Three sides: a, b, c	A (2), B and c (1)
Three angles: A, B, C	a (3), b and c (1)
Two sides and the included angle: a, b, C	B (4), A and c (1)
Two angles and the included side: A, B, c	b (5), a and C (1)
Two sides and the angle opposite one of them: a, b, B	A (1), c (5), C (1)
Two angles and the side opposite one of them: A, B, b	a (1), C (4), c (1)

C. HYPERBOLIC TRIGONOMETRY

8. Hyperbolic functions [1]

Definitions of hyperbolic functions. The *hyperbolic sine* (abbreviation: sinh), the *hyperbolic cosine* (cosh) and the *hyperbolic tangent* (tanh) are defined by the formulas:

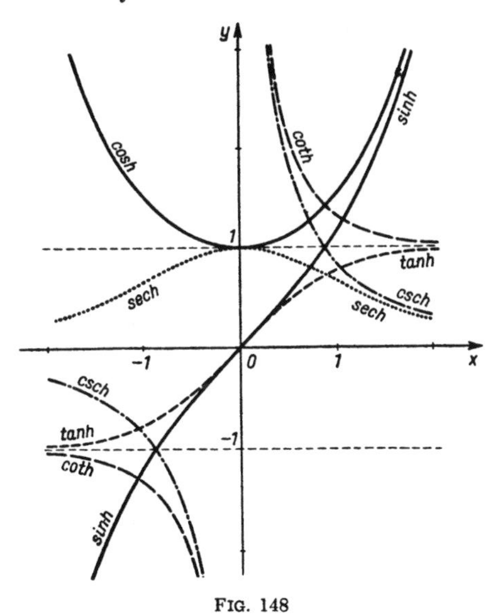

FIG. 148

[1] Elementary information about hyperbolic functions analogous to those concerning trigonometric functions is collected under this title.

$$\sinh x = \frac{e^x - e^{-x}}{2}, \quad \cosh x = \frac{e^x + e^{-x}}{2}, \quad \tanh x = \frac{e^x - e^{-x}}{e^x + e^{-x}}.$$

See below for a geometric definition of hyperbolic functions analogous to the definition of trigonometric functions (sine, cosine, tangent) (pp. 233–234).

The *hyperbolic cotangent, secant* and *cosecant* are defined as the reciprocals of the hyperbolic tangent, cosine and sine:

$$\coth x = \frac{1}{\tanh x} = \frac{e^x + e^{-x}}{e^x - e^{-x}},$$

$$\operatorname{sech} x = \frac{1}{\cosh x} = \frac{2}{e^x + e^{-x}}, \quad \operatorname{cosech} x = \frac{1}{\sinh x} = \frac{2}{e^x - e^{-x}}.$$

Variation of hyperbolic functions is illustrated by the graphs in Fig. 148. See also p. 156.

For the tables of hyperbolic functions see pp. 61–65.

9. Fundamental formulas of hyperbolic trigonometry

Formulas for the hyperbolic functions are analogous to those for trigonometric functions (pp. 216–218)(¹).

Functions of the same argument:

$$\cosh^2 x - \sinh^2 x = 1, \quad \operatorname{sech}^2 x + \tanh^2 x = 1, \quad \coth^2 x - \operatorname{cosech}^2 x = 1,$$

$$\tanh x \coth x = 1, \quad \frac{\sinh x}{\cosh x} = \tanh x, \quad \frac{\cosh x}{\sinh x} = \coth x.$$

Expression of one function by another of the same argument:

$$\sinh x = \sqrt{\cosh^2 x - 1} = \frac{\tanh x}{\sqrt{1 - \tanh^2 x}} = \frac{1}{\sqrt{\coth^2 x - 1}}$$

$$= \frac{\sqrt{1 - \operatorname{sech}^2 x}}{\operatorname{sech} x} = \frac{1}{\operatorname{cosech} x},$$

$$\cosh x = \sqrt{\sinh^2 x + 1} = \frac{1}{\sqrt{1 - \tanh^2 x}} = \frac{\coth x}{\sqrt{\coth^2 x - 1}}$$

$$= \frac{1}{\operatorname{sech} x} = \frac{\sqrt{1 + \operatorname{cosech}^2 x}}{\operatorname{cosech} x},$$

(¹) These relations can be deduced from the corresponding formulas for the trigonometric functions by means of a simple rule, see p. 232.

$$\tanh x = \frac{\sinh x}{\sqrt{\sinh^2 x + 1}} = \frac{\sqrt{\cosh^2 x - 1}}{\cosh x} = \frac{1}{\coth x}$$

$$= \sqrt{1 - \operatorname{sech}^2 x} = \frac{1}{\sqrt{1 + \operatorname{cosech}^2 x}},$$

$$\coth x = \frac{\sqrt{\sinh^2 x + 1}}{\sinh x} = \frac{\cosh x}{\sqrt{\cosh^2 x - 1}} = \frac{1}{\tanh x}$$

$$= \frac{1}{\sqrt{1 - \operatorname{sech}^2 x}} = \sqrt{\operatorname{cosech}^2 x + 1}\,.$$

Functions of a sum and difference of two arguments:

$$\sinh (x \pm y) = \sinh x \cosh y \pm \cosh x \sinh y,$$

$$\cosh (x \pm y) = \cosh x \cosh y \pm \sinh x \sinh y,$$

$$\tanh (x \pm y = \frac{\tanh x \pm \tanh y}{1 \pm \tanh x \tanh y},$$

$$\coth (x \pm y) = \frac{1 \pm \coth x \coth y}{\coth x \pm \coth y}\,.$$

Functions of a double argument:

$$\sinh 2x = 2 \sinh x \cosh x,$$

$$\cosh 2x = \sinh^2 x + \cosh^2 x,$$

$$\tanh 2x = \frac{2 \tanh x}{1 + \tanh^2 x}, \qquad \coth 2x = \frac{1 + \coth^2 x}{2 \coth x}\,.$$

De Moivre's formula (see p. 217):

$$(\cosh x \pm \sinh x)^n = \cosh nx \pm \sinh nx.$$

Functions of a half argument:

$$\sinh \tfrac{1}{2} x = \varepsilon \sqrt{\tfrac{1}{2}(\cosh x - 1)}, \qquad \tanh \tfrac{1}{2} x = \frac{\cosh x - 1}{\sinh x} = \frac{\sinh x}{\cosh x + 1},$$

$$\cosh \tfrac{1}{2} x = \sqrt{\tfrac{1}{2}(\cosh x + 1)}, \qquad \coth \tfrac{1}{2} x = \frac{\sinh x}{\cosh x - 1} = \frac{\cosh x + 1}{\sinh x},$$

where $\varepsilon = +1$, if $x > 0$ and $\varepsilon = -1$, if $x < 0$.

Sum and difference of functions:

$$\sinh x \pm \sinh y = 2 \sinh \tfrac{1}{2}(x \pm y) \cosh \tfrac{1}{2}(x \mp y),$$

$$\cosh x + \cosh y = 2 \cosh \tfrac{1}{2}(x + y) \cosh \tfrac{1}{2}(x - y),$$

$$\cosh x - \cosh y = 2 \sinh \tfrac{1}{2}(x + y) \sinh \tfrac{1}{2}(x - y),$$

$$\tanh x + \tanh y = \frac{\sinh (x + y)}{\cosh x \cosh y},$$

$$\tanh x - \tanh y = \frac{\sinh (x - y)}{\cosh x \cosh y}.$$

Relations between hyperbolic and trigonometric functions [1]:

$$\sin z = -i \sinh iz, \qquad \cos z = \cosh iz,$$

$$\sinh z = -i \sin iz, \qquad \cosh z = \cos iz,$$

$$\tan z = -i \tanh iz, \qquad \cot z = i \coth iz,$$

$$\tanh z = -i \tan iz, \qquad \coth z = i \cot iz.$$

Every relation between hyperbolic functions of the variable x or ax (but not $ax + b$) can be deduced from the corresponding relation for trigonometric functions (pp. 217–218) by replacing $\sin \alpha$ by $i \sinh x$ and $\cos \alpha$ by $\cosh x$. For example, $\cos^2 \alpha + \sin^2 \alpha = 1$, $\cosh^2 x + i^2 \sinh^2 x = 1$ or $\cosh^2 x - \sinh^2 x = 1$; $\sin 2\alpha = 2 \sin \alpha \cos \alpha$, $i \sinh 2x = 2i \sinh x \cosh x$ or $\sinh 2x = 2 \sinh x \cosh x$ and so on.

10. Inverse hyperbolic functions

Definitions. The *inverse hyperbolic functions* (*"area-functions"*) of the variable x are defined by the equalities

$y = \text{ar} \sinh x$ (*area-sine*), if $x = \sinh y$,

$y = \text{ar} \cosh x$ (*area-cosine*), if $x = \cosh y$,

$y = \text{ar} \tanh x$ (*area-tangent*), if $x = \tanh y$,

$y = \text{ar} \coth x$ (*area-cotangent*), if $x = \coth y$.

The origin of the names of the inverse hyperbolic functions is that they represent the *area* of a sector of hyperbola (see below).

Expressions by means of logarithms. According to the formulas on pp. 230–231, the inverse hyperbolic functions can be expressed by means of logarithms as follows:

$$\text{ar} \sinh x = \ln \left(x + \sqrt{x^2 + 1}\right),$$

$$\text{ar} \cosh x = \pm \ln \left(x + \sqrt{x^2 - 1}\right) \qquad (x > 1),$$

[1] For the functions of a complex variable see pp. 592–595.

$$\operatorname{ar\,tanh} x = \frac{1}{2} \ln \frac{1+x}{1-x} \qquad (|x| < 1),$$

$$\operatorname{ar\,coth} x = \frac{1}{2} \ln \frac{x+1}{x-1} \qquad (|x| > 1).$$

For the graphs of the inverse hyperbolic functions see pp. 115, 116.

Expressions of one function by another:

$$\operatorname{ar\,sinh} x = \varepsilon \operatorname{ar\,cosh} \sqrt{x^2+1} = \operatorname{ar\,tanh} \frac{x}{\sqrt{x^2+1}} = \operatorname{ar\,coth} \frac{\sqrt{x^2+1}}{x},$$

$$\operatorname{ar\,cosh} x = \varepsilon \operatorname{ar\,sinh} \sqrt{x^2-1} = \varepsilon \operatorname{ar\,tanh} \frac{\sqrt{x^2-1}}{x} = \varepsilon \operatorname{ar\,coth} \frac{x}{\sqrt{x^2-1}},$$

$$\operatorname{ar\,tanh} x = \operatorname{ar\,sinh} \frac{x}{\sqrt{1-x^2}} = \varepsilon \operatorname{ar\,cosh} \frac{1}{\sqrt{1-x^2}} = \operatorname{ar\,coth} \frac{1}{x},$$

$$\operatorname{ar\,coth} x = \operatorname{ar\,sinh} \frac{1}{\sqrt{x^2-1}} = \varepsilon \operatorname{ar\,cosh} \frac{x}{\sqrt{x^2-1}} = \operatorname{ar\,tanh} \frac{1}{x},$$

where $\varepsilon = +1$, if $x > 0$ and $\varepsilon = -1$, if $x < 0$.

Certain relations between inverse hyperbolic functions:

$$\operatorname{ar\,sinh} x \pm \operatorname{ar\,sinh} y = \operatorname{ar\,sinh} \left(x\sqrt{1+y^2} \pm y\sqrt{1+x^2}\right),$$

$$\operatorname{ar\,cosh} x \pm \operatorname{ar\,cosh} y = \operatorname{ar\,cosh} \left(xy \pm \sqrt{(x^2-1)(y^2-1)}\right),$$

$$\operatorname{ar\,tanh} x \pm \operatorname{ar\,tanh} y = \operatorname{ar\,tanh} \frac{x \pm y}{1 \pm xy}.$$

11. Geometric definition of hyperbolic functions

The functions $\sin \alpha$, $\cos \alpha$, $\tan \alpha$ have been defined in a trigonometric circle (p. 213) as the lengths of the segments BC, OB, AD (if $R = 1$), where the argument α is the central angle AOC. The argument α can be regarded as the area x (lined in Fig. 149) of the sector COK with the central angle 2α, for $x = \frac{1}{2}R^2 \cdot 2\alpha = \alpha$ ($R = 1$, α is measured in radians). Therefore, $\sin x = BC$, $\cos x = OB$, $\tan x = AD$.

If we consider the analogous areas not in the circle $x^2 + y^2 = 1$, but in the equilateral hyperbola $x^2 - y^2 = 1$ (only in its right-hand branch), and denote the area of the sector COK (lined in Fig. 150) by x, then the hyperbolic functions can be defined as follows: $\sinh x = BC$, $\cosh x = OB$, $\tanh x = AD$.

The area x can be evaluated by aid of the integral calculus (see p. 466); thus we obtain

$$x = \ln\left(BC + \sqrt{BC^2 + 1}\right) = \ln\left(OB + \sqrt{OB^2 - 1}\right) = \frac{1}{2}\ln\frac{1 + AD}{1 - AD},$$

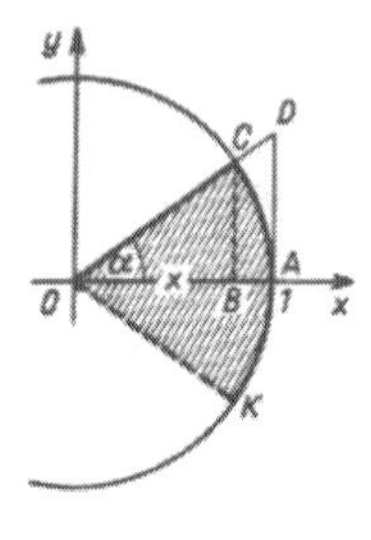

FIG. 149

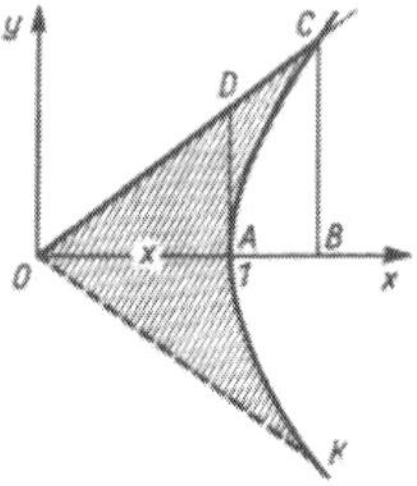

FIG. 150

whence we obtain the following expressions of hyperbolic functions by means of exponential functions (these formulas are also taken as definitions of hyperbolic functions):

$$BC = \frac{e^x - e^{-x}}{2} = \sinh x, \quad OB = \frac{e^x + e^{-x}}{2} = \cosh x,$$

$$AD = \frac{e^x - e^{-x}}{e^x + e^{-x}} = \tanh x.$$

PART THREE

ANALYTIC AND DIFFERENTIAL GEOMETRY

I. ANALYTIC GEOMETRY

A. GEOMETRY IN THE PLANE

1. Fundamental concepts and formulas

Coordinates. The position of an arbitrary point in the plane can be determined by aid of a coordinate system. The numbers which determine the position of a point are called its *coordinates*. The most frequently used are Cartesian rectangular coordinate system and polar coordinate system.

Cartesian rectangular coordinates of a point P (Fig. 151) are the distances, expressed in a certain unit of measure and taken with definite signs, of the point P from two perpendicular straight lines called the *coordinate axes*. The point O of intersection of the coordinate axes is called the *coordinate origin*. The horizontal axis is usually called the *axis of abscissae* (the *x axis*) and the vertical one—the *axis of ordinates* (the *y axis*). The coordinate axes are furnished with a certain positive direction; usually, the x axis is directed to the right and the y axis upwards. The signs of the coordinates of P depend on the half-axis on which the projection of P falls (see the scheme in Fig. 152). The coordinates x and y are called, respectively, the *abscissa* and the *ordinate* of P. The symbol $P(a, b)$ is used to mean the point with the abscissa a and the ordinate b.

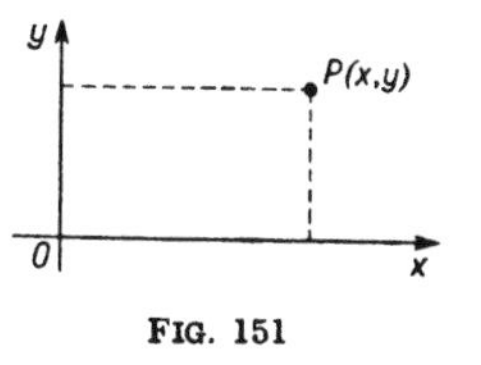

FIG. 151

	I	II	III	IV
x	+	−	−	+
y	+	+	−	−

FIG. 152

The *polar coordinates* of a point P (Fig. 153) are the *radius vector* ϱ, i.e., the distance of P from a fixed point O (the *pole* or the *origin*) and the *polar angle* φ, i.e., the angle between the radius vector OP and a fixed line passing through the pole (the *polar axis*). The polar angle φ is positive when measured counterclockwise and negative when measured clockwise.

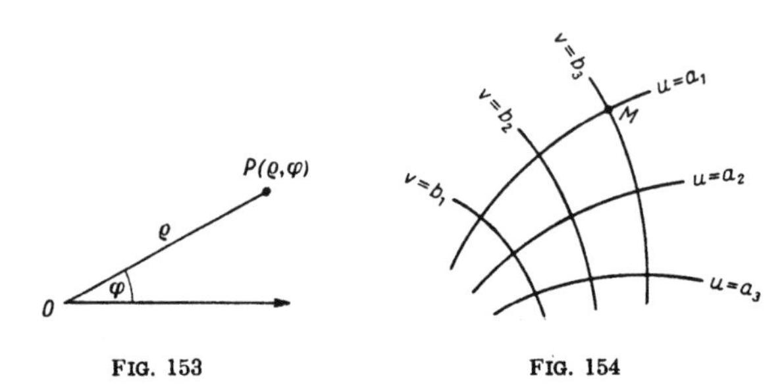

FIG. 153 FIG. 154

Curvilinear coordinates. A more general coordinate system is a curvilinear system introduced as follows. There are given two families of lines in the plane (*coordinate lines*) such that either of them depends on one parameter and that, through every point of the plane, there passes exactly one line of either family. Then the values of parameters corresponding to these lines at the point P are called the *curvilinear coordinates* of P. Thus the point M in Fig. 154 has the curvilinear coordinates $u = a_1$, $v = b_3$. The coordinate lines in the Cartesian coordinate system are straight lines parallel to the axes and in the polar coordinate system—the circles with the centre at the pole and the rays issuing from the pole.

Transformations of coordinates. In passing from one coordinate system to another, the coordinate change as follows:

Parallel translation of Cartesian coordinate axes (Fig. 155) (x, y—the old coordinates, x', y'—the new ones, a, b—coordinates of the new origin):

$$x = x' + a, \qquad y = y' + b,$$

$$x' = x - a, \qquad y' = y - b.$$

Rotation of the axes through the angle φ (1) (Fig. 156):

$$x = x' \cos\varphi - y' \sin\varphi, \qquad y = x' \sin\varphi + y' \cos\varphi,$$

$$x' = x \cos\varphi + y \sin\varphi, \qquad y' = -x \sin\varphi + y \cos\varphi.$$

(1) The angle φ is positive, when the axes rotate counterclockwise.

In a general case, a transformation can be decomposed into a parallel translation and a rotation of the axes.

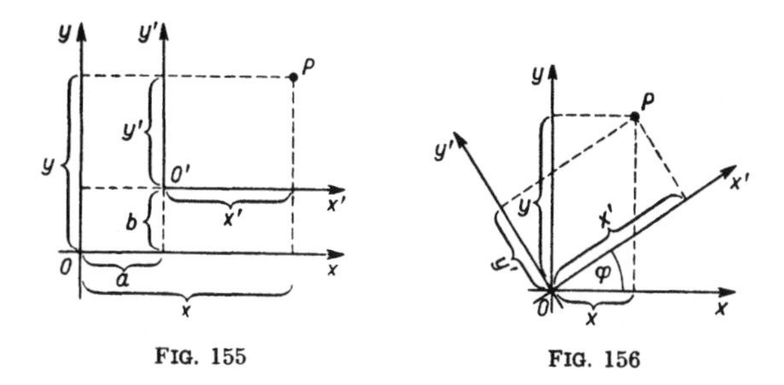

FIG. 155 FIG. 156

Transition from Cartesian coordinates to polar coordinates and conversely. If the origin of Cartesian coordinate system is taken as the pole and the x axis as the polar axis, then (Fig. 157)

$$x = \varrho \cos \varphi, \qquad y = \varrho \sin \varphi,$$

$$\varrho = \sqrt{x^2 + y^2}, \qquad \varphi = \operatorname{arc} \tan \frac{y}{x} = \operatorname{arc} \sin \frac{y}{\varrho}.$$

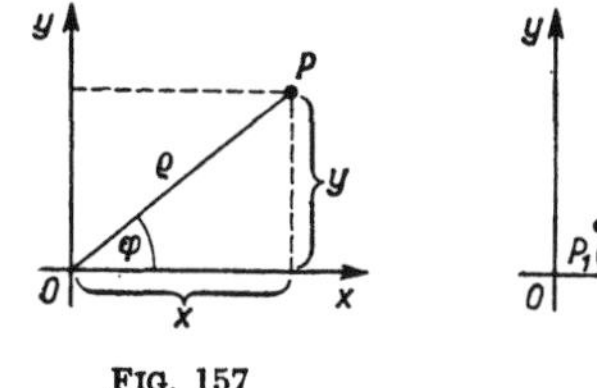

FIG. 157

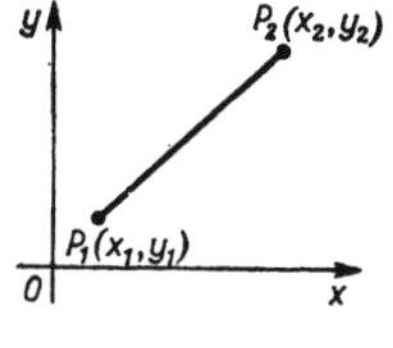

FIG. 158

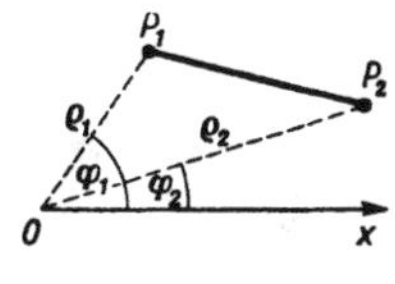

FIG. 159

Distance of two points $P_1(x_1, y_1)$ and $P_2(x_2, y_2)$ (Fig. 158):

$$d = \sqrt{(x_2 - x_1)^2 + (y_2 - y_1)^2}.$$

If polar coordinates are given: $P_1(\varrho_1, \varphi_1)$, $P_2(\varrho_2, \varphi_2)$ (Fig. 159), then

$$d = \sqrt{\varrho_1^2 + \varrho_2^2 - 2\varrho_1 \varrho_2 \cos(\varphi_2 - \varphi_1)}.$$

Division of a segment in a given ratio. Coordinates of a point P such that $\dfrac{P_1P}{PP_2} = \dfrac{m}{n} = \lambda$ (Fig. 160) are given by the formulas

$$(*) \qquad x = \frac{nx_1 + mx_2}{n + m} = \frac{x_1 + \lambda x_2}{1 + \lambda}, \qquad y = \frac{ny_1 + my_2}{n + m} = \frac{y_1 + \lambda y_2}{1 + \lambda}.$$

Coordinates of the *centre* (*midpoint*) of the segment P_1P_2:

$$x = \frac{x_1 + x_2}{2}, \qquad y = \frac{y_1 + y_2}{2}.$$

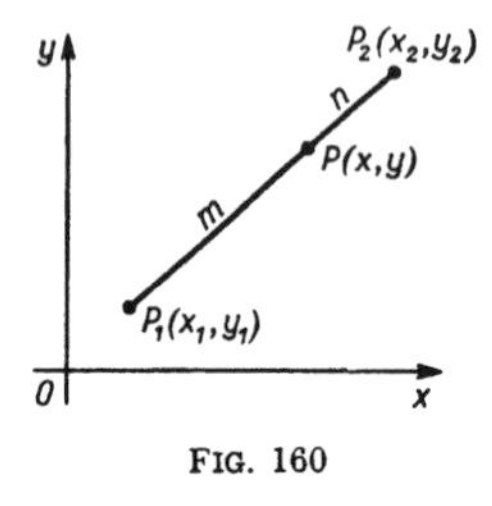

FIG. 160

If the segments P_1P and PP_2 are taken with a positive or negative sign according to whether their directions are the same as the direction of P_1P_2 or opposite, then, for $\lambda < 0$, the formulas (*) give coordinates of an exterior division of the segments P_1P_2 in the given ratio. For example, if the point P is such that P_2 is the midpoint of P_1P, then $\lambda = \dfrac{P_1P}{PP_2} = -2$.

Coordinates of the centre of gravity of a system of particles $M_i(x_i, y_i)$ with the masses m_i $(i = 1, 2, \ldots, n)$ are given by the formulas

$$x = \frac{\sum\limits_{i=1}^{n} m_i x_i}{\sum\limits_{i=1}^{n} m_i}, \qquad y = \frac{\sum\limits_{i=1}^{n} m_i y_i}{\sum\limits_{i=1}^{n} m_i}.$$

Area of a triangle with vertices $P_1(x_1, y_1)$, $P_2(x_2, y_2)$ and $P_3(x_3, y_3)$ (Fig. 161)

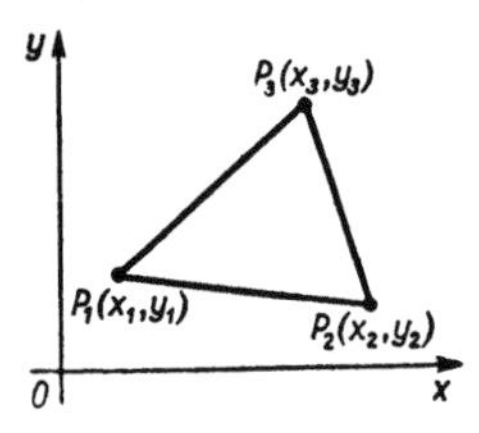

FIG. 161

$$S = \frac{1}{2}\begin{vmatrix} x_1 & y_1 & 1 \\ x_2 & y_2 & 1 \\ x_3 & y_3 & 1 \end{vmatrix} = \frac{1}{2}\left(x_1(y_2 - y_3) + x_2(y_3 - y_1) + x_3(y_1 - y_2)\right)$$

$$= \frac{1}{2}\left((x_1 - x_2)(y_1 + y_2) + (x_2 - x_3)(y_2 + y_3) + (x_3 - x_1)(y_3 + y_1)\right).$$

Three points lie on the same line, if

$$\begin{vmatrix} x_1 & y_1 & 1 \\ x_2 & y_2 & 1 \\ x_3 & y_3 & 1 \end{vmatrix} = 0.$$

Area of a polygon with vertices $P_1(x_1, y_1), P_2(x_2, y_2), \ldots, P_n(x_n, y_n)$:

$$S = \tfrac{1}{2}\big((x_1 - x_2)(y_1 + y_2) + (x_2 - x_3)(y_2 + y_3) + \ldots + (x_n - x_1)(y_n + y_1)\big).$$

The area of a triangle or of a polygon obtained from these formulas will be positive, if the vertices are numbered counter-clockwise, and negative, if they are numbered clockwise.

Equation of a curve. To every equation $F(x, y) = 0$ connecting the coordinates x and y, there corresponds a curve such that the coordinates of every point P lying on this curve satisfy the given equation and, conversely, if the coordinates of a point satisfy the equation, then this point lies on the curve [1]. If $F(x, y)$ is a polynomial, then the curve $F(x, y) = 0$ is called *algebraic*; in this case, the degree of the polynomial $F(x, y)$ (see p. 182) is called the *degree* of the curve. If the equation can not be reduced to the form $F(x, y) = 0$, where $F(x, y)$ is a polynomial, then the curve is called *transcendental*.

One can also consider equations of curves in other coordinate systems. In the following sections, the Cartesian rectangular coordinates are considered, unless otherwise is stated.

2. Straight line

Equation of a straight line. Every linear equation with respect to the coordinates represents a straight line and, conversely, the equation of an arbitrary straight line is of the first degree.

General equation of a straight line:

$$Ax + By + C = 0.$$

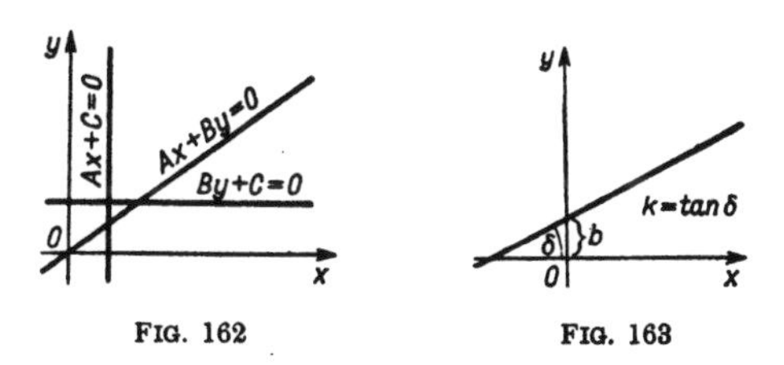

FIG. 162 FIG. 163

If $A = 0$ (Fig. 162), then the line is parallel to the x axis. If $B = 0$, then the line is parallel to the y axis. If $C = 0$, then the line passes through the origin. Equation of any straight line which

[1] It can happen that the given equation $F(x, y) = 0$ is not satisfied by any real point of the plane (for example, $x^2 + y^2 + 1 = 0$, $y = \ln(1 - x^2 - \cosh x)$). Then we conditionally say that the given equation represents an imaginary curve.

is not parallel to the y axis (Fig. 163) can be reduced to the form

$$y = kx + b.$$

k is the *slope* and is equal to $\tan \delta$, where δ is the angle between the line and the positive direction of the x axis; b is the intercept on the y axis (with a corresponding sign).

Equation of a straight line passing through a given point $P_1(x_1, y_1)$ *in a given direction* (Fig. 164):

$$y - y_1 = k(x - x_1), \quad \text{where} \quad k = \tan \delta.$$

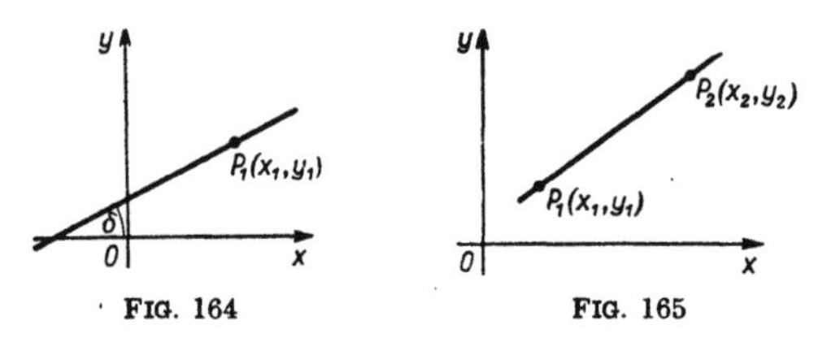

FIG. 164 FIG. 165

Equation of straight line passing through two given points $P_1(x_1, y_1)$ and $P_2(x_2, y_2)$ (Fig. 165):

$$\frac{y - y_1}{y_2 - y_1} = \frac{x - x_1}{x_2 - x_1}.$$

Intercept equation of a straight line. If a straight line intersects the coordinate axes in the points $A(a, 0)$ and $B(0, b)$ (Fig. 166), then its equation is

$$\frac{x}{a} + \frac{y}{b} = 1.$$

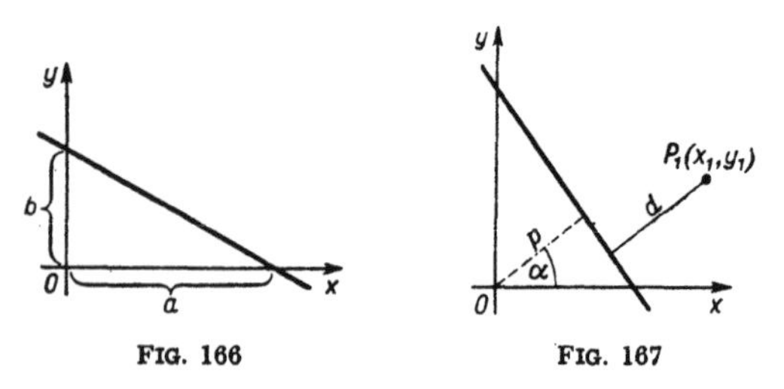

FIG. 166 FIG. 167

Normal equation of a straight line:

$$x \cos \alpha + y \sin \alpha - p = 0,$$

where p is the distance of the line from the origin, and α is the angle between the perpendicular to the line at the origin and the x axis $(p > 0, 0 < \alpha < 2\pi)$, (Fig. 167). Normal equation of a line can be obtained from the general one by multiplying it by the factor $\mu = \pm \frac{1}{\sqrt{A^2+B^2}}$. The sign of μ should be opposite to the sign of C.

Distance of a point $P_1(x_1, y_1)$ from a straight line (Fig. 167):

$$d = x_1 \cos \alpha + y_1 \sin \alpha - p$$

is the result of substitution of coordinates of the given point to the left member of the normal equation of the line. In this formula, $d > 0$, if P_1 and the origin lie on the same side of the given line, and $d < 0$ otherwise.

Intersection of straight lines. Coordinates (x_0, y_0) of a point of intersection of two straight lines are obtained by simultaneous solution of their equations. If the straight lines are given by the equations $A_1x + B_1y + C_1 = 0$ and $A_2x + B_2y + C_2 = 0$, then

$$x_0 = \begin{vmatrix} B_1 & C_1 \\ B_2 & C_2 \end{vmatrix} : \begin{vmatrix} A_1 & B_1 \\ A_2 & B_2 \end{vmatrix}, \quad y_0 = \begin{vmatrix} C_1 & A_1 \\ C_2 & A_2 \end{vmatrix} : \begin{vmatrix} A_1 & B_1 \\ A_2 & B_2 \end{vmatrix}.$$

If

$$\begin{vmatrix} A_1 & B_1 \\ A_2 & B_2 \end{vmatrix} = 0,$$

then the lines are parallel; in particular,

if

$$\frac{A_1}{A_2} = \frac{B_1}{B_2} = \frac{C_1}{C_2},$$

then the lines coincide.

A third line $A_3x + B_3y + C_3 = 0$ passes through the point of intersection of the two first lines (Fig. 168), if

$$\begin{vmatrix} A_1 & B_1 & C_1 \\ A_2 & B_2 & C_2 \\ A_3 & B_3 & C_3 \end{vmatrix} = 0.$$

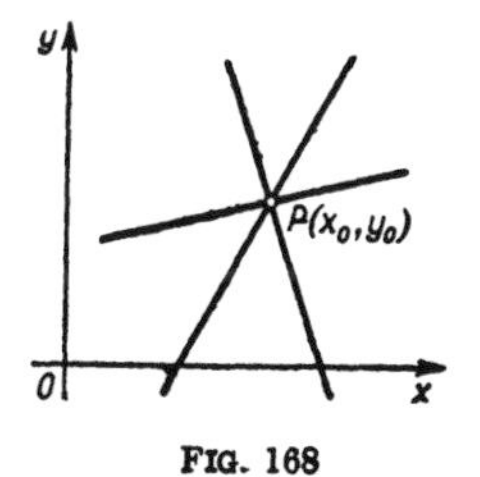

FIG. 168

Equation of any straight line passing through the point of intersection of two lines (*equation of straight lines*):

$$(A_1x + B_1y + C_1) + \lambda(A_2x + B_2y + C_2) = 0.$$

This equation gives all the lines of the bundle, when λ varies from $-\infty$ to $+\infty$. If the equations of the given lines are in the normal form, we obtain, for $\lambda = \pm 1$, the *equations of bisectors of the angles* formed by the given lines (Fig. 169).

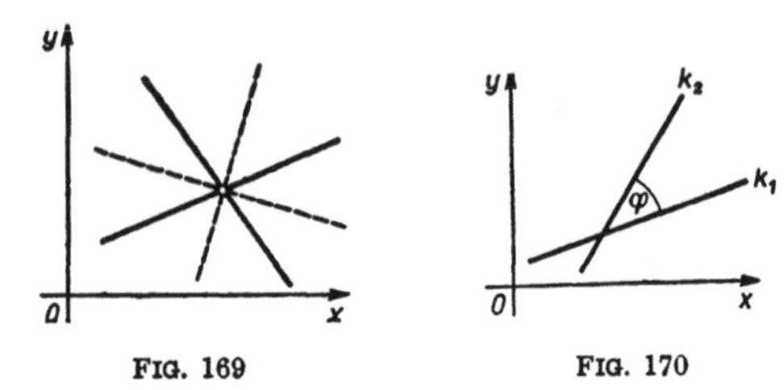

FIG. 169 FIG. 170

Angle between two lines. The angle φ between two straight lines (Fig. 170) can be determined as follows.

If the equations of the lines are given in the general form

$$A_1x + B_1y + C_1 = 0, \quad A_2x + B_2y + C_2 = 0,$$

then

$$\tan\varphi = \frac{A_1B_2 - A_2B_1}{A_1A_2 + B_1B_2},$$

$$\cos\varphi = \frac{A_1A_2 + B_1B_2}{\sqrt{A_1^2 + B_1^2}\sqrt{A_2^2 + B_2^2}}, \quad \sin\varphi = \frac{A_1B_2 - A_2B_1}{\sqrt{A_1^2 + B_1^2}\sqrt{A_2^2 + B_2^2}}.$$

If the slopes k_1 and k_2 are given, then

$$\tan\varphi = \frac{k_2 - k_1}{1 + k_1k_2},$$

$$\cos\varphi = \frac{1 + k_1k_2}{\sqrt{1 + k_1^2}\sqrt{1 + k_2^2}}, \quad \sin\varphi = \frac{k_2 - k_1}{\sqrt{1 + k_1^2}\sqrt{1 + k_2^2}}$$

(the angle φ is measured from the first line to the second one counterclockwise).

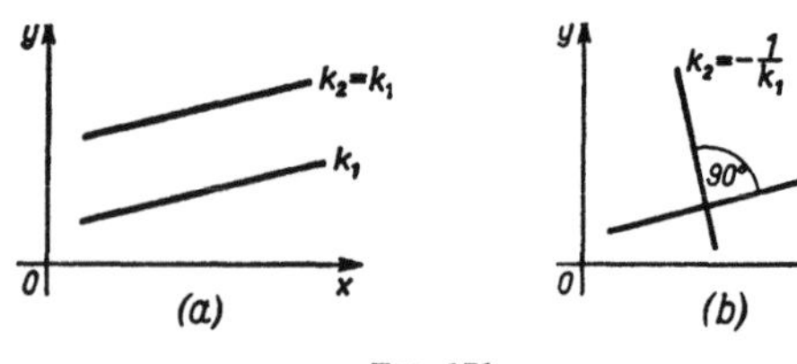

FIG. 171

The *straight lines are parallel* (Fig. 171a), if

$$\frac{A_1}{A_2} = \frac{B_1}{B_2} \quad \text{or} \quad k_1 = k_2.$$

The *straght lines are perpendicular* (Fig. 171b), if

$$A_1A_2 + B_1B_2 = 0 \quad \text{or} \quad k_2 = -\frac{1}{k_1}.$$

Equation of a straight line in polar coordinates (Fig. 172):

$$\varrho = \frac{p}{\cos(\varphi - \alpha)},$$

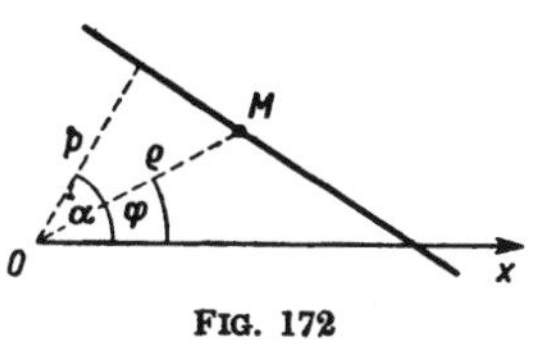

FIG. 172

where p is the distance of the line from the pole and α is the angle between the polar axis and the perpendicular from the pole to the line.

3. Circle

Equation in Cartesian coordinates. Equation of a circle of the radius R with the centre at the origin (Fig. 173a):

$$x^2 + y^2 = R^2.$$

Equation of a circle of the radius R with the centre $C(x_0, y_0)$ (Fig. 173b):

$$(x - x_0)^2 + (y - y_0)^2 = R^2.$$

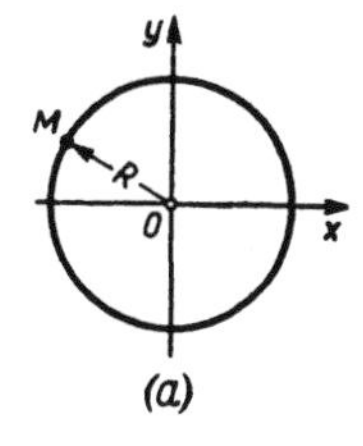

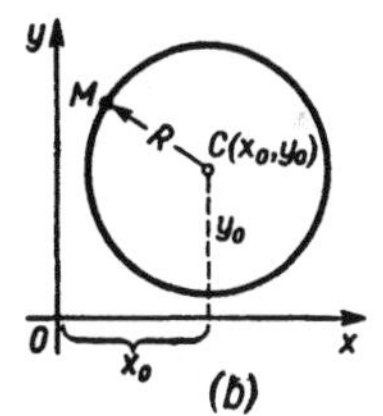

FIG. 173

A general equation of the second degree $ax^2 + 2bxy + cy^2 + 2dx + 2ey + f = 0$ represents a circle if and only if $b = 0$ and $a = c$. In this case the equation can be reduced to the form

$$x^2 + y^2 + 2mx + 2ny + q = 0.$$

Then the radius $R = \sqrt{m^2 + n^2 - q}$ and the coordinates of the centre $x_0 = -m$, $y_0 = -n$. If $q > m^2 + n^2$ then the equation does not represent any real curve; if $q = m^2 + n^2$, then the equation is satisfied by a single point $M(x_0, y_0)$.

Equation of a circle in the parametric form:

$$x = x_0 + R\cos t, \qquad y = y_0 + R\sin t,$$

where t is the angle between the moving radius and positive direction of the x axis (Fig. 174).

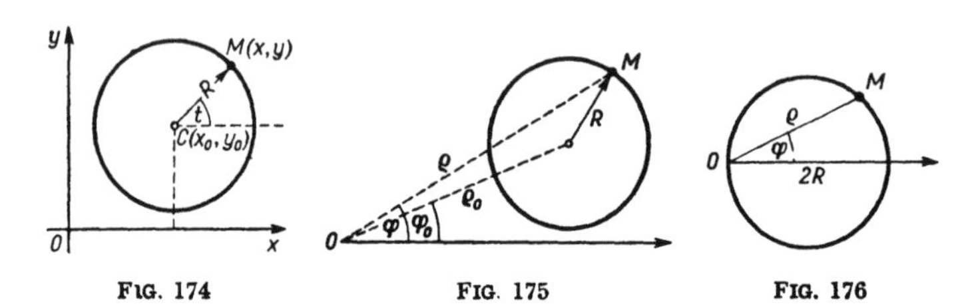

FIG. 174 FIG. 175 FIG. 176

Equation in polar coordinates. A general equation (Fig. 175) $\varrho^2 - 2\varrho\varrho_0\cos(\varphi - \varphi_0) + \varrho_0^2 = R^2$. If the centre lies on the polar axis and the circle passes through the origin (Fig. 176), then the equation has the form $\varrho = 2R\cos\varphi$.

4. Ellipse

Elements of an ellipse (Fig. 177): The *major axis* AB $(=2a)$, the *minor axis* CD $(=2b)$, the *vertices* A, B, C, D, the *centre* O, the *foci* F_1 and F_2 (the points lying on the major axis on both sides

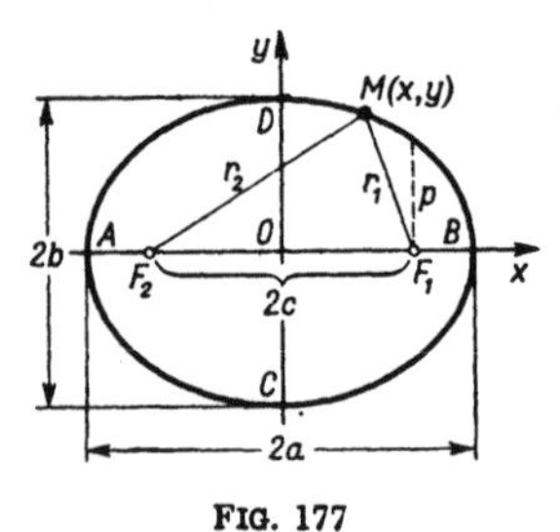

FIG. 177

of the centre at the distance $c = \sqrt{a^2 - b^2}$ from the centre), the *eccentricity* $e = \dfrac{c}{a}$ $(e < 1)$, the *semi-latus rectum* $p = \dfrac{b^2}{a}$ (one half of the chord passing through a focus parallel to the minor axis).

Equation of an ellipse. The *canonical equation* (if the coordinate axes coincide with the axes of the ellipse):

$$\frac{x^2}{a^2}+\frac{y^2}{b^2}=1, \qquad a>b.$$

Parametric form:

$$x=a\cos t, \qquad y=b\sin t.$$

For the *polar equation* see p. 256.

Focal property of the ellipse (definition of an ellipse). An ellipse is the locus of a point the sum of whose distances from two fixed points (the foci) is constant ($=2a$). Each of these distances (the focal radius vector of a point of the ellipse with the abscissa x) is expressed by the formula [1]

$$r_1=MF_1=a-ex, \qquad r_2=MF_2=a+ex, \qquad r_1+r_2=2a.$$

Directrices of an ellipse are straight lines parallel to the minor axis at the distance $d=\frac{a}{e}$ from it (Fig. 178). For every point $M(x, y)$ of an ellipse,

$$\frac{r_1}{d_1}=\frac{r_2}{d_2}=e;$$

this property can be taken as a definition of an ellipse (see p. 253).

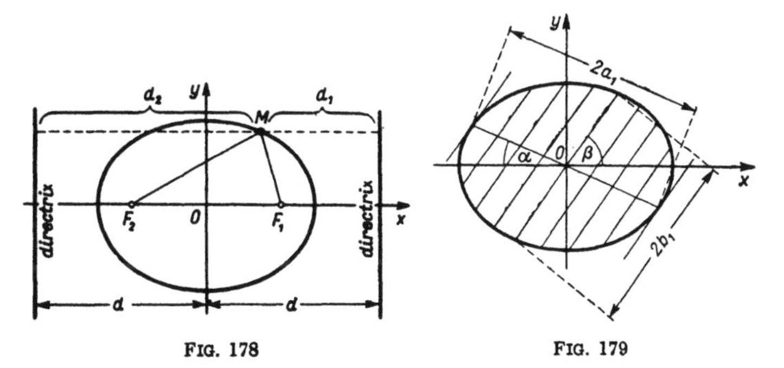

FIG. 178 FIG. 179

Diameters of the ellipse, i.e., the chords through its centre, are bisected by the centre (Fig. 179). The locus of midpoints of chords parallel to one diameter of the ellipse is the diameter *conjugate* to the given diameter. If k and k' are slopes of a pair of conjugate

---

(1) In the following formulas containing coordinates, the ellipse is assumed to be given by its canonical equation.

diameters, then $kk' = -\frac{b^2}{a^2}$. If $2a_1$ and $2b_1$ are the lengths of conjugate diameters and α and β are acute angles between the diameters and the major axis ($k = -\tan\alpha$, $k' = \tan\beta$), then

$$a_1 b_1 \sin(\alpha + \beta) = ab$$

and

$$a_1^2 + b_1^2 = a^2 + b^2$$

(*Apollonius' theorem*).

Tangent line at $M(x_0, y_0)$ has the equation

$$\frac{xx_0}{a^2} + \frac{yy_0}{b^2} = 1.$$

The normal and the tangent to an ellipse are bisectors of the interior and exterior angles, respectively, between the focal radius vectors of the point of contact (Fig. 180). The straight line $Ax + By + C = 0$ is tangent to the ellipse, if $A^2a^2 + B^2b^2 + C^2 = 0$.

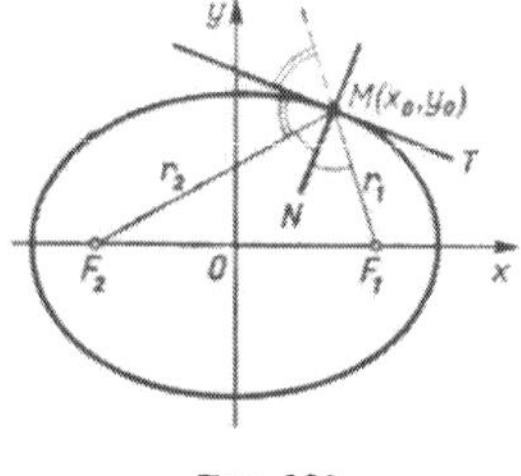

FIG. 180

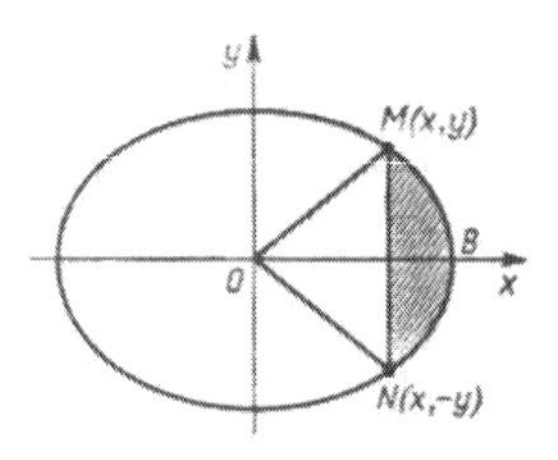

FIG. 181

Radius of curvature at $M(x_0, y_0)$ (see Fig. 180):

$$R = a^2b^2\left(\frac{x_0^2}{a^4} + \frac{y_0^2}{b^4}\right)^{3/2} = \frac{(r_1 r_2)^{3/2}}{ab} = \frac{p}{\sin^3 u},$$

where u is the angle between the tangent and one of the radius vectors of the point of contact. At the vertices A and B (Fig. 177), $R = \frac{b^2}{a} = p$. For the vertices C and D, $R = \frac{a^2}{b}$.

Area $S = \pi ab$. Area of a sector $BOM = \frac{1}{2} ab \arccos\frac{x}{a}$ (Fig. 181). Area of a segment $MBN = ab \arccos\frac{x}{a} - xy$.

Perimeter of the ellipse:

$$L = 4a\,E(e) = 2\pi a\left[1 - \left(\frac{1}{2}\right)^2 e^2 - \left(\frac{1\cdot 3}{2\cdot 4}\right)^2 \frac{e^4}{3} - \left(\frac{1\cdot 3\cdot 5}{2\cdot 4\cdot 6}\right)^2 \frac{e^6}{5} - \dots\right],$$

where $E(e) = E(e, \frac{1}{2}\pi)$ is the complete elliptic integral of the second kind (see p. 407). If $\dfrac{a-b}{a+b} = \lambda$, then

$$L = \pi(a+b)\left(1 + \frac{\lambda^2}{4} + \frac{\lambda^4}{64} + \frac{\lambda^6}{256} + \frac{25\lambda^8}{16384} + \dots\right).$$

Approximate formulas:

$$L = \pi\left[1.5\,(a+b) - \sqrt{ab}\right], \quad L = \pi(a+b)\,\frac{64 - 3\lambda^4}{64 - 16\lambda^2}.$$

5. Hyperbola

Elements of a hyperbola (Fig. 182): The *transverse (real) axis* $AB\ (=2a)$, the *vertices* A, B, the *centre* O, the *foci* F_1 and F_2 (the points lying on the transversal axis on both sides of the centre at

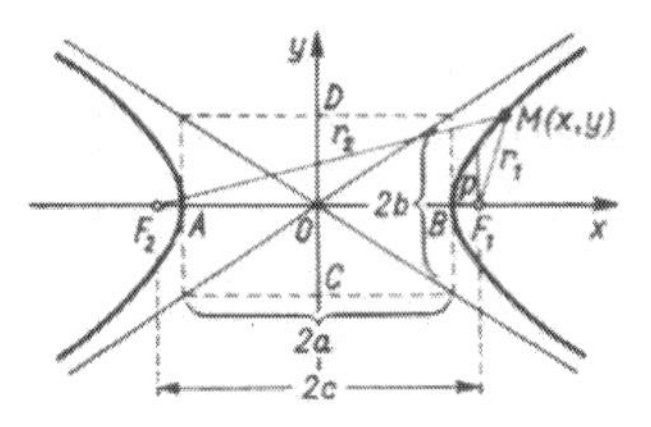

FIG. 182

the distance c (greater than a) from the centre), the *conjugate (imaginary) axis* $CD\ (=2b = 2\sqrt{c^2 - a^2})$, the *semi-latus rectum* p (one half of the chord passing through a focus and perpendicular to the transverse axis, $p = \dfrac{b^2}{a}$); the *eccentricity* $e = \dfrac{c}{a} > 1$.

Equation of a hyperbola. The *canonical equation* (if the x axis coincides with the transverse axis of hyperbola):

$$\frac{x^2}{a^2} - \frac{y^2}{b^2} = 1.$$

Parametric form:

$$x = a \cosh t, \quad y = b \sinh t \quad \text{or} \quad x = a \sec t, \quad y = b \tan t.$$

In *polar coordinates*: see p. 256.

Focal property of a hyperbola (definition of a hyperbola). A hyperbola is the locus of a point the difference of whose distances from two fixed points (the foci) is constant ($=2a$). The points such that $r_1 - r_2 = 2a$ belong to one *branch* of the hyperbola (the left one in Fig. 182); the points such that $r_2 - r_1 = 2a$ belong to the other branch (the right one). Each of these distances (the focal radius vector of a point of hyperbola with the abscissa x) is expressed by the formula [1] $r_1 = \pm (ex - a)$, $r_2 = \pm (ex + a)$ (the upper sign for the points of the right branch, the lower sign for the points of the left branch); $r_2 - r_1 = \pm 2a$.

Directrices of the hyperbola are straight lines parallel to the transverse axis at the distance $d = \frac{a}{e}$ from the centre (Fig. 183). For a point $M(x, y)$ of the hyperbola $\frac{r_1}{d_1} = \frac{r_2}{d_2} = e$ (this property can be taken as a definition of a hyperbola, see p. 256).

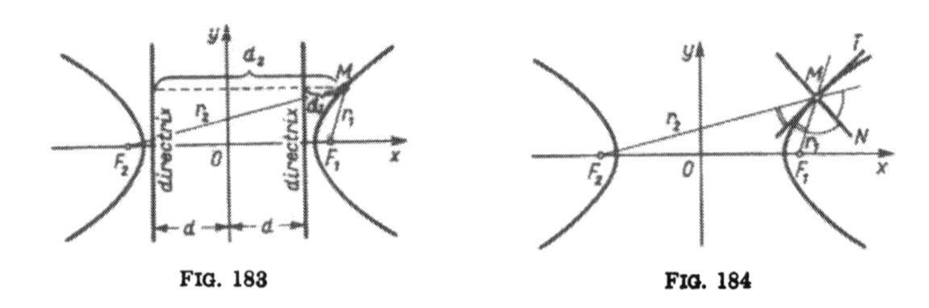

FIG. 183

FIG. 184

Tangent line to hyperbola at $M(x_0, y_0)$ has the equation

$$\frac{xx_0}{a^2} - \frac{yy_0}{b^2} = 1.$$

The tangent and the normal lines to a hyperbola are bisectors of the interior and exterior angles, respectively, between the focal radius vectors of the point of contact (Fig. 184). The line $Ax + By + C = 0$ is tangent to the hyperbola if $A^2a^2 - B^2b^2 = C^2$.

---

[1] In the formulas containing coordinates, the hyperbola is assumed to be given by its canonical equation.

Asymptotes of hyperbola (Fig. 185) are straight lines approached by the branches of a hyperbola, when the coordinates tend to infinity (for a general definition of asymptotes see p. 290). The slope of asymptotes $k = \pm \tan \delta = \pm \frac{b}{a}$. Equation of both asymptotes: $y = \pm \frac{b}{a} x$.

A point of contact M bisects the segment of a tangent TT_1 between the asymptotes: $TM = MT_1$. Area of the triangle TOT_1 between the tangents and the asymptotes is equal to ab (for every point M). If the lines MF and MG pass through a point M of hyperbola parallel to the asymptotes, then the area $OFMG = \frac{1}{4}(a^2 + b^2) = \frac{1}{4}c^2$.

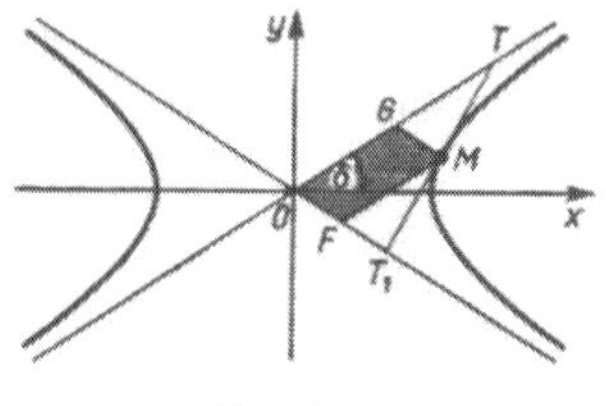

FIG. 185

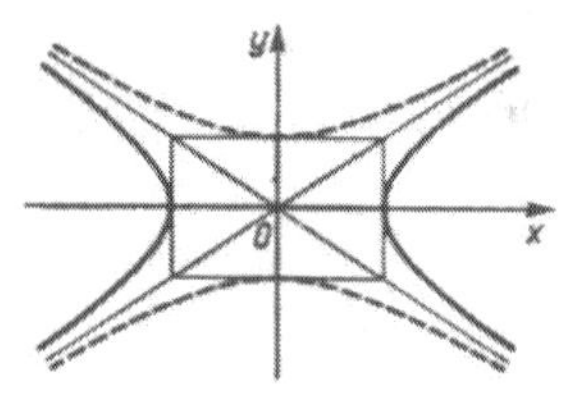

FIG. 186

Conjugate hyperbolas (Fig. 186)

$$\frac{x^2}{a^2} - \frac{y^2}{b^2} = 1 \quad \text{and} \quad \frac{y^2}{b^2} - \frac{x^2}{a^2} = 1$$

(the second one is drawn in Fig. 186 by the dotted line) have common asymptotes. The transverse axis of one of them is the conjugate axis of the other and conversely.

Diameters are chords of a given hyperbola and of its conjugate passing through the centre; they are bisected by the centre.

Two diameters with the slopes k and k' are called *conjugate*, if $kk' = \frac{b^2}{a^2}$. Each of two conjugate diameters bisects the chords (of the given hyperbola and of its conjugate) parallel to the other [1] (Fig. 187). If the lengths of two conjugate diameters are $2a_1$ and $2b_1$ and α and β are the acute angles between the diameters and the transverse axis ($\alpha > \beta$), then $a_1^2 - b_1^2 = a^2 - b^2$, $ab = a_1 b_1 \sin(\alpha - \beta)$.

---

[1] Only one of two conjugate diameters intersects the hyperbola (this one for which $|k| < b/a$). The chord obtained here is a diameter in a more restricted sense; it is bisected by the centre.

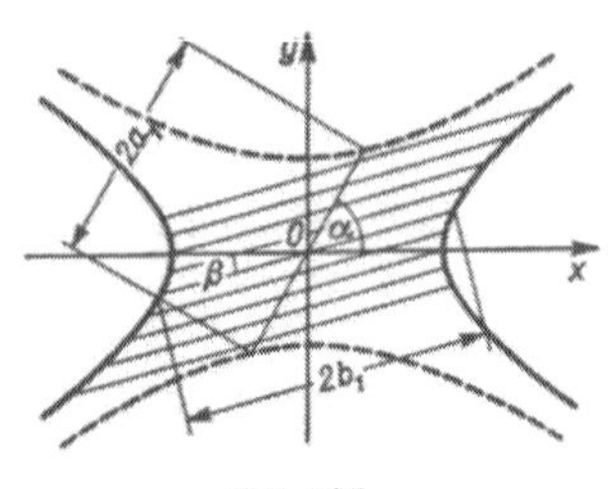

FIG. 187

Radius of curvature of hyperbola at the point $M(x_0, y_0)$ (for notation see p. 247):

$$R = a^2b^2\left(\frac{x_0^2}{a^4} + \frac{y_0^2}{b^4}\right)^{3/2} = \frac{(r_1 r_2)^{3/2}}{ab} = \frac{p}{\sin^3 u},$$

where u is the angle between the tangent and a radius vector of the point of contact. At the vertices A and B (Fig. 182): $R = p = b^2/a$.

Area of a segment of hyperbola (Fig. 188):

$$AMN = xy - ab\ln\left(\frac{x}{a} + \frac{y}{b}\right) = xy - ab \operatorname{ar\,cosh}\frac{x}{a}.$$

Area of $OAMG = \frac{1}{4}ab + \frac{1}{2}ab\ln\frac{2OG}{c}$ (MG is parallel to an asymptote).

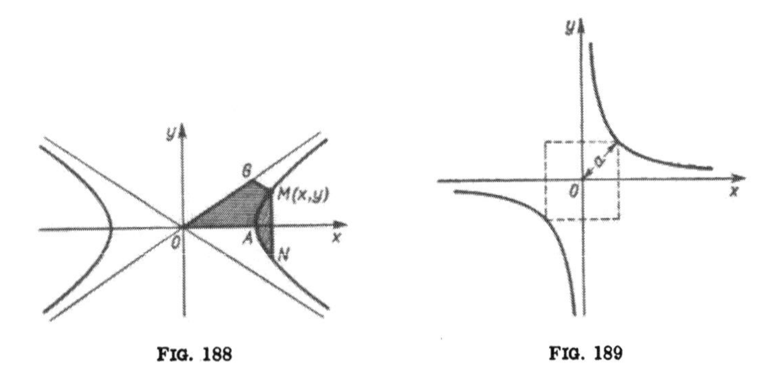

FIG. 188 FIG. 189

Equilateral hyperbola is a hyperbola whose axes are equal: $a = b$. Its equation $x^2 - y^2 = a^2$. Asymptotes of an equilateral hyperbola are perpendicular. Equation of an equilateral hyperbola referred to its axes (Fig. 189): $xy = \frac{1}{2}a^2$.

6. Parabola

Elements of a parabola (Fig. 190): *Axis* of the parabola is the x axis, *focus* F (the point lying on the axis at the distance $\frac{1}{2}p$ from the vertex), *directrix* NN' (the straight line perpendicular to the axis at the distance $\frac{1}{2}p$ from the vertex opposite the focus), *semi-latus rectum* (or *parameter*) p (the distance from the focus to the directrix or one half of the chord passing through the focus perpendicular to the axis). The *eccentricity* of the parabola is equal to 1 (see p. 256).

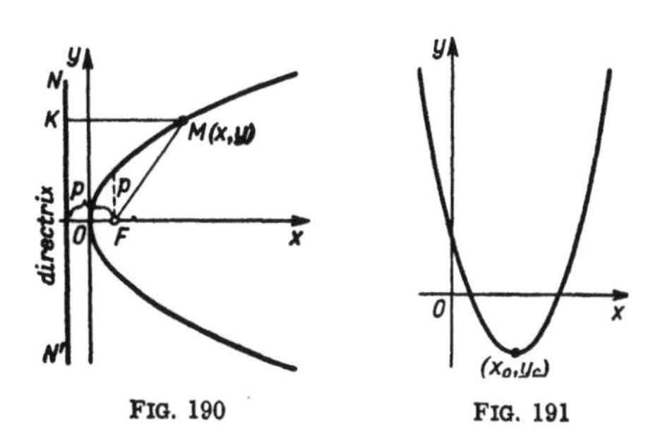

FIG. 190 FIG. 191

Equation of the parabola. Canonical equation $y^2 = 2px$ (if the parabola has the x axis as its axis and its vertex is at the origin turned to the left [1]). For the *equation of the parabola in polar coordinates* see p. 256. *Equation of a parabola with a vertical axis* (Fig. 191):

$$y = ax^2 + bx + c;$$

the parameter of this parabola: $p = \dfrac{1}{2|a|}$. For $a>0$, the parabola is turned upwards [1] and for $a<0$, it is turned downwards [1]. Coordinates of the vertex:

$$x_0 = -\frac{b}{2a}, \quad y_0 = \frac{4ac - b^2}{4a}.$$

Fundamental property (definition of a parabola): A parabola is the locus of a point $M(x, y)$ whose distance from a fixed point (the focus) is equal to its distance from a fixed straight line (the directrix) (Fig. 190): $MF = MK = x + \frac{1}{2}p$ [2]; MF is the focal radius vector of a point of the parabola.

---

[1] The x axis is assumed to be directed to the right and the y axis—upwards.

[2] In the formulas containing coordinates, the parabola is assumed to be given by its canonical equation.

Diameter of the parabola is any straight line parallel to its axis. A diameter bisects the chords parallel to the line tangent to the parabola at its point of intersection with the diameter (Fig. 192). If the slope of these chords is k, then the equation of the diameter is $y = \frac{k}{p}$.

Tangent to the parabola (Fig. 193) at the point $M(x_0, y_0)$ has the equation $yy_0 = p(x + x_0)$. The tangent and the normal lines to a parabola bisect the angles between the focal radius vector and the diameter passing through the point of contact. The tangent to a parabola at the vertex (the y axis) bisects the segment of any tangent to the parabola contained between the point of contact and the axis of the parabola (the x axis): $TS = SM$; $TF = FM$; $TO = OP = x_0$. The line $y = kx + b$ is tangent to the parabola, if $p = 2bk$.

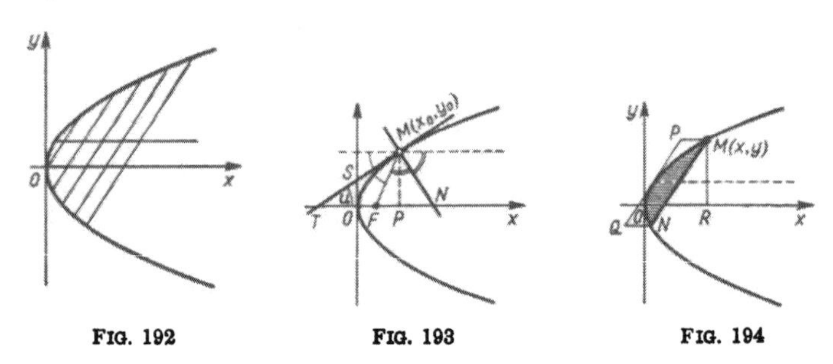

FIG. 192 FIG. 193 FIG. 194

Radius of curvature of the parabola at $M(x_1, y_1)$:

$$R = \frac{(p + 2x_1)^{3/2}}{\sqrt{p}} = \frac{p}{\sin^3 u} = \frac{n^3}{p^2},$$

where n is the length of the normal line MN (Fig. 193). At the vertex $R = p$.

Area of a segment of parabola $MON = \frac{2}{3}$ times the area of $PQNM$ (Fig. 194). Area $OMR = \frac{2}{3}xy$.

Length of arc of parabola from the vertex O to $M(x, y)$:

$$OM = \frac{p}{2}\left[\sqrt{\frac{2x}{p}\left(1 + \frac{2x}{p}\right)} + \ln\left(\sqrt{\frac{2x}{p}} + \sqrt{1 + \frac{2x}{p}}\right)\right]$$

$$= \sqrt{x\left(x + \frac{p}{2}\right)} + \frac{p}{2} \operatorname{ar\,sinh} \sqrt{\frac{2x}{p}}.$$

Approximately, for small values of $\frac{x}{y}$:

$$OM \approx y\left[1 + \frac{2}{3}\left(\frac{x}{y}\right)^2 - \frac{2}{5}\left(\frac{x}{y}\right)^4\right].$$

7. Curves of the second degree (conic sections)

General equation of curves of the second degree

$$ax^2 + 2bxy + cy^2 + 2dx + 2ey + f = 0$$

can represent an ellipse (in particular, a circle), a hyperbola, a parabola or a pair of straight lines (a *degenerate* curve of the second degree).

Invariants of a curve of the second degree:

$$\Delta = \begin{vmatrix} a & b & d \\ b & c & e \\ d & e & f \end{vmatrix}, \qquad \delta = \begin{vmatrix} a & b \\ b & c \end{vmatrix} = ac - b^2, \qquad S = a + c$$

remain unchanged in translating the origin and rotating the coördinate axes, i.e., if, after a transformation of coordinates, the equation of the curve has the form

$$a'x'^2 + 2b'x'y' + c'y'^2 + 2d'x' + 2e'y' + f' = 0,$$

then the values of Δ, δ, S computed in the new coordinates are the same as previously.

Form of a curve. The table on pp. 254, 255 enables us to determine the form of any curve given by an equation of the second degree and to reduce it to the canonical form.

General properties of curves of the second degree. *Conic sections.* A section of a cone of revolution by a plane is a curve called a conic section. If the plane of the section does not pass through the vertex of the cone, the section is a hyperbola, a parabola or an ellipse according to that whether the plane is parallel to two, one or none of the generators of the cone. A section of a cone by a plane passing through the vertex is a degenerate curve of the second degree ($\Delta = 0$, see the table on pp. 254, 255). Two parallel straight lines are obtained, when the cone degenerates to a cylinder (the vertex of the cone tends to infinity).

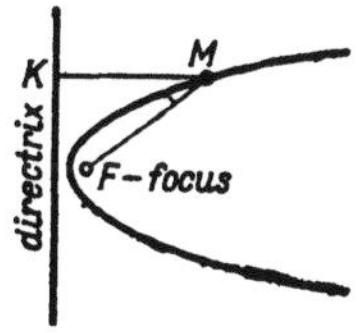

FIG. 195

Reduction of equations of curves of the

<table>
<tr><th colspan="3">Invariants δ and Δ</th><th>Form of the curve</th></tr>
<tr><td rowspan="4">Central curves $\delta\neq 0$</td><td rowspan="2">$\delta>0$</td><td>$\Delta\neq 0$</td><td>Ellipse
(a) $\Delta\cdot S<0$, real ellipse,
(b) $\Delta\cdot S>0$, imaginary ellipse ([2])</td></tr>
<tr><td>$\Delta=0$</td><td>A pair of two imaginary ([2]) straight lines having one real point in common</td></tr>
<tr><td rowspan="2">$\delta<0$</td><td>$\Delta\neq 0$</td><td>Hyperbola</td></tr>
<tr><td>$\Delta=0$</td><td>Two intersecting straight lines</td></tr>
<tr><td colspan="2" rowspan="2">Parabolic curves $\delta=0$ ([3])</td><td>$\Delta\neq 0$</td><td>Parabola</td></tr>
<tr><td>$\Delta=0$</td><td>A pair of straight lines, parallel, if $d^2-af>0$, coincident, if $d^2-af=0$, imaginary, if $d^2-af<0$</td></tr>
</table>

([1]) For notation see p. 253.

([2]) See footnote on p. 239.

([3]) In the case $\delta=0$, it is assumed that neither of the coefficients a, b, c is equal to zero. If two coefficients (a and b or b and c) are zero, then the transformation reduces

second degree to the canonical form [1]

The necessary transformation	Equation of the curve after transformation
(1) Translation of the origin to the centre of the curve with coordinates $x_0 = \frac{be - cd}{\delta}$, $y_0 = \frac{bd - ae}{\delta}$ (2) Rotation of the coordinate axes through the angle α determined by the equation $\tan 2\alpha = \frac{2b}{a - c}$. The sign of $\sin 2\alpha$ should coincide with the sign of $2b$. The slope of the new x' axis is $k = \frac{c - a + \sqrt{(c-a)^2 + 4b^2}}{2b}$	$a'x'^2 + c'y'^2 + \frac{\Delta}{\delta} = 0$, where a' and c' are the roots of the quadratic equation $u^2 - Su + \delta = 0$, i.e., $a' = \frac{a + c + \sqrt{(a-c)^2 + 4b^2}}{2}$, $c' = \frac{a + c - \sqrt{(a-c)^2 + 4b^2}}{2}$
(1) Translation of the origin to the vertex of the parabola with the coordinates x_0, y_0 determined by the equations $ax_0 + by_0 + \frac{ad + be}{S} = 0$, $\left(d + \frac{dc - be}{S}\right) x_0 + \left(e + \frac{ae - bd}{S}\right) y_0 + f = 0$ (2) Rotation of the axes through the angle α determined by the equation $\tan \alpha = -\frac{a}{b}$; the sign of $\sin \alpha$ should be opposite to the sign of a.	$y'^2 = 2px'$; $p = \frac{ae - bd}{S\sqrt{a^2 + b^2}}$
Rotation of the axes through the angle α determined by the equation $\tan \alpha = -\frac{a}{b}$; the sign of $\sin \alpha$ should be opposite to the sign of a.	$Sy'^2 + 2\frac{ad + be}{\sqrt{a^2 + b^2}} y' + f = 0$, $(y' - y'_0)(y' - y'_1) = 0$

to a parallel translation of the axes; the equation $cy^2 + 2dx + 2ey + f = 0$ reduces to the form $(y - y_0)^2 = 2p(x - x_0)$, and the equation $ax^2 + 2dx + 2ey + f = 0$—to the form $(x - x_0)^2 = 2p(y - y_0)$.

Definition of a conic section by means of a focus and a directrix. The locus of a point M (Fig. 195) such that the ratio of the distance of M from a fixed point F (the focus) to the distance of M from a fixed line (the directrix) is a constant value e is a curve of the second degree with the eccentricity e. If $e < 1$, the curve is an ellipse, if $e = 1$—a parabola, if $e > 1$—a hyperbola.

Determining of a curve by five points. Every five points determine a unique curve of the second degree. If at least three of these points lie on one straight line, the curve is degenerated.

Polar equation. Curves of the second degree are represented in polar coordinates by the equation

$$\varrho = \frac{p}{1 + e\cos\varphi}$$

where p is the parameter (the semi-latus rectum), and e is the eccentricity. The pole is at a focus and the polar axis is directed towards the nearest vertex. For a hyperbola, this equation represents only one branch.

B. GEOMETRY IN SPACE

8. Fundamental concepts and formulas

Coordinates. The position of an arbitrary point P in the space can be determined by aid of a coordinate system. The most frequently used are

(1) Cartesian rectangular,
(2) Cylindrical and
(3) Spherical coordinate systems.

Cartesian rectangular coordinates of a point P are the distances [1], taken with definite sign, of P from three mutually perpendicular coordinate planes, i.e., the projections of the radius vector $\boldsymbol{r}$ of P (see p. 614) on three mutually perpendicular *coordinate axes*. According to the mutual location of the positive directions of the coordinate axes, two kinds of coordinate systems are possible: the *right-handed* (Fig. 196a) and the *left-handed* (Fig. 196b) coordinate system. We shall use right-handed systems (the formulas do not depend on the kind of a coordinate system). The point of intersection of the coordinate axes is called the

---

[1] Expressed in a certain units of measure.

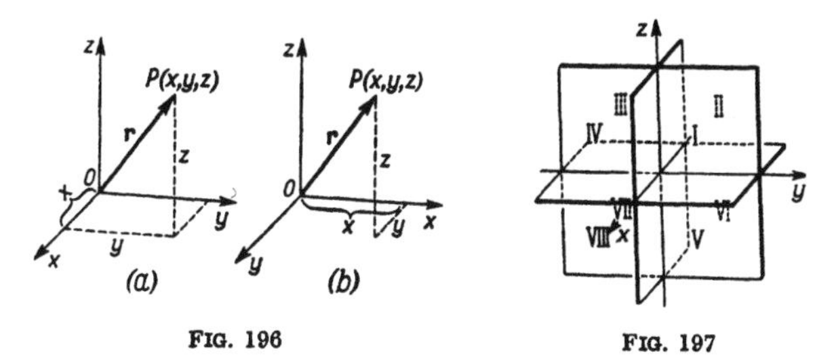

FIG. 196

FIG. 197

coordinate origin. Every point in the space has three coordinates: x, y, z. The symbol $P(a, b, c)$ is used to mean the point P with the coordinates $x = a$, $y = b$, $z = c$. The sign of coordinates depend on the octant containing the given point (Fig. 197):

Octant	*I*	*II*	*III*	*IV*	*V*	*VI*	*VII*	*VIII*
x	+	−	−	+	+	−	−	+
y	+	+	−	−	+	+	−	−
z	+	+	+	+	−	−	−	−

A coordinate surface in the space is a surface one coordinate of whose points is constant; a *coordinate line* is a line two coordinates of whose points are constant. The coordinate surfaces intersect in coordinate lines. In the Cartesian rectangular coordinate system, the coordinate surfaces are planes parallel to the coordinate planes and the coordinate lines are straight lines parallel to the axes.

A more general coordinate system is a *curvilinear coordinate system* introduced as follows. There are given three families of coordinate surfaces such that, through every point of the space, there passes exactly one surface of such family. The position of a point is then determined by the values of parameters of the coordinate surfaces passing through this point. The most frequently used curvilinear coordinates are cylindrical and spherical coordinates described below.

Cylindrical coordinates (Fig. 198): ϱ and φ are polar coordinates of the projection of the point P onto the base plane (usually the xy plane), and the third coordinate, z, is the distance of P from the base plane.

Coordinate surface of cylindrical coordinates are planes perpendicular to the z axis ($z = \text{const}$), half-planes bounded by the z axis ($\varphi = \text{const}$) and cylindrical surfaces whose axis is the

z axis ($\varrho = \text{const}$). Coordinate lines are lines of intersection of these surfaces.

Relation between cylindrical and Cartesian coordinates:

$$x = \varrho \cos \varphi, \qquad y = \varrho \sin \varphi, \qquad z = z;$$

$$\varrho = \sqrt{x^2 + y^2}, \qquad \varphi = \arctan \frac{y}{x} = \arcsin \frac{y}{\varrho}.$$

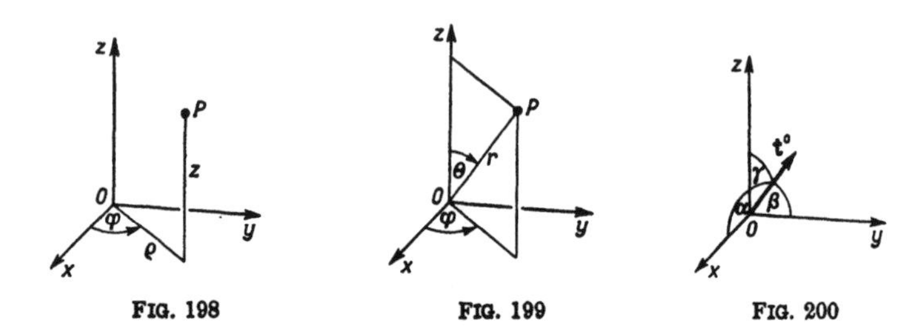

FIG. 198 FIG. 199 FIG. 200

Spherical coordinates: r is the length of the *radius vector* $\mathbf{r}$ φ is the *longitude* (the angle between the positive part of the x axis and the projection of the radius vector onto the xy plane) and θ is the *polar distance* (the angle between the positive part of the z axis and the radius vector). Positive directions of the angles φ and θ are shown in Fig. 199. Taking the values of coordinates from the intervals $0 < r < \infty$, $-\pi < \varphi < \pi$, $0 < \theta < \pi$ we obtain uniquely all the points of the space.

Coordinate surfaces: spheres with the centre at the origin ($r = \text{const}$), half-planes bounded by the z axis ($\varphi = \text{const}$) and cones whose common axis is the z axis and whose common vertex is the origin ($\theta = \text{const}$). Coordinate lines are intersections of these surfaces.

Relation between the spherical and Cartesian coordinates:

$$x = r \sin \theta \cos \varphi, \qquad y = r \sin \theta \sin \varphi, \qquad z = r \cos \theta;$$

$$r = \sqrt{x^2 + y^2 + z^2}, \qquad \varphi = \arctan \frac{y}{x}, \qquad \theta = \arctan \frac{\sqrt{x^2 + y^2}}{z}.$$

A direction in the space is uniquely determined by a *unit vector* $\mathbf{t}^0$ (see p. 614) or by its coordinates, i.e., the cosines of the angles (Fig. 200) which the direction makes with the positive directions of the coordinate axes (*direction cosines*):

$$l = \cos \alpha, \qquad m = \cos \beta, \qquad n = \cos \gamma, \qquad l^2 + m^2 + n^2 = 1.$$

The angle between two given directions with the direction cosines l_1, m_1, n_1 and l_2, m_2, n_2:

$$\cos\varphi = l_1l_2 + m_1m_2 + n_1n_2.$$

Two directions are perpendicular, if $l_1l_2 + m_1m_2 + n_1n_2 = 0$.

Transformation of rectangular coordinates. Notation: x, y, z — old coordinates, x', y', z' — new coordinates, a, b, c—coordinates of the new origin in the old coordinates.

Parallel translation (Fig. 201):

$$x = x' + a, \quad y = y' + b, \quad z = z' + c;$$

$$x' = x - a, \quad y' = y - b, \quad z' = z - c.$$

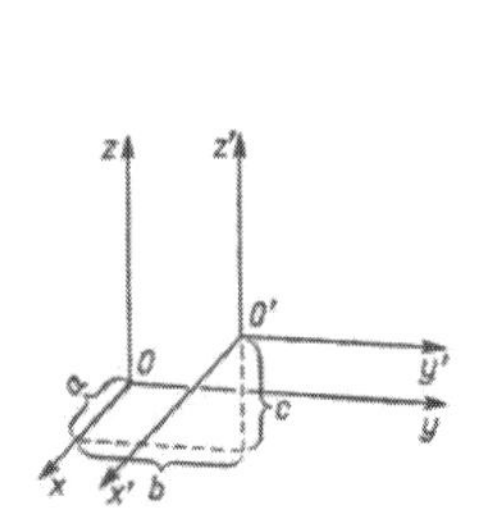

FIG. 201

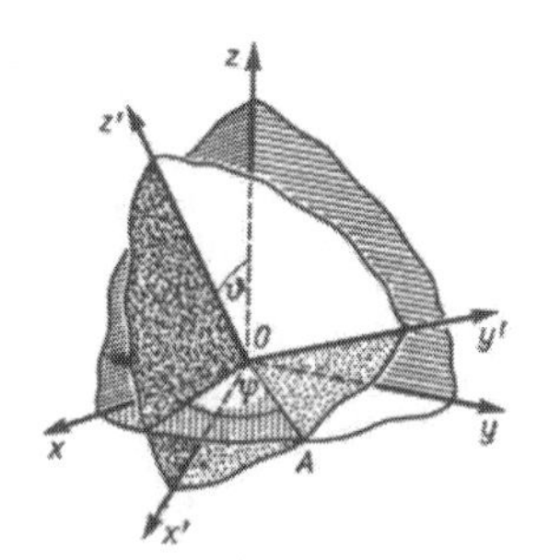

FIG. 202

Rotation of the axes (Fig. 202). If the direction cosines of the new axes are denoted as follows

Old axes	Cosines of the new axes with respect to the old axes		
	x'	y'	z'
x	l_1	l_2	l_3
y	m_1	m_2	m_3
z	n_1	n_2	n_3

then

$$x = l_1x' + l_2y' + l_3z', \quad y = m_1x' + m_2y' + m_3z', \quad z = n_1x' + n_2y' + n_3z';$$

$$x' = l_1x + m_1y + n_1z, \quad y' = l_2x + m_2y + n_2z, \quad z' = l_3x + m_3y + n_3z.$$

Determinant of the transformation:

$$\Delta = \begin{vmatrix} l_1 & l_2 & l_3 \\ m_1 & m_2 & m_3 \\ n_1 & n_2 & n_3 \end{vmatrix}.$$

Properties of the determinant of a transformation:

(1) $\Delta = \pm 1$; $\Delta = +1$, if a left-handed system is transformed into a left-handed one, or a right-handed one into a right-handed one, and $\Delta = -1$, if a left-handed system is transformed into a right-handed one or conversely.

(2) The sum of squares of each row or column is equal to 1.

(3) The sum of products of the corresponding elements of two rows or two columns is equal to 0.

(4) Each element is equal to the product of its algebraic complement (see p. 172) times the value of Δ (equal to $+1$ or -1).

Euler's angles. The position of the new coordinate system with respect to the old one can be determined by means of three angles introduced by Euler (Fig. 202):

(1) The *angle ϑ of nutation* contained between the positive directions of the z and z' axes $(0 \leqslant \vartheta < \pi)$.

(2) The *angle ψ of precession* contained between the x axis and the line OA of intersection of the planes xy and $x'y'$; the positive direction of the line OA is chosen so that the system of axes OA, Oz, Oz' has the same orientation as the system of coordinate axes x, y, z [1]. The angle ψ is directed from the x axis to the y axis $(0 \leqslant \psi < 2\pi)$.

(3) The *angle φ of proper rotation* contained between OA and the x' axis; the angle φ is directed from the x' axis to the y' axis $(0 \leqslant \varphi < 2\pi)$.

Denoting

$$\cos\vartheta = c_1, \qquad \cos\psi = c_2, \qquad \cos\varphi = c_3,$$
$$\sin\vartheta = s_1, \qquad \sin\psi = s_2, \qquad \sin\varphi = s_3,$$

we have

$$\begin{aligned} l_1 &= c_2c_3 - c_1s_2s_3, & m_1 &= s_2c_3 + c_1c_2s_3, & n_1 &= s_1s_3, \\ l_2 &= -c_2s_3 - c_1s_2c_3, & m_2 &= -s_2s_3 + c_1c_2c_3, & n_2 &= s_1c_3, \\ l_3 &= s_1s_2, & m_3 &= -s_1c_2, & n_3 &= c_1. \end{aligned}$$

Distance of two points $P_1(x_1, y_1, z_1)$ and $P_2(x_2, y_2, z_2)$ (Fig. 203) is equal to

[1] For the orientation of a triple of axes see p. 617.

$$d = \sqrt{(x_2 - x_1)^2 + (y_2 - y_1)^2 + (z_2 - z_1)^2}.$$

Direction cosines of the segment P_1P_2:

$$\cos\alpha = \frac{x_2 - x_1}{d}, \quad \cos\beta = \frac{y_2 - y_1}{d}, \quad \cos\gamma = \frac{z_2 - z_1}{d}.$$

Coordinates of the point P such that

$$\frac{P_1P}{PP_2} = \frac{m}{n} = \lambda$$

are

$$x = \frac{nx_1 + mx_2}{n+m} = \frac{x_1 + \lambda x_2}{1+\lambda}, \quad y = \frac{ny_1 + my_2}{n+m} = \frac{y_1 + \lambda y_2}{1+\lambda},$$

$$z = \frac{nz_1 + mz_2}{n+m} = \frac{z_1 + \lambda z_2}{1+\lambda}.$$

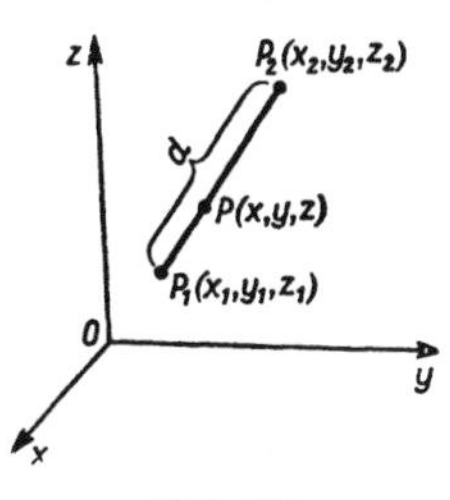

FIG. 203

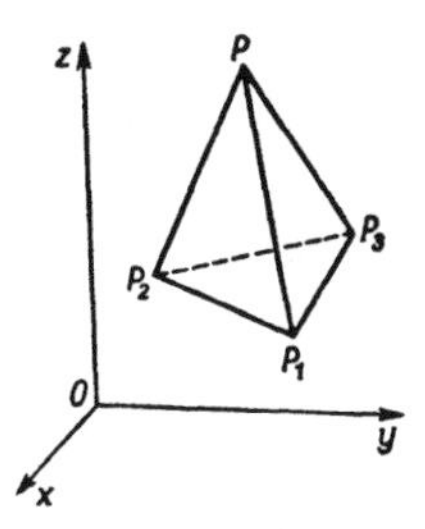

FIG. 204

Coordinates of the centre (midpoint) of the segment P_1P_2:

$$x = \frac{x_1 + x_2}{2}, \quad y = \frac{y_1 + y_2}{2}, \quad z = \frac{z_1 + z_2}{2}.$$

Coordinates of the centre of gravity of a system of particles $M_i(x_i, y_i, z_i)$ with the masses m_i $(i = 1, 2, \ldots, n)$ are given by the formulas

$$\bar{x} = \frac{\sum_{i=1}^{n} m_i x_i}{\sum_{i=1}^{n} m_i}, \quad \bar{y} = \frac{\sum_{i=1}^{n} m_i y_i}{\sum_{i=1}^{n} m_i}, \quad \bar{z} = \frac{\sum_{i=1}^{n} m_i z_i}{\sum_{i=1}^{n} m_i}.$$

Volume of a tetrahedron with the vertices $P(x, y, z)$, $P_1(x_1, y_1, z_1)$, $P_2(x_2, y_2, z_2)$, $P_3(x_3, y_3, z_3)$ (Fig. 204):

$$V = \frac{1}{6}\begin{vmatrix} x & y & z & 1 \\ x_1 & y_1 & z_1 & 1 \\ x_2 & y_2 & z_2 & 1 \\ x_3 & y_3 & z_3 & 1 \end{vmatrix} = \frac{1}{6}\begin{vmatrix} x-x_1 & y-y_1 & z-z_1 \\ x-x_2 & y-y_2 & z-z_2 \\ x-x_3 & y-y_3 & z-z_3 \end{vmatrix}.$$

The volume evaluated by this formula is positive, if the orientation of the triple $\overline{PP_1}, \overline{PP_2}, \overline{PP_3}$ of vectors coincide with that of the coordinate system (see p. 617) and is negative otherwise.

Four points P, P_1, P_2, P_3 lie in the same plane, if

$$\begin{vmatrix} x & y & z & 1 \\ x_1 & y_1 & z_1 & 1 \\ x_2 & y_2 & z_2 & 1 \\ x_3 & y_3 & z_3 & 1 \end{vmatrix} = 0.$$

Equation of a surface. To each equation $F(x, y, z) = 0$ there corresponds a certain surface such that the coordinates of an arbitrary point P lying on this surface satisfy this equation and, conversely, every point whose coordinates satisfy this equation lies on the surface. The equation $F(x, y, z) = 0$ is called the *equation of this surface.*

Equation of a *cylindrical surface* (p. 217) whose generators are parallel to the x axis (resp. to the y axis or to the z axis) does not involve the variable x (resp. y or z): $F(y, z) = 0$ (resp. $F(x, z) = 0$ or $F(x, y) = 0$). The equation $F(y, z) = 0$ represents, in the yz plane, the line of intersection of the surface with this plane.

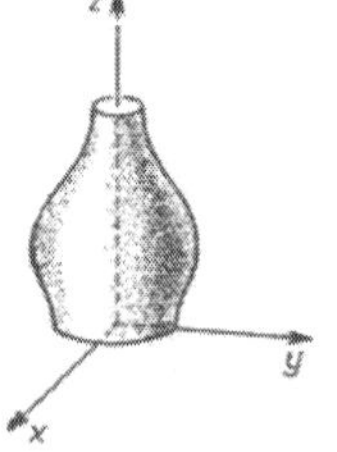

FIG. 205

A cylindrical surface whose direction cosines are equal (or proportional) to the numbers l, m, n has the equation $F(nx - lz, ny - mz) = 0$.

A surface generated by revolution of the curve $z = f(x)$ lying in the xz plane about the z axis (Fig. 205) has the equation $z = f(\sqrt{x^2 + y^2})$. Equations of surfaces of revolution about another coordinate axes are analogous.

Equation of a *conical surface* (see p. 208) with the vertex at the origin is of the form $F(x, y, z) = 0$, where $F(x, y, z) = 0$ is a homogenous function (see p. 344) with respect to x, y, z.

Equation of a line in the space. A line in the space can be given by three parametric equations: $x = \varphi_1(t)$, $y = \varphi_2(t)$, $z = \varphi_3(t)$. To each value of the parameter t there corresponds a definite point of the line (the parameter t may not have a definite geometric significance). A curve in the space can also be given by two equations $F_1(x, y, z) = 0$, $F_2(x, y, z) = 0$. Each of these equations represents a surface; the points whose coordinates satisfy both equations lie on the line of intersection of these two surfaces. Every equation $F_1 + \lambda F_2 = 0$, for an arbitrary λ, represents a surface passing through the line of intersection and can replace either of the originally given equations.

9. Plane and straight line in space

Equation of a plane. Every linear equation with respect to the coordinates represents a plane and, conversely, an equation of an arbitrary plane is of the first degree.

General equation of a plane: $Ax + By + Cz + D = 0$; in the vector form: $\boldsymbol{r}\boldsymbol{N} + D = 0$ (see p. 616 and p. 621). The vector $\boldsymbol{N}(A, B, C)$ (Fig. 206) is perpendicular to the plane and has the direction cosines

$$\cos\alpha = \frac{A}{\sqrt{A^2 + B^2 + C^2}}, \quad \cos\beta = \frac{B}{\sqrt{A^2 + B^2 + C^2}},$$

$$\cos\gamma = \frac{C}{\sqrt{A^2 + B^2 + C^2}}.$$

If $D = 0$, then the plane passes through the origin; if $A = 0$ (resp. $B = 0$ or $C = 0$), then the plane is parallel to the x axis (resp. to the y axis or to the z axis); if $A = B = 0$ (resp. $A = C = 0$ or $B = C = 0$), then the plane is parallel to the xy plane (resp. to the xz plane or to the yz plane).

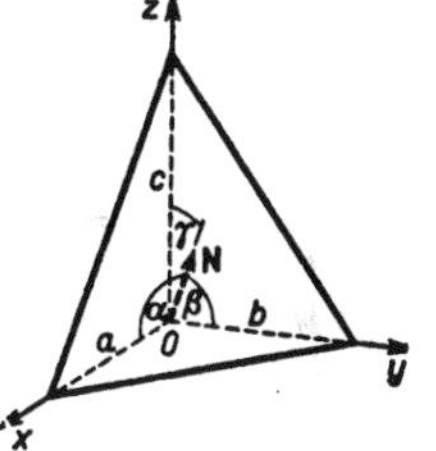

FIG. 206

Normal equation of a plane: $x\cos\alpha + y\cos\beta + z\cos\gamma - p = 0$; in the vector form $\boldsymbol{r}\boldsymbol{N}^0 - p = 0$, where $\boldsymbol{N}^0$ is a unit vector and p is the distance of the plane from the origin. The normal equation of a plane can be obtained from the general equation by multiplying it by the factor

$$\pm\mu = \frac{1}{N} = \frac{1}{\sqrt{A^2 + B^2 + C^2}}.$$

Intercept equation of a plane:

$$\frac{x}{a} + \frac{y}{b} + \frac{z}{c} = 1,$$

the plane intersects the coordinate axes at the points $A(a, 0, 0)$, $B(0, b, 0)$, $C(0, 0, c)$ (Fig. 206).

Equation of the plane passing

(a) *through three points* $P_1(x_1, y_1, z_1)$, $P_2(x_2, y_2, z_2)$, $P_3(x_3, y_3, z_3)$:

$$\begin{vmatrix} x-x_1 & y-y_1 & z-z_1 \\ x_2-x_1 & y_2-y_1 & z_2-z_1 \\ x_3-x_1 & y_3-y_1 & z_3-z_1 \end{vmatrix} = 0, \quad \text{or} \quad \begin{vmatrix} x & y & z & 1 \\ x_1 & y_1 & z_1 & 1 \\ x_2 & y_2 & z_2 & 1 \\ x_3 & y_3 & z_3 & 1 \end{vmatrix} = 0;$$

in the vector form $(\boldsymbol{r}-\boldsymbol{r}_1)(\boldsymbol{r}_2-\boldsymbol{r}_1)(\boldsymbol{r}_3-\boldsymbol{r}_1) = 0$ (1).

(b) *through two points* $P_1(x_1, y_1, z_1)$, $P_2(x_2, y_2, z_2)$ and parallel to a straight line with the direction vector $\boldsymbol{R}(l, m, n)$:

$$\begin{vmatrix} x-x_1 & y-y_1 & z-z_1 \\ x_2-x_1 & y_2-y_1 & z_2-z_1 \\ l & m & n \end{vmatrix} = 0, \quad \text{or} \quad \begin{vmatrix} x & y & z & 1 \\ x_1 & y_1 & z_1 & 1 \\ x_2 & y_2 & z_2 & 1 \\ l & m & n & 0 \end{vmatrix} = 0;$$

in the vector form $(\boldsymbol{r}-\boldsymbol{r}_1)(\boldsymbol{r}_2-\boldsymbol{r}_1)\boldsymbol{R} = 0$ (1).

(c) *through one point* $P_1(x_1, y_1, z_1)$ and parallel to two straight lines with the direction vectors $\boldsymbol{R}_1(l_1, m_1, n_1)$, $\boldsymbol{R}_2(l_2, m_2, n_2)$:

$$\begin{vmatrix} x-x_1 & y-y_1 & z-z_1 \\ l_1 & m_1 & n_1 \\ l_2 & m_2 & n_2 \end{vmatrix} = 0, \quad \text{or} \quad \begin{vmatrix} x & y & z & 1 \\ x_1 & y_1 & z_1 & 1 \\ l_1 & m_1 & n_1 & 0 \\ l_2 & m_2 & n_2 & 0 \end{vmatrix} = 0;$$

in the vector form $(\boldsymbol{r}-\boldsymbol{r}_1)\boldsymbol{R}_1\boldsymbol{R}_2 = 0$ (2).

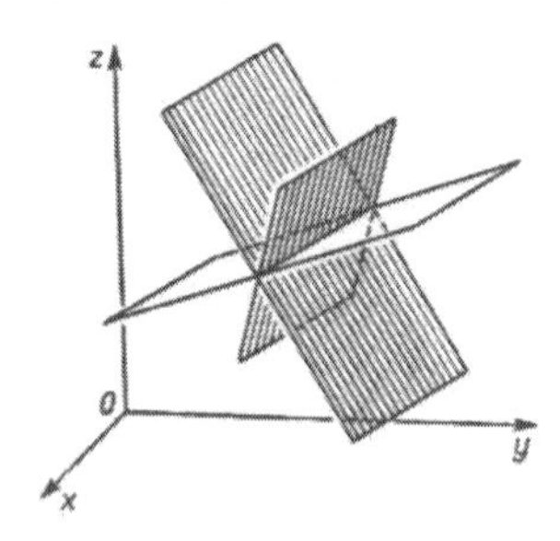

FIG. 207

(d) *through one point* $P_1(x_1, y_1, z_1)$ and perpendicular to a straight line with the direction vector $\boldsymbol{N}(A, B, C)$:

$$A(x-x_1) + B(y-y_1) + C(z-z_1) = 0;$$

in the vector form $(\boldsymbol{r}-\boldsymbol{r}_1)\boldsymbol{N} = 0$ (2).

---

(1) For the triple (box) product of vectors see p. 618.

(2) For the scalar product of vectors see p. 616.

(e) *through the line* of intersection of two planes

$$A_1x + B_1y + C_1z + D_1 = 0 \quad \text{and} \quad A_2x + B_2y + C_2z + D_2 = 0:$$

$$A_1x + B_1y + C_1z + D_1 + \lambda(A_2x + B_2y + C_2z + D_2) = 0$$

(*equation of a bundle of planes*, Fig. 207). When λ varies from $-\infty$ to $+\infty$, we obtain all planes of the bundle. If the equations of the planes are given in the normal form, then for $\lambda = \pm 1$ we obtain the planes bisecting the angles between two given planes.

The angle between two planes, see p. 269.

The point of intersection of three planes, see p. 268.

Distance between two parallel planes [1] $Ax + By + Cz + D_1 = 0$ and $Ax + By + Cz + D_2 = 0$:

$$\delta = \frac{|D_1 - D_2|}{\sqrt{A^2 + B^2 + C^2}}.$$

Distance of the point $M(a, b, c)$ **from the plane** is equal to the result of substitution of the coordinates a, b, c to the left member of the normal equation ($x \cos\alpha + y \cos\beta + z \cos\gamma - p = 0$) of the plane [2]:

$$\delta = a \cos\alpha + b \cos\beta + c \cos\gamma - p.$$

If the point M and the origin lie on different sides of the plane, then $\delta > 0$; otherwise, $\delta < 0$.

Equations of a straight line in the space. A straight line in the space as a intersection of two planes is represented analytically in the form of a system of two linear equations.

General equations of a straight line:

$$(1) \quad A_1x + B_1y + C_1z + D_1 = 0, \quad A_2x + B_2y + C_2z + D_2 = 0,$$

$$A_1^2 + B_1^2 + C_1^2 \neq 0 \text{ and } A_2^2 + B_2^2 + C_2^2 \neq 0,$$

in the vector form [3]

$$\boldsymbol{r}\boldsymbol{N}_1 + D_1 = 0, \quad \boldsymbol{r}\boldsymbol{N}_2 + D_2 = 0.$$

Equation of a straight line in two projecting planes:

$$y = kx + a, \quad z = hx + b;$$

---

[1] A condition for two planes to be parallel is given on p. 269.

[2] For reduction of a general equation of a plane to the normal form see p. 263.

[3] For the scalar product of vectors see p. 616.

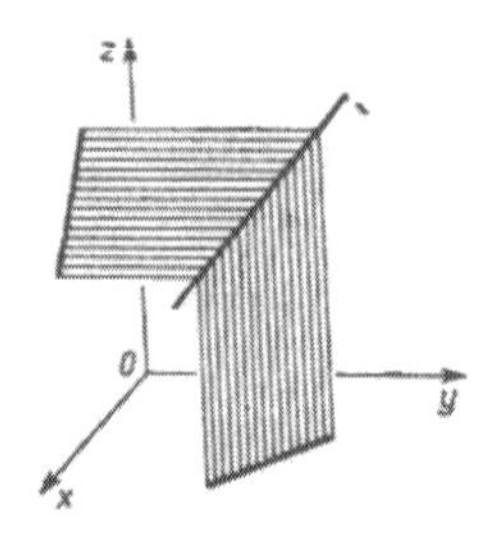

FIG. 208

each of these equations represents a plane projecting the straight line on the xy and xz plane, respectively (Fig. 208). This form of equations is not applicable to planes parallel to the yz plane; in this case we consider the projections of the line on another pair of coordinate planes.

Equations of the straight line passing

(a) through a given point $P_1(x_1, y_1, z_1)$ and parallel to the direction vector $\boldsymbol{R}(l, m, n)$ (Fig. 209):

$$\frac{x - x_1}{l} = \frac{y - y_1}{m} = \frac{z - z_1}{n}; \tag{2}$$

in the vector form $(\boldsymbol{r} - \boldsymbol{r}_1) \times \boldsymbol{R} = 0$ [1].

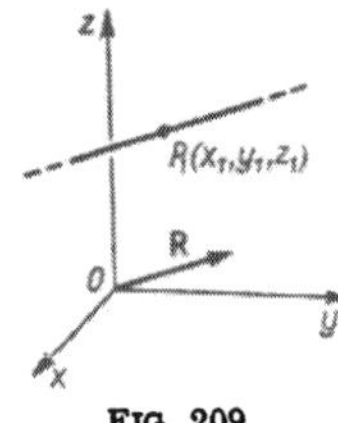

FIG. 209

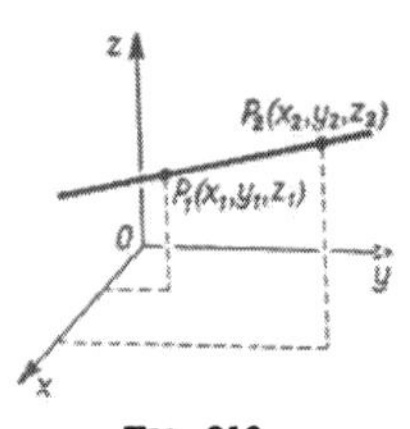

FIG. 210

Parametric equations

$$x = x_1 + lt, \quad y = y_1 + mt, \quad z = z_1 + nt;$$

in the vector form $\boldsymbol{r} = \boldsymbol{r}_1 + \boldsymbol{R}t$.

To obtain the "canonical form" (2) from (1), we set

$$l = \begin{vmatrix} B_1 & C_1 \\ B_2 & C_2 \end{vmatrix}, \quad m = \begin{vmatrix} C_1 & A_1 \\ C_2 & A_2 \end{vmatrix}, \quad n = \begin{vmatrix} A_1 & B_1 \\ A_2 & B_2 \end{vmatrix},$$

in the vector form $\boldsymbol{R} = \boldsymbol{N}_1 \times \boldsymbol{N}_2$ [1]; the numbers x_1, y_1, z_1 are chosen so as to satisfy the equations (1).

(b) through two given points $P_1(x_1, y_1, z_1)$ and $P_2(x_2, y_2, z_2)$ (Fig. 210):

$$\frac{x - x_1}{x_2 - x_1} = \frac{y - y_1}{y_2 - y_1} = \frac{z - z_1}{z_2 - z_1};$$

in the vector form $(\boldsymbol{r} - \boldsymbol{r}_1) \times (\boldsymbol{r}_2 - \boldsymbol{r}_1) = 0$ [1].

---

[1] For products of vectors, see p. 616.

(c) through a given point $P_1(x_1, y_1, z_1)$ and perpendicular to the plane $Ax + By + Cz + D = 0$ or $\boldsymbol{r}\boldsymbol{N} + D = 0$ (¹) (Fig. 211):

$$\frac{x - x_1}{A} = \frac{y - y_1}{B} = \frac{z - z_1}{C};$$

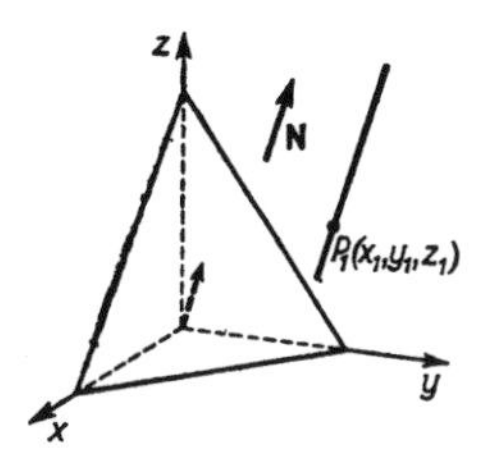

FIG. 211

in the vector form $(\boldsymbol{r} - \boldsymbol{r}_1) \times \boldsymbol{N} = 0$ (¹).

Distance of the point $M(a, b, c)$ **from a straight line** given by the equations (2) is expressed by the formula

$$\delta^2 = \frac{[(a - x_1)\, m - (b - y_1)\, l]^2 + [(b - y_1)\, n - (c - z_1)\, m]^2}{l^2 + m^2 + n^2} + \frac{[(c - z_1)\, l - (a - x_1)\, n]^2}{l^2 + m^2 + n^2}.$$

Distance between two straight lines given by the equations in the canonical form

$$\frac{x - x_1}{l_1} = \frac{y - y_1}{m_1} = \frac{z - z_1}{n_1},$$

$$\frac{x - x_2}{l_2} = \frac{y - y_2}{m_2} = \frac{z - z_2}{n_2}$$

is expressed by the formula

$$\delta = \frac{\begin{vmatrix} x_1 - x_2 & y_1 - y_2 & z_1 - z_2 \\ l_1 & m_1 & n_1 \\ l_2 & m_2 & n_2 \end{vmatrix}}{\sqrt{\begin{vmatrix} l_1 & m_1 \\ l_2 & m_2 \end{vmatrix}^2 + \begin{vmatrix} m_1 & n_1 \\ m_2 & n_2 \end{vmatrix}^2 + \begin{vmatrix} n_1 & l_1 \\ n_2 & l_2 \end{vmatrix}^2}}.$$

Two given lines intersect if and only if the numerator of this expression is zero.

(¹) For products of vectors, see p. 616.

Points of intersection of planes and straight lines

	Equations of planes and lines	Coordinates of points of intersection	Remarks
Three planes	$A_1x + B_1y + C_1z + D_1 = 0$, $A_2x + B_2y + C_2z + D_2 = 0$, $A_3x + B_3y + C_3z + D_3 = 0$	$\bar{x} = \frac{-\Delta_x}{\Delta}$, $\bar{y} = \frac{-\Delta_y}{\Delta}$, $\bar{z} = \frac{-\Delta_z}{\Delta}$ where $\Delta = \begin{vmatrix} A_1 & B_1 & C_1 \\ A_2 & B_2 & C_2 \\ A_3 & B_3 & C_3 \end{vmatrix}$, $\Delta_x = \begin{vmatrix} D_1 & B_1 & C_1 \\ D_2 & B_2 & C_2 \\ D_3 & B_3 & C_3 \end{vmatrix}$, $\Delta_y = \begin{vmatrix} A_1 & D_1 & C_1 \\ A_2 & D_2 & C_2 \\ A_3 & D_3 & C_3 \end{vmatrix}$, $\Delta_z = \begin{vmatrix} A_1 & B_1 & D_1 \\ A_2 & B_2 & D_2 \\ A_3 & B_3 & D_3 \end{vmatrix}$	Three planes intersect in one point, if $\Delta \neq 0$; if $\Delta = 0$ and at least one minor of the second order is different from zero, then the planes are parallel to a certain direction; if all minors are zero, then the planes pass through a common line
Four planes	$A_1x + B_1y + C_1z + D_1 = 0$, $A_2x + B_2y + C_2z + D_2 = 0$, $A_3x + B_3y + C_3z + D_3 = 0$, $A_4x + B_4y + C_4z + D_4 = 0$	We find a point of intersection of any three of the four planes (see below). In this case ($\delta = 0$) one of the equations follows from three remaining ones	Four planes intersect in one point only if $\delta = \begin{vmatrix} A_1 & B_1 & C_1 & D_1 \\ A_2 & B_2 & C_2 & D_2 \\ A_3 & B_3 & C_3 & D_3 \\ A_4 & B_4 & C_4 & D_4 \end{vmatrix} = 0$
A plane and a line	(1) $Ax + By + Cz + D = 0$, $\frac{x - x_1}{l} = \frac{y - y_1}{m} = \frac{z - z_1}{n}$, (2) $Ax + By + Cz + D = 0$, $y = kx + a$, $z = hx + b$	(1) $\bar{x} = x_1 - l\varrho$, $\bar{y} = y_1 - m\varrho$, $\bar{z} = z_1 - n\varrho$, where $\varrho = \frac{Ax_1 + By_1 + Cz_1 + D}{Al + Bm + Cn}$; (2) $\bar{x} = -\frac{Ba + Cb + D}{A + Bk + Ch}$, $\bar{y} = k\bar{x} + a$, $\bar{z} = h\bar{x} + b$	If $Al + Bm + Cn = 0$ or $A + Bk + Ch = 0$, then the line is parallel to the plane, if, moreover, $Ax_1 + By_1 + Cz_1 + D = 0$ or $Ba + Cb + D = 0$, then the line lies on the plane
Two lines	$y = k_1x + a_1$, $z = h_1x + b_1$; $y = k_2x + a_2$, $z = h_2x + b_2$	$\bar{x} = \frac{a_2 - a_1}{k_1 - k_2} = \frac{b_2 - b_1}{h_1 - h_2}$, $\bar{y} = \frac{k_1a_2 - k_2a_1}{k_1 - k_2}$, $\bar{z} = \frac{h_1b_2 - h_2b_1}{h_1 - h_2}$	These formulas determine a point of intersection provided that $(a_1 - a_2)(h_1 - h_2) = (b_1 - b_2)(k_1 - k_2)$, otherwise, the lines do not intersect (see p. 267).

Angle between planes and straight lines

Angle between	Equations of planes and lines	Formula for the angle
two planes	$A_1x + B_1y + C_1z + D_1 = 0,$ $A_2x + B_2y + C_2z + D_2 = 0$	$\cos\varphi = \dfrac{A_1A_2 + B_1B_2 + C_1C_2}{\sqrt{(A_1^2 + B_1^2 + C_1^2)(A_2^2 + B_2^2 + C_2^2)}}$
in the vector form	$\boldsymbol{r}\boldsymbol{N}_1 + D_1 = 0,$ $\boldsymbol{r}\boldsymbol{N}_2 + D_2 = 0$	$\cos\varphi = \dfrac{\boldsymbol{N}_1\boldsymbol{N}_2}{N_1N_2}$
two lines	$\dfrac{x - x_1}{l_1} = \dfrac{y - y_1}{m_1} = \dfrac{z - z_1}{n_1},$ $\dfrac{x - x_2}{l_2} = \dfrac{y - y_2}{m_2} = \dfrac{z - z_2}{n_2}$	$\cos\varphi = \dfrac{l_1l_2 + m_1m_2 + n_1n_2}{\sqrt{(l_1^2 + m_1^2 + n_1^2)(l_2^2 + m_2^2 + n_2^2)}}$
in the vector form	$(\boldsymbol{r} - \boldsymbol{r}_1) \times \boldsymbol{R}_1 = 0,$ $(\boldsymbol{r} - \boldsymbol{r}_2) \times \boldsymbol{R}_2 = 0$	$\cos\varphi = \dfrac{\boldsymbol{R}_1\boldsymbol{R}_2}{R_1R_2}$
a line and a plane	$\dfrac{x - x_1}{l} = \dfrac{y - y_1}{m} = \dfrac{z - z_1}{n},$ $Ax + By + Cz + D = 0$	$\sin\varphi = \dfrac{Al + Bm + Cn}{\sqrt{(A^2 + B^2 + C^2)(l^2 + m^2 + n^2)}}$
in the vector form	$(\boldsymbol{r} - \boldsymbol{r}_1) \times \boldsymbol{R} = 0,$ $\boldsymbol{r}\boldsymbol{N} + D = 0$	$\sin\varphi = \dfrac{\boldsymbol{R}\boldsymbol{N}}{RN}$

Conditions for parallelness (notation as above):

(a) *Two planes:* $\dfrac{A_1}{A_2} = \dfrac{B_1}{B_2} = \dfrac{C_1}{C_2}$ or $\boldsymbol{N}_1 \times \boldsymbol{N}_2 = 0.$

(b) *Two lines:* $\dfrac{l_1}{l_2} = \dfrac{m_1}{m_2} = \dfrac{n_1}{n_2}$ or $\boldsymbol{R}_1 \times \boldsymbol{R}_2 = 0.$

(c) *A line and a plane:* $Al + Bm + Cn = 0$ or $\boldsymbol{R}\boldsymbol{N} = 0.$

Conditions for perpendicularity (notation as above):

(a) *Two planes:* $A_1A_2 + B_1B_2 + C_1C_2 = 0$ or $\boldsymbol{N}_1\boldsymbol{N}_2 = 0.$

(b) *Two lines:* $l_1l_2 + m_1m_2 + n_1n_2 = 0$ or $\boldsymbol{R}_1\boldsymbol{R}_2 = 0.$

(c) *A line and a plane:* $\dfrac{A}{l} = \dfrac{B}{m} = \dfrac{C}{n}$ or $\boldsymbol{N} \times \boldsymbol{R} = 0.$

10. Surfaces of the second degree (canonical equations)([1])

Central surfaces. The following equations are given in the *canonical form*: the *centre* of the surface (the point bisecting all chords passing through it) lies at the origin and the coordinate axes are the axes of symmetry of the surface. Then the coordinate planes are planes of symmetry.

Ellipsoid (Fig. 212):

$$\frac{x^2}{a^2}+\frac{y^2}{b^2}+\frac{z^2}{c^2}=1,$$

where a, b, c are *semiaxes* of the ellipsoid.

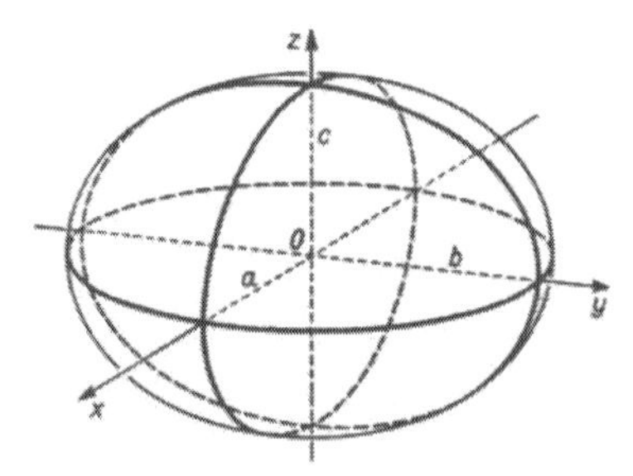

FIG. 212

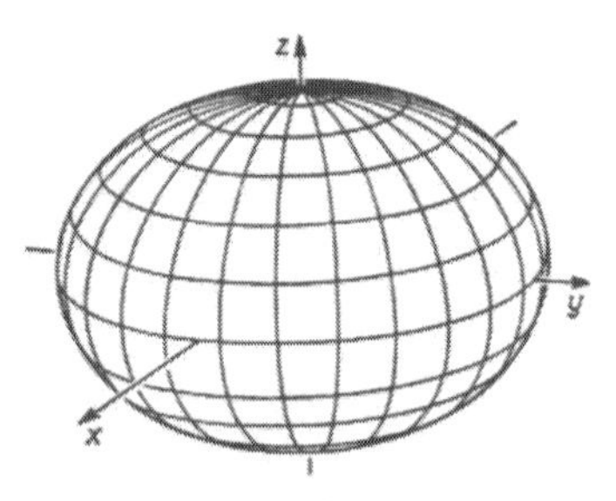

FIG. 213

If $a=b>c$, then we have a *flattened ellipsoid of revolution* or a *flattened spheroid* (Fig. 213). It is obtained by revolving an ellipse

$$\frac{x^2}{a^2}+\frac{z^2}{c^2}=1$$

([1]) For general equations of surfaces of the second degree, see p. 275.

lying in the xz plane, about its minor axis. If $a = b < c$, then we have a *lengthened ellipsoid of revolution* or a *lengthened spheroid* (Fig. 214) obtained by revolving the ellipse

$$\frac{x^2}{a^2} + \frac{z^2}{c^2} = 1,$$

lying in the xz plane about its major axis. If $a = b = c$, we obtain a *sphere*

$$x^2 + y^2 + z^2 = a^2.$$

Any plane section of an ellipsoid is an ellipse (in a particular case a circle). Volume of the ellipsoid is equal to $\frac{4}{3}\pi abc$.

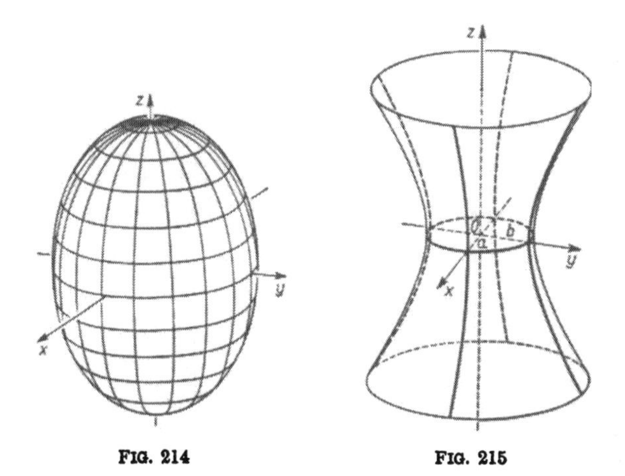

FIG. 214 FIG. 215

Hyperboloid of one sheet (Fig. 215):

$$\frac{x^2}{a^2} + \frac{y^2}{b^2} - \frac{z^2}{c^2} = 1,$$

where a and b are the *real semiaxes* and c is the *imaginary semiaxis*.
For *linear generators* see p. 273.

Hyperboloid of two sheets (Fig. 216):

$$\frac{x^2}{a^2} + \frac{y^2}{b^2} - \frac{z^2}{c^2} = -1,$$

where c is the *real semiaxis* and a and b are the *imaginary semiaxes*.

For both types of hyperboloids, the plane sections parallel to the z axis are hyperbolas (in the case of a hyperboloid of one

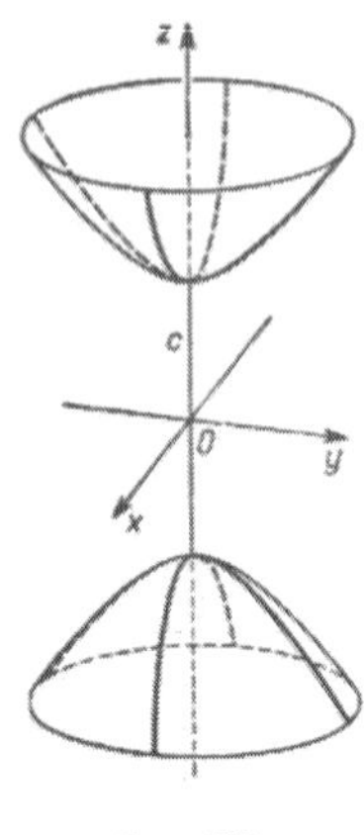

FIG. 216

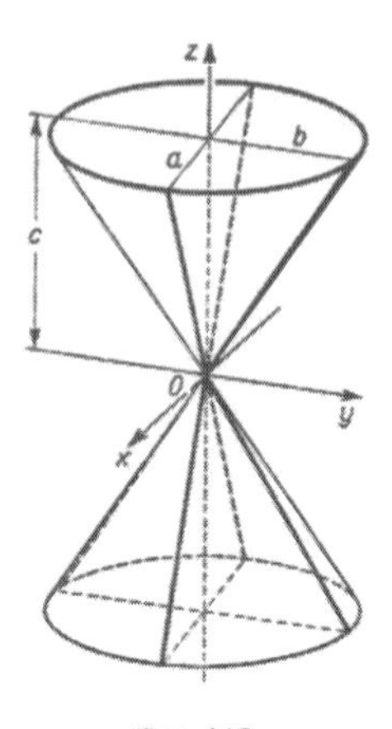

FIG. 217

sheet, it can be a pair of intersecting lines) and the plane sections parallel to the xy plane are ellipses.

If $a = b$, then the hyperboloid can be obtained by revolving a hyperbola with the semiaxes a and c about the axis $2c$: a hyperboloid of one sheet is a result of revolving the hyperbola about the real (*transverse*) axis and a hyperboloid of two sheets—a result of revolving the hyperbola about the imaginary (*conjugate*) axis.

Cone (Fig. 217):

$$\frac{x^2}{a^2} + \frac{y^2}{b^2} - \frac{z^2}{c^2} = 0.$$

It has the *vertex* at the origin and its *directing curve* (see p. 208) can be taken as an ellipse with the semiaxes a and b lying in the plane perpendicular to the z axis at the distance c from the origin. This is also an asymptotical cone of two hyperboloids

$$\frac{x^2}{a^2} + \frac{y^2}{b^2} - \frac{z^2}{c^2} = \pm 1,$$

i.e., a point of every generator of the cone approaches both hyperboloids, when its coordinates tend to infinity (Fig. 218). If $a = b$, then we have a *cone of revolution* (see p. 208).

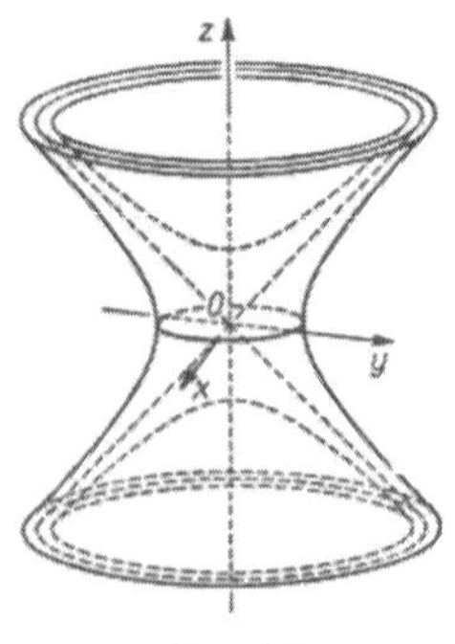

FIG. 218

Paraboloids. Paraboloids do not have a centre. In the following equations, the *vertex* of the paraboloid lies at the origin, the z axis is an *axis of symmetry* and the xz and yz planes are *planes of symmetry*.

Elliptic paraboloid (Fig. 219):

$$z = \frac{x^2}{a^2} + \frac{y^2}{b^2}.$$

Plane sections parallel to the z axis are parabolas; sections parallel to the xy plane are ellipses. If $a = b$, then we have a *paraboloid of revolution* obtained by revolving the parabola $z = x^2/a^2$ lying in the xz plane about its axis.

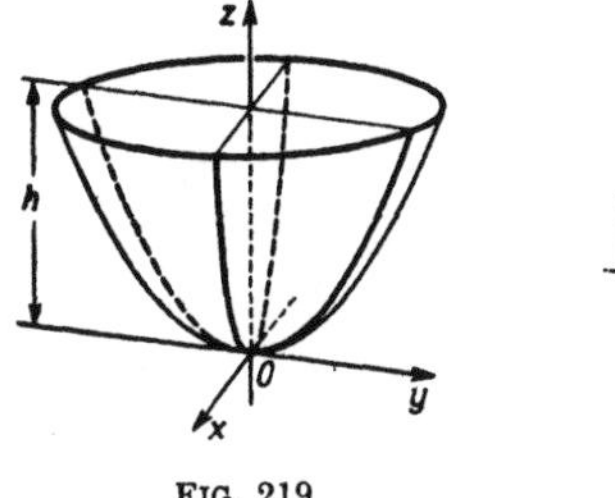

FIG. 219

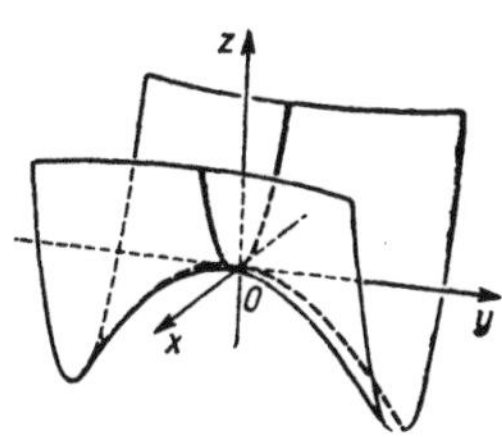

FIG. 220

Volume of the part of the paraboloid cut off by the plane perpendicular to its axis at the altitude h is equal to $\frac{1}{2}\pi abh$, i.e., one half of volume of an elliptic cylinder with the same base and altitude.

Hyperbolic paraboloid (Fig. 220):

$$z = \frac{x^2}{a^2} - \frac{y^2}{b^2}.$$

Sections parallel to the yz plane are all equal parabolas; sections parallel to the xz plane are also all equal parabolas; sections parallel to the xy plane are hyperbolas or a pair of intersecting straight lines.

A rectilinear generator of a surface is a straight line lying wholly on the surface, as, for examples, the generators of a cone or a cylinder.

A hyperboloid of one sheet (Fig. 221)

$$\frac{x^2}{a^2} + \frac{y^2}{b^2} - \frac{z^2}{c^2} = 1$$

has two families of generators:

$$\frac{x}{a}+\frac{z}{c}=u\left(1+\frac{y}{b}\right),\quad u\left(\frac{x}{a}-\frac{z}{c}\right)=1-\frac{y}{b},$$

$$\frac{x}{a}+\frac{z}{c}=v\left(1-\frac{y}{b}\right),\quad v\left(\frac{x}{a}-\frac{z}{c}\right)=1+\frac{y}{b},$$

where u and v are arbitrary numbers.

A hyperbolic paraboloid (Fig. 222)

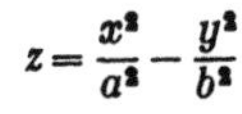

$$z=\frac{x^2}{a^2}-\frac{y^2}{b^2}$$

has also two families of generators:

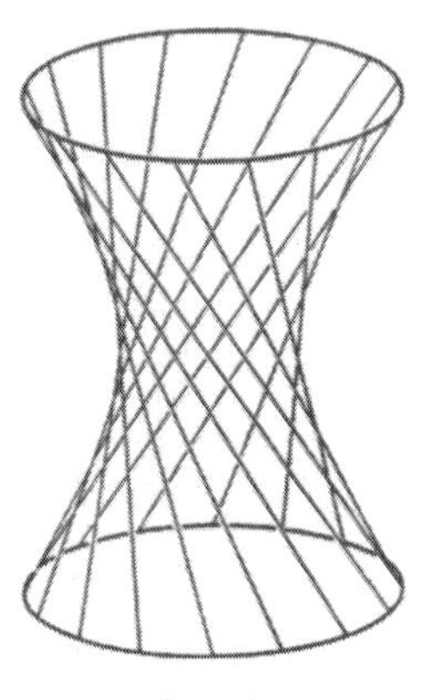

FIG. 221

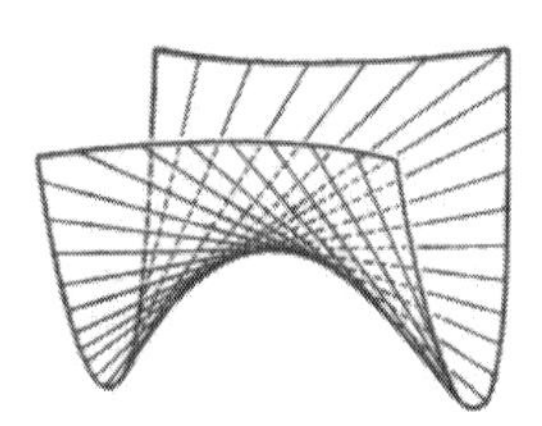

FIG. 222

$$\frac{x}{a}+\frac{y}{b}=u,\quad u\left(\frac{x}{a}-\frac{y}{b}\right)=z,$$

$$\frac{x}{a}-\frac{y}{b}=v,\quad v\left(\frac{x}{a}+\frac{y}{b}\right)=z,$$

where u and v are arbitrary numbers. Through each point of these surfaces there pass two generators: one of either family (Figs. 221 and 222 show only one family).

Cylinders. *Elliptic cylinder* (Fig. 223):

$$\frac{x^2}{a^2}+\frac{y^2}{b^2}=1,$$

hyperbolic cylinder (Fig. 224):

$$\frac{x^2}{a^2}-\frac{y^2}{b^2}=1,$$

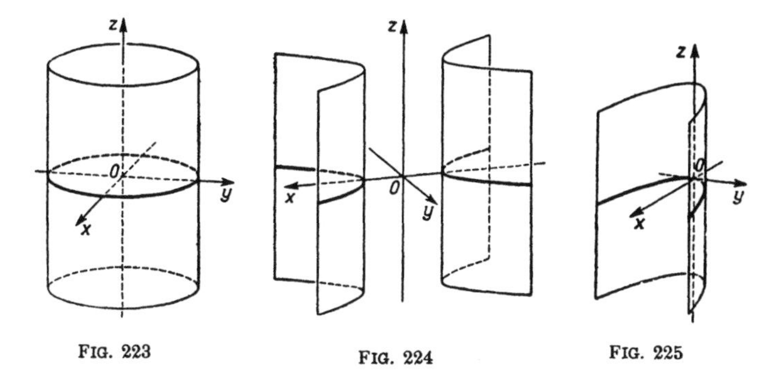

FIG. 223 FIG. 224 FIG. 225

parabolic cylinder (Fig. 225):

$$y^2 = 2px.$$

11. Surfaces of the second degree (general theory)

General equation of a surface of the second degree.

$$a_{11}x^2 + a_{22}y^2 + a_{33}z^2 + 2a_{12}xy + 2a_{23}yz + 2a_{31}zx + 2a_{14}x +$$
$$+ 2a_{24}y + 2a_{34}z + a_{44} = 0.$$

Invariants of a surface of the second degree [1]:

$$\Delta = \begin{vmatrix} a_{11} & a_{12} & a_{13} & a_{14} \\ a_{21} & a_{22} & a_{23} & a_{24} \\ a_{31} & a_{32} & a_{33} & a_{34} \\ a_{41} & a_{42} & a_{43} & a_{44} \end{vmatrix}, \qquad \delta = \begin{vmatrix} a_{11} & a_{12} & a_{13} \\ a_{21} & a_{22} & a_{23} \\ a_{31} & a_{32} & a_{33} \end{vmatrix},$$

$$S = a_{11} + a_{22} + a_{33}, \qquad T = a_{22}a_{33} + a_{33}a_{11} + a_{11}a_{22} - a_{23}^2 - a_{31}^2 - a_{12}^2.$$

These quantities remain unchanged while translating the origin and rotating the coordinate axes.

The shape of a surface of the second degree given by an equation can be determined according to the signs of its invariants Δ, δ, S and T, using the following table. The table shows also the canonical equations to which the given equation of a surface can be reduced by a transformation of coordinates. The equation of so called imaginary surfaces is not satisfied by any real point except in two cases: the vertex of an imaginary cone and the line of intersection of two imaginary planes.

[1] We assume here $a_{ik} = a_{ki}$.

I. $\delta \neq 0$ (central surfaces)

	$S\delta > 0,\ T > 0$	$S\delta$ and T are not both > 0
$\Delta < 0$	Ellipsoid $\frac{x^2}{a^2}+\frac{y^2}{b^2}+\frac{z^2}{c^2}=1$	Hyperboloid of two sheets $\frac{x^2}{a^2}+\frac{y^2}{b^2}-\frac{z^2}{c^2}=-1$
$\Delta > 0$	Imaginary ellipsoid $\frac{x^2}{a^2}+\frac{y^2}{b^2}+\frac{z^2}{c^2}=-1$	Hyperboloid of one sheet $\frac{x^2}{a^2}+\frac{y^2}{b^2}-\frac{z^2}{c^2}=1$
$\Delta = 0$	Imaginary cone (with a real vertex) $\frac{x^2}{a^2}+\frac{y^2}{b^2}+\frac{z^2}{c^2}=0$	Cone $\frac{x^2}{a^2}+\frac{y^2}{b^2}-\frac{z^2}{c^2}=0$

II. $\delta = 0$ (paraboloids, cylinders and pairs of planes)

	$T > 0$	$T < 0$
$\Delta \neq 0$	$\Delta < 0$ Elliptic paraboloid $\frac{x^2}{a^2}+\frac{y^2}{b^2}=\pm z$	$\Delta > 0$ Hyperbolic paraboloid $\frac{x^2}{a^2}-\frac{y^2}{b^2}=\pm z$
$\Delta = 0$	A cylindrical surface whose generating curve is a curve of the second degree. According to the shape of this curve (see pp. 254, 255), we obtain various types of cylinders: imaginary or real elliptic cylinder (if $T > 0$); hyperbolic cylinder (if $T < 0$); parabolic cylinder (if $T = 0$), provided that the surface is not degenerate, i.e., does not reduce to a pair of real or imaginary planes or to one plane. Condition for degeneracy: $\begin{vmatrix} a_{11} & a_{12} & a_{14} \\ a_{21} & a_{22} & a_{24} \\ a_{41} & a_{42} & a_{44} \end{vmatrix} + \begin{vmatrix} a_{11} & a_{13} & a_{14} \\ a_{31} & a_{33} & a_{34} \\ a_{41} & a_{43} & a_{44} \end{vmatrix} + \begin{vmatrix} a_{22} & a_{23} & a_{24} \\ a_{32} & a_{33} & a_{34} \\ a_{42} & a_{43} & a_{44} \end{vmatrix} = 0.$	

II. DIFFERENTIAL GEOMETRY

In differential geometry we investigate plane or space curves and surfaces by using the methods of differential calculus. Therefore we assume that the functions involved in the equations are continuous and have continuous derivatives up to a certain order which is needed in the considered problem [1]. In dealing with geometrical objects given by their equations, we distinguish those properties which *depend* on the choice of a coordinate system (as, for example, points of intersection of the curve with the coordinate axes, the slope of a tangent line, maxima and minima) and *invariant properties* which are not disturbed by transformations of coordinates and which therefore depend only on the curve or surface itself (as, for example, points of inflection, vertices or curvature of a curve). On the other hand, we distinguish the *local properties* which concern only small parts of a curve or a surface (e.g., curvature, linear element of a surface) and the properties of a curve or surface in the whole (e.g., number of vertices, length of a closed curve).

A. PLANE CURVES

1. Ways in which a curve can be defined

Equation of a curve [2]. A plane curve can be defined analytically in one of the following forms:

In Cartesian coordinates:

(1) *implicit form* $F(x, y) = 0$,

(2) *explicit form* $y = f(x)$,

(3) *parametric form* $x = x(t)$, $y = y(t)$.

---

[1] This condition may fail only at certain separate points of a curve or surface; in this case we have points of a special type (as, for example, a discontinuity or a bend of a curve). For such points see pp. 285, 306.

[2] For the general notion of an equation of a line see p. 239.

In polar coordinates:

(4) $\varrho = f(\varphi)$.

Positive direction of a curve. The *positive direction* of a curve given in the form (3) is that in which a point $M(x(t), y(t))$ of the curve moves, when the parameter t increases. If the curve is given in the form (2), then the abscissa x can be taken as a parameter: $x = x$, $y = f(x)$, and the positive direction of the curve corresponds

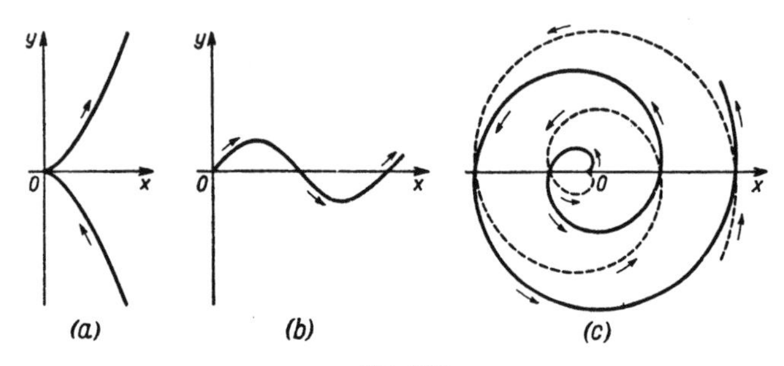

FIG. 226

to the positive direction of the x axis (i.e., from the left to the right). If the curve is given in the form (4), then the angle φ can be taken as a parameter: $x = f(\varphi)\cos\varphi$, $y = f(\varphi)\sin\varphi$ and the positive direction of the curve corresponds to the increasing φ (i.e., counterclockwise).

Examples (Fig. 226). (a) $x = t^2$, $y = t^3$. (b) $y = \sin x$. (c) $\varrho = a\varphi$.

2. Local elements of a curve

In this section we denote by M a variable point of the curve determined by: the value of x, in the form (2), the value of t, in the form (3) and the value of φ in the form (4); N denotes a point infinitely near to M determined respectively by the values $x + dx$, $t + dt$ and $\varphi + d\varphi$.

Differential of an arc. If s is the length of the curve from a certain fixed point A to M then an infinitely small increment $\Delta s = \breve{MN}$ of length is approximately expressed by the differential [1] ds of arc:

---

[1] For the differential, see pp. 362–364.

$$\Delta s \approx ds = \sqrt{1 + \left(\frac{dy}{dx}\right)^2}\, dx, \quad \text{for the form (2),}$$

$$= \sqrt{x_t'^2 + y_t'^2}\, dt, \quad \text{for the form (3),}$$

$$= \sqrt{\varrho^2 + \varrho'^2}\, d\varphi, \quad \text{for the form (4).}$$

Examples.

(1) $y = \sin x$, $ds = \sqrt{1 + \cos^2 x}\, dx$,

(2) $x = t^2$, $y = t^3$, $ds = |t| \sqrt{4 + 9t^2}\, dt$,

(3) $\varrho = a\varphi$, $ds = a\sqrt{1 + \varphi^2}\, d\varphi$.

Tangent and normal line. A *tangent line* at the point M is the limit position of a chord MN, when $N \to M$. A *normal line* is the straight line passing through M and perpendicular to the tangent (Fig. 227).

Equations of the tangent and the normal (x, y are coordinates of the point M of the curve; X, Y are running coordinates of points of the tangent or the normal; derivatives are taken at the point M).

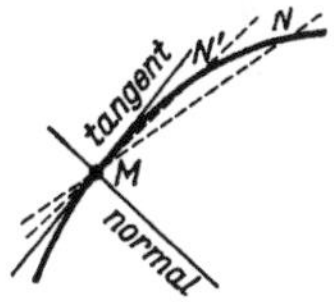

FIG. 227

Form of the curve	Equation of the tangent	Equation of the normal
(1)	$\frac{\partial F}{\partial x}(X - x) + \frac{\partial F}{\partial y}(Y - y) = 0$	$\frac{X - x}{\frac{\partial F}{\partial x}} = \frac{Y - y}{\frac{\partial F}{\partial y}}$
(2)	$Y - y = \frac{dy}{dx}(X - x)$	$Y - y = -\frac{1}{\frac{dy}{dx}}(X - x)$
(3)	$\frac{Y - y}{y_t'} = \frac{X - x}{x_t'}$	$x_t'(X - x) + y_t'(Y - y) = 0$

Examples. Find the equations of the tangent and the normal:

(1) For the circle $x^2 + y^2 = 25$ at the point $M(3,4)$. Equation of the tangent $2x(X - x) + 2y(Y - y) = 0$ or, by the equation of the circle, $Xx + Yy = 25$; at the point M: $3X + 4Y = 25$. Equation of the normal: $\frac{X - x}{2x} = \frac{Y - y}{2y}$ or $Y = \frac{y}{x} X$; at M: $Y = \frac{4}{3} X$.

(2) For the sine curve $y = \sin x$ at $O\,(0,0)$. Equation of the tangent: $Y - \sin x = (X - x)\cos x$ or $Y = X\cos x + \sin x - x\cos x$; at the point O: $Y = X$. Equation of the normal: $Y - \sin x$

$= -\frac{1}{\cos x}(X - x)$ or $Y = -X \sec x + \sin x + x \sec x$; at the point O: $Y = -X$.

(3) For the curve $x = t^2$, $y = t^3$ at the point $M(4, -8)$, $t = -2$. Equation of the tangent: $\frac{Y - t^3}{3t^2} = \frac{X - t^2}{2t}$ or $Y = \frac{3}{2}tX - \frac{1}{2}t^3$; at the point $M(4, -8)$: $Y = -3X + 4$. Equation of the normal $2t(X - t^2) + 3t^2(Y - t^3) = 0$ or $2X + 3tY = t^2(2 + 3t^2)$; at M: $X - 3Y = 28$.

Positive direction of the tangent and normal for a curve given in the form (2), (3) or (4) (see pp. 277, 278), is determi- as follows. Positive direction of the tangent coincides with that of the curve at the point of contact (see p. 278). Positive direction of the normal is obtained from that of the tangent by rotating the tangent counterclockwise about the point of contact through the angle 90°. The point M divides the tangent and the normal into positive and negative half lines (Fig. 228).

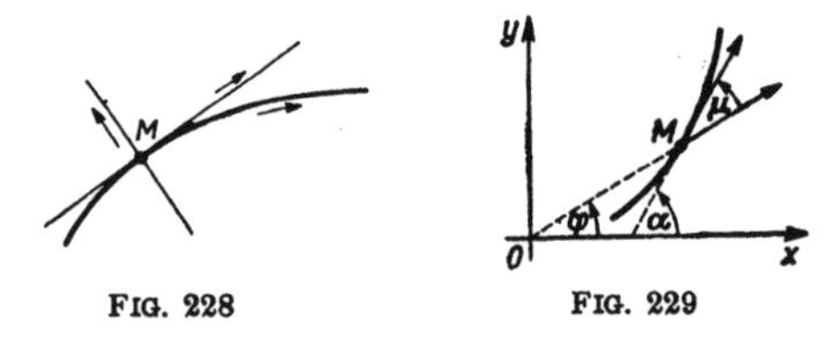

FIG. 228 FIG. 229

The slope of the tangent is determined by the angle α between the positive direction of the x axis and positive direction of the tangent or (if the curve is given in polar coordinates) by the angle μ between the positive direction of the radius vector $OM = \varrho$ and the positive direction of the tangent (Fig. 229). The angles α and μ can be found from the formulas (ds is computed from the formulas on p. 279):

$$\tan \alpha = \frac{dy}{dx}, \qquad \cos \alpha = \frac{dx}{ds}, \qquad \sin \alpha = \frac{dy}{ds},$$

$$\tan \mu = \frac{\varrho}{\left(\frac{d\varrho}{d\varphi}\right)}, \qquad \cos \mu = \frac{d\varrho}{ds}, \qquad \sin \mu = \varrho \frac{d\varphi}{ds}.$$

Examples.

(1) $y = \sin x$; then

$$\tan \alpha = \cos x,$$

$$\cos \alpha = \frac{1}{\sqrt{1 + \cos^2 x}}, \qquad \sin \alpha = \frac{\cos x}{\sqrt{1 + \cos^2 x}}.$$

(2) $x = t^2$, $y = t^3$; then $\tan\alpha = \frac{3}{2}t$ and

$$\cos\alpha = \frac{2}{\sqrt{4+9t^2}}, \qquad \sin\alpha = \frac{3t}{\sqrt{4+9t^2}};$$

$$\cos\alpha = -\frac{2}{\sqrt{4+9t^2}}, \qquad \sin\alpha = -\frac{3t}{\sqrt{4+9t^2}}.$$

(3) $\varrho = a\varphi$; then $\tan\mu = \varphi$ and

$$\cos\mu = \frac{1}{\sqrt{1+\varphi^2}}, \qquad \sin\mu = \frac{\varphi}{\sqrt{1+\varphi^2}}.$$

Segments of the tangent and normal; subtangent and subnormal (Fig. 230).

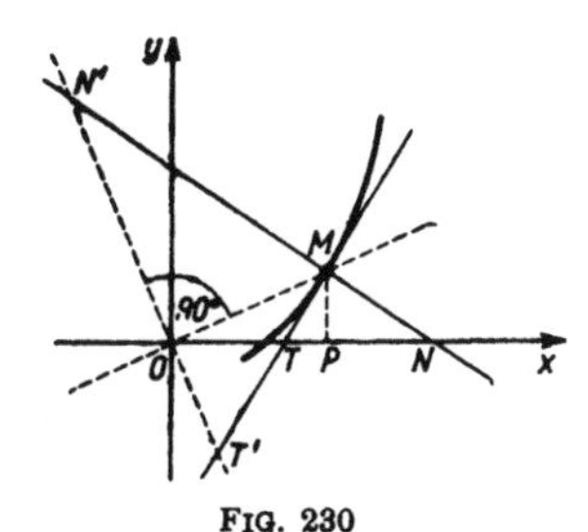

FIG. 230

(a) In Cartesian coordinates, for the form (2) and (3) of the curve (see p. 277):

Segment of the tangent $\quad MT = \left|\frac{y}{y'}\sqrt{1+y'^2}\right|,$

Segment of the normal $\quad MN = \left|y\sqrt{1+y'^2}\right|,$

Subtangent $\quad PT = \left|\frac{y}{y'}\right|,$

Subnormal $\quad PN = |yy'|.$

(b) In polar coordinates, for the form (4) of the curve (see p. 278):

Segment of the polar tangent $\quad MT' = \left|\frac{\varrho}{\varrho'}\sqrt{\varrho^2+\varrho'^2}\right|,$

Segment of the polar normal $\quad MN' = \left|\sqrt{\varrho^2+\varrho'^2}\right|,$

Polar subtangent $\quad OT' = \left|\frac{\varrho^2}{\varrho'}\right|,$

Polar subnormal $\quad ON' = |\varrho'|.$

Examples.

(1) $y = \cosh x$; $y' = \sinh x$, $\sqrt{1+y'^2} = \cosh x$; $MT = |\cosh x \coth x|$, $MN = |\cosh^2 x|$, $PT = |\coth x|$, $PN = |\sinh x \cosh x|$.

(2) $\varrho = a\varphi$; $\varrho' = a$, $\sqrt{\varrho^2 + \varrho'^2} = a\sqrt{1+\varphi^2}$; $MT' = \left|a\varphi\sqrt{1+\varphi^2}\right|$, $MN' = \left|a\sqrt{1+\varphi^2}\right|$, $OT' = |a\varphi^2|$, $ON' = a$.

Angle between two curves. The angle between two curves Γ_1 and Γ_2 intersecting in the point M is understood to be the angle β between the tangents to these curves at the point M (Fig. 231).

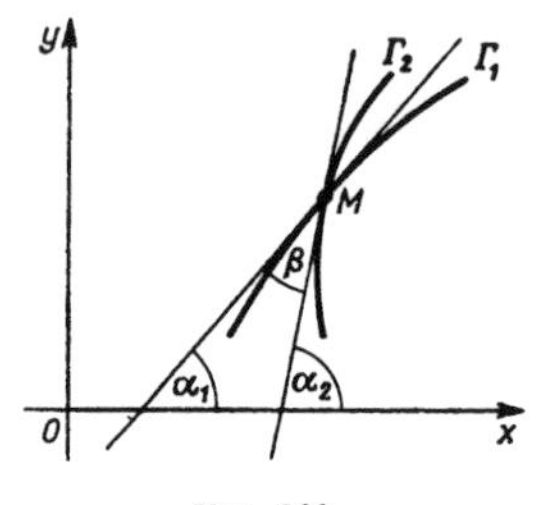

FIG. 231

Evaluation of the angle β reduces to determining an angle between two straight lines (see p. 240) with the slopes

$$k_1 = \tan\alpha_1 = \left(\frac{df_1}{dx}\right)_M, \qquad k_2 = \tan\alpha_2 = \left(\frac{df_2}{dx}\right)_M,$$

where $y = f_1(x)$ is the equation of Γ_1 and $y = f_2(x)$ is the equation of Γ_2; the derivatives are taken at the point M.

Example. Determine the angle between the parabolas $y = \sqrt{x}$ and $y = x^2$ at the point $M(1,1)$.

$$\tan\alpha_1 = \left(\frac{d\sqrt{x}}{dx}\right)_{x=1} = \frac{1}{2}, \qquad \tan\alpha_2 = \left(\frac{d(x^2)}{dx}\right)_{x=1} = 2,$$

$$\tan\beta = \frac{\tan\alpha_2 - \tan\alpha_1}{1+\tan\alpha_1\tan\alpha_2} = \frac{3}{4}.$$

Concavity and convexity of a curve. Given a point M of a curve $y=f(x)$, it is possible to determine (except for the case when M is a point of inflection p. 285) whether in a small neighbourhood of M, the convexity of the curve is directed upwards or downwards: if $f''(x) < 0$ at the point M, then the curve is *upwards convex* [1] (the point M_1 in Fig. 232); if $f''(x) > 0$, then the

[1] More precisely, it is convex in the positive direction of the y axis.

curve is *downwards convex* (the point M_2); if, $f''(x) = 0$, then some further considerations are necessary (see p. 286, the points of inflection).

Example. $y = x^3$ (see Fig. 6b, p. 99). $y'' = 6x$; for $x < 0$, the curve is upwards convex and for $x > 0$ it is downwards convex.

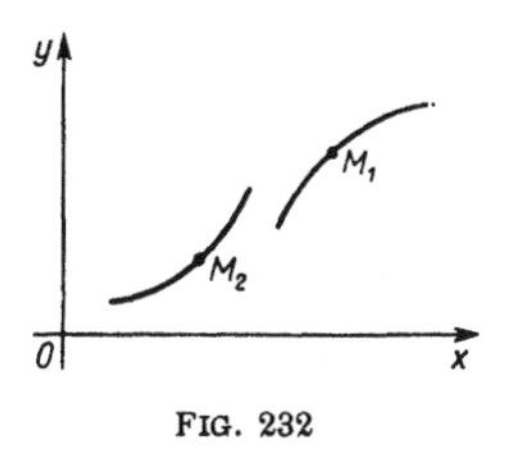

FIG. 232

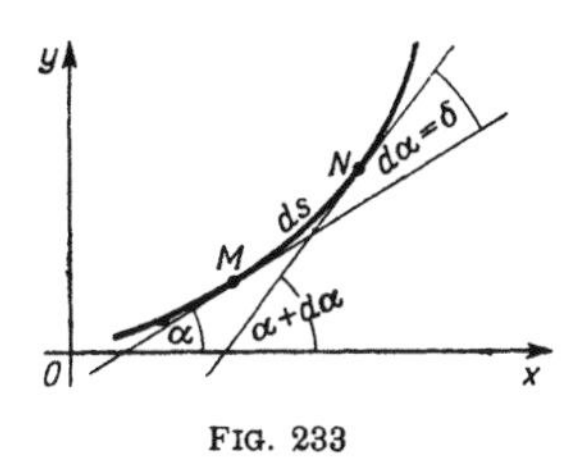

FIG. 233

Curvature and radius of curvature. The *curvature* K of a curve at a point M is the limit of the ratio of the angle δ contained between the positive directions of the tangents at the point M and N (Fig. 233) to the length of the arc $\breve{MN}$, when $\breve{MN} \to 0$:

$$K = \lim_{MN \to 0} \frac{\delta}{\breve{MN}}.$$

The curvature is positive or negative according to the sign of this limit. If $K > 0$, then the centre of curvature lies on the positive half of the normal line (see p. 280) (i.e., the curve is convex in the negative direction of its normal); if $K < 0$, the convexity of the curve is directed otherwise.

The curvature is commonly defined as a positive number; then we take the absolute value of the above limit.

Radius of curvature at the point M of the curve is the reciprocal of the curvature: $R = 1/K$. If the line is more curved, then its curvature is greater and its radius of curvature is less. The curvature of a circle with the radius a is constant and equal to $K = 1/a$ and the radius of curvature is $R = a$. For a straight line $K = 0$, $R = \infty$. In general, the curvature is different for various points of the curve.

Formulas for K and R. Putting $\delta = d\alpha$, $\breve{MN} = ds$ (Fig. 233) we have

$$K = \frac{d\alpha}{ds}, \qquad R = \frac{ds}{d\alpha}. \tag{*}$$

If the curve is given in the form (1), (2), (3) or (4) (see p. 277–278), then K and R can be computed from the formulas:

(**)

For a curve given in the form (2):

$$K = \frac{\dfrac{d^2y}{dx^2}}{\left[1+\left(\dfrac{dy}{dx}\right)^2\right]^{3/2}}, \quad R = \frac{\left[1+\left(\dfrac{dy}{dx}\right)^2\right]^{3/2}}{\dfrac{d^2y}{dx^2}}.$$

For a curve given in the form (3):

$$K = \frac{\begin{vmatrix} x'_t & y'_t \\ x''_t & y''_t \end{vmatrix}}{(x_t'^2 + y_t'^2)^{3/2}}, \quad R = \frac{(x_t'^2 + y_t'^2)^{3/2}}{\begin{vmatrix} x'_t & y'_t \\ x''_t & y''_t \end{vmatrix}}.$$

For a curve given in the form (1):

$$K = \frac{\begin{vmatrix} F''_{xx} & F''_{xy} & F'_x \\ F''_{yx} & F''_{yy} & F'_y \\ F'_x & F'_y & 0 \end{vmatrix}}{(F_x'^2 + F_y'^2)^{3/2}}, \quad R = \frac{(F_x'^2 + F_y'^2)^{3/2}}{\begin{vmatrix} F''_{xx} & F''_{xy} & F'_x \\ F''_{yx} & F''_{yy} & F'_y \\ F'_x & F'_y & 0 \end{vmatrix}}.$$

For a curve given in the form (4):

$$K = \frac{\varrho^2 + 2\varrho'^2 - \varrho\varrho''}{(\varrho^2 + \varrho'^2)^{3/2}}, \quad R = \frac{(\varrho^2 + \varrho'^2)^{3/2}}{\varrho^2 + 2\varrho'^2 - \varrho\varrho''}.$$

Examples.

(1) $y = \cosh x$; $\quad K = \dfrac{1}{\cosh^2 x}$.

(2) $x = t^2$, $y = t^3$; $K = \dfrac{6}{t(4 + 9t^2)^{3/2}}$.

(3) $y^2 - x^2 = a^2$; $K = \dfrac{a^2}{(x^2 + y^2)^{3/2}}$.

(4) $\varrho = a\varphi$; $K = \dfrac{1}{a} \cdot \dfrac{\varphi^2 + 2}{(\varphi^2 + 1)^{3/2}}$.

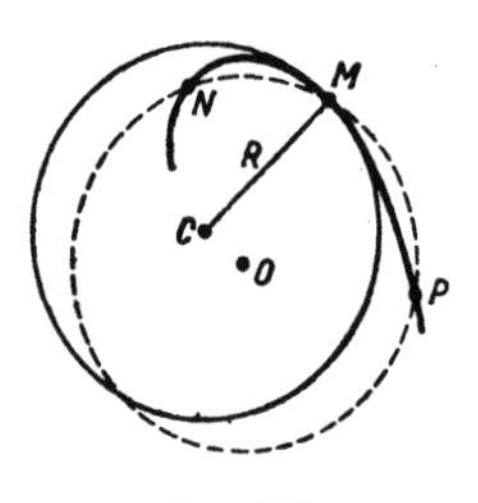

FIG. 234

Circle of curvature and centre of curvature. *Circle of curvature* of a curve at the point M is the limiting position of a circle passing through M and through two points N and P lying in the neighbourhood of M (Fig. 234), when $N \to M$ and $P \to M$. Radius of the circle of curvature is equal to the *radius of curvature* and is expressed by the formulas (**). Centre of the circle of curvature is called the *centre of curvature*. It lies on the normal at the point M in the direction of concavity of the curve. Coordinates (x_C, y_C) of the centre of curvature are given by the formulas;

(**) For a curve given in the form (2) (see p. 277):

$$x_C = x - \frac{\frac{dy}{dx}\left[1+\left(\frac{dy}{dx}\right)^2\right]}{\frac{d^2y}{dx^2}}, \qquad y_C = y + \frac{1+\left(\frac{dy}{dx}\right)^2}{\frac{d^2y}{dx^2}}.$$

For a curve given in the form (3):

$$x_C = x - \frac{y'_t(x'^2_t + y'^2_t)}{\begin{vmatrix} x'_t & y'_t \\ x''_t & y''_t \end{vmatrix}}, \qquad y_C = y + \frac{x'_t(x'^2_t + y'^2_t)}{\begin{vmatrix} x'_t & y'_t \\ x''_t & y''_t \end{vmatrix}}.$$

For a curve given in the form (4):

$$x_C = \varrho \cos\varphi - \frac{(\varrho^2+\varrho'^2)(\varrho\cos\varphi + \varrho'\sin\varphi)}{\varrho^2 + 2\varrho'^2 - \varrho\varrho''},$$

$$y_C = \varrho \sin\varphi - \frac{(\varrho^2+\varrho'^2)(\varrho\sin\varphi - \varrho'\cos\varphi)}{\varrho^2 + 2\varrho'^2 - \varrho\varrho''}.$$

For a curve given in the form (1):

$$x_C = x + \frac{F'_x(F'^2_x + F'^2_y)}{\begin{vmatrix} F''_{xx} & F''_{xy} & F'_x \\ F''_{yx} & F''_{yy} & F'_y \\ F'_x & F'_y & 0 \end{vmatrix}}, \qquad y_C = y + \frac{F'_y(F'^2_x + F'^2_y)}{\begin{vmatrix} F''_{xx} & F''_{xy} & F'_x \\ F''_{yx} & F''_{yy} & F'_y \\ F'_x & F'_y & 0 \end{vmatrix}}.$$

These formulas can be written in the form

$$x_C = x - R\sin\alpha, \qquad y_C = y + R\cos\alpha$$

or

$$x_C = x - R\frac{dy}{ds}, \qquad y_C = y + R\frac{dx}{ds}$$

(Fig. 235), where R is given by the formulas (**) (see p. 284).

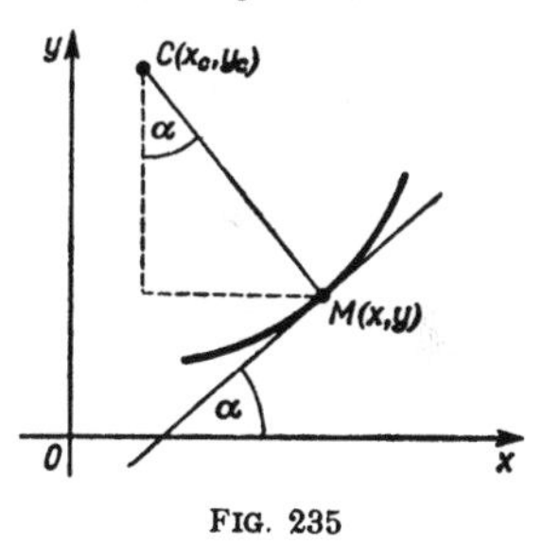

FIG. 235

3. Points of special types [1]

Point of inflection. A *point of inflection* is a point of a curve in which the direction of convexity reverses into the opposite direction (Fig. 236). In a small neighbourhood of a point of inflection, the curve does not lie wholly on one side of the tangent but intersects it; then curvature $K = 0$ and radius of curvature $R = \infty$.

[1] We shall discuss here only those points which are invariant with respect to transformations of coordinates. Maxima and minima are discussed on pp. 379–383.

Determining points of inflection.

For a curve given in the form (2) (p. 277): $y = f(x)$.

A necessary condition for a point of inflection is that the second derivative $f''(x)$ is equal to zero, provided that it exists. To determine the points of inflection in which the second derivative $f''(x)$ exists [1], we find all roots $x_1, x_2, \ldots$ of the equation $f''(x) = 0$ and substitute each root x_i succesively to the next derivatives. If $f'''(x_i) \neq 0$, then x_i is the abscissa of a point of inflection; if $f'''(x_i) = 0$ and $f^{IV}(x_i) \neq 0$, then x_i is not a point of inflection, and so on. If the first non-zero derivative at x_i is of an odd order, then x_i is a point of inflection; if the first non-zero derivative is of an even order, then x_i is not a point of inflection. If the point in question is not a point of inflection (hence the first non-zero derivative is of an even order k), then the curve is upwards convex, if $f^{(k)}(x_i) < 0$ and is downwards convex, if $f^{(k)}(x_i) > 0$.

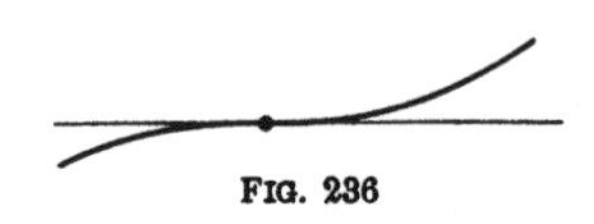

FIG. 236

Examples.

$$(1)\ y = \frac{1}{1 + x^2}; \quad y''(x) = -2\frac{1 - 3x^2}{(1 + x^2)^3}, \quad x_1 = -\frac{1}{\sqrt{3}},$$

$$x_2 = \frac{1}{\sqrt{3}}, \quad f'''(x) = 24x\frac{1 - x^2}{(1 + x^2)^4}, \quad f'''(x_1) \neq 0, \quad f'''(x_2) \neq 0.$$

Points of inflection:

$$A\left(-\frac{1}{\sqrt{3}}, \frac{3}{4}\right), \quad B\left(\frac{1}{\sqrt{3}}, \frac{3}{4}\right).$$

$$(2)\ y = x^4; f''(x) = 12x^2, x_1 = 0, f'''(x) = 24x, f'''(x_1) = 0,$$

$f^{IV}(x) = 24$; no point of inflection.

The question whether a given root x_i is a point of inflection can also be answered immediately by examining change of sign of the second derivative $f''(x)$ in passing through x_i. If the sign of the second derivative changes into the opposite one, then the convexity of the curve changes direction (see pp. 282 and 283) and we have a point of inflection. This method can also be applied in the case when $y'' = \infty$.

Example. $y = x^{5/3}$; $y' = \frac{5}{3}x^{2/3}$, $y'' = \frac{10}{9}x^{-1/3}$; for $x = 0$, $y'' = \infty$. In passing from negative to positive x, the second derivative changes its sign from "−" to "+". Hence the curve has a point of inflection for $x = 0$.

---

[1] For determining the points of inflection in which $f''(x)$ does not exist (for example, becomes infinite) see below.

Practically, if the points of inflection are seen from the graph of a function (for example, between a maximum and a minimum of a function having a continuous derivative), it is sufficient to find only x_i and disregard the higher derivatives.

Other forms of equation of a curve. The above necessary condition $f''(x) = 0$ for a point of inflection, can be written for other form of equation of a curve as follows:

For the parametric form (3) (see p. 277):

$$\begin{vmatrix} x_t' & y_t' \\ x_t'' & y_t'' \end{vmatrix} = 0.$$

For a curve given in polar coordinates (4):

$$\varrho^2 + 2\varrho'^2 - \varrho\varrho'' = 0.$$

For the implicite form (1), we solve the system of equations:

$$F(x, y) = 0 \qquad \text{and} \qquad \begin{vmatrix} F''_{xx} & F''_{xy} & F'_x \\ F''_{yx} & F''_{yy} & F'_y \\ F'_x & F'_y & 0 \end{vmatrix} = 0.$$

The solutions are coordinates of points of inflection.

Examples.

(1) $x = a(t - \frac{1}{2}\sin t)$, $\quad y = a(1 - \frac{1}{2}\cos t)$ (a shortened cycloid, see p. 123);

$$\begin{vmatrix} x_t' & y_t' \\ x_t'' & y_t'' \end{vmatrix} = \frac{a^2}{4}\begin{vmatrix} 2-\cos t & \sin t \\ \sin t & \cos t \end{vmatrix} = \frac{a^2}{4}(2\cos t - 1),$$

$$\cos t = \tfrac{1}{2}, \qquad t = \pm\tfrac{1}{3}\pi + 2k\pi.$$

There are infinitely many points of inflection, for $t = \pm\frac{1}{3}\pi + 2k\pi$.

(2) $\varrho = \dfrac{1}{\sqrt{\varphi}}$; $\quad \varrho^2 + 2\varrho'^2 - \varrho\varrho'' = \dfrac{1}{\varphi} + \dfrac{1}{2\varphi^3} - \dfrac{3}{4\varphi^3} = \dfrac{1}{4\varphi^3}(4\varphi^2 - 1).$

A point of inflection corresponds to $\varphi = \frac{1}{2}$.

(3) $x^2 - y^2 = a^2$ (a hyperbola);

$$\begin{vmatrix} F'' & . & . \\ . & . & . \\ . & . & . \end{vmatrix} = \begin{vmatrix} 2 & 0 & 2x \\ 0 & -2 & -2y \\ 2x & -2y & 0 \end{vmatrix} = 8(x^2 - y^2).$$

The equations $x^2 - y^2 = a^2$ and $8(x^2 - y^2) = 0$ are inconsistent, hence the parabola has no points of inflection.

Vertices of a curve are points of maximum or minimum of its curvature (the points where the line is the most or the least

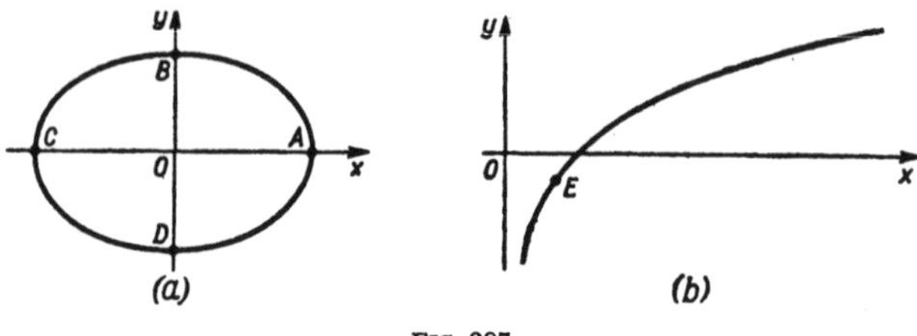

FIG. 287

curved); e.g., an ellipse has four vertices A, B, C, D (Fig. 237a) a logarithmic curve has one vertex E (Fig. 237b). Determining of vertices reduces to determining of extremes of the curvature K given by formulas (**) on p. 284 or extremes of the radius of curvature $R = 1/K$ according to which calculation is more simple.

Singular points. This name comprises points of various types.

(a) *Branch points,* in which the curve intersects with itself (Fig. 238a); (b) *Isolated points* which are separated from the

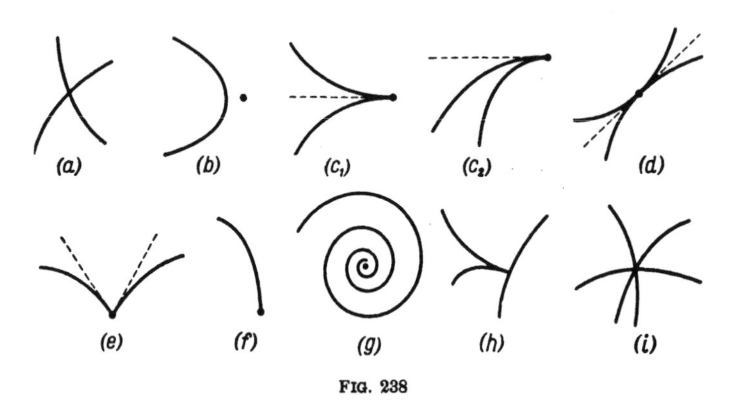

FIG. 238

rest of the curve but whose coordinates satisfy the equation of the curve (Fig. 238b); (c) *Cusps* in which the direction of the curve reverses; we distinguish points of return of the first kind (Fig. 238c_1) and of the second kind (Fig. 238c_2) according to the position of the tangent with respect to both branches; (d) *Points of self-contact* in which the curve is tangent to itself; (e) *Bend points* in which the curve changes its direction by a "jump"; in contrast to a cusp, the tangent lines to both branches of the curve at a bend point are different (see Fig. 238e); (f) *Points of stopping* at which the curve ends (Fig. 238f); (g) *Asymptotical points* which are encircled by the curve infinitely many times and approached arbitrarily near (Fig. 238g). Certain combinations of singular points of these types can also occur (Fig. 238h, i).

Determining singular points of types (e), (f), (g). These singularities can occur only for transcendental curves ([1]). Bend points occur when the derivative dy/dx has a finite jump, e.g., at the origin for the curve $y = \frac{x}{1+e^{1/x}}$ (see Fig. 284c on p. 361). Points of stopping correspond to finite discontinuities of the function $y = f(x)$, as, for example, the points (1,0) and (1,1) for the curve $y = \frac{1}{1+e^{1/(x-1)}}$ (see Fig. 272 on p. 329). Asymptotical points are easiest to detection for curves given in polar coordinates $\varrho = f(\varphi)$; if $\lim \varrho = 0$, when $\varphi \to +\infty$ or $\varphi \to -\infty$, then the pole is an asymptotical point; for example, in the logarithmic spiral $\varrho = ae^{k\varphi}$ (see Fig. 60 on p. 128)

Determining singular points of types (a), (b), (c), (d) and (h), (i) (called *multiple points*). We consider the curve in the form $F(x, y) = 0$. A point $A(x_1, y_1)$ whose coordinates satisfy simultaneously three equations $F = 0$, $F'_x = 0$, $F'_y = 0$, is a *double point*, if at least one of the second derivatives $F''_{xx}, F''_{xy}, F''_{yy}$ is different from zero; if these derivatives are also equal to zero, then the point is a *threefold* or a *multiple point*.

The type of a double point depends on the sign of the determinant

$$\Delta = \begin{vmatrix} F''_{xx} & F''_{xy} \\ F''_{yx} & F''_{yy} \end{vmatrix}_{\binom{x=x_1}{y=y_1}} .$$

(1) If $\Delta < 0$, then A is a branch point the slopes of the tangent lines at which are equal to the roots of the equation

$$F''_{yy}\, k^2 + 2F''_{xy}\, k + F''_{xx} = 0.$$

(2) If $\Delta > 0$, then A is an isolated point.

(3) If $\Delta = 0$, then A is a point of return or of self-contact. Then the slope of the tangent lines is

$$\tan\alpha = -\frac{F''_{xy}}{F''_{yy}}.$$

For detailed examining the multiplicity of a point, when $\Delta > 0$, we translate this point to the origin and turn the coordinate axes so that the x axis coincide with the tangent to the curve at the point A; then we recognize the type of the point from the equation of the curve.

([1]) See p. 239.

Examples.

(1) $F(x, y) = (x^2+y^2)^2-2a^2(x^2-y^2) = 0$ (a lemniscate, see Fig. 51 on p. 123), $F'_x = 4x(x^2+y^2-a^2)$, $F'_y = 4y(x^2+y^2+a^2)$; the equations $F'_x = 0$, $F'_y = 0$ have three solutions $(0,0)$, $(a, 0)$ and $(-a, 0)$, but only the first one satisfies the equation $F = 0$. Substituting $(0,0)$ to the second derivatives, we have $F''_{xx} = -4a^2$, $F''_{xy} = 0$, $F''_{yy} = 4a^2$, $\Delta = -16a^4 < 0$, hence the origin is a branch point; the slopes of the tangents are $\tan\alpha = \pm 1$; the tangents: $y = x$ and $y = -x$.

(2) $F(x, y) = x^3+y^3-x^2-y^2 = 0$; $F'_x = x(3x-2)$, $F'_y = y(3y-2)$; among the four points $(0,0)$, $(0,\frac{2}{3})$, $(\frac{2}{3}, 0)$, $(\frac{2}{3}, \frac{2}{3})$, only the first one lies on the curve; $(F''_{xx})_0 = -2$, $(F''_{xy})_0 = 0$, $(F''_{yy})_0 = -2$, $\Delta = 4 > 0$; hence the origin is an isolated point.

(3) $F(x, y) = (y-x^2)^2-x^5 = 0$. The equations $F'_x = 0$, $F'_y = 0$ has a single solution $(0,0)$ satisfying the equation $F = 0$. $\Delta = 0$, $\tan\alpha = 0$. In this case we have a cusp of the second kind which is evident from the equation of the curve in the explicit form $y = x^2(1 \pm \sqrt{x})$; y does not exist for $x < 0$, and for small values of $x > 0$, the two values of y are positive (the tangent at the origin is horizontal).

Case of an algebraic curve $F(x, y) = 0$. If the equation does not contain free terms or terms of the first degree, then the origin is a double point. Equations of the tangents can be immediately obtained by setting the terms of the second degree equal to zero. For example for the lemniscate (see above, example 1) equations of the tangents are $x^2-y^2 = 0$ or $y = \pm x$. If the equation does not contain terms of the third degree, then the origin is a threefold point and so on.

4. Asymptotes

General case. A part of a curve which can be indefinitely prolonged, i.e., an infinite branch of a curve, can have an asymptote; an *asymptote* is a straight line approached by the curve from one side (Fig. 239a) or from both sides (Fig. 239b). To find asymptotes of a curve given in parametric form $x = x(t)$, $y = y(t)$, we determine the values t_i of t for which $x(t) \to \infty$ or $y(t) \to \infty$.

FIG 239

If $x(t_i)=\infty$ but $y(t_i)=a\neq\infty$ then the horizontal line $y=a$ is an asymptote.

If $y(t_i)=\infty$ but $x(t_i)=a\neq\infty$ then the vertical line $x=a$ is an asymptote.

If $x(t_i)=\infty$ and $y(t_i)=\infty$ then we evaluate two limits:

$$k=\lim_{t\to t_i}\frac{y(t)}{x(t)}\quad\text{and}\quad b=\lim_{t\to t_i}\big(y(t)-kx(t)\big);$$

if the both limits exist, then the line $y=kx+b$ is an asymptote.

If the curve is given in the form $y=f(x)$, then vertical asymptotes are sought as points of discontinuity of the function $f(x)$ (see pp. 334–338) and horizontal asymptotes are of the form $y=kx+b$, where

$$k=\lim_{x\to\infty}\frac{f(x)}{x},\qquad b=\lim_{x\to\infty}(f(x)-kx).$$

Example. $x=\dfrac{m}{\cos t}$, $y=n(\tan t-t)$, $t_1=\frac{1}{2}\pi$, $t_2=-\frac{1}{2}\pi$ etc.

$$x(t_1)=y(t_1)=\infty,\ k=\lim_{t\to\frac{1}{2}\pi}\frac{n}{m}(\sin t-t\cos t)=\frac{n}{m},$$

$$b=\lim_{t\to\frac{1}{2}\pi}\left[n\,(\tan t-t)-\frac{n}{m}\cdot\frac{m}{\cos t}\right]=n\lim_{t\to\frac{1}{2}\pi}\frac{\sin t-t\cos t-1}{\cos t}=-\frac{n\pi}{2},$$

$$y=\frac{n}{m}x-\frac{n\pi}{2},\qquad \frac{x}{m}-\frac{y}{n}=\frac{\pi}{2}.$$

Similarly, the second asymptote is $\dfrac{x}{n}+\dfrac{y}{m}=\dfrac{\pi}{2}$ and so on.

Case of an algebraic curve $F(x,y)=0$. The function $F(x,y)$ is a polynomial with respect to x and y. We select the terms of $F(x,y)$ of a highest degree [1]. Let $\Phi(x,y)$ be the sum of the highest terms of $F(x,y)$. We solve the equation $\Phi(x,y)=0$ with respect to x and y:

$$x=\varphi(y),\qquad y=\psi(x).$$

The values $y_1=a$ for which $x=\infty$ give a horizontal asymptote; the values $x_1=b$ for which $y=\infty$ give a vertical asymptote. To find sloping asymptotes, we substitute the expression $y=kx+b$ to $F(x,y)$ and arrange the obtained polynomial according to powers of x:

$$F(x,kx+b)=f_1(k,b)x^m+f_2(k,b)x^{m-1}+\dots$$

---

[1] The *degree* of a term Ax^my^n is the sum $m+n$ of the exponents. E.g., the term $3x^2y^3$ is of degree 5, the term $2y^2$ is of degree 2; the highest terms of the polynomial x^3+y^3-3xy are x^3 and y^3 (both of degree 3).

We put the coefficients f_1 and f_2 at the highest powers of x equal to zero

$$f_1(k, b) = 0, \quad f_2(k, b) = 0.$$

If these equation have a solution, then k and b determine an asymptote $y = kx + b$.

Example. $x^3 + y^3 - 3axy = 0$ (Descartes' leaf, see Fig. 44 on p. 118), $F(x, kx + b) = (1 + k^3)x^3 + 3(k^2 b - ka)x^2 + \ldots$; $1 + k^3 = 0$, $k^2 b - ka = 0$; these equations have the solution $k = -1$, $b = -a$; equation of an asymptote: $y = -x - a$.

5. General examining of a curve by its equation

We examine a curve to determine the form of variation of a function $y = f(x)$ or to determine the shape of a curve given analytically by one of formulas (1), (2), (3) or (4) of pp. 277, 278.

Plotting the graph of functions given in the form $y = f(x)$.

(1) Find the domain of existence (p. 322).

(2) Examine symmetry of the curve with respect to the y axis and with respect to the origin, i.e., whether the function is even or odd (p. 327).

(3) Determine the behaviour of the function at infinity by evaluating the limits $\lim_{x \to -\infty} f(x)$ and $\lim_{x \to +\infty} f(x)$ (p. 329).

(4) Find the points of discontinuity and their type (pp. 334–338).

(5) Find the intersection of the curve with the y axis by calculating $f(0)$ and the intersection with the x axis by solving the equation $f(x) = 0$ (for solution of algebraic and transcendental equation in general form see p. 169).

(6) Find the maxima and minima and then the domain of increase and decrease of the function (see p. 379).

(7) Find the points of inflection (pp. 285, 286) and then the domains where the curve is upwards or downwards convex (pp. 282, 283); at the points of inflection find the slopes of tangent lines.

Having known these, we sketch the graph and then plot it precisely at these points, where it is necessary.

Example. Plot the graph of the function

$$y = \frac{2x^2 + 3x - 4}{x^2}.$$

(1) The function exists for every x except $x = 0$.

(2) No symmetry.

(3) $y \to 2$ when $x \to \pm\infty$ so that if $x \to -\infty$, then $y = 2 - 0$ and if $x \to +\infty$, then $y = 2 + 0$.

(4) An infinite discontinuity for $x = 0$ (from $-\infty$ to $+\infty$, since y is negative for small x).

(5) $f(0) = \infty$. The equation $2x^2 + 3x - 4 = 0$ has the roots $x_{1,2} = \frac{-3 \pm \sqrt{41}}{4}$; hence the intersections with the x axis are $x_1 \approx 0.85$, $x_2 \approx -2.35$.

(6) Point of maximum $x = \frac{8}{3} \approx 2.66$, $y \approx 2.56$.

(7) Point of inflection $x = 4$, $y = 2.5$, $\tan\alpha = -\frac{1}{16}$.

Moreover, after plotting these data, we find

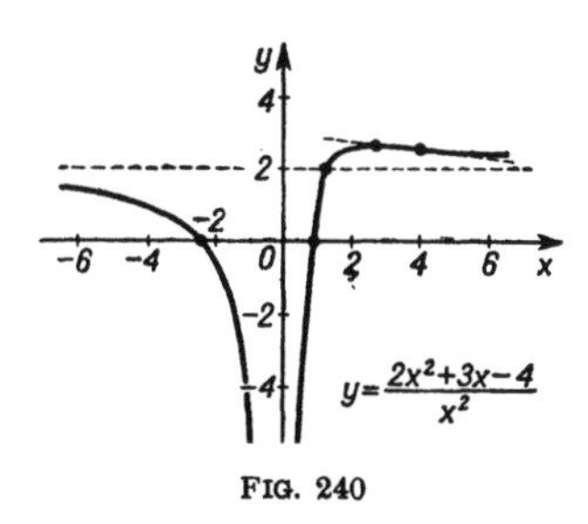

FIG. 240

(8) Intersection of the curve with its asymptote: $x = \frac{4}{3} \approx 1.33$, $y = 2$.

The curve is shown in Fig. 240.

Plotting the graphs of curves given in implicit form $F(x, y) = 0$. The general rules are inapplicable in this case, for they often lead to long computations. It is useful to find the following data:

(1) Intersections with the coordinate axes.

(2) Symmetry with respect to the axes and the origin (replacing x by $-x$ and y by $-y$).

(3) Maxima and minima with respect to the x axis (p. 379) and to the y axis, by analogous formulas obtained by interchanging the variables x and y.

(4) Points of inflection (p. 285) with slopes of tangents.

(5) Singular points (pp. 288, 289).

(6) Vertices of the curve (p. 287) and the circles of curvature (p. 284) at the vertices; arcs of the circles of curvature will approximate the curve in neighbourhoods of the vertices.

(7) All asymptotes (p. 290) and the position of branches of the curve with respect to the asymptotes.

6. Evolutes and involutes

The evolute of a given curve is the curve generated by the centres of curvature (p. 284) of all points of the curve. The evolute is the envelope (p. 295) of normals to the given curve. Parametric equations of the evolute are given by the formulas (∵) on p. 285 (equation for the centre of curvature with x_C and y_C the running coordinates of the evolute). If the parameter (x, t or φ) can be eliminated, then the rectangular equation of the evolute connecting X and Y can be found.

Example. Find the evolute of the parabola $y = x^2$ (Fig. 241). We have

$$X = x - \frac{2x(1+4x^2)}{2} = -4x^3, \qquad Y = x^2 + \frac{1+4x^2}{2} = \frac{1+6x^2}{2},$$

whence $Y = \frac{1}{2} + 3(\frac{1}{4}X)^{2/3}$, where X and Y are the running coordinates of the evolute.

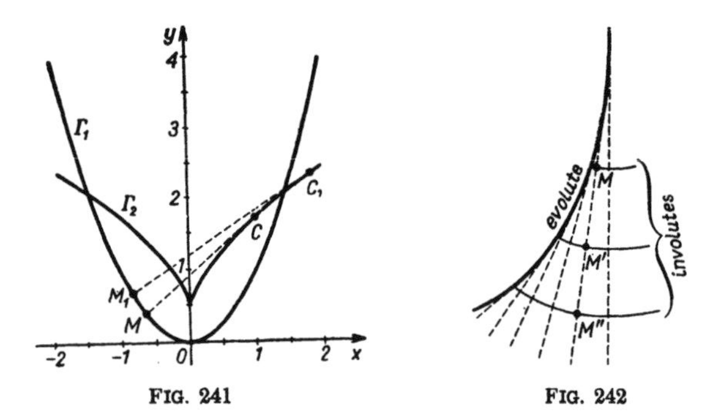

FIG. 241 FIG. 242

The involute Γ_1 of a curve Γ_2 is a curve whose evolute is Γ_2. A normal MC of the involute is a tangent of the evolute and length of arc $\breve{CC_1}$ of the evolute is equal to the increment of radius of curvature of the involute (Fig. 241):

$$\breve{CC_1} = M_1C_1 - MC.$$

These properties show that the involute Γ_1 can be regarded as the curve traced by an end of a taut string unwound from its evolute Γ_2. A given evolute has a family of involutes each of which corresponds to the initial length of the string (Fig. 242). The equation of the involute can be obtained by integration of the system of differential equations corresponding to its evolute. For the equation of the involute of a circle see p. 129.

7. Envelope of a family of curves

Characteristic points. Let

$$(\bullet) \qquad F(x, y, \alpha) = 0$$

be the equation of a family of curves depending on one parameter α. Every two infinitely near curves of this family corresponding to the values of parameter α and $\alpha + \Delta\alpha$ have the points K of the nearest approach. Such a point is either a point of intersection of the curves (α) and $(\alpha + \Delta\alpha)$ or a point of the curve (α) whose distance from the curve $(\alpha + \Delta\alpha)$ along the normal is an infinitesimal of a higher order with respect to $\Delta\alpha$ (Fig. 243a and b). When $\Delta\alpha \to 0$, then the curve $(\alpha + \Delta\alpha)$ tends to (α) and the point K in some cases tends to a limiting position called the *characteristic point*. Singular points of the curve (α) are always characteristic points.

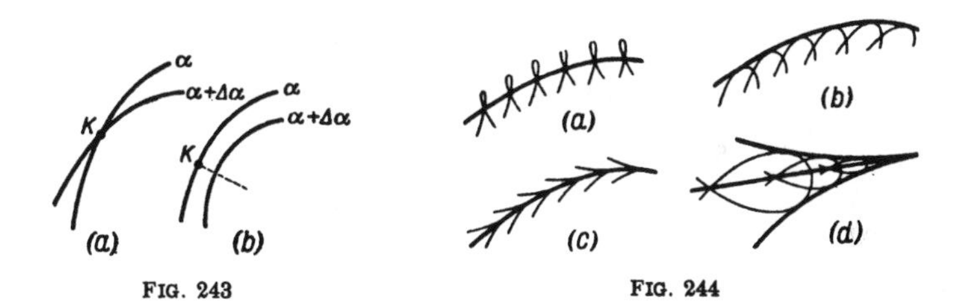

FIG. 243 FIG. 244

Characteristics. The locus of characteristic points for all curves of the family $(\bullet)$ forms a curve (or a set of curves) called the *characteristic* of the family; it consists either of singular points of the given family (Fig. 244a) or is an envelope of the family (Fig. 244b), i.e., is tangent to each curve of the family. Certain combinations of these two cases can also occur (Fig. 244c, d).

The equation of the envelope (and of the characteristic, in the general case) of the family $F(x, y, \alpha) = 0$ can be obtained by eliminating α from the equations $F = 0$, $\frac{\partial F}{\partial \alpha} = 0$.

Example. Find the envelope of the family of straight lines

AB if the end points of the segments AB of a fixed length l slide on the coordinate axes (Fig. 245a).

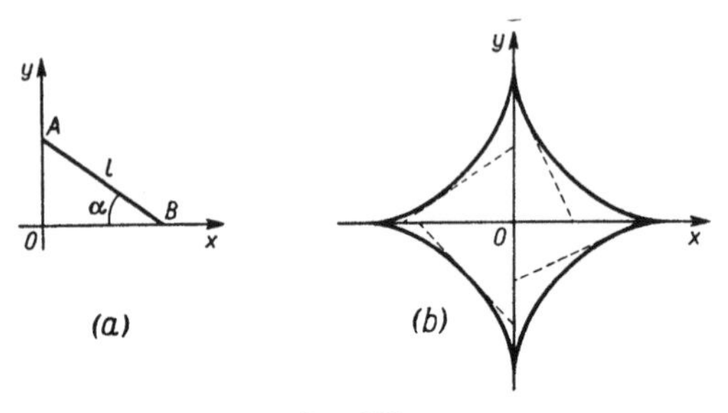

FIG. 245

Equation of the family

$$\frac{x}{l \sin \alpha} + \frac{y}{l \cos \alpha} = 1,$$

or

$$F \equiv x \cos \alpha + y \sin \alpha - l \sin \alpha \cos \alpha = 0,$$

$$\frac{\partial F}{\partial \alpha} = - x \sin \alpha + y \cos \alpha - l \cos^2 \alpha + l \sin^2 \alpha = 0.$$

Eliminating α from these equations, we obtain $x^{2/3} + y^{2/3} = l^{2/3}$ as the equation of the envelope, i.e., the equation of an astroid (Fig. 245b; see also p. 126).

B. SPACE CURVES

8. Ways in which a curve can be defined

Coordinate equations. A curve in the space (or a skew curve) can be defined in one of the following forms:

(a) As an intersection of two surfaces

$$(1) \qquad F(x, y, z) = 0, \qquad \Phi(x, y, z) = 0.$$

(b) In parametric form

$$(2) \qquad x = x(t), \qquad y = y(t), \qquad z = z(t),$$

where t is an arbitrary parameter, in particular, $t = x$, y or z,

(c) In parametric form

$$x = x(s), \quad y = y(s), \quad z = z(s), \tag{3}$$

where s is the length of arc from a fixed point A to the running point M, i.e.,

$$s = \int_{t_0}^{t} \sqrt{\left(\frac{dx}{dt}\right)^2 + \left(\frac{dy}{dt}\right)^2 + \left(\frac{dz}{dt}\right)^2}\, dt.$$

Vector equation. Denoting by $\boldsymbol{r}$ the radius vector of a point of the curve (see p. 614), we can write equation (2) in the form

$$\boldsymbol{r} = \boldsymbol{r}(t), \quad \text{where} \quad \boldsymbol{r}(t) = x(t)\boldsymbol{i} + y(t)\boldsymbol{j} + z(t)\boldsymbol{k}, \tag{2a}$$

and equation (3) in the form

$$\boldsymbol{r} = \boldsymbol{r}(s), \quad \text{where} \quad \boldsymbol{r}(s) = x(s)\boldsymbol{i} + y(s)\boldsymbol{j} + z(s)\boldsymbol{k}. \tag{3a}$$

Positive direction of a curve given by equation (2) or (2a) corresponds to the increasing parameter t, and for a curve given by equation (3) or (3a)—to the direction in which the length of arc of the curve is measured.

9. Moving trihedral

Definitions. At every point M of a curve (except singular points), 3 mutually perpendicular straight lines and 3 mutually

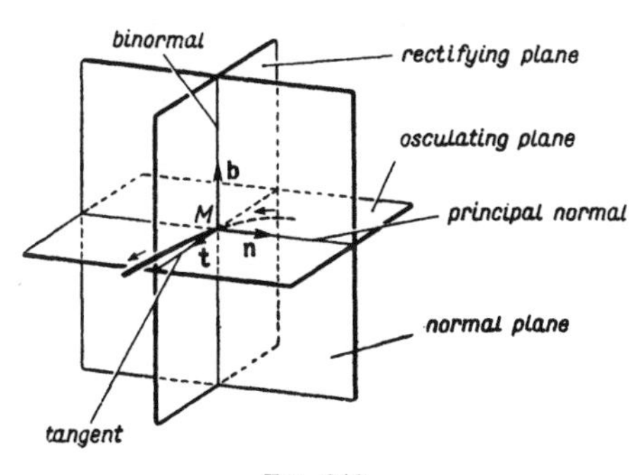

FIG. 246

perpendicular planes intersecting in M are defined (Fig. 246):

(1) The *tangent*—the limiting position of a secant MN, when $N \to M$ (see Fig. 227 on p. 279).

(2)The *normal plane*, perpendicular to the tangent. All straight lines passing through M and lying in this plane are called *normals* to the curve at M.

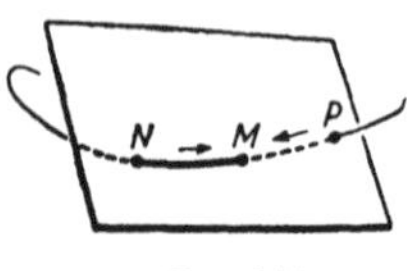

Fig. 247

(3) The *osculating plane*—the limiting position of a plane passing through three near points M, N, P of the curve, when $N \to M$ and $P \to M$ (Fig. 247). The osculating plane contains the tangent.

(4) The *principal normal* — the intersection of the normal and osculating planes (i.e., the normal line lying in the osculating plane).

(5) The *binormal*—the line perpendicular to the osculating plane.

(6) The *rectifying plane*, containing the tangent and the binormal.

Positive directions are defined on lines (1), (4) and (5) as follows: positive direction of the tangent is defined by that of the curve and is determined by the unit vector $\boldsymbol{t}$; positive direction of the principal normal is that where the curve is concave and is determined by the unit vector $\boldsymbol{n}$; positive direction of the binormal is defined by the unit vector $\boldsymbol{b} = \boldsymbol{t} \times \boldsymbol{n}$ ($\boldsymbol{t}$, $\boldsymbol{n}$ and $\boldsymbol{b}$ should constitute a right handed triple, see p. 617). Three vectors $\boldsymbol{t}$, $\boldsymbol{n}$ and $\boldsymbol{b}$ together with the planes spanned by them form the *moving trihedral* of the space curve.

Position of a curve with respect to the moving trihedral. In points of a general type, the curve lies on one side of the rectifying plane and intersects the normal and osculating planes (Fig. 248a).

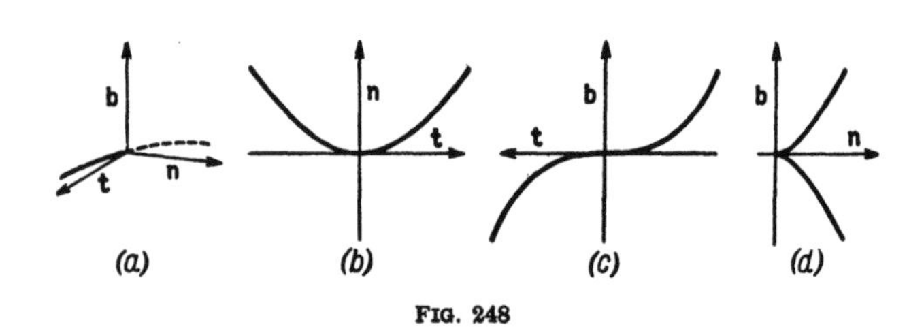

Fig. 248

Projections of a small neighbourhood of the point M of the curve on the faces of the trihedral have, approximately, the forms of the following curves:

projection on the osculating plane—the form of a parabola (Fig. 248b), projection of the rectifying plane—the form of a cubical parabola (Fig. 248c),

projection on the normal plane—the form of a semicubical parabola (Fig. 248d).

If curvature and torsion (see below) of the curve at the point M are zero or if the point is a singular point ($x'(t) = y'(t) = z'(t) = 0$), then the curve has a different position.

Equations of elements of the moving trihedral.

(a) For a curve given in the form (1) (p. 296).

Tangent:

$$\frac{X - x}{\begin{vmatrix} \frac{\partial F}{\partial y} & \frac{\partial F}{\partial z} \\ \frac{\partial \Phi}{\partial y} & \frac{\partial \Phi}{\partial z} \end{vmatrix}} = \frac{Y - y}{\begin{vmatrix} \frac{\partial F}{\partial z} & \frac{\partial F}{\partial x} \\ \frac{\partial \Phi}{\partial z} & \frac{\partial \Phi}{\partial x} \end{vmatrix}} = \frac{Z - z}{\begin{vmatrix} \frac{\partial F}{\partial x} & \frac{\partial F}{\partial y} \\ \frac{\partial \Phi}{\partial x} & \frac{\partial \Phi}{\partial y} \end{vmatrix}}$$

Normal plane:

$$\begin{vmatrix} X - x & Y - y & Z - z \\ \frac{\partial F}{\partial x} & \frac{\partial F}{\partial y} & \frac{\partial F}{\partial z} \\ \frac{\partial \Phi}{\partial x} & \frac{\partial \Phi}{\partial y} & \frac{\partial \Phi}{\partial z} \end{vmatrix} = 0,$$

where x, y, z are coordinates of the point M and X, Y, Z are running coordinates of the tangent line or normal plane. The partial derivatives are taken at M.

(b) For a curve given in the form (2) (p. 296) or (2a) (p. 297).

In the following formulas, x, y, z and $\boldsymbol{r}$ are coordinates and radius vector of the point M of the curve and X, Y, Z and $\boldsymbol{R}$ are running coordinates and radius vector of the corresponding element of the trihedral. The derivatives with respect to t are taken at the point M.

Coordinate equation	Vector equation
Tangent	
$\frac{X - x}{x'} = \frac{Y - y}{y'} = \frac{Z - z}{z'}$	$\boldsymbol{R} = \boldsymbol{r} + \lambda \frac{d\boldsymbol{r}}{dt}$
Normal plane	
$x'(X - x) + y'(Y - y) + z'(Z - z) = 0$	$(\boldsymbol{R} - \boldsymbol{r})\frac{d\boldsymbol{r}}{dt} = 0$

Coordinate equation	Vector equation
Osculating plane	
$\begin{vmatrix} X-x & Y-y & Z-z \\ x' & y' & z' \\ x'' & y'' & z'' \end{vmatrix} = 0$	$(\boldsymbol{R}-\boldsymbol{r})\frac{d\boldsymbol{r}}{dt}\cdot\frac{d^2\boldsymbol{r}}{dt^2}=0$
Binormal	
$\frac{X-x}{\begin{vmatrix} y' & z' \\ y'' & z'' \end{vmatrix}} = \frac{Y-y}{\begin{vmatrix} z' & x' \\ z'' & x'' \end{vmatrix}} = \frac{Z-z}{\begin{vmatrix} x' & y' \\ x'' & y'' \end{vmatrix}}$	$\boldsymbol{R}=\boldsymbol{r}+\lambda\left(\frac{d\boldsymbol{r}}{dt}\times\frac{d^2\boldsymbol{r}}{dt^2}\right)$
Rectifying plane	
$\begin{vmatrix} X-x & Y-y & Z-z \\ x' & y' & z' \\ l & m & n \end{vmatrix} = 0,$ where $l = y'z''-y''z'$, $m = z'x''-z''x'$, $n = x'y''-x''y'$	$(\boldsymbol{R}-\boldsymbol{r})\frac{d\boldsymbol{r}}{dt}\left(\frac{d\boldsymbol{r}}{dt}\times\frac{d^2\boldsymbol{r}}{dt^2}\right)=0$
Principal normal	
$\frac{X-x}{\begin{vmatrix} y' & z' \\ m & n \end{vmatrix}} = \frac{Y-y}{\begin{vmatrix} z' & x' \\ n & l \end{vmatrix}} = \frac{Z-z}{\begin{vmatrix} x' & y' \\ l & m \end{vmatrix}}$	$\boldsymbol{R}=\boldsymbol{r}+\lambda\frac{d\boldsymbol{r}}{dt}\times\left(\frac{d\boldsymbol{r}}{dt}\times\frac{d^2\boldsymbol{r}}{dt^2}\right)$

(c) For a curve given in the form (3) or (3a) (p. 297).

If the length of arc s is taken as a parameter, then the equation of the tangent, normal plane, osculating plane and binormal can be taken from the above (t should be replaced by s) and the equations of the principal normal and the rectifying plane can be simplified as follows:

Coordinate equation	Vector equation
Principal normal	
$\frac{X-x}{x''} = \frac{Y-y}{y''} = \frac{Z-z}{z''}$	$\boldsymbol{R}=\boldsymbol{r}+\lambda\frac{d^2\boldsymbol{r}}{ds^2}$
Rectifying plane	
$x''(X-x)+y''(Y-y)+z''(Z-z)=0$	$(\boldsymbol{R}-\boldsymbol{r})\frac{d^2\boldsymbol{r}}{ds^2}=0$

10. Curvature and torsion

Curvature. The *curvature* of a curve at the point M is a number which characterizes the deviation of the curve (in a small neighbourhood containing M) from a straight line. Precisely (Fig. 249)

$$K = \lim_{\breve{MN}\to 0}\left|\frac{\Delta t}{\breve{MN}}\right| = \left|\frac{dt}{ds}\right|.$$

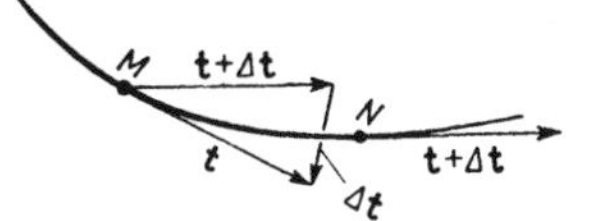

FIG. 249

Radius of curvature: $\varrho = 1/K$. For space curves, K and ϱ are always positive.

Formulas for K and ϱ:

(a) For a curve in the form (3) (see p. 297):

$$(\bullet)\qquad K = \left|\frac{d^2\mathbf{r}}{ds^2}\right| = \sqrt{x''^2 + y''^2 + z''^2}$$

(derivatives with respect to s).

(b) For a curve in the form (2) (see p. 296):

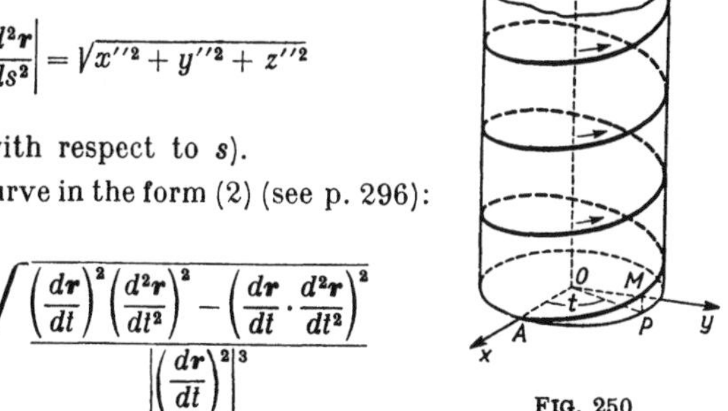

FIG. 250

$$(\bullet\bullet)\quad K = \sqrt{\frac{\left(\frac{d\mathbf{r}}{dt}\right)^2\left(\frac{d^2\mathbf{r}}{dt^2}\right)^2 - \left(\frac{d\mathbf{r}}{dt}\cdot\frac{d^2\mathbf{r}}{dt^2}\right)^2}{\left|\left(\frac{d\mathbf{r}}{dt}\right)^2\right|^3}}$$

$$= \sqrt{\frac{(x'^2 + y'^2 + z'^2)(x''^2 + y''^2 + z''^2) - (x'x'' + y'y'' + z'z'')^2}{(x'^2 + y'^2 + z'^2)^3}}$$

derivatives with respect to t).

Example. Find the curvature of the *circular helix* (or *screw thread*) (Fig. 250): $x = a\cos t$, $y = a\sin t$, $z = bt$ [1]. Replacing the parameter t by $s = t\sqrt{a^2 + b^2}$, we obtain

$$x = a\cos\frac{s}{\sqrt{a^2 + b^2}},\qquad y = a\sin\frac{s}{\sqrt{a^2 + b^2}},\qquad z = \frac{bs}{\sqrt{a^2 + b^2}},$$

and, from the formula $(\bullet)$

---

[1] The helix defined by these equations and shown in Fig. 250 is called *right handed*. An observer looking at the helix down its axis (the z axis) sees the line ascending counterclockwise. A helix symmetric to the right handed helix with respect to a plane is *left handed*.

$$K = \frac{a}{a^2 + b^2} = \text{const}, \qquad \varrho = \frac{a^2 + b^2}{a} = \text{const}.$$

The same result can be obtained without changing the parameter from the formula (**).

Torsion. The *torsion* of a curve at the point M is a number which characterizes the deviation of the curve (in a small neighbourhood containing M) from a plane curve. Precisely (Fig. 251).

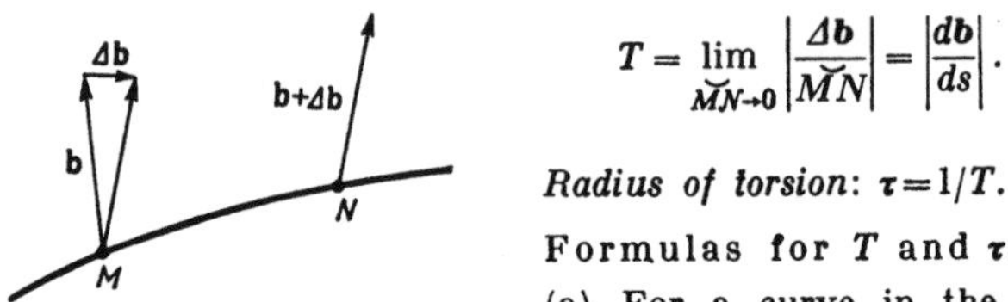

FIG. 251

$$T = \lim_{\overset{\smile}{MN}\to 0} \left|\frac{\Delta \boldsymbol{b}}{\overset{\smile}{MN}}\right| = \left|\frac{d\boldsymbol{b}}{ds}\right|.$$

Radius of torsion: $\tau = 1/T$.

Formulas for T and τ:

(a) For a curve in the form (3) (see p. 297):

$$(\ast\ast\ast) \qquad T = \frac{1}{\tau} = \varrho^2 \left(\frac{d\boldsymbol{r}}{ds} \cdot \frac{d^2\boldsymbol{r}}{ds^2} \cdot \frac{d^3\boldsymbol{r}}{ds^3}\right) = \frac{\begin{vmatrix} x' & y' & z' \\ x'' & y'' & z'' \\ x''' & y''' & z''' \end{vmatrix}}{(x''^2 + y''^2 + z''^2)}$$

(derivatives with respect to s).

(b) For a curve in the form (2):

$$(\ast\ast\ast\ast) \qquad T = \frac{1}{\tau} = \varrho^2 \frac{\frac{d\boldsymbol{r}}{dt} \cdot \frac{d^2\boldsymbol{r}}{dt^2} \cdot \frac{d^3\boldsymbol{r}}{dt^3}}{\left|\left(\frac{d\boldsymbol{r}}{dt}\right)^2\right|^3} = \varrho^2 \frac{\begin{vmatrix} x' & y' & z' \\ x'' & y'' & z'' \\ x''' & y''' & z''' \end{vmatrix}}{(x'^2 + y'^2 + z'^2)^3}$$

(ϱ computed from the formulas (*) or (**)).

The torsion evaluated from the formulas $(\ast\ast\ast)$ or $(\ast\ast\ast\ast)$ can be positive or negative. If $T > 0$, then, for an observer situated on the principal normal parallel to the binormal (see Fig. 246), the curve ascends from the right to the left like in a right handed screw. If $T < 0$, then, from the point of view of this observer, the curve ascendes from the left to the right.

Example. The torsion of the circular helix is constant. For the right handed helix it is equal to

$$T = \left(\frac{a^2 + b^2}{a}\right)^2 \frac{\begin{vmatrix} -a\sin t & a\cos t & b \\ -a\cos t & -a\sin t & 0 \\ a\sin t & -a\cos t & 0 \end{vmatrix}}{[(-a\sin t)^2 + (a\cos t)^2 + b^2]^3} = \frac{b}{a^2 + b^2}, \qquad \tau = \frac{a^2 + b^2}{b}.$$

For the left handed helix the torsion is negative:

$$T = -\frac{b}{a^2 + b^2}.$$

Serre-Frenet's formulas. The derivatives of the vectors $\boldsymbol{t}$, $\boldsymbol{n}$ and $\boldsymbol{b}$ can be computed from the following Serre-Frenet's formulas:

$$\frac{d\boldsymbol{t}}{ds} = \frac{\boldsymbol{n}}{\varrho}, \quad \frac{d\boldsymbol{n}}{ds} = \frac{\boldsymbol{t}}{\varrho} - \frac{\boldsymbol{b}}{\tau}, \quad \frac{d\boldsymbol{b}}{ds} = -\frac{\boldsymbol{n}}{\tau},$$

where ϱ is the radius of curvature and τ the radius of torsion.

C. SURFACES

11. Ways in which a surface can be defined

Equation of a surface. A surface can be defined in one of the following forms:

(a) *Implicit form*

(1) $$F(x, y, z) = 0.$$

(b) *Explicit form*

(2) $$z = f(x, y).$$

(c) *Parametric form*

(3) $$x = x(u, v), \quad y = y(u, v), \quad z = z(u, v).$$

(d) *Vector form*

(3a) $$\boldsymbol{r} = \boldsymbol{r}(u, v) \quad \text{or} \quad \boldsymbol{r} = x(u, v)\,\boldsymbol{i} + y(u, v)\,\boldsymbol{j} + z(u, v)\,\boldsymbol{k}.$$

For various parameters u and v we obtain the radius vector and coordinates of various points of the surface. Eliminating the parameters u and v from (3), we obtain formula (1). Formula (2) is a particular case of (3), when $u = x$ and $v = y$.

Example. Equation of the sphere

(1) $$x^2 + y^2 + z^2 - a^2 = 0$$

or

(3) $$x = a \cos u \sin v, \quad y = a \sin u \sin v, \quad z = a \cos v$$

or

(3a) $$\boldsymbol{r} = a\,(\cos u \sin v\,\boldsymbol{i} + \sin u \sin v\,\boldsymbol{j} + \cos v\,\boldsymbol{k}).$$

Curvilinear coordinates on a surface. If the surface is given in the form (3) or (3a), then, for a fixed parameter $v = v_0$, the point $\boldsymbol{r}(x, y, z)$ traces a curve $\boldsymbol{r} = \boldsymbol{r}(u, v_0)$ on the surface. For various values $v = v_1$, $v = v_2$, ... we obtain a family of curves on the surface. Since only the parameter u varies along each of these lines, they are called *u-lines* (Fig. 252). Similarly, for a fixed parameter u_0,

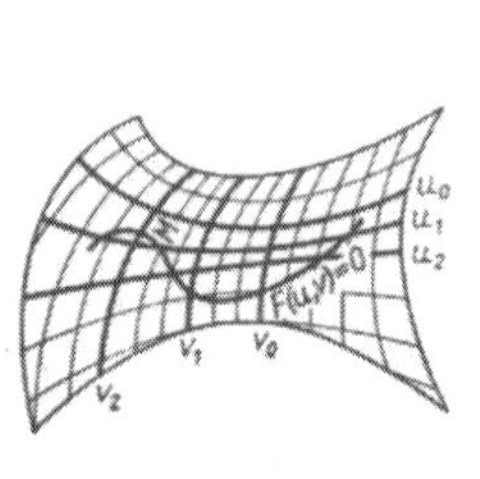

FIG. 252

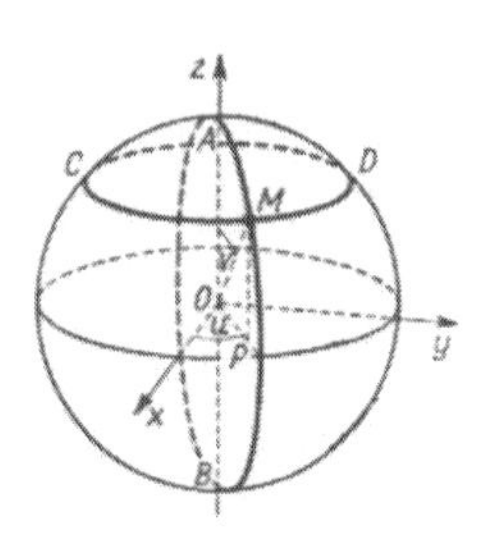

FIG. 253

the point $r(u_0, v)$ traces a second curve. For various values $u = u_1$, $u = u_2$, ... we obtain the second family of lines called the *v-lines* ($u =$ const). Thus we obtain a net of curves on the surface (3) called the *coordinate lines*. The numbers $u = u_i$ and $v = v_k$ are called *curvilinear* or *Gaussian coordinates* of a point M of the surface. In the case (2) the coordinate lines are sections of the surface $z = f(x, y)$ by planes $x =$ const and $y =$ const.

Every equation connecting the curvilinear coordinates u and v, e.g., $F(u, v) = 0$ or $u = u(t)$, $v = v(t)$ determines a curve on the surface.

Example. In parametric representation of the spherical surface (see previous example), the parameter u is the longitude of M ($u = \sphericalangle OxOP$) and v is the polar distance of M ($v = \sphericalangle OzOM$). The u-lines are parallels and the v-lines are meridians (Fig. 253).

12. Tangent plane and normal

Definitions. The tangent lines to various curves lying on the surface and passing through a given point $M(\boldsymbol{r})$ or $M(x, y, z)$ will lie in one plane called the *tangent plane* to the surface at M. (The exceptions from this rule are so-called conical points discussed below.) The straight line perpendicular to the tangent plane at M is called the *normal* to the surface at M (Fig. 254). The vectors $\boldsymbol{r}_1 = \frac{\partial \boldsymbol{r}}{\partial u}$ and $\boldsymbol{r}_2 = \frac{\partial \boldsymbol{r}}{\partial v}$ are tangent to the u-line and v-line, respectively, and lie in the tangent plane at M. The vector product $\mathbf{r}_1 \times \mathbf{r}_2$ is parallel to the normal at M and its unit vector $\mathbf{N}^0 = \dfrac{\mathbf{r}_1 \times \mathbf{r}_2}{|\mathbf{r}_1 \times \mathbf{r}_2|}$ is called the *unit normal vector*. The direction of $\mathbf{N}^0$ depends on the ordering of the parameters u and v.

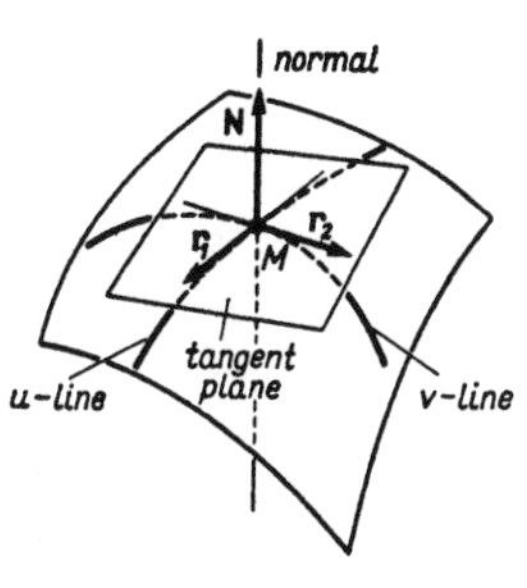

FIG. 254

Equations of the tangent plane and normal line to a surface

Equation of the surface (see p. 303)	Equation of the tangent plane	Equation of the normal line
(1)	$\frac{\partial F}{\partial x}(X-x)+\frac{\partial F}{\partial y}(Y-y)+\frac{\partial F}{\partial z}(Z-z)=0$	$\frac{X-x}{\frac{\partial F}{\partial x}}=\frac{Y-y}{\frac{\partial F}{\partial y}}=\frac{Z-z}{\frac{\partial F}{\partial z}}$
(2)	$Z-z=p(X-x)+q(Y-y)$	$\frac{X-x}{p}=\frac{Y-y}{q}=\frac{Z-z}{-1}$
(3)	$\begin{vmatrix} X-x & Y-y & Z-z \\ \frac{\partial x}{\partial u} & \frac{\partial y}{\partial u} & \frac{\partial z}{\partial u} \\ \frac{\partial x}{\partial v} & \frac{\partial y}{\partial v} & \frac{\partial z}{\partial v}\end{vmatrix}=0$	$\frac{X-x}{\begin{vmatrix}\frac{\partial y}{\partial u} & \frac{\partial z}{\partial u}\\ \frac{\partial y}{\partial v} & \frac{\partial z}{\partial v}\end{vmatrix}}=\frac{Y-y}{\begin{vmatrix}\frac{\partial z}{\partial u} & \frac{\partial x}{\partial u}\\ \frac{\partial z}{\partial v} & \frac{\partial x}{\partial v}\end{vmatrix}}=\frac{Z-z}{\begin{vmatrix}\frac{\partial x}{\partial u} & \frac{\partial y}{\partial u}\\ \frac{\partial x}{\partial v} & \frac{\partial y}{\partial v}\end{vmatrix}}$
(3a)	$(\mathbf{R}-\mathbf{r})\mathbf{r}_1\mathbf{r}_2=0$ or $(\mathbf{R}-\mathbf{r})\mathbf{N}=0$	$\mathbf{R}=\mathbf{r}+\lambda(\mathbf{r}_1\times\mathbf{r}_2)$ or $\mathbf{R}=\mathbf{r}+\lambda\mathbf{N}$

In this table $x, y, z, \mathbf{r}$ denote coordinates and radius vector of a point M of the surface; $X, Y, Z, \mathbf{R}$ denote running coordinates and radius vector of a point of the tangent plane or the normal to the surface. The derivatives are taken at the point M; $p = \dfrac{\partial z}{\partial x}$, $q = \dfrac{\partial z}{\partial y}$.

Example. For the sphere $x^2+y^2+z^2-a^2=0$, the tangent plane:

$$2x(X-x)+2y(Y-y)+2z(Z-z)=0 \quad \text{or} \quad xX+yY+zZ-a^2=0;$$

the normal:

$$\frac{X-x}{2x}=\frac{Y-y}{2y}=\frac{Z-z}{2z} \quad \text{or} \quad \frac{X}{x}=\frac{Y}{y}=\frac{Z}{z}.$$

For the sphere

$$x=a\cos u\sin v, \qquad y=a\sin u\sin v, \qquad z=a\cos v,$$

the tangent plane:

$$X\cos u\sin v+Y\sin u\sin v+Z\cos v=a,$$

the normal:

$$\frac{X}{\cos u\sin v}=\frac{Y}{\sin u\sin v}=\frac{Z}{\cos v}.$$

Singular (conical) points of a surface. If the coordinates $x=x_1$, $y=y_1$, $z=z_1$ of a point $M(x_1, y_1, z_1)$ of a surface given in the form (1) (see p. 303) satisfy the equations

$$\frac{\partial F}{\partial x}=\frac{\partial F}{\partial y}=\frac{\partial F}{\partial z}=F(x, y, z)=0,$$

then the point $M(x_1, y_1, z_1)$ is called a *singular* (or *conical*) *point* of the surface. In this case, the tangent lines to curves in the surface at the point M do not lie in a plane but generate a cone of the second degree with the equation

$$\frac{\partial^2 F}{dx^2}(X-x)+\frac{\partial^2 F}{\partial y^2}(Y-y)+\frac{\partial^2 F}{\partial z^2}(Z-z)+2\frac{\partial^2 F}{\partial x\partial y}(X-x)(Y-y)+$$

$$+2\frac{\partial^2 F}{\partial y\,\partial z}(Y-y)(Z-z)+2\frac{\partial^2 F}{\partial z\,\partial x}(Z-z)(X-x)=0,$$

where the derivatives are taken at M. If all six partial derivatives are equal to zero, then the singular point is more complicated (the tangents generate a cone of the third or a higher degree).

13. Linear element of a surface

Differential of arc. Let a surface be given in the form (3) or (3a) (see p. 303). If $M(u, v)$ is a point of the surface and $N(u+du, v+dv)$ is a point near to M, then the length of arc MN

is equal approximately to the *differential of arc* or the *linear element of the surface* according to the formula

$$\text{(I)} \qquad ds^2 = E\,du^2 + 2F\,du dv + G\,dv^2,$$

where

$$E = \mathbf{r}_1^2 = \left(\frac{\partial x}{\partial u}\right)^2 + \left(\frac{\partial y}{\partial u}\right)^2 + \left(\frac{\partial z}{\partial u}\right)^2, \quad G = \mathbf{r}_2^2 = \left(\frac{\partial x}{\partial v}\right)^2 + \left(\frac{\partial y}{\partial v}\right)^2 + \left(\frac{\partial z}{\partial v}\right)^2,$$

$$F = \mathbf{r}_1\mathbf{r}_2 = \frac{\partial x}{\partial u}\cdot\frac{\partial x}{\partial v} + \frac{\partial y}{\partial u}\cdot\frac{\partial y}{\partial v} + \frac{\partial z}{\partial u}\cdot\frac{\partial z}{\partial v}.$$

The right member of formula (I) is also called the *first fundamental form* of a surface given in the form (3) (p. 303). The coefficients E, F, G depend on a point of the surface.

Example. For the sphere $\mathbf{r} = a(\cos u \sin v\,\mathbf{i} + \sin u \sin v\,\mathbf{j} + \cos v\,\mathbf{k})$,

$$E = a^2 \sin^2 v, \qquad F = 0, \qquad G = a^2.$$

The first fundamental form: $ds^2 = a^2(\sin^2 v\,du^2 + dv^2)$.

For a surface given in the form (2) (p. 303):

$$E = 1 + p^2, \quad F = pq, \quad G = 1 + q^2, \quad \text{where} \quad p = \frac{\partial z}{\partial x}, \quad q = \frac{\partial z}{\partial y}.$$

Measurements on a surface. *Length of arc* of a curve $u = u(t)$, $v = v(t)$ on a surface, for $t_0 < t < t_1$ can be computed from the formula

$$(*) \qquad L = \int_{t_0}^{t_1} ds = \int_{t_0}^{t_1} \sqrt{E\left(\frac{du}{dt}\right)^2 + 2F\frac{du}{dt}\cdot\frac{dv}{dt} + G\left(\frac{dv}{dt}\right)^2}\,dt.$$

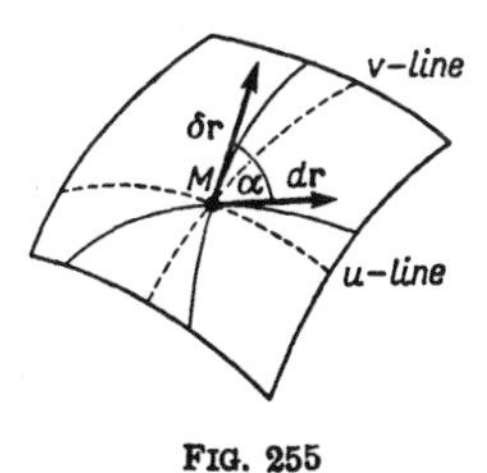

FIG. 255

Angle α between two curves (i.e., between their tangents) intersecting at M and having the direction vectors $d\mathbf{r}\{du, dv\}$ and $\delta\mathbf{r}\{\delta u, \delta v\}$ (Fig. 255) can be computed from the formula

$$(\cdot\cdot)\qquad \cos\alpha = \frac{d\mathbf{r}\,\delta\mathbf{r}}{\sqrt{(d\mathbf{r})^2(\delta\mathbf{r})^2}}$$

$$= \frac{E\,du\delta u + F(du\delta v + dv\delta u) + G\,dv\delta v}{\sqrt{E\,du^2 + 2F\,dudv + G\,dv^2}\sqrt{E\,\delta u^2 + 2F\,\delta u\delta v + G\,\delta v^2}},$$

where the coefficients E, F, G should be taken for the point M. In particular, the curves are orthogonal, if the numerator of formula $(\cdot\cdot)$ is zero. A condition for the coordinate lines $v = \text{const}$ $(dv = 0)$ and $u = \text{const}$ $(\delta u = 0)$ to be orthogonal is $F = 0$.

Area of a surface S bounded by a curve can be computed as a double integral

$$S = \int_{(S)} dS,$$

where

$$(\because)\qquad dS = \sqrt{EG - F^2}\,dudv.$$

Thus knowing the coefficients E, F, G of the first fundamental form, we can compute lengths of lines, angles and areas on the surface from the formulas $(\cdot)$, $(\cdot\cdot)$, $(\because)$, i.e., the first fundamental form determines the metric on the surface.

Applicability of surfaces. If a surface is deformed by bending, without stretching, compression or tearing, then its metric remains unchanged, i.e., the first fundamental form is the same. Two surfaces having the same first fundamental form are called applicable; they can be applied, or deformed, one onto the other by bending.

14. Curvature of a surface

Curvature of a line on a surface. The radii ϱ of curvature of various curves Γ lying on a surface and passing through a point M are related as follows:

(1) Radius of curvature ϱ of a curve Γ is equal to the radius of curvature of the curve C which is the section of the surface by the osculating plane to Γ at the point M (Fig. 256a).

(2) Radius of curvature of any plane section C is equal to

$$(\text{M})\qquad \varrho = R\cos(\mathbf{n}, \mathbf{N}),$$

where R is the radius of curvature of the normal section (C_{norm}) passing through the normal vector $\mathbf{N}$ and having the same tangent PQ as the plane section C; $(\mathbf{n}, \mathbf{N})$ is the angle between the principal normal $\mathbf{n}$ to the curve and the normal vector $\mathbf{N}$ to the surface

(*Meusnier's theorem*, Fig. 256b). In formula (M), the sign should be positive, if the vector $\boldsymbol{N}$ is directed towards the concavity of the curve C_{norm}, and negative otherwise.

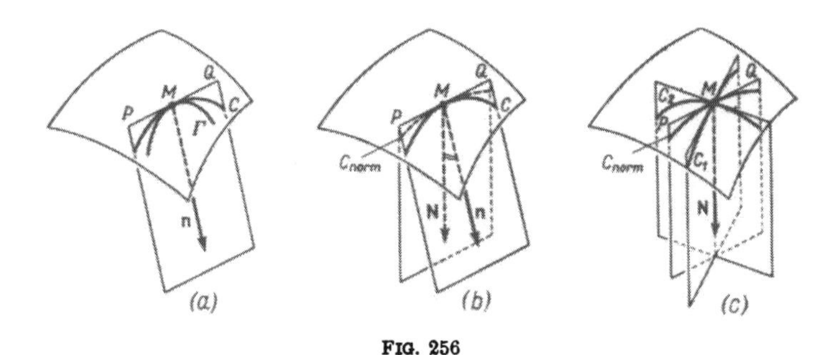

FIG. 256

(3) Curvature of any normal section C_{norm} is given by *Euler's formula*

(E) $$\frac{1}{R} = \frac{\cos^2\alpha}{R_1} + \frac{\sin^2\alpha}{R_2},$$

where R_1 and R_2 are the principal radii of curvature, i.e., the maximal and minimal values of R, for principal normal sections C_1 and C_2 (Fig. 256c). The signs at R, R_1, and R_2 are defined as in formula (M).

Principal radii of curvature. If the surface is given by the equation $z = f(x, y)$, then the *principal radii of curvature* R_1 and R_2 are roots of the equation

(A) $$(rt - s^2)R^2 + h[2pqs - (1 + p^2)t - (1 + q^2)r]R + h^4 = 0,$$

where

$$p = \frac{\partial z}{\partial x}, \quad q = \frac{\partial z}{\partial y}, \quad r = \frac{\partial^2 z}{\partial x^2}, \quad s = \frac{\partial^2 z}{\partial x \partial y}, \quad t = \frac{\partial^2 z}{\partial y^2}, \quad h = \sqrt{1 + p^2 + q^2}.$$

The planes of principal normal sections C_1 and C_2 are perpendicular and their directions are determined by the value of dy/dx from the quadratic equation

(B) $$[tpq - s(1 + q^2)]\left(\frac{dy}{dx}\right)^2 + [t(1 + p^2) - r(1 + q^2)]\frac{dy}{dx} + \\ + [s(1 + p^2) - rpq] = 0.$$

If the surface is given in parametric form $\boldsymbol{r}=\boldsymbol{r}(u, v)$, then the equations corresponding to (A) and (B) have the form

$$\text{(A}'\text{)} \quad (DD''-D'^2)\,R^2-(ED''-2FD'+GD)\,R+(EG-F^2)=0,$$

$$\text{(B}'\text{)} \quad (GD'-FD'')\left(\frac{dv}{du}\right)^2+(GD-ED'')\frac{dv}{du}+(FD-ED')=0,$$

where D, D', D'' are the coefficients of the *second fundamental form* defined by the equalities

$$D=\boldsymbol{r}_{11}\boldsymbol{N}=\frac{d}{\sqrt{EG-F^2}}, \quad D'=\boldsymbol{r}_{12}\boldsymbol{N}=\frac{d'}{\sqrt{EG-F^2}},$$

$$D''=\boldsymbol{r}_{22}\boldsymbol{N}=\frac{d''}{\sqrt{EG-F^2}};$$

the vectors $\boldsymbol{r}_{11}, \boldsymbol{r}_{12}, \boldsymbol{r}_{22}$ are second partial derivatives of the radius vector $\boldsymbol{r}$ with respect to the parameters u and v, and the numerators d, d', d'' are:

$$d=\begin{vmatrix} \frac{\partial^2 x}{\partial u^2} & \frac{\partial^2 y}{\partial u^2} & \frac{\partial^2 z}{\partial u^2} \\ \frac{\partial x}{\partial u} & \frac{\partial y}{\partial u} & \frac{\partial z}{\partial u} \\ \frac{\partial x}{\partial v} & \frac{\partial y}{\partial v} & \frac{\partial z}{\partial v} \end{vmatrix}, \quad d'=\begin{vmatrix} \frac{\partial^2 x}{\partial u\partial v} & \frac{\partial^2 y}{\partial u\partial v} & \frac{\partial^2 z}{\partial u\partial v} \\ \frac{\partial x}{\partial u} & \frac{\partial y}{\partial u} & \frac{\partial z}{\partial u} \\ \frac{\partial x}{\partial v} & \frac{\partial y}{\partial v} & \frac{\partial z}{\partial v} \end{vmatrix}, \quad d''=\begin{vmatrix} \frac{\partial^2 x}{\partial v^2} & \frac{\partial^2 y}{\partial v^2} & \frac{\partial^2 z}{\partial v^2} \\ \frac{\partial x}{\partial u} & \frac{\partial y}{\partial u} & \frac{\partial z}{\partial u} \\ \frac{\partial x}{\partial v} & \frac{\partial y}{\partial v} & \frac{\partial z}{\partial v} \end{vmatrix}$$

The curves which are everywhere tangent to the direction of a principal normal section are called the *lines of curvature.* Their equation can be obtained by integration of differential equations (B) or (B′).

Classification of points of a surface. If the principal radii of curvature R_1 and R_2 (p. 309) at a given point M are of the same sign, then the principal normal sections are concave in the same direction. In this case, the surface lies on one side of the tangent plane; such a point is called *elliptic* (Fig. 257a); the analytic condition for an elliptic point: $DD''-D'^2>0$. In particular, if $R_1=R_2$, then the point is called *umbilical.* In an umbilical point, the curvature of all normal sections is constant.

If the principal radii of curvature R_1 and R_2 are of different signs, then the principal normal sections are concave in different directions. In this case, the surface intersects the tangent plane and has the form of a saddle; such a point is called *hyperbolic* (Fig. 257b); the analytic condition: $DD''-D'^2<0$.

If one of the principal radii of curvature, R_1 or R_2, is infinitely great, then one of the principal normal sections has a point of

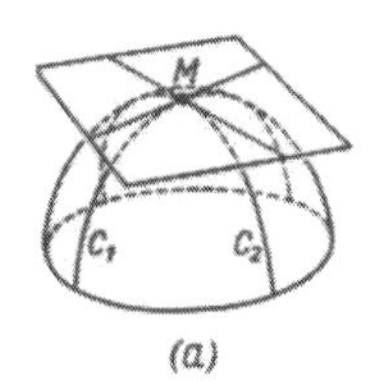

(a)

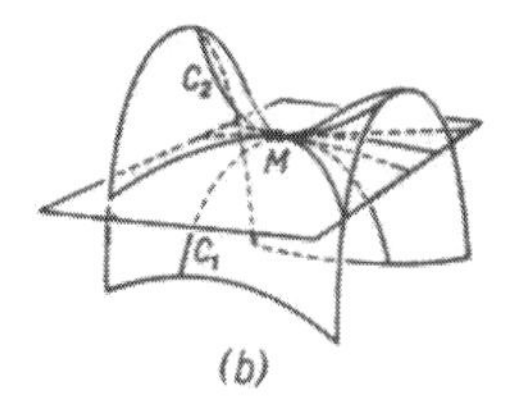

(b)

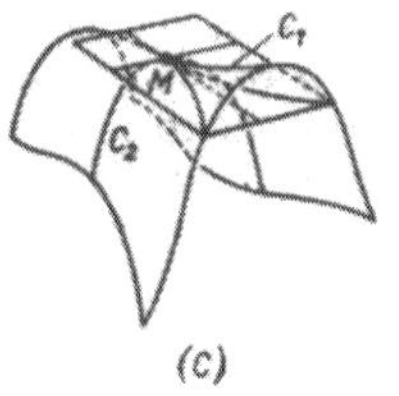

(c)

FIG. 257

inflection or is a straight line; such a point is called *parabolic* (Fig. 257c); the analytic condition: $DD'' - D'^2 = 0$.

Examples. All points of the ellipsoid are elliptic, of the hyperboloid of one sheet—hyperbolic, and of the cylinder—parabolic.

Curvature of a surface. The expression

$$H = \frac{1}{2}\left(\frac{1}{R_1} + \frac{1}{R_2}\right)$$

is called the *mean curvature* of the surface at the point M and the expression

$$K = \frac{1}{R_1 R_2}$$

is called the *Gaussian* or *total curvature*.

Example. For the circular cylinder of radius a, the mean curvature $H = 1/2a$ and Gaussian curvature $K = 0$.

In elliptic points $K > 0$, in hyperbolic points $K < 0$, in parabolic points $K = 0$.

If the surface is given by the equation $z = f(x, y)$, then formulas for H and K have the form

$$H = \frac{r(1+q^2) - 2pqs + t(1+p^2)}{2(1+p^2+q^2)^{3/2}},$$

$$K = \frac{rt - s^2}{(1+p^2+q^2)^2} \text{ [1]}.$$

The surfaces whose mean curvature is everywhere equal to zero (i.e. $R_1 = -R_2$) are called *minimal surfaces*. The surfaces whose

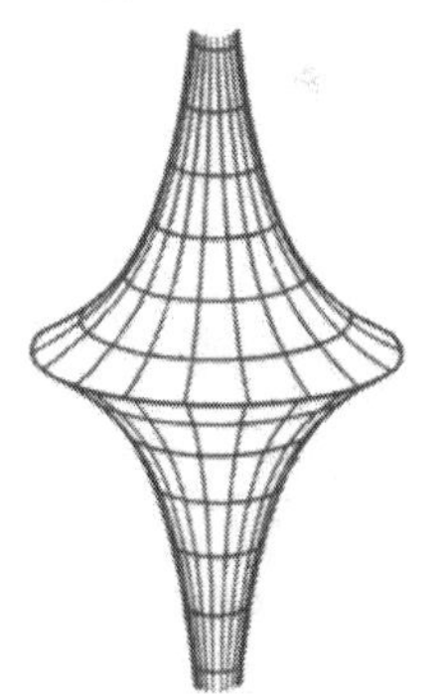

FIG. 258

[1] For notation p, q, r, s, t, see p. 309.

Gaussian curvature is everywhere constant are called *surfaces of a constant curvature*. The simplest example of a surface with a constant positive Gaussian curvature K is the sphere; a surface with a constant negative Gaussian curvature is the pseudosphere, i.e., the surface obtained by revolving the tractrix (see p. 131) about its asymptote (Fig. 258).

15. Ruled and developable surfaces

A surface is said to be a *ruled surface* if it can be generated by a motion of a straight line; if a surface can be developed into the plane, then it is called *developable*. The simplest examples of developable surfaces are cylindrical and conical surfaces (see pp. 207, 208). There exist ruled surfaces which are not developable, e.g., hyperboloid of one sheet and hyperbolic paraboloid (see pp. 271, 273). The Gaussian curvature of a developable surface is everywhere equal to zero.

A condition that a surface given by the equation $z = f(x, y)$ is developable is

$$rt - s^2 = 0$$ [1].

16. Geodesic lines on a surface

Concept of geodesic lines. Given a point $M(u, v)$ of a surface and a direction dv/du, there exists a line on the surface passing through the given point in the given direction. It is called a *geodesic line* or a *geodesic* and plays on the surface the role of a straight line on the plane as follows.

(1) If a particle is forced to remain on the surface but is free from external forces, then it will move along a geodesic.

(2) An elastic string tauten on a surface admits a form of a geodesic.

(3) The shortest distance between two points of a surface is given by a geodesic.

Definition. A *geodesic* is a curve on the surface whose principal normal coincides at every point with the normal to the surface.

Example. Geodesics on a circular cylinder are helixes.

---

[1] For notation p, q, r, s, t, see p. 309.

Equation of geodesics. If the surface is given in the form $z = f(x, y)$, then the differential equation for geodesics is

$$(1 + p^2 + q^2)\frac{d^2y}{dx^2} = pt\left(\frac{dy}{dx}\right)^3 + (2ps - qt)\left(\frac{dy}{dx}\right)^2 + (pr - 2qs)\frac{dy}{dx} - qr\ (^1).$$

The equation for geodesics of a surface given in the form (3) (p. 303) has a more complicated form.

---

(1) For notation p, q, r, s, t, see p. 303.

PART FOUR

FOUNDATIONS OF MATHEMATICAL ANALYSIS

I. INTRODUCTION TO ANALYSIS

1. Real numbers

Rational numbers. All integral and fractional numbers (positive, negative and zero) are called *rational*. Rational numbers form an infinite set with the following properties:

(1) This set is *ordered*, i.e., for every two rational numbers it is possible to indicate which of them is greater.

(2) This set is *everywhere dense*, i.e., between every two rational numbers a and b $(a < b)$ there exists at least one rational number c $(a < c < b)$; hence there exists also an infinite set of rational numbers between a and b.

(3) The arithmetic operations (addition, subtraction, multiplication and division) with two arbitrary rational numbers are always performable and give as the result a rational number. Division by zero is an exception; it is *not* allowed. The expression $a/0$ has no definite meaning, for there exists no definite number b satisfying the equation $b \cdot 0 = a$ (if $a = 0$, then b can be arbitrary, and if $a \neq 0$, then b does not exist) [1].

(4) Every rational number can be represented by a decimal fraction (which may be finite or recurring).

Geometric representation of rational numbers. If a point O of a straight line xx (Fig. 259) is taken as the *origin*, one of its directions is taken as positive (*orientation*) and if a *unit segment* l is chosen, then to every rational number a there corresponds a definite point of the straight line with the *coordinate* a (a rational

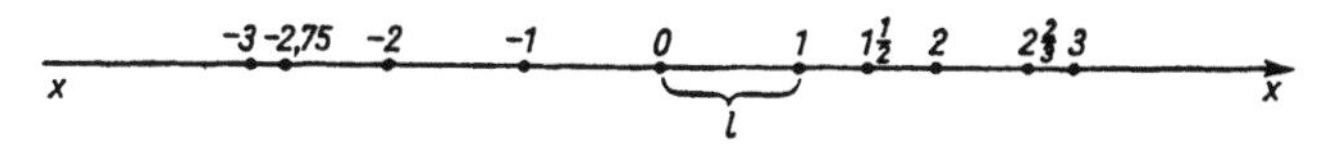

FIG. 259

[1] The equality $a/0 = \infty$ (*infinity*) which is often used does not mean that this division is performable (infinity is not a number); it is simply an abbreviation for the statement: "if the divisor tends to zero, then the absolute value of the quotient increases beyond bound".

point). The line xx is called a *number scale*. According to property (2) of rational numbers, there exists an infinite set of rational points between every two rational points.

Irrational numbers. The set of rational numbers is not sufficient for mathematical analysis. Although it is everywhere dense, it does not fill up the whole number scale. For example, if a diagonal AB of a square with a side equal to 1 is placed on a number scale so that the point A coincides with the origin, then the point K corresponding to B on the number scale has no rational coordinate (Fig. 260). Introduction of *irrational numbers* enables us to assign a definite number to every point of the number scale; thus the set of all numbers becomes continuous.

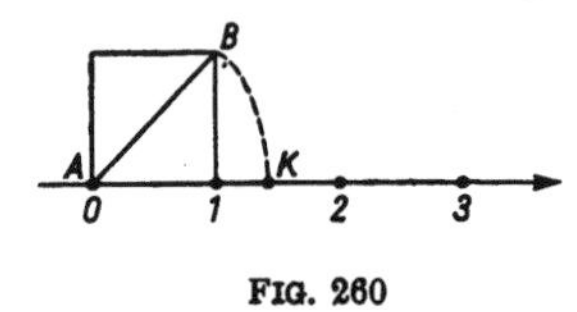

FIG. 260

An exact definition of irrational numbers is introduced in more extensive courses of mathematical analysis. The points of the number scale which correspond to the irrational numbers fill all the gaps between the rational numbers. Each irrational number can be represented as an infinite non-recurring decimal fraction.

In particular, the non-integral roots of algebraic equations of the form $x^n + a_1x^{n-1} + a_2x^{n-2} + \ldots + a_{n+1}x + a_n = 0$ (with integral coefficients) belong to the irrational numbers; for example, the equation $x^3 - 9x + 4 = 0$ has irrational roots (see p. 162); such numbers are called *algebraic*. The simplest algebraic numbers are the roots of the binomial equations $x^n - a = 0$, that is to say are the numbers of the form $\sqrt[n]{a}$, if they are not rational (for example, $\sqrt{2} = 1.414\ldots$, $\sqrt[3]{10} = 2.154\ldots$). The irrational numbers that are not algebraic are called *transcendental*; they include, in particular, the numbers $\pi = 3.141593\ldots$, $e = 2.718282\ldots$, decimal logarithms of integers (except of those of the form 10^n), most of the values of the trigonometric functions for the angles whose degree measure is expressed by an integer.

Real numbers. All rational and irrational numbers are called *real*. The following are the fundamental properties of real numbers:

(1) The set of real numbers is *ordered* (see p. 315).

(2) It is *everywhere dense* (see p. 315).

(3) It is *continuous*, i.e. (in contrast to the set of rational numbers) every point of the number scale has a real coordinate.

(4) The arithmetic operations with real numbers are always performable (with except of division by zero, see p. 315) and give as result a real number. Raising to a power and inverse opera-

tions are also performable in the system of real numbers (every real positive number has a root of an arbitrary degree and also a logarithm to an arbitrary base (different from 1)).

A further generalization of the notion of real numbers in mathematical analysis is complex numbers (see p. 585).

2. Sequences and their limits

Sequences. An infinite ordered set of numbers

$$a_1, a_2, a_3, \ldots, a_n, \ldots$$

is called a *sequence of numbers* (¹). The numbers of the sequence are said to be its *terms*. The terms of a sequence are not necessarily different.

A sequence is considered to be defined if a principle for forming it is given, i.e., if a rule for obtaining an arbitrary term of the sequence is known. In many cases, it is possible to give a formula for the general term a_n of the sequence.

Examples: (1) $a_n = n$, (2) $a_n = 4 + 3(n-1)$, (3) $a_n = 3(-\tfrac{1}{2})^{n-1}$, (4) $a_n = (-1)^{n+1}$, (5) $a_n = 3 - \dfrac{1}{2^{n-2}}$, (6) $a_n = 3\tfrac{1}{3} - \tfrac{1}{3}\cdot 10^{-(n-1)/2}$ when n is odd and $a_n = 3\tfrac{1}{3} + \tfrac{2}{3}\cdot 10^{-\frac{1}{2}n+1}$, when n is even, (7) $a_n = 1/n$, (8) $a_n = (-1)^{n+1}n$, (9) $a_n = -\tfrac{1}{2}(n+1)$ when n is odd and $a_n = 0$, when n is even, (10) $a_n = 3 - \dfrac{1}{2^{\frac{1}{2}n-\frac{3}{2}}}$, when n is odd and $a_n = 13 - \dfrac{1}{2^{\frac{1}{2}n-2}}$, when n is even.

The first terms of these sequences are the following:

(1) 1, 2, 3, 4, 5, ... (*natural sequence*),
(2) 4, 7, 10, 13, 16, ... (*arithmetical progression*),
(3) $3, -\tfrac{3}{2}, \tfrac{3}{4}, -\tfrac{3}{8}, \tfrac{3}{16}, \ldots$ (*geometrical progression*),
(4) $1, -1, 1, -1, 1, \ldots,$
(5) $1, 2, 2\tfrac{1}{2}, 2\tfrac{3}{4}, 2\tfrac{7}{8}, \ldots,$
(6) 3, 4, 3.3, 3.4, 3.33, 3.34, 3.333, 3.334, ...,
(7) $1, \tfrac{1}{2}, \tfrac{1}{3}, \tfrac{1}{4}, \tfrac{1}{5}, \ldots,$
(8) $1, -2, 3, -4, 5, -6, \ldots,$
(9) $-1, 0, -2, 0, -3, 0, -4, 0, \ldots,$
(10) $1, 11, 2, 12, 2\tfrac{1}{2}, 12\tfrac{1}{2}, 2\tfrac{3}{4}, 12\tfrac{3}{4}, \ldots$

Limit of a sequence. If for a given sequence $a_1, a_2, \ldots, a_n$ there exists a number A such that the numbers a_n approach A arbi-

(¹) We consider here only infinite sequences.

trarily near, when n increases, then the number A is called a *limit* of the sequence [1]. Notation:

$$A = \lim_{n \to \infty} a_n.$$

Precise formulation: $A = \lim_{n \to \infty} a_n$, if, for any preassigned positive number ε, however small, a number a_N of the given sequence can be indicated, such that all the numbers a_n beyond a_N (i.e., for $n > N$) differ from A, in the absolute value, less than ε:

$$|a_n - A| < \varepsilon \quad (n > N).$$

Among the examples (1)–(10), the sequences (3), (5), (6) and (7) have limits:

$$(3) \lim_{n \to \infty} a_n = 0, \quad (5) \lim_{n \to \infty} a_n = 3, \quad (6) \lim_{n \to \infty} a_n = 3\tfrac{1}{3}, \quad (7) \lim_{n \to \infty} a_n = 0.$$

Geometric significance. If the terms of a sequence with a limit A are marked by points of a number scale, then all the

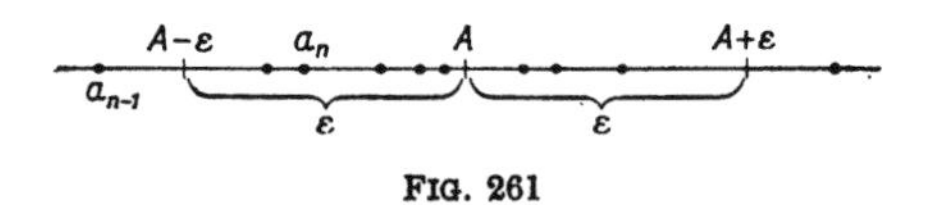

FIG. 261

points a_n, beyond a_N, are contained inside the interval bounded by the points $A - \varepsilon$ and $A + \varepsilon$ (Fig. 261).

Infinite limit. If the absolute value of a_n increases infinitely, when n increases, then a limit does not exist; this case is denoted by the symbol

$$\lim_{n \to \infty} a_n = \infty \quad \text{("the limit is equal to the infinity").}$$

Exact formulation: $\lim_{n \to \infty} a_n = \infty$, if, for any preassigned positive number K, however great, a number N can be indicated such that all the numbers a_n, for $n > N$, are, in the absolute value, greater than K:

$$|a_n| > K \quad (n > N).$$

If, moreover, all the numbers a_n $(n > N)$ are positive, then we write $\lim_{n \to \infty} a_n = +\infty$; if the numbers a_n are all negative, then we write $\lim_{n \to \infty} a_n = -\infty$.

[1] The numbers a_n may be equal to A for certain values of n.

The sequences (1), (2) and (8) from the examples (1)–(10) have infinite limits and in the cases (1) and (2) $\lim_{n\to\infty} a_n = +\infty$.

Monotone sequences. A sequence $a_1, a_2, \ldots, a_n, \ldots$ is said to be *increasing*, if

(1) $$a_1 < a_2 < a_3 < \ldots < a_n < \ldots,$$

decreasing, if

(2) $$a_1 > a_2 > a_3 > \ldots > a_n > \ldots,$$

non-decreasing, if

(3) $$a_1 \leq a_2 \leq a_3 \leq \ldots \leq a_n \leq \ldots$$

and *non-increasing*, if

(4) $$a_1 \geq a_2 \geq a_3 \geq \ldots \geq a_n \geq \ldots$$

Sequences of the form (1), (2), (3), (4) are called *monotone*; the sequences (1) and (3) are called *monotone increasing* the sequences (2) and (4)—*monotone decreasing.*

The sequences (1) and (2), in contrast to the sequences (3) and (4), are sometimes called *strictly monotone.* The points of the number scale representing the terms of a monotone sequence proceed (in the order of indices) in one direction, and, for the sequences (3) and (4), some consecutive terms may coincide. From the examples (1)–(10) on p. 317, the sequences (1), (2), (5) are monotone increasing and the sequence (7) is strictly decreasing.

Bounded sequences. If, for a given sequence, there exists a positive number K, such that the absolute value of all the terms of the sequence is less than K ($|a_n| < K$), then the sequence is said to be *bounded*; if such a number does not exist, then the sequence is said to be *unbounded.* The following sequences, from the examples (1)–(10) on p. 317, are bounded: (3) ($K = 4$), (4) ($K = 2$), (5) ($K = 3$), (6) ($K = 5$), (7) ($K = 2$), (10) ($K = 13$).

Fundamental theorems on limits of sequences.

(1) A sequence can have at most one limit.

(2) A sequence having a finite limit is bounded; a sequence having an infinite limit is unbounded.

(3) A monotone bounded sequence has a finite limit; if this sequence is monotone increasing, then $\lim_{n\to\infty} a_n \geq a_n$; if it is monotone decreasing, then $\lim_{n\to\infty} a_n \leq a_n$.

(4) A monotone unbounded sequence has an infinite limit; if this sequence is monotone increasing, then $\lim_{n\to\infty} a_n = +\infty$; if it is monotone decreasing, then $\lim_{n\to\infty} a_n = -\infty$.

(5) A necessary and sufficient condition for the existence of a limit of a sequence: In order that a sequence $a_1, a_2, \ldots, a_n, \ldots$ has a limit it is necessary and sufficient that for any preassigned positive number ε, however small, a term a_N of the sequence can be found, such that any two numbers of the sequence, beyond a_N, differ by less than ε, i.e.,

$$|a_i - a_j| < \varepsilon \quad \text{for} \quad i > N \text{ and } j > N \text{ (}^1\text{)}.$$

For other properties and for evaluation of limits see pp. 330–333 ("Limit of a function").

3. Functions of one variable (²)

Definition. A variable quantity y is said to be a *function* of a variable quantity x, (x is called an *argument*, or the *independent variable*), if, for a given value of x, the quantity y assumes a certain definite value (*single-valued functions*, for example, $y = x^2$) or a number of definite values (*multiple-valued functions*, for example, $y = \pm\sqrt{x}$ is a two-valued function). The symbols $f(x)$, $F(x)$, $\varphi(x)$, ... denote different functions of the variable x; $f(a)$ denotes the value of $f(x)$ assumed for $x = a$; for example, if $f(x) = x^2 + 2x - 5$, then $f(3) = 3^2 + 2 \cdot 3 - 5 = 10$.

The set of all values of x for which the function is definite is called the *domain of existence* (*or of definition*) of the function. Functions with a connected domain of existance are mostly considered. A domain of real numbers is said to be *connected*, if (1) it contains more than one point, (2) it has no gaps, i.e., all the numbers lying between any two numbers of the domain also belong to the domain. A connected domain may be *unbounded* on both sides (i.e., contain all the points of the straight line), it may be *bounded from the left* or *from the right* (i.e., contain all the numbers greater, or, respectively, less than a certain given number) or it may be *bounded from both sides* (i.e., contain all the numbers lying between two given numbers). A connected domain of real numbers is also called an *interval* with the end points a and b ($a < b$; a may be

(¹) A sequence satisfying this condition is said to be a *fundamental sequence*.

(²) We consider here only functions of a real variable. For functions of a complex variable see pp. 590–605.

equal to $-\infty$ and b may be equal to $+\infty$). An end point a or b of the interval is said to be *open* if it does not belong to the interval and it is said to be *closed,* if it does belong to it (the end points $-\infty$ and $+\infty$ are regarded as open).

An interval is denoted by its end points taken in brackets; an open end point is denoted by a round bracket and a closed end point by a square bracket. An interval with two open end points is called *open,* with one open end point—*semi-open* and with two closed end points—*closed* (see Fig. 262).

Name of the interval	Bounds of the domain	Notation for the interval	Graphical notation
Open	$a < x < b$	(a, b)	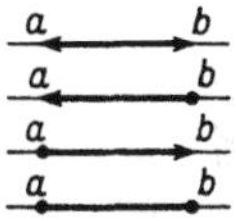
Semi-open	$a < x \leqslant b$	$(a, b]$	
	$a \leqslant x < b$	$[a, b)$	
Closed	$a \leqslant x \leqslant b$	$[a, b]$	
Unbounded intervals	$-\infty < x < +\infty$	$(-\infty, +\infty)$	
	$-\infty < x < +b$	$(-\infty, b)$	
	$-\infty < x \leqslant b$	$(-\infty, b]$	
	$a < x < +\infty$	$(a, +\infty)$	
	$a \leqslant x < +\infty$	$[a, +\infty)$	

FIG. 262

Functions whose domain of existence is a finite set or an infinite sequence of separate numbers are often considered. Sometimes the sequence of all natural numbers (the natural sequence) is taken as a domain of existence; the values of such a function can be ordered into a sequence

$$f(1),\ f(2),\ f(3),\ldots,\ f(n),\ldots$$

(a function with an integral argument).

Domains of existence in the form of a union of intervals and of separate numbers are also considered.

Ways of defining a function. A function can be defined in different ways, for example, by means of tables giving its values, by a graph or by one or several formulas (for different interval of existence).

Examples of functions defined by several formulas or by a graph (Fig. 263)[1]:

(1) The arrows mean that the end point in the arrowhead does not belong to the graph.

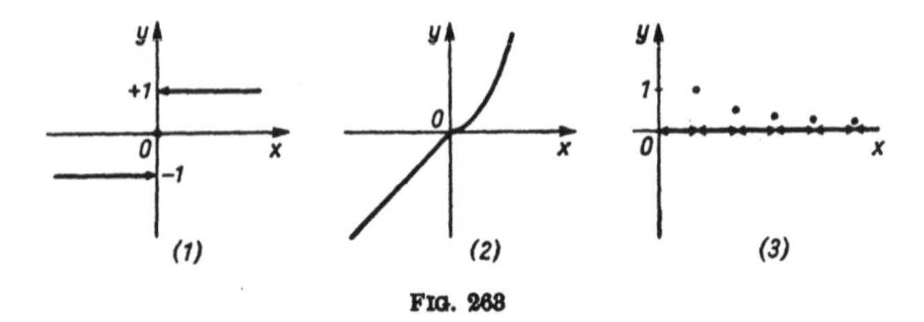

FIG. 268

$$(1)\ y = \begin{cases} -1 & \text{for } x < 0, \\ 0 & \text{for } x = 0, \\ +1 & \text{for } x > 0. \end{cases} \qquad (2)\ y = \begin{cases} x & \text{for } x < 0, \\ x^2 & \text{for } x > 0. \end{cases}$$

$$(3)\ y = \begin{cases} \dfrac{1}{n} & \text{when } n \text{ is a positive integer,} \\ 0 & \text{for all remaining numbers.} \end{cases}$$

Domain of existence of an analytic expression. Most of the functions considered in mathematical analysis are defined by a single formula. The domain of existence of such a function includes all the values of the argument for which the given analytic expression has a definite meaning, i.e., it has a finite real value. Such a domain is said to be the *domain of existence of an analytic expression*. If there are no additional restrictions, the domain of existence of a function defined by a single formula is understood to be the domain of existence of this analytic expression. In particular, the domain of existence does not include the values of variables, for which (1) the function assumes complex values (2) "becomes infinite" (see p. 336, the types of discontinuities of functions), (3) is undetermined (see pp. 331–333, evaluation of indeterminate forms).

Examples: (1) $y = \sqrt{1 - x^2}$; domain of existence $-1 < x < 1$; (2) $y = \log \cos x$; domain of existence $-\frac{1}{2}\pi < x < +\frac{1}{2}\pi$, $\frac{3}{2}\pi < x < \frac{5}{2}\pi$, ..., $\frac{1}{2}(4n - 1)\pi < x < \frac{1}{2}(4n + 1)\pi$, ..., n is integral.

Fundamental forms of analytic definitions of functions. A functions may be defined

(1) in the *explicit form*, when an expression for y by means of x is given ($y = f(x)$),

(2) in the *implicit form*, when y is related to x by an equation ($F(x, y) = 0$, for example $x^2 + y^2 - 1 = 0$, or $x^y - xy = 0$),

(3) in the *parametric form* ($x = \varphi(t)$, $y = \psi(t)$), when both variables x and y are given as function of a certain variable t, for example, $x = a \cos t$, $y = a \sin t$.

Inverse functions. Two functions $y = f(x)$ and $y = \varphi(x)$ are said to be *inverse* each to another, if every pair of values a, b satisfying $b = f(a)$, satisfies also $a = \varphi(b)$, and, conversely, every pair satisfying $a = \varphi(b)$ satisfies $b = f(a)$. One of two inverse functions (no matter which) may be taken as *simple* and then the other one is called *inverse*.

Examples of inverse functions (Fig. 264): (1) $y = x^2$ and $y = \pm\sqrt{x}$, (2) $y = e^x$ and $y = \ln x$, (3) $y = \sin x$ and $y = \arcsin x$.

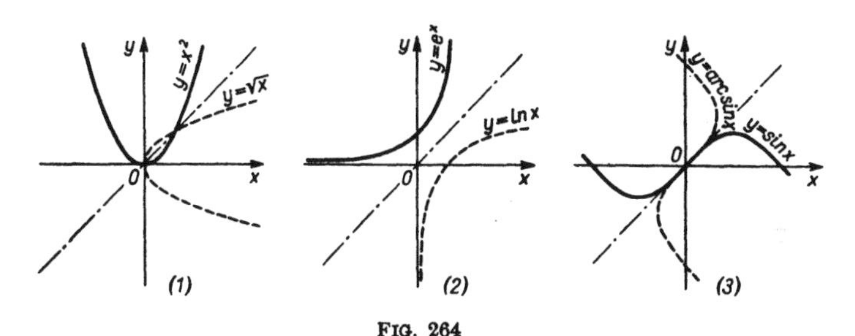

Fig. 264

To obtain an inverse function from a given one $y = f(x)$, we have to interchange the positions of the argument and of the function; the expression $x = \varphi(y)$ defines a function inverse to $y = f(x)$ in an implicit form. Solving the equation $x = \varphi(y)$ with respect to y we obtain the inverse function $y = \varphi(x)$ in the explicit form.

The graphs of two functions inverse each to another are symmetric with respect to the bisector of the coordinate system (see Fig. 264).

Elementary functions [1] are those defined by formulas containing a finite number of algebraic or trigonometric operations performed with the argument, the function and certain constants. (These operations include addition, subtraction, powers, roots, exponential and logarithmic operations, taking a trigonometric or inverse trigonometric function.) The elementary functions can be divided into *algebraic* and *transcendental* ones.

The variables x and y in the algebraic functions are connected by an algebraic equation of the form

$$\sum_{i=0}^{k} a_i x^n y^m = 0,$$

[1] The tables of elementary functions are given on pp. 19–83 and their graphs on pp. 96–116.

for example, $3xy^3 - 4xy + x^3 - 1 = 0$. If this equation can be solved with respect to y, we get one of the following simple types of algebraic functions:

(1) *Polynomial* (an *integral function*): it contains only addition, subtraction and multiplication: $y = a_0x^n + a_1x^{n-1} + \ldots + a_n$. In particular, $y = a$ (*constant function*), $y = ax + b$ (*linear function*), $y = ax^2 + bx + c$ (*quadratic function*) are polynomials.

(2) *Rational* (*fractional*) *function*: A rational function is one which can always be expressed as a quotient of two polynomials:

$$y = \frac{a_0x^n + a_1x^{n-1} + \ldots + a_n}{b_0x^m + b_1x^{m-1} + \ldots + b_m}$$

in particular, $y = \dfrac{ax+b}{cx+d}$ is called the *homographic transformation*.

(3) *Irrational function*: it contains roots, besides the operations mentioned in (2). For example, $y = \sqrt{2x+3}$, $y = \sqrt[3]{(x^2-1)\sqrt{x}}$.

Transcendental functions are those in which the argument and the function cannot be connected by an algebraic equation of the form $\sum a_i x^m y^n = 0$. The simplest of them are elementary transcendental functions, i.e., the following:

(1) *Exponential functions*: the variable x or an algebraic function of x is the exponent of a power (for example, $y = e^x$, $y = a^x$, $y = 2^{3x^2-5x}$).

(2) *Logarithmic functions*: the variable x or an algebraic function of x is under a logarithm sign (for example, $y = \ln x$, $y = \log x$, $y = \log_2(5x^2 - 3x)$).

(3) *Trigonometric functions*: the variable x or an algebraic function of x is under a sign of sin, cos, tan, cot, sec, cosec (for example, $y = \sin x$, $y = \cos(2x+3)$, $y = \tan\sqrt{x}$) [1].

(4) *Inverse trigonometric functions*: the variable x or an algebraic function of x is under the sign of arc sin, arc cos, etc. (for example, $y = \arcsin x$, $y = \arccos\sqrt{1-x}$).

Various combinations of elementary and transcendental

---

[1] An argument x of a trigonometric function $\sin x$, $\cos x$, $\tan x$, ... in mathematical analysis is understood to be an arbitrary number and not an angle or an arc of a circle (as in elementary trigonometry). Trigonometric functions can be defined purely analytically, without geometric notions; for example, $\sin x$ may be defined by its expansion into a power series (see p. 388) or as a solution of the differential equation $d^2y/dx^2 + y = 0$ with the initial condition: $x = 0$, $y = 0$, $dy/dx = 1$. In this sense, the argument of a trigonometric function is equal numerically to its circular measure expressed in radians. Thus to evaluate trigonometric functions, trigonometric tables can be used, but the radian measure of angle should be changed into degree.

functions in which one function is an argument of the other lead to *composite functions*, for example, $y = \ln \sin x$, $y = \frac{\ln x + \sqrt{\arcsin x}}{x^2 + 5e^x}$. Such combinations of elementary functions taken in a finite number also give elementary functions.

Non-elementary functions. Functions which are not elementary can be defined in various ways, by describing a correspondence between the argument and the function. The following ways of defining non-elementary functions are used in mathematical analysis [1]:

(1) By means of several mathematical formulas (see p. 322).

(2) By passing to a limit; in particular:

(a) by means of infinite series or products (see p. 355),

(b) by means of definite integrals (with one or two variable limits) which cannot be expressed in terms of elementary functions (see p. 395).

(3) By differential equations whose solution cannot be expressed in quadratures.

(4) By functional equations.

Some non-elementary functions are of theoretical importance; for these tables are arranged, graphs are constructed and their properties are investigated. Such functions are called *special*; they often have their own names and notations.

Examples of non-elementary functions.

(1) *Integral part of the number* x: y is equal to the greatest integer not exceeding x. Notation $y = E(x)$. The graph is in Fig. 265.

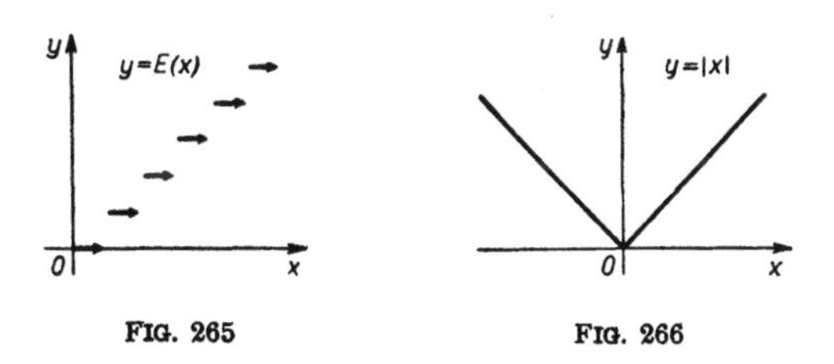

FIG. 265 FIG. 266

(2) *Absolute value of* x: $y = \begin{cases} -x & \text{for } x < 0, \\ x & \text{for } x > 0. \end{cases}$ Notation $y = |x|$. The graph in Fig. 266.

---

[1] Sometimes one function can be defined in many ways.

(3) The *sign of* x: $y = \begin{cases} -1 \text{ for } x < 0, \\ 0 \text{ for } x = 0, \\ 1 \text{ for } x > 0. \end{cases}$

Notation $y = \operatorname{sgn} x$. For the graph see Fig. 263, (1) on p. 322.

(4) $y = \lim\limits_{n\to\infty} \dfrac{1}{1+x^{2n}}$ or $y = \begin{cases} 1 \text{ for } |x| < 1, \\ \frac{1}{2} \text{ for } |x| = 1, \\ 0 \text{ for } |x| > 1. \end{cases}$

The graph is in Fig. 267.

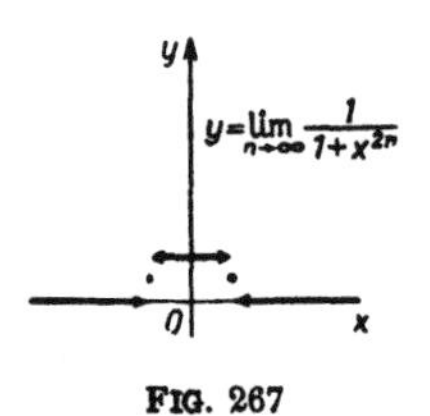

FIG. 267

(5) $y = \displaystyle\int_0^x \frac{\sin x}{x}\,dx$ or $y = x - \dfrac{x^3}{3\cdot 3!} + \dfrac{x^5}{5\cdot 5!} - \dfrac{x^7}{7\cdot 7!} + \cdots$

Notation $y = \operatorname{Si} x$ ("the *integral sine*", see p. 434).

(6) $y = \displaystyle\int_0^\infty e^{-t}t^{x-1}dt$ or $y = \lim\limits_{n\to\infty} \dfrac{n!\,n^{x-1}}{x(x+1)(x+2)\dots(x+n-1)}$.

Notation $y = \Gamma(x)$ ("the *Gamma function*", see p. 191).

(7) Solution of Bessel's equation $x^2y'' + xy' + (x^2 - n^2)y = 0$ with certain boundary conditions ("*Bessel's function*", see p. 549).

Certain types of functions.

(1) *Monotone functions.* A function $f(x)$ is said to be *monotone increasing* (Fig. 268a) or *monotone decreasing* (Fig. 268b), if the

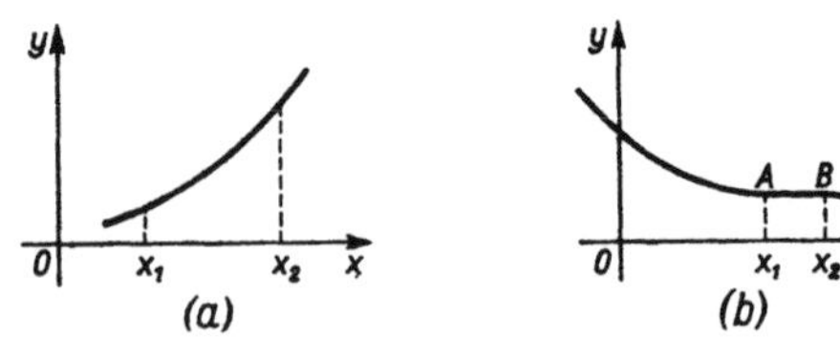

FIG. 268

condition $x_1 < x_2$ implies $f(x_1) < f(x_2)$ or $f(x_1) > f(x_2)$, respectively. For example, $y = \ln x$, $y = e^{-x}$.

If the function $f(x)$ satisfies this condition only in a certain domain of x (an interval, for example), then $f(x)$ is said to be *monotone* in this domain ([1]).

(2) *Bounded functions.* A function is called *bounded from above,* if the values it takes do not exceed a certain number. It is called *bounded from below,* if it is always greater than a certain number. A function satisfying both conditions is said to be *bounded.*

Examples. The function $y = 1 - x^2$ is bounded from above $(y < 1)$; $y = e^x$ is bounded from below $(y > 0)$; $y = \sin x$ is bounded $(-1 < y < +1)$; $y = \dfrac{4}{1+x^2}$ is bounded $(0 < y < 4)$.

(3) *Even functions.* A function is called *even* if it satisfies the condition $f(-x) = f(+x)$ (Fig. 269a); for example, $y = \cos x$, $y = x^4 - 3x^2 + 1$.

(4) *Odd functions.* A function is called *odd* if it satisfies the condition $f(-x) = -f(+x)$ (Fig. 269b); for example, $y = \sin x$, $y = x^3 - x$.

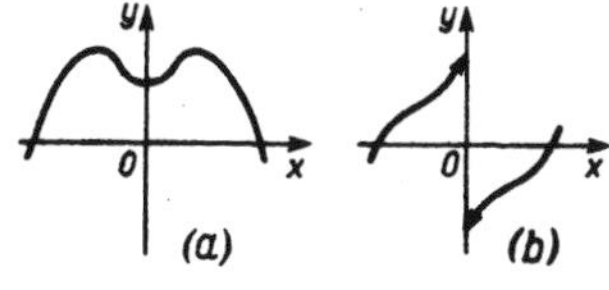

FIG. 269

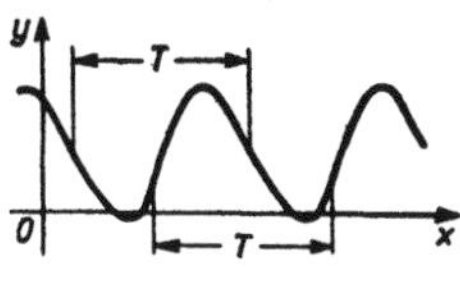

FIG. 270

(5) *Periodic functions.* A function $f(x)$ is called *periodic* if it satisfies the condition $f(x + T) = f(x)$. The least number T satisfying this condition is called the *period* of the function $f(x)$ (Fig. 270).

4. Limit of a function

The notion of *limit* is considered here only for functions of two types: (1) for functions of an integral argument (see p. 321) and (2) for functions with a connected domain of existence (see p. 320) ([2]).

([1]) The monotone functions defined as above are sometimes called *monotone in the wider sense.* A function satisfying the condition $f(x_1) < f(x_2)$ or $f(x_1) > f(x_2)$ (a sharp inequality), for every $x_1 < x_2$, is said to be *strictly monotone.* The function in Fig. 268a is strictly monotone increasing and the function in Fig. 268b is monotone in the wider sense (it is constant in the interval AB).

([2]) The notion of a limit can also be introduced for functions defined in more complicated domains.

Limit of a function of an integral argument. A limit of a function $y=f(x)$, where $x=1, 2, 3, \ldots, n, \ldots$ is defined only for $x\to\infty$; it is the limit of the sequence of numbers [1] $f(1), f(2), f(3), \ldots, f(n), \ldots$:

$$A=\lim_{x\to\infty} f(x)=\lim_{n\to\infty} f(n).$$

Examples. (1) $\lim\limits_{n\to\infty}\frac{1}{n}=0$. (2) $\lim\limits_{n\to\infty}\left(1+\frac{1}{n}\right)^n=e$.

Limit of a function defined in a connected domain.

Definition. The function $y=f(x)$ has the *limit* A for $x=a$:

$$A=\lim_{x\to a} f(x),$$

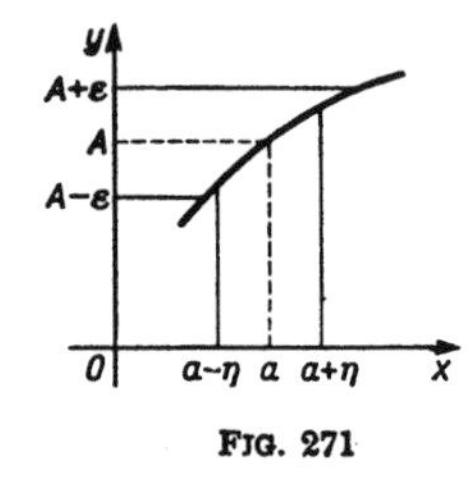

FIG. 271

if $f(x)$ approaches the number A arbitrarily near, when x approaches a. The function $f(x)$ may not assume the value A for $x=a$; it need not even be defined for $x=a$.

Precisely: $A=\lim\limits_{x\to a} f(x)$ if, for an arbitrarily small positive number ε, a positive number η can be found such that, for each value of x from the interval $a-\eta<x<a+\eta$ [2] (except, perhaps, the value $x=a$), the corresponding values of $f(x)$ lie in the interval $A-\varepsilon<f(x)<A+\varepsilon$ (Fig. 271).

Criteria for the existence of a limit.

(1) *Reducing to a limit of a sequence.* A function $f(x)$ has the limit A for $x=a$ if, for an arbitrary sequence $(x_1, x_2, \ldots, x_n, \ldots)$ of values of x, from the given domain, tending to a, the sequence of the corresponding values of the function $(f(x_1), f(x_2), \ldots, f(x_n), \ldots)$ has the limit A. The number A is a common limit for all such sequences.

(2) *Cauchy's criterion.* A necessary and sufficient condition that the function $f(x)$ has a limit is that, for arbitrary x_1 and x_2, from the given domain of existence, which are sufficiently near to a, the corresponding values of the function, $f(x_1)$, $f(x_2)$ are arbitrarily near.

Precisely: A necessary and sufficient condition that the function $f(x)$ have a limit for $x=a$ is that, for an arbitrary choice of $\varepsilon>0$, we can always find a number $\eta>0$ such that for arbitrary

---

[1] See p. 317.

[2] If a is a boundary point of the considered domain, then this double inequality should be replaced by a simple one: $a-\eta<x$ or $x<a+\eta$.

x_1 and x_2 from the given domain of existence satisfying the conditions $|x_1 - a| < \eta$ and $|x_2 - a| < \eta$, we have

$$\left|f(x_1) - f(x_2)\right| < \varepsilon.$$

An infinite limit of a function is a notion analogous to that of an infinite limit of a sequence (p. 318). The symbol $\lim_{x \to a} f(x) = \infty$ (the limit is equal to infinity) is used to mean that the limit of the function does not exist for $x = a$, but, when x approaches a, the absolute value of the function increases without bound.

Precisely: $\lim_{a \to x} f(x) = \infty$ if, for an arbitrary choice of a positive number K, we can find a number η such that, for every x from the interval $a - \eta < x < a + \eta$, the corresponding values of $f(x)$ are absolutely greater than K:

$$\left|f(x)\right| > K.$$

If, moreover, the function $f(x)$ is positive in the interval $a - \eta < x < a + \eta$, then we write $\lim_{x \to a} f(x) = +\infty$; if the function is negative, then $\lim_{x \to a} f(x) = -\infty$.

Left-hand and right-hand limits of a function. The number A is a *left-hand limit* of the function $f(x)$ for $x = a$, if the values of $f(x)$ approach A for increasing values of x tending to a. Notation: $A = f(a - 0)$. Analogously, A is a *right-hand limit* of $f(x)$, if the values of $f(x)$ approach A for decreasing values of x tending to a. Notation: $A = f(a + 0)$. For example, the function $f(x) = \dfrac{1}{1 + e^{1/(x-1)}}$ has, for $x \to 1$, different left-hand and right-hand limits: $f(1 - 0) = 1$, $f(1 + 0) = 0$ (Fig. 272).

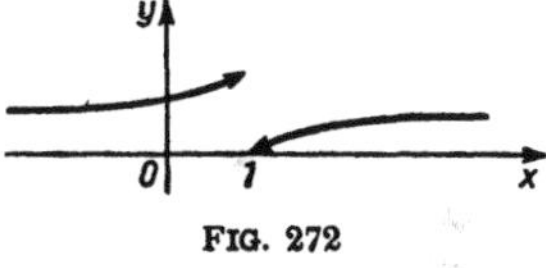

FIG. 272

Limit of a function at infinity. The number A is called a *limit* of the function $y = f(x)$ for $x \to +\infty$:

$$A = \lim_{x \to +\infty} f(x),$$

if, for an arbitrary choice of a positive number ε, we can find a number N such that, for any $x > N$, the values of $f(x)$ will lie in the interval $A - \varepsilon < f(x) < A + \varepsilon$. Similarly,

$$A = \lim_{x \to -\infty} f(x),$$

if, given a number $\varepsilon > 0$, we can find a number $-N$ such that, for any $x < -N$, the values of $f(x)$ will lie in the interval $A - \varepsilon < f(x) < A + \varepsilon$. For example,

$$\lim_{x\to+\infty} \frac{x+1}{x} = 1, \quad \lim_{x\to-\infty} \frac{x+1}{x} = 1, \quad \lim_{x\to-\infty} e^x = 0.$$

In both cases, if the absolute value of $f(x)$ infinitely increases, then the limit of $f(x)$ at infinity does not exist; we then write

$$\lim_{x\to+\infty} f(x) = \infty, \quad \lim_{x\to-\infty} f(x) = \infty.$$

If, at the same time, the function $f(x)$ remains positive, then we write $\lim_{x\to+\infty} f(x) = +\infty$ or $\lim_{x\to-\infty} f(x) = +\infty$; if $f(x)$ remains negative, then we write $\lim_{x\to-\infty} f(x) = -\infty$ or $\lim_{x\to+\infty} f(x) = -\infty$.

For example,

$$\lim_{x\to+\infty} \frac{x^3-1}{x^2} = +\infty, \quad \lim_{x\to-\infty} \frac{x^3-1}{x^2} = -\infty,$$

$$\lim_{x\to+\infty} \frac{1-x^3}{x^2} = -\infty, \quad \lim_{x\to-\infty} \frac{1-x^3}{x^2} = +\infty.$$

Main theorems on limits of functions.

(1) The limit of a constant is equal to this constant: $\lim A = A$.

(2) The limit of a sum (difference) of a finite number of functions is equal to the sum (difference) of limits of the corresponding functions:

$$\lim_{x\to a} \left(f(x) + \varphi(x) - \psi(x)\right) = \lim_{x\to a} f(x) + \lim_{x\to a} \varphi(x) - \lim_{x\to a} \psi(x).$$

(3) The limit of a product of a finite number of functions is equal to the product of limits of the corresponding functions:

$$\lim_{x\to a} \left(f(x)\varphi(x)\psi(x)\right) = \lim_{x\to a} f(x) \lim_{x\to a} \varphi(x) \lim_{x\to a} \psi(x).$$

(4) The limit of a quotient of two functions:

$$\lim_{x\to a} \frac{f(x)}{\varphi(x)} = \frac{\lim\limits_{x\to a} f(x)}{\lim\limits_{x\to a} \varphi(x)},$$

provided that $\lim_{x\to a} \varphi(x) \neq 0$.

(5) If the function $f(x)$ lies between two other functions $\varphi(x)$ and $\psi(x)$, i.e., $\varphi(x) < f(x) < \psi(x)$ and $\lim_{x\to a} \varphi(x) = A$, $\lim_{x\to a} \psi(x) = A$, then

$$\lim_{x\to a} f(x) = A.$$

(6) A monotone function defined in a connected domain has a (finite or infinite) limit for every value of x (finite or infinite). A bounded monotone function has a finite limit for any value of x.

Certain important limits.

(1) The number $e = \lim_{x\to\infty}\left(1+\frac{1}{x}\right)^x = e = 2.71828\ldots$ (irrational number). For the table of quantities related to e, see p. 19. The number e is used as a base for natural logarithms (see p. 156).

(2) The number C: $\lim_{n\to\infty}\left(1+\frac{1}{2}+\frac{1}{3}+\ldots+\frac{1}{n}-\ln n\right) = C = 0.5772\ldots$ (*Euler's constant*).

(3) $\lim_{x\to 0}\frac{\sin x}{x} = 1$, if x is length of arc or an angle expressed in radians.

Evaluation of limits. For evaluation of limits, we use the theorems given above and apply the following methods:

(1) We transform a function to a form convenient for evaluation of limit.

Examples.

$$\lim_{x\to 1}\frac{x^3-1}{x-1} = \lim_{x\to 1}(x^2+x+1) = 3;$$

$$\lim_{x\to 0}\frac{\sqrt{1+x}-1}{x} = \lim_{x\to 0}\frac{(\sqrt{1+x}-1)(\sqrt{1+x}+1)}{x(\sqrt{1+x}+1)} = \lim_{x\to 0}\frac{1}{\sqrt{1+x}+1} = \frac{1}{2};$$

$$\lim_{x\to 0}\frac{\sin 2x}{x} = \lim_{x\to 0}\frac{2(\sin 2x)}{2x} = 2\lim_{2x\to 0}\frac{\sin 2x}{2x} = 2.$$

(2) In cases leading to "indeterminate forms" $0/0$, ∞/∞, $\infty\cdot 0$, $\infty-\infty$, 0^0, ∞^0, 1^∞, we apply l'Hôpital's rule:

(a) Indeterminate forms $0/0$ or ∞/∞. If $f(x) = \varphi(x)/\psi(x)$ where the functions $\varphi(x)$ and $\psi(x)$ are defined in an interval containing the point a [1] and have finite derivatives in this interval ($\psi'(x) \neq 0$) and if

$$\lim_{x\to a}\varphi(x) = 0 \quad \text{and} \quad \lim_{x\to a}\psi(x) = 0 \quad \text{(indeterminate form } 0/0)$$

or

$$\lim_{x\to a}\varphi(x) = \infty \quad \text{and} \quad \lim_{x\to a}\psi(x) = \infty \quad \text{(indeterminate form } \infty/\infty),$$

[1] At the point a the functions $\varphi(x)$ and $\psi(x)$ need not be defined.

then

$$\lim_{x \to a} f(x) = \lim \frac{\varphi'(x)}{\psi'(x)},$$

provided that this limit exists or is equal to ∞ (*l'Hôpital's rule*).

In the case when $\lim\limits_{x \to a} \dfrac{\varphi'(x)}{\psi'(x)}$ is also an indeterminate form $0/0$ or ∞/∞, then we apply l'Hôpital's rule again and so on.

Example.

$$\lim_{x \to 0} \frac{\ln \sin 2x}{\ln \sin x} = \lim_{x \to 0} \frac{\dfrac{2 \cos 2x}{\sin 2x}}{\dfrac{\cos x}{\sin x}} = \lim \frac{2 \tan x}{\tan 2x}$$

$$= \lim_{x \to 0} \frac{\dfrac{2}{\cos^2 x}}{\dfrac{2}{\cos^2 2x}} = \lim_{x \to 0} \frac{\cos^2 2x}{\cos^2 x} = 1.$$

(b) Indeterminate form $0 \cdot \infty$. If $f(x) = \varphi(x) \cdot \psi(x)$ (under the same condition as in case (a)) and $\lim\limits_{x \to a} \varphi(x) = 0$, $\lim\limits_{x \to a} \psi(x) = \infty$ (indeterminate form $0 \cdot \infty$), then transforming the function to the form $f(x) = \dfrac{\varphi(x)}{1/\psi(x)}$ or $f(x) = \dfrac{\psi(x)}{1/\varphi(x)}$ we reduce the problem to the case (a).

Example.

$$\lim_{x \to \pi/2} (\pi - 2x) \tan x = \lim_{x \to \pi/2} \frac{\pi - 2x}{\cot x} = \lim_{x \to \pi/2} \frac{-2}{-1/\sin^2 x} = 2.$$

(c) Indeterminate form $\infty - \infty$. If $f(x) = \varphi(x) - \psi(x)$ and $\lim\limits_{x \to a} \varphi(x) = \infty$ and $\lim\limits_{x \to a} \psi(x) = \infty$ (indeterminate form $\infty - \infty$), then, to find the limit $\lim\limits_{x \to a} f(x)$, we transform algebraically the difference $\varphi(x) - \psi(x)$ to the form $0/0$ or ∞/∞. This can be done in various ways, e.g.

$$\varphi - \psi = \left(\frac{1}{\psi} - \frac{1}{\varphi}\right) : \frac{1}{\varphi\psi}.$$

Example.

$$\lim_{x \to 1} \left(\frac{x}{x-1} - \frac{1}{\ln x}\right) = \lim_{x \to 1} \left(\frac{x \ln x - x + 1}{x \ln x - \ln x}\right)$$

(indeterminate form 0/0). Applying l'Hôpital's rule twice, we obtain

$$\lim_{x\to1}\left(\frac{x\ln x-x+1}{x\ln x-\ln x}\right)=\lim_{x\to1}\left(\frac{\ln x}{\ln x+1-\frac{1}{x}}\right)=\lim_{x\to1}\left(\frac{\frac{1}{x}}{\frac{1}{x}+\frac{1}{x^2}}\right)=\frac{1}{2}.$$

(d) Indeterminate forms 0^0, ∞^0, 1^∞. If $f(x)=\varphi(x)^{\psi(x)}$ and $\lim_{x\to a}\varphi(x)=0$, $\lim_{x\to a}\psi(x)=0$, then we first find the limit A of the function $\ln f(x)=\psi(x)\ln\varphi(x)$ which is of the form $0\cdot\infty$ (case (b)) and then evaluate e^A.

Example. $\lim_{x\to0}x^x=X$, $\ln x^x=x\ln x$, $\lim_{x\to0}x\ln x=\lim_{x\to0}\frac{\ln x}{1/x}$ $=\lim_{x\to0}(-x)=0$, $\ln X=0$, $X=e^0=1$. Therefore, $\lim_{x\to0}x^x=1$.

In the cases ∞^0 and 1^∞ the procedure is analogous.

(3) To evaluate a limit of a function, we can also apply, besides l'Hôpital's rule, the expansion of the function into Taylor's series. For example,

$$\lim_{x\to0}\frac{x-\sin x}{x^3}=\lim_{x\to0}\frac{x-\left(x-\frac{x^3}{3!}+\frac{x^5}{5!}-\dots\right)}{x^3}=\lim_{x\to0}\left(\frac{1}{3!}-\frac{x^2}{5!}+\dots\right)=\frac{1}{6}.$$

5. Infinitesimals

Definition. A function α of one variable x is called an *infinitesimal* for $x\to a$, if its limit is zero ($\lim_{x\to a}\alpha=0$). If $\alpha=c$ (constant) and $\lim\alpha=0$, then $c=0$ [1], i.e., the only constant infinitesimal is zero.

If a function A of the variable x has an infinite limit for $x\to a$ (see p. 329), then it is called an *infinitely great quantity* for $x\to a$.

Main properties. If α, β, γ, ..., are infinitesimals and a is a finite quantity (i.e., not tending to zero or to infinity), then:

(1) the sum or difference $\alpha\pm\beta\pm\gamma\pm\dots$ is an infinitesimal (the number of summands is finite);

(2) the product $\varphi\cdot\beta$ or $\alpha\cdot a$ is an infinitesimal;

(3) the quotient α/a is an infinitesimal (if $a\neq0$);

(4) the quotient α/β can be an infinitesimal, a finite or infinitely great quantity, or a quantity without a limit.

Examples. (1) $\alpha=\sin x$, $\beta=1-\cos x$, $\gamma=x^2$. For $x\to0$, α, β and γ are infinitesimals. Since

[1] The limit of a constant function is equal to this constant.

$$\lim_{x\to 0}\frac{\beta}{\alpha}=\lim_{x\to 0}\frac{1-\cos x}{\sin x}=\lim_{x\to 0}\frac{\sin x}{\cos x}=0 \text{ (l'Hôpital's rule)},$$

$$\lim_{x\to 0}\frac{\beta}{\gamma}=\lim_{x\to 0}\frac{1-\cos x}{x^2}=\lim_{x\to 0}\frac{\sin x}{2x}=\lim_{x\to 0}\frac{\cos x}{2}=\frac{1}{2},$$

$$\lim_{x\to 0}\frac{\alpha}{\gamma}=\lim_{x\to 0}\frac{\sin x}{x^2}=\lim_{x\to 0}\frac{\cos x}{2x}=\infty,$$

therefore β/α is an infinitesimal, β/γ—a finite quantity, α/γ is infinitely great.

(2) $\alpha=1/n$, $\beta=(-1)^n/n$, where n is a natural number. When $n\to\infty$, then α and β are infinitesimals, but the limit $\lim\limits_{n\to\infty}\frac{\alpha}{\beta}$ $=\lim\limits_{n\to\infty}(-1)^n$ does not exist.

Order of infinitesimals. Two infinitesimals α and β are said to be of *the same order*, if their ratio α/β is a finite quantity. If α/β is an infinitesimal, then α is said to be of *higher order* than β. If γ/α is infinitely great, then α/γ is an infinitesimal and α is of higher order than γ.

Example. The quantities $\beta=1-\cos x$ and $\gamma=x^2$ are of the same order; β and γ are of higher order than $\alpha=\sin x$.

An infinitesimal α is said to be of the m-th *order* with respect to another infinitesimal β, if the order of α is equal to that of the infinitesimal β^m.

Example. The infinitesimals $\sin x$ and $1-\cos x$ (for $x\to 0$) are, respectively, of the first and second order with respect to the infinitesimal x.

Two infinitesimals are called *equivalent*, if the limit of their ratio is equal to 1.

Examples. The infinitesimals x and $\sin x$ are equivalent (for $x\to 0$) and the infinitesimals x^2 and $1-\cos x$ are not equivalent.

In evaluating a limit of a quotient of two infinitesimals, each of them can be replaced by an equivalent infinitesimal without changing the limit.

6. Continuity and points of discontinuity of functions

The concept of continuity and discontinuity. Most of the functions considered in mathematical analysis are *continuous*, i.e., small variations of the argument x induce small variation of the function y, and the graph of such a function is a continuous line.

For some values of x, the function may fail to be continuous and the graph becomes disconnected; thus the function has a *discontinuity*. The values of the argument corresponding to discontinuities of the function are called *points of discontinuity*. Fig. 273 shows

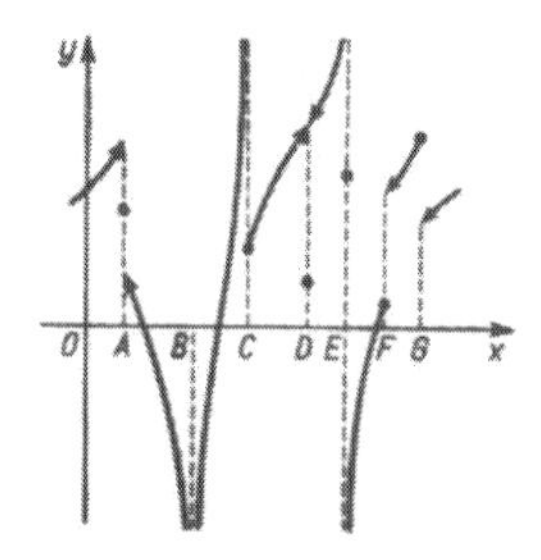

FIG. 273

a graph of a function continuous everywhere except the points A, B, C, D, E, F, G of discontinuity (the letters denote the projections of the points) (1).

Definition. The function $y = f(x)$ is said to be *continuous* for $x = a$, if (1) it is defined for $x = a$, (2) the limit $\lim_{a \leftarrow x} f(x)$ exists and is equal to $f(a)$ (2).

If the function $f(x)$ is defined and continuous for all values of x in the interval from a to b, then it is said to be *continuous in this interval*. A function defined and continuous for all points of the number scale is called *continuous everywhere*.

At the points $x = a$ (lying inside the domain of existence or on its boundary) where the function is not defined or the value $f(a)$ is not equal to $\lim f(x)$, the function is *discontinuous*; such points are called *points of discontinuity* of the function (3).

If the function $f(x)$ is continuous for all points of a certain interval except a finite number of points at which the function $f(x)$ admits a finite jump (see below), then $f(x)$ is said to be *piecewise continuous*; its graph is then composed of a finite number of simple arcs.

---

(1) An arrow in the figure denotes that the point in the arrowhead does not belong to the graph; a heavy point denotes a point belonging to the graph.

(2) The second condition can be replaced by the following equivalent one: for any infinitesimal α, the difference $\beta = f(a+\alpha) - f(a)$ is an infinitesimal, i.e., infinitesimal increments of the argument induce infinitesimal increments of the function.

(3) If a function is defined only on one side of an argument $x = a$ (as, for example, $+\sqrt{x}$ for $x = 0$ or $\arccos x$ for $x = 1$), then it is said *to break off*.

Some frequently occurring types of discontinuity at $x = a$.

(1) *Infinite discontinuity* (the function "becomes infinite"): the function has an infinite left-hand or right-hand limit or both infinite limits, as at the points B, C, E in Fig. 273 (this is the most frequently occurring case).

Examples. $f(x) = \tan x$, $f(\frac{1}{2}\pi - 0) = +\infty$, $f(\frac{1}{2}\pi + 0) = -\infty$ [1] (see the graph on p. 111). Discontinuity of type E in Fig. 273.

$f(x) = 1/(x-1)^2$; $f(1-0) = +\infty$, $f(1+0) = +\infty$. Discontinuity of type B in Fig. 273.

$f(x) = e^{1/(x-1)}$; $f(1-0) = 0$, $f(1+0) = \infty$. Discontinuity of type C in Fig. 273, but the function $f(x)$ is here not defined for $x = 1$.

(2) *Finite discontinuity*; when x passes through the point a, the function jumps from one finite value to another (the points A, F, G in Fig. 273). The exact value of $f(x)$ for $x = a$ may either be not defined (the point G) or coincide with one of the limits $f(a-0)$ or $f(a+0)$ (the point F) or be different from either of them (the point A).

Examples. $f(x) = \dfrac{1}{1 + e^{1/(x-1)}}$, $f(1-0) = 1$, $f(1+0) = 0$ (see Fig. 272, p. 329).

$f(x) = E(x)$ (Fig. 265, p. 325), $f(a-0) = a-1$, $f(a+0) = a$.

$f(x) = \lim\limits_{n\to\infty} \dfrac{1}{1+x^{2n}}$ (Fig. 267, p. 326), $f(1-0) = 1$, $f(1+0) = 0$, $f(1) = \frac{1}{2}$.

(3) *Removable discontinuity*: $\lim\limits_{x\to a} f(x)$ exists, $f(a-0) = f(a+0)$, but for $x = a$ the function is either not defined or its value $f(a) \neq \lim\limits_{x\to a} f(x)$ (the point D in Fig. 273). Putting $f(a) = \lim\limits_{x\to a} f(x)$, we add the point $x = a$, $y = f(a)$ and remove the discontinuity.

Various cases of indeterminate forms examined by aid of l'Hôpital's rule (see pp. 331–333) and leading to a finite limit represent removable discontinuities.

Example. $f(x) = \dfrac{\sqrt{1+x} - 1}{x}$ for $x = 0$ is an indeterminate form $0/0$; $\lim\limits_{x\to 0} f(x) = \frac{1}{2}$; the function

$$f(x) = \begin{cases} \dfrac{\sqrt{1+x}-1}{x} & \text{for} \quad x \neq 0, \\ \dfrac{1}{2} & \text{for} \quad x = 0 \end{cases}$$

becomes continuous.

---

[1] For the notation $f(a-0)$, $f(a+0)$, see p. 329.

Continuity and discontinuities of elementary functions. All elementary functions are continuous in their domains of existence. The points of discontinuity do not belong to the domain of existence. For a detailed study of plotting the graph of an elementary function see p. 292; for the graphs of the simplest functions see pp. 96–116. We give here only some general information about discontinuities of elementary functions.

Integral functions (polynomials) are continuous everywhere (for all points of the real line).

Rational functions $\frac{P(x)}{Q(x)}$ ($P(x)$ and $Q(x)$ are polynomials) are continuous everywhere except those values of x for which $Q(x)=0$ but $P(x)\neq 0$; if $x=a$ is a root of the denominator, but $P(a)\neq 0$, then the function has an infinite discontinuity for $x=a$. If a is a root of both the numerator and denominator, then the function has an infinite discontinuity only in the case when the multiplicity of the root in the denominator is greater than that in the numerator; otherwise the discontinuity is removable.

Irrational functions. Roots (with an integral index) from integral functions are continuous for all values of x from the domain of existence; they may break off at ends of their intervals of existence (as, for example, a root of an even index at a boundary point between positive and negative values of the function under the radical). Roots from rational functions are discontinuous at the same values of x at which the function under the root sign is discontinuous.

Trigonometric functions. $\sin x$ and $\cos x$ are continuous everywhere. $\tan x$ and $\sec x$ have infinite discontinuities for $x=\frac{1}{2}(2n+1)\pi$, and $\cot x$ and $\operatorname{cosec} x$ have infinite discontinuities for $x=n\pi$ (n is an integer).

Inverse trigonometric functions. $\operatorname{arc\,tan} x$ and $\operatorname{arc\,cot} x$ are continuous everywhere; $\arcsin x$ and $\arccos x$ break off at the ends of the interval of definition $(-1< x<+1)$.

Exponential function e^x or $a^x (a>0)$ is continuous everywhere.

Logarithmic function $\log x$ (to any positive base) is continuous for all positive x and breaks off at $x=0$; the right-hand limit $\lim\limits_{x\to 0}\log x=-\infty$.

In the case of a composite elementary function, the discontinuities can be detected by examining the involved simple functions.

Example. Determine discontinuities of the function

$$y = \frac{e^{1/(x-2)}}{x \sin \sqrt[3]{1-x}}.$$

The exponent $1/(x-2)$ has a discontinuity at $x=2$ and

$$\lim_{x\to 2-0} e^{1/(x-2)} = 0 \quad \text{and} \quad \lim_{x\to 2+0} e^{1/(x-2)} = \infty.$$

The denominator of $f(x)$ is zero at $x=0$ and at the points x for which $\sin \sqrt[3]{1-x} = 0$. These are the roots of the equation $\sqrt[3]{1-x} = n\pi$, whence $x = 1 - n^3\pi^3$, where n is an arbitrary integer. For any of these values the numerator is different from zero, hence the function has infinite discontinuities of the type of the point E in Fig. 273 at the points: $x=0$, $x=1$, $x = 1 \pm \pi^3$, $x = 1 \pm 8\pi^3$, $x = 1 \pm 27\pi^3, \ldots$

Properties of continuous functions.

(1) Passing through zero (*Cauchy's theorem*). If a function $f(x)$ is defined and continuous in a closed interval $[a, b]$ and the values $f(a)$ and $f(b)$ at the end points have different signs, then there exsists at least one value c between a and b for which $f(x)$ is equal to zero:

$$f(c) = 0 \qquad (a < c < b).$$

Geometric significance: a continuous curve passing from one side of the x axis to the other must intersect it.

(2) Mean value theorem. If a function $f(x)$ is defined and continuous in a connected domain and assumes different values A and B at two points a and b $(a < b)$ of this domain:

$$f(a) = A, \qquad f(b) = B \qquad (A \neq B),$$

then, for any number C lying between A and B, there exists at least one point c between a and b such that

$$f(c) = C \qquad (a<c<b; \quad A<C<B \text{ or } A>C>B),$$

i.e., the function passes through all intermediate values between A and B.

(3) Existence of the inverse function [1]. If a function $f(x)$ is defined in a connected domain I, is continuous and strictly monotone (increasing or decreasing, see p. 326) in the domain I, then, for the given function, there exists the inverse function $\varphi(x)$ defined in the domain II of values assumed by $f(x)$ (Fig. 274).

(4) Theorem that a continuous function defined in

---

[1] See p. 323.

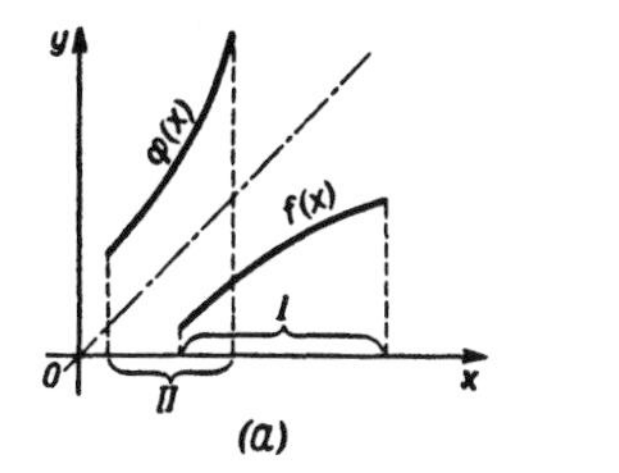

FIG. 274

a closed interval is bounded. If a function $f(x)$ is defined and continuous in a closed interval $[a, b]$, then there exist two numbers m and M such that $m < f(x) < M$ for $a < x < b$.

(5) Greatest and least value theorem. If a function $f(x)$ is defined and continuous in a closed interval $[a, b]$, then it assumes in this interval a greatest value at least once and a least value at least once, i.e., there exist in the interval at least one point c and at least one point d such that

$$f(c) > f(x) \quad \text{and} \quad f(d) < f(x) \quad \text{for every } x \text{ such that } a < x < b.$$

The difference between the greatest and the least value of the function in a given interval is called the *oscillation* of the function in this interval [1].

(6) A function continuous in a closed interval is also uniformly continuous in this interval (see below).

Uniform continuity. A function $y = f(x)$ is said to be *uniformly continuous* in a given domain of definition, if, for every positive number ε, there exists a corresponding positive number η such that for every pair of points x_1, x_2 in the domain whose distance apart is less than η, the difference of the corresponding values $f(x_1)$ and $f(x_2)$ of the function is less than ε:

$$|f(x_1) - f(x_2)| < \varepsilon \quad \text{for} \quad |x_1 - x_2| < \eta.$$

Uniform continuity means that the difference η of the arguments which provides that the difference of the corresponding values of the function is less than ε is independent of a particular part of the domain of definition of the function.

A function continuous in a given domain is *not* necessarily uniformly continuous.

---

[1] The concept of oscillation of a function can be extended to functions without a greatest and least value.

7. Functions of several variables

Definition. A variable value u is said to be a *function of n variable values* $x, y, z, \ldots, t$ (called the *arguments*) if, for given values of these variables, the variable u assumes a definite value (a *single-valued function*) or a certain number of definite values (a *multiple-valued function*).

Notation: a function of two variables $u = f(x, y)$; of three variables: $u = F(x, y, z)$, a function of n variables: $u = \varphi(x, y, z, \ldots, t)$. A system of n numbers representing values of the variables is called a *system of arguments* (¹).

Examples. A function of two variables: $u = f(x, y) = xy^2$; for the arguments $x = 2$, $y = 3$, the function assumes the values $f(2, 3) = 2 \cdot 3^2 = 18$. A function of four variables: $u = \varphi(x, y, z, t) = x \ln(y - zt)$; for the arguments $x = 3, y = 4, z = 3, t = 1$, the function assumes the value $\varphi(3, 4, 3, 1) = 3 \ln(4 - 3 \cdot 1) = 0$.

Geometric representation.

Representation of a system of arguments. A system of values of two variables x, y can be represented as a point P of the plane with Cartesian coordinates x, y (see p. 235). A system of values of three variables x, y, z can be represented as a point P of the space with the coordinates x, y, z. Such a representation is impossible for systems of four or more variables; however, we call a system of n variables $x, y, z, \ldots, t$ a *point of the n-dimensional space* with coordinates $x, y, z, \ldots, t$, by analogy with the previous cases. In this sense, the system (3, 4, 3, 1), from the preceeding example, is a point of the 4 dimensional space with coordinates $x = 3, y = 4, z = 3, t = 1$. Thus functions of several variables are sometimes called *functions of a point* (see p. 625).

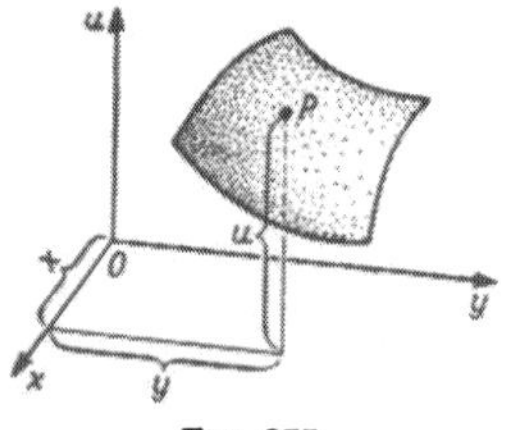

FIG. 275

Geometric representation of a function $u = f(x, y)$ of two variables. Just as the graph of a function of one variable is a curve, a function of two variables represents a surface in the space with the equation $u - f(x, y) = 0$ (Fig. 275, see p. 262). For example, the function $u = 1 - \frac{1}{2}x - \frac{1}{3}y$ represents a plane (see p. 263), the function $u = \frac{1}{2}x^2 + \frac{1}{4}y^2$ represents an elliptic paraboloid (see p. 273), the function $u = \sqrt{16 - x^2 - y^2}$ represents a hemisphere and so on (²) (Fig. 276).

(¹) Such a system can be regarded as a current point of an n-dimensional space.

(²) Functions of three or more variables cannot be represented geometrically. But, by analogy with three dimensional space, we introduce also the concept of a hypersurface in the n-dimensional space.

The *domain of existence* (or of definition) of a function is the set of systems of the arguments (i.e., the set of points) for which the function assumes a definite value. Such domains can have various forms; functions with connected domains of existence are often considered.

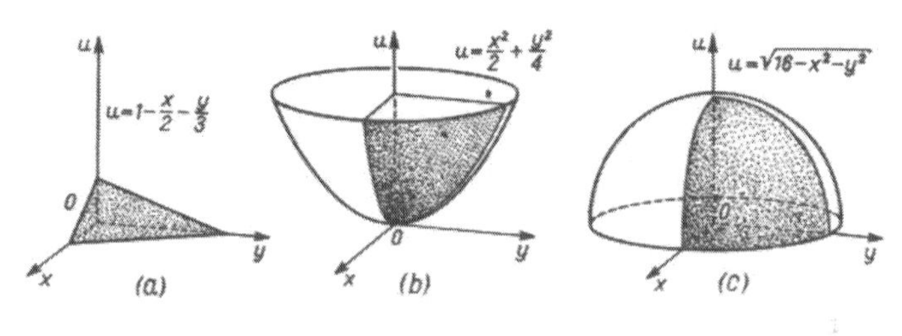

FIG. 276

Connected domains of two variables. Domains such as shown in Fig. 277 are called *simply connected* [1] (also the whole plane is simply connected). If in the interior of a given part of the plane there is a point or a simply connected bounded domain which does not belong to the domain of definition of the function,

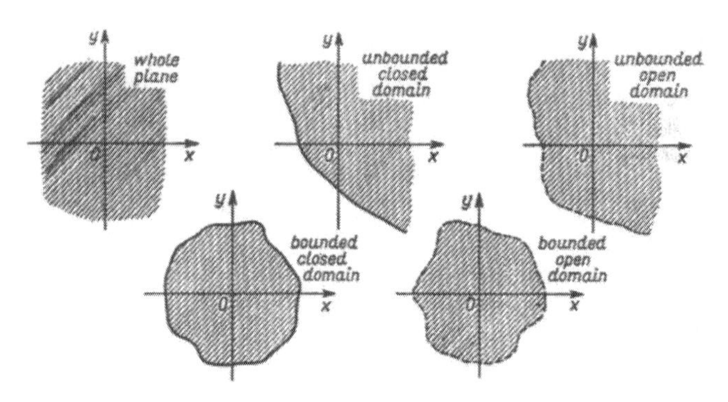

FIG. 277

[1] Fig. 277 shows some simplest examples of connected domains of two variables with their names. The domains are lined; if a boundary belongs to the domain, it is drawn by a continuous line; if a boundary does not belong to the domain, it is drawn by a dotted line.

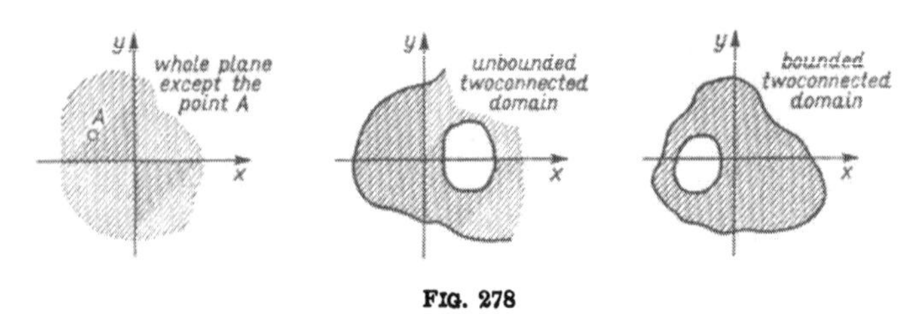

FIG. 278

then this part of the plane is called *two-connected*. Fig. 278 shows examples of two-connected domains. Examples of multi-connected domains are shown in Fig. 279.

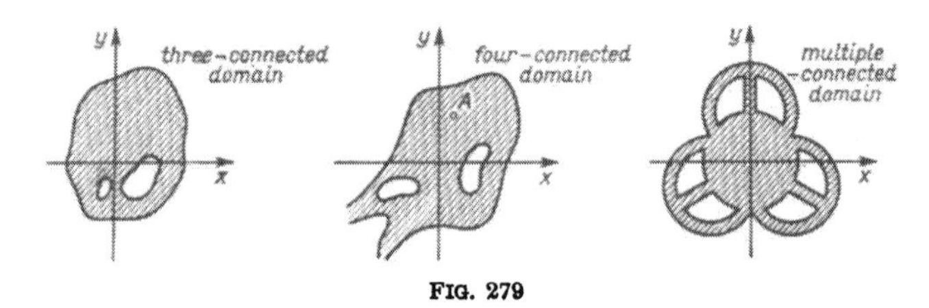

FIG. 279

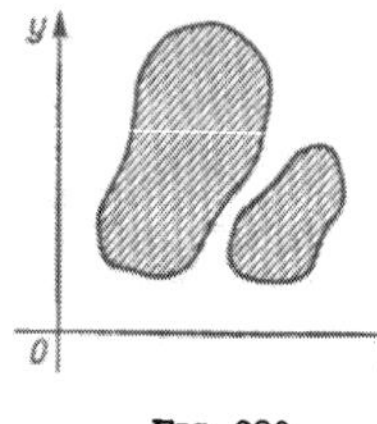

FIG. 280

A disconnected domain is shown in Fig. 280.

Connected domains of three variables (some simplest cases): the whole space or a part of it bounded by one or several surfaces whose points may or may not belong to the domain; the names of such domains are analogous to those given in Figs. 277–279 for two variables. A similar geometrical concept of a domain can be introduced in spaces of dimension greater than 3.

Ways of defining a function.

Definition by a table. A function of two variables can be defined by a table (as, for example, the table of elliptic integrals on p. 91). The values of arguments in such a table are given usually in the upper row and in the left column and the values of the function lie in the intersection of the corresponding rows and columns. A table of this type is called a *table of two entries.*

Definition by formulas. A function of several variables can be defined by one or several formulas.

Examples:

(1) $u = xy^2$, (2) $u = x \ln (y - zt)$,

$$(3)\qquad u=\begin{cases} x+y & \text{for } x>0,\ y>0,\\ x-y & \text{for } x>0,\ y<0,\\ -x+y & \text{for } x<0,\ y>0,\\ -x-y & \text{for } x<0,\ y<0.\end{cases}$$

This function can also be written in the form $u=|x|+|y|$.

Domain of definition of an analytic expression (domain of existence of a function). In mathematical analysis we are mainly concerned with functions defined by a single formula and then the domain of definition of such a function includes all these systems of arguments for which the given analytic expression has a definite sense, i.e., assumes a finite real value. Such a domain is called the *domain of definition of an analytic expression*. The domain of definition (or of existence) of a function defined by a single formula is usually understood to be the domain of definition of this analytic expression.

Examples. (1) $u=x^2+y^2$; the domain of definition consists of all values of x and y, i.e., is the whole plane.

(2) $u=\dfrac{1}{\sqrt{16-x^2-y^2}}$; the domain of definition consists of the systems x, y satisfying the inequality $x^2+y^2<16$, i.e., of the interior points of a circle (Fig. 281a).

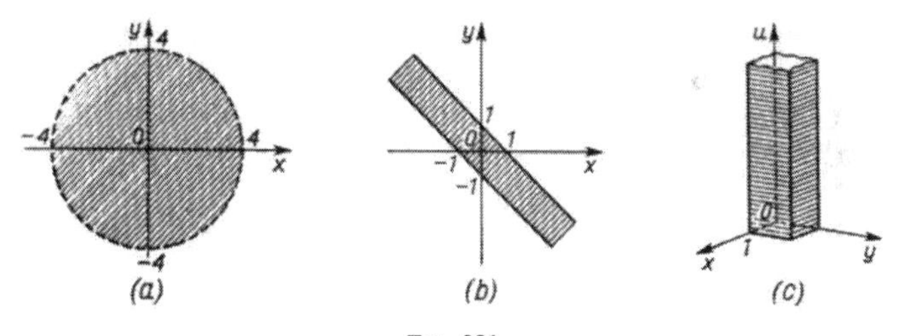

Fig. 281

(3) $u=\arcsin(x+y)$; the domain of existence consists of the values of x, y satisfying the inequality $-1<x+y<+1$; this is a closed domain on the plane contained between two parallel straight lines $x+y=-1$ and $x+y=+1$ (Fig. 281b).

(4) $u=\arcsin(2x-1)+\sqrt{1-y^2}+\sqrt{y}+\ln z$; the domain of definition consists of points x, y, z satisfying the inequalities $0<x<1$, $0<y<1$, $z>0$, i.e., of all points of the space lying above the square with the side equal to 1 and lying on the xy plane (Fig. 281c).

Fundamental forms of an analytic definition of a function. A function of several variables is defined *explicitly*, when an expression of it by the arguments of the form $u = f(x, y, z, \ldots, t)$ is given. It is defined in the *implicit* form, when the function and the arguments are related by an equation $F(x, y, z, \ldots, t, u) = 0$. It is defined *parametrically*, when n arguments and the function are expressed in terms of other n variables (the parameters): $x = \varphi(r, s)$, $y = \psi(r, s)$, $u = \chi(r, s)$ (a function of two variables); $x = \varphi(r, s, t)$, $y = \psi(r, s, t)$, $z = \chi(r, s, t)$, $u = \varkappa(r, s, t)$ (a function of three variables) and so on.

Homogeneous functions. A function $f(x, y, z, \ldots, t)$ of several variables is called *homogeneous*, if it satisfies the condition

$$f(\lambda x, \lambda y, \lambda z, \ldots, \lambda t) = \lambda^m f(x, y, z, \ldots, t),$$

for an arbitrary number $\lambda \neq 0$. The number m is called the *degree of homogeneity*. For example, the function $u = x^2 - 3xy + y^2 + x\sqrt{xy + x^3/y}$ is homogeneous of degree 2; the function $u = \frac{x + z}{2x - 3y}$ is homogeneous of degree 0.

For a homogeneous function $f(x, y, z, \ldots, t)$ we have Euler's theorem:

$$x\frac{\partial f}{\partial x} + y\frac{\partial f}{\partial y} + \ldots + t\frac{\partial f}{\partial t} = nf(x, y, \ldots, t).$$

Dependence of functions of several variables. Two single-valued functions $u = f(x, y)$, $v = \varphi(x, y)$ of two variables defined in a certain domain are said to be *dependent* one on another, if one of them can be expressed as a function of the other: $u = F(v)$, i.e., if the identity

$$f(x, y) \equiv F(\varphi(x, y)) \quad \text{or} \quad \Phi(f, \varphi) = 0$$

holds for every point of the domain; otherwise they are called *independent*. For example, the functions $u = (x^2 + y^2)^2$ and $v = \sqrt{x^2 + y^2}$ defined in the domain $x^2 + y^2 > 0$ are dependent, for $u = v^4$.

Similarly, m functions $u_1, u_2, \ldots, u_m$ of n variables $x_1, x_2, \ldots, x_n$ defined in a certain domain are called *dependent*, if one of them can be expressed as a function of the remaining ones, i.e., the identity

$$u_i \equiv F(u_1, u_2, \ldots, u_{i-1}, u_{i+1}, \ldots, u_m) \quad \text{or} \quad \Phi(u_1, u_2, \ldots, u_m) = 0$$

holds for every point of the domain; otherwise they are called *independent*. For example, three functions of n variables:

$$u = x_1 + x_2 + \ldots + x_n, \qquad v = x_1^2 + x_2^2 + \ldots + x_n^2,$$

$$w = x_1 x_2 + x_1 x_3 + \ldots + x_1 x_n + x_2 x_3 + \ldots + x_{n-1} x_n$$

defined in the n-dimensional space are dependent, for $v = u^2 - 2w$.

Analytic condition of independence of two functions $u = f(x, y)$ and $v = \varphi(x, y)$: the Jacobian of these functions, i.e., the determinant

$$\begin{vmatrix} \frac{\partial f}{\partial x} & \frac{\partial f}{\partial y} \\ \frac{\partial \varphi}{\partial x} & \frac{\partial \varphi}{\partial y} \end{vmatrix} \quad \text{denoted by} \quad \frac{\partial(f, \varphi)}{\partial(x, y)} \quad \text{or} \quad \frac{\partial(u, v)}{\partial(x, y)},$$

is not equal identically to zero in the considered domain. This condition can be generalized to the case of n functions of the same number n of unknowns: $u_1 = f_1(x_1, x_2, \ldots, x_n), \ldots, u_n = f_n(x_1, x_2, \ldots, x_n)$:

$$\begin{vmatrix} \frac{\partial f_1}{\partial x_1} & \frac{\partial f_1}{\partial x_2} & \cdots & \frac{\partial f_1}{\partial x_n} \\ \frac{\partial f_2}{\partial x_1} & \frac{\partial f_2}{\partial x_2} & \cdots & \frac{\partial f_2}{\partial x_n} \\ \cdots & \cdots & \cdots & \cdots \\ \frac{\partial f_n}{\partial x_1} & \frac{\partial f_n}{\partial x_2} & \cdots & \frac{\partial f_n}{\partial x_n} \end{vmatrix} \equiv \frac{\partial(f_1, f_2, \ldots, f_n)}{\partial(x_1, x_2, \ldots, x_n)} \neq 0.$$

In the case when the number m of functions $u_1, u_2, \ldots, u_m$ is less than the number n of variables, the functions are independent, if at least one determinant of order m of the matrix

$$\begin{bmatrix} \frac{\partial u_1}{\partial x_1} & \frac{\partial u_1}{\partial x_2} & \cdots & \frac{\partial u_1}{\partial x_n} \\ \frac{\partial u_2}{\partial x_1} & \frac{\partial u_2}{\partial x_2} & \cdots & \frac{\partial u_2}{\partial x_n} \\ \cdots & \cdots & \cdots & \cdots \\ \frac{\partial u_m}{\partial x_1} & \frac{\partial u_m}{\partial x_2} & \cdots & \frac{\partial u_m}{\partial x_n} \end{bmatrix}$$

is different from zero. The number of independent functions is, in this case, equal to the rank μ [1] of this matrix. The independent functions are those whose derivatives form the determinant of order μ not equal identically to zero.

If $m > n$, then the number of independent functions cannot be greater than n.

---

[1] For the rank of a matrix, see p. 176.

Limit of a function of several variables [1]. A function $u = f(x, y)$ has the *limit* A for $x = a$ and $y = b$ (notation: $A = \lim\limits_{\substack{x \to a \\ y \to b}} f(x, y)$) if the function approaches arbitrarily near the number A, when x approaches a and y approaches b in an arbitrary way. At the point $P(a, b)$ itself (i.e., for $x = a$ and $y = b$) the function may be not defined or, if it is defined, it may not assume the value A.

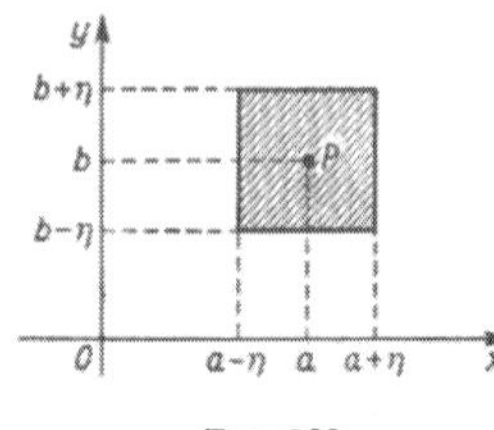

FIG. 282

Precisely: $A = \lim\limits_{\substack{x \to a \\ y \to b}} f(x, y)$, if, for a prescribed positive number ε, however small, we can find a positive number η such that, for two independent values x and y taken from the intervals $a - \eta < x < a + \eta$, $b - \eta < y < b + \eta$ (Fig. 282), the corresponding values $f(x, y)$ of the function will lie in the interval

$$A - \varepsilon < f(x, y) < A + \varepsilon.$$

The concept of a limit of a function $f(x, y, z, \ldots, t)$ of many variables can be introduced similarly.

Iterated limits. If, for a function $f(x, y)$ of two variables, we first find the limit $\lim\limits_{x \to a} f(x, y)$ (for a constant y) and then, for the obtained expression which is a function of y, we find the limit for $y \to b$, we obtain a number

$$B = \lim_{y \to b} \left(\lim_{x \to a} f(x, y) \right)$$

which is called an *iterated limit*. Changing the order of limits, we obtain another iterated limit

$$C = \lim_{x \to a} \left(\lim_{y \to b} f(x, y) \right).$$

In general, $B \neq C$ (even if the two limits exist). For example, if $f(x, y) = \dfrac{x^2 - y^2 + x^3 + y^3}{x^2 + y^2}$, then for $x \to 0$, $y \to 0$, we have $B = -1$, and $C = +1$.

If the function $f(x, y)$ has a limit $A = \lim\limits_{\substack{x \to a \\ y \to b}} f(x, y)$, then $B = C = A$. But the equality $B = C$ does not necessarily imply the existence of the limit A.

---

[1] Only functions defined in a connected domain are cons dered here.

Continuous functions of several variables.

Definition. A function $u = f(x, y)$ of two variables is called *continuous* for $x = a$, $y = b$ (i.e., at the point $P(a, b)$), if

(1) the point $P(a, b)$ belongs to the domain of definition of the function $f(x, y)$,

(2) $\lim\limits_{\substack{x \to a \\ y \to b}} f(x, y)$ exists and is equal to $f(a, b)$; otherwise the function is said to have a *discontinuity*. If the function is defined and continuous at every point of a given connected domain, then it is called continuous in this domain.

Continuity of functions of many variables is defined similarly.

Uniform continuity of a function of several variables defined in a connected domain can be defined in the same way as for a function of one variable (see p. 339). Thus the function $f(x, y)$ of two variables is uniformly continuous in a given domain if, for every positive number ε, there exists a positive number η such that, for any two points $P_1(x_1, y_1, z_1)$ and $P_2(x_2, y_2, z_2)$ satisfying the conditions $|x_1 - x_2| < \eta$, $|y_1 - y_2| < \eta$, the corresponding values of the function differ less than ε:

$$|f(x_1, y_1) - f(x_2, y_2)| < \varepsilon.$$

A function continuous in a given domain is *not* necessarily uniformly continuous.

Properties of continuous functions of several variables.

(1) Passing through zero (*Cauchy's theorem*). If the function $f(x, y)$ is defined and continuous in a connected domain and assumes different signs at two different points $P_1(x_1, y_1,)$ and $P_2(x_2, y_2)$ of this domain, then there exists at least one point $P_3(x_3, y_3)$ in this domain at which the function assumes the value zero:

$$f(x_3, y_3) = 0, \quad \text{if} \quad f(x_1, y_1) > 0 \quad \text{and} \quad f(x_2, y_2) < 0.$$

(2) Mean value theorem. If the function $f(x, y)$ is defined and continuous in a connected domain and assumes two different values $A = f(x_1, y_1)$ and $B = f(x_2, y_2)$ at two points $P_1(x_1, y_1)$ and $P_2(x_2, y_2)$ of this domain, then, for an arbitrarily given number C lying between A and B, there exists at least one point $P_3(x_3, y_3)$ of the given domain such that

$$f(x_3, y_3) = C \quad (A < C < B \text{ or } B < C < A).$$

(3) Theorem that a continuous function defined in

a closed bounded domain is bounded. A function $f(x, y)$ continuous in a closed bounded domain is bounded, i.e., there exist two numbers m and M such that, for every point $P(x, y)$ of the domain,

$$m < f(x, y) < M.$$

(4) Greatest and least value theorem. A function $f(x, y)$ continuous in a closed bounded domain assumes a greatest value at least once and a least value at least once, i.e., there exists a point $P'(x', y')$ and a point $P''(x'', y'')$ in the given domain such that

$$f(x', y') > f(x, y) > f(x'', y''),$$

for every point of the domain.

(5) A function continuous in a closed bounded domain is uniformly continuous [1].

8. Series of numbers

Definition. An expression of the form

$$a_1 + a_2 + \ldots + a_n + \ldots,$$

where the numbers $a_1, a_2, \ldots, a_n, \ldots$ form an infinite sequence, is called an *infinite series of numbers*. The sums $S_1 = a_1$, $S_2 = a_1 + a_2$, $\ldots$, $S_n = a_1 + a_2 + \ldots + a_n$ are called the *partial sums* of the series and a_n is called the *general term*. If the sequence $S_1, S_2, \ldots, S_n, \ldots$ has a limit $S = \lim_{n\to\infty} S_n$, when $n\to\infty$, then the series is said to be *convergent* and the number S is called its *sum* (notation: $\sum_{n=1}^{\infty} a_n = S$). If this limit does not exist, then the series diverges; in this case, S_n can increase infinitely ($\lim_{n\to\infty} S_n = \infty$, $\sum_{n=1}^{\infty} a_n = \infty$) or oscillate. Thus a necessary and sufficient condition that the series is convergent is that the sequence $S_1, S_2, \ldots, S_n \ldots$ is convergent (see p. 320).

Examples. The series

$$(1) \qquad 1 + \frac{1}{2} + \frac{1}{4} + \frac{1}{8} + \ldots + \frac{1}{2^n} + \ldots$$

is convergent (a geometric progression).

---

[1] See p. 347.

The series

(2) $$1+1+1+\ldots+1+\ldots,$$

(3) $$1+\frac{1}{2}+\frac{1}{3}+\ldots+\frac{1}{n}+\ldots \quad \text{(harmonic series)},$$

(4) $$1-1+1-\ldots+(-1)^{n-1}+\ldots$$

are divergent. For series (2) and (3), $\lim_{n\to\infty} S_n = \infty$; series (4) is oscillating.

The *remainder* of a convergent series $a_1+a_2+\ldots+a_n+\ldots$ is the difference between its sum S and the partial sum S_n:

$$R_n = S - S_n = a_{n+1} + a_{n+2} + \ldots + a_{n+m} + \ldots$$

Main theorems on convergence of series.

(1) The convergence or divergence of a series is not changed by inserting a finite number of terms at the begining or by removing a finite number of initial terms.

(2) If the terms of a convergent series are multiplied by the same factor c, then the series remains convergent and its sum is multiplied by c.

(3) Convergent series can be added or subtracted term by term: if the series $a_1+a_2+\ldots+a_n+\ldots$ is convergent to the sum S_1 and the series $b_1+b_2+\ldots+b_n+\ldots$ is convergent to the sum S_2 then the series $(a_1\pm b_1)+(a_2\pm b_2)+\ldots+(a_n\pm b_n)+\ldots$ is convergent to the sum $S_1\pm S_2$.

A necessary condition for the convergence of a series is that the general term a_n tends to zero for $n\to\infty$: $\lim a_n = 0$. This condition is not sufficient; for example, in the harmonic series (3), $\lim_{n\to\infty} a_n = 0$, but $\lim_{n\to\infty} S_n = \infty$.

Comparison test for series with positive terms. If two series

(A) $$a_1+a_2+\ldots+a_n+\ldots$$

(B) $$b_1+b_2+\ldots+b_n+\ldots$$

have positive terms and, for every n beyond a certain place, $a_n > b_n$, then the convergence of series (A) implies the convergence of the series (B) and the divergence of (B) implies the divergence of (A).

Examples. The series

(5) $$1+\frac{1}{2^2}+\frac{1}{3^3}+\ldots+\frac{1}{n^n}+\ldots$$

is convergent, since its terms, for $n > 2$, are less than the corresponding terms of (1): $1/n^n < 1/2^{n-1}$ $(n > 2)$ and series (1) is convergent.

The series

$$1 + \frac{1}{\sqrt{2}} + \frac{1}{\sqrt{3}} + \ldots + \frac{1}{\sqrt{n}} + \ldots \tag{6}$$

is divergent, since, for $n > 2$, its terms are greater than those of (3), $1/\sqrt{n} > 1/n$ $(n > 2$,, and (3) is divergent.

Tests of convergence for a series with positive terms.

D'Alembert ratio test. If, for the series $a_1 + a_2 + \ldots + a_n + \ldots$, from a certain index onward, the ratio a_{n+1}/a_n is less than a certain number $q < 1$, then the series converges. If all ratios, beyond a certain index, are greater than a certain number $Q > 1$, then the series diverges.

Corollary. If $\lim\limits_{n\to\infty} \frac{a_{n+1}}{a_n} = \varrho$, then the series converges for $\varrho < 1$, and diverges for $\varrho > 1$. For $\varrho = 1$, the test fails: the series may either converge or diverge.

Examples. (1) For the series

$$\frac{1}{2} + \frac{2}{2^2} + \frac{3}{2^3} + \ldots + \frac{n}{2^n} + \ldots \tag{7}$$

$\varrho = \lim\limits_{n\to\infty}\left(\frac{n+1}{2^{n+1}} : \frac{n}{2^n}\right) = \lim\limits_{n\to\infty}\frac{1+1/n}{2} = \frac{1}{2} < 1$, hence series (7) converges.

(2) For the series

$$\frac{2}{1^2} + \frac{3}{2^2} + \frac{4}{3^2} + \ldots + \frac{n+1}{n^2} + \ldots \tag{8}$$

$\varrho = \lim\limits_{n\to\infty}\left(\frac{n+2}{(n+2)^2} : \frac{n+1}{n^2}\right) = 1$, hence d'Alembert test gives no answer.

Cauchy's root test. If, for the series $a_1 + a_2 + \ldots + a_n + \ldots$, all numbers $\sqrt[n]{a_n}$, from a certain index onwards, are less than a certain number $q < 1$, then the series converges. If all these numbers, beyond a certain index, are greater than a certain number $Q > 1$, then the series diverges.

Corollary. If $\lim\limits_{n\to\infty} \sqrt[n]{a_n} = \varrho$, then the series converges for $\varrho < 1$, and diverges for $\varrho > 1$. For $\varrho = 1$, the test fails: the series may either converge or diverge. For example, for the series

$$\frac{1}{2}+\left(\frac{2}{3}\right)^4+\left(\frac{3}{4}\right)^9+\ldots+\left(\frac{n}{n+1}\right)^{n^2}+\ldots \tag{9}$$

$\varrho=\lim_{n\to\infty}\sqrt[n]{\left(\frac{n}{n+1}\right)^{n^2}}=\lim_{n\to\infty}\left(\frac{1}{1+1/n}\right)^n=\frac{1}{e}<1$; hence the series converges.

Integral test. A series with the general term $a_n=f(n)$ converges, if $f(x)$ is a monotone decreasing function and the improper integral $\int_c^\infty f(x)dx$ (see p. 471) converges; if this integral diverges, then the series with the general term $f(n)$ diverges. The lower limit c of the integral can be chosen arbitrarily so that the function $f(x)$ is defined and continuous in the interval $c<x<\infty$. Thus for series (8), we have

$$f(x)=\frac{x+1}{x^2},\qquad \int_c^\infty \frac{x+1}{x^2}dx=\left[\ln x-\frac{1}{x}\right]_c^\infty=\infty;$$

the integral is divergent, hence series (8) is divergent.

Absolute and conditional convergence. In examining a series

$$a_1+a_2+\ldots+a_n+\ldots \tag{A}$$

whose terms may have various signs, it is convenient to consider the series

$$|a_1|+|a_2|+\ldots+|a_n|+\ldots, \tag{B}$$

composed of absolute values of the terms of (A). If the series (B) converges, then the series (A) also converges. In this case the series (A) is said to be *absolutely convergent*. If, however, (B) diverges, then (A) may either converge or diverge; if (A) converge, then it is said to be conditionally convergent. For example, the series

$$\frac{\sin\alpha}{2}+\frac{\sin 2\alpha}{2^2}+\ldots+\frac{\sin n\alpha}{2^n}+\ldots, \tag{10}$$

where α is a fixed number, is absolutely convergent, since the series with general term $\left|\frac{\sin n\alpha}{2^n}\right|$ is convergent; this is evident by comparing it with the series (1):

$$\left|\frac{\sin n\alpha}{2^n}\right|<\frac{1}{2^n}.$$

The series

$$1-\frac{1}{2}+\frac{1}{3}-\ldots+(-1)^{n-1}\frac{1}{n}+\ldots \tag{11}$$

is conditionally convergent (see below, Leibnitz theorem), since series (3) with general term $|a_n| = 1/n$ is divergent.

Properties of absolutely convergent series.

(1) The terms of an absolutely convergent series can be arbitrarily rearranged without change the sum of the series. By a suitable rearrangement of the terms of a conditionally convergent series (in which an *infinite* number of terms is rearranged), the sum of the series can be changed so that it can be made to be equal to an arbitrary number (*Riemann's theorem*) or even to diverge.

(2) Absolutely convergent series can be not only added and subtracted term by term (see p. 349), but also multiplied as polynomials; the result of such multiplication is a series obtained, for example, as follows:

$$(a_1+a_2+\ldots+a_n+\ldots)(b_1+b_2+\ldots+b_n+\ldots)$$
$$=\underbrace{a_1b_1}+\underbrace{a_2b_1+a_1b_2}+\underbrace{a_3b_1+a_2b_2+a_1b_3}+\ldots+$$
$$+\underbrace{a_nb_1+a_{n-1}b_2+\ldots+a_1b_n}+\ldots$$

If $\sum a_n = S_a$ and $\sum b_n = S_b$, then the sum of the series obtained from multiplication is equal to S_aS_b [1].

Alternating series. A series $a_1 - a_2 + a_3 - \ldots + (-1)^{n-1} a_n + \ldots$, where $a_1, a_2, \ldots, a_n, \ldots$ are positive, is called *alternating*. An alternating series satisfying the conditions

$$\lim_{n\to\infty} a_n = 0 \tag{1}$$

and

$$a_1 > a_2 > a_3 > \ldots > a_n > \ldots \tag{2}$$

is convergent (*Leibnitz theorem*). For example, series (11) is convergent.

Estimation of the remainder of an alternating series. If we confine ourselves to n initial terms in a convergent alternating series, then the remainder $R_n = S - S_n$ has the same

---

(1) If two series $a_1 + a_2 + \ldots + a_n + \ldots$ and $b_1 + b_2 + \ldots + b_n + \ldots$ are convergent and at least one of them is absolutely convergent, then the series obtained from multiplication is convergent though not necessarily absolutely convergent.

sign as the first term neglected and is numerically less than it:

$$|S - S_n| < |a_{n+1}|.$$

For example, the remainder of the series (1) whose sum is ln 2 satisfies the inequality $|\ln 2 - S_n| < \frac{1}{n+1}$.

Table of sums of certain series of numbers

1. $1 + \frac{1}{1!} + \frac{1}{2!} + \frac{1}{3!} + \ldots + \frac{1}{n!} + \ldots = e,$

2. $1 - \frac{1}{1!} + \frac{1}{2!} - \frac{1}{3!} + \ldots + (-1)^n \frac{1}{n!} \pm \ldots = \frac{1}{e},$

3. $1 - \frac{1}{2} + \frac{1}{3} - \frac{1}{4} + \ldots + (-1)^{n-1} \frac{1}{n} \pm \ldots = \ln 2.$

4. $1 + \frac{1}{2} + \frac{1}{4} + \frac{1}{8} + \ldots + \frac{1}{2^n} + \ldots = 2,$

5. $1 - \frac{1}{2} + \frac{1}{4} - \frac{1}{8} + \ldots + (-1)^{n-1} \frac{1}{2^n} \pm \ldots = \frac{2}{3},$

6. $1 - \frac{1}{3} + \frac{1}{5} - \frac{1}{7} + \frac{1}{9} - \ldots + (-1)^{n-1} \frac{1}{2n-1} \pm \ldots = \frac{\pi}{4},$

7. $\frac{1}{1 \cdot 2} + \frac{1}{2 \cdot 3} + \frac{1}{3 \cdot 4} + \ldots + \frac{1}{n(n+1)} + \ldots = 1,$

8. $\frac{1}{1 \cdot 3} + \frac{1}{3 \cdot 5} + \frac{1}{5 \cdot 7} + \ldots + \frac{1}{(2n-1)(2n+1)} + \ldots = \frac{1}{2},$

9. $\frac{1}{1 \cdot 3} + \frac{1}{2 \cdot 4} + \frac{1}{3 \cdot 5} + \ldots + \frac{1}{(n-1)(n+1)} + \ldots = \frac{3}{4},$

10. $\frac{1}{3 \cdot 5} + \frac{1}{7 \cdot 9} + \frac{1}{11 \cdot 13} + \ldots + \frac{1}{(4n-1)(4n+1)} + \ldots = \frac{1}{2} - \frac{\pi}{8},$

11. $\frac{1}{1 \cdot 2 \cdot 3} + \frac{1}{2 \cdot 3 \cdot 4} + \ldots + \frac{1}{n(n+1)(n+2)} + \ldots = \frac{1}{4},$

12. $\frac{1}{1 \cdot 2 \ldots l} + \frac{1}{2 \cdot 3 \ldots (l+1)} + \ldots + \frac{1}{n \ldots (n+l-1)} + \ldots$

$$= \frac{1}{(l-1)(l-1)!},$$

13. $1 + \frac{1}{2^2} + \frac{1}{3^2} + \frac{1}{4^2} + \ldots + \frac{1}{n^2} + \ldots = \frac{\pi^2}{6},$

14. $1 - \frac{1}{2^2} + \frac{1}{3^2} - \frac{1}{4^2} + \ldots + (-1)^{n-1} \frac{1}{n^2} \pm \ldots = \frac{\pi^2}{12},$

15. $\frac{1}{1^2}+\frac{1}{3^2}+\frac{1}{5^2}+\ldots+\frac{1}{(2n+1)^2}+\ldots=\frac{\pi^2}{8}$,

16. $1+\frac{1}{2^4}+\frac{1}{3^4}+\ldots+\frac{1}{n^4}+\ldots=\frac{\pi^4}{90}$,

17. $1-\frac{1}{2^4}+\frac{1}{3^4}-\ldots+(-1)^{n-1}\frac{1}{n^4}\pm\ldots=\frac{7\pi^4}{720}$,

18. $\frac{1}{1^4}+\frac{1}{3^4}+\frac{1}{5^4}+\ldots+\frac{1}{(2n+1)^4}+\ldots=\frac{\pi^4}{96}$.

Bernoulli's numbers B_k

19. $1+\frac{1}{2^{2k}}+\frac{1}{3^{2k}}+\frac{1}{4^{2k}}+\ldots+\frac{1}{n^{2k}}+\ldots=\frac{\pi^{2k}2^{2k-1}}{(2k)!}B_k$,

20. $1-\frac{1}{2^{2k}}+\frac{1}{3^{2k}}-\frac{1}{4^{2k}}+\ldots+(-1)^{n-1}\frac{1}{n^{2k}}\pm\ldots=\frac{\pi^{2k}(2^{2k}-1)}{(2k)!}B_k$,

21. $1+\frac{1}{3^{2k}}+\frac{1}{5^{2k}}+\frac{1}{7^{2k}}+\ldots+\frac{1}{(2n-1)^{2k}}+\ldots=\frac{\pi^{2k}(2^{2k}-1)}{2\cdot(2k)!}B_k$.

Table of initial Bernoulli's numbers

k	B_k	k	B_k	k	B_k	k	B_k
1	$\frac{1}{6}$	4	$\frac{1}{30}$	7	$\frac{7}{6}$	10	$\frac{174611}{330}$
2	$\frac{1}{30}$	5	$\frac{5}{66}$	8	$\frac{3617}{510}$	11	$\frac{854513}{138}$
3	$\frac{1}{42}$	6	$\frac{691}{2730}$	9	$\frac{43867}{798}$		

Euler's numbers E_k

22. $1-\frac{1}{3^{2k+1}}+\frac{1}{5^{2k+1}}-\frac{1}{7^{2k+1}}+\ldots+(-1)^{n-1}\frac{1}{(2n-1)^{2k+1}}\pm\ldots$

$$=\frac{\pi^{2k+1}}{2^{2k+2}(2k)!}E_k.$$

Table of initial Euler's numbers

k	E_k	k	E_k
1	1	5	50 521
2	5	6	2 702 765
3	61	7	199 360 981
4	1385		

Remark. Some authors use another notation for Bernoulli's and Euler's numbers:

$$B_1 = -\frac{1}{2},\ B_2 = \frac{1}{6},\ B_3 = 0,\ B_4 = -\frac{1}{30},\ B_5 = 0,$$

$$B_6 = \frac{1}{42},\ B_7 = 0,\ B_8 = -\frac{1}{30},\ \ldots,$$

$$E_1 = 0,\ E_2 = -1,\ E_3 = 0,\ E_4 = 5,\ E_5 = 0,\ E_6 = -61,$$

$$E_7 = 0,\ E_8 = 1385,\ \ldots$$

9. Series of functions

Definitions. A series composed of functions of a common variable x:

$$f_1(x) + f_2(x) + \ldots + f_n(x) + \ldots = \sum_{n=1}^{\infty} f_n(x), \tag{1}$$

is called a *series of functions*. The values $x = a$ belonging to a common domain of definition of the functions $f_1(x), f_2(x), \ldots, f_n(x), \ldots$ and such that the series of numbers

$$f_1(a) + f_2(a) + \ldots + f_n(a) + \ldots$$

is convergent (i.e., there exists a limit of partial sums $S_n(a)$): $\lim_{n\to\infty} S_n(a) = \lim_{n\to\infty} \sum_{n=1}^{\infty} f_n(a) = S(a)$ form the *domain of convergence* of the series of functions (1). The function $S(x)$ is called the *sum* of the series of functions (1); then we say that the series (1) is convergent to the function $S(x)$. The sum of n initial terms of series (1) is called a *partial sum*: $S_n(x) = f_1(x) + f_2(x) + \ldots + f_n(x)$. The difference between the sum $S(x)$ of a convergent series and the partial sum $S_n(x)$ is called the *remainder* of series (1) and denoted by $R_n(x)$:

$$R_n(x) = S(x) - S_n(x) = f_{n+1}(x) + f_{n+2}(x) + \ldots + f_{n+m}(x) + \ldots$$

Uniform and non-uniform convergence of a series. According to the definition of limit of a sequence (p. 317), the series (1) is convergent in a given domain, if, for any preassigned positive number $\varepsilon > 0$, however small, there exists an integer N such that $|S(x) - S_n(x)| < \varepsilon$ for $n > N$. Thus, for a series of functions, there can be two possibilities:

(1) There exists a common number N, for all values of x in the given domain of convergence of the series. In this case,

the series (1) is said to be *uniformly convergent* in the given domain.

(2) There is no such number N common for all x in the given domain of convergence, i.e., for every n, we can find a point x in the domain of convergence such that $|S(x) - S_n(x)| > \varepsilon$; in this case, the series is said to be *non-uniformly convergent*.

Examples. (1) The series

$$(\bullet) \qquad 1 + \frac{x}{1!} + \frac{x^2}{2!} + \ldots + \frac{x^n}{n!} + \ldots$$

is convergent for every value x; its sum is equal to e^x (see p. 389). This series is uniformly convergent in an arbitrary bounded domain of x. In fact, if $|x| < a$, then

$$|S(x) - S_n(x)| = \left| \frac{x^{n+1}}{(n+1)!} e^{\theta x} \right| {}^{(1)} < \frac{a^{n+1}}{(n+1)!} e^a,$$

but, for sufficiently great n (independent of x), $\frac{a^{n+1}}{(n+1)!} e^a$ can be made less than ε, for $(n+1)!$ increases faster than a^{n+1}. However, the series converges non-uniformly in the domain of all real numbers: for any fixed n, it is always possible to find an x such that $\left| \frac{x^{n+1}}{(n+1)!} e^{\theta x} \right|$ is greater than any preassigned ε.

(2) The series $x + x(1-x) + x(1-x)^2 + \ldots + x(1-x)^n + \ldots$ converges for every value of x from 0 to 1, since $\varrho = \lim\limits_{n \to \infty} \left| \frac{a_{n+1}}{a_n} \right| = |1-x| < 1$ for $x > 0$ (and $S = 0$, for $x = 0$). But this convergence is not uniform: $S(x) - S_n(x) = x\big((1-x)^{n+1} + (1-x)^{n+2} + \ldots\big) = (1-x)^{n+1}$ and for any preassigned n, it is possible to find such a small x, that $(1-x)^{n+1}$ can be made arbitrarily near to 1, i.e., it will not be less than ε. However, the series converges uniformly in the interval $a < x < 1$, where a is positive and less than 1.

Weierstrass' test of uniform convergence. If there exists a convergent series of numbers

$$(1) \qquad c_1 + c_2 + \ldots + c_n + \ldots$$

such that the terms of the series of functions

$$(2) \qquad f_1(x) + f_2(x) + \ldots + f_n(x) + \ldots$$

satisfy, in the given domain, the condition

$$|f_n(x)| < c_n,$$

then the series (2) is uniformly convergent in the given domain.

---

(1) From a formula for the remainder of a Maclaurin's series, see p. 386.

In this case, the series (1) is called the *majorant* of (2).

Examples. The series $\sum_{n=1}^{\infty} a_n \cos nx$, $\sum_{n=1}^{\infty} a_n \sin nx$ are uniformly convergent in any domain, provided that the series $\sum_{n=1}^{\infty} |a_n|$ is uniformly convergent, since $|a_n \cos nx| < |a_n|$ and $|a_n \sin nx| < |a_n|$.

Properties of uniformly convergent series.

(1) If the functions $f_1(x), f_2(x), \ldots, f_n(x), \ldots$ are continuous in a certain domain of x and the series $f_1(x)+f_2(x)+\ldots+f_n(x)+\ldots$ converges uniformly in this domain, then its sum $S(x)$ is also a continuous function in this domain. If, however, the series does not converge uniformly in a bounded domain, then $S(x)$ can be discontinuous; thus, the sum in example (2) is discontinuous: $S(x)$ is equal to zero for $x = 0$ and is equal to 1 for $x > 0$; the function e^x in example (•) is continuous; the series converges non-uniformly in the whole infinite domain, but converges uniformly in every finite domain.

(2) An uniformly convergent series can be integrated term by term and the sum of the series of integrals is equal to the integral of the sum of the given series.

A power series is a series of functions of the form

$$\text{(A)} \qquad a_0 + a_1x + a_2x^2 + \ldots + a_nx^n + \ldots$$

or of the form

$$\text{(B)} \qquad a_0 + a_1(x-a) + a_2(x-a)^2 + \ldots + a_n(x-a)^n + \ldots$$

where a_i are constant coefficients.

Main properties of power series:

(1) The series (A) is absolutely convergent for all values of x less in the absolute value than a certain number ϱ ($|x| < \varrho$), called the *radius of convergence* of the power series. The series (B) is absolutely convergent for all values of x satisfying the inequality $|x-a| < \varrho$. The radius of convergence is determined by the formulas

$$\frac{1}{\varrho} = \lim_{n\to\infty} \frac{|a_{n+1}|}{|a_n|}, \qquad \frac{1}{\varrho} = \lim_{n\to\infty} \sqrt[n]{|a_n|}\ (^1).$$

At the end-points of the domain of convergence (the points $x = \pm\varrho$ for the series (A) and the points $x = a+\varrho$, $x = a-\varrho$ for the series (B)), the series can converge or diverge.

---

(1) If these limits do not exist, then lim sup should be taken.

(2) If the series (A) converges for a certain positive value $x = x_1$, then it converges uniformly inside the interval $(-x_1 + \varepsilon, x_1)$ (*Abel's theorem*).

Example. The series $1 + \frac{x}{1} + \frac{x^2}{2} + \ldots + \frac{x^n}{n} + \ldots$ has $\frac{1}{\varrho} = \lim_{n\to\infty} \frac{n+1}{n} = 1$, i.e., $\varrho = 1$, and it converges absolutely in the interval $-1 < x < +1$; for $x = -1$ the series converges conditionally (comp. series (11) on p. 352) and for $x = 1$ series diverges (comp. series (3) on p. 349). By Abel's theorem, the series converges uniformly in the interval $[-x_1, +x_1]$, where x_1 is an arbitrary positive number, less than 1.

Table of initial terms of certain powers of a power series

$$S = a + bx + cx^2 + dx^3 + ex^4 + fx^5 + \ldots,$$

$$S^2 = a^2 + 2abx + (b^2 + 2ac)x^2 + 2(ad + bc)x^3 + (c^2 + 2ae + 2bd)x^4 + 2(af + be + cd)x^5 + \ldots,$$

$$\sqrt{S} = S^{1/2} = a^{1/2}\left[1 + \frac{1}{2}\frac{b}{a}x + \left(\frac{1}{2}\frac{c}{a} - \frac{1}{8}\frac{b^2}{a^2}\right)x^2 + \left(\frac{1}{2}\frac{d}{a} - \frac{1}{4}\frac{bc}{a^2} + \frac{1}{16}\frac{b^3}{a^3}\right)x^3 + \left(\frac{1}{2}\frac{e}{a} - \frac{1}{4}\frac{bd}{a^2} - \frac{1}{8}\frac{c^2}{a^2} + \frac{3}{16}\frac{b^2c}{a^3} - \frac{5}{128}\frac{b^4}{a^4}\right)x^4 + \ldots\right],$$

$$\frac{1}{\sqrt{S}} = S^{-1/2} = a^{-1/2}\left[1 - \frac{1}{2}\frac{b}{a}x + \left(\frac{3}{8}\frac{b^2}{a^2} - \frac{1}{2}\frac{c}{a}\right)x^2 + \left(\frac{3}{4}\frac{bc}{a^2} - \frac{1}{2}\frac{d}{a} - \frac{5}{16}\frac{b^3}{a^3}\right)x^3 + \left(\frac{3}{4}\frac{bd}{a^2} + \frac{3}{8}\frac{c^2}{a^2} - \frac{1}{2}\frac{e}{a} - \frac{15}{16}\frac{b^2c}{a^3} + \frac{35}{128}\frac{b^4}{a^4}\right)x^4 + \ldots\right],$$

$$\frac{1}{S} = S^{-1} = a^{-1}\left[1 - \frac{b}{a}x + \left(\frac{b^2}{a^2} - \frac{c}{a}\right)x^2 + \left(\frac{2bc}{a^2} - \frac{d}{a} - \frac{b^3}{a^3}\right)x^3 + \left(\frac{2bd}{a^2} + \frac{c^2}{a^2} - \frac{e}{a} - 3\frac{b^2c}{a^3} + \frac{b^4}{a^4}\right)x^4 + \ldots\right],$$

$$\frac{1}{S^2} = S^{-2} = a^{-2}\left[1 - 2\frac{b}{a}x + \left(3\frac{b^2}{a^2} - 2\frac{c}{a}\right)x^2 + \left(6\frac{bc}{a^2} - 2\frac{d}{a} - 4\frac{b^3}{a^3}\right)x^3 + \left(6\frac{bd}{a^2} + 3\frac{c^2}{a^2} - 2\frac{e}{a} - 12\frac{b^2c}{a^3} + 5\frac{b^4}{a^4}\right)x^4 + \ldots\right].$$

Inverse of a power series. Given a series

$$y = f(x) = ax + bx^2 + cx^3 + dx^4 + ex^5 + fx^6 + \ldots \qquad (a \neq 0),$$

the expansion of the function inverse to $f(x)$:

$$x = \varphi(y) = Ay + By^2 + Cy^3 + Dy^4 + Ey^5 + Fy^6 + \dots$$

has the coefficients

$$A = \frac{1}{a}, \quad B = -\frac{b}{a^3}, \quad C = \frac{1}{a^5}(2b^2 - ac), \quad D = \frac{1}{a^7}(5abc - a^2d - 5b^3),$$

$$E = \frac{1}{a^9}(6a^2bd + 3a^2c^2 + 14b^4 - a^3e - 21ab^2c),$$

$$F = \frac{1}{a^{11}}(7a^3be + 7a^3cd + 84ab^3c - a^4f - 28a^2b^2d - 28a^2bc^2 - 42b^5).$$

For expansions of functions into power series see p. 385 and for expansions into trigonometric series see p. 727.

II. DIFFERENTIAL CALCULUS

1. Fundamental concepts

Derivative of a function of one variable [1]. The *derivative* of a function $y = f(x)$ (denoted by y', $y^{\bullet}$, Dy, $\frac{dy}{dx}$, $f'(x)$, $Df(x)$, $\frac{df(x)}{dx}$) is equal, for a given value of x, to the limit of the ratio of the increment Δy of the function to the increment Δx of the argument, when Δx tends to zero:

$$f'(x) = \lim_{\Delta x \to 0} \frac{f(x + \Delta x) - f(x)}{\Delta x}. \tag{1}$$

Geometric significance of the derivative. If $y = f(x)$ is represented by a graph in Cartesian coordinates (Fig. 283), then $f'(x) = \tan \alpha$, where α is the angle between the x axis and the tangent line to the curve in the given point (the angle is measured counterclockwise) [2].

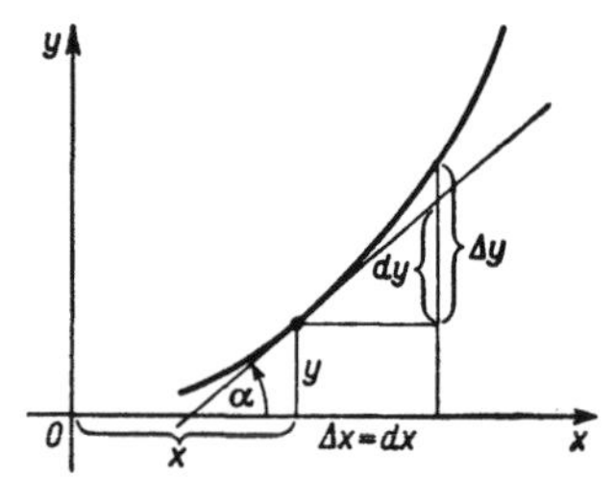

FIG. 283

Existence of the derivative. The derivative $f'(x)$ exists for these values of the argument x for which: (1) the function

[1] Only single-valued functions will be considered in this chapter.

[2] The formula $f'(x)$ holds only in the case when a common unit of measure is chosen on the x and y axes.

$f(x)$ is defined and continuous, (2) the considered ratio has a finite limit (1). If, for a value x_1, the derivative does not exist, then either there is no definite tangent line at the corresponding point of the graph of the function, or the tangent is perpendicular to the x axis. In the latter case, the limit (1) is infinite; we then write (inaccurately) $f'(x_1) = \infty$ (the derivative becomes infinite).

Examples when the derivative does not exist.

(1) $f(x) = \sqrt[3]{x}$, $f'(x) = \frac{1}{3\sqrt[3]{x^2}}$; $f'(0) = \infty$, the derivative becomes infinite at the point 0 (Fig. 284a).

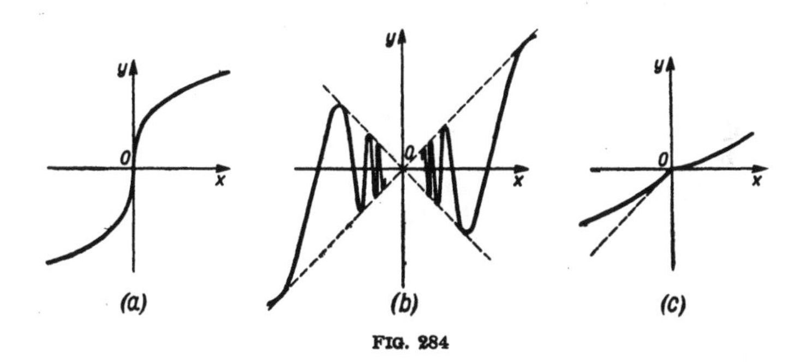

FIG. 284

(2) $f(x) = x^2 \sin \frac{1}{x}$; the limit (1) does not exist at $x = 0$ (Fig. 284b).

(3) $f(x) = \frac{x}{1 + e^{1/x}}$; the limit (1) does not exist at $x = 0$, but the left-hand limit $f'(-0) = 1$ and the right-hand limit $f'(+0) = 0$; in this case, the curve has a *bend* (Fig. 284c).

Left-hand and right-hand limits. If, for a given $x = a$, the limit (1) does not exist, but left-hand and right-hand limits exist (as in example (3), Fig. 284c), then these limits are called the *left-hand* and *right-hand derivatives*, respectively. Geometric significance of such derivatives: $f'(a-0) = \tan \alpha_1$, $f'(a+0) = \tan \alpha_2$ (Fig. 285); the curve has a bend.

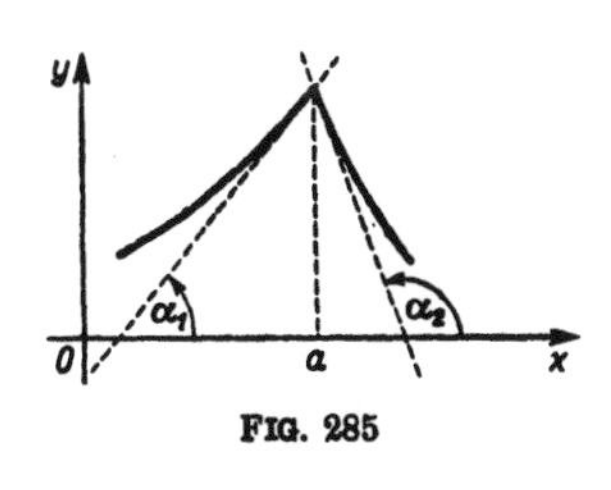

FIG. 285

The elementary functions possess derivatives in the whole domain of definition except for isolated points, where the cases of the types mentioned above can occur (Fig. 284a, b, c).

Partial derivative. The *partial derivative* of a function $u = f(x, y, z, \ldots, t)$ of several variables with respect to one of them, say x (notation: $\frac{\partial u}{\partial x}$, u'_x, $\frac{\partial f}{\partial x}$, f'_x) is defined by the formula

$$\frac{\partial u}{\partial x} = \lim_{\Delta x \to 0} \frac{f(x + \Delta x, y, z, \ldots, t) - f(x, y, z, \ldots, t)}{\Delta x};$$

in this case, only an increment of one variable is considered. A function of n variables has n *partial derivatives of the first order*: $\frac{\partial u}{\partial x}, \frac{\partial u}{\partial y}, \frac{\partial u}{\partial z}, \ldots, \frac{\partial u}{\partial t}$. Partial derivatives can be evaluated by using the rules of differentiation of functions of one variable (see pp. 366, 367), so that the remaining variables are regarded as constants. For example,

$$u = \frac{x^2 y}{z}, \quad \frac{\partial u}{\partial x} = \frac{2xy}{z}, \quad \frac{\partial u}{\partial y} = \frac{x^2}{z}, \quad \frac{\partial u}{\partial z} = -\frac{x^2 y}{z^2}.$$

Geometric significance of a partial derivative of a function of two variables. If $u = f(x, y)$ is represented by a surface in Cartesian coordinates, then $\frac{\partial u}{\partial x} = \tan \alpha$, where α is the angle between the straight line tangent to the surface and parallel to the xu plane and the positive direction of the x axis. Similarly, $\frac{\partial u}{\partial y} = \tan \beta$ (Fig. 286; in this case both the angles α and β are positive).

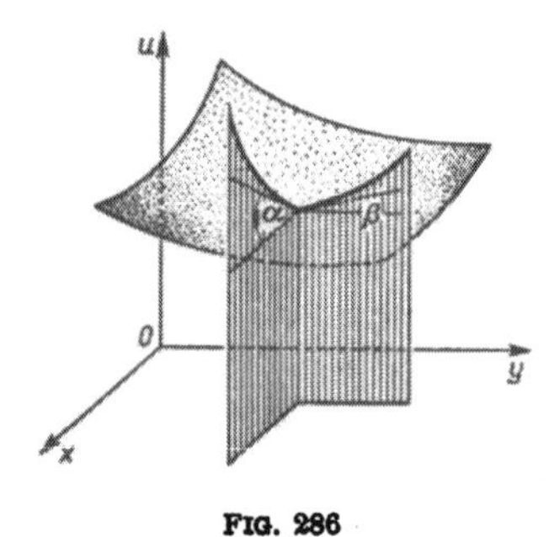

FIG. 286

For the *directional derivative* and *space derivative*, see field theory pp. 632 and 649.

The differentials of variable quantities x, y, and so on (notation dx, dy and so on) are defined according to whether the quantity is an independent variable or a function. The differential of an independent variable x is its increment which may have an arbitrary value ($dx = \Delta x$). The differential of a function $y = f(x)$ for a given value of x and a given differential dx of the argument dx is defined as the product

$$dy = f'(x)\,dx.$$

Geometric significance of the differential. If the function is represented by a curve in Cartesian coordinates, then dy is the increment of the ordinate of the tangent line to the curve at the point x corresponding to the given increment dx (Fig. 283).

Fundamental properties of the differential.

(1) *Invariance*: the equation $dy = f'(x)\,dx$ holds independently of whether x is an independent variable or a function of a new variable t.

(2) If dx is an infinitesimal, then dy and Δy are equivalent infinitesimals ($\lim \frac{\Delta y}{dy} = 1$) and their difference is an infinitesimal of a higher order than that of $dx, dy, \Delta y$. This property enables us to replace, for small increments, the increment of the function by the corresponding differential; this can be applied in approximate computation (see p. 133) as well as in the differential and integral calculus.

The partial differential of a function $u = f(x, y, z, \ldots, t)$ with respect to one of them, say x (notation $d_x u$ or $d_x f$), is defined by the equation

$$d_x u = \frac{\partial u}{\partial x}\,dx.$$

Differentiable function and total differential. A function $u = f(x, y, \ldots, t)$ is said to be *differentiable* at the point $M_0(x_0, y_0, \ldots, t_0)$, if the total increment of the function between the point M_0 and an infinitely near point $M(x_0 + dx, y_0 + dy, \ldots, t_0 + dt)$ (where $dx, dy, \ldots, dt$ are infinitesimals):

$$\Delta u = f(x_0 + dx, y_0 + dy, \ldots, t_0 + dt) - f(x_0, y_0, \ldots, t_0)$$

differs from the sum of its partial derivatives with respect to all variables

$$(1) \qquad \left(\frac{\partial u}{\partial x}\,dx + \frac{\partial u}{\partial y}\,dy + \ldots + \frac{\partial u}{\partial t}\,dt\right)_{x_0, y_0, \ldots, t_0}$$

by an infinitesimal of a higher order than the distance

$$M_0 M = \sqrt{dx^2 + dy^2 + \ldots + dt^2}.$$

If u is a differentiable function, then the expression (1) is called its *total differential* and is denoted by du:

$$du = \frac{\partial u}{\partial x}dx + \frac{\partial u}{\partial y}dy + \ldots + \frac{\partial u}{\partial t}dt \quad (2)$$ [1].

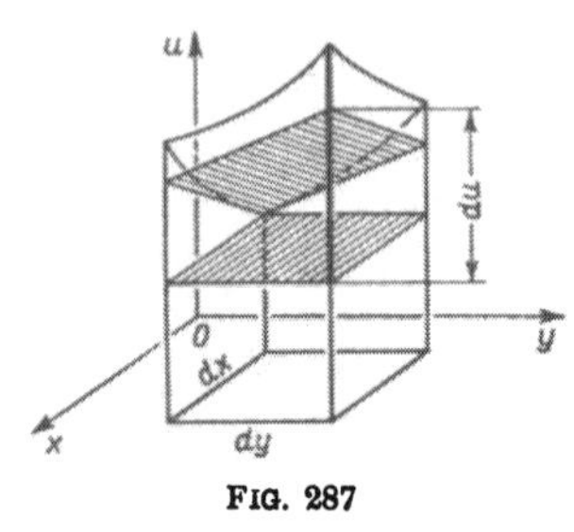

FIG. 287

Every continuous function of several variables having partial derivatives with respect to all variables at a given point is differentiable at this point. However, merely the existence of the partial derivatives of the function with respect to all variables does not imply that the function is differentiable.

Geometric significance of the total differential du of a function $u = f(x, y)$ represented by a surface in Cartesian coordinates (Fig. 287): du is equal to the increment of the u coordinate of the tangent plane at the given point, when x and y increase by dx and dy.

A fundamental property of the total differential is analogous to that of the differential of a function of one variable [2]: the expression (2) is invariant with respect to the involved variables.

Linearity of differential expressions. Differentials of variables related by a functional relation (an equation in a finite form) are always related by a linear relation (a differential equation of the first order). This holds for differentials of independent variables as well as for (partial or total) differentials of functions of one or several variables.

To *differentiate* an equation means to obtain a differential equation from the given finite equation. In a narrow sense, to *differentiate* means simply to find the derivative or a differential of a function.

Higher derivatives and differentials. The second derivative of a function $y = f(x)$ of one variable (notation: y'', $D^2y, y^{\cdot\cdot}$, $\frac{d^2y}{dx^2}$ [3], $f''(x)$, $D^2f(x)$, $\frac{d^2f(x)}{dx^2}$ [3]) is the derivative of the derivative: $f''(x)$

(1) For the total differential in vector form see p. 633, theory of field.

(2) See p. 363.

(3) This notation is convenient only in the case when x is an independent variable. It is inconvenient, for example, when $x = \varphi(v)$; see p. 373, change of variables.

$= \frac{d}{dx} f'(x)$. Higher derivatives are defined by successive differentiation (notation: y''', $\frac{d^3y}{dx^3}$ [1], D^4y, $f^{(IV)}(x)$, $D^nf(x)$, $\frac{d^nf(x)}{dx^n}$ [1]).

Second partial derivatives of a function $u = f(x, y, z)$ may be taken with respect to the same variable, as the first derivative $\left(\frac{\partial^2 u}{\partial x^2}, \frac{\partial^2 u}{\partial y^2}, \dots\right)$ or with respect to another $\left(\frac{\partial^2 u}{\partial x \partial y}, \frac{\partial^2 u}{\partial x \partial z}, \dots\right)$; in the latter case, the *mixed derivatives* are equal $\left(\frac{\partial^2 u}{\partial x \partial y} = \frac{\partial^2 u}{\partial y \partial x}\right)$ only if they are continuous for given x and y. Similarly for the higher partial derivatives of an arbitrary order $\left(\frac{\partial^3 u}{\partial x^3}, \frac{\partial^3 u}{\partial x \partial y^2}, \frac{\partial^3 u}{\partial x \partial y \partial z}, \dots\right)$.

The *second differential* of a function of one variable $y = f(x)$ is denoted by d^2y, $d^2f(x)$, and represents the differential of the first differential: $d^2y = d(dy) = f''(x)\,dx^2$ [1]. Higher differentials are defined similarly: $d^3y = d(d^2y) = f'''(x)\,dx^3$ [1], d^ny.

Second total differential of a function $u = f(x, y)$:

$$d^2u = d(du) = \frac{\partial^2 u}{\partial x^2} dx^2 + 2\frac{\partial^2 u}{\partial x \partial y} dx\,dy + \frac{\partial^2 u}{\partial y^2} dy^2, \text{ [1]}$$

$$= \left(\frac{\partial}{\partial x} dx + \frac{\partial}{\partial y} dy\right)^2 u;$$

d^nu can be evaluated by the symbolical formula

$$d^nu = \left(\frac{\partial}{\partial x} dx + \frac{\partial}{\partial y} dy\right)^n u \text{ [2]}.$$

For functions $u = f(x, y, z, \dots)$ of several variables,

$$d^nu = \left(\frac{\partial}{\partial x} dx + \frac{\partial}{\partial y} dy + \frac{\partial}{\partial z} dz + \dots\right)^n u \text{ [2]}.$$

2. Technique of differentiation

General indications. By using the rules of differentiation and the table of derivatives given below it is possible to find the derivative of any elementary function. The function obtained is also an elementary one. The most essential is the rule for differentia-

[1] This notation is convenient only in the case, when x is an independent variable. It is inconvenient when $x = \varphi(v)$ (see p. 373, change of variables).

[2] When x and y are functions of new variables, the formulas are more complicated; see pp. 375, 376.

tion of composite functions (the *chain rule*). We explain it by the following example:

Example: $y = e^{\tan\sqrt{x}}; \frac{dy}{dx} = e^{\tan\sqrt{x}} \cdot \frac{d(\tan\sqrt{x})}{dx} = e^{\tan\sqrt{x}} \cdot \frac{1 \cdot d\sqrt{x}}{\cos^2\sqrt{x}\,dx}$

$$= e^{\tan\sqrt{x}} \cdot \frac{1}{\cos^2\sqrt{x}} \cdot \frac{1}{2\sqrt{x}} = \frac{e^{\tan\sqrt{x}}}{2\sqrt{x}\cos^2\sqrt{x}}.$$

Before differentiation, it is convenient to express the function, if that is possible, as a sum by removing the brackets, separating the integral part of a quotient (pp. 149, 150), evaluating logarithms (p. 157) etc.

Examples.

(1) $y = \frac{2 - 3\sqrt{t} + 4\sqrt[3]{t} + t^2}{t} = \frac{2}{t} - 3t^{-1/2} + 4t^{-2/3} + t,$

$$\frac{dy}{dt} = -2t^{-2} + \frac{3}{2}t^{-3/2} - \frac{8}{3}t^{-5/3} + 1.$$

(2) $y = \ln\sqrt{\frac{x^2+1}{x^2-1}} = \frac{1}{2}\ln(x^2+1) - \frac{1}{2}\ln(x^2-1),$

$$\frac{dy}{dx} = \frac{1}{2}\left(\frac{2x}{x^2+1}\right) - \frac{1}{2}\left(\frac{2x}{x^2-1}\right) = -\frac{2x}{x^4-1}.$$

Fundamental rules of differentiation. Notation: $u, v, w, \ldots$— functions of an independent variable x; $u', v', w', \ldots$—derivative of these functions with respect to x.

(1) The derivative (or differential) of an algebraic sum of two or several functions is equal to the algebraic sum of derivatives (or differentials) of these functions:

$$(u + v - w + \ldots + t)' = u' + v' - w' + \ldots + t';$$

$$d(u + v - w + \ldots + t) = du + dv - dw + \ldots + dt.$$

(2) The derivative (or differential) of a product of n functions is the sum of n summands each of whose is obtained from the given product by successive replacing one of the factors by its derivative (or differential): for two functions:

$$(uv)' = u'v + uv', \quad d(uv) = v\,du + u\,dv,$$

for three functions:

$$(uvw)' = u'vw + uv'w + uvw', \quad d(uvw) = vw\,du + uw\,dv + uv\,dw,$$

and so on.

To find the derivative of a product of several functions, we sometimes compute first their *logarithmic derivative* (i.e., the derivative of the logarithm of the given function, $(\ln|y|)' = y'/y$). For example,

$$y = \sqrt{x^3 e^{4x} \sin x}, \quad \ln y = \tfrac{1}{2}(3\ln|x| + 4x + \ln|\sin x|),$$

$$\frac{d\ln y}{dx} = \frac{y'}{y} = \frac{1}{2}\left(\frac{3}{x} + 4 + \cot x\right), \quad y' = \left(\frac{3}{2x} + 2 + \frac{1}{2}\cot x\right)\sqrt{x^3 e^{4x} \sin x}.$$

This rule can be applied in differentiation of a function of the form u^v. For example, if $y = (2x+1)^{3x}$, then

$$\ln y = 3x \ln(2x+1), \quad \frac{y'}{y} = 3\left(\frac{2x}{2x+1} + \ln(2x+1)\right),$$

$$y' = 3\left(\frac{2x}{2x+1} + \ln(2x+1)\right)y = 3\left(\frac{2x}{2x+1} + \ln(2x+1)\right)(2x+1)^{3x}.$$

(3) The derivative (or differential) of a function with a constant factor. A constant factor can be brought out from under the sign of the derivative (or differential):

$$(cu)' = cu', \quad d(cu) = c\,du.$$

(4) The derivative (or differential) of a fraction can be computed from the formula

$$\left(\frac{u}{v}\right)' = \frac{vu' - uv'}{v^2}, \quad d\left(\frac{u}{v}\right) = \frac{v\,du - u\,dv}{v^2}.$$

(5) The derivative of a composite function ("the *chain rule*"). If $y = f(u)$ and $u = \varphi(x)$, then

$$\frac{dy}{dx} = f'(u)\,\varphi'(x);$$

if $y = f(u)$, $u = \varphi(t)$, $t = \psi(x)$, then

$$\frac{dy}{dx} = f'(u)\,\varphi'(t)\,\psi'(x).$$

If the "chain" is composed of a greater number of functions, the procedure is analogous.

Table of derivatives of elementary functions

Function	Derivative	Function	Derivative
constant	0	$\frac{1}{x}$	$-\frac{1}{x^2}$
x	1		
x^n	nx^{n-1}	$\frac{1}{x^n}$	$-\frac{n}{x^{n+1}}$

Table of derivatives of elementary functions

Function	Derivative	Function	Derivative
$\sqrt{x}$	$\frac{1}{2\sqrt{x}}$	$\operatorname{arc}\tan x$	$\frac{1}{1+x^2}$
$\sqrt[n]{x}$	$\frac{1}{n\sqrt[n]{x^{n-1}}}$	$\operatorname{arc}\cot x$	$-\frac{1}{1+x^2}$
e^x	e^x	$\operatorname{arc}\sec x$	$\frac{1}{x\sqrt{x^2-1}}$
a^x	$a^x \ln a$	$\operatorname{arc}\operatorname{cosec} x$	$-\frac{1}{x\sqrt{x^2-1}}$
$\ln \lvert x\rvert$	$\frac{1}{x}$	$\sinh x$	$\cosh x$
$\log_a \lvert x\rvert$	$\frac{1}{x}\log_a e = \frac{1}{x\ln a}$	$\cosh x$	$\sinh x$
$\log \lvert x\rvert$	$\frac{1}{x}\log e \approx \frac{0.4343}{x}$	$\tanh x$	$\frac{1}{\cosh^2 x}$
$\sin x$	$\cos x$	$\coth x$	$-\frac{1}{\sinh^2 x}$
$\cos x$	$-\sin x$	$\operatorname{ar}\sinh x$	$\frac{1}{\sqrt{x^2+1}}$
$\tan x$	$\frac{1}{\cos^2 x} = \sec^2 x = 1 + \tan^2 x$	$\operatorname{ar}\cosh x$	$\frac{1}{\sqrt{x^2-1}}$
$\cot x$	$-\frac{1}{\sin^2 x} = -\operatorname{cosec}^2 x = -(1+\cot^2 x)$	$\operatorname{ar}\tanh x$	$\frac{1}{1-x^2}$
$\sec x$	$\frac{\sin x}{\cos^2 x} = \tan x \sec x$	$\operatorname{ar}\coth x$	$-\frac{1}{1-x^2}$
$\operatorname{cosec} x$	$-\frac{\cos x}{\sin^2 x} = -\cot x \operatorname{cosec} x$		
$\operatorname{arc}\sin x$	$\frac{1}{\sqrt{1-x^2}}$		
$\operatorname{arc}\cos x$	$-\frac{1}{\sqrt{1-x^2}}$		

Higher derivatives of the simplest functions

Function	n-th derivative
x^m [1]	$m(m-1)(m-2)\dots(m-n+1)x^{m-n}$
$\ln \lvert x\rvert$	$(-1)^{n-1}(n-1)!\frac{1}{x^n}$
$\log_a \lvert x\rvert$	$(-1)^{n-1}\frac{(n-1)!}{\ln a}\cdot\frac{1}{x^n}$
e^{kx}	$k^n e^{kx}$
a^x	$(\ln a)^n a^x$
a^{kx}	$(k \ln a)^n a^{kx}$

[1] If m is a natural number and $n > m$, then the n-th derivative is zero.

Higher derivatives of the simplest functions

Function	n-th derivative
$\sin x$	$\sin (x + n \cdot \frac{1}{2}\pi)$
$\cos x$	$\cos (x + n \cdot \frac{1}{2}\pi)$
$\sin kx$	$k^n \sin (kx + n \cdot \frac{1}{2}\pi)$
$\cos kx$	$k^n \cos (kx + n \cdot \frac{1}{2}\pi)$
$\sinh x$	$\sinh x$, when n is even, $\cosh x$, when n is odd
$\cosh x$	$\cosh x$, when n is even, $\sinh x$, when n is odd

(6) The n-th derivative of a product of two functions (*Leibnitz's formula*):

$$D^n(uv) = uD^nv \binom{n}{1} DuD^{n-1}v + \binom{n}{2} D^2uD^{n-2}v + \ldots +$$

$$+ \binom{n}{k} D^kuD^{n-k}v + \ldots + D^nu \cdot v$$

or, putting $D^0u = u$, $D^0v = v$,

$$D^n(uv) = \sum_{m=0}^{n} C_n^m D^m u D^{n-m} v.$$

This is analogous to Newton's formula, see p. 193.

Derivative of a composite function of several variables. Case of one independent variable:

$$u = f(x, y, \ldots, z),$$

where $x = \varphi(t)$, $y = \psi(t)$, $\ldots$, $z = \chi(t)$:

$$\frac{du}{dt} = \frac{\partial u}{\partial x} \cdot \frac{dx}{dt} + \frac{\partial u}{\partial y} \cdot \frac{dy}{dt} + \ldots + \frac{\partial u}{\partial z} \cdot \frac{dz}{dt}. \tag{1}$$

Case of several independent variables:

$$u = f(x, y, \ldots, t),$$

where $x = \varphi(\xi, \eta, \ldots, \tau)$, $y = \psi(\xi, \eta, \ldots, \tau)$, $\ldots$, $t = \chi(\xi, \eta, \ldots, \tau)$:

$$\begin{aligned} \frac{\partial u}{\partial \xi} &= \frac{\partial u}{\partial x} \cdot \frac{\partial x}{\partial \xi} + \frac{\partial u}{\partial y} \cdot \frac{\partial y}{\partial \xi} + \ldots + \frac{\partial u}{\partial t} \cdot \frac{\partial t}{\partial \xi}, \\ \frac{\partial u}{\partial \eta} &= \frac{\partial u}{\partial x} \cdot \frac{\partial x}{\partial \eta} + \frac{\partial u}{\partial y} \cdot \frac{\partial y}{\partial \eta} + \ldots + \frac{\partial u}{\partial t} \cdot \frac{\partial t}{\partial \eta}, \\ &\ldots\ldots\ldots\ldots\ldots\ldots\ldots\ldots\ldots \\ \frac{\partial u}{\partial \tau} &= \frac{\partial u}{\partial x} \cdot \frac{\partial x}{\partial \tau} + \frac{\partial u}{\partial y} \cdot \frac{\partial y}{\partial \tau} + \ldots + \frac{\partial u}{\partial t} \cdot \frac{\partial t}{\partial \tau}. \end{aligned} \tag{2}$$

Differentiation of an implicit function.

(1) A function $y = f(x)$ of one variable given by the equation

$$\text{(A)} \qquad F(x, y) = 0.$$

Differentiating (A) with respect to x according to (1), we obtain

$$\text{(B)} \qquad F'_x + F'_y y' = 0,$$

whence

$$y' = -\frac{F'_x}{F'_y}.$$

Differentiating formula (B) with respect to x by the same formula (1), we have

$$\text{(C)} \qquad F''_{xx} + 2F''_{xy}y' + F''_{yy}(y')^2 + F'_y y'' = 0,$$

hence, by formula (B),

$$\text{(D)} \qquad y'' = \frac{2F'_x F'_y F''_{xy} - (F'_y)^2 F''_{xx} - (F' \;)^2 F''_{yy}}{(F'_y)^3}.$$

In the same way we obtain

$$\text{(E)} \qquad F'''_{xxx} + 3F'''_{xxy}y' + 3F'''_{xyy}(y')^2 + F'''_{yyy}(y')^3 +$$
$$+ 3F''_{xy}y'' + 3F''_{yy}y'y'' + F'_y y''' = 0,$$

whence, by (B) and (D), we can determine y''' and so on.

(2) A function $u = f(x, y, z, \ldots, t)$ of several variables given by the equation $F(x, y, \ldots, t, u) = 0$. By aid of formulas (2), the partial derivatives can be determined as follows:

$$\frac{\partial u}{\partial x} = -\frac{F'_x}{F'_u}, \quad \frac{\partial u}{\partial y} = -\frac{F'_y}{F'_u}, \quad \ldots, \quad \frac{\partial u}{dt} = -\frac{F'_t}{F'_u}.$$

Higher partial derivatives can be determined in the same way.

(3) Two functions $y = f(x)$, $z = \varphi(x)$ given by the system of equations

$$\text{(A)} \qquad F(x, y, z) = 0, \qquad \Phi(x, y, z) = 0.$$

Differentiating (A) according to (1), we have

$$\text{(B)} \qquad F'_x + F'_y y' + F'_z z' = 0, \qquad \Phi'_x + \Phi'_y y' + \Phi'_z z' = 0,$$

whence

$$y' = \frac{F'_z\Phi'_x - \Phi'_z F'_x}{F'_y\Phi'_z - F'_z\Phi'_y}, \qquad z' = \frac{F'_x\Phi'_y - F'_y\Phi'_x}{F'_y\Phi'_z - F'_z\Phi'_y}.$$

Having found y' and z', we find the second derivatives in the same way by differentiating (B), and so on.

(4) *n* functions of one variable: $y=f(x)$, $z=\varphi(x)$, ..., $t=\psi(x)$ defined by the system of equations

(A) $F(x,y,z,\dots,t)=0, \quad \Phi(x,y,z,\dots,t)=0, \quad \dots, \quad \Psi(x,y,z,\dots,t)=0.$

Differentiating (A) according to formula (1), we obtain

$$\text{(B)} \qquad \begin{array}{l} F'_x+F'_y y'+F'_z z'+\dots+F'_t t'=0, \\ \Phi'_x+\Phi'_y y'+\Phi'_z z'+\dots+\Phi'_t t'=0, \\ \dots\dots\dots\dots\dots\dots\dots\dots \\ \Psi'_x+\Psi'_y y'+\Psi'_z z'+\dots+\Psi'_t t'=0. \end{array}$$

If

$$\begin{vmatrix} F'_y & F'_z & \dots & F'_t \\ \Phi'_y & \Phi'_z & \dots & \Phi'_t \\ \dots & \dots & \dots & \dots \\ \Psi'_y & \Psi'_z & \dots & \Psi'_t \end{vmatrix} \neq 0,$$

then, solving the system (B) with respect to $y', z', \dots, t'$, we find the first derivatives. Higher derivatives can be found in the same way.

(5) Two functions of two variables $u=f(x,y)$, $v=\varphi(x,y)$ defined by a system of two equations

$$\text{(A)} \qquad F(x,y,u,v)=0, \qquad \Phi(x,y,u,v)=0.$$

Differentiating equations (A) with respect to x and y according to formulas (2) on p. 369, we obtain

$$(\mathrm{B_x}) \qquad \begin{array}{l} \dfrac{\partial F}{\partial x}+\dfrac{\partial F}{\partial u}\cdot\dfrac{\partial u}{\partial x}+\dfrac{\partial F}{\partial v}\cdot\dfrac{\partial v}{\partial x}=0, \\[2ex] \dfrac{\partial \Phi}{\partial x}+\dfrac{\partial \Phi}{\partial u}\cdot\dfrac{\partial u}{\partial x}+\dfrac{\partial \Phi}{\partial v}\cdot\dfrac{\partial v}{\partial x}=0, \end{array}$$

$$(\mathrm{B_y}) \qquad \begin{array}{l} \dfrac{\partial F}{\partial y}+\dfrac{\partial F}{\partial u}\cdot\dfrac{\partial u}{\partial y}+\dfrac{\partial F}{\partial v}\cdot\dfrac{\partial v}{\partial y}=0, \\[2ex] \dfrac{\partial \Phi}{\partial y}+\dfrac{\partial \Phi}{\partial u}\cdot\dfrac{\partial u}{\partial y}+\dfrac{\partial \Phi}{\partial v}\cdot\dfrac{\partial v}{\partial y}=0. \end{array}$$

Solving the system $(\mathrm{B_x})$ with respect to $\dfrac{\partial u}{\partial x}, \dfrac{\partial v}{\partial x}$ and the system $(\mathrm{B_y})$ with respect to $\dfrac{\partial u}{\partial y}, \dfrac{\partial v}{\partial y}$, we obtain the first partial derivatives. Higher derivatives can be found similarly.

(6) *n* functions of *m* variables defined by a system of *n* equations. Partial derivatives of the first and higher order can be found in an analogous way.

Derivatives of a function $y = f(x)$ given in parametric form $x = x(t)$, $y = y(t)$ (the dashes denote derivatives with respect to t):

$$\frac{dy}{dx} = \frac{y'}{x'}, \quad \frac{d^2y}{dx^2} = \frac{x'y'' - y'x''}{x'^3},$$

$$\frac{d^3y}{dx^3} = \frac{x'(x'y''' - y'x''') - 3x''(x'y'' - y'x'')}{x'^5}, \ldots$$

Derivative of the inverse function. If $y = f(x)$ is the function inverse to $x = \varphi(y)$, then its derivatives are given by the formulas

$$\frac{dy}{dx} = \frac{1}{\varphi'(y)}, \quad \frac{d^2y}{dx^2} = -\frac{\varphi''(y)}{[\varphi'(y)]^3}, \quad \frac{d^3y}{dx^3} = \frac{3[\varphi''(y)]^2 - \varphi'(y)\varphi'''(y)}{[\varphi'(y)]^5}, \ldots,$$

$\varphi'(y) \neq 0$.

For example, $y = \arcsin x$ is inverse to $x = \sin y$ and

$$\frac{dy}{dx} = \frac{1}{(\sin y)'} = \frac{1}{\cos y} = \frac{1}{\sqrt{1 - \sin^2 y}} = \frac{1}{\sqrt{1 - x^2}}.$$

Graphical differentiation. Let $y = f(x)$ be a differentiable function represented by its graph Γ in the interval $a < x < b$. The graph Γ' of its derivative can be plotted approximately in the following way.

Preliminary step: construction of the tangent. A construction "by eye" is rather inexact. If, however, the direction of the tangent is known (MN in Fig. 288), the point of contact can be found more exactly as follows. Draw two chords M_1N_1 and M_2N_2 in the direction MN near each to another; draw a straight line PQ through the midpoints R_1 and R_2 of the chords; the tangent at the point A of intersection of PQ with the curve has (approximately) the direction MN. The accuracy of the construction can be verified by drawing a third chord: it should be bisected by the line PQ.

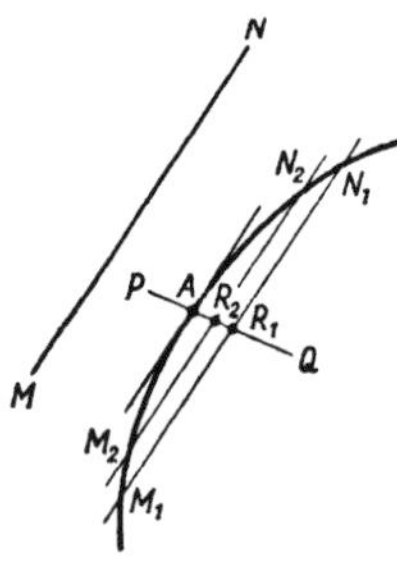

FIG. 288

Plotting of the graph of the derivative. (1) Choosing several directions l_1, l_2, ... of tangents to the curve so that they correspond approximately to the given segment of the curve, we determine the points of contact in the way described above (the tangent lines need not be constructed).

(2) We select an arbitrary point P ("the pole") on the negative part of the x axis; the segment $PQ = a$ should be greater, the steeper is the curve.

(3) We draw the straight lines PB_1, PB_2, ... from the pole P parallel to the direction l_1, l_2, ... to intersect the y axis at the points B_1, B_2, ...

(4) We draw the horizontal lines B_1C_1, B_2C_2, ... from the points B_1, B_2, ... to intersect the corresponding ordinates of the points A_1, A_2, ... at C_1, C_2, ...

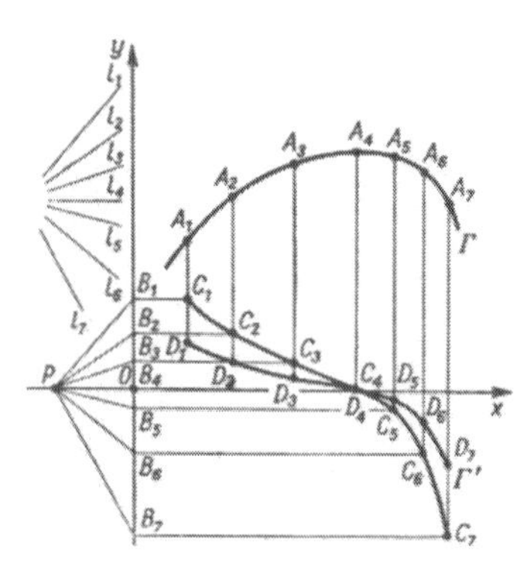

FIG. 289

(5) We join the points C_1, C_2, ... by a smooth curve; its equation is $y = af'(x)$ and for $a = 1$ we obtain the graph of the derivative.

(6) If $a \neq 1$, we find the points D_1, D_2, ... whose ordinates are equal to the ordinates of C_1, C_2, ... divided by a ($a = PO$, Fig. 289). The curve Γ' is then the graph of the derivative.

3. Change of variables in differential expressions

Functions of one variable. If $y = f(x)$ and an expression

$$H = F\left(x, y, \frac{dy}{dx}, \frac{d^2y}{dx^2}, \frac{d^3y}{dx^3}, \dots\right)$$

involving the argument, the function and its derivatives is given, then, in case of a change of variables, the derivatives can be found from the following formulas.

(1) If the argument x is replaced by t related to x by the formula $x = \varphi(t)$, then

$$\frac{dy}{dx}=\frac{1}{\varphi'(t)}\cdot\frac{dy}{dt},$$

$$\frac{d^2y}{dx^2}=\frac{1}{[\varphi'(t)]^3}\left(\varphi'(t)\frac{d^2y}{dt^2}-\varphi''(t)\frac{dy}{dt}\right),$$

$$\frac{d^3y}{dx^3}=\frac{1}{[\varphi'(t)]^5}\left([\varphi'(t)]^2\frac{d^3y}{dt^3}-3\varphi'(t)\varphi''(t)\frac{d^2y}{dt^2}+\right.$$

$$\left.+\left[3[\varphi''(t)]^2-\varphi'(t)\varphi'''(t)\right]\frac{dy}{dt}\right)^{(1)},$$

$\varphi'(t)\neq 0$.

(2) In the case, when the function y is replaced by a function u related to y by the equation $y=\varphi(u)$, then

$$\frac{dy}{dx}=\varphi'(u)\frac{du}{dx},$$

$$\frac{d^2y}{dx^2}=\varphi'(u)\frac{d^2u}{dx^2}+\varphi''(u)\left(\frac{du}{dx}\right)^2,$$

$$\frac{d^3y}{dx^3}=\varphi'(u)\frac{d^3u}{dx^3}+3\varphi''(u)\frac{du}{dx}\cdot\frac{d^2u}{dx^2}+\varphi'''(u)\left(\frac{du}{dx}\right)^3.$$

(3) In the case, when x and y are replaced by t and u related to x and y by the equations $x=\varphi(t,u)$, $y=\psi(t,u)$:

$$\frac{dy}{dx}=\frac{\dfrac{\partial\psi}{\partial t}+\dfrac{\partial\psi}{\partial u}\cdot\dfrac{du}{dt}}{\dfrac{\partial\varphi}{\partial t}+\dfrac{\partial\varphi}{\partial u}\cdot\dfrac{du}{dt}},$$

$$\frac{d^2y}{dx^2}=\frac{d}{dx}\left(\frac{dy}{dx}\right)=\frac{d}{dx}\left(\frac{\dfrac{\partial\psi}{\partial t}+\dfrac{\partial\psi}{\partial u}\cdot\dfrac{du}{dt}}{\dfrac{\partial\varphi}{\partial t}+\dfrac{\partial\varphi}{\partial u}\cdot\dfrac{du}{dt}}\right)$$

$$=\frac{1}{\dfrac{\partial\varphi}{\partial t}+\dfrac{\partial\varphi}{\partial u}\cdot\dfrac{du}{dt}}\left[\frac{d}{dt}\left(\frac{\dfrac{\partial\psi}{\partial t}+\dfrac{\partial\psi}{\partial u}\cdot\dfrac{du}{dt}}{\dfrac{\partial\varphi}{\partial t}+\dfrac{\partial\varphi}{\partial u}\cdot\dfrac{du}{dt}}\right)\right]$$

$$=\frac{1}{B}\cdot\frac{d}{dt}\left(\frac{A}{B}\right)=\frac{1}{B^3}\left(B\frac{dA}{dt}-A\frac{dB}{dt}\right),$$

---

(1) If the formula of transformation is given in implicit form $\Phi(x,t)=0$, then the derivatives $\frac{dy}{dx}$, $\frac{d^2y}{dx^2}$, $\frac{d^3y}{dx^3}$ are computed by the same formula, but $\varphi'(t)$, $\varphi''(t)$, $\varphi'''(t)$ are computed as derivatives of an implicit function; if the expression involves the variable x, then x should be eliminated by means of the equation $\Phi(x,t)=0$.

where $A=\frac{\partial\psi}{\partial t}+\frac{\partial\psi}{\partial u}\cdot\frac{du}{dt}$, $\quad B=\frac{\partial\varphi}{\partial t}+\frac{\partial\varphi}{\partial u}\cdot\frac{du}{dt}$.

$\frac{d^3y}{dx^3}$ can be found similarly.

For example, for transformation of Cartesian coordinates to polar coordinates by the formulas $x=\varrho\cos\varphi$, $y=\varrho\sin\varphi$, we have

$$\frac{dy}{dx}=\frac{\varrho'\sin\varphi+\varrho\cos\varphi}{\varrho'\cos\varphi-\varrho\sin\varphi},\quad \frac{d^2y}{dx^2}=\frac{\varrho^2+2\varrho'^2-\varrho\varrho''}{(\varrho'\cos\varphi-\varrho\sin\varphi)^3}.$$

Functions of two variables. If $\omega=f(x, y)$ and an expression

$$H=F\left(x, y, \omega, \frac{\partial\omega}{\partial x}, \frac{\partial\omega}{\partial y}, \frac{\partial^2\omega}{\partial x^2}, \frac{\partial^2\omega}{\partial x\partial y}, \frac{\partial^2\omega}{\partial y^2}, \ldots\right)$$

containing the variables, the function and its partial derivatives is given, then, in case when the variables x and y are changed into u and v related to x and y by the equations $x=\varphi(u, v)$, $y=\psi(u, v)$, the partial derivatives $\frac{\partial\omega}{\partial x}$, $\frac{\partial\omega}{\partial y}$ can be found from the following system of equations:

$$\frac{\partial\omega}{\partial u}=\frac{\partial\omega}{\partial x}\cdot\frac{\partial\varphi}{\partial u}+\frac{\partial\omega}{\partial y}\cdot\frac{\partial\psi}{\partial u},\quad \frac{\partial\omega}{\partial v}=\frac{\partial\omega}{\partial x}\cdot\frac{\partial\varphi}{\partial v}+\frac{\partial\omega}{\partial y}\cdot\frac{\partial\psi}{\partial v}.$$

Hence

$$\frac{\partial\omega}{\partial x}=A\frac{\partial\omega}{\partial u}+B\frac{\partial\omega}{\partial v},\quad \frac{\partial\omega}{\partial y}=C\frac{\partial\omega}{\partial u}+D\frac{\partial\omega}{\partial v},$$

where A, B, C, D are functions of u, v. Second partial derivatives are obtained from the same formulas, by applying them not to the function ω, but to $\frac{\partial\omega}{\partial x}$ and $\frac{\partial\omega}{\partial y}$; for example,

$$\frac{\partial^2\omega}{\partial x^2}=\frac{\partial}{\partial x}\left(\frac{\partial\omega}{\partial x}\right)=\frac{\partial}{\partial x}\left(A\frac{\partial\omega}{\partial u}+B\frac{\partial\omega}{\partial v}\right)$$

$$=A\left(A\frac{\partial^2\omega}{\partial u^2}+B\frac{\partial^2\omega}{\partial u\partial v}+\frac{\partial A}{\partial u}\cdot\frac{\partial\omega}{\partial u}+\frac{\partial B}{\partial u}\cdot\frac{\partial\omega}{\partial v}\right)+$$

$$+B\left(A\frac{\partial^2\omega}{\partial u\partial v}+B\frac{\partial^2\omega}{\partial v^2}+\frac{\partial A}{\partial v}\cdot\frac{\partial\omega}{\partial u}+\frac{\partial B}{\partial v}\cdot\frac{\partial\omega}{\partial v}\right).$$

Higher partial derivatives can be obtained similarly.

Example. Express Laplacian [1]

---

[1] See p. 645.

$$\Delta\omega = \frac{\partial^2\omega}{\partial x^2} + \frac{\partial^2\omega}{\partial y^2}$$

in the polar coordinates $x = \varrho\cos\varphi$, $y = \varrho\sin\varphi$.

$$\frac{\partial\omega}{\partial\varrho} = \frac{\partial\omega}{\partial x}\cos\varphi + \frac{\partial\omega}{\partial y}\sin\varphi, \qquad \frac{\partial\omega}{\partial\varphi} = -\frac{\partial\omega}{\partial x}\varrho\sin\varphi + \frac{\partial\omega}{\partial y}\varrho\cos\varphi;$$

whence

$$\frac{\partial\omega}{\partial x} = \cos\varphi\frac{\partial\omega}{\partial\varrho} - \frac{\sin\varphi}{\varrho}\cdot\frac{\partial\omega}{\partial\varphi}, \qquad \frac{\partial\omega}{\partial y} = \sin\varphi\frac{\partial\omega}{\partial\varrho} + \frac{\cos\varphi}{\varrho}\cdot\frac{\partial\omega}{\partial\varphi};$$

$$\frac{\partial^2\omega}{\partial x^2} = \cos\varphi\frac{\partial}{\partial\varphi}\left(\cos\varphi\frac{\partial\omega}{\partial\varrho} - \frac{\sin\varphi}{\varrho}\cdot\frac{\partial\omega}{\partial\varphi}\right) -$$

$$-\frac{\sin\varphi}{\varrho}\cdot\frac{\partial}{\partial\varphi}\left(\cos\varphi\frac{\partial\omega}{\partial\varrho} - \frac{\sin\varphi}{\varrho}\cdot\frac{\partial\omega}{\partial\varphi}\right)$$

Calculating similarly $\frac{\partial^2\omega}{\partial y^2}$, we obtain

$$\Delta\omega = \frac{\partial^2\omega}{\partial\varrho^2} + \frac{1}{\varrho^2}\cdot\frac{\partial^2\omega}{\partial\varphi^2} + \frac{1}{\varrho}\cdot\frac{\partial\omega}{\partial\varrho}.$$

In the case of functions of several variables the transformation formulas can be obtained in a similar way.

4. Main theorems of differential calculus

A condition of monotony of a function. If the function $y = f(x)$ is defined and continuous in a certain connected domain and has a derivative at each interior point of this domain (¹), then a necessary and sufficient condition that $f(x)$ is monotone throughout the given domain is that

$f'(x) \geq 0$, for a monotone increasing function,

$f'(x) \leq 0$, for a monotone decreasing function (²).

Geometric significance. The graph of a monotone increasing function is a curve which, when running from the left to the right, does not drop anywhere (it rises or runs horizontally), (Fig. 290a); the tangent to the curve either forms an acute angle

(¹) I.e., at the points which are not end points of the interval.

(²) This condition is valid for functions monotonely increasing in the wider sense (see p. 326). In order that the function increase or decrease strictly monotonely, another condition should be added: that the derivative $f'(x)$ does not identically vanish in any interval contained in the domain of existence. This condition is not fulfilled, for example, on the segment BC in Fig. 290b.

with the positive direction of the x axis or is parallel to it. A decreasing function has a graph which does not rise anywhere (Fig. 290b) ([1]).

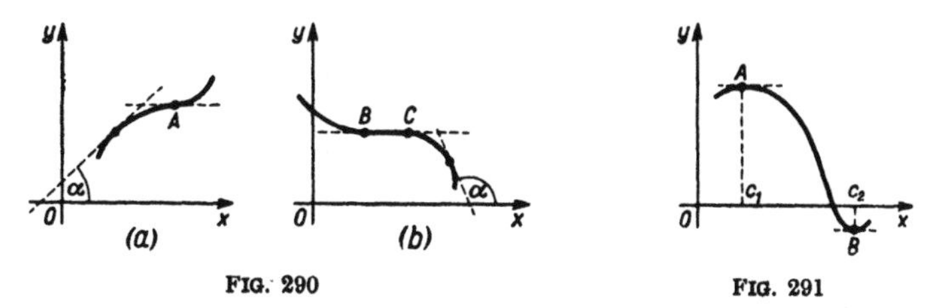

FIG. 290

FIG. 291

Fermat's theorem. If a function $y = f(x)$ defined in a connected domain assumes a greatest or a least value at an interior ([2]) point $x = c$, i.e.,

$$f(c) < f(x) \quad \text{or} \quad f(c) > f(x),$$

and possesses a finite derivative at the points c, then this derivative is zero:

$$f'(c) = 0.$$

Geometric significance. At the points A and B of the graph of the function satisfying the condition of the theorem, the tangent is parallel to the x axis (Fig. 291).

Fermat's theorem gives only a necessary condition for a greatest and a least value of the function. It is not sufficient: at the point A in Fig. 290a, $f'(x) = 0$, but the function does not assume a greatest or least value at this point.

The assumption of Fermat's theorem that the derivative is finite is essential: the function in Fig. 292d assumes a greatest value at the point E, but the derivative is not zero.

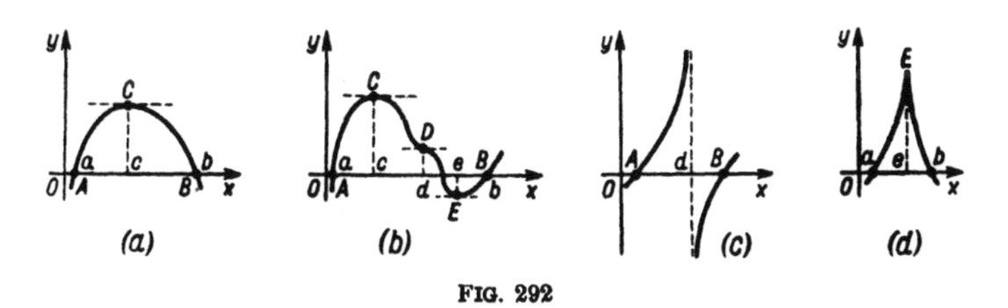

FIG. 292

([1]) In the case of a strictly monotone function, the tangent can be parallel to the x axis only at isolated points (e.g., the point A, Fig. 290a), but not in a whole interval (BC in Fig. 290b).

([2]) I.e., a point which is not an end-point of the interval.

Rolle's theorem. If a function $y = f(x)$ is continuous in the closed interval $[a, b]$, has a continuous derivative in this interval and vanishes at the end points of the interval:

$$f(a) = 0, \quad f(b) = 0 \quad (a < b),$$

then there exists at least one point c between a and b such that

$$f'(c) = 0, \quad (a < c < b).$$

Geometric significance. If the graph of the function $y = f(x)$ is a continuous curve whose tangent varies continuously and if the curve intersects the x axis at the points A and B, then there exists at least one point C between A and B such that the tangent at C is parallel to the x axis (Fig. 292a).

There can be more such points (the points C, D, E in Fig. 292b). The assumptions that the function and its derivative are continuous is essential: the function in Fig. 292c has a discontinuity at $x=d$; the derivative of the function in Fig. 292d has a discontinuity at $x = e$. In either case, a point C with $f'(x) = 0$ does not exist.

Lagrange's (mean value) theorem. If a function $y = f(x)$ is continuous in the closed interval $[a, b]$ and has a continuous derivative in this interval, then there exists at least one point c between a and b such that

$$\frac{f(b) - f(a)}{b - a} = f'(c) \quad (a < c < b).$$

This can be rewritten (by putting $b = a + h$ and denoting by θ a certain number contained between 0 and 1) as follows

$$f(a + h) = f(a) + hf'(a + \theta h) \quad (0 < \theta < 1).$$

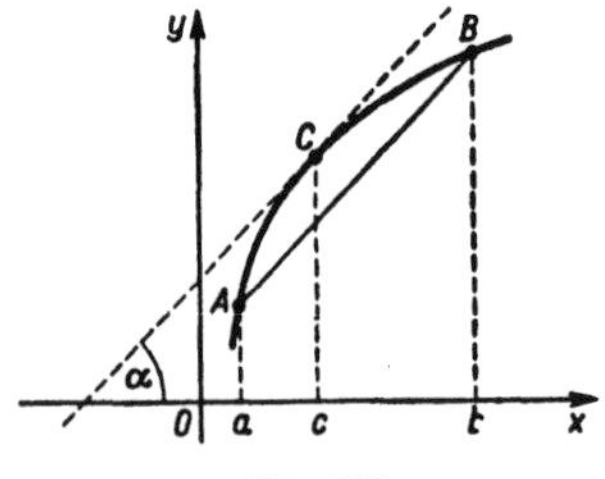

FIG. 293

Geometric significance. If the tangent line to a continuou curve $y = f(x)$ (Fig. 293) varies continuously in the interval AB then there exists a point C between A and B such that the tangen at C is parallel to the chord AB.

There can be more points with this property; the assumption that the function and its derivative are continuous is essential (examples analogous to those in Fig. 292b, c, d can be easily constructed).

Taylor's theorem (a generalization of Lagrange's theorem). If the function $y = f(x)$ is continuous in the interval $[a, a+h]$ [1] and has continuous derivatives from the first up to the n-th inclusively, then we have Taylor's formula

$$f(a+h) = f(a) + \frac{h}{1!}f'(a) + \frac{h^2}{2!}f''(a) + \ldots + \\ + \frac{h^{(n-1)}}{(n-1)!}f^{(n-1)}(a) + \frac{h^n}{n!}f^{(n)}(a+\theta h),$$

where θ is a certain number contained between 0 and 1 $(0 < \theta < 1)$.

Cauchy's theorem. If two functions $y = f(x)$ and $y = \varphi(x)$ defined in a closed interval $[a, b]$ are continuous and have continuous derivatives in this interval and, in addition, $\varphi'(x)$ does not vanish in the interval, then there exists a number c between a and b such that

$$\frac{f(b)-f(a)}{\varphi(b)-\varphi(a)} = \frac{f'(c)}{\varphi'(c)} \quad (a < c < b).$$

Geometric significance of Cauchy's theorem is similar to that of Lagrange's theorem: if the curve in Fig. 293 is considered as given in parametric form $x = \varphi(t)$, $y = f(t)$, where the point A corresponds to $t = a$ and the point B to $t = b$, then, for the point C,

$$\tan\alpha = \frac{f(b)-f(a)}{\varphi(b)-\varphi(a)} = \frac{f'(c)}{\varphi'(c)}.$$

5. Finding maxima and minima

Function of one variable. Definition. A *maximum* (M) or a *minimum* (m) [2] of a function $y = f(x)$ is its value $f(x_0)$ such that the following inequalities hold

$$f(x_0 + h) < f(x_0) \quad \text{(for a maximum)},$$

---

[1] h can be here positive as well as negative.

[2] In mathematical analysis maxima and minima have a common name of "turning values" or "extrema".

and

$$f(x_0 + h) > f(x_0) \quad \text{(for a minimum)},$$

for arbitrary small values of h, positive or negative. Thus the value $f(x_0)$ at a maximum (minimum) is greater (less) than all neighbouring values of the function.

A necessary condition for a maximum or minimum of a continuous function. A maximum or minimum of a continuous function can occur only at the points where the derivative either is equal to zero or does not exists (in particular, is infinite).

Geometric significance. At points of the graph corresponding to a maximum or a minimum, the tangent either is parallel to the x axis (Fig. 294a) or is parallel to the y axis (Fig. 294b) or does not exist (Fig. 294c).

This condition is not sufficient: it is satisfied at the points

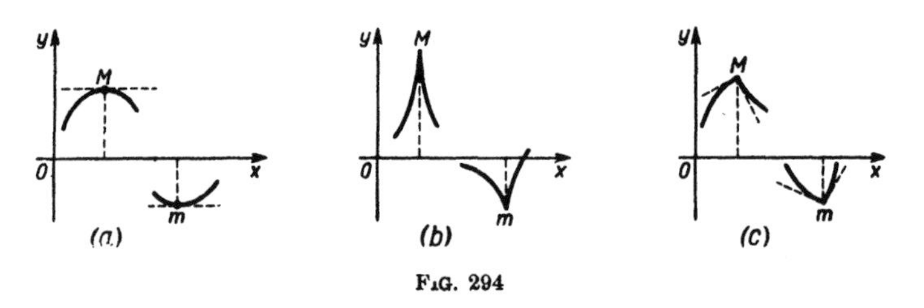

FIG. 294

A, B, C in Fig. 295, but these points are not points of maximum or minimum.

Maxima and minima of a continuous function alternate: a minimum lies between two successive maxima and a maximum between two minima.

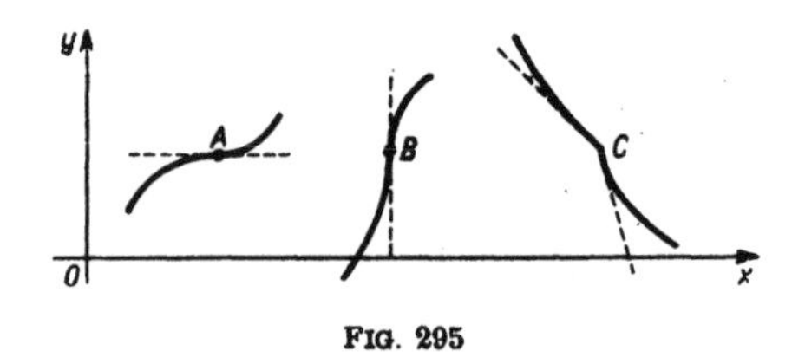

FIG. 295

Finding maxima and minima of a continuous function given in explicit form $y = f(x)$ with a continuous derivative. First we find the points satisfying the condition $f'(x) = 0$

(*stationary points*): we compute the derivative and find all real roots $x_1, x_2, \ldots, x_n$ of the equation $f'(x) = 0$. Then we examine each root, for example, x_1, in one of the following ways:

(1) *Method of comparison of signs*. We determine the sign of $f'(x)$ for $\tilde{x}$ a little less and $\tilde{\tilde{x}}$ a little greater than x_1 (precisely: for the values $\tilde{x}$ and $\tilde{\tilde{x}}$ lying on both sides of x_1 so that there are no roots of the equation $f'(x) = 0$ between $\tilde{x}$ and x_1 or between x_1 and $\tilde{\tilde{x}}$). If the derivative $f'(x)$ changes sign from "+" to "−" (Fig. 296a), then there is a maximum of $f(x)$ at $x = x_1$; if the sign

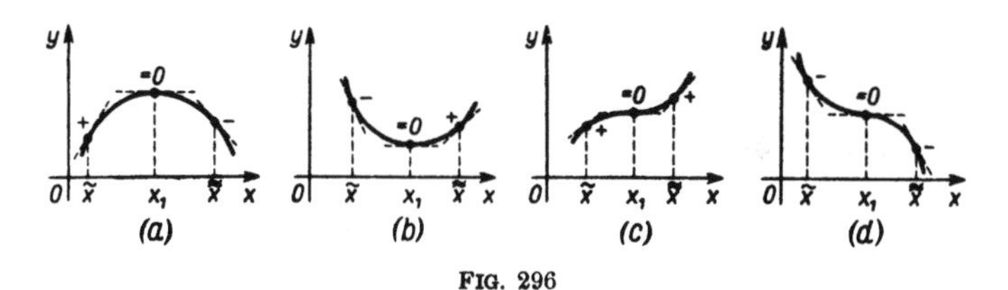

FIG. 296

of $f'(x)$ changes from "−" to "+" (Fig. 296b), then the function has a minimum at $x = x_1$. If the derivative does not change sign (Fig. 296c and d), then there is neither a maximum nor a minimum at $x = x_1$ and we have a point of inflection with the tangent parallel to the x axis.

(2) *Method of higher derivatives*. This method can be applied, if the function has higher derivatives at $x = x_1$. We substitute each root x_1 to the second derivative. If $f''(x_1) < 0$, then the function has a maximum; if $f''(x_1) > 0$, then the function has a minimum. If $f''(x_1) = 0$, then we substitute x_1 to the third derivative. If, in this case, $f'''(x_1) \neq 0$, then the function has neither a maximum nor a minimum at $x = x_1$ (a point of inflection). If $f'''(x_1) = 0$, we find the fourth derivative an so on.

General rule. If the lowest order of a derivative non-vanishing at the point $x = x_1$ is even, then the function $f(x)$ has a maximum or a minimum at $x = x_1$ according as this derivative is negative or positive. If the order of the first non-vanishing derivative is odd, then there is neither a maximum nor a minimum at $x = x_1$.

The method of comparison of signs can also be applied for those values of the function where the derivative does not exist (see Fig. 294b and c and Fig. 295).

To find the greatest and the least value of the function in a given interval $a < x < b$, we determine all the maxima and minima lying inside the interval and also the values of the function at the end points of the interval, at points of discontinuity of the function and of its derivative. Then we find the greatest and the least of these values.

Examples of determining the greatest and the least values.

(a) $y = e^{-x^2}$ in the interval $[-1, +1]$. The greatest value at $x = 0$ (a maximum, Fig. 297a).

(b) $y = x^3 - x^2$ in the interval $[-1, +2]$. The greatest value at $x = +2$ (the right end point of the interval, Fig. 297b).

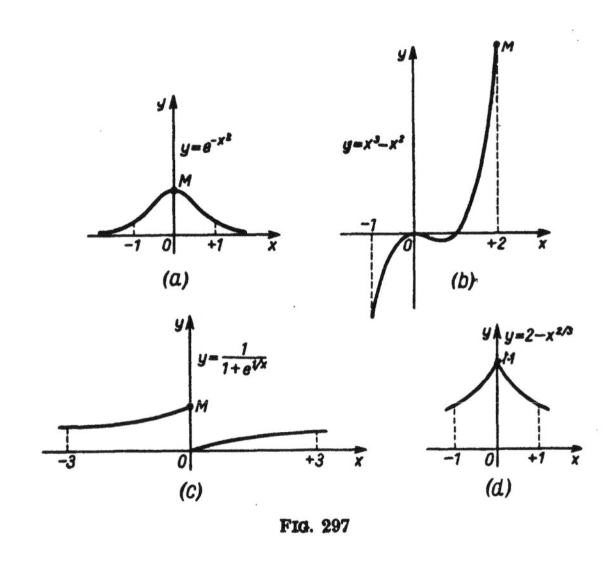

FIG. 297

(c) $y = \dfrac{1}{1 + e^{1/x}}$ in the interval $[-3, +3]$. The greatest value at $x = 0$ (a discontinuity of the function, if we assume $y = 1$ for $x = 0$; Fig. 297c).

(d) $y = 2 - x^{2/3}$ in the interval $[-1, +1]$. The greatest value at $x = 0$ (a maximum, the derivative is infinite, Fig. 297d).

Finding maxima and minima of a function given in implicit form. To find maxima and minima of a function $y = f(x)$ given in implicit form by the equation $F(x, y) = 0$ where F, F'_x and F'_y are continuous, we proceed as follows. We solve the system of equations $F(x, y) = 0$ and $F'_x(x, y) = 0$ and substitute the obtained solutions $(x_1, y_1), (x_2, y_2), \ldots$ to F'_y and F''_{xx}.

If, at the point (x_i, y_i), F'_y and F''_{xx} have different signs, then the function $y = f(x)$ has a minimum at $x = x_i$; if F'_y and F''_{xx} are of the same sign, then the function $y = f(x)$ has a maximum; if one of the expressions F'_y or F''_{xx} is zero, then further examining is more complicated.

Functions of several variables. Definition. A function $u = f(x, y, \ldots, t)$ has a maximum or a minimum for the system $x_0, y_0, \ldots, t_0$ ("at the point $P_0(x_0, y_0, \ldots, t_0)$") if there exists a number ε such that the domain

$$x_0 - \varepsilon < x < x_0 + \varepsilon, \quad y_0 - \varepsilon < y < y_0 + \varepsilon, \quad \ldots, \quad t_0 - \varepsilon < t < t_0 + \varepsilon$$

is contained in the domain of definition of the function and, for every point $(x, y, \ldots, t)$ of this domain except for the point $(x_0, y_0, \ldots, t_0)$, the following inequalities hold:

$$f(x, y, \ldots, t) < f(x_0, y_0, \ldots, t_0) \quad \text{(for a maximum)},$$

and

$$f(x, y, \ldots, t) > f(x_0, y_0, \ldots, t_0) \quad \text{(for a minimum)}.$$

Using the concept of multi-dimensional space (see p. 340), we can say that the function u assumes at a point of maximum (or minimum) a value greater (respectively, less) than the values in neighbouring points.

Geometric significance of a maximum or minimum of a function of two variables represented by a surface in Cartesian coordinates (see p. 262). At a point A of maximum (or minimum), the third coordinate u is greater (or less) than the coordinate u of any point sufficiently near to A. See Fig. 298: (a) maximum, (b) minimum.

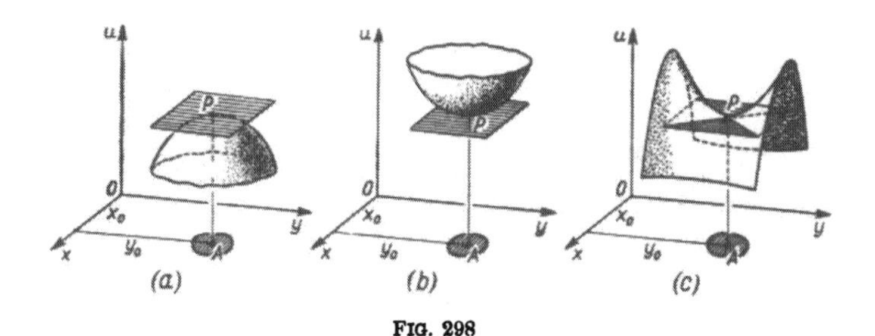

FIG. 298

If the surface has a tangent plane at a point P of maximum or minimum, then this plane is parallel to the xy plane (Fig. 298a, b). This condition is necessary for a maximum or a minimum, but not sufficient: the surface in Fig. 298c has a horizontal tangent plane at P, but neither a maximum nor a minimum (P is a saddle point).

Finding maxima and minima of a function $u = f(x, y)$ of two variables. We solve the system of equations

$$f'_x = 0, \quad f'_y = 0$$

and substitute the obtained solutions (x_1, y_1), (x_2, y_2), ... for the second partial derivatives $A = \frac{\partial^2 f}{\partial x^2}$, $B = \frac{\partial^2 f}{\partial x \partial y}$, $C = \frac{\partial^2 f}{\partial y^2}$. We form the expression

$$\Delta = \begin{vmatrix} A & B \\ B & C \end{vmatrix} = AC - B^2 = [f''_{xx} f''_{yy} - (f''_{xy})^2]_{x=x_1, y=y_1}.$$

If $\Delta > 0$, then the function $f(x, y)$ has a maximum at the point (x_1, y_1), when $f''_{xx} < 0$, and a minimum, when $f''_{xx} > 0$. If $\Delta < 0$, then the function has neither a maximum nor a minimum. If $\Delta = 0$, then examining is more complicated.

Finding maxima and minima of functions of n variables. A necessary, but not sufficient condition for a maximum or minimum of a differentiable function $u = f(x, y, \ldots, t)$ of n variables at a point $(x, y, \ldots, t)$ is the simultaneous vanishing of all first partial derivatives at this point:

$$\text{(A)} \qquad f'_x = 0, \quad f'_y = 0, \quad \ldots, \quad f'_t = 0.$$

Sufficient conditions are, in the general case, more complicated. Practically, to determine, whether a solution $x_1, y_1, \ldots, t_1$ of equations (A) is a maximum or a minimum, we examine the function at the points near to $(x_1, y_1, \ldots, t_1)$.

Relative maxima and minima (Lagrange's method). To find maxima and minima of a function $u = f(x, y, \ldots, t)$ of n variables which are not independent but connected by k additional conditions ($k < n$):

$$\varphi(x, y, \ldots, t) = 0, \quad \psi(x, y, \ldots, t) = 0, \quad \ldots, \quad \chi(x, y, \ldots, t) = 0,$$

we introduce undetermined coefficients $\lambda, \mu, \ldots, \varkappa$ and consider the following function of $n + k$ variables $x, y, \ldots, t, \lambda, \mu, \ldots, \varkappa$:

$$F(x, y, \ldots, t, \lambda, \mu, \ldots, \varkappa)$$
$$\equiv f(x, y, \ldots, t) + \lambda\varphi(x, y, \ldots, t) + \mu\psi(x, y, \ldots, t) + \ldots + \varkappa\chi(x, y, \ldots, t).$$

A necessary condition for a maximum or minimum of the function F is the system (A) of $n + k$ equations for the function F with unknowns $x, y, \ldots, t, \lambda, \mu, \ldots, \varkappa$:

$$\varphi = 0, \quad \psi = 0, \quad \ldots, \quad \chi = 0, \quad F'_x = 0, \quad F'_y = 0, \quad \ldots, \quad F'_t = 0.$$

A system $(x_1, y_1, \ldots, t_1)$ satisfying these equations can give a maximum or a minimum for the function f. This condition is necessary, but not sufficient.

For example, for a function $u = f(x, y)$, if $\varphi(x, y) = 0$, a point of maximum or minimum can be determined from three equations

$$\varphi(x, y) = 0, \qquad \frac{\partial}{\partial x}\big(f(x, y) + \lambda\varphi(x, y)\big) = 0,$$

$$\frac{\partial}{\partial y}\big(f(x, y) + \lambda\varphi(x, y)\big) = 0$$

with the unknowns x, y, λ.

6. Expansion of a function into a power series

Taylor's series for a function of one variable. A function $y = f(x)$, continuous and having all derivatives at $x = a$, can be, in many cases, expressed as a power series (see p. 357) obtained from Taylor's formula (see p. 379):

$$\text{(T)} \quad f(x) = f(a) + \frac{x-a}{1!}f'(a) + \frac{(x-a)^2}{2!}f''(a) + \ldots + \frac{(x-a)^n}{n!}f^{(n)}(a) + \ldots$$

(*Taylor's series*).

Formula (T) holds for each value of x for which the remainder [1] $f(x) - S_n = R_n$ tends to zero, when $n \to \infty$.

Expressions for the remainder:

$$R_n = \frac{(x-a)^{n+1}}{(n+1)!}f^{(n+1)}(\xi) \qquad (\xi \text{ lies between } a \text{ and } x),$$

$$R_n = \frac{1}{n!}\int_a^x (x-t)^n f^{(n+1)}(t)\, dt.$$

Another form of Taylor's series:

$$f(a+h) = f(a) + \frac{h}{1!}f'(a) + \frac{h^2}{2!}f''(a) + \ldots + \frac{h^n}{n!}f^{(n)}(a) + \ldots,$$

---

[1] This concept of the remainder of a Taylor series does not coincide with that of the remainder of a series as introduced on p. 355. Both concepts are the same only in cases for which the formula (T) is true.

where the remainder

$$R_n = \frac{h^{n+1}}{(n+1)!} f^{(n+1)}(a+\theta h) = \frac{1}{n!} \int_0^h (h-t)^n f^{(n+1)}(a+t)\, dt.$$

Maclaurin's series—a formula for expansion of the function $f(x)$ in powers of x; it is obtained from Taylor's series for $a = 0$:

$$\text{(M)} \qquad f(x) = f(0) + \frac{x}{1!} f'(0) + \frac{x^2}{2!} f''(0) + \ldots + \frac{x^n}{n!} f^{(n)}(0) + \ldots$$

The *remainder:*

$$R_n = \frac{x^{n+1}}{(n+1)!} f^{(n+1)}(\theta x) = \frac{1}{n!} \int_0^x (x-t)^n f^{(n+1)}(t)\, dt.$$

$(0 < \theta < 1)$.

The convergence of Taylor's or Maclaurin's series is verified by examining its remainder or by calculating its radius of convergence (p. 357); in the latter case, the series may turn out to converge, without having its sum $S(x)$ equal to $f(x)$.

Taylor's series for a function of two variables:

$$f(x+h, y+k) = f(x, y) + \left\{\frac{\partial f(x, y)}{\partial x} h + \frac{\partial f(x, y)}{\partial y} k\right\} +$$

$$+ \frac{1}{2}\left\{\frac{\partial^2 f(x, y)}{\partial x^2} h^2 + 2\frac{\partial^2 f(x, y)}{\partial x \partial y} hk + \frac{\partial^2 f(x, y)}{\partial y^2} k^2\right\} +$$

$$+ \frac{1}{6}\left\{\ldots\right\} + \ldots + \frac{1}{n!}\left\{\ldots\right\} + R_n$$

or (in the symbolic form)

$$f(x+h, y+k) = f(x, y) + \frac{1}{1!}\left(\frac{\partial}{\partial x} h + \frac{\partial}{\partial y} k\right) f(x, y) +$$

$$+ \frac{1}{2!}\left(\frac{\partial}{\partial x} h + \frac{\partial}{\partial y} k\right)^2 f(x, y) + \frac{1}{3!}\left(\frac{\partial}{\partial x} h + \frac{\partial}{\partial y} k\right)^3 f(x, y) + \ldots$$

$$+ \frac{1}{n!}\left(\frac{\partial}{\partial x} h + \frac{\partial}{\partial y} k\right)^n f(x, y) + R_n,$$

where

$$R_n = \frac{1}{(n+1)!}\left(\frac{\partial}{\partial x} h + \frac{\partial}{\partial y} k\right)^{n+1} f(x + \theta_1 h, y + \theta_2 k) \quad \begin{pmatrix} 0 < \theta_1 < 1 \\ 0 < \theta_2 < 1 \end{pmatrix}.$$

Taylor's series for functions of several variables can be written in an analogous form:

$$f(x+h, y+k, \ldots, t+l)$$

$$= f(x, y, \ldots, t) + \sum_{i=1}^{n} \frac{1}{i!}\left(\frac{\partial}{\partial x}h + \frac{\partial}{\partial y}k + \ldots + \frac{\partial}{\partial t}l\right)^{i} f(x, y, \ldots, t) + R_n,$$

where

$$R_n = \frac{1}{(n+1)!}\left(\frac{\partial}{\partial x}h + \frac{\partial}{\partial y}k + \ldots + \frac{\partial}{\partial t}l\right)^{n+1} f(x+\theta_1 h, y+\theta_2 k, \ldots, t+\theta_m l)$$

$$(0 < \theta_i < 1).$$

Table of expansion of certain functions into power series

Function	Expansion into series	Domain of convergence
	Algebraic functions	
	Binomial series	
$(a \pm x)^m$	By a transformation of the binomial to the form $a^m(1 \pm x/a)^m$ we reduce it to the following series:	$\|x\| \leqslant a$, when $m > 0$; $\|x\| < a$, when $m < 0$
	Binomial series with a positive exponent ([1])	
$(1 \pm x)^m$ $(m > 0)$	$1 \pm mx + \frac{m(m-1)}{2!}x^2 \pm \frac{m(m-1)(m-2)}{3!}x^3 + \ldots$	
	$\ldots + (\pm 1)^n \frac{m(m-1)\ldots(m-n+1)}{n!}x^n + \ldots$	$\|x\| \leqslant 1$
$(1 \pm x)^{1/4}$	$1 \pm \frac{1}{4}x - \frac{1\cdot 3}{4\cdot 8}x^2 \pm \frac{1\cdot 3\cdot 7}{4\cdot 8\cdot 12}x^3 - \frac{1\cdot 3\cdot 7\cdot 11}{4\cdot 8\cdot 12\cdot 16}x^4 \pm \ldots$	$\|x\| \leqslant 1$
$(1 \pm x)^{1/3}$	$1 \pm \frac{1}{3}x - \frac{1\cdot 2}{3\cdot 6}x^2 \pm \frac{1\cdot 2\cdot 5}{3\cdot 6\cdot 9}x^3 - \frac{1\cdot 2\cdot 5\cdot 8}{3\cdot 6\cdot 9\cdot 12}x^4 \pm \ldots$	$\|x\| \leqslant 1$
$(1 \pm x)^{1/2}$	$1 \pm \frac{1}{2}x - \frac{1\cdot 1}{2\cdot 4}x^2 \pm \frac{1\cdot 1\cdot 3}{2\cdot 4\cdot 6}x^3 - \frac{1\cdot 1\cdot 3\cdot 5}{2\cdot 4\cdot 6\cdot 8}x^4 \pm \ldots$	$\|x\| \leqslant 1$
$(1 \pm x)^{3/2}$	$1 \pm \frac{3}{2}x + \frac{3\cdot 1}{2\cdot 4}x^2 \mp \frac{3\cdot 1\cdot 1}{2\cdot 4\cdot 6}x^3 + \frac{3\cdot 1\cdot 1\cdot 3}{2\cdot 4\cdot 6\cdot 8}x^4 \mp \ldots$	$\|x\| \leqslant 1$
$(1 \pm x)^{5/2}$	$1 \pm \frac{5}{2}x + \frac{5\cdot 3}{2\cdot 4}x^2 \pm \frac{5\cdot 3\cdot 1}{2\cdot 4\cdot 6}x^3 - \frac{5\cdot 3\cdot 1\cdot 1}{2\cdot 4\cdot 6\cdot 8}x^4 \mp \ldots$	$\|x\| \leqslant 1$

([1]) If m is a natural number, the series is finite and has $m+1$ terms with coefficients of the form $\frac{m(m-1)\cdots(m-n+1)}{n!} = \binom{m}{n}$; a table of binomial coefficients is given on p. 193.

Table of expansion of certain functions into power series (cont.)

Function	Expansion into series	Domain of convergence
	Binomial series with a negative exponent	
$(1 \pm x)^{-m}$ $(m > 0)$	$1 \mp mx + \frac{m(m+1)}{2!}x^2 \mp \frac{m(m+1)(m+2)}{3!}x^3 + \ldots$	
	$\ldots + (\pm 1)^n \frac{m(m+1)\ldots(m+n-1)}{n!}x^n \pm \ldots$	$\lvert x\rvert < 1$
$(1 \pm x)^{-1/4}$	$1 \mp \frac{1}{4}x + \frac{1\cdot 5}{4\cdot 8}x^2 \mp \frac{1\cdot 5\cdot 9}{4\cdot 8\cdot 12}x^3 + \frac{1\cdot 5\cdot 9\cdot 13}{4\cdot 8\cdot 12\cdot 16}x^4 \mp \ldots$	$\lvert x\rvert < 1$
$(1 \pm x)^{-1/3}$	$1 \mp \frac{1}{3}x + \frac{1\cdot 4}{3\cdot 6}x^2 \mp \frac{1\cdot 4\cdot 7}{3\cdot 6\cdot 9}x^3 + \frac{1\cdot 4\cdot 7\cdot 10}{3\cdot 6\cdot 9\cdot 12}x^4 \mp \ldots$	$\lvert x\rvert < 1$
$(1 \pm x)^{-1/2}$	$1 \mp \frac{1}{2}x + \frac{1\cdot 3}{2\cdot 4}x^2 \mp \frac{1\cdot 3\cdot 5}{2\cdot 4\cdot 6}x^3 + \frac{1\cdot 3\cdot 5\cdot 7}{2\cdot 4\cdot 6\cdot 8}x^4 \mp \ldots$	$\lvert x\rvert < 1$
$(1 \pm x)^{-1}$	$1 \mp x + x^2 \mp x^3 + x^4 \mp \ldots$	$\lvert x\rvert < 1$
$(1 \pm x)^{-3/2}$	$1 \mp \frac{3}{2}x + \frac{3\cdot 5}{2\cdot 4}x^2 \mp \frac{3\cdot 5\cdot 7}{2\cdot 4\cdot 6}x^3 + \frac{3\cdot 5\cdot 7\cdot 9}{2\cdot 4\cdot 6\cdot 8}x^4 \mp \ldots$	$\lvert x\rvert < 1$
$(1 \pm x)^{-2}$	$1 \mp 2x + 3x^2 \mp 4x^3 + 5x^4 \mp \ldots$	$\lvert x\rvert < 1$
$(1 \pm x)^{-5/2}$	$1 \mp \frac{5}{2}x + \frac{5\cdot 7}{2\cdot 4}x^2 \mp \frac{5\cdot 7\cdot 9}{2\cdot 4\cdot 6}x^3 + \frac{5\cdot 7\cdot 9\cdot 11}{2\cdot 4\cdot 6\cdot 8}x^4 \mp \ldots$	$\lvert x\rvert < 1$
$(1 \pm x)^{-3}$	$1 \mp \frac{1}{1\cdot 2}(2\cdot 3x \mp 3\cdot 4x^2 + 4\cdot 5x^3 \mp 5\cdot 6x^4 + \ldots)$	$\lvert x\rvert < 1$
$(1 \pm x)^{-4}$	$1 \mp \frac{1}{1\cdot 2\cdot 3}(2\cdot 3\cdot 4x \mp 3\cdot 4\cdot 5x^2 +$	
	$+ 4\cdot 5\cdot 6x^3 \mp 5\cdot 6\cdot 7x^4 + \ldots)$	$\lvert x\rvert < 1$
$(1 \pm x)^{-5}$	$1 \mp \frac{1}{1\cdot 2\cdot 3\cdot 4}(2\cdot 3\cdot 4\cdot 5x \mp 3\cdot 4\cdot 5\cdot 6x^2 +$	
	$+ 4\cdot 5\cdot 6\cdot 7x^3 \mp 5\cdot 6\cdot 7\cdot 8x^4 + \ldots)$	$\lvert x\rvert < 1$
	Trigonometric functions	
$\sin x$	$x - \frac{x^3}{3!} + \frac{x^5}{5!} - \ldots + (-1)^n \frac{x^{2n+1}}{(2n+1)!} \pm \ldots$	$\lvert x\rvert < \infty$
$\sin(x + a)$	$\sin a + x\cos a - \frac{x^2 \sin a}{2!} - \frac{x^3 \cos a}{3!} +$	
	$+ \frac{x^4 \sin a}{4!} + \ldots + \frac{x^n \sin\left(a + \frac{n\pi}{2}\right)}{n!} \pm \ldots$	$\lvert x\rvert < \infty$
$\cos x$	$1 - \frac{x^2}{2!} + \frac{x^4}{4!} - \frac{x^6}{6!} + \ldots + (-1)^n \frac{x^{2n}}{(2n)!} \pm \ldots$	$\lvert x\rvert < \infty$
$\cos(x + a)$	$\cos a - x\sin a - \frac{x^2 \cos a}{2!} + \frac{x^3 \sin a}{3!} +$	
	$+ \frac{x^4 \cos a}{4!} - \ldots + \frac{x^n \cos\left(a + \frac{n\pi}{2}\right)}{n!} \pm \ldots$	$\lvert x\rvert < \infty$

Table of expansion of certain functions into power serier (cont.)

Function	Expansion into series	Domain of convergence
$\tan x$	$x+\frac{1}{3}x^3+\frac{2}{15}x^5+\frac{17}{315}x^7+\frac{62}{2835}x^9+\ldots+\frac{2^{2n}(2^{2n}-1)B_n}{(2n)!}x^{2n-1}+\ldots$ [1]	$\lvert x\rvert<\frac{\pi}{2}$
$\cot x$	$\frac{1}{x}-\left[\frac{x}{3}+\frac{x^3}{45}+\frac{2x^5}{945}+\frac{x^7}{4725}+\ldots+\frac{2^{2n}B_n}{(2n)!}x^{2n-1}+\ldots\right]$ [1]	$0<\lvert x\rvert<\pi$
$\sec x$	$1+\frac{1}{2}x^2+\frac{5}{24}x^4+\frac{61}{720}x^6+\frac{277}{8064}x^8+\ldots+\frac{E_n}{(2n)!}x^{2n}+\ldots$ [2]	$\lvert x\rvert<\frac{\pi}{2}$
$\operatorname{cosec} x$	$\frac{1}{x}+\frac{1}{6}x+\frac{7}{360}x^3+\frac{31}{15120}x^5+\frac{127}{604800}x^7+\ldots+\frac{2(2^{2n-1}-1)}{(2n)!}B_n x^{2n-1}$ [1]	$0<\lvert x\rvert<\pi$
	Exponential functions	
e^x	$1+\frac{x}{1!}+\frac{x^2}{2!}+\frac{x^3}{3!}+\ldots+\frac{x^n}{n!}+\ldots$	$\lvert x\rvert<\infty$
$a^x=e^{x\ln a}$	$1+\frac{x\ln a}{1!}+\frac{(x\ln a)^2}{2!}+\frac{(x\ln a)^3}{3!}+\ldots+\frac{(x\ln a)^n}{n!}+\ldots$	$\lvert x\rvert<\infty$
$\frac{x}{e^x-1}$	$1-\frac{x}{2}+\frac{B_1x^2}{2!}-\frac{B_2x^4}{4!}+\frac{B_3x^6}{6!}-\ldots+(-1)^{n+1}\frac{B_nx^{2n}}{(2n)!}\pm\ldots$ [1]	$\lvert x\rvert<2\pi$
	Logarithmic functions	
$\ln x$	$2\left[\frac{x-1}{x+1}+\frac{(x-1)^3}{3(x+1)^3}+\frac{(x-1)^5}{5(x+1)^5}+\ldots+\frac{(x-1)^{2n+1}}{(2n+1)(x+1)^{2n+1}}+\ldots\right]$	$x>0$
$\ln x$	$(x-1)-\frac{(x-1)^2}{2}+\frac{(x-1)^3}{3}-\frac{(x-1)^4}{4}+\ldots+(-1)^{n+1}\frac{(x-1)^n}{n}\pm\ldots$	$0<x\leqslant 2$
$\ln x$	$\frac{x-1}{x}+\frac{(x-1)^2}{2x^2}+\frac{(x-1)^3}{3x^3}+\ldots+\frac{(x-1)^n}{nx^n}+\ldots$	$x>\frac{1}{2}$

[1] B_n denotes Bernoulli's numbers (see p. 354).
[2] E_n denotes Euler's numbers (see p. 354).

Table of expansion of certain functions into power series (cont.)

Function	Expansion into series	Domain of convergence
$\ln(1+x)$	$x-\frac{x^2}{2}+\frac{x^3}{3}-\frac{x^4}{4}+\ldots+(-1)^{n+1}\frac{x^n}{n}\pm\ldots$	$-1<x\leqslant 1$
$\ln(1-x)$	$-\left[x+\frac{x^2}{2}+\frac{x^3}{3}+\frac{x^4}{4}+\frac{x^5}{5}+\ldots+\frac{x^n}{n}+\ldots\right]$	$-1\leqslant x<1$
$\ln\left(\frac{1+x}{1-x}\right)$ $=2\operatorname{ar}\tanh x$	$2\left[x+\frac{x^3}{3}+\frac{x^5}{5}+\frac{x^7}{7}+\ldots+\frac{x^{2n+1}}{2n+1}+\ldots\right]$	$\|x\|<1$
$\ln\left(\frac{x+1}{x-1}\right)$ $=2\operatorname{ar}\coth x$	$2\left[\frac{1}{x}+\frac{1}{3x^3}+\frac{1}{5x^5}+\frac{1}{7x^7}+\ldots+\right.$ $\left.+\frac{1}{(2n+1)x^{2n+1}}+\ldots\right]$	$\|x\|>1$
$\ln\|\sin x\|$	$\ln\|x\|-\frac{x^2}{6}-\frac{x^4}{180}-\frac{x^6}{2835}-\ldots-$ $-\frac{2^{2n-1}B_n x^{2n}}{n(2n)!}-\ldots$ [1]	$0<\|x\|<\pi$
$\ln\cos x$	$-\frac{x^2}{2}-\frac{x^4}{12}-\frac{x^6}{45}-\frac{17x^8}{2520}-\ldots-$ $-\frac{2^{2n-1}(2^{2n}-1)B_n x^{2n}}{n(2n)!}-\ldots$ [1]	$\|x\|<\frac{\pi}{2}$
$\ln\|\tan x\|$	$\ln\|x\|+\frac{1}{3}x^2+\frac{7}{90}x^4+\frac{62}{2835}x^6+\ldots+$ $+\frac{2^{2n}(2^{2n-1}-1)B_n}{n(2n)!}x^{2n}+\ldots$ [1]	$0<\|x\|<\frac{\pi}{2}$
	Inverse trigonometric functions	
$\arcsin x$	$x+\frac{x^3}{2\cdot 3}+\frac{1\cdot 3x^5}{2\cdot 4\cdot 5}+\frac{1\cdot 3\cdot 5x^7}{2\cdot 4\cdot 6\cdot 7}+\ldots+$ $+\frac{1\cdot 3\cdot 5\ldots(2n-1)x^{2n+1}}{2\cdot 4\cdot 6\ldots(2n)(2n+1)}+\ldots$	$\|x\|<1$
$\arccos x$	$\frac{\pi}{2}-\left[x+\frac{x^3}{2\cdot 3}+\frac{1\cdot 3x^5}{2\cdot 4\cdot 5}+\frac{1\cdot 3\cdot 5x^7}{2\cdot 4\cdot 6\cdot 7}+\ldots+\right.$ $\left.+\frac{1\cdot 3\cdot 5\ldots(2n-1)x^{2n+1}}{2\cdot 4\cdot 6\ldots(2n)(2n+1)}+\ldots\right]$	$\|x\|<1$

[1] B_n denotes Bernoulli's numbers (see p. 354).

Table of expansion of certain functions into power series (cont.)

Function	Expansion into series	Domain of convergence
arc tan x	$x - \frac{x^3}{3} + \frac{x^5}{5} - \frac{x^7}{7} + \ldots + (-1)^n \frac{x^{2n+1}}{2n+1} \pm \ldots$	$\|x\| < 1$
	$= \pm \frac{\pi}{2} - \frac{1}{x} + \frac{1}{3x^3} - \frac{1}{5x^5} + \frac{1}{7x^7} - \ldots +$	
	$+ (-1)^{n+1} \cdot \frac{1}{(2n+1)\, x^{2n+1}} \pm \ldots$ (1)	$\|x\| > 1$
arc cot x	$\frac{\pi}{2} - \left[x - \frac{x^3}{3} + \frac{x^5}{5} - \frac{x^7}{7} + \ldots + \right.$	
	$\left. + (-1)^n \frac{x^{2n+1}}{2n+1} \pm \ldots \right]$	$\|x\| < 1$
	Hyperbolic functions	
sinh x	$x + \frac{x^3}{3!} + \frac{x^5}{5!} + \frac{x^7}{7!} + \ldots + \frac{x^{2n+1}}{(2n+1)!} + \ldots$	$\|x\| < \infty$
cosh x	$1 + \frac{x^2}{2!} + \frac{x^4}{4!} + \frac{x^6}{6!} + \ldots + \frac{x^{2n}}{(2n)!} + \ldots$	$\|x\| < \infty$
tanh x	$x - \frac{1}{3} x^3 + \frac{2}{15} x^5 - \frac{17}{315} x^7 + \frac{62}{2835} x^9 - \ldots +$	
	$+ \frac{(-1)^{n+1} 2^{2n} (2^{2n} - 1)}{(2n)!} B_n x^{2n-1} \pm \ldots$ (2)	$\|x\| < \frac{\pi}{2}$
coth x	$\frac{1}{x} + \frac{x}{3} - \frac{x^3}{45} + \frac{2x^5}{945} - \frac{x^7}{4725} + \ldots +$	
	$+ \frac{(-1)^{n+1} 2^{2n}}{(2n)!} B_n x^{2n-1} \pm \ldots$ (2)	$0 < \|x\| < \pi$
sech x	$1 - \frac{1}{2!} x^2 + \frac{5}{4!} x^4 - \frac{61}{6!} x^6 + \frac{1385}{8!} x^8 - \ldots +$	
	$+ \frac{(-1)^n}{(2n)!} E_n x^{2n} \pm \ldots$ (3)	$\|x\| < \frac{\pi}{2}$
cosech x	$\frac{1}{x} - \frac{x}{6} + \frac{7x^3}{360} - \frac{31x^5}{15120} + \ldots +$	
	$+ \frac{2(-1)^n (2^{2n-1} - 1)}{(2n)!} B_n x^{2n-1} + \ldots$ (2)	$0 < \|x\| < \pi$

(1) The first term $\frac{1}{2}\pi$ is taken with the sign "+", for $x > 1$ and with the sign "−", for $x < 1$.

(2) B_n denotes Bernoulli's numbers (see p. 354).

(3) E_n denotes Euler's numbers (see p. 354).

Table of expansion of certain functions into power series (cont.)

Function	Expansion into series	Domain of convergence
	Inverse hyperbolic functions	
ar sinh x	$x-\frac{1}{2\cdot 3}x^3+\frac{1\cdot 3}{2\cdot 4\cdot 5}x^5-\frac{1\cdot 3\cdot 5}{2\cdot 4\cdot 6\cdot 7}x^7+\ldots+$	
	$+(-1)^n\cdot\frac{1\cdot 3\cdot 5\,(2n-1)}{2\cdot 4\cdot 6\ldots 2n(2n+1)}x^{2n+1}\pm\ldots$	$\lvert x\rvert<1$
ar cosh x	$\pm\left[\ln(2x)-\frac{1}{2\cdot 2x^2}-\frac{1\cdot 3}{2\cdot 4\cdot 4x^4}-\frac{1\cdot 3\cdot 5}{2\cdot 4\cdot 6\cdot 6x^6}-\ldots\right]$	$x>1$
ar tanh x	$x+\frac{x^3}{3}+\frac{x^5}{5}+\frac{x^7}{7}+\ldots+\frac{x^{2n+1}}{2n+1}+\ldots$	$\lvert x\rvert<1$
ar coth x	$\frac{1}{x}+\frac{1}{3x^3}+\frac{1}{5x^5}+\frac{1}{7x^7}+\ldots+\frac{1}{(2n+1)\,x^{2n+1}}+\ldots$	$\lvert x\rvert>1$

III. INTEGRAL CALCULUS

A. INDEFINITE INTEGRALS

1. Fundamental concepts and theorems

Primitive function. Given a function $y = f(x)$ of one variable defined in a closed domain, a function $F(x)$ defined in the same domain [1] whose derivative is equal to $f(x)$ (or, that is to say, whose differential is equal to $f(x)dx$) is called a *primitive function* of $f(x)$:

$$F'(x) = f(x) \qquad \text{or} \qquad dF(x) = f(x)dx.$$

A given function $f(x)$ has infinitely many primitive functions; the difference between two primitive functions $F_1(x)$ and $F_2(x)$ of $f(x)$ is constant. The graphs of all primitive functions $F_1(x)$, $F_2(x)$, $F_3(x)$, ... of a given function represent the same curve and can be obtained one from another by a parallel translation along the y axis (Fig. 299).

Geometric significance of a primitive function. If the function $f(x)$ is represented in Cartesian coordinates by a curve

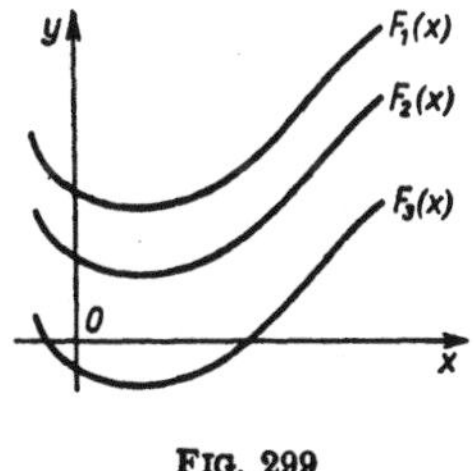

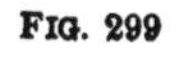

FIG. 299

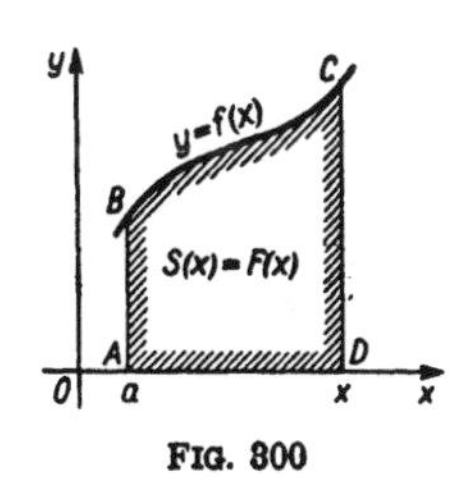

FIG. 300

---

(1) In certain cases the domain of definition of a primitive function $F(x)$ is wider than that of the given function $f(x)$. If the function $f(x)$ is defined in a connected domain except for isolated points of discontinuity $x_1, x_2, \ldots, x_n$, then the domain of definition of its primitive function may contain these points (see p. 394).

(Fig. 300), then the value of its primitive function is equal to the area $S(x)$ bounded by the curve $y = f(x)$, the x axis, a fixed ordinate AB (corresponding to $x = a$) and a variable ordinate CD (corresponding to the abscissa x). By an arbitrary choice of the constant a, we obtain various primitive functions.

The area $S(x)$ should be understood here in the algebraic sense [1].

Theorem on existence of a primitive function. Every function continuous in a given domain possesses a primitive function which is also continuous in this domain. A function discontinuous at certain separate values of x possesses a primitive function which is either continuous or discontinuous at these values of x [2].

Examples.

$$1.\ f(x) = \frac{1}{\sqrt[3]{x^2}},\ F(x) = 3\sqrt[3]{x}.\quad 2.\ f(x) = \frac{1}{x^2},\ F(x) = -\frac{1}{x}.$$

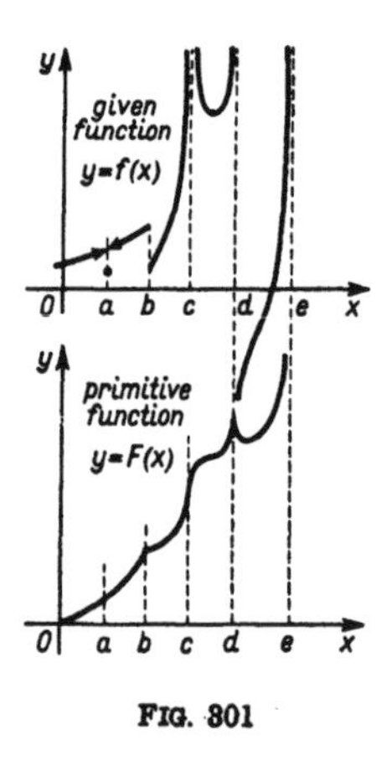

FIG. 301

In both examples, the function $f(x)$ has a discontinuity at $x = 0$; the function $F(x) = 3\sqrt[3]{x}$ is continuous, but $F(x) = -1/x$ has also a discontinuity at $x = 0$.

For the behaviour of the graph of a primitive function $F(x)$ at various points of discontinuity of a given function $f(x)$, see Fig. 301. In the case of a removable discontinuity (a) or a finite discontinuity (b) of $f(x)$, the primitive function is continuous; in the case of an infinite discontinuity of $f(x)$, the primitive function may be continuous (the curve $F(x)$ can have a point of inflection (c) or a point of return (d) with a vertical tangent) or discontinuous (e). For an analytical test, see pp. 474–478.

Indefinite integral. The general expression $F(x) + C$ for all primitive functions of a given function $f(x)$ is called the *indefinite integral* of the function $f(x)$ or of its differential $f(x)dx$. Notation:

$$F(x) + C = \int f(x)dx$$

($\int$—the integral sign, $f(x)$—the integrand).

---

[1] The area of $ABCD = \int_a^x f(x)\,dx$, see p. 455.

[2] See footnote on p. 393.

A definite integral (see p. 457) with an arbitrary fixed lower limit and a variable upper limit can always be taken as a primitive function.

Integrals of elementary functions are not necessarily elementary ones. Examples of evaluation of integrals (examples of the integration) of simple functions leading to elementary functions are given on pp. 395–411. The results of integration are collected the table on pp. 411–453 [1].

If an integral cannot be expressed in terms of elementary functions, then, in cases of theoretical importance or practical application, the values of this function are collected in tables; such functions (for a fixed constant of integration) have often special names. For example,

$$\int_0^x \frac{dx}{\ln x} = \mathrm{Li}\, x \quad \text{(the integral logarithm)},$$

$$\int_0^{\sin\varphi} \frac{dx}{\sqrt{(1-x^2)(1-k^2x^2)}} = F(k, \varphi) \quad \text{(elliptic integral of the 1st kind) [2].}$$

A function whose integral is not expressible in terms of elementary functions or whose integration is very complicated is expanded into a series (see p. 385) and then (when it converges uniformly) integrated term by term. For an approximate integration, the function can be replaced by a polynomial (see pp. 757, 758)[3].

2. General rules of integration

Fundamental integrals. The table on pp. 396, 397 gives formulas of integration obtained as inverse to the corresponding formulas of differentiation (p. 386). We try to reduce a given integral to those simple integrals by means of algebraic or trigonometric transformations or by application of the rules of integration.

Fundamental rules of integration are the properties of indefinite integrals which enable us to transform the integral of a given function to the integrals of other functions:

---

[1] The name "primitive function" will be used in the sequel as a synonym of "integral", but, in the tables, the constant C of integration will be omitted for the sake of brevity.

[2] For tables of elliptic integrals see p. 407.

[3] For the graphical integration (construction of the graph of a primitive function from that of the given function) see pp. 463, 464.

(1) A constant factor can be brought out from under the integral sign:

$$\int af(x)\,dx = a\int f(x)\,dx.$$

(2) Integral of the sum (difference) is equal to the sum (difference) of integrals:

$$\int (u+v-w)dx = \int u\,dx + \int v\,dx - \int w\,dx \text{ (}^1\text{)}.$$

(3) Integration by substitution: if $x = \varphi(t)$, then

$$\int f(x)dx = \int f(\varphi(t))\varphi'(t)dt.$$

(4) Integration "by parts":

$$\int u\,dv = uv - \int v\,du \text{ (}^2\text{)}.$$

Table of fundamental integrals
(the constants of integration are omitted here and in the following tables)

Power functions	Exponential functions
$\int x^n\,dx = \frac{x^{n+1}}{n+1}\quad (n \neq -1),$	$\int e^x\,dx = e^x,$
$\int \frac{dx}{x} = \ln\lvert x\rvert.$	$\int a^x\,dx = \frac{a^x}{\ln a},\ a > 0,\ a \neq 1.$
Trigonometric functions	**Hyperbolic functions**
$\int \sin x\,dx = -\cos x,$	$\int \sinh x\,dx = \cosh x,$
$\int \cos x\,dx = \sin x,$	$\int \cosh x\,dx = \sinh x,$
$\int \tan x\,dx = -\ln\lvert\cos x\rvert,$	$\int \tanh x\,dx = \ln\cosh x,$
$\int \cot x\,dx = \ln\lvert\sin x\rvert,$	$\int \coth x\,dx = \ln\lvert\sinh x\rvert,$
$\int \frac{dx}{\cos^2 x} = \tan x,$	$\int \frac{dx}{\cosh^2 x} = \tanh x,$
$\int \frac{dx}{\sin^2 x} = -\cot x.$	$\int \frac{dx}{\sinh^2 x} = -\coth x.$

(1) u, v, w are functions of x.
(2) u and v are functions of x.

Table of fundamental integrals
(the constants of integration are omitted here and in a following tables)

Rational functions $(a > 0)$	Irrational functions $(a > 0)$
$\int \frac{dx}{a^2 + x^2} = \frac{1}{a} \operatorname{arc tan} \frac{x}{a}$,	$\int \frac{dx}{\sqrt{a^2 - x^2}} = \operatorname{arc sin} \frac{x}{a}$, $a > \lvert x \rvert$,
$\int \frac{dx}{a^2 - x^2} = \frac{1}{a} \operatorname{ar tanh} \frac{x}{a} = \frac{1}{2a} \ln \frac{a+x}{a-x}$, $\lvert x \rvert < a$;	$\int \frac{dx}{\sqrt{a^2 + x^2}} = \operatorname{ar sinh} \frac{x}{a} = \ln (x + \sqrt{x^2 + a^2})$,
$\int \frac{dx}{x^2 - a^2} = -\frac{1}{a} \operatorname{ar coth} \frac{x}{a} = \frac{1}{2a} \ln \frac{x-a}{x+a}$, $\lvert x \rvert > a$.	$\int \frac{dx}{\sqrt{x^2 - a^2}} = \operatorname{ar cosh} \frac{x}{a} = \ln \lvert x + \sqrt{x^2 - a^2} \rvert$.

General hints concerning evaluation of integrals. There is no general rule of integration; the technique of integration can be acquired only by experience. Examples of integration of elementary functions will be developed in the following sections. On pp. 411–453 there is given a table of integrals, where we can try to find a given integral or an integral similar to it.

Some general methods leading to evaluation of integrals are the following:

(1) By means of algebraic or trigonometric transformations we express the integrand as the sum of certain functions and decompose the integral into the sum of integrals.

Examples:

$$\int (x+3)^2 (x^2+1)\, dx = \int (x^4 + 6x^3 + 10x^2 + 6x + 9)\, dx = \frac{x^5}{5} + \frac{3}{2}x^4 + \frac{10}{3}x^3 + 3x^2 + 9x + C;$$

$$\int \sin 2x \cos x\, dx = \int \tfrac{1}{2}(\sin 3x + \sin x) dx = -\tfrac{1}{6}\cos 3x - \tfrac{1}{2}\cos x + C.$$

(2) If $\int f(x)dx = F(x)$ is known (for instance, from the tables), then

$$\int f(ax)dx = \frac{1}{a}F(ax) + C, \qquad \int f(x+b)dx = F(x+b) + C,$$

$$\int f(ax+b)dx = \frac{1}{a}F(ax+b) + C.$$

Examples:

$$\int \sin ax\,dx = -\frac{1}{a}\cos ax + C, \quad \int e^{ax+b}\,dx = \frac{1}{a}e^{ax+b} + C,$$

$$\int \frac{dx}{1+(x+a)^2} = \text{arc tan}\,(x+a) + C.$$

(3) If the integrand is a quotient whose numerator is the differential of the denominator, then the integral is equal to the logarithm of the denominator:.

$$\int \frac{f'(x)}{f(x)}\,dx = \int \frac{df(x)}{f(x)} = \ln|f(x)| + C.$$

Example:

$$\int \frac{2x+3}{x^2+3x-5}\,dx = \ln|x^2+3x-5| + C.$$

3. Integration of rational functions

Integrals of rational functions can always be expressed by means of elementary functions.

General rules. A *polynomial* (*integral function*) can be integrated directly:

$$\int (a_0x^n + a_1x^{n-1} + \ldots + a_{n-1}x + a_n)\,dx$$

$$= \frac{a_0}{n+1}x^{n+1} + \frac{a_1}{n}x^n + \ldots + \frac{a_{n-1}}{2}x^2 + a_nx + C.$$

Rational function $\int \frac{Q(x)}{P(x)}\,dx$ (where $Q(x)$ and $P(x)$ are two polynomials of degrees, respectively, m and n) can be transformed to a form convenient to integration in the following way: (1) we cancel the quotient so that $Q(x)$ and $P(x)$ have no common factor; (2) if $m > n$, we separate the integral part of $\frac{Q(x)}{P(x)}$ by a division (see p. 150); we get a polynomial which can be integrated directly and the remainder which is a proper fraction with $m < n$; (3) we decompose the denominator $P(x)$ into a product of linear or quadratic factors (see p. 165):

$$P(x) = a_0(x-\alpha)^k(x-\beta)^l\ldots(x^2+px+q)^r(x^2+p'x+q')^s\ldots,$$

where

$$\frac{p^2}{4} - q < 0, \quad \frac{p'^2}{4} - q' < 0, \quad \ldots$$

(4) we bring out the coefficient a_0 from under the integral sign; (5) a proper fraction obtained in this way, with the denominator decomposed into simple factors, can be resolved into a sum of partial fractions (see pp. 151, 152) which can be easily integrated. The following four cases can then occur:

(1) All the roots of the denominator are real and simple:

$$P(x) = (x-\alpha)(x-\beta)\ldots(x-\lambda).$$

The resolution has the form

$$\frac{Q(x)}{P(x)} = \frac{A}{x-\alpha} + \frac{B}{x-\beta} + \ldots + \frac{L}{x-\lambda},$$

where

$$A = \frac{Q(\alpha)}{P'(\alpha)}, \quad B = \frac{Q(\beta)}{P'(\beta)}, \quad \ldots, \quad L = \frac{Q(\lambda)}{P'(\lambda)}\ [1].$$

The integration proceeds according to the formula

$$\int \frac{A\,dx}{x-\alpha} = A \ln|x-\alpha| \quad \text{and so on.}$$

Example:

$$I = \int \frac{(2x+3)\,dx}{x^3+x^2-2x}; \quad \frac{2x+3}{x(x-1)(x+2)} = \frac{A}{x} + \frac{B}{x-1} + \frac{C}{x+2};$$

$$A = \frac{Q(0)}{P'(0)} = \left(\frac{2x+3}{3x^2+2x-2}\right)_{x=0} = -\frac{3}{2},$$

$$B = \left(\frac{2x+3}{3x^2+2x-2}\right)_{x=1} = \frac{5}{3}, \quad C = \left(\frac{2x+3}{3x^2+2x-2}\right)_{x=-2} = -\frac{1}{6},$$

$$I = \int\left(\frac{-\frac{3}{2}}{x} + \frac{\frac{5}{3}}{x-1} + \frac{-\frac{1}{6}}{x+2}\right)dx = \ln\left|\frac{(x-1)^{5/3}}{x^{3/2}(x+2)^{1/6}}\right| + C.$$

(2) All the roots of denominator are real, but multiple roots occur:

$$P(x) = (x-\alpha)^l (x-\beta)^m \ldots$$

---

[1] The numbers $A, B, \ldots, L$ can also be obtained by the method of undetermined coefficients (see p. 151).

Resolution has the form

$$\frac{Q(x)}{P(x)} = \frac{A_1}{(x-\alpha)} + \frac{A_2}{(x-\alpha)^2} + \dots + \frac{A_l}{(x-\alpha)^l} + \\ + \frac{B_1}{x-\beta} + \frac{B_2}{(x-\beta)^2} + \dots + \frac{B_m}{(x-\beta)^m} + \dots$$

The constants $A_1, A_2, \dots, A_l, B_1, B_2, \dots, B_m, \dots$ can be calculated by using the method of undetermined coefficients (see p. 152) and integration proceeds according to the formulas

$$\int \frac{A_1 dx}{x-\alpha} = A_1 \ln |x-\alpha|,$$

$$\int \frac{A_k dx}{(x-\alpha)^k} = -\frac{A_k}{(k-1)(x-\alpha)^{k-1}}, \quad k > 1$$

Example:

$$I = \int \frac{x^3+1}{x(x-1)^3} dx; \quad \frac{x^3+1}{x(x-1)^3} = \frac{A}{x} + \frac{B_1}{x-1} + \frac{B_2}{(x-1)^2} + \frac{B_3}{(x-1)^3}.$$

Method of undetermined coefficients leads to the equations

$$A + B_1 = 1, \quad -3A - 2B_1 + B_2 = 0,$$
$$3A + B_1 - B_2 + B_3 = 0, \quad -A = 1,$$

whence $A = -1$, $B_1 = 2$, $B_2 = 1$, $B_3 = 2$;

$$\int \left(-\frac{1}{x} + \frac{2}{x-1} + \frac{1}{(x-1)^2} + \frac{2}{(x-1)^3}\right) dx = -\ln|x| + 2\ln|x-1| - \\ - \frac{1}{x-1} - \frac{1}{(x-1)^2} + C = \ln\frac{(x-1)^2}{|x|} - \frac{x}{(x-1)^2} + C.$$

(3) Some roots of the denominator are complex and simple:

$$P(x) = (x-\alpha)^l (x-\beta)^m \dots (x^2+px+q)(x^2+p'x+q') \dots,$$

where

$$\frac{p^2}{4} < q, \quad \frac{p'^2}{4} < q', \quad \dots$$

Resolution has the form

$$\frac{Q(x)}{P(x)} = \frac{A_1}{x-\alpha} + \frac{A_2}{(x-\alpha)^2} + \dots + \frac{A_l}{(x-\alpha)^l} + \frac{B_1}{x-\beta} + \frac{B_2}{(x-\beta)^2} + \dots + \\ + \frac{B_m}{(x-\beta)^m} + \dots + \frac{Cx+D}{x^2+px+q} + \frac{Ex+F}{x^2+p'x+q'} + \dots$$

The constants are evaluated by the method of undetermined coefficients (see p. 152).

The expression $\frac{Cx+D}{x^2+px+q}$ is integrated according to the formula

$$\int \frac{(Cx+D)\,dx}{x^2+px+q} = \frac{C}{2}\ln|x^2+px+q| + \frac{D-\frac{1}{2}Cp}{\sqrt{q-\frac{1}{4}p^2}}\arctan\frac{x+\frac{1}{2}p}{\sqrt{q-\frac{1}{4}p^2}}.$$

Example:

$$I = \int \frac{4dx}{x^3+4x}; \qquad \frac{4}{x^3+4x} = \frac{A}{x} + \frac{Cx+D}{x^2+4}.$$

The method of undetermined coefficients leads to the equations

$$A + C = 0, \qquad D = 0, \qquad 4A = 4,$$

whence $A = 1$, $C = -1$, $D = 0$.

$$I = \int \left(\frac{1}{x} - \frac{x}{x^2+4}\right) dx = \ln|x| - \frac{1}{2}\ln(x^2+4) + \ln C_1 = \ln\frac{C_1|x|}{\sqrt{x^2+4}}$$

(in this case, the term containing arc tan does not occur).

(4) Denominator has complex multiple roots:

$$P(x) = (x-\alpha)^k (x-\beta)^l \ldots (x^2+px+q)^m (x^2+p'x+q')^n \ldots$$

Resolution has the form

$$\frac{Q(x)}{P(x)} = \frac{A_1}{x-\alpha} + \frac{A_2}{(x-\alpha)^2} + \ldots + \frac{B_1}{x-\beta} + \frac{B_2}{(x-\beta)^2} + \ldots + \frac{B_l}{(x-\beta)_l} +$$

$$+ \frac{C_1x+D_1}{x^2+px+q} + \frac{C_2x+D_2}{(x^2+px+q)^2} + \ldots + \frac{C_mx+D_m}{(x^2+px+q)^m} +$$

$$+ \frac{E_1x+F_1}{x^2+p'x+q'} + \frac{E_2x+F_2}{(x^2+p'x+q')^2} + \ldots + \frac{E_nx+F_n}{(x^2+p'x+q')^n} + \ldots$$

The constants are evaluated by using the method of undetermined coefficients (see p. 152).

The expression $\frac{C_mx+D_m}{(x^2+px+q)^m}$ is integrated as follows. We transform the numerator

$$C_mx + D_m = \frac{C_m}{2}(2x+p) + \left(D_m - \frac{C_mp}{2}\right).$$

We decompose the integral into two sumands. The first one can be integrated at once:

$$\int \frac{C_m}{2}\cdot\frac{(2x+p)\,dx}{(x^2+px+q)^m} = -\frac{C_m}{2(m-1)}\cdot\frac{1}{(x^2+px+q)^{m-1}},$$

and the second one (without the coefficient)—by formula for lowering of the exponent:

$$(1)\quad \int \frac{dx}{(x^2+px+q)^m} = \frac{x+\frac{1}{2}p}{2(m-1)(q-\frac{1}{4}p^2)(x^2+px+q)^{m-1}} + \\ + \frac{2m-3}{2(m-1)(q-\frac{1}{4}p^2)}\int \frac{dx}{(x^2+px+q)^{m-1}}.$$

Example:

$$I = \int \frac{2x^2+2x+13}{(x-2)(x^2+1)^2}\,dx,$$

$$\frac{2x^2+2x+13}{(x-2)(x^2+1)^2} = \frac{A}{x-2} + \frac{C_1x+D_1}{x^2+1} + \frac{C_2x+D_2}{(x^2+1)^2}.$$

The method of undetermined coefficients leads to the equations

$$A+C_1=0,\ -2C_1+D_1=0,\quad 2A+C_1-2D_1+C_2=2,$$

$$-2C_1+D_1-2C_2+D_2=2,\qquad A-2D_1-2D_2=13,$$

whence

$$A=1,\quad C_1=-1,\quad D_1=-2,\quad C_2=-3,\quad D_2=-4$$

and

$$I = \int\left(\frac{1}{x-2} - \frac{x+2}{x^2+1} - \frac{3x+4}{(x^2+1)^2}\right)dx = \ln|x-2| -$$

$$-\left(\tfrac{1}{2}\ln(x^2+1) + 2\arctan x\right) - \left(-\frac{3}{2(x^2+1)} + \int \frac{4\,dx}{(x^2+1)^2}\right).$$

But, by formula (1),

$$\int \frac{dx}{(x^2+1)^2} = \frac{x}{2(x^2+1)} - \frac{1}{2}\int\frac{dx}{x^2+1} = \frac{x}{2(x^2+1)} - \frac{1}{2}\arctan x.$$

Therefore, we have finally

$$I = \frac{3-4x}{2(x^2+1)} + \frac{1}{2}\ln\frac{(x-2)^2}{x^2+1} - 4\arctan x + C.$$

Separation of the rational part of an integral (*method of Ostrogradsky*). The integral of a rational function is an elementary function composed of a *rational part* (i.e., a certain algebraic fraction) and a *transcendental part* (involving logarithms and inverse tangents). The rational part occurs only in the second and fourth of the considered cases, i.e., only when the denominator of the integrand has multiple (real or complex) roots. The rational part can be found without integration by aid of the *method of Ostrogradsky* and the evaluation of the integral can be reduced to the cases when the denominator has only simple roots. This can be done as follows:

The denominator $P(x)$ of the integrand $\frac{Q(x)}{P(x)}$ ($\frac{Q(x)}{P(x)}$ is a proper fraction such that $Q(x)$ and $P(x)$ have no common factors, see p. 399) has the form

$$P(x) = (x-\alpha)^k (x-\beta)^l \dots (x^2+px+q)^m (x^2+p'x+q')^n \dots$$

This can be written as a product of two factors $P_1(x)$ and $P_2(x)$, where $P_2(x)$ is the product of all factors occuring in $P(x)$ taken with the power 1:

$$P_2(x) = (x-\alpha)(x-\beta)\dots(x^2+px+q)(x^2+p'x+q')\dots,$$

hence

$$P_1(x) = (x-\alpha)^{k-1}(x-\beta)^{l-1}\dots(x^2+px+q)^{m-1}(x^2+p'x+q')^{n-1}\dots\ ^{(1)}$$

The given integral can be written in the form

$$\text{(A)} \qquad \int \frac{Q(x)}{P(x)}\,dx = \frac{Q_1(x)}{P_1(x)} + \int \frac{Q_2(x)}{P_2(x)}\,dx$$

(*Ostrogradsky's formula*), where $P(x), P_1(x), P_2(x)$ are known polynomials of degrees, respectively, r, s, and t and $Q(x)$ is a polynomial of degree at most $r-1$ and $Q_1(x)$ and $Q_2(x)$ are unknown polynomials of degrees, respectively, at most $s-1$ and $t-1$:

$$Q_1(x) = ax^{s-1} + bx^{s-2} + \dots + d, \qquad Q_2(x) = ex^{t-1} + fx^{t-2} + \dots + h.$$

By integrating (A) we obtain

$$\text{(B)} \qquad \frac{Q(x)}{P(x)} = \left[\frac{Q_1(x)}{P_1(x)}\right]' + \frac{Q_2(x)}{P_2(x)}.$$

The unknown coefficients of the polynomials $Q_1(x)$ and $Q_2(x)$ can be determined from equation (B) by the method of undetermined coefficients.

Once the polynomials $Q_1(x)$, $P_1(x)$, $Q_2(x)$ and $P_2(x)$ are known,

---

(1) The polynomials $P_1(x)$ and $P_2(x)$ can be found easily, if the factors of $P(x)$ are known, i.e., if all the roots of the equation $P(x) = 0$ are determined. But $P_1(x)$ and $P_2(x)$ can also be found without solving this equation: it is sufficient to integrate the polynomial $P(x)$ and find the highest common factor of $P(x)$ and $P'(x)$ (see p. 149), which is equal to $P_1(x)$, while $P_2(x) = \frac{P(x)}{P_1(x)}$.

the given integral is reduced to the integral $\int \frac{Q_2(x)}{P_2(x)}\,dx$ where the denominator of the integrand has no multiple roots.

Example.

$$\int \frac{x^4 + x^3 + 4x^2 + 3x + 2}{(x+1)^2\,(x^2+1)^2}\,dx,$$

we have here

$$P_1 = P_2 = (x+1)(x^2+1) = x^3 + x^2 + x + 1,$$
$$P = (x^3 + x^2 + x + 1)^2,$$
$$Q = x^4 + x^3 + 4x^2 + 3x + 2,$$
$$Q_1 = ax^2 + bx + c,$$
$$Q_2 = ex^2 + fx + g.$$

By formula (B)

$$\frac{x^4 + x^3 + 4x^2 + 3x + 2}{(x^3 + x^2 + x + 1)^2} = \left(\frac{ax^2 + bx + c}{x^3 + x^2 + x + 1}\right)' + \frac{ex^2 + fx + g}{x^3 + x^2 + x + 1},$$

hence

$$x^4 + x^3 + 4x^2 + 3x + 2 = (2ax + b)(x^3 + x^2 + x + 1) -$$
$$-(ax^2 + bx + c)(3x^2 + 2x + 1) + (ex^2 + fx + g)(x^3 + x^2 + x + 1).$$

Comparing the coefficients of equal powers of x on both sides, we obtain a system of equations for a, b, c, e, f, g:

$$1.\ e = 0,\quad 2.\ -a + f = 1,$$
$$3.\ -2b + f + g = 1,\quad 4.\ a - b - 3c + f + g = 4,$$
$$5.\ 2a - 2c + f + g = 3,\quad 6.\ b - c + g = 2.$$

The coefficient e has been omitted in equations 2–6. Hence

$$a = -\tfrac{1}{4},\ b = \tfrac{1}{4},\ c = -1,\ e = 0,\ f = \tfrac{3}{4},\ g = \tfrac{3}{4}.$$

Therefore

$$\int \frac{x^4 + x^3 + 4x^2 + 3x + 2}{(x+1)^2\,(x^2+1)^2}\,dx$$
$$= -\frac{1}{4}\cdot\frac{x^2 - x + 4}{x^3 + x^2 + x + 1} + \frac{3}{4}\int \frac{x+1}{(x+1)(x^2+1)}\,dx$$

The last integral is equal to arc tan x.

For tables of integrals of rational functions see pp. 411–420

4. Integration of irrational functions

Irrational functions are not always integrable in terms of elementary functions. In simplest cases, integrals of irrational functions can be reduced to those of rational functions by means of the following substitutions:

Integral ([1])	Substitution
$\int R\left(x, \sqrt[n]{\frac{ax+b}{cx+e}}\right) dx,$	$\sqrt[n]{\frac{ax+b}{cx+e}} = t,$
$\int R\left(x, \sqrt[n]{\frac{ax+b}{cx+e}}, \sqrt[m]{\frac{ax+b}{cx+e}}, \dots\right) dx,$	$\sqrt[r]{\frac{ax+b}{cx+e}} = t,$ where r is the least common multiple of $n, m, \dots$
$\int R\left(x, \sqrt{ax^2+bx+c}\right) dx,$	One of Euler's substitutions:
1° Case $a > 0$ ([2]),	$\sqrt{ax^2+bx+c} = t - \sqrt{a}\,x,$
2° Case $c > 0$,	$\sqrt{ax^2+bx+c} = xt + \sqrt{c},$
3° Case, when the trinomial has two different real roots: $ax^2+bx+c = a(x-\alpha)(x-\beta)$	$\sqrt{ax^2+bx+c} = t(x-\alpha)$

([1]) The symbol R denotes a rational function of the expression to which it refers.

([2]) If $a < 0$ and the trinomial ax^2+bx+c has two complex roots, then the function under the integral sign does not exist, since $\sqrt{ax^2+bx+c}$ is an imaginary number for any real value of x.

The integral $\int R\left(x, \sqrt{ax^2+bx+c}\right)dx$ can also be reduced to one of the following forms

$$\int R\left(x, \sqrt{x^2+\alpha^2}\right)dx, \qquad \int R\left(x, \sqrt{x^2-\alpha^2}\right)dx, \qquad \int R\left(x, \sqrt{\alpha^2-x^2}\right)dx,$$

for the quadratic ax^2+bx+c can always be represented as a sum or difference of two squares.

Examples.

(1) $4x^2+16x+17 = 4\,(x^2+4x+4+\frac{1}{4})$

$$= 4[(x+2)^2+(\tfrac{1}{2})^2] = 4[x_1^2+(\tfrac{1}{2})^2], \text{ where } x_1 = x+2.$$

(2) $x^2+3x+1 = x^2+3x+\frac{9}{4}-\frac{5}{4} = \left(x+\frac{3}{2}\right)^2 - \left(\frac{\sqrt{5}}{2}\right)^2$

$$= x_1^2 - \left(\frac{\sqrt{5}}{2}\right)^2, \text{ where } x_1 = x+\tfrac{3}{2}.$$

(3) $-x^2+2x=1-x^2+2x-1=1^2-(x-1)^2=1^2-x_1^2,$

where $x_1=x-1.$

Such integrals can be evaluated by the following substitutions:

Integral	Substitution
$\int R(x, \sqrt{x^2+\alpha^2})\,dx$	$x=\alpha\sinh t$ or $x=\alpha\tan t$
$\int R(x, \sqrt{x^2-\alpha^2})\,dx$	$x=\alpha\cosh t$ or $x=\alpha\sec t$
$\int R(x, \sqrt{\alpha^2-x^2})\,dx$	$x=\alpha\sin t$ or $x=\alpha\cos t$

These substitutions lead to integrals of rational expressions involving trigonometric or hyperbolic functions (see pp. 408 or 411).

Integration of binomial differentials. An expression

$$x^m(a+bx^n)^p dx,$$

where a and b are arbitrary real numbers and m, n, p are arbitrary (positive or negative) rational numbers is called a *binomial differential.*

Tchebyscheff's theorem. The integral

$$\int x^m(a+bx^n)^p dx \tag{1}$$

can be expressed in terms of elementary functions only in the following three cases:

(1) p is an integer. We expand $(a+bx^n)^p$ according to Newton's formula (see p. 193). Then the integrand is a sum of terms of the form cx^k which can be easily integrated.

(2) $\dfrac{m+1}{n}$ is an integer. We reduce (1) to an integral of a rational function by the substitution $t=\sqrt[r]{a+bx^n}$, where r is the denominator of p.

(3) $\dfrac{m+1}{n}+p$ is an integer. We reduce (1) to an integral of a rational function by the substitution $t=\sqrt[r]{\dfrac{a+bx^n}{x^n}}$, where r is the denominator of p.

Examples.

$$\int \frac{\sqrt[3]{1+\sqrt[4]{x}}}{\sqrt{x}}\,dx=\int x^{-1/2}(1+x^{1/4})^{1/3}\,dx, \tag{1}$$

$$m=-\frac{1}{2},\quad n=\frac{1}{4},\quad p=\frac{1}{3},\quad \frac{m+1}{n}=2$$

(case 2). Substitution

$$t=\sqrt[3]{1+\sqrt[4]{x}},\quad x=(t^3-1)^4,\quad dx=12t^2(t^3-1)^3\,dt,$$

$$\int\frac{\sqrt[3]{1+\sqrt[4]{x}}}{\sqrt{x}}\,dx=12\int(t^6-t^3)\,dt=\frac{3}{7}t^4(4t^3-7)+C.$$

(2)
$$\int\frac{x^3\,dx}{\sqrt[4]{1+x^3}}=\int x^3(1+x^3)^{-1/4}\,dx,$$

$$m=3,\quad n=3,\quad p=-\frac{1}{4};\quad \frac{m+1}{n}=\frac{4}{3},\quad \frac{m+1}{n}+p=\frac{13}{12};$$

neither of the conditions (1), (2), (3) is fulfilled, hence the integral is not an elementary function.

Elliptic integrals. Integrals such as

(A)
$$\int R\left(x,\sqrt{ax^3+bx^2+cx+\partial}\right)dx,$$
$$\int R\left(x,\sqrt{ax^4+bx^3+cx^2+\partial x+e}\right)dx$$

cannot, in general, be expressed in terms of elementary functions. In cases when these integrals are not elementary functions, they are called *elliptic integrals* ([1]).

Integrals of the type (A) which are not expressable in terms of elementary functions can be reduced by means of transformation to elementary functions and to integrals of the following three types:

(B)
$$\int\frac{dt}{\sqrt{(1-t^2)(1-k^2t^2)}},\quad \int\frac{t^2\,dt}{\sqrt{(1-t^2)(1-k^2t)^2}},$$
$$\int\frac{dt}{(1+ht^2)\sqrt{(1-t^2)(1-k^2t^2)}},$$

where $0<k<1$.

By the substitution $t=\sin\varphi$ $(0<\varphi<\frac{1}{2}\pi)$, the integrals (B) can be reduced to the following form of Legendre:

$$\int\frac{d\varphi}{\sqrt{1-k^2\sin^2\varphi}}\quad \textit{(elliptic integral of the 1st kind)},$$

$$\int\sqrt{1-k^2\sin^2\varphi}\,d\varphi\quad \textit{(elliptic integral of the 2nd kind)},$$

$$\int\frac{d\varphi}{(1+h\sin^2\varphi)\sqrt{1-k^2\sin^2\varphi}}\quad \textit{(elliptic integral of the 3rd kind)}.$$

([1]) In cases, when the integrals (A) are expressible in terms of elementary functions, they are called *pseudo elliptic integrals*.

The corresponding definite integrals with the lower limit of integration equal to zero are denoted as follows:

(I) $$\int_0^{\varphi} \frac{d\psi}{\sqrt{1-k^2\sin^2\psi}} = F(k,\varphi),$$

(II) $$\int_0^{\varphi} \sqrt{1-k^2\sin^2\psi}\,d\psi = E(k,\varphi),$$

(III) $$\int_0^{\varphi} \frac{d\psi}{(1+h\sin^2\psi)\sqrt{1-k^2\sin^2\psi}} = \Pi(h,k,\varphi),$$

where $k < 1$.

These are *incomplete elliptic integrals of the first, second and third kind respectively*. When $\varphi = \frac{1}{2}\pi$, (I) and (II) are called *complete elliptic integrals* and denoted by

$$\mathrm{K} = F\left(k,\frac{\pi}{2}\right) = \int_0^{\pi/2} \frac{d\psi}{\sqrt{1-k^2\sin^2\psi}}, \quad \mathrm{E} = E\left(k,\frac{\pi}{2}\right) = \int_0^{\pi/2} \sqrt{1-k^2\sin^2\psi}\,d\psi.$$

For tables of values of incomplete and complete elliptic integrals of the first and second kind see pp. 91–93.

Tables of integrals of irrational functions—see pp. 420–433.

5. Integration of trigonometric functions

An integral of the form

(A) $$\int R(\sin x, \cos x)\,dx \text{ }[1]$$

can always be reduced to an integral of a rational function by means of the universal substitution given below; in particular cases it can also be evaluated in simpler ways.

Universal substitution for the integral (A):

$$t = \tan\tfrac{1}{2}x, \quad \text{hence } dx = \frac{2\,dt}{1+t^2}, \quad \sin x = \frac{2t}{1+t^2}, \quad \cos x = \frac{1-t^2}{1+t^2}.$$

---

[1] The symbol R denotes a rational function of the expressions to which it refers.

For example,

$$\int \frac{1+\sin x}{\sin x(1+\cos x)}\,dx = \int \frac{\left(1+\frac{2t}{1+t^2}\right)\frac{2\,dt}{1+t^2}}{\frac{2t}{1+t^2}\left(1+\frac{1-t^2}{1+t^2}\right)}$$

$$= \frac{1}{2}\int \left(t+2+\frac{1}{t}\right)dt = \tfrac{1}{4}t^2 + t + \tfrac{1}{2}\ln|t| + C$$

$$= \tfrac{1}{4}\tan^2 \tfrac{1}{2}x + \tan \tfrac{1}{2}x + \tfrac{1}{2}\ln|\tan \tfrac{1}{2}x| + C.$$

If $\sin x$ and $\cos x$ occur in the integrand of (A) only in powers with an integral exponent, then the integral can be reduced to a rational function more directly by the substitution $t = \tan x$.

Simplified methods for certain frequently occurring cases:

(1) $\int R(\sin x)\cos x\,dx$. Substitution $t = \sin x$, $\cos x\,dx = dt$.

(2) $\int R(\cos x)\sin x\,dx$. Substitution $t = \cos x$, $\sin x\,dx = -dt$.

(3) $\int \sin^n x\,dx$. If n is odd $(n = 2m+1)$, then

$$\int \sin^{2m+1} x\,dx = \int (1-\cos^2 x)^m \sin x\,dx = -\int (1-t^2)^m dt,$$

where $t = \cos x$.

If n is even $(n = 2m)$, then

$$\int \sin^{2m} x\,dx = \int [\tfrac{1}{2}(1-\cos 2x)]^m dx$$

$$= \frac{1}{2^{m+1}}\int (1-\cos t)^m dt, \quad \text{where} \quad t = 2x.$$

The exponent has been lowered two times. Now we open the parentheses and integrate term by term (see below, case (4)).

(4) $\int \cos^n x\,dx$. If n is odd $(n = 2m+1)$, then

$$\int \cos^{2m+1} x\,dx = \int (1-\sin^2 x)^m \cos x\,dx = \int (1-t^2)^m\,dt,$$

where $t = \sin x$.

If n is even, $(n = 2m)$, then

$$\int \cos^{2m} x\,dx = \int [\tfrac{1}{2}(1+\cos 2x)]^m dx$$

$$= \frac{1}{2^{m+1}}\int (1+\cos t)^m dt, \quad \text{where} \quad t = 2x.$$

We open the parentheses and integrate term by term.

(5) $\int \sin^n x \cos^m x\,dx$ can be reduced to cases (1) or (2), if at least one of the indices m or n is odd.

Examples.

$$\int \sin^2 x \cos^5 x\,dx = \int \sin^2 x(1-\sin^2 x)^2 \cos x\,dx$$
$$= \int t^2(1-t^2)^2 dt, \qquad \text{where } t = \sin x,$$
$$\int \frac{\sin x}{\sqrt{\cos x}}\,dx = -\int \frac{dt}{\sqrt{t}}, \qquad \text{where} \quad t = \cos x.$$

If both m and n are even, then the exponents can be lowered two times, like in cases (3) and (4). We use the formulas

$$\sin x \cos x = \frac{\sin 2x}{2}, \qquad \sin^2 x = \frac{1-\cos 2x}{2}, \qquad \cos^2 x = \frac{1+\cos 2x}{2}.$$

Example.

$$\int \sin^2 x \cos^4 x\,dx = \int (\sin x \cos x)^2 \cos^2 x\,dx$$
$$= \tfrac{1}{8}\int \sin^2 2x\,(1+\cos 2x)\,dx$$
$$= \tfrac{1}{8}\int \sin^2 2x \cos 2x\,dx + \tfrac{1}{16}\int (1-\cos 4x)\,dx$$
$$= \tfrac{1}{48}\sin^3 2x + \tfrac{1}{16}x - \tfrac{1}{64}\sin 4x + C.$$

(6) $\int \tan^n x\,dx = \int \tan^{n-2} x\,(\sec^2 x - 1)\,dx$

$$= \int \tan^{n-2} x\,d\tan x - \int \tan^{n-2} x\,dx$$
$$= \frac{\tan^{n-1} x}{n-1} - \int \tan^{n-2} x\,dx$$

and so on. Repeating this process, we reduce the integral to $\int dx = x + C$, if n is even or to $\int \tan x\,dx = -\ln|\cos x| + C$, if n is odd.

(7) $\int \cot^n x\,dx$ can be integrated like case (6).

Tables of integrals of trigonometric functions—see pp. 433–446.

6. Integration of other transcendental functions

Exponential functions. Integrals of the form

$$\int R(e^{mx}, e^{nx}, \ldots, e^{px})\,dx,$$

where $m, n, \ldots, p$ are rational numbers, can be reduced, by the substitution $t = e^x$, to the integral $\int \frac{1}{t} R(t^m, t^n, \ldots, t^p)\,dt$ which can be reduced to an integral of a rational function (see p. 389) by the substitution $z = \sqrt[r]{t}$, where r is the least common multiple of the denominators of $m, n, \ldots, p$.

Hyperbolic functions. Integrals involving $\sinh x$, $\cosh x$, $\tanh x$, $\coth x$, can be computed by expressing the hyperbolic functions in terms of the exponential ones (see p. 230). The most frequently occurring cases $\int \sinh^n x\,dx$, $\int \cosh^n x\,dx$, $\int \sinh^n x \cosh^m x\,dx$ can be integrated by rules similar to those applicable to trigonometric functions (pp. 408–409).

Application of integration by parts. Functions involving logarithms, inverse trigonometric and inverse hyperbolic functions, the products of x^m times $\ln x$, e^{ax}, $\sin ax$ or $\cos ax$ are mostly integrated by using (once or several times) the formula for integration by parts (p. 396). In certain cases, repeated application of integration by parts leads to the original integral and then its evaluation reduces to a solution of an algebraic equation. For example, the integrals $\int e^{ax} \sin bx\,dx$, $\int e^{ax} \cos bx\,dx$ can be evaluated in this way (we integrate them twice by parts, taking for u, in both cases, a function of one type—the exponential or trigonometric one).

The cases $\int P(x)e^{ax}\,dx$, $\int P(x) \sin bx\,dx$, $\int P(x) \cos bx\,dx$, where $P(x)$ is a polynomial, are also integrated by parts.

Tables of integrals of transcendental functions are given on pp. 446–453.

7. Tables of indefinite integrals

General hints

(1) The constant of integration is always omitted except for cases when the integral can be represented in various forms with various constants.

(2) In cases when the function has been written in the form of a power series, it cannot be expressed in terms of elementary functions.

Integrals of rational functions

Integrals involving $ax+b$, where $a \neq 0$

Notation: $X = ax + b$.

(1) $\displaystyle\int X^n\,dx = \frac{1}{a(n+1)} X^{n+1}$, $n \neq -1$ (when $n = -1$, see (2)).

(2) $\int \frac{dx}{X} = \frac{1}{a} \ln |X|.$

(3) $\int x X^n dx = \frac{1}{a^2 (n+2)} X^{n+2} - \frac{b}{a^2 (n+1)} X^{n+1},$

$n \neq -1, -2$ (when $n = -1, -2$, see (5) and (6)).

(4) $\int x^m X^n dx = \frac{1}{a^{m+1}} \int (X-b)^m X^n dX$; this formula can be applied when $m < n$ or when m is an integer and n a fraction. In these cases we expand $(X-b)^m$ according to Newton's formula, p. 193; $n \neq -1, -2, \ldots, -m$.

(5) $\int \frac{x\,dx}{X} = \frac{x}{a} - \frac{b}{a^2} \ln |X|.$

(6) $\int \frac{x\,dx}{X^2} = \frac{b}{a^2 X} + \frac{1}{a^2} \ln |X|.$

(7) $\int \frac{x\,dx}{X^3} = \frac{1}{a^2}\left(-\frac{1}{X} + \frac{b}{2X^2}\right).$

(8) $\int \frac{x\,dx}{X^n} = \frac{1}{a^2}\left(\frac{-1}{(n-2) X^{n-2}} + \frac{b}{(n-1) X^{n-1}}\right), \quad n \neq 1, 2.$

(9) $\int \frac{x^2\,dx}{X} = \frac{1}{a^3}\left(\frac{1}{2} X^2 - 2bX + b^2 \ln |X|\right).$

(10) $\int \frac{x^2\,dx}{X^2} = \frac{1}{a^3}\left(X - 2b \ln |X| - \frac{b^2}{X}\right).$

(11) $\int \frac{x^2\,dx}{X^3} = \frac{1}{a^3}\left(\ln |X| + \frac{2b}{X} - \frac{b^2}{2X^2}\right).$

(12) $\int \frac{x^2\,dx}{X^n} = \frac{1}{a^3}\left(\frac{-1}{(n-3) X^{n-3}} + \frac{2b}{(n-2) X^{n-2}} - \frac{b^2}{(n-1) X^{n-1}}\right),$

$n \neq 1, 2, 3.$

(13) $\int \frac{x^3\,dx}{X} = \frac{1}{a^4}\left(\frac{X^3}{3} - \frac{3bX^2}{2} + 3b^2 X - b^3 \ln |X|\right).$

(14) $\int \frac{x^3\,dx}{X^2} = \frac{1}{a^4}\left(\frac{X^2}{2} - 3bX + 3b^2 \ln |X| + \frac{b^3}{X}\right).$

(15) $\int \frac{x^3\,dx}{X^3} = \frac{1}{a^4}\left(X - 3b \ln |X| - \frac{3b^2}{X} + \frac{b^3}{2X^2}\right).$

(16) $\int \frac{x^3\,dx}{X^4} = \frac{1}{a^4}\left(\ln |X| + \frac{3b}{X} - \frac{3b^2}{2X^2} + \frac{b^3}{3X^3}\right).$

(17) $$\int \frac{x^3\,dx}{X^n} = \frac{1}{a^4}\left(\frac{-1}{(n-4)X^{n-4}} + \frac{3b}{(n-3)X^{n-3}} - \frac{3b^2}{(n-2)X^{n-2}} + \right.$$
$$\left. + \frac{b^3}{(n-1)X^{n-1}}\right), \quad n \neq 1, 2, 3, 4.$$

(18) $$\int \frac{dx}{xX} = -\frac{1}{b}\ln\left|\frac{X}{x}\right|.$$

(19) $$\int \frac{dx}{xX^2} = -\frac{1}{b^2}\left(\ln\left|\frac{X}{x}\right| + \frac{ax}{X}\right).$$

(20) $$\int \frac{dx}{xX^3} = -\frac{1}{b^3}\left(\ln\left|\frac{X}{x}\right| + \frac{2ax}{X} - \frac{a^2x^2}{2X^2}\right).$$

(21) $$\int \frac{dx}{xX^n} = -\frac{1}{b^n}\left[\ln\left|\frac{X}{x}\right| - \sum_{i=1}^{n-1}\binom{n-1}{i}\frac{(-a)^i x^i}{iX^i}\right], \quad n > 1.$$

(22) $$\int \frac{dx}{x^2X} = -\frac{1}{bx} + \frac{a}{b^2}\ln\left|\frac{X}{x}\right|.$$

(23) $$\int \frac{dx}{x^2X^2} = -a\left(\frac{1}{b^2X} + \frac{1}{ab^2x} - \frac{2}{b^3}\ln\left|\frac{X}{x}\right|\right).$$

(24) $$\int \frac{dx}{x^2X^3} = -a\left(\frac{1}{2b^2X^2} + \frac{2}{b^3X} + \frac{1}{ab^3x} - \frac{3}{b^4}\ln\left|\frac{X}{x}\right|\right).$$

(25) $$\int \frac{dx}{x^2X^n} = -\frac{1}{b^{n+1}}\left[-\sum_{i=2}^{n}\binom{n}{i}\frac{(-a)^i x^{i-1}}{(i-1)X^{i-1}} + \frac{X}{x} - na\ln\left|\frac{X}{x}\right|\right],$$
$$n > 2.$$

(26) $$\int \frac{dx}{x^3X} = -\frac{1}{b^3}\left(a^2\ln\left|\frac{X}{x}\right| - \frac{2aX}{x} + \frac{X^2}{2x^2}\right).$$

(27) $$\int \frac{dx}{x^3X^2} = -\frac{1}{b^4}\left(3a^2\ln\left|\frac{X}{x}\right| + \frac{a^3x}{X} + \frac{X^2}{2x^2} - \frac{3aX}{x}\right).$$

(28) $$\int \frac{dx}{x^3X^3} = -\frac{1}{b^5}\left(6a^2\ln\left|\frac{X}{x}\right| + \frac{4a^3x}{X} - \frac{a^4x^2}{2X^2} + \frac{X^2}{2x^2} - \frac{4aX}{x}\right).$$

(29) $$\int \frac{dx}{x^3X^n} = -\frac{1}{b^{n+2}}\left[-\sum_{i=3}^{n+1}\binom{n+1}{i}\frac{(-a)^i x^{i-2}}{(i-2)X^{i-2}} + \frac{a^2X^2}{2x^2} - \right.$$
$$\left. - \frac{(n+1)aX}{x} + \frac{n(n+1)a^2}{2}\ln\left|\frac{X}{x}\right|\right], \quad n > 3.$$

(30) $$\int \frac{dx}{x^m X^n} = -\frac{1}{b^{m+n-1}} \sum_{i=0}^{m+n-2} \binom{m+n-2}{i} \frac{X^{m-i-1}(-a)^i}{(m-i-1)x^{m-i-1}};$$

if the denominator of one of the summands is zero, then we replace this term by

$$\binom{m+n-2}{n-1}(-a)^{m-1} \ln\left|\frac{X}{x}\right|.$$

Notation: $\Delta = bf - ag$.

(31) $$\int \frac{ax+b}{fx+g}\,dx = \frac{ax}{f} + \frac{\Delta}{f^2} \ln|fx+g|.$$

(32) $$\int \frac{dx}{(ax+b)(fx+g)} = \frac{1}{\Delta} \ln\left|\frac{fx+g}{ax+b}\right|, \quad \Delta \neq 0.$$

(33) $$\int \frac{x\,dx}{(ax+b)(fx+g)} = \frac{1}{\Delta}\left(\frac{b}{a}\ln|ax+b| - \frac{g}{f}\ln|fx+g|\right), \quad \Delta \neq 0.$$

(34) $$\int \frac{dx}{(ax+b)^2(fx+g)} = \frac{1}{\Delta}\left(\frac{1}{ax+b} + \frac{f}{\Delta}\ln\left|\frac{fx+g}{ax+b}\right|\right), \quad \Delta \neq 0.$$

(35) $$\int \frac{x\,dx}{(a+x)(b+x)^2} = \frac{b}{(a-b)(b+x)} - \frac{a}{(a-b)^2}\ln\left|\frac{a+x}{b+x}\right|, \quad a \neq b.$$

(36) $$\int \frac{x^2\,dx}{(a+x)(b+x)^2} = \frac{b^2}{(b-a)(b+x)} + \frac{a^2}{(b-a)^2}\ln|a+x| + \frac{b^2-2ab}{(b-a)^2}\ln|b+x|, \quad a \neq b.$$

(37) $$\int \frac{dx}{(a+x)^2(b+x)^2} = \frac{-1}{(a-b)^2}\left(\frac{1}{a+x} + \frac{1}{b+x}\right) + \frac{2}{(a-b)^3}\ln\left|\frac{a+x}{b+x}\right|, \quad a \neq b.$$

(38) $$\int \frac{x\,dx}{(a+x)^2(b+x)^2} = \frac{1}{(a-b)^2}\left(\frac{a}{a+x} + \frac{b}{b+x}\right) + \frac{a+b}{(a-b)^3}\ln\left|\frac{a+x}{b+x}\right|, \quad a \neq b.$$

(39) $$\int \frac{x^2\,dx}{(a+x)^2(b+x)^2} = \frac{-1}{(a-b)^2}\left(\frac{a^2}{a+x} + \frac{b^2}{b+x}\right) + \frac{2ab}{(a-b)^3}\ln\left|\frac{a+x}{b+x}\right|, \quad a \neq b.$$

Integrals involving ax^2+bx+c, where $a \neq 0$

Notation: $X = ax^2+bx+c$, $\Delta = 4ac-b^2$.

(40) $$\int \frac{dx}{X} = \frac{2}{\sqrt{\Delta}} \operatorname{arc\,tan} \frac{2ax+b}{\sqrt{\Delta}}, \quad \Delta > 0,$$

$$= -\frac{2}{\sqrt{-\Delta}} \operatorname{ar\,tanh} \frac{2ax+b}{\sqrt{-\Delta}} = \frac{1}{\sqrt{-\Delta}} \ln \left| \frac{2ax+b-\sqrt{-\Delta}}{2ax+b+\sqrt{-\Delta}} \right|, \quad \Delta < 0.$$

(41) $$\int \frac{dx}{X^2} = \frac{2ax+b}{\Delta X} + \frac{2a}{\Delta} \int \frac{dx}{X} \quad \text{(see (40))}.$$

(42) $$\int \frac{dx}{X^3} = \frac{2ax+b}{\Delta} \left(\frac{1}{2X^2} + \frac{3a}{\Delta X} \right) + \frac{6a^2}{\Delta^2} \int \frac{dx}{X} \quad \text{(see (40))}.$$

(43) $$\int \frac{dx}{X^n} = \frac{2ax+b}{(n-1)\Delta X^{n-1}} + \frac{(2n-3)\,2a}{(n-1)\Delta} \int \frac{dx}{X^{n-1}}.$$

(44) $$\int \frac{x\,dx}{X} = \frac{1}{2a} \ln |X| - \frac{b}{2a} \int \frac{dx}{X} \quad \text{(see (40))}.$$

(45) $$\int \frac{x\,dx}{X^2} = -\frac{bx+2c}{\Delta X} - \frac{b}{\Delta} \int \frac{dx}{X} \quad \text{(see (40))}.$$

(46) $$\int \frac{x\,dx}{X^n} = -\frac{bx+2c}{(n-1)\Delta X^{n-1}} - \frac{b(2n-3)}{(n-1)\Delta} \int \frac{dx}{X^{n-1}}.$$

(47) $$\int \frac{x^2\,dx}{X} = \frac{x}{a} - \frac{b}{2a^2} \ln |X| + \frac{b^2-2ac}{2a^2} \int \frac{dx}{X} \quad \text{(see (40))}.$$

(48) $$\int \frac{x^2\,dx}{X^2} = \frac{(b^2-2ac)\,x+bc}{a\Delta X} + \frac{2c}{\Delta} \int \frac{dx}{X} \quad \text{(see (40))}.$$

(49) $$\int \frac{x^2\,dx}{X^n} = \frac{-x}{(2n-3)\,aX^{n-1}} + \frac{c}{(2n-3)\,a} \int \frac{dx}{X^n} -$$

$$- \frac{(n-2)\,b}{(2n-3)\,a} \int \frac{x\,dx}{X^n} \quad \text{(see (43) and (46))}.$$

(50) $$\int \frac{x^m\,dx}{X^n} = -\frac{x^{m-1}}{(2n-m-1)\,aX^{n-1}} +$$

$$+ \frac{(m-1)\,c}{(2n-m-1)\,a} \int \frac{x^{m-2}\,dx}{X^n} - \frac{(n-m)\,b}{(2n-m-1)\,a} \int \frac{x^{m-1}\,dx}{X^n},$$

$m \neq 2n-1$ (when $m = 2n-1$, see (51)).

(51) $$\int \frac{x^{2n-1}\,dx}{X^n} = \frac{1}{a} \int \frac{x^{2n-3}\,dx}{X^{n-1}} - \frac{c}{a} \int \frac{x^{2n-3}\,dx}{X^n} - \frac{b}{a} \int \frac{x^{2n-2}\,dx}{X^n}.$$

(52) $$\int \frac{dx}{xX} = \frac{1}{2c} \ln \frac{x^2}{|X|} - \frac{b}{2c} \int \frac{dx}{X} \quad \text{(see (40))}.$$

(53) $$\int \frac{dx}{xX^n} = \frac{1}{2c(n-1)X^{n-1}} - \frac{b}{2c} \int \frac{dx}{X^n} + \frac{1}{c} \int \frac{dx}{xX^{n-1}}.$$

(54) $$\int \frac{dx}{x^2 X} = \frac{b}{2c^2} \ln \frac{|X|}{x^2} - \frac{1}{cx} + \left(\frac{b^2}{2c^2} - \frac{a}{c}\right) \int \frac{dx}{X} \quad \text{(see (40))}.$$

(55) $$\int \frac{dx}{x^m X^n} = -\frac{1}{(m-1)\,cx^{m-1}X^{n-1}} -$$

$$-\frac{(2n+m-3)\,a}{(m-1)\,c} \int \frac{dx}{x^{m-2}X^n} - \frac{(n+m-2)\,b}{(m-1)c} \int \frac{dx}{x^{m-1}X^n}, \quad m > 1.$$

(56) $$\int \frac{dx}{(fx+g)X} = \frac{1}{2(cf^2 - gbf + g^2 a)} \left(f \ln \frac{(fx+g)^2}{|X|}\right) +$$

$$+ \frac{2ga - bf}{2(cf^2 - gbf + g^2 a)} \int \frac{dx}{X} \quad \text{(see(40))}.$$

Integrals involving $a^2 \pm x^2$, where $a \neq 0$

Assume $a > 0$ and denote

$$X = a^2 \pm x^2, \quad Y = \begin{cases} \arctan \dfrac{x}{a}, & \text{when } X = a^2 + x^2, \\[2ex] \operatorname{ar\,tanh} \dfrac{x}{a} = \frac{1}{2} \ln \left|\dfrac{a+x}{a-x}\right|, & \text{when } X = a^2 - x^2, \\[2ex] \operatorname{ar\,coth} \dfrac{x}{a} = \frac{1}{2} \ln \left|\dfrac{x+a}{x-a}\right|, & \text{when } X = a^2 - x^2 \text{ and } |x| > a. \end{cases}$$

If a double sign in a formula is used, then the upper sign corresponds to the case when $X = a^2 + x^2$ and the lower sign to the case when $X = a^2 - x^2$.

(57) $$\int \frac{dx}{X} = \frac{1}{a} Y.$$

(58) $$\int \frac{dx}{X^2} = \frac{x}{2a^2 X} + \frac{1}{2a^3} Y.$$

(59) $$\int \frac{dx}{X^3} = \frac{x}{4a^2 X^2} + \frac{3x}{8a^4 X} + \frac{3}{8a^5} Y.$$

(60) $$\int \frac{dx}{X^{n+1}} = \frac{x}{2na^2 X^n} + \frac{2n-1}{2na^2} \int \frac{dx}{X^n}.$$

(61) $$\int \frac{x\,dx}{X} = \pm \frac{1}{2} \ln |X|.$$

(62) $$\int \frac{x\,dx}{X^2} = \mp \frac{1}{2X}.$$

(63) $$\int \frac{x\,dx}{X^3} = \mp \frac{1}{4X^2}.$$

(64) $$\int \frac{x\,dx}{X^{n+1}} = \mp \frac{1}{2n\,X^n}, \quad n \neq 0.$$

(65) $$\int \frac{x^2\,dx}{X} = \pm x \mp aY.$$

(66) $$\int \frac{x^2\,dx}{X^2} = \mp \frac{x}{2X} \pm \frac{1}{2a}\,Y.$$

(67) $$\int \frac{x^2\,dx}{X^3} = \mp \frac{x}{4X^2} \pm \frac{x}{8a^2X} \pm \frac{1}{8a^3}\,Y.$$

(68) $$\int \frac{x^2\,dx}{X^{n+1}} = \mp \frac{x}{2n\,X^n} \pm \frac{1}{2n}\int \frac{dx}{X^n}, \quad n \neq 0.$$

(69) $$\int \frac{x^3\,dx}{X} = \pm \frac{x^2}{2} - \frac{a^2}{2}\ln|X|.$$

(70) $$\int \frac{x^3\,dx}{X^2} = \frac{a^2}{2X} + \frac{1}{2}\ln|X|.$$

(71) $$\int \frac{x^3\,dx}{X^3} = -\frac{1}{2X} + \frac{a^2}{4X^2}.$$

(72) $$\int \frac{x^3\,dx}{X^{n+1}} = -\frac{1}{2(n-1)\,X^{n-1}} + \frac{a^2}{2n\,X^n}, \quad n > 1.$$

(73) $$\int \frac{dx}{xX} = \frac{1}{2a^2}\ln\frac{x^2}{|X|}.$$

(74) $$\int \frac{dx}{xX^2} = \frac{1}{2a^2X} + \frac{1}{2a^4}\ln\frac{x^2}{|X|}.$$

(75) $$\int \frac{dx}{xX^3} = \frac{1}{4a^2X^2} + \frac{1}{2a^4X} + \frac{1}{2a^6}\ln\frac{x^2}{|X|}.$$

(76) $$\int \frac{dx}{x^2X} = -\frac{1}{a^2x} \mp \frac{1}{a^3}\,Y.$$

(77) $$\int \frac{dx}{x^2X^2} = -\frac{1}{a^4x} \mp \frac{x}{2a^4X} \mp \frac{3}{2a^5}\,Y.$$

(78) $$\int \frac{dx}{x^2X^3} = -\frac{1}{a^6x} \mp \frac{x}{4a^4X^2} \mp \frac{7x}{8a^6X} \mp \frac{15}{8a^7}Y.$$

(79) $$\int \frac{dx}{x^3 X} = -\frac{1}{2a^2 x^2} \mp \frac{1}{2a^4} \ln \frac{x^2}{|X|}.$$

(80) $$\int \frac{dx}{x^3 X^2} = -\frac{1}{2a^4 x^2} \mp \frac{1}{2a^4 X} \mp \frac{1}{a^6} \ln \frac{x^2}{|X|}.$$

(81) $$\int \frac{dx}{x^3 X^3} = -\frac{1}{2a^6 x^2} \mp \frac{1}{a^6 X} \mp \frac{1}{4a^4 X^2} \mp \frac{3}{2a^8} \ln \frac{x^2}{|X|}.$$

(82) $$\int \frac{dx}{(b+cx) X} = \frac{1}{a^2 c^2 \pm b^2} \left(c \ln |b+cx| - \frac{c}{2} \ln |X| \pm \frac{b}{a} Y \right).$$

Integrals involving $a^3 \pm x^3$, where $a \neq 0$

Notation: $X = a^3 \pm x^3$. If a double sign in a formula is used, then the upper sign corresponds to the case when $X = a^3 + x^3$, and the lower sign to the case when $X = a^3 - x^3$.

(83) $$\int \frac{dx}{X} = \pm \frac{1}{6a^2} \ln \frac{(a \pm x)^2}{|a^2 \mp ax + x^2|} + \frac{1}{a^2 \sqrt{3}} \operatorname{arc\,tan} \frac{2x \mp a}{a\sqrt{3}}.$$

(84) $$\int \frac{dx}{X^2} = \frac{x}{3a^3 X} + \frac{2}{3a^3} \int \frac{dx}{X} \quad \text{(see (83))}.$$

(85) $$\int \frac{x\,dx}{X} = \frac{1}{6a} \ln \frac{|a^2 \mp ax + x^2|}{(a \pm x)^2} \pm \frac{1}{a\sqrt{3}} \operatorname{arc\,tan} \frac{2x \mp a}{a\sqrt{3}}.$$

(86) $$\int \frac{x\,dx}{X^2} = \frac{x^2}{3a^3 X} + \frac{1}{3a^3} \int \frac{x\,dx}{X} \quad \text{(see (85))}.$$

(87) $$\int \frac{x^2\,dx}{X} = \pm \frac{1}{3} \ln |X|.$$

(88) $$\int \frac{x^2\,dx}{X^2} = \mp \frac{1}{3X}.$$

(89) $$\int \frac{x^3\,dx}{X} = \pm x \mp a^3 \int \frac{dx}{X} \quad \text{(see (83))}.$$

(90) $$\int \frac{x^3\,dx}{X^2} = \mp \frac{x}{3X} \pm \frac{1}{3} \int \frac{dx}{X} \quad \text{(see (83))}.$$

(91) $$\int \frac{dx}{xX} = \frac{1}{3a^3} \ln \left| \frac{x^3}{X} \right|.$$

(92) $$\int \frac{dx}{xX^2} = \frac{1}{3a^3 X} + \frac{1}{3a^6} \ln \left| \frac{x^3}{X} \right|.$$

(93) $$\int \frac{dx}{x^2 X} = -\frac{1}{a^3 x} \mp \frac{1}{a^3} \int \frac{x\,dx}{X} \quad \text{(see (85))}.$$

(94) $$\int \frac{dx}{x^2X^2} = -\frac{1}{a^6x} \mp \frac{x^3}{3a^6X} \mp \frac{4}{3a^6}\int \frac{x\,dx}{X} \quad \text{(see (85))}.$$

(95) $$\int \frac{dx}{x^3X} = -\frac{1}{2a^3x^2} \mp \frac{1}{a^3}\int \frac{dx}{X} \quad \text{(see (83))}.$$

(96) $$\int \frac{dx}{x^3X^2} = -\frac{1}{2a^6x^2} \mp \frac{x}{3a^6X} \mp \frac{5}{3a^6}\int \frac{dx}{X} \quad \text{(see (83))}.$$

Integrals involving $a^4 + x^4$, where $a \neq 0$

(97) $$\int \frac{dx}{a^4+x^4} = \frac{1}{4a^3\sqrt{2}} \ln\left|\frac{x^2+ax\sqrt{2}+a^2}{x^2-ax\sqrt{2}+a^2}\right| + \frac{1}{2a^3\sqrt{2}} \arctan \frac{ax\sqrt{2}}{a^2-x^2}.$$

(98) $$\int \frac{x\,dx}{a^4+x^4} = \frac{1}{2a^2} \arctan \frac{x^2}{a^2}.$$

(99) $$\int \frac{x^2\,dx}{a^4+x^4} = -\frac{1}{4a\sqrt{2}} \ln\left|\frac{x^2+ax\sqrt{2}+a^2}{x^2-ax\sqrt{2}+a^2}\right| + \frac{1}{2a\sqrt{2}} \arctan \frac{ax\sqrt{2}}{a^2-x^2}.$$

(100) $$\int \frac{x^3\,dx}{a^4+x^4} = \frac{1}{4} \ln (a^4+x^4).$$

Integrals involving $a^4 - x^4$, where $a \neq 0$

(101) $$\int \frac{dx}{a^4-x^4} = \frac{1}{4a^3} \ln\left|\frac{a+x}{a-x}\right| + \frac{1}{2a^3} \arctan \frac{x}{a}.$$

(102) $$\int \frac{x\,dx}{a^4-x^4} = \frac{1}{4a^2} \ln \frac{a^2+x^2}{|a^2-x^2|}.$$

(103) $$\int \frac{x^2\,dx}{a^4-x^4} = \frac{1}{4a} \ln\left|\frac{a+x}{a-x}\right| - \frac{1}{2a} \arctan \frac{x}{a}.$$

(104) $$\int \frac{x^3\,dx}{a^4-x^4} = -\frac{1}{4} \ln |a^4-x^4|.$$

Certain cases of resolution of a fraction into partial fractions

(105) $$\frac{1}{(a+bx)(f+gx)} \equiv \frac{1}{fb-ag}\left(\frac{b}{a+bx} - \frac{g}{f+gx}\right).$$

(106) $$\frac{1}{(x+a)(x+b)(x+c)} \equiv \frac{A}{x+a} + \frac{B}{x+b} + \frac{C}{x+c},$$

where $A=\frac{1}{(b-a)(c-a)}$, $B=\frac{1}{(a-b)(c-b)}$, $C=\frac{1}{(a-c)(b-c)}$.

(107) $$\frac{1}{(x+a)(x+b)(x+c)(x+d)} \equiv \frac{A}{x+a}+\frac{B}{x+b}+\frac{C}{x+c}+\frac{D}{x+d},$$

where $A=\frac{1}{(b-a)(c-a)(d-a)}$, $B=\frac{1}{(a-b)(c-b)(d-b)}$, etc.

(108) $$\frac{1}{(a+bx^2)(f+gx^2)} \equiv \frac{1}{fb-ag}\left(\frac{b}{a+bx^2}-\frac{g}{f+gx^2}\right).$$

Integrals of irrational functions

Integrals involving $\sqrt{x}$ and $a^2 \pm b^2x$, where $a \neq 0$

Notation:

$$X=a^2 \pm b^2 x, \qquad Y=\begin{cases} \arctan \frac{b\sqrt{x}}{a}, & \text{when } x=a^2+b^2x, \\ \frac{1}{2}\ln\left|\frac{a+b\sqrt{x}}{a-b\sqrt{x}}\right|, & \text{when } x=a^2-b^2x. \end{cases}$$

If a double sign is used in a formula, then the upper sign corresponds to the case when $X=a^2+b^2x$ and the lower sign to the case when $X=a^2-b^2x$.

(109) $$\int \frac{\sqrt{x}\,dx}{X}=\pm\frac{2\sqrt{x}}{b^2}\mp\frac{2a}{b^3}Y.$$

(110) $$\int \frac{\sqrt{x^3}\,dx}{X}=\pm\frac{2}{3}\cdot\frac{\sqrt{x^3}}{b^2}-\frac{2a^2\sqrt{x}}{b^4}+\frac{2a^3}{b^5}Y.$$

(111) $$\int \frac{\sqrt{x}\,dx}{X^2}=\mp\frac{\sqrt{x}}{b^2X}+\frac{1}{ab^3}Y.$$

(112) $$\int \frac{\sqrt{x^3}\,dx}{X^2}=\pm\frac{2\sqrt{x^3}}{b^2X}+\frac{3a^2\sqrt{x}}{b^4X}-\frac{3a}{b^5}Y.$$

(113) $$\int \frac{dx}{X\sqrt{x}}=\frac{2}{ab}Y.$$

(114) $$\int \frac{dx}{X\sqrt{x^3}}=-\frac{2}{a^2\sqrt{x}}\mp\frac{2b}{a^3}Y.$$

(115) $\int \frac{dx}{X^2\sqrt{x}} = \frac{\sqrt{x}}{a^2 X} + \frac{1}{a^3 b} Y.$

(116) $\int \frac{dx}{X^2\sqrt{x^3}} = -\frac{2}{a^2 X\sqrt{x}} \mp \frac{3b^2\sqrt{x}}{a^4 X} \mp \frac{3b}{a^5} Y.$

Other integrals involving $\sqrt{x}$

(117) $\int \frac{\sqrt{x}\,dx}{a^4 + x^2}$

$$= -\frac{1}{2a\sqrt{2}} \ln\left|\frac{x + a\sqrt{2x} + a^2}{x - a\sqrt{2x} + a^2}\right| + \frac{1}{a\sqrt{2}} \arctan \frac{a\sqrt{2x}}{a^2 - x}, \quad a \neq 0.$$

(118) $\int \frac{dx}{(a^4 + x^2)\sqrt{x}}$

$$= \frac{1}{2a^3\sqrt{2}} \ln\left|\frac{x + a\sqrt{2x} + a^2}{x - a\sqrt{2x} + a^2}\right| + \frac{1}{a^3\sqrt{2}} \arctan \frac{a\sqrt{2x}}{a^2 - x}, \quad a \neq 0.$$

(119) $\int \frac{\sqrt{x}\,dx}{a^4 - x^2} = \frac{1}{2a} \ln\left|\frac{a + \sqrt{x}}{a - \sqrt{x}}\right| - \frac{1}{a} \arctan \frac{\sqrt{x}}{a}, \quad a \neq 0.$

(120) $\int \frac{dx}{(a^4 - x^2)\sqrt{x}} = \frac{1}{2a^3} \ln\left|\frac{a + \sqrt{x}}{a - \sqrt{x}}\right| + \frac{1}{a^3} \arctan \frac{\sqrt{x}}{a}, \quad a \neq 0.$

Integrals involving $\sqrt{ax + b}$, where $a \neq 0$

Notation: $X = ax + b$.

(121) $\int \sqrt{X}\,dx = \frac{2}{3a}\sqrt{X^3}.$

(122) $\int x\sqrt{X}\,dx = \frac{2(3ax - 2b)\sqrt{X^3}}{15a^2}.$

(123) $\int x^2\sqrt{X}\,dx = \frac{2(15a^2x^2 - 12abx + 8b^2)\sqrt{X^3}}{105a^3}.$

(124) $\int \frac{dx}{\sqrt{X}} = \frac{2\sqrt{X}}{a}.$

(125) $\int \frac{x\,dx}{\sqrt{X}} = \frac{2(ax - 2b)}{3a^2}\sqrt{X}.$

(126) $\int \frac{x^2\,dx}{\sqrt{X}} = \frac{2(3a^2x^2 - 4abx + 8b^2)\sqrt{X}}{15a^3}.$

(127) $$\int \frac{dx}{x\sqrt{X}} = -\frac{2}{\sqrt{b}} \operatorname{ar\,tanh} \sqrt{\frac{X}{b}} = \frac{1}{\sqrt{b}} \ln \left| \frac{\sqrt{X}-\sqrt{b}}{\sqrt{X}+\sqrt{b}} \right|, \quad b > 0,$$

$$= \frac{2}{\sqrt{-b}} \arctan \sqrt{\frac{X}{-b}}, \quad b < 0.$$

(128) $$\int \frac{\sqrt{X}}{x} dx = 2\sqrt{X} + b \int \frac{dx}{x\sqrt{X}} \quad \text{(see (127))}.$$

(129) $$\int \frac{dx}{x^2\sqrt{X}} = -\frac{\sqrt{X}}{bx} - \frac{a}{2b} \int \frac{dx}{x\sqrt{X}} \quad \text{(see (127))}.$$

(130) $$\int \frac{\sqrt{X}}{x^2} dx = -\frac{\sqrt{X}}{x} + \frac{a}{2} \int \frac{dx}{x\sqrt{X}} \quad \text{(see (127))}.$$

(131) $$\int \frac{dx}{x^n\sqrt{X}} = -\frac{\sqrt{X}}{(n-1)\,bx^{n-1}} - \frac{(2n-3)\,a}{(2n-2)\,b} \int \frac{dx}{x^{n-1}\sqrt{X}}.$$

(132) $$\int \sqrt{X^3}\, dx = \frac{2\sqrt{X^5}}{5a}.$$

(133) $$\int x\sqrt{X^3}\, dx = \frac{2}{35a^2}\left(5\sqrt{X^7} - 7b\sqrt{X^5}\right).$$

(134) $$\int x^2\sqrt{X^3}\, dx = \frac{2}{a^3}\left(\frac{\sqrt{X^9}}{9} - \frac{2b\sqrt{X^7}}{7} + \frac{b^2\sqrt{X^5}}{5}\right).$$

(135) $$\int \frac{\sqrt{X^3}}{x} dx = \frac{2\sqrt{X^3}}{3} + 2b\sqrt{X} + b^2 \int \frac{dx}{x\sqrt{X}} \quad \text{(see (127))}.$$

(136) $$\int \frac{x\,dx}{\sqrt{X^3}} = \frac{2}{a^2}\left(\sqrt{X} + \frac{b}{\sqrt{X}}\right).$$

(137) $$\int \frac{x^2\,dx}{\sqrt{X^3}} = \frac{2}{a^3}\left(\frac{\sqrt{X^3}}{3} - 2b\sqrt{X} - \frac{b^2}{\sqrt{X}}\right).$$

(138) $$\int \frac{dx}{x\sqrt{X^3}} = \frac{2}{b\sqrt{X}} + \frac{1}{b} \int \frac{dx}{x\sqrt{X}} \quad \text{(see (127))}.$$

(139) $$\int \frac{dx}{x^2\sqrt{X^3}} = -\frac{1}{bx\sqrt{X}} - \frac{3a}{b^2\sqrt{X}} - \frac{3a}{2b^2} \int \frac{dx}{x\sqrt{X}} \quad \text{(see (127))}.$$

(140) $$\int X^{\pm n/2}\, dx = \frac{2X^{(2\pm n)/2}}{a(2 \pm n)}.$$

(141) $$\int xX^{\pm n/2}\, dx = \frac{2}{a^2}\left(\frac{X^{(4\pm n)/2}}{4 \pm n} - \frac{bX^{(2\pm n)/2}}{2 \pm n}\right).$$

(142) $$\int x^2 X^{\pm n/2} dx = \frac{2}{a^3}\left(\frac{X^{(6\pm n)/2}}{6\pm n} - \frac{2b X^{(4\pm n)/2}}{4\pm n} + \frac{b^2 X^{(2\pm n)/2}}{2\pm n}\right).$$

(143) $$\int \frac{X^{n/2} dx}{x} = \frac{2X^{n/2}}{n} + b\int \frac{X^{(n-2)/2}}{x} dx.$$

(144) $$\int \frac{dx}{xX^{n/2}} = \frac{2}{(n-2)\,bX^{(n-2)/2}} + \frac{1}{b}\int \frac{dx}{xX^{(n-2)/2}}.$$

(145) $$\int \frac{dx}{x^2 X^{n/2}} = -\frac{1}{bxX^{(n-2)/2}} - \frac{na}{2b}\int \frac{dx}{xX^{n/2}}.$$

Integrals involving $\sqrt{ax+b}$ and $\sqrt{fx+g}$, where $a \neq 0$ and $f \neq 0$

Notation:

$$X = ax+b, \qquad Y = fx+g, \qquad \Delta = bf - ag.$$

(146) $$\int \frac{dx}{\sqrt{XY}} = \frac{2}{\sqrt{-af}} \operatorname{arc\,tan} \sqrt{-\frac{fX}{aY}}, \quad af < 0,$$

$$= \frac{2}{\sqrt{af}} \operatorname{ar\,tanh} \sqrt{\frac{fX}{aY}} = \frac{1}{2\sqrt{af}} \ln\left|\sqrt{aY} + \sqrt{fX}\right|, \quad af > 0.$$

(147) $$\int \frac{x\,dx}{\sqrt{XY}} = \frac{\sqrt{XY}}{af} - \frac{ag+bf}{2af}\int \frac{dx}{\sqrt{XY}} \quad \text{(see (146))}.$$

(148) $$\int \frac{dx}{\sqrt{X}\sqrt{Y^3}} = -\frac{2\sqrt{X}}{\Delta\sqrt{Y}}.$$

(149) $$\int \frac{dx}{Y\sqrt{X}} = \frac{2}{\sqrt{-\Delta f}} \operatorname{arc\,tan} \frac{f\sqrt{X}}{\sqrt{-\Delta f}}, \quad \Delta f < 0,$$

$$= \frac{1}{\sqrt{\Delta f}} \ln\left|\frac{f\sqrt{X} - \sqrt{\Delta f}}{f\sqrt{X} + \sqrt{\Delta f}}\right|, \quad \Delta f > 0.$$

(150) $$\int \sqrt{XY}\,dx = \frac{\Delta + 2aY}{4af}\sqrt{XY} - \frac{\Delta^2}{8af}\int \frac{dx}{\sqrt{XY}} \quad \text{(see (146))}.$$

(151) $$\int \sqrt{\frac{Y}{X}}\,dx = \frac{1}{a}\sqrt{XY} - \frac{\Delta}{2a}\int \frac{dx}{\sqrt{XY}} \quad \text{(see (146))}.$$

(152) $$\int \frac{\sqrt{X}\,dx}{Y} = \frac{2\sqrt{X}}{f} + \frac{\Delta}{f}\int \frac{dx}{Y\sqrt{X}} \quad \text{(see (149))}.$$

(153) $$\int \frac{Y^n\,dx}{\sqrt{X}} = \frac{2}{(2n+1)\,a}\left(\sqrt{X}\,Y^n - n\Delta \int \frac{Y^{n-1}\,dx}{\sqrt{X}}\right).$$

(154) $$\int \frac{dx}{\sqrt{X}Y^n} = -\frac{1}{(n-1)\Delta}\left[\frac{\sqrt{X}}{Y^{n-1}} + \left(n-\frac{3}{2}\right)a\int\frac{dx}{\sqrt{X}\,Y^{n-1}}\right].$$

(155) $$\int \sqrt{X}\,Y^n\,dx = \frac{1}{(2n+3)f}\left(2\sqrt{X}\,Y^{n+1} + \Delta\int\frac{Y^n\,dx}{\sqrt{X}}\right) \text{(see (153))}.$$

(156) $$\int \frac{\sqrt{X}\,dx}{Y^n} = \frac{1}{(n-1)f}\left(-\frac{\sqrt{X}}{Y^{n-1}} + \frac{a}{2}\int\frac{dx}{\sqrt{X}\,Y^{n-1}}\right).$$

Integrals involving $\sqrt{a^2-x^2}$, where $a \neq 0$

Notation: $X = a^2 - x^2$.

(157) $$\int \sqrt{X}\,dx = \frac{1}{2}\left(x\sqrt{X} + a^2 \arcsin\frac{x}{a}\right).$$

(158) $$\int x\sqrt{X}\,dx = -\frac{1}{3}\sqrt{X^3}.$$

(159) $$\int x^2\sqrt{X}\,dx = -\frac{x}{4}\sqrt{X^3} + \frac{a^2}{8}\left(x\sqrt{X} + a^2\arcsin\frac{x}{a}\right).$$

(160) $$\int x^3\sqrt{X}\,dx = \frac{\sqrt{X^5}}{5} - a^2\frac{\sqrt{X^3}}{3}.$$

(161) $$\int \frac{\sqrt{X}}{x}\,dx = \sqrt{X} - a\ln\left|\frac{a+\sqrt{X}}{x}\right|.$$

(162) $$\int \frac{\sqrt{X}}{x^2}\,dx = -\frac{\sqrt{X}}{x} - \arcsin\frac{x}{a}.$$

(163) $$\int \frac{\sqrt{X}}{x^3}\,dx = -\frac{\sqrt{X}}{2x^2} + \frac{1}{2a}\ln\left|\frac{a+\sqrt{X}}{x}\right|.$$

(164) $$\int \frac{dx}{\sqrt{X}} = \arcsin\frac{x}{a}.$$

(165) $$\int \frac{x\,dx}{\sqrt{X}} = -\sqrt{X}.$$

(166) $$\int \frac{x^2\,dx}{\sqrt{X}} = -\frac{x}{2}\sqrt{X} + \frac{a^2}{2}\arcsin\frac{x}{a}.$$

(167) $$\int \frac{x^3\,dx}{\sqrt{X}} = \frac{\sqrt{X^3}}{3} - a^2\sqrt{X}.$$

(168) $$\int \frac{dx}{x\sqrt{X}} = -\frac{1}{a}\ln\left|\frac{a+\sqrt{X}}{x}\right|.$$

(169) $$\int \frac{dx}{x^2\sqrt{X}} = -\frac{\sqrt{X}}{a^2 x}.$$

(170) $$\int \frac{dx}{x^3\sqrt{X}} = -\frac{\sqrt{X}}{2a^2x^2} - \frac{1}{2a^3}\ln\left|\frac{a+\sqrt{X}}{x}\right|.$$

(171) $$\int \sqrt{X^3}\,dx = \frac{1}{4}\left(x\sqrt{X^3} + \frac{3a^2x}{2}\sqrt{X} + \frac{3a^4}{2}\arcsin\frac{x}{a}\right).$$

(172) $$\int x\sqrt{X^3}\,dx = -\frac{1}{5}\sqrt{X^5}.$$

(173) $$\int x^2\sqrt{X^3}\,dx = -\frac{x\sqrt{X^5}}{6} + \frac{a^2x\sqrt{X^3}}{24} + \frac{a^4x\sqrt{X}}{16} + \frac{a^6}{16}\arcsin\frac{x}{a}.$$

(174) $$\int x^3\sqrt{X^3}\,dx = \frac{\sqrt{X^7}}{7} - \frac{a^2\sqrt{X^5}}{5}.$$

(175) $$\int \frac{\sqrt{X^3}}{x}\,dx = \frac{\sqrt{X^3}}{3} + a^2\sqrt{X} - a^3\ln\left|\frac{a+\sqrt{X}}{x}\right|.$$

(176) $$\int \frac{\sqrt{X^3}}{x^2}\,dx = -\frac{\sqrt{X^3}}{x} - \frac{3}{2}x\sqrt{X} - \frac{3}{2}a^2\arcsin\frac{x}{a}.$$

(177) $$\int \frac{\sqrt{X^3}}{x^3}\,dx = -\frac{\sqrt{X^3}}{2x^2} - \frac{3\sqrt{X}}{2} + \frac{3a}{2}\ln\left|\frac{a+\sqrt{X}}{x}\right|.$$

(178) $$\int \frac{dx}{\sqrt{X^3}} = \frac{x}{a^2\sqrt{X}}.$$

(179) $$\int \frac{x\,dx}{\sqrt{X^3}} = \frac{1}{\sqrt{X}}.$$

(180) $$\int \frac{x^2\,dx}{\sqrt{X^3}} = \frac{x}{\sqrt{X}} - \arcsin\frac{x}{a}.$$

(181) $$\int \frac{x^3\,dx}{\sqrt{X^3}} = \sqrt{X} + \frac{a^2}{\sqrt{X}}.$$

(182) $$\int \frac{dx}{x\sqrt{X^3}} = \frac{1}{a^2\sqrt{X}} - \frac{1}{a^3}\ln\left|\frac{a+\sqrt{X}}{x}\right|.$$

(183) $$\int \frac{dx}{x^2\sqrt{X^3}} = \frac{1}{a^4}\left(-\frac{\sqrt{X}}{x} + \frac{x}{\sqrt{X}}\right).$$

(184) $$\int \frac{dx}{x^3\sqrt{X^3}} = -\frac{1}{2a^2x^2\sqrt{X}} + \frac{3}{2a^4\sqrt{X}} - \frac{3}{2a^5}\ln\left|\frac{a+\sqrt{X}}{x}\right|.$$

Integrals involving $\sqrt{x^2+a^2}$, where $a \neq 0$

Notation: $X = x^2 + a^2$.

(185) $$\int \sqrt{X}\,dx = \frac{1}{2}\left(x\sqrt{X} + a^2 \operatorname{ar\,sinh}\frac{x}{a}\right) + C$$
$$= \tfrac{1}{2}\left(x\sqrt{X} + a^2 \ln\left|x+\sqrt{X}\right|\right) + C_1.$$

(186) $$\int x\sqrt{X}\,dx = \frac{1}{3}\sqrt{X^3}.$$

(187) $$\int x^2\sqrt{X}\,dx = \frac{x}{4}\sqrt{X^3} - \frac{a^2}{8}\left(x\sqrt{X} + a^2 \operatorname{ar\,sinh}\frac{x}{a}\right) + C$$
$$= \frac{x}{4}\sqrt{X^3} - \frac{a^2}{8}\left(x\sqrt{X} + a^2 \ln\left|x+\sqrt{X}\right|\right) + C_1.$$

(188) $$\int x^3\sqrt{X}\,dx = \frac{\sqrt{X^5}}{5} - \frac{a^2\sqrt{X^3}}{3}.$$

(189) $$\int \frac{\sqrt{X}}{x}\,dx = \sqrt{X} - a\ln\left|\frac{a+\sqrt{X}}{x}\right|.$$

(190) $$\int \frac{\sqrt{X}}{x^2}\,dx = -\frac{\sqrt{X}}{x} + \operatorname{ar\,sinh}\frac{x}{a} + C$$
$$= -\frac{\sqrt{X}}{x} + \ln\left|x+\sqrt{X}\right| + C_1.$$

(191) $$\int \frac{\sqrt{X}}{x^3}\,dx = -\frac{\sqrt{X}}{2x^2} - \frac{1}{2a}\ln\left|\frac{a+\sqrt{X}}{x}\right|.$$

(192) $$\int \frac{dx}{\sqrt{X}} = \operatorname{ar\,sinh}\frac{x}{a} + C = \ln\left|x+\sqrt{X}\right| + C_1.$$

(193) $$\int \frac{x\,dx}{\sqrt{X}} = \sqrt{X}.$$

(194) $$\int \frac{x^2\,dx}{\sqrt{X}} = \frac{x}{2}\sqrt{X} - \frac{a^2}{2}\operatorname{ar\,sinh}\frac{x}{a} + C$$
$$= \frac{x}{2}\sqrt{X} - \frac{a^2}{2}\ln\left|x+\sqrt{X}\right| + C_1.$$

(195) $$\int \frac{x^3\,dx}{\sqrt{X}} = \frac{\sqrt{X^3}}{3} - a^2\sqrt{X}.$$

(196) $$\int \frac{dx}{x\sqrt{X}} = -\frac{1}{a}\ln\left|\frac{a+\sqrt{X}}{x}\right|.$$

(197) $$\int \frac{dx}{x^2\sqrt{X}} = -\frac{\sqrt{X}}{a^2 x}.$$

(198) $$\int \frac{dx}{x^3\sqrt{X}} = -\frac{\sqrt{X}}{2a^2x^2} + \frac{1}{2a^3}\ln\left|\frac{a+\sqrt{X}}{x}\right|.$$

(199) $$\int \sqrt{X^3}\,dx = \frac{1}{4}\left(x\sqrt{X^3} + \frac{3a^2x}{2}\sqrt{X} + \frac{3a^4}{2}\operatorname{ar\,sinh}\frac{x}{a}\right) + C$$
$$= \frac{1}{4}\left(x\sqrt{X^3} + \frac{3a^2x}{2}\sqrt{X} + \frac{3a^4}{2}\ln|x+\sqrt{X}|\right) + C_1.$$

(200) $$\int x\sqrt{X^3}\,dx = \frac{1}{5}\sqrt{X^5}.$$

(201) $$\int x^2\sqrt{X^3}\,dx$$
$$= \frac{x\sqrt{X^5}}{6} - \frac{a^2x\sqrt{X^3}}{24} - \frac{a^4x\sqrt{X}}{16} - \frac{a^6}{16}\operatorname{ar\,sinh}\frac{x}{a} + C$$
$$= \frac{x\sqrt{X^5}}{6} + \frac{a^2x\sqrt{X^3}}{24} - \frac{a^4x\sqrt{X}}{16} - \frac{a^6}{16}\ln|x+\sqrt{X}| + C_1.$$

(202) $$\int x^3\sqrt{X^3}\,dx = \frac{\sqrt{X^7}}{7} - \frac{a^2\sqrt{X^5}}{5}.$$

(203) $$\int \frac{\sqrt{X^3}}{x}\,dx = \frac{\sqrt{X^3}}{3} + a^2\sqrt{X} - a^3\ln\left|\frac{a+\sqrt{X}}{x}\right|.$$

(204) $$\int \frac{\sqrt{X^3}}{x^2}\,dx = -\frac{\sqrt{X^3}}{x} + \frac{3}{2}x\sqrt{X} + \frac{3}{2}a^2\operatorname{ar\,sinh}\frac{x}{a} + C$$
$$= -\frac{\sqrt{X^3}}{x} + \frac{3}{2}x\sqrt{X} + \frac{3}{2}a^2\ln|x+\sqrt{X}| + C_1.$$

(205) $$\int \frac{\sqrt{X^3}}{x^3}\,dx = -\frac{\sqrt{X^3}}{2x^2} + \frac{3}{2}\sqrt{X} - \frac{3}{2}a\ln\left|\frac{a+\sqrt{X}}{x}\right|.$$

(206) $$\int \frac{dx}{\sqrt{X^3}} = \frac{x}{a^2\sqrt{X}}.$$

(207) $$\int \frac{x\,dx}{\sqrt{X^3}} = -\frac{1}{\sqrt{X}}.$$

(208) $$\int \frac{x^2\,dx}{\sqrt{X^3}} = -\frac{x}{\sqrt{X}} + \operatorname{ar\,sinh}\frac{x}{a} + C$$
$$= -\frac{x}{\sqrt{X}} + \ln|x+\sqrt{X}| + C_1.$$

(209) $$\int \frac{x^3\,dx}{\sqrt{X^3}} = \sqrt{X} + \frac{a^2}{\sqrt{X}}.$$

(210) $$\int \frac{dx}{x\sqrt{X^3}} = \frac{1}{a^2\sqrt{X}} - \frac{1}{a^3}\ln\left|\frac{a+\sqrt{X}}{x}\right|.$$

(211) $$\int \frac{dx}{x^2\sqrt{X^3}} = -\frac{1}{a^4}\left(\frac{\sqrt{X}}{x} + \frac{x}{\sqrt{X}}\right).$$

(212) $$\int \frac{dx}{x^3\sqrt{X^3}} = -\frac{1}{2a^2x^2\sqrt{X}} - \frac{3}{2a^4\sqrt{X}} + \frac{3}{2a^5}\ln\left|\frac{a+\sqrt{X}}{x}\right|.$$

Integrals involving $\sqrt{x^2-a^2}$, where $a \neq 0$

Notation: $X = x^2 - a^2$.

(213) $$\int \sqrt{X}\,dx = \frac{1}{2}\left(x\sqrt{X} - a^2 \operatorname{ar\,cosh}\frac{x}{a}\right) + C$$
$$= \tfrac{1}{2}\left(x\sqrt{X} - a^2 \ln\left|x+\sqrt{X}\right|\right) + C_1.$$

(214) $$\int x\sqrt{X}\,dx = \frac{1}{3}\sqrt{X^3}.$$

(215) $$\int x^2\sqrt{X}\,dx = \frac{x}{4}\sqrt{X^3} + \frac{a^2}{8}\left(x\sqrt{X} - a^2 \operatorname{ar\,cosh}\frac{x}{a}\right) + C$$
$$= \frac{x}{4}\sqrt{X^3} + \frac{a^2}{8}\left(x\sqrt{X} - a^2\ln\left|x+\sqrt{X}\right|\right) + C_1.$$

(216) $$\int x^3\sqrt{X}\,dx = \frac{\sqrt{X^5}}{5} + \frac{a^2\sqrt{X^3}}{3}.$$

(217) $$\int \frac{\sqrt{X}}{x}\,dx = \sqrt{X} - a \arccos\frac{a}{x}.$$

(218) $$\int \frac{\sqrt{X}}{x^2}\,dx = -\frac{\sqrt{X}}{x} + \operatorname{ar\,cosh}\frac{x}{a} + C$$
$$= -\frac{\sqrt{X}}{x} + \ln\left|x+\sqrt{X}\right| + C_1.$$

(219) $$\int \frac{\sqrt{X}}{x^3}\,dx = -\frac{\sqrt{X}}{2x^2} + \frac{1}{2a}\arccos\frac{a}{x}.$$

(220) $$\int \frac{dx}{\sqrt{X}} = \operatorname{ar\,cosh}\frac{x}{a} + C = \ln\left|x+\sqrt{X}\right| + C_1.$$

(221) $$\int \frac{x\,dx}{\sqrt{X}} = \sqrt{X}.$$

(222) $\int \frac{x^2\,dx}{\sqrt{X}} = \frac{x}{2}\sqrt{X} + \frac{a^2}{2}\operatorname{ar\,cosh}\frac{x}{a} + C$

$= \frac{x}{2}\sqrt{X} + \frac{a^2}{2}\ln|x+\sqrt{X}| + C_1.$

(223) $\int \frac{x^3\,dx}{\sqrt{X}} = \frac{\sqrt{X^3}}{3} + a^2\sqrt{X}.$

(224) $\int \frac{dx}{x\sqrt{X}} = \frac{1}{a}\arccos\frac{a}{x}.$

(225) $\int \frac{dx}{x^2\sqrt{X}} = \frac{\sqrt{X}}{a^2 x}.$

(226) $\int \frac{dx}{x^3\sqrt{X}} = \frac{\sqrt{X}}{2a^2x^2} + \frac{1}{2a^3}\arccos\frac{a}{x}.$

(227) $\int \sqrt{X^3}\,dx = \frac{1}{4}\left(x\sqrt{X^3} - \frac{3a^2x}{2}\sqrt{X} + \frac{3a^4}{2}\operatorname{ar\,cosh}\frac{x}{a}\right) + C$

$= \frac{1}{4}\left(x\sqrt{X^3} - \frac{3a^2x}{2}\sqrt{X} + \frac{3a^4}{2}\ln|x+\sqrt{X}|\right) + C_1.$

(228) $\int x\sqrt{X^3}\,dx = \frac{1}{5}\sqrt{X^5}.$

(229) $\int x^2\sqrt{X^3}\,dx$

$= \frac{x\sqrt{X^5}}{6} + \frac{a^2x\sqrt{X^3}}{24} - \frac{a^4x\sqrt{X}}{16} + \frac{a^6}{16}\operatorname{ar\,cosh}\frac{x}{a} + C$

$= \frac{x\sqrt{X^5}}{6} + \frac{a^2x\sqrt{X^3}}{24} - \frac{a^4x\sqrt{X}}{16} + \frac{a^6}{16}\ln|x+\sqrt{X}| + C_1.$

(230) $\int x^3\sqrt{X^3}\,dx = \frac{\sqrt{X^7}}{7} + \frac{a^2\sqrt{X^5}}{5}.$

(231) $\int \frac{\sqrt{X^3}}{x}\,dx = \frac{\sqrt{X^3}}{3} - a^2\sqrt{X} + a^3\arccos\frac{a}{x}.$

(232) $\int \frac{\sqrt{X^3}}{x^2}\,dx = -\frac{\sqrt{X^3}}{2} + \frac{3}{2}x\sqrt{X} - \frac{3}{2}a^2\operatorname{ar\,cosh}\frac{x}{a} + C$

$= -\frac{\sqrt{X^3}}{2} + \frac{3}{2}x\sqrt{X} - \frac{3}{2}a^2\ln|x+\sqrt{X}| + C_1.$

(233) $\int \frac{\sqrt{X^3}}{x^3}\,dx = -\frac{\sqrt{X^3}}{2x^2} + \frac{3\sqrt{X}}{2} - \frac{3}{2}a\arccos\frac{a}{x}.$

(234) $$\int \frac{dx}{\sqrt{X^3}} = -\frac{x}{a^2\sqrt{X}}.$$

(235) $$\int \frac{x\,dx}{\sqrt{X^3}} = -\frac{1}{\sqrt{X}}.$$

(236) $$\int \frac{x^2\,dx}{\sqrt{X^3}} = -\frac{x}{\sqrt{X}} + \operatorname{ar\,cosh}\frac{x}{a} + C$$
$$= -\frac{x}{\sqrt{X}} + \ln|x+\sqrt{X}| + C_1.$$

(237) $$\int \frac{x^3\,dx}{\sqrt{X^3}} = \sqrt{X} - \frac{a^2}{\sqrt{X}}.$$

(238) $$\int \frac{dx}{x\sqrt{X^3}} = -\frac{1}{a^2\sqrt{X}} - \frac{1}{a^3}\arccos\frac{a}{x}.$$

(239) $$\int \frac{dx}{x^2\sqrt{X^3}} = -\frac{1}{a^4}\left(\frac{\sqrt{X}}{x} + \frac{x}{\sqrt{X}}\right).$$

(240) $$\int \frac{dx}{x^3\sqrt{X^3}} = \frac{1}{2a^2x^2\sqrt{X}} - \frac{3}{2a^4\sqrt{X}} - \frac{3}{2a^5}\arccos\frac{a}{x}.$$

Integrals involving ax^2+bx+c, where $a \neq 0$

Notation: $X = ax^2+bx+c,\ \Delta = 4ac - b^2,\ k = \frac{4a}{\Delta}.$

(241) $$\int \frac{dx}{\sqrt{X}} = \frac{1}{\sqrt{a}}\ln|2\sqrt{aX}+2ax+b| + C, \quad a>0,$$
$$= \frac{1}{\sqrt{a}}\operatorname{ar\,sinh}\frac{2ax+b}{\sqrt{\Delta}} + C_1, \quad a>0, \Delta>0,$$
$$= \frac{1}{\sqrt{a}}\ln|2ax+b|, \quad a>0, \Delta=0,$$
$$= -\frac{1}{\sqrt{-a}}\arcsin\frac{2ax+b}{\sqrt{-\Delta}}, \quad a<0, \Delta<0.$$

(242) $$\int \frac{dx}{X\sqrt{X}} = \frac{2(2ax+b)}{\Delta\sqrt{X}}.$$

(243) $$\int \frac{dx}{X^2\sqrt{X}} = \frac{2(2ax+b)}{3\Delta\sqrt{X}}\left(\frac{1}{X} + 2k\right).$$

(244) $$\int \frac{dx}{X^{(2n+1)/2}} = \frac{2(2ax+b)}{(2n-1)\Delta X^{(2n-1)/2}} + \frac{2k(n-1)}{2n-1}\int \frac{dx}{X^{(2n-1)/2}}.$$

(245) $$\int \sqrt{X}\,dx = \frac{(2ax+b)\sqrt{X}}{4a} + \frac{1}{2k}\int \frac{dx}{\sqrt{X}} \quad \text{(see (241))}.$$

(246) $$\int X\sqrt{X}\,dx = \frac{(2ax+b)\sqrt{X}}{8a}\left(X + \frac{3}{2k}\right) + \frac{3}{8k^2}\int \frac{dx}{\sqrt{X}} \text{ (see (241))}.$$

(247) $$\int X^2\sqrt{X}\,dx$$

$$= \frac{(2ax+b)\sqrt{X}}{12a}\left(X^2 + \frac{5X}{4k} + \frac{15}{8k^2}\right) + \frac{5}{16k^3}\int \frac{dx}{\sqrt{X}} \quad \text{(see (241))}.$$

(248) $$\int X^{(2n+1)/2}\,dx = \frac{(2ax+b)\,X^{(2n+1)/2}}{4a(n+1)} + \frac{2n+1}{2k(n+1)}\int X^{(2n-1)/2}dx.$$

(249) $$\int \frac{x\,dx}{\sqrt{X}} = \frac{\sqrt{X}}{a} - \frac{b}{2a}\int \frac{dx}{\sqrt{X}} \quad \text{(see (241))}.$$

(250) $$\int \frac{x\,dx}{X\sqrt{X}} = -\frac{2(bx+2c)}{\Delta\sqrt{X}}.$$

(251) $$\int \frac{x\,dx}{X^{(2n+1)/2}} = -\frac{1}{(2n-1)\,aX^{(2n-1)/2}} - \frac{b}{2a}\int \frac{dx}{X^{(2n+1)/2}} \text{ (see (244))}.$$

(252) $$\int \frac{x^2\,dx}{\sqrt{X}} = \left(\frac{x}{2a} - \frac{3b}{4a^2}\right)\sqrt{X} + \frac{3b^2-4ac}{8a^2}\int \frac{dx}{\sqrt{X}} \quad \text{(see (241))}.$$

(253) $$\int \frac{x^2\,dx}{X\sqrt{X}} = \frac{(2b^2-4ac)\,x+2bc}{a\Delta\sqrt{X}} + \frac{1}{a}\int \frac{dx}{\sqrt{X}} \quad \text{(see (241))}.$$

(254) $$\int x\sqrt{X}\,dx = \frac{X\sqrt{X}}{3a} - \frac{b(2ax+b)}{8a^2}\sqrt{X} - \frac{b}{4ak}\int \frac{dx}{\sqrt{X}} \text{ (see (241))}.$$

(255) $$\int xX\sqrt{X}\,dx = \frac{X^2\sqrt{X}}{5a} - \frac{b}{2a}\int X\sqrt{X}\,dx \quad \text{(see (246))}.$$

(256) $$\int xX^{(2n+1)/2}\,dx = \frac{X^{(2n+3)/2}}{(2n+3)\,a} - \frac{b}{2a}\int X^{(2n+1)/2}\,dx \quad \text{(see (248))}.$$

(257) $$\int x^2\sqrt{X}\,dx = \left(x - \frac{5b}{6a}\right)\frac{X\sqrt{X}}{4a} + \frac{5b^2-4ac}{16a^2}\int \sqrt{X}\,dx \text{ (see (245))}.$$

(258) $$\int \frac{dx}{x\sqrt{X}} = -\frac{1}{\sqrt{c}}\ln\left|\frac{2\sqrt{cX}}{x} + \frac{2c}{x} + b\right| + C, \quad c>0,$$

$$= -\frac{1}{\sqrt{c}}\,\text{ar sinh}\,\frac{bx+2c}{x\sqrt{\Delta}} + C_1, \quad c>0, \Delta>0,$$

$$= -\frac{1}{\sqrt{c}}\ln\left|\frac{bx+2c}{x}\right|, \quad c>0, \Delta=0,$$

$$= \frac{1}{\sqrt{-c}}\,\text{arc sin}\,\frac{bx+2c}{x\sqrt{-\Delta}}, \quad c<0, \Delta<0.$$

(259) $$\int \frac{dx}{x^2\sqrt{X}} = -\frac{\sqrt{X}}{cx} - \frac{b}{2c}\int \frac{dx}{x\sqrt{X}} \quad \text{(see 258).}$$

(260) $$\int \frac{\sqrt{X}\,dx}{x} = \sqrt{X} + \frac{b}{2}\int \frac{dx}{\sqrt{X}} + c\int \frac{dx}{x\sqrt{X}} \quad \text{(see (241) and (258)).}$$

(261) $$\int \frac{\sqrt{X}\,dx}{x^2} = -\frac{\sqrt{X}}{x} + a\int \frac{dx}{\sqrt{X}} + \frac{b}{2}\int \frac{dx}{x\sqrt{X}} \quad \text{(see (241) and (258)).}$$

(262) $$\int \frac{x^{(2n+1)/2}}{x}\,dx = \frac{X^{(2n+1)/2}}{2n+1} + \frac{b}{2}\int X^{(2n-1)/2}\,dx + c\int \frac{X^{(2n-1)/2}}{x}\,dx$$

(see (248) and (260))

(263) $$\int \frac{dx}{x\sqrt{ax^2+bx}} = -\frac{2}{bx}\sqrt{ax^2+bx}\,.$$

(264) $$\int \frac{dx}{\sqrt{2ax-x^2}} = \arcsin\frac{x-a}{a}.$$

(265) $$\int \frac{x\,dx}{\sqrt{2ax-x^2}} = -\sqrt{2ax-x^2} + a\arcsin\frac{x-a}{a}.$$

(266) $$\int \sqrt{2ax-x^2}\,dx = \frac{x-a}{2}\sqrt{2ax-x^2} + \frac{a^2}{2}\arcsin\frac{x-a}{a}.$$

(267) $$\int \frac{dx}{(ax^2+b)\sqrt{fx^2+g}} = \frac{1}{\sqrt{b}\sqrt{ag-bf}}\arctan\frac{x\sqrt{ag-bf}}{\sqrt{b}\sqrt{fx^2+g}},$$

$$ag-bf>0,$$

$$= \frac{1}{2\sqrt{b}\sqrt{bf-ag}}\ln\left|\frac{\sqrt{b}\sqrt{fx^2+g}+x\sqrt{bf-ag}}{\sqrt{b}\sqrt{fx^2+g}-x\sqrt{bf-ag}}\right|, \quad ag-bf<0.$$

Integrals involving other irrational expressions

(268) $$\int \sqrt[n]{ax+b}\,dx = \frac{n(ax+b)}{(n+1)a}\sqrt[n]{ax+b}, \quad a \neq 0.$$

(269) $$\int \frac{dx}{\sqrt[n]{ax+b}}\,dx = \frac{n(ax+b)}{(n-1)a}\cdot\frac{1}{\sqrt[n]{ax+b}}, \quad a \neq 0.$$

(270) $$\int \frac{dx}{x\sqrt{x^n+a^2}} = -\frac{2}{na}\ln\left|\frac{a+\sqrt{x^n+a^2}}{\sqrt{x^n}}\right|, \quad a \neq 0.$$

(271) $$\int \frac{dx}{x\sqrt{x^n-a^2}} = \frac{2}{na}\arccos\frac{a}{\sqrt{x^n}}, \quad a \neq 0.$$

(272) $$\int \frac{\sqrt{x}\,dx}{\sqrt{a^3-x^3}} = \frac{2}{3}\arcsin\sqrt{\left(\frac{x}{a}\right)^3}, \quad a \neq 0.$$

Recurrence formulas for the integral of the binomial differential

(273) $\int x^m (ax^n + b)^p \, dx$

$$= \frac{1}{m+np+1}\left(x^{m+1}(ax^n+b)^p + npb\int x^m (ax^n+b)^{p-1}\,dx\right),$$

$$= \frac{1}{bn(p+1)}\left(-x^{m+1}(ax^n+b)^{p+1} + \right.$$

$$\left. + (m+n+np+1)\int x^m (ax^n+b)^{p+1}\,dx\right),$$

$$= \frac{1}{(m+1)\,b}\left(x^{m+1}(ax^n+b)^{p+1} - \right.$$

$$\left. - a(m+n+np+1)\int x^{m+n}(ax^n+b)^p\,dx\right),$$

$$= \frac{1}{a(m+np+1)}\left(x^{m-n+1}(ax^n+b)^{p+1} - \right.$$

$$\left. -(m-n+1)\,b\int x^{m-n}(ax^n+b)^p\,dx\right).$$

Integrals of trigonometric functions [1]

Integrals involving $\sin ax$, where $a \neq 0$

(274) $\int \sin ax\,dx = -\frac{1}{a}\cos ax.$

(275) $\int \sin^2 ax\,dx = \frac{1}{2}x - \frac{1}{4a}\sin 2ax.$

(276) $\int \sin^3 ax\,dx = -\frac{1}{a}\cos ax + \frac{1}{3a}\cos^3 ax.$

(277) $\int \sin^4 ax\,dx = \frac{3}{8}x - \frac{1}{4a}\sin 2ax + \frac{1}{32a}\sin 4ax.$

(278) $\int \sin^n ax\,dx = -\frac{\sin^{n-1} ax\cos ax}{na} + \frac{n-1}{n}\int \sin^{n-2} ax\,dx$

(n is a positive integer).

(279) $\int x\sin ax\,dx = \frac{\sin ax}{a^2} - \frac{x\cos ax}{a}.$

---

[1] Integrals of functions involving $\sin x$ and $\cos x$ together with the hyperbolic functions and the function e^{ax} are given on pp. 448, 449.

(280) $$\int x^2 \sin ax\,dx = \frac{2x}{a^2}\sin ax - \left(\frac{x^2}{a} - \frac{2}{a^3}\right)\cos ax.$$

(281) $$\int x^3 \sin ax\,dx = \left(\frac{3x^2}{a^2} - \frac{6}{a^4}\right)\sin ax - \left(\frac{x^3}{a} - \frac{6x}{a^3}\right)\cos ax.$$

(282) $$\int x^n \sin ax\,dx = -\frac{x^n}{a}\cos ax + \frac{n}{a}\int x^{n-1}\cos ax\,dx, \quad n > 0.$$

(283) $$\int \frac{\sin ax}{x}\,dx = ax - \frac{(ax)^3}{3\cdot 3!} + \frac{(ax)^5}{5\cdot 5!} - \frac{(ax)^7}{7\cdot 7!} + \ldots \text{ [1]}$$

(284) $$\int \frac{\sin ax}{x^2}\,dx = -\frac{\sin ax}{x} + a\int \frac{\cos ax\,dx}{x} \quad \text{(see (322))}.$$

(285) $$\int \frac{\sin ax}{x^n}\,dx = -\frac{1}{n-1}\cdot\frac{\sin ax}{x^{n-1}} + \frac{a}{n-1}\int \frac{\cos ax}{x^{n-1}}\,dx \quad \text{(see (324))}.$$

(286) $$\int \frac{dx}{\sin ax} = \frac{1}{a}\ln\left|\tan\frac{ax}{2}\right|.$$

(287) $$\int \frac{dx}{\sin^2 ax} = -\frac{1}{a}\cot ax.$$

(288) $$\int \frac{dx}{\sin^3 ax} = -\frac{\cos ax}{2a\sin^2 ax} + \frac{1}{2a}\ln\left|\tan\frac{ax}{2}\right|.$$

(289) $$\int \frac{dx}{\sin^n ax} = -\frac{1}{a(n-1)}\cdot\frac{\cos ax}{\sin^{n-1} ax} + \frac{n-2}{n-1}\int \frac{dx}{\sin^{n-2} ax}, \quad n>1.$$

(290) $$\int \frac{x\,dx}{\sin ax} = \frac{1}{a^2}\left(ax + \frac{(ax)^3}{3\cdot 3!} + \frac{7(ax)^5}{3\cdot 5\cdot 5!} + \frac{31(ax)^7}{3\cdot 7\cdot 7!} + \frac{127(ax)^9}{3\cdot 5\cdot 9!} + \right.$$
$$\left. + \ldots + \frac{2(2^{2n-1}-1)}{(2n+1)!}B_n\,(ax)^{2n+1} + \ldots\right) \text{ [2]}.$$

(291) $$\int \frac{x\,dx}{\sin^2 ax} = -\frac{x}{a}\cot ax + \frac{1}{a^2}\ln|\sin ax|.$$

---

[1] The definite integral $\int_0^x \frac{\sin t}{t}\,dt$ is called the *integral sine* and denoted by Si x:

$$\operatorname{Si} x = x - \frac{x^3}{3\cdot 3!} + \frac{x^5}{5\cdot 5!} - \frac{x^7}{7\cdot 7!} + \ldots$$

[2] B_n are Bernoulli's numbers (see p. 354).

(292) $$\int \frac{x\,dx}{\sin^n ax} = -\frac{x\cos ax}{(n-1)a\sin^{n-1} ax} - \frac{1}{(n-1)(n-2)a^2\sin^{n-2} ax} + \frac{n-2}{n-1}\int \frac{x\,dx}{\sin^{n-2} ax}, \quad n > 2.$$

(293) $$\int \frac{dx}{1+\sin ax} = -\frac{1}{a}\tan\left(\frac{\pi}{4} - \frac{ax}{2}\right).$$

(294) $$\int \frac{dx}{1-\sin ax} = \frac{1}{a}\tan\left(\frac{\pi}{4} + \frac{ax}{2}\right).$$

(295) $$\int \frac{x\,dx}{1+\sin ax} = -\frac{x}{a}\tan\left(\frac{\pi}{4} - \frac{ax}{2}\right) + \frac{2}{a^2}\ln\left|\cos\left(\frac{\pi}{4} - \frac{ax}{2}\right)\right|.$$

(296) $$\int \frac{x\,dx}{1-\sin ax} = \frac{x}{a}\cot\left(\frac{\pi}{4} - \frac{ax}{2}\right) + \frac{2}{a^2}\ln\left|\sin\left(\frac{\pi}{4} - \frac{ax}{2}\right)\right|.$$

(297) $$\int \frac{\sin ax\,dx}{1 \pm \sin ax} = \pm x + \frac{1}{a}\tan\left(\frac{\pi}{4} \mp \frac{ax}{2}\right).$$

(298) $$\int \frac{dx}{\sin ax(1 \pm \sin ax)} = \frac{1}{a}\tan\left(\frac{\pi}{4} \mp \frac{ax}{2}\right) + \frac{1}{a}\ln\left|\tan\frac{ax}{2}\right|.$$

(299) $$\int \frac{dx}{(1+\sin ax)^2} = -\frac{1}{2a}\tan\left(\frac{\pi}{4} - \frac{ax}{2}\right) - \frac{1}{6a}\tan^3\left(\frac{\pi}{4} - \frac{ax}{2}\right).$$

(300) $$\int \frac{dx}{(1-\sin ax)^2} = \frac{1}{2a}\cot\left(\frac{\pi}{4} - \frac{ax}{2}\right) + \frac{1}{6a}\cot^3\left(\frac{\pi}{4} - \frac{ax}{2}\right).$$

(301) $$\int \frac{\sin ax\,dx}{(1+\sin ax)^2} = -\frac{1}{2a}\tan\left(\frac{\pi}{4} - \frac{ax}{2}\right) + \frac{1}{6a}\tan^3\left(\frac{\pi}{4} - \frac{ax}{2}\right).$$

(302) $$\int \frac{\sin ax\,dx}{(1-\sin ax)^2} = -\frac{1}{2a}\cot\left(\frac{\pi}{4} - \frac{ax}{2}\right) + \frac{1}{6a}\cot^3\left(\frac{\pi}{4} - \frac{ax}{2}\right).$$

(303) $$\int \frac{dx}{1+\sin^2 ax} = \frac{1}{2\sqrt{2}\,a}\arcsin\left(\frac{3\sin^2 ax - 1}{\sin^2 ax + 1}\right).$$

(304) $$\int \frac{dx}{1-\sin^2 ax} = \int \frac{dx}{\cos^2 ax} = \frac{1}{a}\tan ax.$$

(305) $$\int \sin ax \sin bx\,dx = \frac{\sin(a-b)x}{2(a-b)} - \frac{\sin(a+b)x}{2(a+b)}, \quad |a| \neq |b|$$

(when $|a| = |b|$, see (275)).

(306) $$\int \frac{dx}{b + c\sin ax} = \frac{2}{a\sqrt{b^2 - c^2}} \arctan \frac{b\tan\frac{1}{2}ax + c}{\sqrt{b^2 - c^2}}, \quad b^2 > c^2,$$

$$= \frac{1}{a\sqrt{c^2 - b^2}} \ln\left|\frac{b\tan\frac{1}{2}ax + c - \sqrt{c^2 - b^2}}{b\tan\frac{1}{2}ax + c + \sqrt{c^2 - b^2}}\right|, \quad b^2 < c^2.$$

(307) $$\int \frac{\sin ax\,dx}{b + c\sin ax} = \frac{x}{c} - \frac{b}{c}\int \frac{dx}{b + c\sin ax}, \quad c \neq 0 \text{ (see (306))}.$$

(308) $$\int \frac{dx}{\sin ax\,(b + c\sin ax)} = \frac{1}{ab}\ln\left|\tan\frac{ax}{2}\right| - \frac{c}{b}\int \frac{dx}{b + c\sin ax},$$

$b \neq 0$ (see (306)).

(309) $$\int \frac{dx}{(b + c\sin ax)^2} = \frac{c\cos ax}{a(b^2 - c^2)(b + c\sin ax)} +$$

$$+ \frac{b}{b^2 - c^2}\int \frac{dx}{b + c\sin ax}, \quad |b| \neq |c| \text{ (see (306))}.$$

(310) $$\int \frac{\sin ax}{(b + c\sin ax)^2}\,dx = \frac{b\cos ax}{a(c^2 - b^2)(b + c\sin ax)} +$$

$$+ \frac{c}{c^2 - b^2}\int \frac{dx}{b + c\sin ax}, \quad |b| \neq |c| \text{ (see (306))}.$$

(311) $$\int \frac{dx}{b^2 + c^2\sin^2 ax} = \frac{1}{ab\sqrt{b^2 + c^2}} \arctan \frac{\sqrt{b^2 + c^2}\tan ax}{b}, \quad b > 0.$$

(312) $$\int \frac{dx}{b^2 - c^2\sin^2 ax} = \frac{1}{ab\sqrt{b^2 - c^2}} \arctan \frac{\sqrt{b^2 - c^2}\tan ax}{b},$$

$b^2 > c^2,\ b > 0.$

$$= \frac{1}{2ab\sqrt{c^2 - b^2}} \ln\left|\frac{\sqrt{c^2 - b^2}\tan ax + b}{\sqrt{c^2 - b^2}\tan ax - b}\right|,$$

$c^2 > b^2,\ b > 0.$

Integrals involving $\cos ax$, where $a \neq 0$

(313) $$\int \cos ax\,dx = \frac{1}{a}\sin ax.$$

(314) $$\int \cos^2 ax\,dx = \frac{1}{2}x + \frac{1}{4a}\sin 2ax.$$

(315) $$\int \cos^3 ax\,dx = \frac{1}{a}\sin ax - \frac{1}{3a}\sin^3 ax.$$

(316) $\int \cos^4 ax\,dx = \frac{3}{8}x + \frac{1}{4a}\sin 2ax + \frac{1}{32a}\sin 4ax.$

(317) $\int \cos^n ax\,dx = \frac{\cos^{n-1} ax \sin ax}{na} + \frac{n-1}{n}\int \cos^{n-2} ax\,dx.$

(318) $\int x\cos ax\,dx = \frac{\cos ax}{a^2} + \frac{x\sin ax}{a}.$

(319) $\int x^2\cos ax\,dx = \frac{2x}{a^2}\cos ax + \left(\frac{x^2}{a} - \frac{2}{a^3}\right)\sin ax.$

(320) $\int x^3\cos ax\,dx = \left(\frac{3x^2}{a^2} - \frac{6}{a^4}\right)\cos ax + \left(\frac{x^3}{a} - \frac{6x}{a^3}\right)\sin ax.$

(321) $\int x^n\cos ax\,dx = \frac{x^n\sin ax}{a} - \frac{n}{a}\int x^{n-1}\sin ax\,dx.$

(322) $\int \frac{\cos ax}{x}dx = \ln|ax| - \frac{(ax)^2}{2\cdot 2!} + \frac{(ax)^4}{4\cdot 4!} - \frac{(ax)^6}{6\cdot 6!} + \dots$ [1].

(323) $\int \frac{\cos ax}{x^2}dx = -\frac{\cos ax}{x} - a\int \frac{\sin ax}{x}dx$ (see 283).

(324) $\int \frac{\cos ax}{x^n}dx = -\frac{\cos ax}{(n-1)x^{n-1}} - \frac{a}{n-1}\int \frac{\sin ax}{x^{n-1}}dx,$

$n \neq 1$ (see (285)).

(325) $\int \frac{dx}{\cos ax} = \int \sec ax\,dx = \frac{1}{a}\ln\left|\tan\left(\frac{ax}{2} + \frac{\pi}{4}\right)\right|$

$= \frac{1}{a}\ln|\sec ax + \tan ax|.$

(326) $\int \frac{dx}{\cos^2 ax} = \frac{1}{a}\tan ax.$

(327) $\int \frac{dx}{\cos^3 ax} = \frac{\sin ax}{2a\cos^2 ax} + \frac{1}{2a}\ln\left|\tan\left(\frac{\pi}{4} + \frac{ax}{2}\right)\right|.$

---

[1] The definite integral $\int_x^\infty \frac{\cos t}{t}dt$ is called the *integral cosine* and denoted by $\mathrm{Ci}\,x$:

$$\mathrm{Ci}\,x = C - \ln|x| - \frac{x^2}{2\cdot 2!} + \frac{x^4}{4\cdot 4!} - \frac{x^6}{6\cdot 6!} + \dots,$$

where C is Euler's constant (see p. 381).

(328) $$\int \frac{dx}{\cos^n ax} = \frac{1}{a(n-1)} \cdot \frac{\sin ax}{\cos^{n-1} ax} + \frac{n-2}{n-1} \int \frac{dx}{\cos^{n-2} ax}, \quad n > 1.$$

(329) $$\int \frac{x}{\cos ax} dx = \frac{1}{a^2} \left(\frac{(ax)^2}{2} + \frac{(ax)^4}{4 \cdot 2!} + \frac{5(ax)^6}{6 \cdot 4!} + \frac{61(ax)^8}{8 \cdot 6!} + \right.$$
$$\left. + \frac{1385(ax)^{10}}{10 \cdot 8!} + \ldots + \frac{E_n (ax)^{2n+2}}{(2n+2)(2n!)} + \ldots \right) (^1).$$

(330) $$\int \frac{x}{\cos^2 ax} dx = \frac{x}{a} \tan ax + \frac{1}{a^2} \ln |\cos ax|.$$

(331) $$\int \frac{x}{\cos^n ax} dx = \frac{x \sin ax}{(n-1) a \cos^{n-1} ax} -$$
$$- \frac{1}{(n-1)(n-2) a^2 \cos^{n-2} ax} + \frac{n-2}{n-1} \int \frac{x}{\cos^{n-2} ax} dx, \quad n > 2.$$

(332) $$\int \frac{dx}{1 + \cos ax} = \frac{1}{a} \tan \frac{ax}{2}.$$

(333) $$\int \frac{dx}{1 - \cos ax} = -\frac{1}{a} \cot \frac{ax}{2}.$$

(334) $$\int \frac{x}{1 + \cos ax} dx = \frac{x}{a} \tan \frac{ax}{2} + \frac{2}{a^2} \ln \left| \cos \frac{ax}{2} \right|.$$

(335) $$\int \frac{x}{1 - \cos ax} dx = -\frac{x}{a} \cot \frac{ax}{2} + \frac{2}{a^2} \ln \left| \sin \frac{ax}{2} \right|.$$

(336) $$\int \frac{\cos ax}{1 + \cos ax} dx = x - \frac{1}{a} \tan \frac{ax}{2}.$$

(337) $$\int \frac{\cos ax}{1 - \cos ax} dx = -x - \frac{1}{a} \cot \frac{ax}{2}.$$

(338) $$\int \frac{dx}{\cos ax(1 + \cos ax)} = \frac{1}{a} \ln \left| \tan\left(\frac{\pi}{4} + \frac{ax}{2} \right) \right| - \frac{1}{a} \tan \frac{ax}{2}.$$

(339) $$\int \frac{dx}{\cos ax(1 - \cos ax)} = \frac{1}{a} \ln \left| \tan\left(\frac{\pi}{4} + \frac{ax}{2} \right) \right| - \frac{1}{a} \cot \frac{ax}{2}.$$

(340) $$\int \frac{dx}{(1 + \cos ax)^2} = \frac{1}{2a} \tan \frac{ax}{2} + \frac{1}{6a} \tan^3 \frac{ax}{2}.$$

(341) $$\int \frac{dx}{(1 - \cos ax)^2} = -\frac{1}{2a} \cot \frac{ax}{2} - \frac{1}{6a} \cot^3 \frac{ax}{2}.$$

(¹) E_n are Euler's numbers (see p. 354).

(342) $$\int \frac{\cos ax}{(1+\cos ax)^2}\,dx = \frac{1}{2a}\tan\frac{ax}{2} - \frac{1}{6a}\tan^3\frac{ax}{2}.$$

(343) $$\int \frac{\cos ax}{(1-\cos ax)^2}\,dx = \frac{1}{2a}\cot\frac{ax}{2} - \frac{1}{6a}\cot^3\frac{ax}{2}.$$

(344) $$\int \frac{dx}{1+\cos^2 ax} = \frac{1}{2\sqrt{2}a}\arcsin\left(\frac{1-3\cos^2 ax}{1+\cos^2 ax}\right).$$

(345) $$\int \frac{dx}{1-\cos^2 ax} = \int \frac{dx}{\sin^2 ax} = -\frac{1}{a}\cot ax.$$

(346) $$\int \cos ax \cos bx\,dx = \frac{\sin(a-b)x}{2(a-b)} + \frac{\sin(a+b)x}{2(a+b)},$$

$|a| \neq |b|$ (if $|a| = |b|$, see (314)).

(347) $$\int \frac{dx}{b+c\cos ax} = \frac{2}{a\sqrt{b^2-c^2}}\arctan\frac{(b-c)\tan\frac{1}{2}ax}{\sqrt{b^2-c^2}}, \quad b^2 > c^2,$$

$$= \frac{1}{a\sqrt{c^2-b^2}}\ln\left|\frac{(c-b)\tan\frac{1}{2}ax + \sqrt{c^2-b^2}}{(c-b)\tan\frac{1}{2}ax - \sqrt{c^2-b^2}}\right|, \quad b^2 < c^2.$$

(348) $$\int \frac{\cos ax}{b+c\cos ax}\,dx = \frac{x}{c} - \frac{b}{c}\int \frac{dx}{b+c\cos ax}, \quad c \neq 0 \text{ (see (347))}.$$

(349) $$\int \frac{dx}{\cos ax(b+c\cos ax)}$$

$$= \frac{1}{ab}\ln\left|\tan\left(\frac{ax}{2}+\frac{\pi}{4}\right)\right| - \frac{c}{b}\int \frac{dx}{b+c\cos ax} \quad \text{(see (347))}.$$

(350) $$\int \frac{dx}{(b+c\cos ax)^2} = \frac{c\sin ax}{a(c^2-b^2)(b+c\cos ax)} -$$

$$-\frac{b}{c^2-b^2}\int \frac{dx}{b+c\cos ax}, \quad |b| \neq |c| \text{ (see (347))}.$$

(351) $$\int \frac{\cos ax}{(b+c\cos ax)^2}\,dx$$

$$= \frac{b\sin ax}{a(b^2-c^2)(b+c\cos ax)} - \frac{c}{b^2-c^2}\int \frac{dx}{b+c\cos ax}, \quad |b| \neq |c|$$

(see (347)).

(352) $$\int \frac{dx}{b^2+c^2\cos^2 ax} = \frac{1}{ab\sqrt{b^2+c^2}}\arctan\frac{b\tan ax}{\sqrt{b^2+c^2}}, \quad b > 0.$$

(353) $$\int \frac{dx}{b^2 - c^2 \cos^2 ax} = \frac{1}{ab\sqrt{b^2 - c^2}} \arctan \frac{b \tan ax}{\sqrt{b^2 - c^2}},$$

$$b^2 > c^2,\ b > 0,$$

$$= \frac{1}{2ab\sqrt{c^2 - b^2}} \ln \left| \frac{b \tan ax - \sqrt{c^2 - b^2}}{b \tan ax + \sqrt{c^2 - b^2}} \right|,$$

$$c^2 > b^2,\ b > 0.$$

Integrals involving $\sin ax$ and $\cos ax$, where $a \neq 0$

(354) $$\int \sin ax \cos ax \, dx = \frac{1}{2a} \sin^2 ax.$$

(355) $$\int \sin^2 ax \cos^2 ax \, dx = \frac{x}{8} - \frac{\sin 4ax}{32a}.$$

(356) $$\int \sin^n ax \cos ax \, dx = \frac{1}{a(n+1)} \sin^{n+1} ax, \quad n \neq -1.$$

(357) $$\int \sin ax \cos^n ax \, dx = -\frac{1}{a(n+1)} \cos^{n+1} ax, \quad n \neq -1.$$

(358) $$\int \sin^n ax \cos^m ax \, dx$$

$$= -\frac{\sin^{n-1} ax \cos^{m+1} ax}{a(n+m)} + \frac{n-1}{n+m} \int \sin^{n-2} ax \cos^m ax \, dx,$$

$m > 0$ $n > 0$ (lowering the exponent n),

$$= \frac{\sin^{n+1} ax \cos^{m-1} ax}{a(n+m)} + \frac{m-1}{n+m} \int \sin^n ax \cos^{m-2} ax \, dx,$$

$m > 0$, $n > 0$ (lowering the exponent m).

(359) $$\int \frac{dx}{\sin ax \cos ax} = \frac{1}{a} \ln |\tan ax|.$$

(360) $$\int \frac{dx}{\sin^2 ax \cos ax} = \frac{1}{a} \left[\ln \left| \tan \left(\frac{\pi}{4} + \frac{ax}{2} \right) \right| - \frac{1}{\sin ax} \right].$$

(361) $$\int \frac{dx}{\sin ax \cos^2 ax} = \frac{1}{a} \left(\ln \left| \tan \frac{ax}{2} \right| + \frac{1}{\cos ax} \right).$$

(362) $$\int \frac{dx}{\sin^3 ax \cos ax} = \frac{1}{a} \left(\ln |\tan ax| - \frac{1}{2 \sin^2 ax} \right).$$

(363) $$\int \frac{dx}{\sin ax \cos^3 ax} = \frac{1}{a} \left(\ln |\tan ax| + \frac{1}{2 \cos^2 ax} \right).$$

(364) $$\int \frac{dx}{\sin^2 ax \cos^2 ax} = -\frac{2}{a} \cot 2ax.$$

(365) $$\int \frac{dx}{\sin^2 ax \cos^3 ax} = \frac{1}{a}\left[\frac{\sin ax}{2\cos^2 ax} - \frac{1}{\sin ax} + \frac{3}{2}\ln\left|\tan\left(\frac{\pi}{4} + \frac{ax}{2}\right)\right|\right].$$

(366) $$\int \frac{dx}{\sin^3 ax \cos^2 ax} = \frac{1}{a}\left(\frac{1}{\cos ax} - \frac{\cos ax}{2\sin^2 ax} + \frac{3}{2}\ln\left|\tan\frac{ax}{2}\right|\right).$$

(367) $$\int \frac{dx}{\sin ax \cos^n ax} = \frac{1}{a(n-1)\cos^{n-1} ax} + \int \frac{dx}{\sin ax \cos^{n-2} ax},$$
$n \neq 1$, (see (361), (363)).

(368) $$\int \frac{dx}{\sin^n ax \cos ax} = -\frac{1}{a(n-1)\sin^{n-1} ax} + \int \frac{dx}{\sin^{n-2} ax \cos ax},$$
$n \neq 1$ (see (360), (362)).

(369) $$\int \frac{dx}{\sin^n ax \cos^m ax}$$
$$= -\frac{1}{a(n-1)} \cdot \frac{1}{\sin^{n-1} ax \cos^{m-1} ax} + \frac{n+m-2}{n-1}\int \frac{dx}{\sin^{n-2} ax \cos^m ax},$$
$m > 0$, $n > 1$ (lowering the exponent n),
$$= \frac{1}{a(m-1)} \cdot \frac{1}{\sin^{n-1} ax \cos^{m-1} ax} + \frac{n+m-2}{m-1}\int \frac{dx}{\sin^n ax \cos^{m-2} ax},$$
$m > 1$, $n > 0$ (lowering the exponent m).

(370) $$\int \frac{\sin ax}{\cos^2 ax}\,dx = \frac{1}{a\cos ax} = \frac{1}{a}\sec ax.$$

(371) $$\int \frac{\sin ax}{\cos^3 ax}\,dx = \frac{1}{2a\cos^2 ax} + C = \frac{1}{2a}\tan^2 ax + C_1.$$

(372) $$\int \frac{\sin ax}{\cos^n ax}\,dx = \frac{1}{a(n-1)\cos^{n-1} ax}.$$

(373) $$\int \frac{\sin^2 ax}{\cos ax}\,dx = -\frac{1}{a}\sin ax + \frac{1}{a}\ln\left|\tan\left(\frac{\pi}{4} + \frac{ax}{2}\right)\right|.$$

(374) $$\int \frac{\sin^2 ax}{\cos^3 ax}\,dx = \frac{1}{a}\left[\frac{\sin ax}{2\cos^2 ax} - \frac{1}{2}\ln\left|\tan\left(\frac{\pi}{4} + \frac{ax}{2}\right)\right|\right].$$

(375) $$\int \frac{\sin^2 ax}{\cos^n ax}\,dx = \frac{\sin ax}{a(n-1)\cos^{n-1} ax} - \frac{1}{n-1}\int \frac{dx}{\cos^{n-2} ax},$$

$n \neq 1$ (see (325), (326), (328)).

(376) $$\int \frac{\sin^3 ax}{\cos ax}\,dx = -\frac{1}{a}\left(\frac{\sin^2 ax}{2} + \ln|\cos ax|\right).$$

(377) $$\int \frac{\sin^3 ax}{\cos^2 ax}\,dx = \frac{1}{a}\left(\cos ax + \frac{1}{\cos ax}\right).$$

(378) $$\int \frac{\sin^3 ax}{\cos^n ax}\,dx = \frac{1}{a}\left(\frac{1}{(n-1)\cos^{n-1} ax} - \frac{1}{(n-3)\cos^{n-3} ax}\right),$$

$n \neq 1,\ n \neq 3.$

(379) $$\int \frac{\sin^n ax}{\cos ax}\,dx = -\frac{\sin^{n-1} ax}{a(n-1)} + \int \frac{\sin^{n-2} ax}{\cos ax}\,dx, \quad n \neq 1.$$

(380) $$\int \frac{\sin^n ax}{\cos^m ax}\,dx$$

$$= \frac{\sin^{n+1} ax}{a(m-1)\cos^{m-1} ax} - \frac{n-m+2}{m-1}\int \frac{\sin^n ax}{\cos^{m-2} ax}\,dx, \quad m \neq 1,$$

$$= -\frac{\sin^{n-1} ax}{a(n-m)\cos^{m-1} ax} + \frac{n-1}{n-m}\int \frac{\sin^{n-2} ax}{\cos^m ax}\,dx, \quad m \neq n,$$

$$= \frac{\sin^{n-1} ax}{a(m-1)\cos^{m-1} ax} - \frac{n-1}{m-1}\int \frac{\sin^{n-1} ax}{\cos^{m-2} ax}\,dx, \quad m \neq 1.$$

(381) $$\int \frac{\cos ax}{\sin^2 ax}\,dx = -\frac{1}{a \sin ax} = -\frac{1}{a}\operatorname{cosec} ax.$$

(382) $$\int \frac{\cos ax}{\sin^3 ax}\,dx = -\frac{1}{2a \sin^2 ax} + C = -\frac{\cot^2 ax}{2a} + C_1.$$

(383) $$\int \frac{\cos ax}{\sin^n ax}\,dx = -\frac{1}{a(n-1)\sin^{n-1} ax}.$$

(384) $$\int \frac{\cos^2 ax}{\sin ax}\,dx = \frac{1}{a}\left(\cos ax + \ln\left|\tan\frac{ax}{2}\right|\right).$$

(385) $$\int \frac{\cos^2 ax}{\sin^3 ax}\,dx = -\frac{1}{2a}\left(\frac{\cos ax}{\sin^2 ax} - \ln\left|\tan\frac{ax}{2}\right|\right).$$

(386) $$\int \frac{\cos^2 ax}{\sin^n ax}\,dx = -\frac{1}{(n-1)}\left(\frac{\cos ax}{a \sin^{n-1} ax} + \int \frac{dx}{\sin^{n-2} ax}\right),$$

$n \neq 1$ (see (289)).

(387) $$\int \frac{\cos^3 ax}{\sin ax} dx = \frac{1}{a}\left(\frac{\cos^2 ax}{2} + \ln|\sin ax|\right).$$

(388) $$\int \frac{\cos^3 ax}{\sin^2 ax} dx = -\frac{1}{a}\left(\sin ax + \frac{1}{\sin ax}\right).$$

(389) $$\int \frac{\cos^3 ax}{\sin^n ax} dx = \frac{1}{a}\left(\frac{1}{(n-3)\sin^{n-3} ax} - \frac{1}{(n-1)\sin^{n-1} ax}\right),$$
$$n \neq 1, 3.$$

(390) $$\int \frac{\cos^n ax}{\sin ax} dx = \frac{\cos^{n-1} ax}{a(n-1)} + \int \frac{\cos^{n-2} ax}{\sin ax} dx, \quad n \neq 1.$$

(391) $$\int \frac{\cos^n ax}{\sin^m ax} dx$$
$$= -\frac{\cos^{n+1} ax}{a(m-1)\sin^{m-1} ax} - \frac{n-m+2}{m-1}\int \frac{\cos^n ax}{\sin^{m-2} ax} dx, \quad m \neq 1,$$
$$= \frac{\cos^{n-1} ax}{a(n-m)\sin^{m-1} ax} + \frac{n-1}{n-m}\int \frac{\cos^{n-2} ax}{\sin^m ax} dx, \quad m \neq n,$$
$$= -\frac{\cos^{n-1} ax}{a(m-1)\sin^{m-1} ax} - \frac{n-1}{m-1}\int \frac{\cos^{n-2} ax}{\sin^{m-2} ax} dx, \quad m \neq 1.$$

(392) $$\int \frac{dx}{\sin ax(1 \pm \cos ax)} = \pm \frac{1}{2a(1 \pm \cos ax)} + \frac{1}{2a}\ln\left|\tan\frac{ax}{2}\right|.$$

(393) $$\int \frac{dx}{\cos ax(1 \pm \sin ax)}$$
$$= \mp \frac{1}{2a(1 \pm \sin ax)} + \frac{1}{2a}\ln\left|\tan\left(\frac{\pi}{4} + \frac{ax}{2}\right)\right|.$$

(394) $$\int \frac{\sin ax}{\cos ax(1 \pm \cos ax)} dx = \frac{1}{a}\ln\left|\frac{1 \pm \cos ax}{\cos ax}\right|.$$

(395) $$\int \frac{\cos ax}{\sin ax(1 \pm \sin ax)} dx = -\frac{1}{a}\ln\left|\frac{1 \pm \sin ax}{\sin ax}\right|.$$

(396) $$\int \frac{\sin ax}{\cos ax(1 \pm \sin ax)} dx$$
$$= \frac{1}{2a(1 \pm \sin ax)} \pm \frac{1}{2a}\ln\left|\tan\left(\frac{\pi}{4} + \frac{ax}{2}\right)\right|.$$

(397) $$\int \frac{\cos ax}{\sin ax(1 \pm \cos ax)} dx = -\frac{1}{2a(1 \pm \cos ax)} \pm \frac{1}{2a}\ln\left|\tan\frac{ax}{2}\right|.$$

(398) $\int \frac{\sin ax}{\sin ax \pm \cos ax} dx = \frac{x}{2} \mp \frac{1}{2a} \ln |\sin ax \pm \cos ax|.$

(399) $\int \frac{\cos ax}{\sin ax \pm \cos ax} dx = \pm \frac{x}{2} + \frac{1}{2a} \ln |\sin ax \pm \cos ax|.$

(400) $\int \frac{dx}{\sin ax \pm \cos ax} = \frac{1}{a\sqrt{2}} \ln \left|\tan\left(\frac{ax}{2} \pm \frac{\pi}{8}\right)\right|.$

(401) $\int \frac{dx}{1 + \cos ax \pm \sin ax} = \pm \frac{1}{a} \ln \left|1 \pm \tan \frac{ax}{2}\right|.$

(402) $\int \frac{dx}{b \sin ax + c \cos ax} = \frac{1}{a\sqrt{b^2 + c^2}} \ln \left|\tan \frac{ax + \theta}{2}\right|, \; b \neq 0, \; c \neq 0$

(notation: $\sin \theta = \frac{c}{\sqrt{b^2 + c^2}}, \; \tan \theta = \frac{c}{b}$).

(403) $\int \frac{\sin ax}{b + c \cos ax} dx = -\frac{1}{ac} \ln |b + c \cos ax|.$

(404) $\int \frac{\cos ax}{b + c \sin ax} dx = \frac{1}{ac} \ln |b + c \sin ax|.$

(405) $\int \frac{dx}{b + c \cos ax + f \sin ax} = \int \frac{d(x + \theta/a)}{b + \sqrt{c^2 + f^2} \sin (ax + \theta)},$

$c \neq 0, \; f \neq 0.$

(notation: $\sin \theta = \frac{c}{\sqrt{c^2 + f^2}}, \; \tan \theta = \frac{c}{f}$, see (306))

(406) $\int \frac{dx}{b^2 \cos^2 ax + c^2 \sin^2 ax} = \frac{1}{abc} \arctan\left(\frac{c}{b} \tan ax\right),$

$b \neq 0, \; c \neq 0.$

(407) $\int \frac{dx}{b^2 \cos^2 ax - c^2 \sin^2 ax} = \frac{1}{2abc} \ln \left|\frac{c \tan ax + b}{c \tan ax - b}\right|, \; b \neq 0, \; c \neq 0.$

(408) $\int \sin ax \cos bx \, dx = -\frac{\cos(a + b)x}{2(a + b)} - \frac{\cos(a - b)x}{2(a - b)},$

$a^2 \neq b^2$ (if $a = b$, see (354)).

Integrals involving $\tan ax$, where $a \neq 0$

(409) $\int \tan ax \, dx = -\frac{1}{a} \ln |\cos ax|.$

(410) $\int \tan^2 ax \, dx = \frac{\tan ax}{a} - x.$

(411) $$\int \tan^3 ax\,dx = \frac{1}{2a}\tan^2 ax + \frac{1}{a}\ln|\cos ax|.$$

(412) $$\int \tan^n ax\,dx = \frac{1}{a(n-1)}\tan^{n-1} ax - \int \tan^{n-2} ax\,dx,\quad n>3.$$

(413) $$\int x\tan ax\,dx = \frac{ax^3}{3} + \frac{a^3x^5}{15} + \frac{2a^5x^7}{105} + \frac{17a^7x^9}{2835} + \ldots + \frac{2^{2n}(2^{2n}-1)B_n a^{2n-1}x^{2n+1}}{(2n+1)!} + \ldots\ (^1).$$

(414) $$\int \frac{\tan ax}{x}\,dx = ax + \frac{(ax)^3}{9} + \frac{2(ax)^5}{75} + \frac{17(ax)^7}{2205} + \ldots + \frac{2^{2n}(2^{2n}-1)B_n(ax)^{2n-1}}{(2n-1)(2n)!} + \ldots\ (^1).$$

(415) $$\int \frac{\tan^n ax}{\cos^2 ax}\,dx = \frac{1}{a(n+1)}\tan^{n+1} ax,\quad n\neq -1.$$

(416) $$\int \frac{dx}{\tan ax \pm 1} = \pm\frac{x}{2} + \frac{1}{2a}\ln|\sin ax \pm \cos ax|.$$

(417) $$\int \frac{\tan ax}{\tan ax \pm 1}\,dx = \frac{x}{2} \mp \frac{1}{2a}\ln|\sin ax \pm \cos ax|.$$

Integrals involving $\cot ax$, where $a\neq 0$

(418) $$\int \cot ax\,dx = \frac{1}{a}\ln|\sin ax|.$$

(419) $$\int \cot^2 ax\,dx = -\frac{\cot ax}{a} - x.$$

(420) $$\int \cot^3 ax\,dx = -\frac{1}{2a}\cot^2 ax - \frac{1}{a}\ln|\sin ax|.$$

(421) $$\int \cot^n ax\,dx = -\frac{1}{a(n-1)}\cot^{n-1} ax - \int \cot^{n-2} ax\,dx,\quad n>3.$$

(422) $$\int x\cot ax\,dx = \frac{x}{a} - \frac{ax^3}{9} - \frac{a^3x^5}{225} - \ldots - \frac{2^{2n}B_n a^{2n-1}x^{2n+1}}{(2n+1)!} - \ldots\ (^1).$$

---

(1) B_n are Bernoulli's numbers (see p. 354).

(423) $\int \frac{\cot ax}{x}\,dx$

$$= -\frac{1}{ax} - \frac{ax}{3} - \frac{(ax)^3}{135} - \frac{2(ax)^5}{4725} - \ldots - \frac{2^{2n} B_n (ax)^{2n-1}}{(2n-1)(2n)!} - \ldots[1].$$

(424) $\int \frac{\cot^n ax}{\sin^2 ax}\,dx = -\frac{1}{a(n+1)}\cot^{n+1} ax, \quad n \neq -1.$

(425) $\int \frac{dx}{1 \pm \cot ax} = \int \frac{\tan ax}{\tan ax \pm 1}\,dx$ (see (417)).

Integrals of other transcendental functions

Integrals of hyperbolic functions

In integrals (426)–(446) we assume $a \neq 0$.

(426) $\int \sinh ax\,dx = \frac{1}{a}\cosh ax.$

(427) $\int \cosh ax\,dx = \frac{1}{a}\sinh ax.$

(428) $\int \sinh^2 ax\,dx = \frac{1}{2a}\sinh ax \cosh ax - \frac{1}{2}x.$

(429) $\int \cosh^2 ax\,dx = \frac{1}{2a}\sinh ax \cosh ax + \frac{1}{2}x.$

(430) $\int \sinh^n ax\,dx$

$$= \frac{1}{an}\sinh^{n-1} ax \cosh ax - \frac{n-1}{n}\int \sinh^{n-2} ax\,dx, \quad n > 0,$$

$$= \frac{1}{a(n+1)}\sinh^{n+1} ax \cosh ax - \frac{n+2}{n+1}\int \sinh^{n+2} ax\,dx,$$

$$n < 0, \quad n \neq -1.$$

(431) $\int \cosh^n ax\,dx$

$$= \frac{1}{an}\sinh ax \cosh^{n-1} ax + \frac{n-1}{n}\int \cosh^{n-2} ax\,dx, \quad n > 0,$$

$$= -\frac{1}{a(n+1)}\sinh ax \cosh^{n+1} ax + \frac{n+2}{n+1}\int \cosh^{n+2} ax\,dx,$$

$$n < 0, \quad n \neq -1.$$

---

[1] B_n are Bernoulli's numbers (see p. 354).

(432) $$\int \frac{dx}{\sinh ax} = \frac{1}{a} \ln \left|\tanh \frac{ax}{2}\right|.$$

(433) $$\int \frac{dx}{\cosh ax} = \frac{2}{a} \arctan e^{ax}.$$

(434) $$\int x \sinh ax\, dx = \frac{1}{a} x \cosh ax - \frac{1}{a^2} \sinh ax.$$

(435) $$\int x \cosh ax\, dx = \frac{1}{a} x \sinh ax - \frac{1}{a^2} \cosh ax.$$

(436) $$\int \tanh ax\, dx = \frac{1}{a} \ln |\cosh ax|.$$

(437) $$\int \coth ax\, dx = \frac{1}{a} \ln |\sinh ax|.$$

(438) $$\int \tanh^2 ax\, dx = x - \frac{\tanh ax}{a}.$$

(439) $$\int \coth^2 ax\, dx = x - \frac{\coth ax}{a}.$$

(440) $$\int \sinh ax \sinh bx\, dx$$
$$= \frac{1}{a^2 - b^2} (a \sinh bx \cosh ax - b \cosh bx \sinh ax), \quad a^2 \neq b^2.$$

(441) $$\int \cosh ax \cosh bx\, dx$$
$$= \frac{1}{a^2 - b^2} (a \sinh ax \cosh bx - b \sinh bx \cosh ax), \quad a^2 \neq b^2.$$

(442) $$\int \cosh ax \sinh bx\, dx$$
$$= \frac{1}{a^2 - b^2} (a \sinh bx \sinh ax - b \cosh bx \cosh ax), \quad a^2 \neq b^2.$$

(443) $$\int \sinh ax \sin ax\, dx = \frac{1}{2a} (\cosh ax \sin ax - \sinh ax \cos ax).$$

(444) $$\int \cosh ax \cos ax\, dx = \frac{1}{2a} (\sinh ax \cos ax + \cosh ax \sin ax).$$

(445) $$\int \sinh ax \cos ax\, dx = \frac{1}{2a} (\cosh ax \cos ax + \sinh ax \sin ax).$$

(446) $$\int \cosh ax \sin ax\, dx = \frac{1}{2a}(\sinh ax \sin ax - \cosh ax \cos ax).$$

Integrals of the exponential functions

In integrals (447)–(464) we assume $a \neq 0$.

(447) $$\int e^{ax}\, dx = \frac{1}{a} e^{ax}.$$

(448) $$\int x e^{ax}\, dx = \frac{e^{ax}}{a^2}(ax - 1).$$

(449) $$\int x^2 e^{ax}\, dx = e^{ax}\left(\frac{x^2}{a} - \frac{2x}{a^2} + \frac{2}{a^3}\right).$$

(450) $$\int x^n e^{ax}\, dx = \frac{1}{a} x^n e^{ax} - \frac{n}{a}\int x^{n-1} e^{ax}\, dx.$$

(451) $$\int \frac{e^{ax}}{x}\, dx = \ln|x| + \frac{ax}{1\cdot 1!} + \frac{(ax)^2}{2\cdot 2!} + \frac{(ax)^3}{3\cdot 3!} + \dots\ (^1).$$

(452) $$\int \frac{e^{ax}}{x^n}\, dx = \frac{1}{n-1}\left(-\frac{e^{ax}}{x^{n-1}} + a\int \frac{e^{ax}}{x^{n-1}}\, dx\right),\quad n \neq 1.$$

(453) $$\int \frac{dx}{1+e^{ax}} = \frac{1}{a}\ln\left|\frac{e^{ax}}{1+e^{ax}}\right|.$$

(454) $$\int \frac{dx}{b+ce^{ax}} = \frac{x}{b} - \frac{1}{ab}\ln|b+ce^{ax}|,\quad b \neq 0.$$

(455) $$\int \frac{e^{ax}}{b+ce^{ax}}\, dx = \frac{1}{ac}\ln|b+ce^{ax}|.$$

(456) $$\int \frac{dx}{be^{ax}+ce^{-ax}} = \frac{1}{a\sqrt{bc}}\arctan\left(e^{ax}\sqrt{\frac{b}{c}}\right),\quad bc > 0,$$

$$= \frac{1}{2a\sqrt{-bc}}\ln\left|\frac{c+e^{ax}\sqrt{-bc}}{c-e^{ax}\sqrt{-bc}}\right|,\quad bc < 0.$$

---

(1) The definite integral $\int_{-\infty}^{x} \frac{e^t}{t}\, dt$ is called the *integral exponential function* and denoted by $\mathrm{Ei}\, x$. When $x < 0$, the integral is divergent for $t = 0$; in this case $\mathrm{Ei}\, x$ should be understood to be the principal value of the improper integral (see p. 475):

$$\int_{-\infty}^{x} \frac{e^t}{t}\, dt = C + \ln|x| + \frac{x}{1\cdot 1!} + \frac{x^2}{2\cdot 2!} + \dots + \frac{x^n}{n\cdot n!} + \dots$$

(C is Euler's constant, see p. 331).

(457) $\displaystyle\int \frac{xe^{ax}}{(1+ax)^2}\,dx = \frac{e^{ax}}{a^2(1+ax)}.$

(458) $\displaystyle\int e^{ax}\ln|x|\,dx = \frac{e^{ax}\ln|x|}{a} - \frac{1}{a}\int \frac{e^{ax}}{x}\,dx$ (see (451)).

(459) $\displaystyle\int e^{ax}\sin bx\,dx = \frac{e^{ax}}{a^2+b^2}(a\sin bx - b\cos bx).$

(460) $\displaystyle\int e^{ax}\cos bx\,dx = \frac{e^{ax}}{a^2+b^2}(a\cos bx + b\sin bx).$

(461) $\displaystyle\int e^{ax}\sin^n x\,dx = \frac{e^{ax}\sin^{n-1}x}{a^2+n^2}(a\sin x - n\cos x) +$

$\displaystyle + \frac{n(n-1)}{a^2+n^2}\int e^{ax}\sin^{n-2}x\,dx$ (see (447), (459)).

(462) $\displaystyle\int e^{ax}\cos^n x\,dx = \frac{e^{ax}\cos^{n-1}x}{a^2+n^2}(a\cos x + n\sin x) +$

$\displaystyle + \frac{n(n-1)}{a^2+n^2}\int e^{ax}\cos^{n-2}x\,dx$ (see (447), (460)).

(463) $\displaystyle\int xe^{ax}\sin bx\,dx = \frac{xe^{ax}}{a^2+b^2}(a\sin bx - b\cos bx) -$

$\displaystyle - \frac{e^{ax}}{(a^2+b^2)^2}[(a^2-b^2)\sin bx - 2ab\cos bx].$

(464) $\displaystyle\int xe^{ax}\cos bx\,dx = \frac{xe^{ax}}{a^2+b^2}(a\cos bx + b\sin bx) -$

$\displaystyle - \frac{e^{ax}}{(a^2+b^2)^2}[(a^2-b^2)\cos bx + 2ab\sin bx].$

Integrals of logarithmic functions

(465) $\displaystyle\int \ln|x|\,dx = x\ln|x| - x.$

(466) $\displaystyle\int (\ln|x|)^2\,dx = x(\ln|x|)^2 - 2x\ln|x| + 2x.$

(467) $\displaystyle\int (\ln|x|)^3\,dx = x(\ln|x|)^3 - 3x(\ln|x|)^2 + 6x\ln|x| - 6x.$

(468) $\displaystyle\int (\ln|x|)^n\,dx = x(\ln|x|)^n - n\int (\ln|x|)^{n-1}\,dx,\ n \neq -1.$

(469) $$\int \frac{dx}{\ln|x|} = \ln|\ln|x|| + \ln|x| + \frac{(\ln|x|)^2}{2\cdot 2!} + \frac{(\ln|x|)^3}{3\cdot 3!} + \dots$$ [1].

(470) $$\int \frac{dx}{(\ln|x|)^n} = -\frac{x}{(n-1)(\ln|x|)^{n-1}} + \frac{1}{n-1}\int \frac{dx}{(\ln|x|)^{n-1}},$$
$n \neq 1$ (see (469)).

(471) $$\int x^m \ln|x|\,dx = x^{m+1}\left(\frac{\ln|x|}{m+1} - \frac{1}{(m+1)^2}\right), \quad m \neq -1.$$

(472) $$\int x^m (\ln|x|)^n\,dx = \frac{x^{m+1}(\ln|x|)^n}{m+1} - \frac{n}{m+1}\int x^m (\ln|x|)^{n-1}\,dx,$$
$m,\ n \neq -1$ (see (470)).

(473) $$\int \frac{(\ln|x|)^n}{x}\,dx = \frac{(\ln|x|)^{n+1}}{n+1}, \quad n \neq -1.$$

(474) $$\int \frac{\ln|x|}{x^m}\,dx = -\frac{\ln|x|}{(m-1)x^{m-1}} - \frac{1}{(m-1)^2 x^{m-1}}, \quad m \neq 1.$$

(475) $$\int \frac{(\ln|x|)^n}{x^m}\,dx = -\frac{(\ln|x|)^n}{(m-1)x^{m-1}} + \frac{n}{m-1}\int \frac{(\ln|x|)^{n-1}}{x^m}\,dx,$$
$m \neq 1$ (see (474)).

(476) $$\int \frac{x^m}{\ln|x|}\,dx = \int \frac{e^{-y}}{y}\,dy, \quad y = -(m+1)\ln|x| \text{ (see (451))}.$$

(477) $$\int \frac{x^m}{(\ln|x|)^n}\,dx = -\frac{x^{m+1}}{(n-1)(\ln|x|)^{n-1}} + \frac{m+1}{n-1}\int \frac{x^m}{(\ln|x|)^{n-1}}\,dx,$$
$n \neq 1$.

(478) $$\int \frac{dx}{x\ln|x|} = \ln|\ln|x||.$$

(479) $$\int \frac{dx}{x^n \ln|x|} = \ln|\ln|x|| - (n-1)\ln|x| + \frac{(n-1)^2(\ln|x|)^2}{2\cdot 2!} - \frac{(n-1)^3(\ln|x|)^3}{3\cdot 3!} + \dots$$

---

[1] The definite integral $\int_0^x \frac{dt}{\ln|t|}$ is called the *integral logarithm* and denoted by $\mathrm{Li}\,x$. If $x > 1$, the integral is divergent for $t = 1$. In this case, $\mathrm{Li}\,x$ should be understood to be the principal value of the improper integral (see p. 475). The integral logarithm is related to the integral exponential function (see p. 448) by $\mathrm{Li}\,x = \mathrm{Ei}(\ln|x|)$.

(480) $$\int \frac{dx}{x(\ln|x|)^n} = \frac{-1}{(n-1)(\ln|x|)^{n-1}}, \quad n \neq 1.$$

(481) $$\int \frac{dx}{x^p(\ln|x|)^n}$$
$$= \frac{-1}{x^{p-1}(n-1)(\ln|x|)^{n-1}} - \frac{p-1}{n-1}\int \frac{dx}{x^p(\ln|x|)^{n-1}}, \quad n \neq 1.$$

(482) $$\int \ln|\sin x|\, dx$$
$$= x\ln|x| - x - \frac{x^3}{18} - \frac{x^5}{900} - \dots - \frac{2^{2n-1}B_n x^{2n+1}}{n(2n+1)!} - \dots \text{ (}^1\text{)}.$$

(483) $$\int \ln|\cos x|\, dx$$
$$= -\frac{x^3}{6} - \frac{x^5}{60} - \frac{x^7}{315} - \dots - \frac{2^{2n-1}(2^{2n}-1)B_n}{n(2n+1)!}x^{2n+1} - \dots \text{ (}^1\text{)}.$$

(484) $$\int \ln|\tan x|\, dx$$
$$= x\ln|x| - x + \frac{x^3}{9} + \frac{7x^5}{450} + \dots + \frac{2^{2n}(2^{2n-1}-1)B_n}{n(2n+1)!}x^{2n+1} + \dots \text{ (}^1\text{)}.$$

(485) $$\int \sin\ln|x|\, dx = \frac{x}{2}(\sin\ln|x| - \cos\ln|x|).$$

(486) $$\int \cos\ln|x|\, dx = \frac{x}{2}(\sin\ln|x| + \cos\ln|x|).$$

(487) $$\int e^{ax}\ln|x|\, dx = \frac{1}{a}e^{ax}\ln|x| - \frac{1}{a}\int \frac{e^{ax}}{x}dx \quad \text{(see (451))}.$$

Integrals of inverse trigonometric functions

In integrals (488)–(511) we assume $a \neq 0$.

(488) $$\int \arcsin\frac{x}{a}\, dx = x\arcsin\frac{x}{a} + \sqrt{a^2 - x^2}.$$

(489) $$\int x\arcsin\frac{x}{a}\, dx = \left(\frac{x^2}{2} - \frac{a^2}{4}\right)\arcsin\frac{x}{a} + \frac{x}{4}\sqrt{a^2 - x^2}.$$

(490) $$\int x^2\arcsin\frac{x}{a}\, dx = \frac{x^3}{3}\arcsin\frac{x}{a} + \frac{1}{9}(x^2 + 2a^2)\sqrt{a^2 - x^2}.$$

---

(1) B_n are Bernoulli's numbers (see p. 354).

(491) $$\int \frac{\arcsin \frac{x}{a}}{x} dx$$

$$= \frac{x}{a} + \frac{1}{2 \cdot 3 \cdot 3} \cdot \frac{x^3}{a^3} + \frac{1 \cdot 3}{2 \cdot 4 \cdot 5 \cdot 5} \cdot \frac{x^5}{a^5} + \frac{1 \cdot 3 \cdot 5}{2 \cdot 4 \cdot 6 \cdot 7 \cdot 7} \cdot \frac{x^7}{a^7} + \cdot$$

(492) $$\int \frac{\arcsin \frac{x}{a}}{x^2} dx = -\frac{1}{x} \arcsin \frac{x}{a} - \frac{1}{a} \ln \left| \frac{a + \sqrt{a^2 - x^2}}{x} \right| .$$

(493) $$\int \arccos \frac{x}{a} dx = x \arccos \frac{x}{a} - \sqrt{a^2 - x^2} .$$

(494) $$\int x \arccos \frac{x}{a} dx = \left(\frac{x^2}{2} - \frac{a^2}{4} \right) \arccos \frac{x}{a} - \frac{x}{4} \sqrt{a^2 - x^2} .$$

(495) $$\int x^2 \arccos \frac{x}{a} dx = \frac{x^3}{3} \arccos \frac{x}{a} - \frac{1}{9} (x^2 + 2a^2) \sqrt{a^2 - x^2} .$$

(496) $$\int \frac{\arccos \frac{x}{a}}{x} dx$$

$$= \frac{\pi}{2} \ln |x| - \frac{x}{a} - \frac{1}{2 \cdot 3 \cdot 3} \cdot \frac{x^3}{a^3} - \frac{1 \cdot 3}{2 \cdot 4 \cdot 5 \cdot 5} \cdot \frac{x^5}{a^5} - \frac{1 \cdot 3 \cdot 5}{2 \cdot 4 \cdot 6 \cdot 7 \cdot 7} \cdot \frac{x^7}{a^7} - \cdots$$

(497) $$\int \frac{\arccos \frac{x}{a}}{x^2} dx = -\frac{1}{x} \arccos \frac{x}{a} + \frac{1}{a} \ln \left| \frac{a + \sqrt{a^2 - x^2}}{x} \right| .$$

(498) $$\int \arctan \frac{x}{a} dx = x \arctan \frac{x}{a} - \frac{a}{2} \ln (a^2 + x^2).$$

(499) $$\int x \arctan \frac{x}{a} dx = \frac{1}{2} (x^2 + a^2) \arctan \frac{x}{a} - \frac{ax}{2} .$$

(500) $$\int x^2 \arctan \frac{x}{a} dx = \frac{x^3}{3} \arctan \frac{x}{a} - \frac{ax^2}{6} + \frac{a^3}{6} \ln (a^2 + x^2).$$

(501) $$\int x^n \arctan \frac{x}{a} dx = \frac{x^{n+1}}{n+1} \arctan \frac{x}{a} - \frac{a}{n+1} \int \frac{x^{n+1}}{a^2 + x^2} dx,$$

$$n \neq -1.$$

(502) $$\int \frac{\arctan \frac{x}{a}}{x} dx = \frac{x}{a} - \frac{x^3}{3^2 a^3} + \frac{x^5}{5^2 a^5} - \frac{x^7}{7^2 a^7} + \ldots, \quad |x| < |a|.$$

(503) $$\int \frac{\arctan \frac{x}{a}}{x^2} dx = -\frac{1}{x} \arctan \frac{x}{a} - \frac{1}{2a} \ln \frac{a^2 + x^2}{x^2} .$$

(504) $$\int \frac{\operatorname{arc\,tan}\frac{x}{a}}{x^n}\,dx$$

$$= -\frac{1}{(n-1)\,x^{n-1}} \operatorname{arc\,tan}\frac{x}{a} + \frac{a}{n-1}\int \frac{dx}{x^{n-1}(a^2+x^2)}, \quad n \neq 1.$$

(505) $$\int \operatorname{arc\,cot}\frac{x}{a}\,dx = x \operatorname{arc\,cot}\frac{x}{a} + \frac{a}{2}\ln(a^2+x^2).$$

(506) $$\int x \operatorname{arc\,cot}\frac{x}{a}\,dx = \frac{1}{2}(x^2+a^2)\operatorname{arc\,cot}\frac{x}{a} + \frac{ax}{2}.$$

(507) $$\int x^2 \operatorname{arc\,cot}\frac{x}{a}\,dx = \frac{x^3}{3}\operatorname{arc\,cot}\frac{x}{a} + \frac{ax^2}{6} - \frac{a^3}{6}\ln(a^2+x^2).$$

(508) $$\int x^n \operatorname{arc\,cot}\frac{x}{a}\,dx = \frac{x^{n+1}}{n+1}\operatorname{arc\,cot}\frac{x}{a} + \frac{a}{n+1}\int \frac{x^{n+1}}{a^2+x^2}\,dx,$$

$$n \neq -1.$$

(509) $$\int \frac{\operatorname{arc\,cot}\frac{x}{a}}{x}\,dx = \frac{\pi}{2}\ln|x| - \frac{x}{a} + \frac{x^3}{3^2a^3} - \frac{x^5}{5^2a^5} + \frac{x^7}{7^2a^7} - \cdots$$

(510) $$\int \frac{\operatorname{arc\,cot}\frac{x}{a}}{x^2}\,dx = -\frac{1}{x}\operatorname{arc\,cot}\frac{x}{a} + \frac{1}{2a}\ln\frac{a^2+x^2}{x^2}.$$

(511) $$\int \frac{\operatorname{arc\,cot}\frac{x}{a}}{x^n}\,dx$$

$$= -\frac{1}{(n-1)\,x^{n-1}}\operatorname{arc\,cot}\frac{x}{a} - \frac{a}{n-1}\int \frac{dx}{x^{n-1}(a^2+x^2)}, \quad n \neq 1$$

Integrals of inverse hyperbolic functions

(512) $$\int \operatorname{ar\,sinh}\frac{x}{a}\,dx = x \operatorname{ar\,sinh}\frac{x}{a} - \sqrt{x^2+a^2}, \quad a \neq 0.$$

(513) $$\int \operatorname{ar\,cosh}\frac{x}{a}\,dx = x \operatorname{ar\,cosh}\frac{x}{a} - \sqrt{x^2-a^2}, \quad a \neq 0.$$

(514) $$\int \operatorname{ar\,tanh}\frac{x}{a}\,dx = x \operatorname{ar\,tanh}\frac{x}{a} + \frac{a}{2}\ln|a^2-x^2|, \quad a \neq 0.$$

(515) $$\int \operatorname{ar\,coth}\frac{x}{a}\,dx = x \operatorname{ar\,coth}\frac{x}{a} + \frac{a}{2}\ln|x^2-a^2|, \quad a \neq 0.$$

B. DEFINITE INTEGRALS

8. Fundamental concepts and theorems

Definition. The *definite integral* between the limits a and b of a function $y=f(x)$ defined in the closed interval $[a, b]$ [1] ($a<b$, for the case A and $a>b$, for the case B) is the number obtained as follows:

(1) we divide the interval $[a, b]$ into n *elementary intervals* by means of arbitrary numbers $x_1, x_2, \ldots, x_{n-1}$ selected so that

$$a = x_0 < x_1 < x_2 < \ldots < x_i < \ldots < x_{n-1} < x_n = b \qquad \text{(case A)}$$

or

$$a = x_0 > x_1 > x_2 > \ldots > x_i > \ldots > x_{n-1} > x_n = b \qquad \text{(case B)};$$

(2) we select an arbitrary point ξ_i inside or on the boundary of each elementary interval $[x_{i-1}, x_i]$ (Fig. 302):

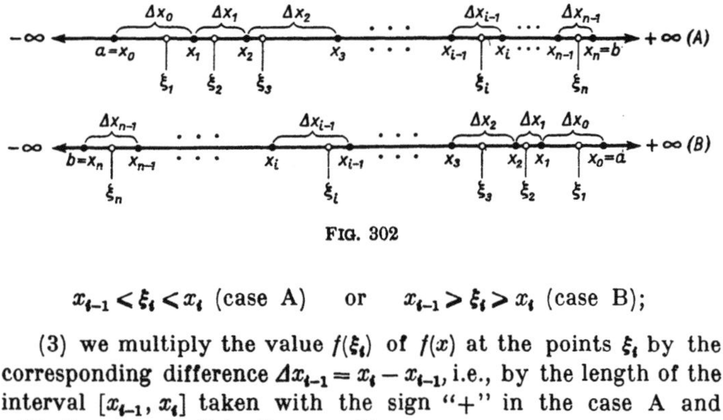

FIG. 302

$$x_{i-1} < \xi_i < x_i \text{ (case A)} \quad \text{or} \quad x_{i-1} > \xi_i > x_i \text{ (case B)};$$

(3) we multiply the value $f(\xi_i)$ of $f(x)$ at the points ξ_i by the corresponding difference $\Delta x_{i-1} = x_i - x_{i-1}$, i.e., by the length of the interval $[x_{i-1}, x_i]$ taken with the sign "+" in the case A and with the sign "−" in the case B;

(4) we add the obtained n products $f(\xi_i)\Delta x_{i-1}$;

(5) we determine the limit of the obtained sum

$$\sum_{i=1}^{n} f(\xi_i)\, \Delta x_{i-1},$$

---

[1] The concept of the definite integral can also be generalized for functions defined in an arbitrary connected interval (an open interval, an interval open on one side, a half-axis or the whole number axis) or in a connected interval except a finite number of separate points. The integrals considered in this more general sense belong to the *improper integrals* (see p. 471–478).

when the length of each elementary interval Δx_{i-1} tends to zero (and, hence, $n \to \infty$).

If this limit exists and is independent of the choice of the points x_i and ξ_i, then it is called the *definite integral*:

$$\int_a^b f(x)\,dx = \lim_{\substack{\Delta x_{i-1} \to 0 \\ n \to \infty}} \sum_{i=1}^{n} f(\xi_i)\,\Delta x_{i-1}. \tag{1}$$

In formula (1), the sign $\int$ is called the *integral sign*, the function $f(x)$—the *integrand* and x—the *variable of integration*. The value of a definite integral depends only on the function $f(x)$ and the limits a and b but not on the variable x which can be denoted by an arbitrary letter. Thus

$$\int_a^b f(x)\,dx = \int_a^b f(t)\,dt = \int_a^b f(z)\,dz.$$

Theorem on existence of the definite integral. Any continuous function in the interval $[a, b]$ has the definite integral, i.e., the limit (1) exists and is independent of the choice of the numbers x_i and ξ_i [1].

Geometric significance of the definite integral of a continuous function. The integral (1) is equal numerically to the area bounded by the graph of the function $y = f(x)$, the x axis and the ordinates

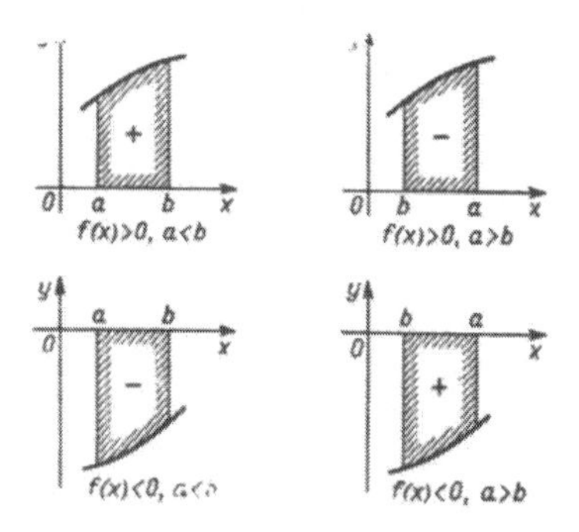

FIG. 303

$f(a)$ and $f(b)$ taken with the sign "+" or "−" as in Fig. 303. If the curve intersects the x axis one or several times inside the

[1] The definite integral exists also for any bounded function with a finite number of points of discontinuity in the interval $[a, b]$. A function for which the definite integral exists in the given interval is called *integrable* in this interval.

interval $[a, b]$, then the integral is equal to the algebraic sum of the areas lying on both sides of the x axis.

The integral between identical limits. By the definition,

$$\int_a^a f(x)\,dx = 0.$$

Fundamental properties of a definite integral are the following:

(1) Theorem on interchanging the limits. After interchanging the limits, the integral changes the sign:

$$\int_a^b f(x)\,dx = -\int_b^a f(x)\,dx.$$

(2) Theorem on decomposition of the interval. For arbitrary numbers a, b, c

$$\int_a^b f(x)\,dx = \int_a^c f(x)\,dx + \int_c^b f(x)\,dx.$$

(3) The integral of an algebraic sum of several functions is equal to the sum of the integrals of these functions:

$$\int_a^b \big(f(x) + \varphi(x) - \psi(x)\big)\,dx = \int_a^b f(x)\,dx + \int_a^b \varphi(x)\,dx - \int_a^b \psi(x)\,dx.$$

(4) A constant factor can be brought out from under the integral sign:

$$\int_a^b cf(x)\,dx = c\int_a^b f(x)\,dx.$$

(5) Mean value theorem. If the function $f(x)$ is continuous in the interval $[a, b]$, then there exists at least one point ξ inside the interval $[a, b]$ ($a < \xi < b$ in the case A and $a > \xi > b$ in the case B, see p. 454) such that

$$\int_a^b f(x)\,dx = (b - a)\,f(\xi).$$

The geometric significance of this theorem is shown in Fig. 304; there exists a point ξ between a and b such that the area of $ABCD$ is equal to the area of the rectangle $AB'C'D$.

Generalized mean value theorem. For the integral of a product of two functions $f(x)$ and $\varphi(x)$ (where $f(x)$ is continuous and $\varphi(x)$ has a constant sign in the interval $[a, b]$), there exists at least one number ξ inside the interval $[a, b]$ such that

$$\int_a^b f(x)\varphi(x)\,dx = f(\xi)\int_a^b \varphi(x)\,dx.$$

(6) Theorem on estimation of the integral. The value of a definite integral is contained between the products of the least and the greatest value of the integrand times the length of the interval of integration:

$$m(b-a) < \int_a^b f(x)\,dx < M(b-a),$$

where m is the least and M is the greatest value of $f(x)$ in the interval $[a, b]$.

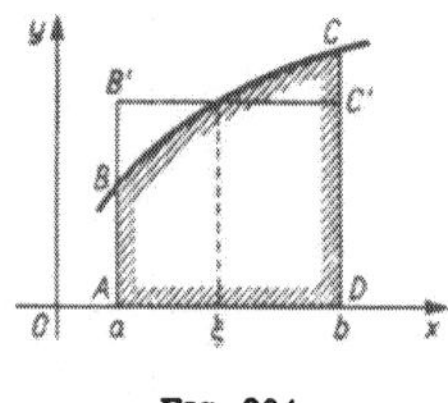

Fig. 304

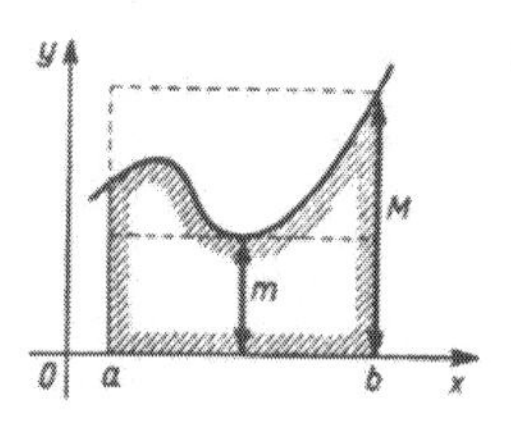

Fig. 305

The significance of this theorem is evident from Fig. 305.

(7) Leibnitz-Newton's theorem. The definite integral $\int_a^x f(t)\,dt$ [1] with a variable upper limit is a continuous function $F(x)$ of x and is a primitive function to the integrand $f(x)$:

$$F'(x) = f(x) \qquad \text{or} \qquad d\int_a^x f(t)\,dt = f(x)\,dx.$$

The significance of this theorem is that the derivative of the variable area $S(x)$ in Fig. 306 is equal to the variable ordinate NM (the area $S(x)$ and the ordinate NM should be taken with the sign "+" or "−", according to Fig. 303 on p. 455).

(8) Fundamental theorem of the integral calculus (expression of the definite integral by an indefinite integral). If

$$\int f(x)\,dx = F(x) + C,$$

[1] In order to avoid misunderstanding, the variable of integration has been denoted by t (see p. 455).

then

$$\int_a^b f(x)\,dx = F(b) - F(a). \tag{2}$$

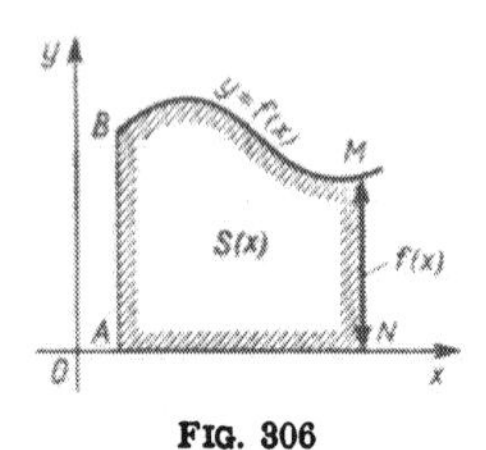

Fig. 306

The right side of the equation (2) is often written by aid of the following symbol of substitution:

$$F(b) - F(a) \equiv [F(x)]_a^b \quad \text{or} \quad F(x)|_a^b.$$

The constant C of integration vanishes after substitution of the limits to $F(x) + C$ and hence it can be omitted in evaluating a definite integral according to formula (2). It follows that equation (2) can be written in the form

$$\int_a^b f(x)\,dx = \left[\int f(x)\,dx\right]_a^b.$$

Equation (2) can also be rewritten in the form of an integral of a differential:

$$\int_a^b dF(x) = F(b) - F(a).$$

9. Evaluation of definite integrals

Fundamental method of evaluation of definite integrals is based on replacing the definite integral by an indefinite integral (see fundamental theorem on p. 457):

$$\int_a^b f(x)dx = \left[\int f(x)\,dx\right]_a^b.$$

In this case, to evaluate the definite integral, a primitive function of $f(x)$ should be found.

Rules of transformations of definite integrals. Definite integrals, likewise indefinite integrals, can be transformed one into another by the following rules:

(1) The substitution rule. By introducing an auxiliary function $x = \varphi(t)$ (where the new variable t is a single-valued function $t = \psi(x)$ of x in the interval $[a, b]$), the integral can be

transformed into the form

$$\int_a^b f(x)\,dx = \int_{\varphi(a)}^{\varphi(b)} f(\varphi(t))\varphi'(t)\,dt.$$

Using this formula we can avoid the inverse substitution in evaluating the indefinite integral.

Example [1].

$$\int_0^a \sqrt{a^2 - x^2}\,dx = \int_{\arcsin 0}^{\arcsin 1} a^2\sqrt{1 - \sin^2 t}\,d\sin t = a^2 \int_0^{\pi/2} \cos^2 t\,dt$$

$$= a^2 \int_0^{\pi/2} \frac{1}{2}(1 + \cos 2t)\,dt = \frac{a^2}{2}\Big[t\Big]_0^{\pi/2} + \frac{a^2}{4}\int_0^{\pi} \cos z\,dz$$

$$= \frac{\pi a^2}{4} + \frac{a^2}{4}\Big[\sin z\Big]_0^{\pi} = \frac{\pi a^2}{4}.$$

(2) Integration by parts. Writing the expression $f(x)\,dx$, in an arbitrary way, in the form $u\,dv$ and finding du (by differentiation) and v (by integration), we can transform the definite integral into the form

$$\int_a^b f(x)\,dx = \int_a^b u\,dv = \Big[uv\Big]_a^b - \int_a^b v\,du.$$

Example:

$$\int_0^1 \underbrace{x}_{u}\,\underbrace{e^x\,dx}_{dv} = \Big[xe^x\Big]_0^1 - \int_0^1 e^x\,dx = e - (e - 1) = 1.$$

Artificial tricks. If the evaluation of a definite integral is very complicated or if the integral cannot be expressed in terms of elementary functions, then its value can sometimes be found by use of artificial tricks. For example, we can use properties of analytic function of a complex variable (see examples on p. 608 and 612) or the theorem on differentiation of an integral with respect to a parameter (see p. 478):

$$\text{(A)} \qquad \frac{d}{dt}\int_a^b f(x, t)\,dx = \int_a^b \frac{\partial f(x, t)}{\partial t}\,dx.$$

[1] In this example, we substitute first $x = \varphi(t) = a \sin t$, whence $t = \psi(x) = \arcsin (x/a)$; the function $\psi(x)$ is single-valued in the interval $[0, a]$ and $\psi(0) = 0$, $\psi(a) = \frac{1}{2}\pi$. Then we substitute $t = \varphi(z) = \frac{1}{2}z$, whence $z = \psi(t) = 2t$; this is a single-valued function in the interval $[0, \frac{1}{2}\pi]$ and $\psi(0) = 0$, $\psi(\frac{1}{2}\pi) = \pi$.

Example. Evaluate

$$I = \int_0^1 \frac{x-1}{\ln|x|}\,dx.$$

We introduce a parameter t and consider the integral

$$F(t) = \int_0^1 \frac{x^t - 1}{\ln|x|}\,dx; \quad F(0) = 0, \quad F(1) = I.$$

Applying formula (A) to the function $F(t)$, we have

$$\frac{dF}{dt} = \int_0^1 \frac{\partial}{\partial t}\left[\frac{x^t-1}{\ln|x|}\right] dt = \int_0^1 \frac{x^t \ln|x|}{\ln|x|}\,dx$$

$$= \int_0^1 x^t\,dx = \left[\frac{1}{t+1}x^{t+1}\right]_0^1 = \frac{1}{t+1}.$$

The integration gives

$$F(t) - F(0) = \int_0^t \frac{dt}{t+1} = [\ln|t+1|]_0^t = \ln|t+1|,$$

whence we obtain the desired integral $I = F(1) = \ln 2$.

Integration by expansion into a series. If the integrand $f(x)$ can be represented in the interval of integration by a uniformly convergent series of functions (see p. 355):

$$f(x) = \varphi_1(x) + \varphi_2(x) + \ldots + \varphi_n(x) + \ldots,$$

then the following equality holds

$$\int f(x)\,dx = \int \varphi_1(x)\,dx + \int \varphi_2(x)\,dx + \ldots + \int \varphi_n(x)\,dx + \ldots$$

hence the definite integral can be represented in the form of a convergent series of numbers

$$\int_a^b f(x)\,dx = \int_a^b \varphi_1(x)\,dx + \int_a^b \varphi_2(x)\,dx + \ldots + \int_a^b \varphi_n(x)\,dx + \ldots$$

If the functions $\varphi_i(x)$ can be easily integrated (as, for example, in the case of an expansion of $f(x)$ into a power series uniformly convergent in the interval $[a, b]$), then the integral $\int_a^b f(x)\,dx$ can be evaluated with an arbitrary accuracy.

Example. Evaluate $I = \int_0^{1/2} e^{-x^2}\,dx$ with an accuracy of 0.0001.

$$e^{-x^2} = 1 - \frac{x^2}{1!} + \frac{x^4}{2!} - \frac{x^6}{3!} + \frac{x^8}{4!} - \dots$$

(see the table on p. 389). This series is uniformly convergent in any finite interval (by Abel's theorem, see p. 358), therefore

$$\int e^{-x^2}\,dx = x\left(1 - \frac{x^2}{1!\cdot 3} + \frac{x^4}{2!\cdot 5} - \frac{x^6}{3!\cdot 7} + \frac{x^8}{4!\cdot 9} - \dots\right),$$

hence

$$I = \int_0^{1/2} e^{-x^2}\,dx = \frac{1}{2}\left(1 - \frac{1}{2^2\cdot 1!\cdot 3} + \frac{1}{2^4\cdot 2!\cdot 5} - \frac{1}{2^6\cdot 3!\cdot 7} + \frac{1}{2^8\cdot 4!\cdot 9} - \dots\right)$$

$$= \frac{1}{2}\left(1 - \frac{1}{12} + \frac{1}{160} - \frac{1}{2688} + \frac{1}{55296} - \dots\right).$$

According to Leibnitz's theorem on alternating series (see p. 352), to compute the integral I with the desired accuracy, we can confine ourselves to four initial terms of the expansion:

$$I \approx \tfrac{1}{2}(1 - 0.08333 + 0.00625 - 0.00037) = \tfrac{1}{2}\cdot 0.92255 = 0.46127,$$

$$\int_0^{1/2} e^{-x^2}\,dx = 0.4613.$$

Approximate methods. The most widely used approximate methods are based on replacing an integral by a finite sum. To compute $\int_a^b y\,dx$, we divide the interval from $a = x_0$ to $b = x_n$ into n equal parts by points $x_0, x_1, \dots, x_n$ and find the values of the integrand y at the points $x_0, x_1, \dots, x_n$. Then we use one of the following three formulas (putting $h = \frac{b-a}{n}$):

(1) Formula of rectangles (Fig 307a):

$$\int_a^b y\,dx \approx h(y_0 + y_1 + \dots + y_{n-1}).$$

(2) Formula of trapezoids (Fig. 307b):

$$\int_a^b y\,dx \approx \frac{h}{2}(y_0 + 2y_1 + 2y_2 + \ldots + 2y_{n-1} + y_n).$$

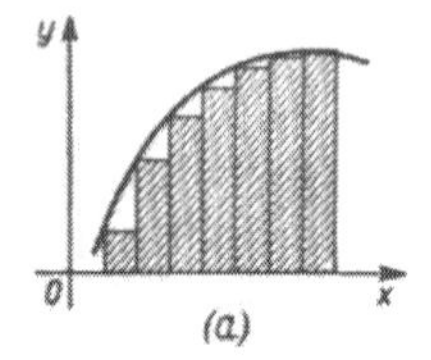

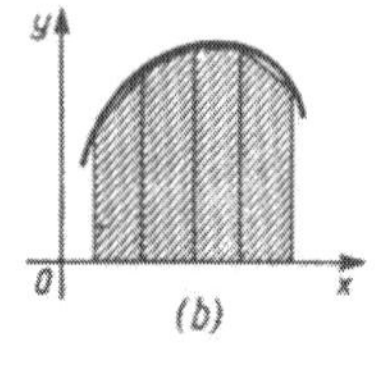

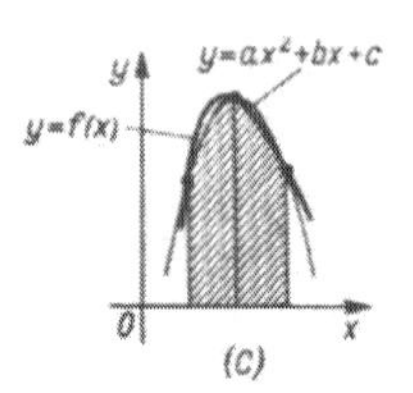

FIG. 307

(3) Formula of parabolas (*Simpson's formula*), for an even n (Fig. 307c):

$$\text{(I)} \quad \int_a^b y\,dx \approx \frac{h}{3}(y_0 + 4y_1 + 2y_2 + 4y_3 + \ldots + 2y_{n-2} + 4y_{n-1} + y_n).$$

All these formulas become more accurate, when n increases. For a fixed n, the second formula is more accurate than the first one and the third one is still more accurate and, therefore, most frequently used. To estimate the error in the result of evaluation of an integral by Simpson's formula (if n is a multiple of 4), we compute an auxiliary sum

$$\text{(II)} \qquad \frac{2h}{3}(y_0 + 4y_2 + 2y_4 + \ldots + 4y_{n-2} + y_n)$$

which represents the same Simpson's formula for strips with $2h$ after discarding the ordinates with odd indices. We can assume that, approximately,

$$\int_a^b y\,dx - \text{(I)} = \frac{\text{(I)} - \text{(II)}}{15}.$$

Many other approximate integration formulas can be obtained by replacing the integrand by an interpolation polynomial (see p. 758). The following formulas are mostly used (for notation see p. 759):

$$\int_a^b y\,dx = 2h\Big[(y_1 + y_3 + \dots + y_{n-1}) +$$
$$+ \tfrac{1}{6}(\Delta^2 y_0 + \Delta^2 y_2 + \dots + \Delta^2 y_{n-2}) -$$
$$- \tfrac{1}{180}(\Delta^4 y_{-1} + \Delta^4 y_1 + \dots + \Delta^4 y_{n-3}) +$$
$$+ \tfrac{1}{1512}(\Delta^6 y_{-2} + \Delta^6 y_0 + \dots + \Delta^6 y_{n-4})\Big],$$

if n is even, and

$$\int_a^b y\,dx = h\Bigg[\left(\frac{y_0}{2} + y_1 + y_2 + \dots + y_{n-1} + \frac{y_n}{2}\right) -$$
$$- \frac{1}{12}\left(\frac{\Delta y_{n-1} + \Delta y_n}{2} - \frac{\Delta y_{-1} + \Delta y_0}{2}\right) +$$
$$+ \frac{11}{720}\left(\frac{\Delta^3 y_{n-2} + \Delta^3 y_{n-1}}{2} - \frac{\Delta^3 y_{-2} + \Delta^3 y_{-1}}{2}\right) -$$
$$- \frac{1}{317}\left(\frac{\Delta^5 y_{n-3} + \Delta^5 y_{n-2}}{2} - \frac{\Delta^5 y_{-3} + \Delta^5 y_{-2}}{2}\right)\Bigg].$$

Usually we do not compute the last summand of these formulas; it is used only to estimate the error of the result.

Graphical integration. If the integrand $y = f(x)$ is represented by a graph AB (Fig. 308), then $\int_a^b f(x)\,dx$ equal to the area M_0ABN can be found graphically as follows:

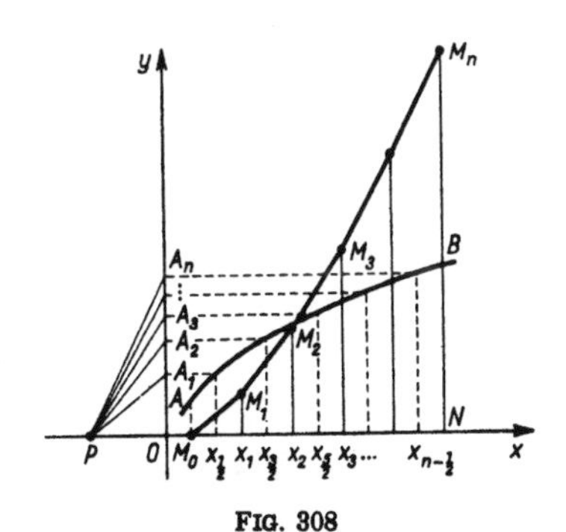

FIG. 308

(1) We divide M_0N into $2n$ equal parts by means of the points

$$x_{1/2},\ x_1,\ x_{3/2},\ x_2,\ \dots,\ x_{n-1},\ x_{n-1/2};$$

for a greater number of points of division, the result will be more accurate;

(2) at the points

$$x_{1/2},\ x_{3/2},\ \ldots,\ x_{n-1/2}$$

of division we erect the ordinates of the curve and lay them off on the y axis (the segments OA_1, OA_2, ..., OA_n);

(3) on the left side of the x axis we lay off a segment OP of an arbitrary length and join it with the points A_1, A_2, ..., A_n;

(4) through the point M_0 we draw the segment M_0M_1 parallel to PA_1 to intersection with the ordinate of x_1, through the point M_1 we draw the segment $M_1M_2 \parallel PA_2$ to intersection with the ordinate of x_2, then we draw the segment $M_2M_3 \parallel PA_3$ and so on till we reach the last ordinate at M_n.

The integral $\int_a^b f(x)\,dx$ is numerically equal to the product of the length of the segments OP times the segment NM_n. By a suitable choice of the segment OP we can obtain the figure of a desired magnitude (to obtain a less graph, we should choose a greater segment OP). If $OP = 1$, then $\int_a^b f(x)\,dx = NM_n$, and the broken line $M_0M_1M_2\ldots M_n$ represents approximately the graph of a primitive function of $f(x)$ (i.e., the indefinite integral $\int f(x)\,dx$).

Evaluation by means of a planimeter. A *planimeter* is a tool which enables us to find the area bounded by a curve and thus to compute the definite integral of a function $y = f(x)$ given by its graph. Planimeters of a special type can evaluate not only $\int y\,dx$ but also $\int y^2\,dx$ and $\int y^3\,dx$.

Integrators. There exist devices which can be used to draw the graph of a primitive function

$$Y = \int_a^x f(t)\,dt$$

after the graph of a given function $y = f(x)$; they are called *integrators*.

10. Applications of definite integrals

A general principle of application of definite integrals to evaluation of geometric, physical and other quantities.

(1) We decompose the evaluated quantity A in a way into a great number n of small quantities: $A = a_1 + a_2 + \ldots + a_n$;

(2) we replace each quantity a_i by a quantity $\tilde{a}_i$ near to a_i that can be evaluated by a known formula; the error $\alpha_i = a_i - \tilde{a}_i$ should be an infinitesimal of a higher order than a_i, i.e., a_i and $\tilde{a}_i$ are equivalent infinitesimals;

(3) we express $\tilde{a}_i$ by a certain variable x chosen so that $\tilde{a}_i$ has the form $f(x_i)\,\Delta x_i$;

(4) we evaluate the desired quantity as the limit of the sum

$$A = \lim_{n\to\infty} \sum_{i=1}^{n} \tilde{a}_i = \lim_{n\to\infty} \sum_{i=1}^{n} f(x_i)\,\Delta x_i = \int_a^b f(x)\,dx,$$

where $a = x_0$ and $b = x_n$ are the end points of the interval of the variable x.

Example. Evaluation of the volume V of a pyramid with base S and altitude H.

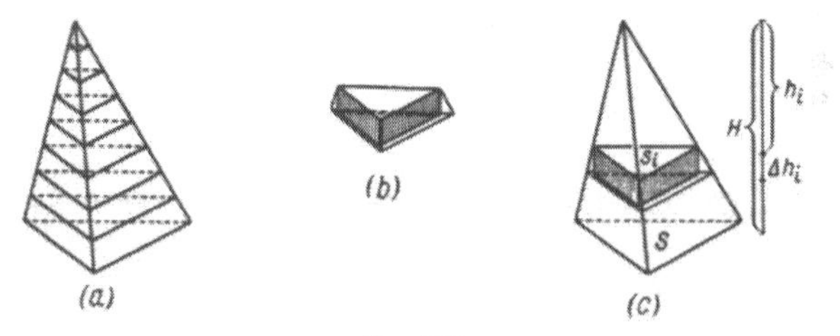

FIG. 309

(1) We decompose the pyramid by plane sections parallel to the base into n frustums (Fig. 309a).

$$V = v_1 + v_2 + \ldots + v_n;$$

(2) we replace each frustum of the pyramid by the prism $\tilde{v}_i$ with the same altitude and with the base equal to the upper base of the frustum (Fig. 309b). The disregarded volume is an infinitesimal of a higher order than v_i;

(3) we write the formula for volume $\tilde{v}_i$ in the form $\tilde{v}_i = S_i \Delta h_i$ where h_i (Fig. 309c) is the distance of the upper base from the vertex of the pyramid or, by the proportion $S_i : S = h_i^2 : H^2$,

$$\tilde{v}_i = \frac{S h_i^2}{H^2} \Delta h_i;$$

(4) we evaluate the desired volume V as the limit of the sum

$$V = \lim_{n\to\infty} \sum_{i=1}^{n} \tilde{v}_i = \lim_{n\to\infty} \sum_{i=1}^{n} \frac{S h_i^2}{H^2} \Delta h_i = \int_0^H \frac{S h^2}{H^2}\, dh = \frac{SH}{3}.$$

Fundamental applications in geometry.

Area. Formula for *area of a curvilinear trapezoid* (Fig. 310a), if the curve is given in the explicit form $y = f(x)$, $a < x < b$, or in

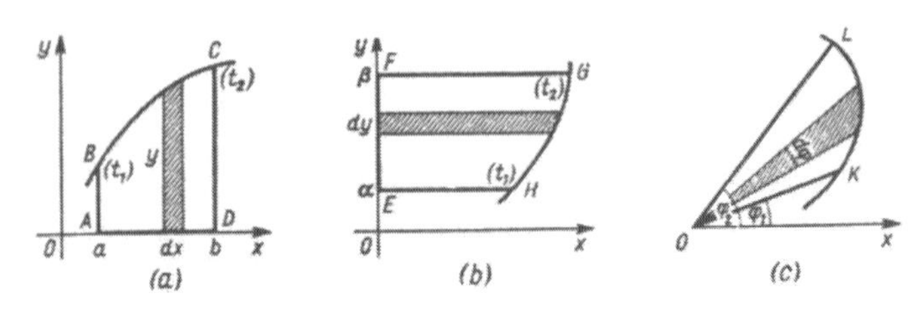

FIG. 310

parametric form $x = \varphi(t)$, $y = \psi(t)$, $t_1 < t < t_2$:

$$S_{ABCD} = \int_a^b f(x)\,dx = \int_{t_1}^{t_2} \psi(t)\,\varphi'(t)\,dt.$$

Formula for area of a curvilinear trapezoid (Fig. 310b), when the curve is given in the form $x = g(y)$, $\alpha < y < \beta$ or in a parametric form $x = \varphi(t)$, $y = \psi(t)$, $t_1 < t < t_2$:

$$S_{EFGH} = \int_\alpha^\beta g(y)\,dy = \int_{t_1}^{t_2} \varphi(t)\,\psi'(t)\,dt.$$

Formula for *area of a curvilinear sector* (Fig. 310c), when the curve is given in polar coordinates by an equation $\varrho = \varrho(\varphi)$, $\varphi_1 < \varphi < \varphi_2$:

$$S_{OKL} = \frac{1}{2}\int_{\varphi_1}^{\varphi_2} \varrho^2\,d\varphi.$$

Areas of more complicated figures can be evaluated by means of the line integral (see p. 492) or the double integral (see p. 504).

Length of arc. Formula for length of arc of a curve (Fig. 311a) given in explicit form $y = f(x)$ or $x = g(y)$ or in parametric form $x = \varphi(t)$, $y = \psi(t)$:

$$L(\breve{AB}) = \int_a^b \sqrt{1 + [f'(x)]^2}\,dx = \int_\alpha^\beta \sqrt{[g'(y)]^2 + 1}\,dy = \int_{t_1}^{t_2} \sqrt{[\varphi'(t)]^2 + [\psi'(t)]^2}\,dt$$

or $L = \int dl$, where dl is the differential of arc: $dl^2 = dx^2 + dy^2$.

If the curve is defined by an equation $\varrho = \varrho(\varphi)$ in polar coordinates

(Fig. 311b), then

$$L(\breve{CD}) = \int_{\varphi_1}^{\varphi_2} \sqrt{\varrho^2 + \left(\frac{d\varrho}{d\varphi}\right)^2}\, d\varphi$$

or $L = \int dl$, where dl is the differential of arc: $dl^2 = \varrho^2 d\varphi^2 + d\varrho^2$.

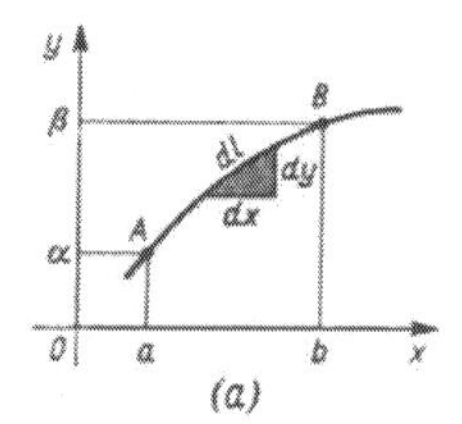

(a) (b)

FIG. 311

Area of a surface. Formula for area of a surface generated by revolution of a curve $y = f(x)$ about the x axis (Fig. 312a):

$$S = 2\pi \int_a^b y\, dl = 2\pi \int_a^b y \sqrt{1 + \left(\frac{dy}{dx}\right)^2}\, dx$$

or by revolution of a curve $x = g(y)$ about the y axis (Fig. 312b):

$$S = 2\pi \int_\alpha^\beta x\, dl = 2\pi \int_\alpha^\beta x \sqrt{\left(\frac{dx}{dy}\right)^2 + 1}\, dy.$$

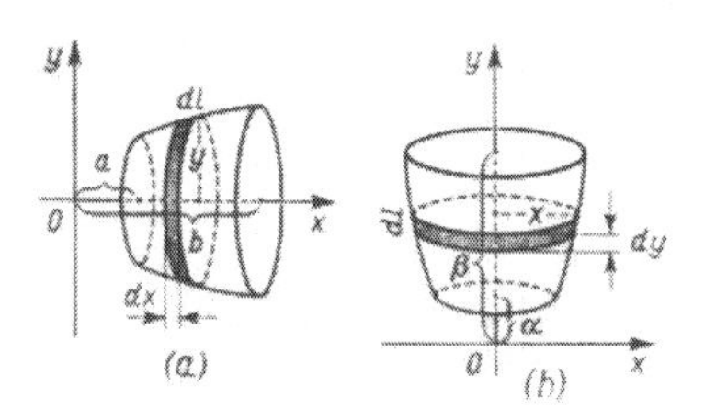

FIG. 312

For areas of more complicated surfaces see pp. 504, 505.

Volume. Formula for the volume interior to a surface of revolution about the x axis (Fig. 312a):

$$V = \pi \int_a^b y^2\, dx;$$

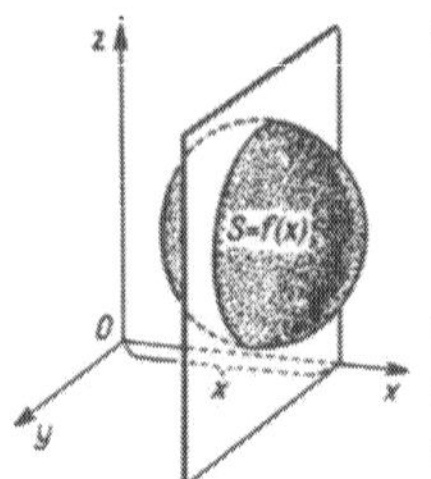

FIG. 313

about the y axis (Fig. 312b):

$$V = \pi \int_{\alpha}^{\beta} x^2\, dy.$$

Formula for the volume of a solid when the area of its plane section perpendicular to the x axis is given as a function $S = f(x)$ of x (Fig. 313):

$$V = \int_{a}^{b} f(x)\, dx.$$

Areas of more complicated solids can be evaluated by means of the double or triple integral (see pp. 496, 504, 505).

Applications to mechanics and physics.

The distance traversed by a point from the initial moment t_0 to the final moment T, when the velocity is depending on the time t by the function $v = f(t)$ is given by the formula

$$S = \int_{t_0}^{T} v\, dt.$$

The work done by the force F along the distance from the point $x = a$ to $x = b$ of the x axis, when the force is a function $F = f(x)$ of x, is given by the formula (Fig. 314):

$$A = \int_{a}^{b} F\, dx \ (^1).$$

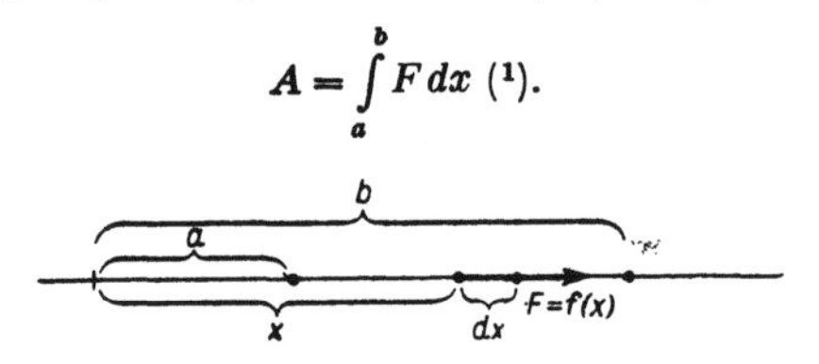

FIG. 314

The pressure exerted by a fluid with specific gravity γ on one side of a vertical plate immersed in the fluid when the distance x of the points of the plate from the level of the fluid varies from a to b is given by the formula (Fig. 315):

$$P = \int_{a}^{b} \gamma x y\, dx,$$

where y is the length of a horizontal section of the plate ($y = f(x)$).

(1) In a general case, when the direction of the force does not coincide with the direction of the motion, the work can be calculated as a line integral (see p. 635).

Moment of inertia.

(1) The moment of inertia of an arc of a homogeneous curve $y=f(x)$, $a<x<b$, with respect to the y axis (Fig. 316a) is given by the formula

$$I_y=\delta\int_a^b x^2\,dl=\delta\int_a^b x^2\sqrt{1+(y')^2}\,dx,$$

where δ is the linear density of the arc.

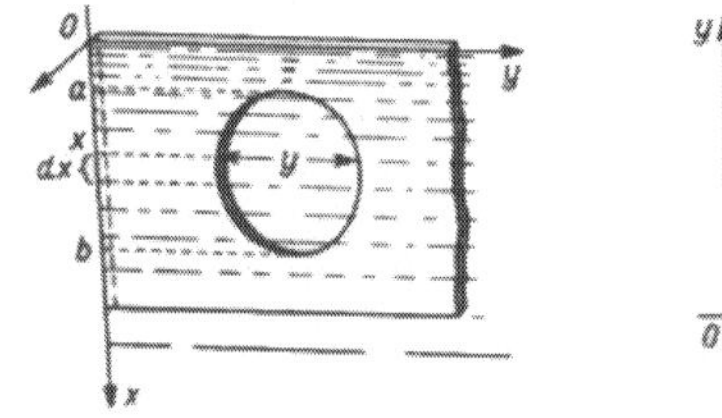

FIG. 315

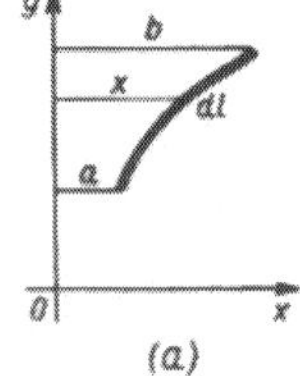

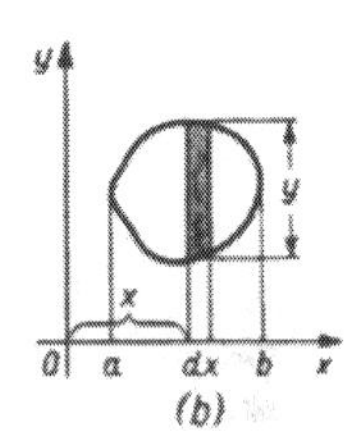

FIG. 316

(2) The moment of inertia of a plane figure (Fig. 316b) with respect to the y axis is given by the formula

$$I_y=\int_a^b x^2 y\,dx,$$

where y is the length of a section parallel to the y axis and δ is the surface density of the figure (see also p. 504).

The centre of gravity C of an arc of a homogeneous plane curve $y=f(x)$, $a<x<b$ (Fig. 317a), has the coordinates

$$x_C=\frac{\int_a^b x\sqrt{1+y'^2}\,dx}{L},\qquad y_C=\frac{\int_a^b y\sqrt{1+y'^2}\,dx}{L},$$

where L is the length of the arc (see p. 466).

The centre of gravity of a closed curve (Fig. 317b):

$$x_C=\frac{\int_a^b x\left(\sqrt{1+|y_1'|^2}+\sqrt{1+|y_2'|^2}\right)dx}{L},$$

$$y_C=\frac{\int_a^b \left(y_1\sqrt{1+|y_1'|^2}+y_2\sqrt{1+|y_2'|^2}\right)dx}{L},$$

where $y_1 = f_1(x)$ and $y_2 = f_2(x)$ are the equations of the upper and lower part of the bounding curve and L is the length of the whole curve.

The first theorem of Guldin. The area of a surface of revolution of a plane curve about an axis lying in this plane and

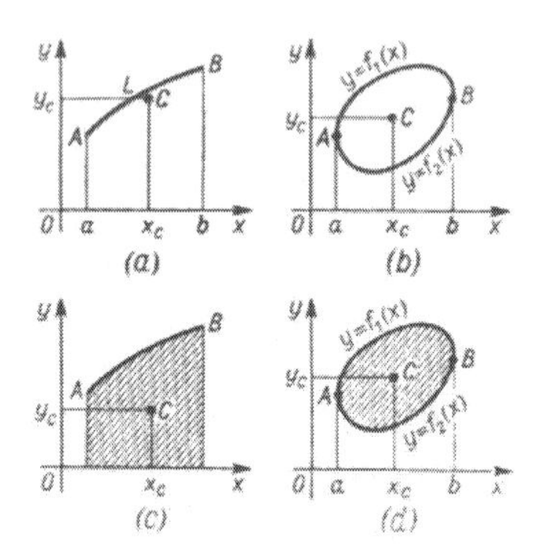

FIG. 317

not intersecting the curve is equal to the length L of the curve multiplied by the distance traversed by the centre of gravity:

$$S_{rev} = L \cdot 2\pi y_C.$$

The centre of gravity C of a homogeneous curvilinear trapezoid (Fig. 317c) has the coordinates

$$x_C = \frac{\int_a^b xy\,dx}{S}, \quad y_C = \frac{\frac{1}{2}\int_a^b y^2\,dx}{S},$$

where S is the area of the trapezoid and $y = f(x)$ is the equation of the curve AB.

The centre of gravity of an arbitrary plane figure (Fig. 317d) has the coordinates

$$x_C = \frac{\int_a^b x(y_1 - y_2)\,dx}{S}, \quad y_C = \frac{\frac{1}{2}\int_a^b (y_1^2 - y_2^2)\,dx}{S},$$

where $y_1 = f_1(x)$ and $y_2 = f_2(x)$ are the equations of the upper and lower part of the bounding curve and S is the area of the figure.

The second theorem of Guldin. The volume interior to

a surface of revolution of a plane figure about an axis lying in this plane and not intersecting the figure is equal to the area S of the figure multiplied by the length of the circumference traversed by the centre of gravity of the figure:

$$V_{rev} = S \cdot 2\pi y_C.$$

For centres of gravity of plane figures and solids see pp. 504, 505 (multiple integrals).

11. Improper integrals

General information. The simplest generalization of the concept of the definite integral (p. 454) are *improper integrals* [1].

Two fundamental types of improper integrals:

(1) Integrals with infinite interval of integration. The domain of definition of the integrand is a closed half axis $[a, \infty)$, $(-\infty, a]$ or the whole axis $(-\infty, \infty)$.

(2) Integrals of discontinuous functions. The integrand is continuous in the interval from a to b except for a certain finite number of points called *singular points*.

Combinations of these two types can also occur.

Integrals with infinite interval of integration.

Definitions. Let the integrand be defined in the half axis $[a, \infty)$. We assume the definition

$$\int_a^{\infty} f(x)\,dx = \lim_{B\to\infty} \int_a^B f(x)\,dx. \tag{1}$$

If this limit exists, then the integral (1) *exists*, i.e., is *convergent* and is called an *improper integral*. If this limit does not exist, then the integral (1) does *not exist*, i.e., is *divergent*. In the case when $\lim_{B\to\infty} \int_a^B f(x)\,dx = \infty$, we write

$$\int_a^{\infty} f(x)\,dx = \infty;$$

the integral is divergent.

The *improper integrals* of functions defined in the half axis $(-\infty, b]$ or in the whole axis $(-\infty, \infty)$ are defined likewise:

---

[1] The concept of an integral can also be generalized to more complicated cases, when the domain of definition of the function (the domain of integration) is the set of values of another function (*Stieltjes' integral*).

$$\int_{-\infty}^{b} f(x)\,dx = \lim_{A\to-\infty} \int_{A}^{b} f(x)\,dx, \tag{2}$$

$$\int_{-\infty}^{\infty} f(x)\,dx = \lim_{\substack{A\to-\infty \\ B\to\infty}} \int_{A}^{B} f(x)\,dx. \tag{3}$$

The numbers A and B tend to infinity independently of one another. If the limit (3) does not exist, but

$$\lim_{A\to\infty} \int_{-A}^{A} f(x)\,dx \tag{4}$$

exists, then the limit (4) is called the *principal value of the improper integral.*

Geometric significance of improper integrals with infinite limits of integration. The integrals (1), (2), (3) are limits of the areas of figures shown in Fig. 318a, b, c.

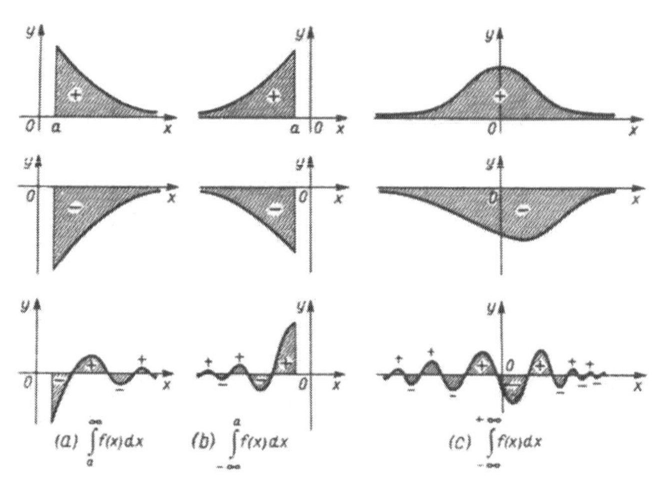

FIG. 318

Examples.

(1) $$\int_{1}^{\infty} \frac{dx}{x} = \lim_{B\to\infty} \int_{1}^{B} \frac{dx}{x} = \lim_{B\to\infty} \ln|B| = \infty$$

(the integral is divergent),

(2) $$\int_2^{\infty} \frac{dx}{x^2} = \lim_{B\to\infty} \int_2^B \frac{dx}{x^2} = \lim_{B\to\infty}\left(\frac{1}{2} - \frac{1}{B}\right) = \frac{1}{2}$$

(the integral is convergent),

(3) $$\int_{-\infty}^{\infty} \frac{dx}{1+x^2} = \lim_{\substack{A\to-\infty \\ B\to\infty}} \int_A^B \frac{dx}{1+x^2} = \lim_{\substack{A\to-\infty \\ B\to\infty}} (\arctan B - \arctan A)$$

$$= \frac{\pi}{2} - \left(-\frac{\pi}{2}\right) = \pi \text{ (the integral is convergent).}$$

If the limits (1), (2), (3) are difficult to find immediately or if we only want to know whether or not the integral is convergent, then we can apply one of the following criteria of convergence.

Sufficient criteria of convergence. We consider here only integrals of type (1). In the case of an integral of type (2), we can change the variables $x = -z$ and reduce the integral to type (1):

$$\int_{-\infty}^{a} f(x)\,dx = \int_{-a}^{\infty} f(-x)\,dx.$$

An integral of type (3) can be decomposed into a sum of two integrals of types (2) and (1):

$$\int_{-\infty}^{\infty} f(x)\,dx = \int_{-\infty}^{c} f(x)\,dx + \int_{c}^{\infty} f(x)\,dx,$$

where c is an arbitrary number.

Criterion 1. If the integral $\int_a^{\infty} |f(x)|\,dx$ exists, then the integral (1) also exists. In this case, the integral (1) is said to be *absolutely convergent* and the function $f(x)$ is said to be *absolutely integrable* in the closed half axis $[a, \infty)$.

Criterion 2. If the functions $f(x)$ and $\varphi(x)$ are positive and satisfy the condition $f(x) < \varphi(x)$ in the interval $a < x < \infty$, then the convergence of the integral $\int_a^{\infty} \varphi(x)\,dx$ implies the convergence of $\int_a^{\infty} f(x)\,dx$ and the divergence of $\int_a^{\infty} f(x)\,dx$ implies that of $\int_a^{\infty} \varphi(x)\,dx$.

In particular, since $\int_a^{\infty} \frac{dx}{x^{\alpha}}$ converges for $\alpha > 1$ (it is equal to

$\left.\frac{1}{(\alpha-1)a^{\alpha-1}}\right)$ and diverges for $\alpha<1$, putting $\varphi(x)=\frac{1}{x^\alpha}$, we obtain the following

Criterion 3. If $f(x)$ is a positive function in the interval $a<x<\infty$ and if there exists a number $\alpha>1$ such that, for sufficiently large x, $f(x)\cdot x^\alpha<c<\infty$, then the integral (1) is convergent; if the function $f(x)$ is positive and there exists a number $\alpha<1$ such that $f(x)\cdot x^\alpha>c>0$, then the integral (1) is divergent.

Example. $\int_0^\infty \frac{x^{3/2}}{1+x^2}\,dx$. Putting $\alpha=1/2$, we obtain $\frac{x^{3/2}}{1+x^2}x^{1/2}$ $=\frac{x^2}{1+x^2}\to 1$. The integral is divergent.

Relation of improper integrals to infinite series. If $x_1, x_2, \ldots, x_n, \ldots$ is an arbitrary infinitely increasing sequence, i.e., if

$$\text{(A)}\qquad a<x_1<x_2<\ldots<x_n<\ldots,\qquad \lim_{n\to\infty}x_n=\infty,$$

and the function $f(x)$ is positive in the interval $a<x<\infty$, then the problem of convergence of the integral (1) can be reduced to that of the series

$$\text{(B)}\qquad \int_a^{x_1}f(x)\,dx+\int_{x_1}^{x_2}f(x)\,dx+\ldots+\int_{x_{n-1}}^{x_n}f(x)\,dx+\ldots$$

If series (B) is convergent, then integral (1) is convergent and is equal to the sum of the series; if series (B) is divergent, then so is integral (1). This enables us to apply criteria of convergence of series to examining convergence of integrals. (The integral criterion of convergence of series (p. 351) was applied to reduce the problem of convergence of series to that of improper integrals.)

Integrals of discontinuous functions.

Definitions. Let $f(x)$ be a function defined in an interval $[a, b)$ open from the right or in the whole closed interval $[a, b]$, but at the point b, $\lim_{x\to b}f(x)=\infty$. In either case we assume the definition

$$\text{(1)}\qquad \int_a^b f(x)\,dx=\lim_{\varepsilon\to 0}\int_a^{b-\varepsilon}f(x)\,dx.$$

If this limit exists, then the integral (1) *exists*, i.e., is *convergent* and is called an *improper integral*. If this limit does not exist, then the integral (1) does *not exist*, i.e., is *divergent*. If $\lim_{\varepsilon \to 0} \int_a^{b-\varepsilon} f(x)\, dx = \infty$, we write

$$\int_a^b f(x)\, dx = \infty;$$

the integral is divergent.

The integral (1) exists, provided that the function $f(x)$ is piecewise continuous and bounded in the interval $[a, b)$. In the sequel we shall assume that the function $f(x)$ is unbounded, i.e., $\lim_{x \to b} f(x) = \infty$.

Improper integrals of functions defined in intervals $(a, b]$ open from the left or in the interval $[a, b]$ with $\lim_{x \to a} f(x) = \infty$ are defined likewise:

$$(2) \qquad \int_a^b f(x)\, dx = \lim_{\varepsilon \to 0} \int_{a+\varepsilon}^b f(x)\, dx.$$

Finally, if the function $f(x)$ is defined in the whole closed interval $[a, b]$ except for one interior point c $(a < c < b)$, i.e., if it is defined in both intervals $[a, c)$ and $(c, b]$ or also at the point c, but $\lim_{x \to c} f(x) = \infty$, then the *improper integral* is defined as follows

$$(3) \qquad \int_a^b f(x)\, dx = \lim_{\varepsilon \to 0} \int_a^{c-\varepsilon} f(x)\, dx + \lim_{\delta \to 0} \int_{c+\delta}^b f(x)\, dx \text{ [1]}.$$

The numbers ε and δ tend to zero independently one of another. If the integral (3) does not exist, but

$$(4) \qquad \lim_{\varepsilon \to 0} \left(\int_a^{c-\varepsilon} f(x)\, dx + \int_{c+\varepsilon}^b f(x)\, dx \right)$$

exists, then the limit (4) is called the *principal value of the improper integral*.

Geometric significance of integrals of discontinuous functions. The integrals (1), (2), (3) represent areas of unbounded figures and have the form as in Fig. 319 (the curves have vertical asymptotes).

---

[1] Similarly as in case (1), we assume in integrals (2) and (3) that $\lim_{x \to a} f(x) = \infty$ or $\lim_{x \to c} f(x) = \infty$.

Examples. (1) $\int\limits_0 \frac{dx}{\sqrt{x}}$; case (2), a singular point $x=0$. We have

$$\int\limits_0^b \frac{dx}{\sqrt{x}} = \lim_{\varepsilon\to 0} \int\limits_\varepsilon^b \frac{dx}{\sqrt{x}} = \lim_{\varepsilon\to 0} (2\sqrt{b} - 2\sqrt{\varepsilon}) = 2\sqrt{b}$$

(the integral is convergent).

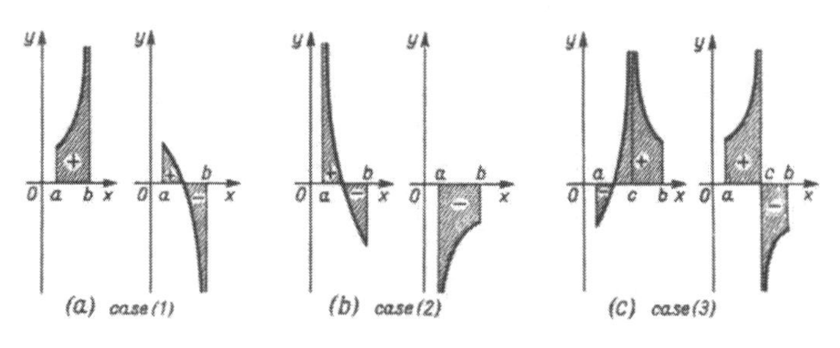

FIG. 319

(2) $\int\limits_0^{\pi/2} \tan x\,dx$; case (1), a singular point $x=\frac{1}{2}\pi$: We have

$$\int\limits_0^{\pi/2} \tan x\,dx = \lim_{\varepsilon\to 0} \int\limits_0^{\pi/2-\varepsilon} \tan x\,dx$$

$$= \lim_{\varepsilon\to 0} \left(\ln\cos 0 - \ln\cos\left(\tfrac{1}{2}\pi - \varepsilon\right)\right) = \infty$$

(the integral is divergent).

(3) $\int\limits_{-1}^{8} \frac{dx}{\sqrt[3]{x}}$; case (3), a singular point at $x=0$. We have

$$\int\limits_{-1}^{8} \frac{dx}{\sqrt[3]{x}} = \lim_{\varepsilon\to 0} \int\limits_{-1}^{-\varepsilon} \frac{dx}{\sqrt[3]{x}} + \lim_{\delta\to 0} \int\limits_{\delta}^{8} \frac{dx}{\sqrt[3]{x}}$$

$$= \lim_{\varepsilon\to 0} \tfrac{3}{2}(\varepsilon^{2/3} - 1) + \lim_{\delta\to 0} \tfrac{3}{2}(4 - \delta^{2/3}) = \tfrac{9}{2}$$

(the integral is convergent).

(4) $\int\limits_{-2}^{2} \frac{2x}{x^2-1}\,dx$; case (3), singular points $x=-1$ and $x=+1$.

We have

$$\int_{-2}^{2} \frac{2x}{x^2-1}\,dx = \lim_{\varepsilon\to 0}\int_{-2}^{-1-\varepsilon} + \lim_{\substack{\delta\to 0\\ \nu\to 0}}\int_{-1+\delta}^{1-\nu} + \lim_{\gamma\to 0}\int_{1+\gamma}^{2}$$

$$= \lim_{\varepsilon\to 0} \ln|x^2-1|\Big|_{-2}^{-1-\varepsilon} + \ldots$$

$$= \lim_{\varepsilon\to 0} [\ln|1+2\varepsilon+\varepsilon^2-1| - \ln 3] + \ldots = \infty$$

(the integral is divergent).

Remarks on application of the fundamental theorem of the integral calculus. In evaluating improper integrals with singular points of type (3), the fundamental theorem of the integral calculus (pp. 457, 458)

$$\int_a^b f(x)\,dx = [F(x)dx]_b^a, \qquad \text{where} \qquad F'(x) = f(x)$$

must not be mechanically applied without regard being paid to singular points inside the interval $[a, b]$ for this can lead to errors. Thus, applying the fundamental theorem to example 4, we obtain

$$\int_{-2}^{+2} \frac{2x}{x^2-1}\,dx = \ln|x^2-1|\Big|_{-2}^{+2} = \ln 3 - \ln 3 = 0,$$

but the integral is divergent.

General rule. The fundamental theorem can be applied provided that the primitive function of $f(x)$ is continuous at the singular point.

This is not the case in example (4): the function $\ln|x^2-1|$ is discontinuous at the points $x=-1$ and $x=+1$; on the other hand, in example (3), the function $\frac{3}{2}x^{2/3}$ is continuous at $x=0$ and hence the fundamental theorem can be applied:

$$\int_{-1}^{8} \frac{dx}{\sqrt[3]{x}} = \frac{3}{2}x^{2/3}\Big|_{-1}^{8} = \frac{3}{2}(8^{2/3} - 1^{2/3}) = \frac{9}{2}.$$

Sufficient criteria for the convergence of integrals of discontinuous functions.

(1) If the integral $\int_a^b |f(x)|\,dx$ exists, then so does the integral

$\int_a^b |f(x)|\, dx$; in this case the integral is said to be *absolutely convergent* and the function $f(x)$ is said to be *absolutely integrable* in the given interval.

(2) If $f(x)$ is a positive function in the interval $[a, b)$ and there exists a number $\alpha < 1$ such that, for x sufficiently near to b, the inequality $f(x)(b-x)^\alpha < c < \infty$ holds, then the integral (1) is convergent. If $f(x)$ is positive and there exists a number $\alpha > 1$ such that, for x sufficiently near to b, the inequality $f(x)(b-x)^\alpha > c > 0$ holds, then the integral (1') is divergent.

12. Integrals depending on a parameter

Definition. The definite integral

$$\int_a^b f(x, y)\, dx = F(y) \tag{1}$$

is a function of the variable y called a *parameter*.

In many cases, the function $F(y)$ is not an elementary function. The integral (1) can be an ordinary integral or an improper integral with an infinite interval of integration or else an integral of a discontinuous function $f(x, y)$.

Example.

$$\Gamma(y) = \int_0^\infty x^{y-1} e^{-x}\, dx \quad \text{(the integral is convergent for } y > 0\text{)}.$$

This is the *Gamma function* or *Euler's integral of the second kind.*

Differentiation under the integral sign. If the function (1) is defined in the interval $c < y < e$ and the function $f(x, y)$ is continuous in the rectangle $a < x < b$, $c < y < e$ and has a continuous partial derivative $\dfrac{\partial f}{\partial y}$ in this domain, then, for every y in the interval $[c, e]$, we have

$$\frac{d}{dy} \int_a^b f(x, y)\, dx = \int_a^b \frac{\partial f(x, y)}{\partial y}\, dx \tag{2}$$

(*differentiation under the integral sign*).

Example. In an arbitrary interval, for $y > 0$,

$$\frac{d}{dy}\int_0^1 \operatorname{arc\,tan}\frac{x}{y}\,dx = \int_0^1 \frac{\partial}{\partial y}\left(\operatorname{arc\,tan}\frac{x}{y}\right)dx$$

$$= -\int_0^1 \frac{x}{x^2+y^2}\,dx = \frac{1}{2}\ln\frac{y^2}{1+y^2}.$$

Proof.

$$\int_0^1 \operatorname{arc\,tan}\frac{x}{y}\,dx = \operatorname{arc\,tan}\frac{1}{y} + \frac{1}{2}y\ln\frac{y^2}{1+y^2},$$

$$\frac{d}{dy}\left(\operatorname{arc\,tan}\frac{1}{y} + \frac{1}{2}y\ln\frac{1}{1+y^2}\right) = \frac{1}{2}\ln\frac{y^2}{1+y^2}.$$

For $y = 0$, the function fails to be continuous; the derivative does not exist.

Extension of formula (2) to the case when the limits of integration also depend on a parameter. If, under the same assumption, the functions $\alpha(y)$ and $\beta(y)$ are defined in the interval $[c, e]$ and have continuous derivatives $\alpha'(y)$ and $\beta'(y)$ and the curves $x = \alpha(y)$, $x = \beta(y)$ lie in the rectangle $a < x < b$, $c < y < e$, then formula (2) can be extended as follows:

$$(2') \quad \frac{d}{dy}\int_{\alpha(y)}^{\beta(y)} f(x, y)\,dx$$

$$= \int_{\alpha(y)}^{\beta(y)} \frac{\partial f(x, y)}{\partial y}\,dx + \beta'(y)\,f(\beta(y), y) - \alpha'(y)\,f(\alpha(y), y).$$

Integration under the integral sign. If the function (1) is defined in the interval $[c, e]$ and the function $f(x, y)$ is continuous in the rectangle $a < x < b$, $c < y < e$, then

$$(3) \qquad \int_c^e\left(\int_a^b f(x, y)\,dx\right)dy = \int_a^b\left(\int_c^e f(x, y)\,dy\right)dx.$$

(*Integration under the integral sign.*)

Examples. (1) $f(x, y) = x^y$ $(0 < x < 1,\ a < y < b,\ a > 0)$. For $a > 0$, the function is continuous; it is discontinuous at $x = 0$, $y = 0$. Hence

$$\int_a^b \Bigl(\int_0^1 x^y\,dx\Bigr)\,dy = \int_0^1 \Bigl(\int_a^b x^y\,dy\Bigr)\,dx.$$

The left side gives $\int_a^b \frac{dy}{1+y} = \ln\frac{1+b}{1+a}$; the right side gives $\int_0^1 \frac{x^b - x^a}{\ln x}\,dx$; the indefinite integral can not be expressed in terms of elementary functions, but the definite integral can be found:

$$\int_0^1 \frac{x^b - x^a}{\ln x}\,dx = \ln\frac{1+b}{1+a} \qquad (0<a<b).$$

(2) $f(x, y) = \frac{y^2 - x^2}{(x^2+y^2)^2}$ $(0 < x < 1,\ 0 < y < 1)$. The function is discontinuous at the point (0, 0) and formula (3) can not be applied. Indeed,

$$\int_0^1 \frac{y^2 - x^2}{(x^2+y^2)^2}\,dx = \frac{x}{x^2+y^2}\Bigg|_{x=0}^{x=1} = \frac{1}{1+y^2}, \quad \int_0^1 \frac{dy}{1+y^2} = \arctan y\Bigg|_0^1 = \frac{\pi}{4}.$$

$$\int_0^1 \frac{y^2 - x^2}{(x^2+y^2)^2}\,dy = -\frac{y}{x^2+y^2}\Bigg|_{y=0}^{y=1} = -\frac{1}{x^2+1},$$

$$-\int_0^1 \frac{dx}{x^2+1} = -\arctan x\Bigg|_0^1 = -\frac{\pi}{4}.$$

13. Tables of certain definite integrals

Integrals of exponential functions

(involved together with algebraic, trigonometric and logarithmic functions).

(1) $\int_0^\infty x^n e^{-ax}\,dx = \frac{\Gamma(n+1)}{a^{n+1}}$, if $a>0$, $n>-1$ [1].

(2) $\int_0^\infty x^n e^{-ax^2}\,dx = \frac{\Gamma\left(\frac{n+1}{2}\right)}{2a^{(n+1)/2}}$, $a>0$, $n>-1$ [1];

[1] For the Gamma function see p. 191; for tables of values of $\Gamma(x)$, see p. 87.

in particular, if n is even ($n = 2k$), then the integral is equal to $\frac{1 \cdot 3 \cdot \ldots \cdot (2k-1)\sqrt{\pi}}{2^{k+1} a^{k+1/2}}$, and if n is odd ($n = 2k+1$), then the integral is equal to $\frac{k!}{2a^{k+1}}$.

$$(3) \quad \int_0^\infty e^{-a^2x^2}\,dx = \frac{\sqrt{\pi}}{2a}, \quad a > 0.$$

$$(4) \quad \int_0^\infty x^2 e^{-a^2x^2}\,dx = \frac{\sqrt{\pi}}{4a^3}, \quad a > 0.$$

$$(5) \quad \int_0^\infty e^{-a^2x^2} \cos bx\,dx = \frac{\sqrt{\pi}}{2a} e^{-b^2/4a^2}, \quad a > 0.$$

$$(6) \quad \int_0^\infty \frac{x}{e^x - 1}\,dx = \frac{\pi^2}{6}.$$

$$(7) \quad \int_0^\infty \frac{x}{e^x + 1}\,dx = \frac{\pi^2}{12}.$$

$$(8) \quad \int_0^\infty \frac{e^{-ax} \sin x}{x}\,dx = \operatorname{arc\,cot} a = \operatorname{arc\,tan} \frac{1}{a}, \quad a > 0.$$

$$(9) \quad \int_0^\infty e^{-x} \ln x\,dx = -C \approx -0.5772 \text{ (}^1\text{)}.$$

Integrals of trigonometric functions
(involved together with algebraic functions)

$$(10) \quad \int_0^{\pi/2} \sin^{2\alpha+1} x \cos^{2\beta+1} x\,dx = \frac{\Gamma(\alpha+1)\,\Gamma(\beta+1)}{2\Gamma(\alpha+\beta+2)}$$

$$= \frac{1}{2} B(\alpha+1, \beta+1) \text{ (}^2\text{)};$$

[1] C is Euler's constant (see p. 331).

[2] $B(x, y) = \frac{\Gamma(x)\,\Gamma(y)}{\Gamma(x+y)}$ is the so-called *Beta function* or *Euler's integral of the first kind* and $\Gamma(x)$ is the *Gamma function* or *Euler's integral of the second kind* (see p. 191).

this formula is true for arbitrary α and β; it can be used to evaluate the integrals

$$\int_0^{\pi/2} \sqrt{\sin x}\, dx, \quad \int_0^{\pi/2} \sqrt[3]{\sin x}\, dx, \quad \int_0^{\pi/2} \frac{dx}{\sqrt[3]{\cos x}} \quad \text{and so on.}$$

If α and β are natural numbers, formula (10) can be written in the form

$$\int_0^{\pi/2} \sin^{2\alpha+1} x \cos^{2\beta+1} x\, dx = \frac{\alpha!\, \beta!}{2(\alpha+\beta+1)!}.$$

(11) $$\int_0^{\infty} \frac{\sin ax}{x}\, dx = \begin{cases} \dfrac{\pi}{2}, & \text{if } a > 0, \\ -\dfrac{\pi}{2}, & \text{if } a < 0. \end{cases}$$

(12) $$\int_0^{\alpha} \frac{\cos ax}{x}\, dx = \infty, \text{ where } \alpha \text{ is arbitrary.}$$

(13) $$\int_0^{\infty} \frac{\tan ax}{x}\, dx = \begin{cases} \dfrac{\pi}{2}, & \text{if } a > 0, \\ -\dfrac{\pi}{2}, & \text{if } a < 0. \end{cases}$$

(14) $$\int_0^{\infty} \frac{\cos ax - \cos bx}{x}\, dx = \ln\left|\frac{b}{a}\right|, \quad a \neq 0, \ b \neq 0.$$

(15) $$\int_0^{\infty} \frac{\sin x \cos ax}{x}\, dx = \begin{cases} \dfrac{\pi}{2}, & \text{if } |a| < 1, \\ \dfrac{\pi}{4}, & \text{if } |a| = 1, \\ 0, & \text{if } |a| > 1. \end{cases}$$

(16) $$\int_0^{\infty} \frac{\sin x}{\sqrt{x}}\, dx = \int_0^{\infty} \frac{\cos x}{\sqrt{x}}\, dx = \sqrt{\frac{\pi}{2}}.$$

(17) $$\int_0^{\infty} \frac{x \sin bx}{a^2 + x^2}\, dx = \pm \frac{\pi}{2} e^{-|ab|}$$

(the sign coincides with the sign of b).

(18) $$\int_0^{\infty} \frac{\cos ax}{1 + x^2}\, dx = \frac{\pi}{2} e^{-|a|}.$$

(19) $$\int_0^\infty \frac{\sin^2 ax}{x^2}\,dx = \frac{\pi}{2}|a|.$$

(20) $$\int_{-\infty}^\infty \sin x^2\,dx = \int_{-\infty}^\infty \cos x^2\,dx = \sqrt{\frac{\pi}{2}}.$$

(21) $$\int_0^{\pi/2} \frac{\sin x}{\sqrt{1-k^2\sin^2 x}}\,dx = \frac{1}{2k}\ln\frac{1+k}{1-k}, \quad |k|<1.$$

(22) $$\int_0^{\pi/2} \frac{\cos x}{\sqrt{1-k^2\sin^2 x}}\,dx = \frac{1}{k}\arcsin k, \quad |k|<1.$$

(23) $$\int_0^{\pi/2} \frac{\sin^2 x}{\sqrt{1-k^2\sin^2 x}}\,dx = \frac{1}{k^2}(\mathrm{K}-\mathrm{E})\,(^1), \quad |k|<1.$$

(24) $$\int_0^{\pi/2} \frac{\cos^2 x}{\sqrt{1-k^2\sin^2 x}}\,dx = \frac{1}{k^2}\big(\mathrm{E}-(1-k^2)\,\mathrm{K}\big)\,(^1), \quad |k|<1.$$

(25) $$\int_0^{\pi} \frac{\cos ax}{1-2b\cos x+b^2}\,dx = \frac{\pi b^a}{1-b^2}, \ a \text{ is a non-negative integer}, \quad |b|<1.$$

Integrals of logarithmic functions
(involved together with algebraic and trigonometric functions)

(26) $$\int_0^1 \ln|\ln x|\,dx = -C = -0.5772\,(^2)$$

(it can be reduced to integral (9)).

(27) $$\int_0^1 \frac{\ln x}{x-1}\,dx = \frac{\pi^2}{6} \quad \text{(it can be reduced to integral (6)).}$$

---

[1] E and K are complete elliptic integrals: $\mathrm{E} = E(k, \frac{1}{2}\pi)$, $\mathrm{K} = F(k, \frac{1}{2}\pi)$ (see p. 408 and the table on pp. 92, 93).

[2] C is Euler's constant (see p. 331).

(28) $\int_0^1 \frac{\ln x}{x+1}\,dx = -\frac{\pi^2}{12}$ (it can be reduced to integral (7)).

(29) $\int_0^1 \frac{\ln x}{x^2-1}\,dx = \frac{\pi^2}{8}.$

(30) $\int_0^1 \frac{\ln(1+x)}{x^2+1}\,dx = \frac{\pi}{8}\ln 2.$

(31) $\int_0^1 \ln\left(\frac{1}{x}\right)^a dx = \Gamma(a+1)$ [1], $-1 < a < \infty.$

(32) $\int_0^{\pi/2} \ln\sin x\,dx = \int_0^{\pi/2} \ln\cos x\,dx = -\frac{\pi}{2}\ln 2.$

(33) $\int_0^{\pi} x\ln\sin x\,dx = -\frac{\pi^2\ln 2}{2}.$

(34) $\int_0^{\pi/2} \sin x\ln\sin x\,dx = \ln 2 - 1.$

(35) $\int_0^{\pi} \ln(a \pm b\cos x)\,dx = \pi\ln\frac{a+\sqrt{a^2-b^2}}{2}, \quad a > b > 0.$

(36) $\int_0^{\pi} \ln(a^2 - 2ab\cos x + b^2)\,dx = \begin{cases} 2\pi\ln a, & a > b > 0, \\ 2\pi\ln b, & b > a > 0. \end{cases}$

(37) $\int_0^{\pi/2} \ln\tan x\,dx = 0.$

(38) $\int_0^{\pi/4} \ln(1+\tan x)\,dx = \frac{\pi}{8}\ln 2.$

---

[1] $\Gamma(x)$ is the Gamma function (see p. 191 and the table on p. 87).

Integrals of algebraic functions

(39) $$\int_0^1 x^{\alpha}(1-x)^{\beta}\,dx = 2\int_0^1 x^{2\alpha+1}(1-x^2)^{\beta}\,dx = \frac{\Gamma(\alpha+1)\,\Gamma(\beta+1)}{\Gamma(\alpha+\beta+2)} = B(\alpha+1,\beta+1)\ ([1])$$

(it can be reduced to integral (10)).

(40) $$\int_0^\infty \frac{dx}{(1+x)\,x^a} = \frac{\pi}{\sin a\pi}, \quad a<1.$$

(41) $$\int_0^\infty \frac{dx}{(1-x)\,x^a} = -\pi\cot a\pi, \quad a<1$$

(42) $$\int_0^\infty \frac{x^{a-1}}{1+x^b}\,dx = \frac{\pi}{b\sin\dfrac{a\pi}{b}}, \quad 0<a<b.$$

(43) $$\int_0^1 \frac{dx}{\sqrt{1-x^a}} = \frac{\sqrt{\pi}\,\Gamma\left(\dfrac{1}{a}\right)}{a\,\Gamma\left(\dfrac{2+a}{2a}\right)}\ ([2]).$$

(44) $$\int_0^1 \frac{dx}{1+2x\cos a+x^2} = \frac{a}{2\sin a}, \quad 0<a<\frac{\pi}{2}.$$

(45) $$\int_0^\infty \frac{dx}{1+2x\cos a+x^2} = \frac{a}{\sin a}, \quad 0<a<\frac{\pi}{2}.$$

C. LINE, MULTIPLE AND SURFACE INTEGRALS

The concept of definite integral can be generalized in various ways. The domain of integration of an ordinary definite integral was a closed interval $[a, b]$ of the real axis. Taking an arc of

---

[1] $B(x, y) = \dfrac{\Gamma(x)\,\Gamma(y)}{\Gamma(x+y)}$ is the Beta function or Euler's integral of the first kind and $\Gamma(x)$ is the Gamma function or Euler's integral of the second kind (see p. 191).

[2] $\Gamma(x)$ is the Gamma function (see p. 191 and the table on p. 87).

a (plane or space) curve as a path of integration, we obtain a line integral; taking a plane region as a domain of integration, we obtain a double integral; taking a curvilinear surface, we obtain a surface integral; finally, if we take a region of space as a domain of integration, we obtain a triple integral.

14. Line integrals of the first type (integrals along an arc of a curve)

Definition. The *line integral of the first type*

$$\int_{(K)} f(x, y)\, ds$$

of a function $u = f(x, y)$ of two variables (defined in a certain connected domain [1]) along an arc $K = \breve{AB}$ of a plane curve defined by its equation (the arc is assumed to lie in the given domain and is called the *path of integration*) is the number obtained as follows (Fig. 320):

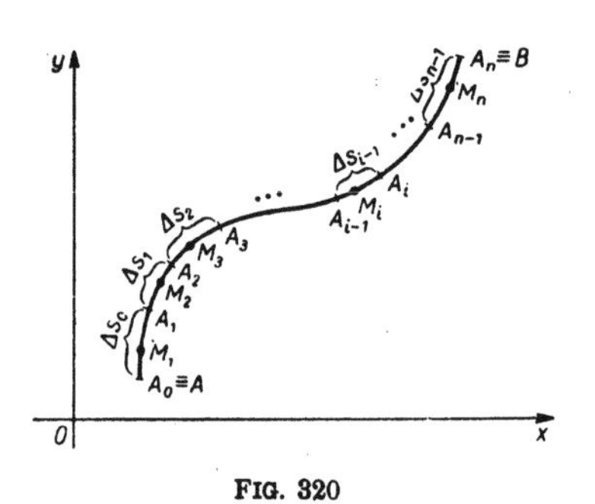

FIG. 320

(1) We divide the arc AB into n *elementary arcs* by means of arbitrary points $A_1, A_2, \ldots, A_{n-1}$ arranged from the initial point $A = A_0$ to the end point $B = A_n$;

(2) inside or on the boundary of each elementary arc $A_{i-1}A_i$ we select an arbitrary point M_i with coordinates ξ_i, η_i;

(3) we multiply the values $f(\xi_i, \eta_i)$ of the function at the points M_i by the lengths $A_{i-1}A_i = \Delta s_{i-1}$ (we take these lengths as positive);

(4) we add all the n products $f(\xi_i, \eta_i)\Delta s_{i-1}$;

---

[1] For connected domains of two variables see p. 341.

(5) we determine the limit of the obtained sum

$$\sum_{i=1}^{n} f(\xi_i, \eta_i) \Delta s_{i-1},$$

when the lengths of all elementary arcs Δs_{i-1} tend to zero (and, consequently, $n \to \infty$).

If this limit exists and is independent of the choice of A_i and M_i, then it is called the *line integral of the first type* of the function $f(x, y)$ along the path K:

$$\text{(A)} \qquad \int_{(K)} f(x, y)\, ds = \lim_{\substack{\Delta s_i \to 0 \\ n \to \infty}} \sum_{i=1}^{n} f(\xi_i, \eta_i) \Delta s_{i-1}.$$

The *line integral of first type* of a function $u = f(x, y, z)$ of three variables along an arc K of a space curve is likewise defined:

$$\text{(B)} \qquad \int_{(K)} f(x, y, z)\, ds = \lim_{\substack{\Delta s_i \to 0 \\ n \to \infty}} \sum_{i=1}^{n} f(\xi_i, \eta_i, \zeta_i) \Delta s_{i-1}.$$

Existence theorem. If the function $f(x, y)$ (resp. $f(x, y, z)$) is continuous and the curve K is continuous and has a tangent which varies continuously, then the line integral (A) (resp., (B)) of the first type exists (i.e., the above limits exist and are independent on the choice of A_i and M).

Evaluation of a line integral of the first type reduces to that of a definite integral. If the path of integration is defined by parametric equations (see pp. 277 and 296, 297) $x = x(t)$, $y = y(t)$ or (for a space curve), $x = x(t)$, $y = y(t)$, $z = z(t)$, then, in case (A):

$$\int_{(K)} f(x, y)\, ds = \int_{t_0}^{T} f[x(t), y(t)] \sqrt{[x'(t)]^2 + [y'(t)]^2}\, dt$$

and in case (B):

$$\int_{(K)} f(x, y, z)\, ds = \int_{t_0}^{T} f[x(t), y(t), z(t)] \sqrt{[x'(t)]^2 + [y'(t)]^2 + [z'(t)]^2}\, dt;$$

t_0 is the value of the parameter t at the point A and T at the point B; the points A and B should be chosen so that $t_0 < T$.

If the path of integration is given in explicit form $y = \varphi(x)$, and, for a space curve in the form $y = \varphi(x)$, $z = \psi(x)$, and if a and b

are the corresponding abscissae of the points A and B, with $a < b$ [1], then, in case (A):

$$\int_{(K)} f(x, y)\,ds = \int_a^b f[x, \varphi(x)]\sqrt{1 + [\varphi'(x)]^2}\,dx$$

and, in case (B):

$$\int_{(K)} f(x, y, z)\,ds = \int_a^b f[x, \varphi(x), \psi(x)]\sqrt{1 + [\varphi'(x)]^2 + [\psi'(x)]^2}\,dx.$$

Applications of the line integrals of the first type.

(1) Length of arc K:

$$L_{(K)} = \int_{(K)} ds.$$

(2) Mass of a non-homogeneous arc K, when δ is a variable linear density ($\delta = f(x, y)$ for a plane curve and $\delta = f(x, y, z)$ for a space curve):

$$M_{(K)} = \int_{(K)} \delta\,ds.$$

15. Line integrals of the second type (Integrals along a projection and integrals in the general form)

Definitions. The *line integral of the second type*

$$(A_x) \qquad \int_{(K)} f(x, y)\,dx$$

or

$$(B_x) \qquad \int_{(K)} f(x, y, z)\,dx$$

of a function $f(x, y)$ of two variables or $f(x, y, z)$ of three variables defined in a connected plane or space domain taken along the projection of an arc $K = \breve{AB}$ of a plane or space curve onto the x axis (the curve is assumed to lie in the domain in question) is the number obtained in the same way as the line integral of the

---

[1] Moreover, we assume that to each point of the projection of K onto the x axis, there corresponds a unique point of K (i.e., a point of the curve is uniquely determined by its projection onto the x axis). If this condition is not satisfied, we decompose the arc K into several parts such that each of them satisfies the condition in question; the line integral along the arc K is regarded as the sum of integrals taken along each of these parts of K.

first type (see pp. 486, 487) except that the step 3 is replaced by the following: we multiply the values $f(\xi_i, \eta_i)$ or $f(\xi_i, \eta_i, \zeta_i)$ of the function not by the lengths $A_{i-1}A_i$ of the arcs but by their projections onto the x axis (Fig. 321):

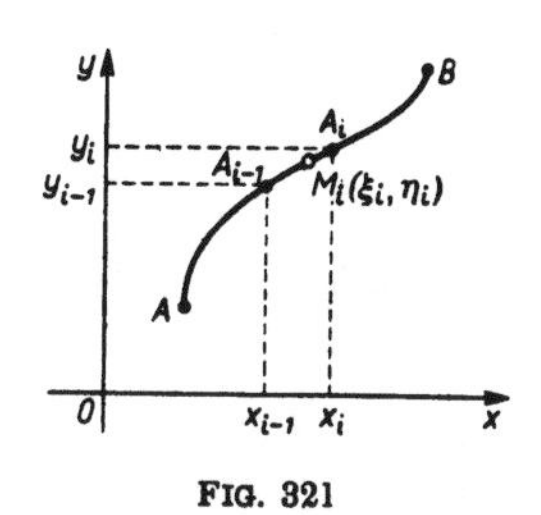

FIG. 321

$$\text{proj}_x \overset{\frown}{A_{i-1}A_i} = x_i - x_{i-1} = \Delta x_{i-1},$$

hence

$$(\mathrm{A}_x) \qquad \int_{(K)} f(x, y)\, dx = \lim_{\substack{\Delta x_{i-1} \to 0 \\ n \to \infty}} \sum_{i=1}^{n} f(\xi_i, \eta_i)\, \Delta x_{i-1},$$

$$(\mathrm{B}_x) \qquad \int_{(K)} f(x, y, z)\, dx = \lim_{\substack{\Delta x_{i-1} \to 0 \\ n \to \infty}} \sum_{i=1}^{n} f(\xi_i, \eta_i, \zeta_i)\, \Delta x_{i-1}.$$

The *line integrals of the second type* along the projection of K onto the y axis, and, in the space, also along the projection onto the z axis are defined likewise:

$$(\mathrm{A}_y) \qquad \int_{(K)} f(x, y)\, dy = \lim_{\substack{\Delta y_{i-1} \to 0 \\ n \to \infty}} \sum_{i=1}^{n} f(\xi_i, \eta_i)\, \Delta y_{i-1},$$

$$(\mathrm{B}_y) \qquad \int_{(K)} f(x, y, z)\, dy = \lim_{\substack{\Delta y_{i-1} \to 0 \\ n \to \infty}} \sum_{i=1}^{n} f(\xi_i, \eta_i, \zeta_i)\, \Delta y_{i-1},$$

$$(\mathrm{B}_z) \qquad \int_{(K)} f(x, y, z)\, dz = \lim_{\substack{\Delta z_{i-1} \to 0 \\ n \to \infty}} \sum_{i=1}^{n} f(\xi_i, \eta_i, \zeta_i)\, \Delta z_{i-1}.$$

Existence theorem. If the function $f(x, y)$ or $f(x, y, z)$ is continuous and the curve K is continuous and has a tangent which varies continuously, then the line integrals of the second type (A_x), (A_y), (B_x), (B_y), (B_z) exist.

Evaluation of line integrals of the second type reduces to that of definite integrals. If the curve is given parametrically (see pp. 277 and 296, 297) $x = x(t), y = y(t)$ (in the plane) or $x = x(t)$, $y = y(t)$, $z = z(t)$ (in the space), then the integrals (A_x), (A_y), (B_x), (B_y), (B_z) can be evaluated by the formulas

$$(A_x) \qquad \int_{(K)} f(x, y)\, dx = \int_{t_0}^{T} f\big(x(t), y(t)\big)\, x'(t)\, dt,$$

$$(A_y) \qquad \int_{(K)} f(x, y)\, dy = \int_{t_0}^{T} f\big(x(t), y(t)\big)\, y'(t)\, dt,$$

$$(B_x) \qquad \int_{(K)} f(x, y, z)\, dx = \int_{t_0}^{T} f\big(x(t), y(t), z(t)\big)\, x'(t)\, dt,$$

$$(B_y) \qquad \int_{(K)} f(x, y, z)\, dy = \int_{t_0}^{T} f\big(x(t), y(t), z(t)\big)\, y'(t)\, dt,$$

$$(B_z) \qquad \int_{(K)} f(x, y, z)\, dz = \int_{t_0}^{T} f\big(x(t), y(t), z(t)\big)\, z'(t)\, dt.$$

In these formulas, t_0 and T denote the values of the parameter t corresponding to the initial point A and the final point B of the path of integration; in contrast to the integrals of the first type, we do not require here that $t_0 < T$. After interchanging the points A and B (i.e., after changing the direction of the path of integration) the integrals change their signs.

If the path of integration is given in explicit form $y = \varphi(x)$, for a plane curve and $y = \varphi(x)$, $z = \psi(x)$ for a space curve, where a and b denote, correspondingly, the abscissas of the initial point A and the final point B (in this case the condition $a < b$ may not be satisfied), then the abscissa x plays the role of the parameter t in the formulas (A_x)–(B_z).

Line integral in the general form [1]. If $P(x, y)$ and $Q(x, y)$ are two functions of two variables or $P(x, y, z)$, $Q(x, y, z)$, $R(x, y, z)$ are three functions of three variables defined in a connected domain and K is an arc of a plane or space curve in this domain, then the line integral in the general form is the sum of the line integrals of the second type along the projections of K onto the coordinate axes, i.e.,

$$\int_{(K)} P\, dx + Q\, dy = \int_{(K)} P\, dx + \int_{(K)} Q\, dy$$

[1] A vectorial exposition of the line integral in the general form and mechanical significance of the line integral is given in the chapter "Field theory", p. 635.

for a plane curve, and

$$\int_{(K)} P\,dx + Q\,dy + R\,dz = \int_{(K)} P\,dx + \int_{(K)} Q\,dy + \int_{(K)} R\,dz$$

for a space curve.

Properties of the line integral.

(1) A line integral along the arc AB of a curve can be decomposed into two line integrals by means of a point C of the curve lying inside or outside of the arc AB:

$$\int_{\breve{AB}} P\,dx + Q\,dy = \int_{\breve{AC}} P\,dx + Q\,dy + \int_{\breve{CB}} P\,dx + Q\,dy;$$

(Fig. 322a, b). Similar formulas hold for an integral along a space curve.

(2) The integral taken along the same path but in the opposite direction has the opposite sign:

$$\int_{\breve{AB}} P\,dx + Q\,dy = -\int_{\breve{BA}} P\,dx + Q\,dy,$$

similar property holds for a space curve.

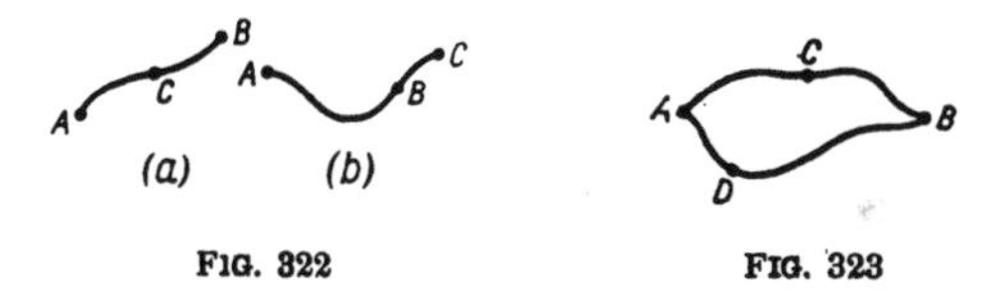

FIG. 322 FIG. 323

(3) In the general case, the line integral depends on the end points A and B as well as on the path of integration:

$$\int_{\breve{ACB}} P\,dx + Q\,dy \neq \int_{\breve{ADB}} P\,dx + Q\,dy$$

Fig. 323); similarly for a space curve.

Examples. (1) $I = \int_{(K)} xy\,dx + yz\,dy + zx\,dz$, where (K) is one coil of the circular helix $x = a\cos t$, $y = a\sin t$, $z = bt$ (see p. 301), from $t_0 = 0$ to $T = 2\pi$:

$$I = \int_0^{2\pi} (-a^3 \sin^2 t \cos t + a^2 bt \sin t \cos t + ab^2 t \cos t)\,dt = -\frac{\pi a^2 b}{2}.$$

(2) $I = \int\limits_{(K)} y^2\,dx + (xy - x^2)\,dy$, where (K) is the arc of the parabola $y^2 = 9x$ from the point $A(0, 0)$ to $B(1, 3)$:

$$I = \int_0^3 \left[\frac{2}{9}y^3 + \left(\frac{y^3}{9} - \frac{y^4}{81}\right)\right] dy = \frac{123}{20}.$$

Circuit integral. A line integral along a closed curve C (the end points A and B of C coincide) is called a *circuit integral* about C and is denoted by

$$\oint\limits_{(C)} P\,dx + Q\,dy \quad \text{or} \quad \oint\limits_{(C)} P\,dx + Q\,dy + R\,dz.$$

In general, a circuit integral is different from zero.

The area of a plane figure bounded by a closed curve C can be evaluated by a circuit integral as follows

$$S = \tfrac{1}{2} \oint\limits_{(C)} (x\,dy - y\,dx),$$

where C taken counterclockwise.

The condition for independence of a line integral of the path of integration (the condition of integrability of a total differential).

Two-dimensional case. The line integral

$$\int P(x, y)\,dx + Q(x, y)\,dy,$$

where P and Q are continuous functions defined in a simply connected domain, taken along a curve lying in the domain depend solely on the initial point A and final point B of the path of integration, but not on the particular choice of the path C, i.e., for any two paths of integration ACB and ADB (Fig. 323) we have the equation

$$\int\limits_{\overset{\frown}{ACB}} P\,dx + Q\,dy = \int\limits_{\overset{\frown}{ADB}} P\,dx + Q\,dy,$$

if, and only if, there exists a function $U(x, y)$ in the domain such that the expression under the integral sign is the total differential $dU(x, y)$ of U:

(1) $$P\,dx + Q\,dy = dU,$$

i.e.,

(2) $$P = \frac{\partial U}{\partial x}, \quad Q = \frac{\partial U}{\partial y}.$$

The function U is then called a *primitive function* [1] of the total differential (1).

A necessary and sufficient condition for the existence of a primitive function of $P\,dx+Q\,dy$ (i.e., for the integrability of the total differential $P\,dx+Q\,dy$) is the equation

$$\frac{\partial P}{\partial y}=\frac{\partial Q}{\partial x} \tag{3}$$

provided that the partial derivatives are continuous.

Three-dimensional case. An analogous theorem holds for the space. The integral

$$\int P(x, y, z)\,dx+Q(x, y, z)\,dy+R(x, y, z)\,dz$$

is independent on the path joining the points A and B if, and only if, there exists a primitive function $U(x, y, z)$ such that

$$P\,dx+Q\,dy+R\,dz=dU, \tag{1'}$$

i.e.,

$$P=\frac{\partial U}{\partial x}, \quad Q=\frac{\partial U}{\partial y}, \quad R=\frac{\partial U}{\partial z}. \tag{2'}$$

A necessary condition of integrability in this case is that the equalities

$$\frac{\partial Q}{\partial z}=\frac{\partial R}{\partial y}, \quad \frac{\partial R}{\partial x}=\frac{\partial P}{\partial z}, \quad \frac{\partial P}{\partial y}=\frac{\partial Q}{\partial x} \tag{3'}$$

shall be simultaneously satisfied, provided that the partial derivatives are continuous.

Evaluation of a primitive function. Provided that condition (3) is satisfied, a primitive function $U(x, y)$ is equal to the line integral

$$U=\int_{\breve{AM}} P\,dx+Q\,dy$$

along an arbitrary path $\breve{AM}$ of integration lying in the domain in which condition (3) is satisfied and joining a fixed point A with the coordinates x_0, y_0 with a variable point M with the running coordinates x, y. In practice, the most convenient path of integration is one of two broken lines AKM or ALM with sides parallel to the coordinate axes, provided that this line lies in the domain,

[1] A primitive function $U(x, y)$ is a potential of the vector field $P\mathbf{i}+Q\mathbf{j}$ (in another notation, it is a potential with the sign "−"), see p. 636.

where condition (3) is satisfied (Fig. 324). This yields the following formulas for evaluation of a primitive function $U(x, y)$ of the total differential $P\,dx + Q\,dy$:

$$(4_1)\qquad U = \int_{\overline{AK}} + \int_{\overline{KM}} + U(x_0, y_0) = \int_{x_0}^{x} P(\xi, y_0)\,d\xi + \int_{y_0}^{y} Q(x, \eta)\,d\eta + C,$$

$$(4_2)\qquad U = \int_{\overline{AL}} + \int_{\overline{LM}} + U(x_0, y_0) = \int_{y_0}^{y} Q(x_0, \eta)\,d\eta + \int_{x_0}^{x} P(\xi, y)\,d\xi + C'.$$

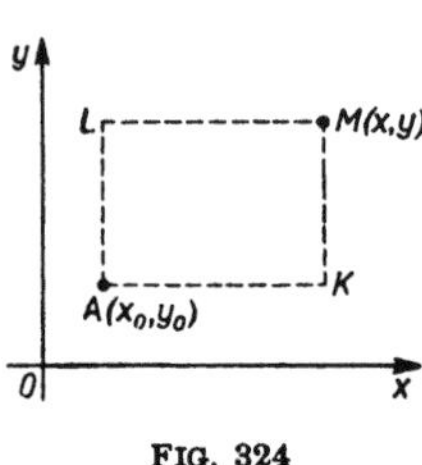

FIG. 324

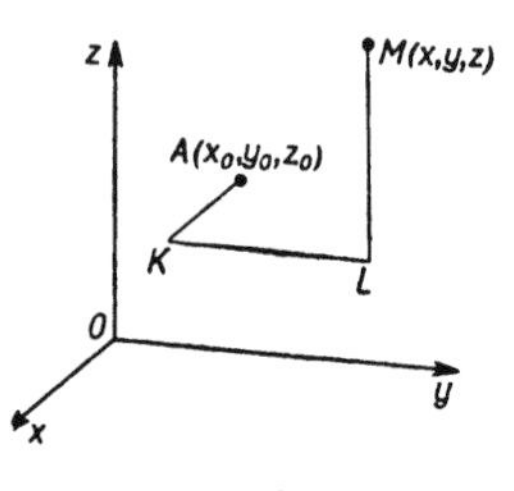

FIG. 325

In the three-dimensional case, if condition (3′) is satisfied, a primitive function $U(x, y, z)$ can be found in an analogous way (Fig. 325):

$$(4')\qquad U = \int_{\overline{AK}} + \int_{\overline{KL}} + \int_{\overline{LM}} + U(x_0, y_0, z_0)$$

$$= \int_{x_0}^{x} P(\xi, y_0, z_0)\,d\xi + \int_{y_0}^{y} Q(x, \eta, z_0)\,d\eta + \int_{z_0}^{z} R(x, y, \zeta)\,d\zeta + C,$$

or according to five similar formulas corresponding to other possible broken lines with sides parallel to the coordinate axes.

Examples. (1) $P\,dx + Q\,dy = -\dfrac{y\,dx}{x^2 + y^2} + \dfrac{x\,dy}{x^2 + y^2}$.

Condition (3) is satisfied: $\dfrac{\partial P}{\partial y} = \dfrac{\partial Q}{\partial x} = \dfrac{y^2 - x^2}{(x^2 + y^2)^2}$. Applying formula (4_2) and taking $x_0 = 0$, $y_0 = 1$ (we cannot take $x_0 = 0$, $y_0 = 0$, for the functions P and Q are discontinuous at this point), we have

$$U = \int_{1}^{y} \frac{0}{0^2 + \eta^2}\,d\eta + \int_{0}^{x} \frac{-y}{\xi^2 + y^2}\,d\xi + U(0,1)$$

$$= -\arctan\frac{x}{y} + C = \arctan\frac{y}{x} + C_1.$$

(2) $P\,dx + Q\,dy + R\,dz$

$$= z\left(\frac{1}{x^2 y} - \frac{1}{x^2 + z^2}\right)dx + \frac{z}{xy^2}dy + \left(\frac{x}{x^2 + z^2} - \frac{1}{xy}\right)dz.$$

Conditions (3′) are satisfied. We apply formula (4′), taking $x_0 = 1,\ y_0 = 1,\ z_0 = 0$:

$$U = \int_1^x 0\,d\xi + \int_1^y 0\,d\eta + \int_0^z \left(\frac{x}{x^2 + \zeta^2} - \frac{1}{xy}\right)d\zeta + C$$

$$= \operatorname{arc\,tan}\frac{z}{x} - \frac{z}{xy} + C.$$

A circuit integral about a closed plane curve (a line integral of a total differential $P\,dx + Q\,dy$ along a closed plane curve) is equal to zero, provided that condition (3) is satisfied and that each of the functions P, Q, $\frac{\partial P}{\partial y}$ and $\frac{\partial Q}{\partial x}$ is defined and continuous at every point inside the curve.

16. Double and triple integrals

Double integral. The *double integral* of a function $u = f(x, y)$ [1] over a domain S is the number denoted by

$$\int_S f(x, y)\,dS \quad \text{or} \quad \iint_S f(x, y)\,dS$$

is the number obtained as follows:

(1) We decompose the domain S (Fig. 326) in an arbitrary way into n elementary domains $\Delta S_1, \Delta S_2, \ldots, \Delta S_n$;

(2) we select an arbitrary point $M_i(x_i, y_i)$ inside or on the boundary of each elementary domain ΔS_i;

(3) we multiply the value $f(x_i, y_i)$ of the function u at M_i by the area dS_i of the corresponding elementary domain ΔS_i;

(4) we add all products $f(x_i, y_i)\,dS_i$;

(5) we determine the limit of this sum

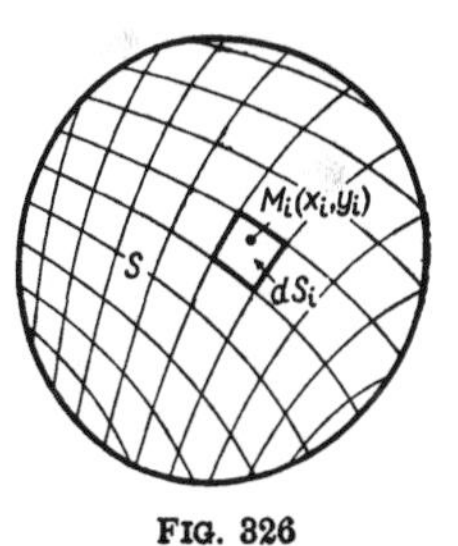

FIG. 326

$$\sum_{i=1}^{n} f(x_i, y_i)\,dS_i,$$

(1) The function u is regarded here as a function of a point (see p. 340) which may be defined not only in Cartesian coordinates.

when each elementary domain dS_i is contracted to a point (¹), and, consequently, their number $n \to \infty$.

If the above limit exists and is independent of the particular decomposition of S into elementary domains or of the choice of the points $M_i(x_i, y_i)$, then it is called the *double integral* of the function u over the domain S called the *domain of integration*:

$$(1) \qquad \int_S f(x, y)\, dS = \lim_{\substack{dS_i \to 0 \\ n \to \infty}} \sum_{i=1}^{n} f(x_i, y_i)\, dS_i.$$

Existence theorem. If the function $f(x, y)$ is continuous in the whole closed domain S (i.e., including all boundary points), then the integral (1) exists.

Geometric significance. The double integral represents the volume of the cylindrical solid (Fig. 327) bounded by (1) the base S on the xy plane, (2) the cylindrical surface whose base curve is the boundary of S and whose generators are lines parallel to the z axis, (3) the surface $u = f(x, y)$. Each summand $f(x_i, y_i)\, dS_i$ of $\sum_{i=1}^{n} f(x_i, y_i)\, dS_i$ represents the volume of a small cylinder with base dS_i and altitude $f(x_i, y_i)$. The volume is taken with the sign "+" or "−" according to whether the corresponding part of the surface $u = f(x, y)$ lies above or below the xy plane. If the surface intersects the xy plane, the volume of the solid is the algebraic sum of the particular positive and negative summands.

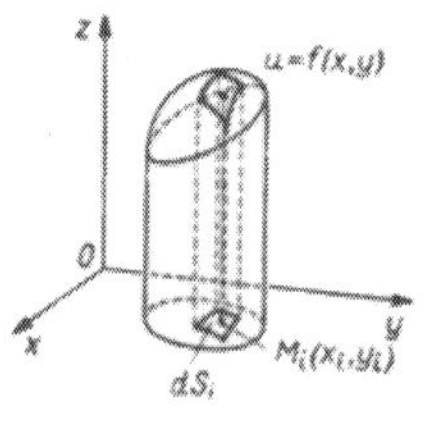

FIG 327

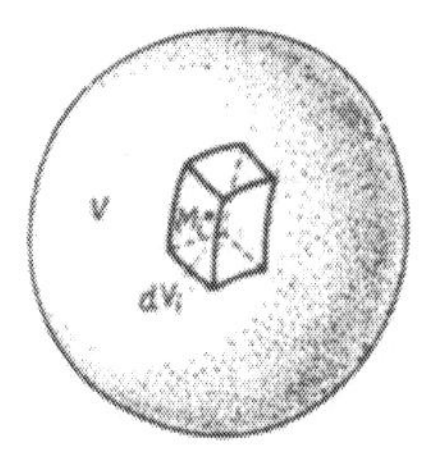

FIG. 328

Triple integral. The *triple integral* of a function $u = f(x, y, z)$ over a space domain V denoted by

$$\int_V f(x, y, z)\, dV \qquad \text{or} \qquad \iiint_V f(x, y, z)\, dV$$

(¹) The requirement that the area dS_i tends to zero is not sufficient. It is necessary to require that the diameter of each ΔS_i, i.e., the distance between the most remote points of ΔS_i tends to zero. For example, if one side of a rectangle tends to zero, its area tends also to zero, but the diameter remains finite.

is defined similarly: we decompose the domain V (Fig. 328) into elementary domains ΔV_i $(i = 1, 2, \ldots, n)$ in an arbitrary way and consider the products of the form $f(x_i, y_i, z_i)\,dV_i$, where $M_i(x_i, y_i, z_i)$ is an arbitrarily selected point inside or on the boundary of ΔV_i and dV_i is the volume of ΔV_i. If there exists a limit of the sum of such products, when each elementary domains is contracted to a point [1], and hence, their number $n \to \infty$, and this limit is independent of the particular choice of the decomposition of V into elementary domains and of the points M_i, then it is called the *triple integral*:

$$\int_V f(x, y, z)\,dV = \lim_{\substack{dV_i \to 0 \\ n \to \infty}} \sum_{i=1}^{n} f(x_i, y_i, z_i)\,dV_i.$$

In an analogous way we define the multiple integrals in the n-dimensional space.

There is an existence theorem for the triple integrals of a continuous function $f(x, y, z)$ of three variables analogous to that for double integral.

17. Evaluation of multiple integrals

Evaluation of a double or triple integral can be reduced to the repeated evaluation of two or three definite integrals. This can be done in many ways according to the choice of a coordinate system.

Double integral.

(1) In Cartesian coordinates. We decompose the domain S of integration into rectangles by means of coordinate lines (Fig.

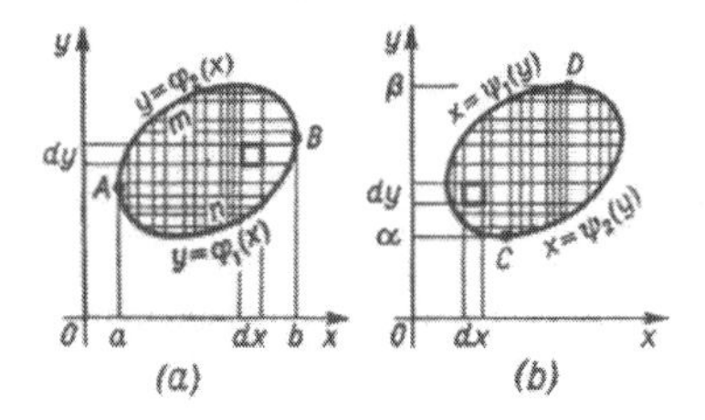

FIG. 329

329a) and first perform the summation in the rectangles of each vertical strip and then over all horizontal strips. Analytically,

[1] In the same sense, as for the double integral, i.e., not only the area of ΔV_i, but also its diameter tends to zero.

$$\int_S f(x, y)\, dS = \int_a^b \Big[\int_{\varphi_1(x)}^{\varphi_2(x)} f(x, y)\, dy\Big]\, dx,$$

where a and b are extreme points on the left and on the right side of the domain S of integration and $y = \varphi_1(x)$ and $y = \varphi_2(x)$ are the curves constituting the lower (AnB) and the upper (AmB) boundary of S in the interval $a < x < b$. The brackets are usually omitted and the above formula is written in the form

$$\int_S f(x, y)\, dS = \int_a^b \int_{\varphi_1(x)}^{\varphi_2(x)} f(x, y)\, dy dx.$$

The interior integral is referred to the variable whose differential is also interior. The product $dydx$ of differentials represents the area dS of an elementary domain in the Cartesian coordinates and in the first integration the variable x is assumed to be constant.

The evaluation of the integral in Cartesian coordinates can also be performed in the reverse order (see Fig. 329b):

$$\int_S f(x, y)\, dS = \int_\alpha^\beta \int_{\psi_1(y)}^{\psi_2(y)} f(x, y)\, dx dy.$$

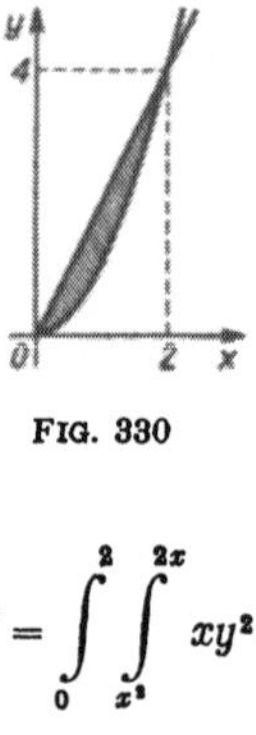

FIG. 330

Example. $I = \int_S xy^2\, dS$ where the domain S is bounded by the parabola $y = x^2$ and the straight line $y = 2x$ (Fig. 330):

$$I = \int_0^2 \int_{x^2}^{2x} xy^2\, dy\, dx = \int_0^2 x\Big[\frac{y^3}{3}\Big]_{x^2}^{2x} dx = \frac{1}{3}\int_0^2 (8x^4 - x^7)\, dx = \frac{32}{5}$$

or else

$$I = \int_0^4 \int_{y/2}^{\sqrt{y}} xy^2\, dx\, dy = \int_0^4 y^2\Big[\frac{x^2}{2}\Big]_{y/2}^{\sqrt{y}} dy = \frac{1}{2}\int_0^4 y^2\Big(y - \frac{y^2}{4}\Big) dy = \frac{32}{5}.$$

(2) In polar coordinates. We decompose the domain S into elementary parts by means of coordinate lines. The elementary domains are bounded by two arcs of circles with the centre at the pole and two segments of rays issuing from the pole (Fig. 331); the integrand should be expressed in polar coordinates $\omega = f(\varrho, \varphi)$. We perform the summation first along each sector and then with respect to all sectors. Analytically,

$$(1)\qquad \int_S f(\varrho, \varphi)\, dS = \int_{\varphi_1}^{\varphi_2} \int_{\varrho_1(\varphi)}^{\varrho_2(\varphi)} f(\varrho, \varphi)\, \varrho\, d\varrho d\varphi,$$

where φ_1 and φ_2 are amplitudes of the extreme radius vectors tangent to S, where $\varphi_1 < \varphi_2$ and $\varrho = \varrho_1(\varphi)$, $\varrho = \varrho_2(\varphi)$ are polar equations of the interior (AmB) and exterior (AnB) boundary of S in the interval $\varphi_1 < \varphi < \varphi_2$. The product $\varrho\, d\varrho d\varphi$ represents the area dS of an elementary domain in polar coordinates. The reverse order of integration is seldom used.

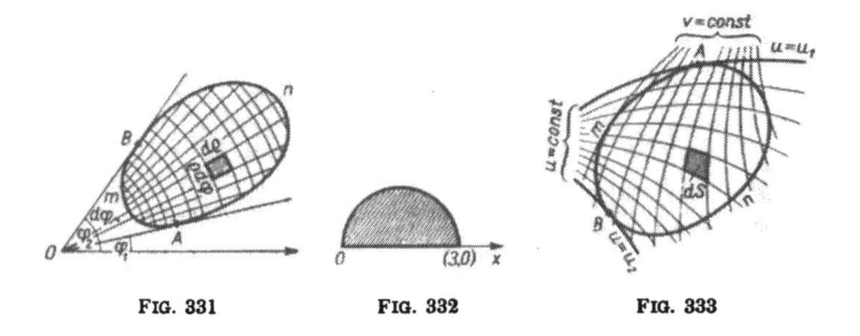

FIG. 331 FIG. 332 FIG. 333

Example. $I = \int_S \varrho \sin^2\varphi\, dS$, where S is the semi-circle $\varrho = 3\cos\varphi$ $0 < \varphi < \frac{1}{2}\pi$ (Fig. 332):

$$I = \int_0^{\pi/2} \int_0^{3\cos\varphi} \varrho \sin^2\varphi\, \varrho\, d\varrho d\varphi = \int_0^{\pi/2} \sin^2\varphi \left[\frac{\varrho^3}{3}\right]_0^{3\cos\varphi} d\varphi$$

$$= 9 \int_0^{\pi/2} \sin^2\varphi \cos^3\varphi\, d\varphi = \frac{6}{5}.$$

(3) In arbitrary curvilinear coordinates u, v defined by the formulas $x = x(u, v)$, $y = y(u, v)$ (see p. 236). We decompose the domain of integration into elementary domains by means of the coordinate lines (Fig. 333), express the integrand in terms of u, v coordinates and perform the summation along one strip (e.g., $v = \text{const}$) and then in all strips. Analytically,

$$(2)\qquad \int_S f(u, v)\, dS = \int_{u_1}^{u_2} \int_{v_1(u)}^{v_2(u)} f(u, v)\, |D|\, dv du,$$

where u_1 and u_2 are coordinates of the extreme coordinate lines tangent to the boundary of S and bounding the domain S and $|D|$ is the absolute value of the Jacobian

$$D = \frac{D(x, y)}{D(u, v)} = \begin{vmatrix} \frac{\partial x}{\partial u} & \frac{\partial x}{\partial v} \\ \frac{\partial y}{\partial u} & \frac{\partial y}{\partial v} \end{vmatrix},$$

$|D|\, dvdu$ represents the area dS of an elementary domain in the curvilinear coordinates u, v.

Formula (1) is a particular case of (2), for, in polar coordinates, $x = \varrho \cos \varphi$, $y = \varrho \sin \varphi$ and Jacobian $D = \varrho$.

We choose the coordinates u, v so that the limits of integration in formula (2) are as simple as possible.

Example. $I = \int\limits_S f(x, y)\, dS$, where S is the domain bounded by the asteroid $x = a \cos^3 t$, $y = a \sin^3 t$.

We introduce curvilinear coordinates $x = u \cos^3 v$, $y = u \sin^3 v$. Coordinate lines $u = c_1$ form a family of similar asteroids $x = c_1 \cos^3 t$, $y = c_1 \sin^3 t$ (Fig. 334); coordinate lines $v = c_2$ are rays $y = kx$, where $k = \tan^3 c_2$.

$$D = \begin{vmatrix} \cos^3 v & -3u \cos^2 v \sin v \\ \sin^3 v & 3u \sin^2 v \cos v \end{vmatrix} = 3u \sin^2 v \cos^2 v,$$

$$I = \int\limits_0^a \int\limits_0^{2\pi} f\big(x(u, v), y(u, v)\big)\, 3u \sin^2 v \cos^2 v\, dvdu.$$

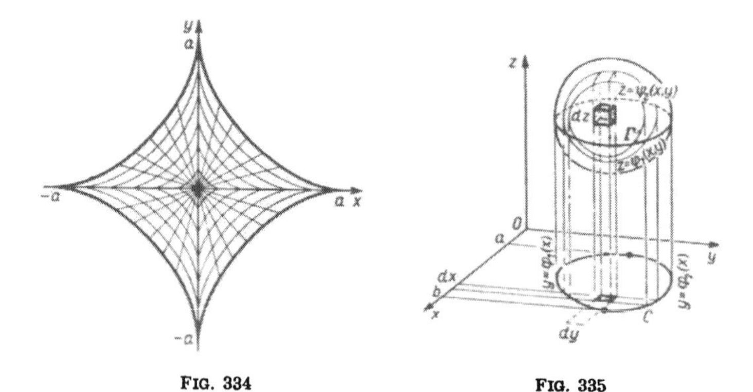

FIG. 334 FIG. 335

Triple integral.

(1) In Cartesian coordinates. We decompose the domain of integration into parallelepipeds by means of coordinate surfaces (in this case, by means of planes) (Fig. 335) and perform the

summation of products $f(x, y, z)\,dV$ first in each vertical column (summation with respect to z), then in all columns of one slice parallel to the xz plane (summation with respect to y) and finally in all such slices (summation with respect to x). Analytically,

$$\int_V f(x, y, z)\,dV = \int_a^b \Big[\int_{\varphi_1(x)}^{\varphi_2(x)} \Big(\int_{\psi_1(x,y)}^{\psi_2(x,y)} f(x, y, z)\,dz\Big)\,dy\Big]\,dx$$

$$= \int_a^b \int_{\varphi_1(x)}^{\varphi_2(x)} \int_{\psi_1(x,y)}^{\psi_2(x,y)} f(x, y, z)\,dz\,dy\,dx.$$

Here $z = \psi_1(x, y)$ and $z = \psi_2(x, y)$ are the equations of the lower and upper part of the surface bounding the domain V of integration; the lower and upper part of the surface are separated by a curve Γ and $y = \varphi_1(x)$, $y = \varphi_2(x)$ are equations of the projection C of Γ onto the xy plane, i.e., of parts of Γ bounded by the extreme abscissae $x = a$ and $x = b$.

Just as in the case of a double integral, the order of integration can be arbitrary; thus a triple integral can be evaluated in six ways.

Example. $I = \int_V (y^2 + z^2)\,dV$, where V is the domain of the pyramid bounded by the coordinate planes and the plane $x + y + z = 1$.

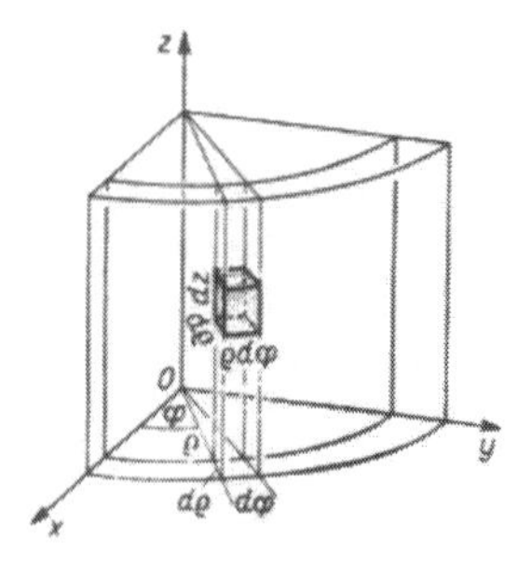

FIG. 336

$$I = \int_0^1 \int_0^{1-x} \int_0^{1-x-y} (y^2 + z^2)\,dz\,dy\,dx = \int_0^1 \Big[\int_0^{1-x} \Big(\int_0^{1-x-y} (y^2 + z^2)\,dz\Big)\,dy\Big]\,dx = \frac{1}{30}.$$

(2) In cylindrical coordinates. We decompose the domain of integration into elementary domains by means of coordinate surfaces $\varrho = \text{const}$, $\varphi = \text{const}$, $z = \text{const}$. The volume of an elementary domain $dV = \varrho\,dz\,d\varrho\,d\varphi$ (Fig. 336); the integrand should be expressed in cylindrical coordinates: $f(\varrho, \varphi, z)$. Then

$$(1) \qquad \int_V f(\varrho, \varphi, z)\,dV = \int_{\varphi_1}^{\varphi_2} \int_{\varrho_1(\varphi)}^{\varrho_2(\varphi)} \int_{z_1(\varrho,\varphi)}^{z_2(\varrho,\varphi)} f(\varrho, \varphi, z)\,\varrho\, dz d\varrho d\varphi.$$

Example. $I = \int_V dV$, where V is the domain bounded by the xy and xz planes, the cylindrical surface $x^2 + y^2 = ax$ and the sphere $x^2 + y^2 + z^2 = a^2$ (Fig. 337) ([1]).

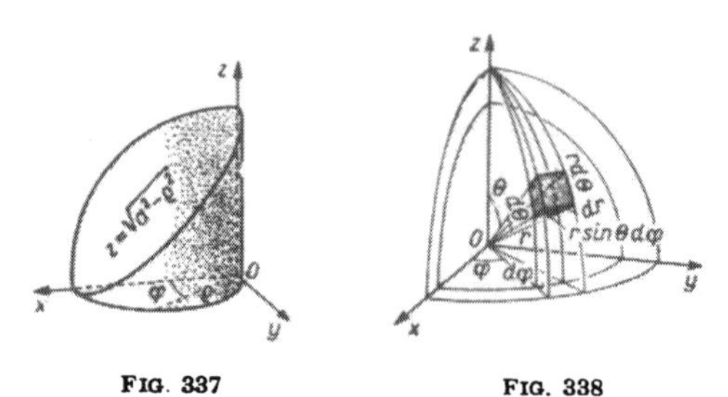

FIG. 337 FIG. 338

We have

$$z_1 = 0,\ z_2 = \sqrt{a^2 - x^2 - y^2} = \sqrt{a^2 - \varrho^2},\ \varrho_1 = 0,\ \varrho_2 = a\cos\varphi,\ \varphi_1 = 0,\ \varphi_2 = \tfrac{1}{2}\pi,$$

$$I = \int_0^{\pi/2} \int_0^{a\cos\varphi} \int_0^{\sqrt{a^2-\varrho^2}} \varrho\, dz d\varrho d\varphi$$

$$= \int_0^{\pi/2} \Big[\int_0^{a\cos\varphi} \Big(\int_0^{\sqrt{a^2-\varrho^2}} dz \Big) \varrho\, d\varrho \Big] d\varphi = \frac{a^3}{18}(3\pi - 4).$$

(3) In spherical coordinates. We decompose the domain into elementary parts by means of the coordinate planes $r =$ const, $\varphi =$ const, $\theta =$ const. The volume of an elementary domain $dV = r^2 \sin\theta\, dr d\theta d\varphi$ (Fig. 338). The function should be expressed in spherical coordinates: $f(r, \varphi, \theta)$.

$$(2) \qquad \int_V f(r, \varphi, \theta)\,dV = \int_{\varphi_1}^{\varphi_2} \int_{\theta_1(\varphi)}^{\theta_2(\varphi)} \int_{r_1(\theta,\varphi)}^{r_2(\theta,\varphi)} f(r, \varphi, \theta) r^2 \sin\theta\, dr d\theta d\varphi.$$

Example. $I = \int_V \frac{\cos\theta}{r^2}\,dV$, where V is the domain interior to

([1]) Since, in this case, $f(x, y, z) \equiv 1$, this integral represents the volume of the solid V.

the cone bounded by the coordinate surface $\varphi=\alpha$ and the plane $z=h$ (Fig. 339).

We have

$$r_1=0,\quad r_2=\frac{h}{\cos\theta},\quad \theta_1=0,\quad \theta_2=\alpha,\quad \varphi_1=0,\quad \varphi_2=2\pi,$$

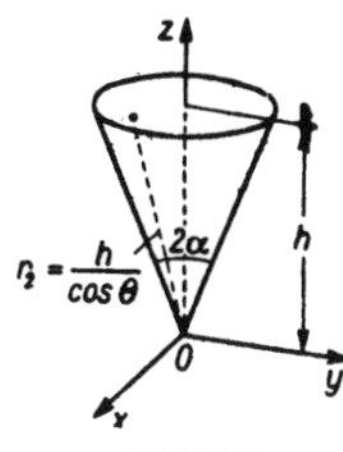

FIG. 339

$$I=\int_0^{2\pi}\int_0^{\alpha}\int_0^{h/\cos\theta}\frac{\cos\theta}{r^2}r^2\sin\theta\, drd\theta d\varphi$$

$$=\int_0^{2\pi}\Big[\int_0^{\alpha}\cos\theta\sin\theta\Big(\int_0^{h/\cos\theta}dr\Big)d\theta\Big]d\varphi=2\pi h(1-\cos\alpha).$$

(4) In arbitrary curvilinear coordinates u, v, w defined by the formulas

$$x=x(u,v,w),\qquad y=y(u,v,w),$$
$$z=z(u,v,w)$$

(see p. 257). We decompose the domain into elementary parts by the coordinate planes $u=\text{const}$, $v=\text{const}$, $w=\text{const}$. The volume of an elementary domain

$$dV=|D|\,dudvdw,\qquad\text{where}\qquad D=\begin{vmatrix}\frac{\partial x}{\partial u}&\frac{\partial x}{\partial v}&\frac{\partial x}{\partial w}\\ \frac{\partial y}{\partial u}&\frac{\partial y}{\partial v}&\frac{\partial y}{\partial w}\\ \frac{\partial z}{\partial u}&\frac{\partial z}{\partial v}&\frac{\partial z}{\partial w}\end{vmatrix}.$$

The integrand should be expressed in terms of u, v, w coordinates.

$$(3)\qquad \int_V f(u,v,w)\,dV=\int_{u_1}^{u_2}\int_{v_1(u)}^{v_2(u)}\int_{w_1(u,v)}^{w_2(u,v)}f(u,v,w)\,|D|\,dwdvdu.$$

Formulas (1) and (2) are particular cases of (3); in cylindrical coordinates, $D=\varrho$, and in spherical coordinates, $D=r^2\sin\theta$.

We choose the curvilinear coordinates so that the limits of integration in formula (3) are as simple as possible.

18. Applications of multiple integrals

Double integrals

Name	Notation	General formula	In Cartesian coordinates	In polar coordinates
Area of a plane figure	S	$\int_S dS$	$\iint dydx$	$\iint \varrho\, d\varrho d\varphi$
Area of a surface ([1])	Σ	$\int_S \frac{dS}{\cos\gamma}$	$\iint \sqrt{1+\left(\frac{\partial z}{\partial x}\right)^2+\left(\frac{\partial z}{\partial y}\right)^2}\, dydx$	$\iint \sqrt{\varrho^2+\varrho^2\left(\frac{\partial z}{\partial \varrho}\right)^2+\left(\frac{\partial z}{\partial \varphi}\right)^2}\, d\varrho d\varphi$
Volume of a cylinder ([2])	V	$\int_S z\, dS$	$\iint z\, dydx$	$\iint z\varrho\, d\varrho d\varphi$
Moment of inertia of a plane figure with respect to the x axis	I_x	$\int_S y^2\, dS$	$\iint y^2\, dydx$	$\iint \varrho^3 \sin^2\varphi\, d\varrho d\varphi$
Moment of inertia of a plane figure with respect to a pole O	I_0	$\int_S \varrho^2 dS$	$\iint (x^2+y^2)\, dydx$	$\iint \varrho^3\, d\varrho d\varphi$
Mass of a plane figure with surface density δ given as a function of a point	M	$\int_S \delta\, dS$	$\iint \delta\, dydx$	$\iint \delta\varrho\, d\varrho d\varphi$
Coordinates of the centre of gravity of a plane homogeneous figure	x_C	$\frac{\int_S x\, dS}{S}$	$\frac{\iint x\, dydx}{\iint dydx}$	$\frac{\iint \varrho^2 \cos\varphi\, d\varrho d\varphi}{\iint \varrho\, d\varrho d\varphi}$
	y_C	$\frac{\int_S y\, dS}{S}$	$\frac{\iint y\, dydx}{\iint dydx}$	$\frac{\iint \varrho^2 \sin\varphi\, d\varrho d\varphi}{\iint \varrho\, d\varrho d\varphi}$

([1]) See p. 506; in this case, S is the projection of the surface onto the xy plane and γ is the angle between the normal to the surface element and the z axis.

([2]) See p. 496.

Triple integrals

Name	Notation	General formula	In Cartesian coordinates	In cylindrical coordinates	In spherical coordinates
Volume of a solid	V	$\int_V dV$	$\iiint dz\,dy\,dx$	$\iiint \varrho\, dz\,d\varrho\,d\varphi$	$\iiint r^2 \sin\theta\, dr\,d\theta\,d\varphi$
Moment of inertia of a solid with respect to the z axis	I_z	$\int_V \varrho^2\, dV$	$\iiint (x^2+y^2)\, dz\,dy\,dx$	$\iiint \varrho^3\, dz\,d\varrho\,d\varphi$	$\iiint r^4 \sin^3\theta\, dr\,d\theta\,d\varphi$
Mass of a solid with density δ given as a function of a point	M	$\int_V \delta\, dV$	$\iiint \delta\, dz\,dy\,dx$	$\iiint \delta\varrho\, dz\,d\varrho\,d\varphi$	$\iiint \delta r^2 \sin\theta\, dr\,d\theta\,d\varphi$
Coordinates of the centre of gravity of a homogeneous solid	x_C	$\dfrac{\int_V x\, dV}{V}$	$\dfrac{\iiint x\, dz\,dy\,dx}{\iiint dz\,dy\,dx}$		
	y_C	$\dfrac{\int_V y\, dV}{V}$	$\dfrac{\iiint y\, dz\,dy\,dx}{\iiint dz\,dy\,dx}$		
	z_C	$\dfrac{\int_V z\, dV}{V}$	$\dfrac{\iiint z\, dz\,dy\,dx}{\iiint dz\,dy\,dx}$		

19. Surface integrals of the first type [1] (Integrals over a surface)

Definition. The *surface integral of the first type* $\int\limits_S f(x, y, z)\,dS$ of a function $u = f(x, y, z)$ of three variables defined in a connected space domain, the integral being taken over a region S of a surface contained in this domain is the number obtained as follows:

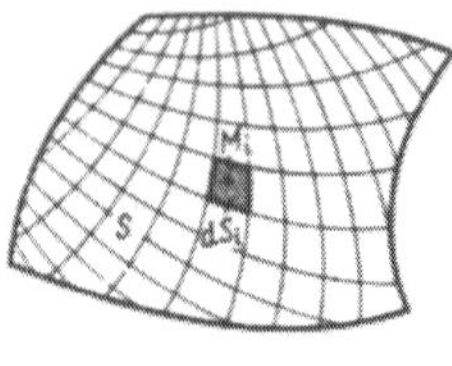

FIG. 340

(1) We decompose the region S (Fig. 340) in an arbitrary way into n elementary regions;

(2) we select an arbitrary point $M_i(x_i, y_i, z_i)$ inside or on the boundary of each elementary region;

(3) we multiply the value $f(x_i, y_i, z_i)$ of the function at M_i by the area dS_i of the corresponding elementary region;

(4) we add the products $f(x_i, y_i, z_i)\,dS_i$ so obtained;

(5) we determine the limit of the sum $\sum\limits_{i=1}^{n} f(x_i, y_i, z_i)\,dS_i$ obtained in this way, when each elementary region is contracted to a point [2], and, hence, $n \to \infty$.

If this limit exists and is independent of the particular decomposition of S into elementary region or of the choice of the points M_i, then it is called the *surface integral of the first type*:

$$\int\limits_S f(x, y, z)\,dS = \lim_{\substack{dS_i \to 0 \\ n \to \infty}} \sum_{i=1}^{\infty} f(x_i, y_i, z_i)\,dS_i.$$

Existence theorem. If the function $f(x, y, z)$ is continuous in the given domain and the functions defining the surface are continuous and have continuous derivatives, then the surface integral of the first type exists.

Evaluation of a surface integral of the first type reduces to the evaluation of a double integral over a plane domain (see pp. 497–500).

---

[1] These integrals are an extension of double integrals (p. 495) just as the line integrals of the first type (p. 486), are an extension of the ordinary definite integrals (p. 454).

[2] In the same sense as in the case of a double integral (see footnote on p. 496).

If the surface S is given in explicit form $z=\varphi(x, y)$, then

$$(1) \qquad \int_S f(x, y, z)\, dS=\iint_{S'} f(x, y, \varphi(x, y)) \sqrt{1+p^2+q^2}\, dx dy,$$

where S' is the projection of S onto the xy plane, $p=\frac{\partial z}{\partial x}$, $q=\frac{\partial z}{\partial y}$ [1].

Since the equation of the normal to the surface $z=\varphi(x, y)$ has the form

$$\frac{X-x}{p}=\frac{Y-y}{q}=\frac{Z-z}{-1}$$

(see p. 305), hence $\frac{1}{\sqrt{1+p^2+q^2}}=\cos\gamma$, where γ is the angle between the normal and the z axis [2]. It follows that equation (1) can be written in the form

$$(2) \qquad \int_S f(x, y, z)\, dS=\iint_{S_{xy}} f(x, y, \varphi(x, y)) \frac{dS_{xy}}{\cos\gamma},$$

where S_{xy} is the projection of S onto the xy plane.

If the surface is given in parametric form $x=x(u, v)$, $y=y(u, v)$ $z=z(u, v)$ (Fig. 341), then

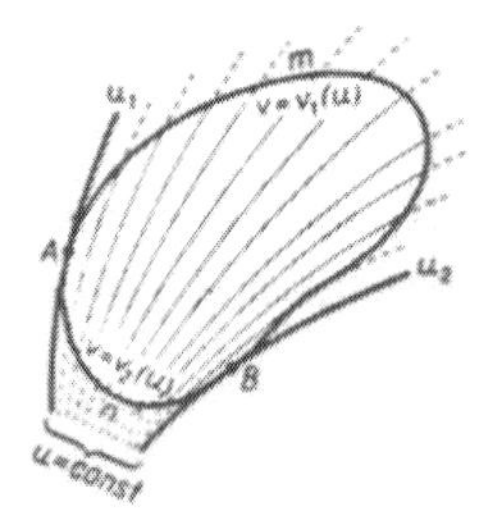

FIG. 341

$$(3) \quad \int_S f(x, y, z)\, dS=\iint_{\Delta} f(x(u, v), y(u, v), z(u, v)) \sqrt{EG-F^2}\, du dv,$$

where E, F and G are defined on p. 307, $\sqrt{EG-F^2}\, du dv=dS$ (area of an elementary region) and Δ is the domain of the para-

(1) We assume here that to every point of the projection S' on the xy plane there corresponds a unique point of S (i.e., that a point of S is uniquely determined by its projection on the xy plane). If the surface S fails to have this property, we decompose it into several parts each of which satisfies this condition and consider the surface integral over S as the sum of the surface integrals taken over all parts of S.

(2) In evaluation of a surface integral of the first type, this angle is always considered as an acute angle; $\cos\gamma > 0$.

meters u, v corresponding to the given surface region S. Integral (3) can be evaluated by a repeated integration as follows

$$(4) \qquad \int_S \Phi(u, v)\, dS = \int_{u_1}^{u_2} \int_{v_1(u)}^{v_2(u)} \Phi(u, v) \sqrt{EG - F^2}\, dv du,$$

where u_1, u_2 are coordinates of the extreme coordinate lines enclosing the region S (Fig. 341), and $v = v_1(u)$, $v = v_2(u)$ are equations of the arcs AmB and AnB of the boundary of S.

Formula (1) is a particular case of (3) for

$$u = x, \quad v = y, \quad E = 1 + p^2, \quad F = pq, \quad G = 1 + q^2.$$

Application of surface integrals of the first type.

(1) Area of a curved surface S:

$$S = \int_S dS$$

(2) Mass of a non-homogeneous curved surface S, if $\delta = f(x, y, z)$ is a variable surface density of S:

$$M_S = \int_S \delta\, dS.$$

20. Surface integrals of the second type (Integrals over a projection)

The concept of an oriented surface. A surface has usually two sides; one of them (no matter which one) can be regarded as *positive* and the other one as *negative* [1]. A surface whose one side has been taken as positive and the other one as negative is said to be *oriented*. On a closed surface without self-intersections enclosing a region of space (as, for example, the surface of a sphere or of an ellipsoid), the exterior side is usually taken as positive and the interior side as negative.

Projection of an oriented surface onto a coordinate plane. A projection of a bounded oriented surface onto a coordinate plane, for example, the xy plane, can be regarded as positive or negative, according to the following rule (Fig. 342). If the surface looked at from the positive direction of the third coordinate axis (in this

[1] There exist surfaces whose two sides cannot be distinguished (e.g., *Möbius band*); we do not consider such surfaces in mathematical analysis.

case, from the positive direction of the z axis, i.e., from up-side) shows its positive side, then the projection $\mathrm{pr}_{xy}\, S$ is regarded as positive (Fig. 342a); otherwise it is regarded as negative (Fig. 342b). If one part of the surface shows its positive side and the other part—its negative side, then the projection $\mathrm{pr}_{xy}\, S$ is regarded as the algebraic sum of positive and negative projections (Fig.

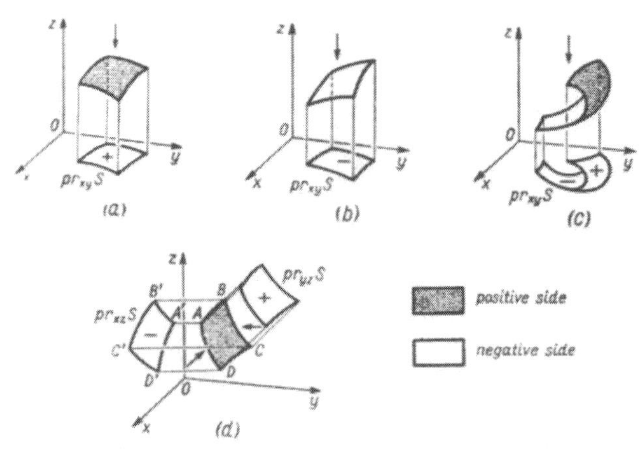

FIG. 342

342c). Fig. 342d illustrates the projections $\mathrm{pr}_{xz}\, S$ and $\mathrm{pr}_{yz}\, S$; one of them is positive and the other one is negative.

The projection of a closed oriented surface onto an arbitrary coordinate plane is equal to zero.

Definition of the surface integral of the second type over the projection onto the xy plane. The surface integral of the second type

$$\int\limits_S f(x, y, z)\, dxdy$$

of a function $f(x, y, z)$ of three variables defined in a connected space domain over the projection of an oriented surface S lying in this domain on the xy plane is the number obtained in the same way as the surface integral of the first type (p. 506) except that the step 3 is replaced by the following one: we multiply the value $f(x_i, y_i, z_i)$ at the point M_i by the projection $\mathrm{pr}_{xy}\, dS_i$ of the area dS_i of the elementary oriented surface region S_i on the xy plane, the projection being taken with their sign (see above):

$$(1_{xy}) \qquad \int_S f(x, y, z)\, dxdy = \lim_{\substack{dS_i \to 0 \\ n \to \infty}} \sum_{i=1}^{n} f(x_i, y_i, z_i)\, \mathrm{pr}_{xy}\, dS_i.$$

The surface integrals over the projection of S onto the yz or zx plane is defined likewise:

$$(1_{yz}) \qquad \int_S f(x, y, z)\, dydz = \lim_{\substack{dS_i \to 0 \\ n \to \infty}} \sum_{i=1}^{n} f(x_i, y_i, z_i)\, \mathrm{pr}_{yz}\, dS_i,$$

$$(1_{zx}) \qquad \int_S f(x, y, z)\, dzdx = \lim_{\substack{dS_i \to 0 \\ n \to \infty}} \sum_{i=1}^{n} f(x_i, y_i, z_i)\, \mathrm{pr}_{zx}\, dS_i.$$

Existence theorem. If the function $f(x, y, z)$ is continuous and the function defining the surface are continuous and have continuous derivatives, then the surface integrals of the second type (1_{xy}), (1_{yz}), (1_{zx}) exist.

Evaluation of surface integrals of the second type can be reduced to evaluation of double integrals. If the surface is given in explicit form $z = \varphi(x, y)$, then integral (1_{xy}) can be evaluated by the formula

$$(2_{xy}) \qquad \int_S f(x, y, z)\, dxdy = \int_{\mathrm{pr}_{xy} S} f\big(x, y, \varphi(x, y)\big)\, dS_{xy},$$

where S_{xy} denotes the projection of S on the xy plane.

Formulas for the surface integrals of $f(x, y, z)$ over the projections of S on the yz and zx planes are analogous:

$$(2_{yz}) \qquad \int_S f(x, y, z)\, dydz = \int_{\mathrm{pr}_{yz} S} f\big(\psi(y, z), y, z\big)\, dS_{yz},$$

where $x = \psi(y, z)$ is the equation of the surface solved with respect to x and $S_{yz} = \mathrm{pr}_{yz} S$, and

$$(2_{zx}) \qquad \int_S f(x, y, z)\, dzdx = \int_{\mathrm{pr}_{zx} S} f\big(x, \chi(z, x), z\big)\, dS_{zx},$$

where $y = \chi(z, x)$ is the equation of the surface solved with respect to y and $S_{zx} = \mathrm{pr}_{zx} S$.

If the orientation of the surface is changed (the positive side is replaced by the negative one and conversely), then the integral over a projection changes its sign.

If the surface is given in parametric form $x = x(u, v)$, $y = y(u, v)$,

$z = z(u, v)$, then integrals (1_{xy}), (1_{yz}), (1_{zx}) are computed from the formulas

$$(3_{xy})\quad \iint_S f(x, y, z)\,dxdy = \iint_\Delta f(x(u, v), y(u, v), z(u, v))\frac{\partial(x, y)}{\partial(u, v)}\,dudv,$$

$$(3_{yz})\quad \iint_S f(x, y, z)\,dydz = \iint_\Delta f(x(u, v), y(u, v), z(u, v))\frac{\partial(y, z)}{\partial(u, v)}\,dudv,$$

$$(3_{zx})\quad \iint_S f(x, y, z)\,dzdx = \iint_\Delta f(x(u, v), y(u, v), z(u, v))\frac{\partial(z, x)}{\partial(u, v)}\,dudv,$$

where $\dfrac{\partial(x,y)}{\partial(u,v)}, \dfrac{\partial(y,z)}{\partial(u,v)}, \dfrac{\partial(z,x)}{\partial(u,v)}$ are Jacobians of pairs of the functions x, y, z with respect to u, v and Δ is the domain of parameters u, v corresponding to the surface S.

Surface integral in general form. If $P(x, y, z)$, $Q(x, y, z)$, $R(x, y, z)$ are three functions of three variables defined in a connected domain and S is an oriented surface contained in this domain, the sum of the integrals of the second type taken over the projection of S on three coordinate planes is called the *surface integral in general form*:

$$\iint_S P\,dydz + Q\,dzdx + R\,dxdy = \iint_S P\,dydz + \iint_S Q\,dzdx + \iint_S R\,dxdy \text{ [1]}.$$

Formula reducing the surface integral to a double integral:

$$\iint_S P\,dydz + Q\,dzdx + R\,dxdy$$
$$= \iint_\Delta \left(P\frac{\partial(y, z)}{\partial(u, v)} + Q\frac{\partial(z, x)}{\partial(u, v)} + R\frac{\partial(x, y)}{\partial(u, v)}\right)dudv,$$

where $\dfrac{\partial(y,z)}{\partial(u,v)}, \ldots$ and Δ are defined as above.

Properties of the surface integrals.

(1) If the surface S is decomposed into a sum of S_1 and S_2, then

$$\iint_S P\,dydz + Q\,dzdx + R\,dxdy$$
$$= \iint_{S_1} P\,dydz + Q\,dzdx + R\,dxdy + \iint_{S_2} P\,dydz + Q\,dzdx + R\,dxdy.$$

---

[1] A vectorial treatment of the surface integral in general form is given in the chapter "Field theory" (p. 625).

(2) If the orientation of the surface is reversed, the integral has its sign changed:

$$\int_{S^+} P\,dydz + Q\,dzdx + R\,dxdy = -\int_{S^-} P\,dydz + Q\,dzdx + R\,dxdy$$

where S^+ and S^- denote the same surface but with opposite orientations.

(3) A surface integral depends, in general, on the line bounding the surface region S as well as on the surface itself. Thus the integrals taken over two surface regions S_1 and S_2 spanned by the same closed curve C (Fig. 343) are, in general, not equal.

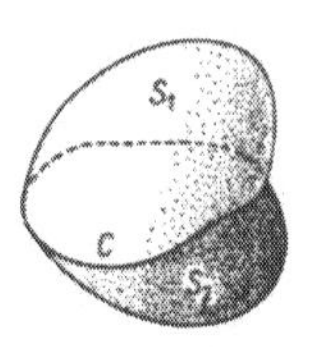

FIG. 343

Volume V of a solid bounded by a closed surface S can be expressed by a surface integral:

$$V = \tfrac{1}{3}\int_S x\,dydz + y\,dzdx + z\,dxdy,$$

where the surface S is oriented so that its exterior side is positive.

21. Formulas of Stokes, Green and Gauss-Ostrogradsky [1]

Stokes's formula (expression of a line integral by a surface integral). If S is an oriented surface bounded by a closed curve K and lying in a connected space domain, and P, Q, R are three functions of three variables x, y, z defined in this domain, then we have *Stokes's formula:*

$$(1)\quad \int_K P\,dx + Q\,dy + R\,dz$$

$$= \int_S \left(\frac{\partial Q}{\partial x} - \frac{\partial P}{\partial y}\right) dxdy + \left(\frac{\partial R}{\partial y} - \frac{\partial Q}{\partial z}\right) dydz + \left(\frac{\partial P}{\partial z} - \frac{\partial R}{\partial x}\right) dzdx \ [2],$$

where the line integral on the left side of (1) is taken along the curve K counterclockwise for an observer looking at the surface from its positive side (Fig. 344).

Green's formula is a particular case of Stokes's formula, for a function of two variables in a plane domain (it expresses a line integral along a plane closed curve by a double integral). If S is

---

[1] A vectorial treatment of these theorems is given in the chapter "Field theory" (pp. 645, 646).

[2] This theorem is valid provided that the functions P, Q, R are continuous and have continuous derivatives of the first order.

a plane domain bounded by a closed curve K, and P, Q are two functions of two variables x, y defined in this domain, then

$$\int_K P\,dx + Q\,dy = \iint_S \left(\frac{\partial Q}{\partial x} - \frac{\partial P}{\partial y}\right) dx\,dy, \tag{2}$$

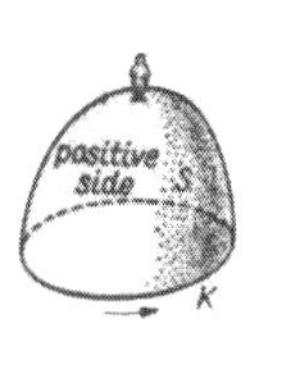

FIG. 344

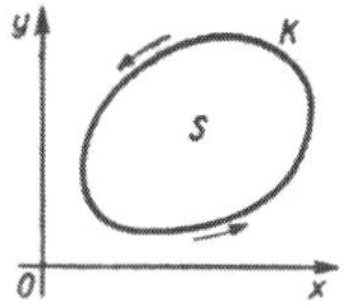

FIG. 345

where the line integral on the left side of (2) is taken along the curve K counterclockwise (Fig. 345).

Gauss-Ostrogradsky formula (expression of a triple integral by a surface integral). If S is a closed oriented surface (whose exterior side is positive) bounding a space domain V, and P, Q, R are three functions of three variables defined in a simply connected domain containing the surface S, then we have the following *Gauss-Ostrogradsky formula*:

$$\iiint_V \left(\frac{\partial P}{\partial x} + \frac{\partial Q}{\partial y} + \frac{\partial R}{\partial z}\right) dV = \iint_S P\,dy\,dz + Q\,dz\,dx + R\,dx\,dy \; [1]. \tag{3}$$

---

[1] This theorem is valid provided that the functions P, Q, R are continuous and have continuous derivatives.

IV. DIFFERENTIAL EQUATIONS

1. General concepts

A differential equation is an equation involving unknown functions, independent variables and derivatives (or differentials) of unknown functions.

Examples.

(1) $\left(\frac{dy}{dx}\right)^2 - xy^5\frac{dy}{dx} + \sin y = 0.$

(2) $x\,d^2y\,dx - dy(dx)^2 = e^y(dy)^3.$

(3) $\frac{\partial^2 z}{\partial x \partial y} = xyz\frac{\partial z}{\partial x}\cdot\frac{\partial z}{\partial y}.$

If the unknown functions depend on one independent variable, then the equation is called *ordinary* (examples (1), (2)). If the unknown functions depend on several independent variables, then the differential equation is called *partial* (example (3)). The *order* of a differential equation is the highest order of derivatives or differentials involved in the equation (equation (1) is of order 1, equations (2) and (3) are of order 2).

An *integral* of a differential equation is one or several equations connecting the unknown functions and the independent variables so that the unknown functions, when determined from these equations and substituted to the given differential equation, satisfy it identically.

Finding the integrals of a differential equation is called its *integration*. An integral expressing explicitly the unknown function by the independent variables is called a *solution* of the differential equation.

Integrals of differential equations may involve certain arbitrary constants or arbitrary functions, hence they are not unique. The functions involved are usually required to satisfy certain imposed additional conditions called *initial* or *boundary conditions* so that the unknown functions and some of their derivatives are required to assume certain preassigned values for some given values of the independent variables. These conditions may be sufficient for the uniqueness of a solution. For example, the ordinary differential equation

$$y^{(n)} = f(x, y, y', \ldots, y^{(n-1)})$$

has (under certain additional assumptions) a unique solution when $y, y', \ldots, y^{(n-1)}$ assume certain preassigned values for $x = a$ (see below, p. 529).

An integral of a differential equation is called ***general***, if any particular integral corresponding to preassigned initial or boundary condition can be obtained from it by a suitable choice of arbitrary constants and arbitrary functions (and ensure the uniqueness of the solution). A differential equation can have *singular solution* which cannot be obtained from a general solution for any values of arbitrary constants and arbitrary functions (see pp. 521, 522).

A. ORDINARY DIFFERENTIAL EQUATIONS

2. Equations of the first order

Theoretical considerations.

Cauchy's existence theorem. If the function $f(x, y)$ is continuous in a neighbourhood of a point (x_0, y_0), i.e., in the domain $|x - x_0| < a$ and $|y - y_0| < b$, then there exists at least one solution of the equation

$$\text{(a)} \qquad y' = f(x, y)$$

defined and continuous in a neighbourhood of x_0 and assuming the value $y = y_0$ for $x = x_0$.

If, moreover, the function $f(x, y)$ satisfies in this domain the following *Lipschitz's condition*

$$|f(x, y_1) - f(x, y_2)| < N|y_1 - y_2|,$$

where N is independent of x, y_1 and y_2, then the solution is unique and is a continuous function with respect to y_0.

The Lipschitz's condition is satisfied provided that the function $f(x, y)$ has a bounded partial derivative $\frac{\partial f}{\partial y}$ in the domain in question (for examples which fail to satisfy the assumptions of Cauchy's theorem see p. 522).

Direction field. If, through a point $M(x, y)$ there passes the graph of a solution $y = \varphi(x)$ of the differential equation $y' = f(x, y)$, then the slope of the tangent to the graph at M (equal to $\frac{dy}{dx}$) can be found directly from the differential equation; it follows that a differential equation determines the slope of the tangent to the graph of its solution at every point. The collection of these directions forms a *direction field* (Fig. 346). A point with a direction assigned to it is called an *element* of the direction field. Integration of a differential equation is equivalent to joining the elements into *integral curves* whose tangents at any point has the direction prescribed for that point by the direction field.

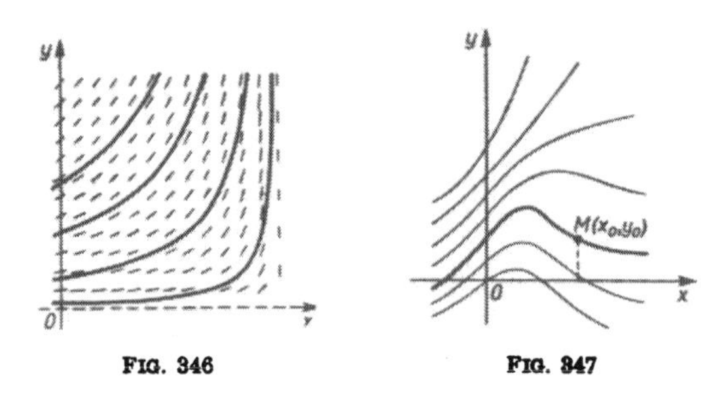

FIG. 346 FIG. 347

Sometimes it is necessary to consider a field in which vertical directions occur; they correspond to infinite values of $f(x, y)$. In such cases we interchange the variables x and y and write the equation in the form

(b) $$\frac{dx}{dy} = \frac{1}{f(x, y)}$$

which is equivalent to the given equation. If the assumption of Cauchy's theorem for the equation (a) or (b) are satisfied in the given domain, then through each point there passes an integral curve (Fig. 347).

The family of integral curves depends on one parameter, i.e., a general integral of a differential equation of the first order involves one arbitrary constant. In order to obtain a particular integral $y = \varphi(x)$ satisfying $y_0 = \varphi(x_0)$, from the general integral $F(x, y, C) = 0$, we determine C from the equation

$$F(x_0, y_0, C) = 0.$$

Fundamental methods of integration.

Separation of variables. If an equation can be reduced to the form

$$M(x)N(y)\,dx + P(x)Q(y)\,dy = 0,$$

then it can be written in the form

$$R(x)\,dx + S(y)\,dy = 0,$$

where the variables x and y are separated. This can be done by dividing both members by $P(x)N(y)$; a general integral has the form

$$\int \frac{M(x)}{P(x)}\,dx + \int \frac{Q(y)}{N(y)}\,dy = C.$$

If the functions $P(x)$ and $N(y)$ vanish for certain values $\bar{x}$ and $\bar{y}$, then the integrals $x = \bar{x}$, $y = \bar{y}$ are also solutions.

Example. $x\,dy + y\,dx = 0$;

$$\int \frac{dy}{y} + \int \frac{dx}{x} = C, \quad \ln y + \ln x = C = \ln c, \quad yx = c.$$

Homogeneous equations. If $M(x, y)$ and $N(x, y)$ are homogeneous functions of their arguments of the same degree (see p. 344), then the variables in the equation $M(x, y)\,dx + N(x, y)\,dy$ will be separated (see above), when we introduce a new variable u by the equation $y = ux$.

Example. $y^2\,dx + x(x - y)\,dy = 0$; $y = ux$; $dy = u\,dx + x\,du$;

$$u^2x^2\,dx + x^2(1 - u)(x\,du + u\,dx) = 0; \quad \frac{dx}{x} + \frac{(1-u)\,du}{u} = 0;$$

$$\ln x + \ln u - u = C = \ln c; \quad ux = ce^u; \quad y = ce^{y/x}.$$

The straight line $x = 0$ is also an integral curve (see above, separation of variables).

Exact equations. An equation of the form

$$(*) \qquad M(x, y)\,dx + N(x, y)\,dy = 0$$

is said to be *exact*, if a function $\Phi(x, y)$ of two variables exists such that

$$M(x, y)\,dx + N(x, y)\,dy \equiv d\Phi(x, y)$$

(see p. 363). If the functions $M(x, y)$ and $N(x, y)$ are continuous and have continuous derivatives of the first order in a simply connected domain, then the equation $\frac{\partial M}{\partial y} = \frac{\partial N}{\partial x}$ is a necessary and sufficient condition that the equation (*) is exact. In this case $\Phi(x, y) = C$ is the general integral of (*). The function $\Phi(x, y)$ can be found from the formula (see p. 494, formula (4_2)):

$$\Phi(x, y) = \int_{x_0}^{x} M(\xi, y)\, d\xi + \int_{y_0}^{y} N(x_0, \eta)\, d\eta,$$

where x_0 and y_0 are arbitrary.

An *integrating factor* is a function $\mu(x, y)$ such that the equation $M(x, y)\,dx + N(x, y)\,dy = 0$ multiplied by $\mu(x, y)$ becomes exact. The function $\mu(x, y)$ satisfies the equation

$$N \frac{\partial \ln \mu}{\partial x} - M \frac{\partial \ln \mu}{\partial y} = \frac{\partial M}{\partial y} - \frac{\partial N}{\partial x}.$$

Any solution of this equation can be taken as an integrating factor.

If μ is a known integrating factor of the differential equation (*) so that equation (*) multiplied by μ changes into an equation $d\Phi(x, y) = 0$, then any integrating factor is of the form $\tilde{\mu} = \mu f(\Phi)$, where f is an arbitrary function.

Example. $(x^2 + y)\,dx - x\,dy = 0$. Equation for an integrating factor:

$$-x \frac{\partial \ln \mu}{\partial x} - (x^2 + y) \frac{\partial \ln \mu}{\partial y} = 2.$$

We seek an integrating factor independent of y. Then

$$x \frac{\partial \ln \mu}{\partial x} = -2, \quad \text{hence} \quad \mu = \frac{1}{x^2}.$$

Multiplying the equation by μ we obtain

$$\left(1 + \frac{y}{x^2}\right) dx - \frac{1}{x}\, dy = 0.$$

Taking $x_0 = 1$, $y_0 = 0$, we find a general integral

$$\Phi(x, y) \equiv \int_1^x \left(1 + \frac{y}{\xi^2}\right) d\xi - \int_0^y d\eta = C, \quad \text{hence} \quad x - \frac{y}{x} = C_1.$$

Linear equation. An equation of the form

$$y' + P(x)y = Q(x),$$

linear in y and y', i.e., involving the unknown function and its derivative to the first degree only, is called a *linear differential equation of the first order*. Such an equation has an integrating factor $\mu = e^{\int P\,dx}$. We find the general integral from the formula

$$(**)\qquad y = e^{-\int P\,dx}\left(\int Q e^{\int P\,dx}\,dx + C\right).$$

Replacing the indefinite integrals in this formula by definite integrals between x_0 and x [1], we obtain the solution which assumes the value C for $x = x_0$.

If any particular solution $y_1(x)$ of a linear differential equation is known, then the general solution of this equation can be found from the formula

$$y = y_1 + Ce^{-\int P\,dx}.$$

If two linearly independent (see p. 532) particular solutions $y_1(x)$ and $y_2(x)$ are known, then the general solution can be found without integration from the formula $y = y_1 + C(y_2 - y_1)$.

Example. Find a solution of the equation $y' - y\tan x = \cos x$ satisfying the initial condition $x_0 = 0$, $y_0 = 0$.

We evaluate $e^{\int_0^x (-\tan x)\,dx} = \cos x$ and obtain from (**)

$$y = \frac{1}{\cos x}\int_0^x \cos^2 x\,dx = \frac{1}{\cos x}\left[\frac{\sin x\cos x + x}{2}\right] = \frac{\sin x}{2} + \frac{x}{2\cos x}.$$

Bernoulli's equation

$$y' + P(x)\,y = Q(x)\,y^n$$

can be reduced to a linear equation by dividing by y^n and introducing a new variable $z = y^{-n+1}$.

Example. $y' - \dfrac{4y}{x} = x\sqrt{y}$.

Here we have $n = \frac{1}{2}$. Dividing by $\sqrt{y}$ and introducing the variable $z = \sqrt{y}$, we obtain $\dfrac{dz}{dx} - \dfrac{2z}{x} = \dfrac{x}{2}$. Now we have $e^{\int P\,dx} = \dfrac{1}{x^2}$, where $P = -\dfrac{2}{x}$. By the formula for a solution of a linear equation,

$$z = x^2\left(\int \frac{x}{2}\cdot\frac{1}{x^2}\,dx + C\right) = x^2\left(\frac{1}{2}\ln|x| + C\right);$$

hence

$$y = x^4\left(\frac{1}{2}\ln|x| + C\right)^2.$$

Riccati's equation

$$y' = P(x)\,y^2 + Q(x)\,y + R(x)$$

---

[1] See p. 394.

cannot, in general, be integrated in *quadratures* (i.e., its solution cannot be reduced to a finite number of integrations). If, however, a particular solution y_1 of Riccati's equation is known, then, by introducing a new variable z by the formula $y = y_1 + 1/z$, it can be reduced to a linear equation. If one more solution y_2 of Riccati's equation is known, then $z_1 = \frac{1}{y_2 - y_1}$ is a particular solution of the linear equation with respect to z, which simplifies the integration of the equation. If three particular solutions y_1, y_2, y_3 of Riccati's equation are known, then its general integral is of the form

$$\frac{y - y_2}{y - y_1} : \frac{y_3 - y_2}{y_3 - y_1} = C.$$

By the change of variables $y = \frac{u}{P(x)} + \beta(x)$, Riccati's equation can be reduced to the canonical form $\frac{du}{dx} = u^2 + R(x)$.

By the substitution $y = -\frac{v'}{P(x)\,v}$, Riccati's equation can be reduced to a linear equation of the second order:

$$Pv'' - (P' + PQ)\,v' + P^2Rv = 0.$$

Example. $y' + y^2 + \frac{1}{x}y - \frac{4}{x^2} = 0.$

We change the variables $y = z + \beta(x)$. The coefficient at the first power of z will be $2\beta(x) + 1/x$, hence it will vanish if we let $\beta(x) = -1/2x$ and we shall obtain $z' + z^2 - 15/4x^2 = 0$. We guess a particular solution to be of the form $z_1 = a/x$. After substituting, we find $a_1 = -\frac{3}{2}$, $a_2 = \frac{5}{2}$, hence we have two particular solutions $z_1 = -3/2x$, $z_2 = 5/2x$. We change the variables again $z = 1/u + z_1 = 1/u - 3/2x$ and obtain $u' + 3u/x = 1$. Using the particular solution $u_1 = \frac{1}{z_2 - z_1} = \frac{x}{4}$ of this equation, we find its general solution $u = \frac{x}{4} + \frac{C}{x^3} = \frac{x^4 + C_1}{4x^3}$, hence

$$y = \frac{1}{u} - \frac{3}{2x} - \frac{1}{2x} = \frac{2x^4 - 2C_1}{x^5 + C_1x}.$$

Equations of the form $F(x, y, y') = 0$ not solvable with respec to y'. If the equation $F(x_0, y_0, p) = 0$, where $p = \frac{dy}{dx}$ has at a certain point $M(x_0, y_0)$ n real roots $p_1, p_2, \ldots, p_n$ and the function $F(x, y, p)$ and its first derivatives are continuous at the point $x = x_0$, $y = y_0$, $p = p_i$ $(i = 1, 2, \ldots, n)$ and $\frac{\partial F}{\partial p} \neq 0$, then through the point $M(x_0, y_0)$ there pass n integral curves.

If the equation $F(x, y, y') = 0$ can be solved with respect to y', then it yields n equations of the form previously considered; solving these equations we obtain n families of integral curves.

If the equation in question can be written in the form $x = \varphi(y, y')$ or $y = \psi(x, y')$, then putting $y' = p$ and regarding p as an auxiliary variable, after differentiation with respect to y or x, we obtain an equation for dp/dy or dp/dx which is solved with respect to the derivative. A solution of this equation together with the original equation determines the desired solution in parametric form.

Example. $x = yy' + y'^2$; $y' = p$, $x = py + p^2$. We differentiate with respect to y substituting $\frac{dx}{dy} = \frac{1}{p}$:

$$\frac{1}{p} = p + (y + 2p)\frac{dp}{dy} \quad \text{or} \quad \frac{dy}{dp} - \frac{py}{1-p^2} = \frac{2p^2}{1-p^2}.$$

This is an equation linear in y. Solving it, we obtain

$$y = -p + \frac{c + \arcsin p}{\sqrt{1-p^2}}$$

which, together with the original equation $x = py + p^2$ gives the desired solution in parametric form.

Lagrange's equation

$$a(y')x + b(y')y + c(y') = 0$$

can be integrated in quadratures in the way shown above. If $a(p) + b(p)p = 0$ for $p = p_0$, then $a(p_0)x + b(p_0)y + c(p_0) = 0$ is a singular integral of Lagrange's equation (see below).

If $a(p) + b(p)p \equiv 0$, then we have so-called *Clairaut's equation* which can always be reduced to the form $y = y'x + f(y')$. Its general solution is of the form $y = Cx + f(C)$. Besides the general solution (which gives geometrically a family of straight lines depending on one parameter), Clairaut's equation has also a singular integral which can be obtained eliminating C from the equation $y = Cx + f(C)$ and from the equation $0 = x + f'(C)$ obtained from the first one by differentiation with respect to C. Geometrically, the singular solution is the envelope of the family of straight lines $y = Cx + f(C)$ (Fig. 348, see p. 522).

Example. $y = xy' + y'^2$.

The general integral $y = Cx + C^2$. Differentiating, we obtain $0 = x + 2C$ and, eliminating C, we get the singular integral $x^2 + 4y = 0$. The integral curves of this differential equation are illustrated in Fig. 348.

Singular integrals. An element (x_0, y_0, y'_0) is called *singu-*

lar, if it satisfies both the equation $F(x, y, y') = 0$ and $\dfrac{\partial F}{\partial y'} = 0$. An integral curve generated by singular elements is called a ***singular integral curve***. The uniqueness of a solution (see Cauchy's theorem, p. 515) fails at each point of a singular curve. Envelopes (see p. 295) of integral curves (Fig. 348) are singular

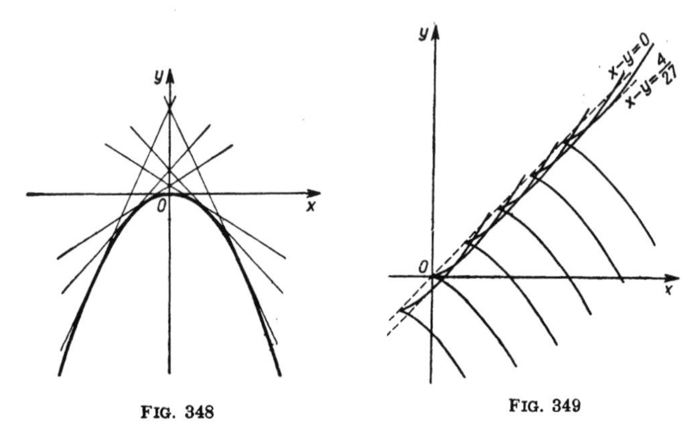

FIG. 348

FIG. 349

integral curves. An equation of a singular curve is a singular integral which, as a rule, cannot be obtained from the general integral for any value of the arbitrary constant. To find a singular integral of the equation $F(x, y, p) = 0$, where $p = y'$, we eliminate p from the equations $F(x, y, p) = 0$ and $\dfrac{\partial F}{\partial p} = 0$. If the obtained relation is an integral of the given differential equation, then it will be a singular integral; the given equation should be first transformed into a form which does not involve multiple-valued functions [1], in particular roots. If the equation of integral curves, that is the general integral, is known, then the envelope of this family, that is the singular solution, can be found by aid of methods of differential geometry (see p. 295).

Examples. (1) $x - y - \frac{4}{9}p^2 + \frac{8}{27}p^3 = 0$. The equation $\dfrac{\partial F}{\partial p} = 0$ has the form $-\frac{8}{9}p + \frac{8}{9}p^2 = 0$. Eliminating p, we obtain (a) $x - y = 0$ or (b) $x - y = \frac{4}{27}$.

The equation (a) is not a solution, the equation (b) is a singular solution. The general solution is $(y - C)^2 = (x - C)^3$. The integral curves (semi-cubical parabolas) and the curves (a) and (b) are shown in Fig. 349.

---

[1] Complex values of functions are also taken into account.

(2) $y' - \ln|x| = 0$. We transform it into the form $e^p - |x| = 0$, since $\ln|x|$ is multiple-valued (see p. 591). $\frac{\partial F}{\partial p} \equiv e^p = 0$. Eliminating p, we obtain the singular integral $x = 0$.

Singular points of a differential equation. The point (0, 0) is a *singular point* of the differential equation

$$\frac{dy}{dx} = \frac{ax + by}{cx + ey}, \quad \text{where} \quad ae - bc \neq 0,$$

for the assumptions of Cauchy's theorem are not fulfilled at this point (see p. 515), while they are fulfilled at any other point sufficiently near to (0, 0) [1]. The behaviour of integral curves in a neighbourhood of such a point depends on the roots of the *characteristic equation*

$$\lambda^2 - (b + c)\lambda + (bc - ae) = 0,$$

as follows:

(1) If the roots of the characteristic equations are real and of the same sign, then the singular point is a *branch point*. The integral curves in a neighbourhood of the singular point pass through this point and, if the roots of the characteristic equation do not coincide, the integral curves, except for one, have a common tangent at this point. If the characteristic equation has a double root, then either all integral curves have a common tangent, or in each direction through the singular point there passes a unique integral curve.

Examples. (1) $\frac{dy}{dx} = \frac{2y}{x}$. The characteristic equation $\lambda^2 - 3\lambda + 2 = 0$ has the roots $\lambda_1 = 2$, $\lambda_2 = 1$. The integral curves $y = Cx^2$ and $x = 0$ [2] (Fig. 350).

(2) $\frac{dy}{dx} = \frac{x + y}{x}$. The characteristic equation $\lambda^2 - 2\lambda + 1 = 0$ has a double root $\lambda = 1$. The integral curves $y = x \ln|x| + Cx$ (Fig. 351).

(3) $\frac{dy}{dx} = \frac{y}{x}$. The characteristic equation $\lambda^2 - 2\lambda + 1 = 0$ has a double root $\lambda = 1$. The integral curves $y = Cx$ (Fig. 352).

(2) If the roots of the characteristic equation are real and of

---

[1] More precisely, the assumptions of Cauchy's theorem are not fulfilled at any point such that $cx + ey = 0$; however, in this case we may interchange the variables and consider the equation

$$\frac{dx}{dy} = \frac{cx + ey}{ax + by}$$

which satisfies these assumptions.

[2] The general solution written in the form $x^2 = C_1 y$ contains also the line $x = 0$.

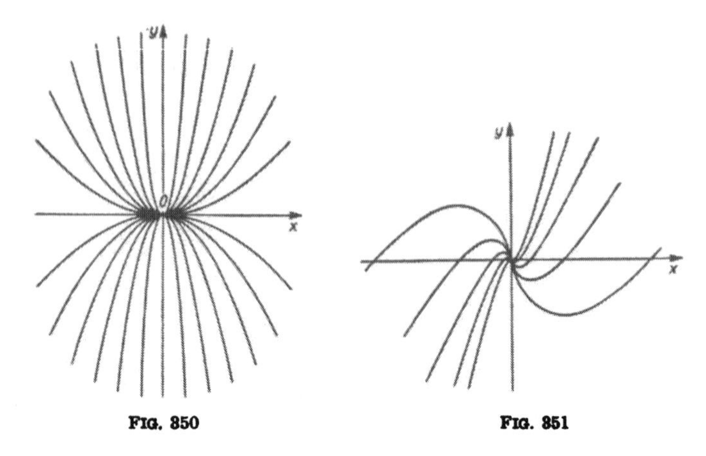

Fig. 350

Fig. 351

different signs, then the singular point is a *saddle point*. Only two of the integral curves lying in a neighbourhood of the singular point pass through this point (they are asymptotes of the remaining integral curves).

Example. $\frac{dy}{dx} = -\frac{y}{x}$. The characteristic equation $\lambda^2 - 1 = 0$ has the roots $\lambda_1 = 1$ and $\lambda_2 = -1$. The integral curves $xy = C$ (Fig. 353); for $C = 0$ we obtain singular integrals $x = 0$ and $y = 0$.

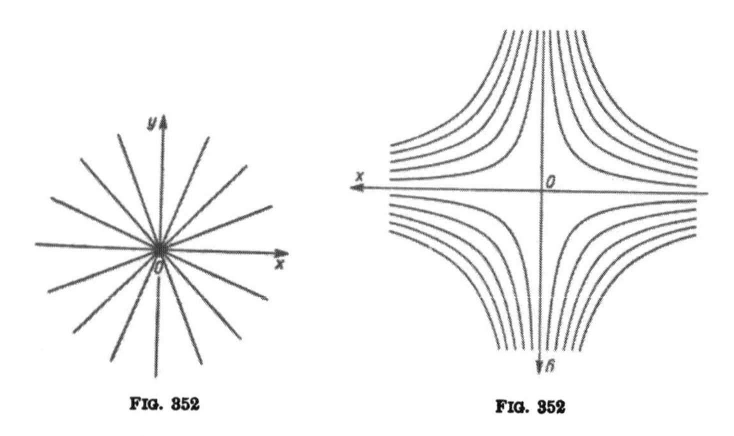

Fig. 352

Fig. 352

(3) If the roots of the characteristic equation are complex conjugate with the real part different from zero, then the singular point is a *focal point*. The curves in the neighbourhood of a focal

point wind on about the singular point; this is an asymptotic point of the integral curves.

Example. $\frac{dy}{dx} = \frac{x+y}{x-y}$. The characteristic equation $\lambda^2 - 2\lambda + + 2 = 0$ has two roots $\lambda_1 = 1 + i$ and $\lambda_2 = 1 - i$. The integral curves have the equation $\varrho = Ce^{\varphi}$ in polar coordinates (Fig. 354).

(4) If the roots of the characteristic equation are imaginary, then the singular point is a *central point*. The integral curves are closed and contain the singular point inside.

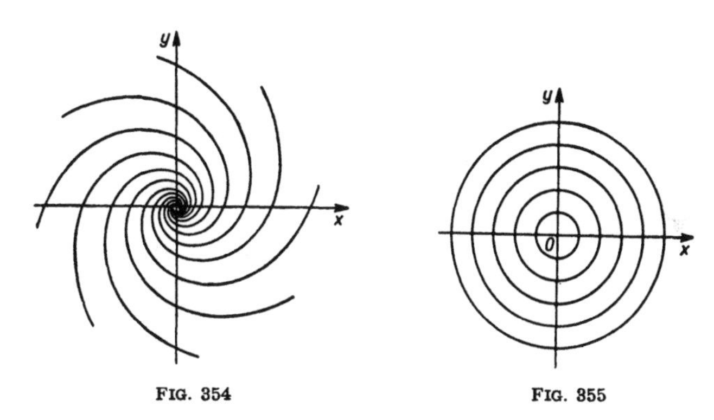

FIG. 354

FIG. 355

Example. $\frac{dy}{dx} = -\frac{x}{y}$. The characteristic equation $\lambda^2 + 1 = 0$ has the roots $\lambda_1 = i$ and $\lambda_2 = -i$. The integral curves $x^2 + y^2 = C$ (Fig. 355).

Singular points of the differential equation

$$\frac{dy}{dx} = \frac{P(x, y)}{Q(x, y)}$$

are the points (x, y) for which $P(x, y) = 0$ and $Q(x, y) = 0$. Assuming that the functions P and Q have continuous derivatives, the given equation can be written in the form

$$\frac{dy}{dx} = \frac{a(x - x_0) + b(y - y_0) + P_1(x, y)}{c(x - x_0) + e(y - y_0) + Q_1(x,y)},$$

where x_0, y_0 are coordinates of a singular point and $P_1(x, y)$ and $Q_1(x, y)$ are infinitesimals of a higher order than the distance of the point (x, y) from (x_0, y_0). Then the type of a singular point of the given equation is the same as that of the first approximate equation obtained by disregarding the terms P_1 and Q_1. Exceptions: (a) if a singular point of the first approximate equation is a central point, then the corresponding singular point of the original equation may be either central or focal; (b) if $ae - bc = 0$, then the type of a singular point may be discovered by examining the terms of higher order.

Approximate methods of solution of differential equations.

Successive approximation method (*Picard's method*). The equation $y' = f(x, y)$ with the initial condition $y = y_0$ for $x = x_0$ can be written in the form

$$y = y_0 + \int_{x_0}^{x} f(x, y)\, dx. \tag{1}$$

Substituting an arbitrary function $y_1(x)$ instead of y to the right member, we obtain a new function y_2 different from y_1, unless y_1 is a solution of the given equation. Substituting y_2 instead of y to the right member of (1), we obtain a function y_3 and so on.

The sequence $y_1, y_2, \ldots, y_n, \ldots$ obtained in this way is convergent in a certain interval containing x_0 to the desired solution of the given equation, provided that the assumptions of Cauchy's theorem are fulfilled (see p. 515). Successive approximation method is sometimes called the *iteration method* (compare p. 170).

Example. $y' = e^x - y^2$ with the initial condition $x_0 = 0$, $y_0 = 0$. We write the equation in the integral form

$$y = \int_0^x (e^x - y^2)\, dx.$$

We apply Picard's method starting from $y_0 = 0$ and obtain successively

$$y_1 = \int_0^x e^x\, dx = e^x - 1,$$

$$y_2 = \int_0^x \left(e^x - (e^x - 1)^2\right) dx = 3e^x - \tfrac{1}{2}e^{2x} - x - \tfrac{5}{2}$$

and so on.

Application of series. If the values $y_0', y_0'', \ldots, y_0^{(n)}$ of all derivatives at the initial point x_0 are known, then the expansion of the solution of the differential equation into Taylor's series (see p. 385) can be written:

$$y = y_0 + (x - x_0)\, y_0' + \frac{(x - x_0)^2}{2!}\, y_0'' + \ldots + \frac{(x - x_0)^n}{n!}\, y_0^{(n)} + \ldots$$

The values of the derivatives can be found by successive differentiating of the given differential equation and substituting the initial conditions. If such infinite differentiation of the differential equation is possible, then the obtained series will be always convergent in a certain neighbourhood of the initial value x_0.

The described method can also be applied to equations of order n.

Practically, it is more convenient to seek a solution in the form of a series with undetermined coefficients which can be found by the fact that the series satisfies the given equation.

Example. $y' = e^x - y^2$ for the initial condition $x_0 = 0$, $y_0 = 0$.

First way. Let

$$y = a_1 x + a_2 x^2 + a_3 x^3 + \ldots + a_n x^n + \ldots$$

Substituting to the equation and using the formula for a square of a series (see p. 358), we obtain

$$a_1 + 2a_2 x + 3a_3 x^2 + \ldots + \left(a_1^2 x^2 + 2a_1 a_2 x^3 + (a_2^2 + 2a_1 a_3)\, x^4 + \ldots\right)$$
$$= 1 + x + \tfrac{1}{2} x^2 + \tfrac{1}{6} x^3 + \ldots$$

Therefore $a_1 = 1$, $2a_2 = 1$, $3a_3 + a_1^2 = \frac{1}{2}$, $4a_4 + 2a_1 a_2 = \frac{1}{6}$ and so on. Solving successively these equations and substituting the coefficients to the series, we obtain $y = x + \frac{1}{2} x^2 - \frac{1}{6} x^3 - \frac{5}{24} x^4 + \ldots$

Second way. Substituting $x = 0$, we obtain $y_0' = 1$. Now $y'' = e^x - 2yy'$; $y_0'' = 1$; $y''' = e^x - 2y'^2 - 2yy''$, $y_0''' = -1$; $y^{(4)} = e^x - 6y'y'' - 2yy'''$, $y_0^{(4)} = -5$ and so on. By Taylor's formula

$$y = x + \frac{x^2}{2!} - \frac{x^3}{3!} - \frac{5x^4}{4!} + \ldots$$

Graphical solution of differential equations is based on the concept of a direction field (see p. 516). An integral curve

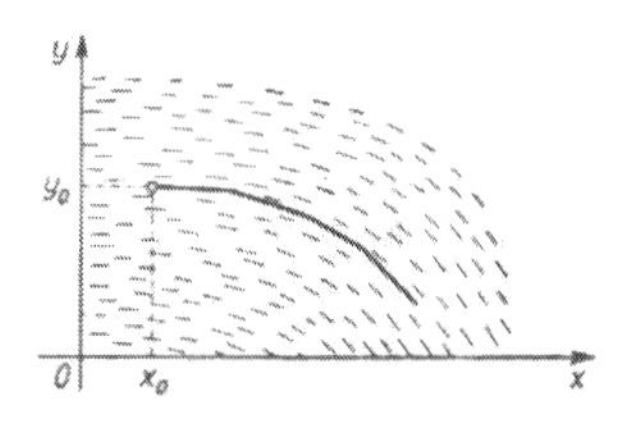

FIG. 356

can be represented approximately by a broken line (Fig. 356) issuing from the initial point and composed of small segments each of which has the direction prescribed by the field at the initial point of the segment; this point coincides with the final point of the preceding segment.

Numerical integration. In numerical integration of the equation $y' = f(x, y)$, we form successively a table of the desired function for the values $x_k = x_0 + kh$ ($k = 1, 2, 3, \ldots$), using an arbitrary formula for approximate integration to compute the differences

$$y_{k+1} - y_k = \int_{x_k}^{x_{k+1}} f(x, y(x))\, dx.$$

The following difference formulas are mostly used (for notation see pp. 758, 759):

(A) $$y_{k+1} - y_k = h\left(f_k + \tfrac{1}{2}\Delta f_{k-1} + \tfrac{5}{12}\Delta^2 f_{k-2} + \tfrac{3}{8}\Delta^3 f_{k-3}\right),$$

(B) $$y_{k+1} - y_k = h\left(f_{k+1} - \tfrac{1}{2}\Delta f_k - \tfrac{1}{12}\Delta^2 f_{k-1} - \tfrac{1}{24}\Delta^3 f_{k-2}\right).$$

Finding a first approximation y_{k+1} from formula (A), we compute f_{k+1} and find a second approximation y_{k+1} from formula (B). A third approximation can be found in the same way, but we usually try to choose the tabular difference h so that a third approximation is unnecessary.

Example. The terms of the series for a solution of the equation $y' = e^x - y^2$ obtained above with the initial condition $x_0 = 0$, $y_0 = 0$ enable us to compute the values of y for $x_1 = 0.1$, $x_2 = 0.2$ and $x_3 = 0.3$ to four decimal figures. To compute further values of y, we form a table as follows (to the step line):

x	y	f	Δf	$\Delta^2 f$	$\Delta^3 f$
0	0.0000	1.0000	942	−147	−35
0.1	0.1048	1.0942	795	−182	−21
0.2	0.2183	1.1737	613	−203	
0.3	0.3389	1.2350	410		
0.4	0.4646	1.2760			

It is worthwhile to test the value y_3 according to formula (B):

$$y_3 = 0.2183 + 0.1(1.2350 - 0.0306 + 0.0015 + 0.0001) = 0.3389.$$

Now, from formula (A), for $x_4 = 0.4$, we obtain

$$y_4 - y_3 = 0.1(1.2350 + 0.0306 - 0.0076 - 0.0013) = 0.1257.$$

Computing the value f_4 by $y_4 = 0.4646$ and extending the table, we find from formula (B)

$$y_4' - y_3 = 0.1(1.2760 - 0.0205 + 0.0017 + 0.0001) = 0.1257.$$

Since the value y_4 is preserved, we make the next step and so on.

Instead of the difference formulas (A) and (B), the following formulas are also in use

(A₁) $$y_{k+1} = y_{k-3} + \tfrac{4}{3}h(2f_k - f_{k-1} + 2f_{k-2}),$$

(B₁) $$y_{k+1} = y_{k-1} + \tfrac{1}{3}h(f_{k+1} + 4f_k + f_{k-1}).$$

The methods described above can also be extended to systems of differential equations.

3. Equations of higher orders and systems of equations

Theoretical considerations.

Existence theorem. Any equation of order n of the form

$$y^{(n)} = F(x, y, y', \ldots, y^{(n-1)})$$

can be reduced to a system of n equations

$$\frac{dy}{dx} = y_1, \quad \frac{dy_1}{dx} = y_2, \quad \ldots, \quad \frac{dy_{n-1}}{dx} = F(x, y, y_1, \ldots, y_{n-1})$$

by introducing new variables $y_1 = y'$, $y_2 = y''$, ..., $y_{n-1} = y^{(n-1)}$.

A more general system of equations

(*) $$\frac{dy_i}{dx} = f_i(x, y_1, y_2, \ldots, y_n), \quad i = 1, 2, \ldots, n$$

has a unique system of solutions $y_i = y_i(x)$ $(i = 1, 2, \ldots, n)$ defined and continuous in an interval $x_0 - h < x < x_0 + h$ and satisfying the initial values $y_i(x_0) = y_i^0$, $i = 1, \ldots, n$, provided that the functions $f_i(x, y_1, y_2, \ldots, y_n)$ are continuous in all variables and satisfy the *Lipschitz's condition*

$$|f_i(x, y_1 + \Delta y_1, y_2 + \Delta y_2, \ldots, y_n + \Delta y_n) - f_i(x, y_1, y_2, \ldots, y_n)| < K(|\Delta y_1| + |\Delta y_2| + \ldots + |\Delta y_n|)$$

for values x, y_i and $y_i + \Delta y_i$ lying in neighbourhood of the initial values, where K is a positive constant.

According to that, the equation $y^{(n)} = f(x, y, y', \ldots, y^{(n-1)})$ has a unique solution continuous together with their derivatives up to the order $n-1$ and satisfying the initial conditions $y = y_0$, $y' = y_0'$, ..., $y^{(n-1)} = y_0^{(n-1)}$, for $x = x_0$, provided that the function $f(x, y, y', \ldots, y^{(n-1)})$ is continuous and satisfies the Lipschitz's condition written above for the functions $f_i(x, y_1, y_2, \ldots, y_n)$.

General solution. The general solution of the equation $y^{(n)} = f(x, y, y', \ldots, y^{(n-1)})$ involves n independent arbitrary constants:

$$y = y(x, C_1, C_2, \ldots, C_n).$$

In the geometrical interpretation, this equation defines a family of integral curves depending on n parameters. Particular integral curves (the graphs of particular solutions) of this family can be obtained by a suitable choice of the arbitrary constants C_1, ..., C_n. If the integral is to satisfy the initial condition given above, then the constants C_1, ..., C_n are determined by the equations

$$y(x_0, C_1, C_2, \ldots, C_n) = y_0,$$

$$\left[\frac{d}{dx} y(x, C_1, C_2, \ldots, C_n)\right]_{x=x_0} = y_0',$$

$$\ldots\ldots\ldots\ldots\ldots\ldots\ldots\ldots$$

$$\left[\frac{d^{n-1}}{dx^{n-1}} y(x, C_1, C_2, \ldots, C_n)\right]_{x=x_0} = y_0^{(n-1)}.$$

If these equations are inconsistent, for any initial values of a certain domain, then the solution is not general in this domain and the arbitrary constants are not independent.

The *general solution* of a system of differential equations (*) involves also n arbitrary constants and can be given in the form solved with respect to the unknown functions:

$$y_1 = F_1(x, C_1, \ldots, C_n),$$

$$y_2 = F_2(x, C_1, \ldots, C_n), \quad \ldots, \quad y_n = F_n(x, C_1, \ldots, C_n)$$

or else in the form solved with respect to the constants:.

$$\varphi_1(x, y_1, \ldots, y_n) = C_1,$$

$$\varphi_2(x, y_1, \ldots, y_n) = C_2, \quad \ldots, \quad \varphi_n(x, y_1, \ldots, y_n) = C_n.$$

In the latter case, each relation $\varphi_i(x, y_1, y_2, \ldots, y_n) = C_i$ defines a *first integral* of the system (*) of equations. A first integral can be defined independently of the general integral as a relation between $x, y_1, y_2, \ldots, y_n$ which becomes a constant value, when $y_1, y_2, \ldots, y_n$ are replaced by any solution of the system of equations. Any first integral of the system (*) satisfies the partial differential equation

$$\frac{\partial\varphi_i}{\partial x} + f_1(x, y_1, \ldots, y_n)\frac{\partial\varphi_i}{\partial y_1} + \ldots + f_n(x, y_1, \ldots, y_n)\frac{\partial\varphi_i}{\partial y_n} = 0,$$

and conversely, each solution $\varphi_i(x, y_1, y_2, \ldots, y_n)$ of this equations is a first integral of the system (*). A system of n independent (see p. 344) first integral of the system (*) of equations constitutes its general integral.

Lowering of the order of an equation. One of the fundamental methods of solving differential equations of order n is a change of variables leading to simpler equations, in particular to equations of a lower degree.

An equation $f(y, y', y'', \ldots, y^{(n)}) = 0$ not involving explicitly the variable x, can be reduced to an equation of order $n-1$ by the substitution $\frac{dy}{dx} = p$, $\frac{d^2y}{dx^2} = p\frac{dp}{dy}$ and so on.

Example. $yy'' - y'^2 = 0$. $p = y'$, $p\frac{dp}{dy} = y''$, hence

$$yp\frac{dp}{dy} - p^2 = 0, \quad y\frac{dp}{dy} - p = 0, \quad p = Cy = \frac{dy}{dx}, \quad y = C_1e^{Cx}$$

(we do not lose a solution when cancelling p, since for $p = 0$ we obtain $y = C_1$ which is included in the obtained general solution for $C = 0$).

An equation $f(x, y', \ldots, y^{(n)}) = 0$ not involving the variable y explicitly can be reduced to an equation of a lower degree by the substitution $y' = p$. If the lowest derivative in the equation is $y^{(k)}$, we substitute $y^{(k)} = p$.

Example. $y'' - xy''' + (y''')^3 = 0$. The substitution $y'' = p$ leads to a Clairaut equation $p - x\frac{dp}{dx} + \left(\frac{dp}{dx}\right)^3 = 0$ with the general solution $p = C_1x + C_1^3$. Hence $y = \frac{1}{6}C_1x^3 - \frac{1}{2}C_1^3x^2 + C_2x + C_3$. The singular solution $p = \frac{2}{9}\sqrt{3}\,x^{3/2}$ of the Clairaut equation gives the singular solution of the original equation:

$$y = \frac{8\sqrt{3}}{315}x^{7/2} + C_1x + C_2.$$

An equation $f(x, y, y', \ldots, y^{(n)})$ where f is a homogeneous function in $y, y', \ldots, y^{(n)}$ (see p. 344) can be reduced to an equation of a lower degree by introducing the new function $z = y'/y$ (that is, $y = e^{\int z\,dx}$).

Example. $yy'' - y'^2 = 0$. $z = \frac{y'}{y}$, $\frac{dz}{dx} = \frac{yy'' - y'^2}{y^2} = 0$, hence $z = C_1$, and

$$\ln y = C_1x + C_2, \quad \text{or} \quad y = Ce^{C_1x}, \quad \text{where} \quad \ln C = C_2.$$

The equation $y^{(n)} = f(x)$. By successive integration we obtain the general solution in the form

$$y = C_1 + C_2x + \ldots + C_nx^{n-1} + \varphi(x),$$

where

$$\varphi(x) = \iint \cdots \int f(x)\,(dx)^n = \frac{1}{(n-1)!}\int_{x_0}^{x} f(t)(x-t)^{n-1}\,dt.$$

In this formula, x_0 is not an additional arbitrary constant. It depends on C_k, since $C_k = \frac{1}{(k-1)!}\, y^{(k-1)}(x_0)$.

Linear equation. *A linear equation* of order n is an equation of the form

$$\text{(L)} \qquad y^{(n)} + a_1 y^{(n-1)} + a_2 y^{(n-2)} + \ldots + a_{n-1} y' + a_n y = F,$$

where a_i $(i = 1, 2, \ldots, n)$ and F (the right member) are functions of x continuous in an interval. If $a_1, a_2, \ldots, a_n$ are constants, then the equation is called *linear equation with constant coefficients.* If $F = 0$, the equation is called *homogeneous*; otherwise it is *non-homogeneous.*

A collection $y_1, y_2, \ldots, y_n$ of solutions of a linear homogeneous differential equation is said to be a *fundamental system of solutions,* if these functions are *linearly independent* in the considered interval, i.e., if no linear combination $C_1y_1 + C_2y_2 + \ldots + C_ny_n$ of them is identically equal to zero in this interval, unless $C_1 = C_2 = \ldots = C_n = 0$. The solutions $y_1, y_2, \ldots, y_n$ of a linear homogeneous equation form a fundamental system of solutions if, and only if, the *Wronskian* of these solutions,

$$W = \begin{vmatrix} y_1 & y_2 & \cdots & y_n \\ y_1' & y_2' & \cdots & y_n' \\ \cdots & \cdots & \cdots & \cdots \\ y_1^{(n-1)} & y_2^{(n-1)} & \cdots & y_n^{(n-1)} \end{vmatrix}$$

is different from zero. For any system of solutions of a linear homogeneous equation, we have *Liouville's formula*

$$W(x) = W(x_0)\, e^{-\int_{x_0}^{x} a_1(x)\,dx}$$

It follows that the Wronskian can be equal to zero only identically, i.e., when $W(x_0)$ is equal to zero. If the functions $y_1, y_2, \ldots, y_n$ constitute a fundamental system of solutions, then $y = C_1y_1 + \ldots + C_ny_n$ is the general solution of the linear homogeneous equation.

If any particular solution y_1 of a linear homogeneous equation is known, then we can lower the order of the equation preserving its linearity by the substitution $u = \frac{d}{dx}\left(\frac{y}{y_1}\right)$.

Superposition theorem. If y_1 and y_2 are solutions of two equations of the form (L) whose right members are, respectively, F_1 and F_2 and whose left members coincide, then the sum $y = y_1 + y_2$ is a solution of the same equation with the right member $F_1 + F_2$.

It follows that in order to obtain the general solution of a linear non-homogeneous equation it is sufficient to add any its particular solution to the general solution of the corresponding homogeneous equation.

Decomposition theorem. If the equation (L) has real coefficients and $F = F_1 + iF_2$, where F_1 and F_2 are real functions, and y_1 and y_2 are solutions of the equation (L), whose right member equals, respectively, F_1 and F_2, then the original equation has the complex solution $y = y_1 + iy_2$.

The solution of a non-homogeneous equation (L) can be found by means of quadratures, if a fundamental system of solutions of the corresponding homogeneous equation is known, by one of the following methods:

The method of variation of constants. Writing the desired solution in the form $y = C_1y_1 + C_2y_2 + \dots + C_ny_n$, we assume that C_1, C_2, ..., C_n are not constants but instead are functions of x. Imposing the following conditions

$$C_1'y_1 + C_2'y_2 + \dots + C_n'y_n = 0,$$
$$C_1'y_1' + C_2'y_2' + \dots + C_n'y_n' = 0,$$
$$\dots\dots\dots\dots\dots\dots\dots$$
$$C_1'y_1^{(n-2)} + C_2'y_2^{(n-2)} + \dots + C_n'y_n^{(n-2)} = 0$$

and substituting y to the equation (L), we obtain

$$C_1'y_1^{(n-1)} + C_2'y_2^{(n-1)} + \dots + C_n'y_n^{(n-1)} = F.$$

Solving this system of linear equations, we find the derivatives C_1', C_2', ..., C_n' and then, by quadratures, we obtain C_1, C_2, ..., C_n.

Cauchy's method. In the general solution $y = C_1y_1 + \dots + C_ny_n$ of a homogeneous equation, we determine such constants that, for $x = x_0$,

$$y = 0, \quad y' = 0, \quad \dots, \quad y^{(n-2)} = 0, \quad y^{(n-1)} = F(a),$$

where a is an arbitrary parameter. Let us denote the obtained solution by $\varphi(x, a)$; then

$$y = \int_{x_0}^{x} \varphi(x, a)\, da$$

will be a particular solution of equation (L); this solution and their derivatives take the value zero for $x = x_0$.

4. Solution of linear differential equations with constant coefficients

Operational notation. Equation (L) (see p. 532) can be written symbolically in the form

$$P_n(D)\,y \equiv (D^n + a_1 D^{n-1} + a_2 D^{n-2} + \ldots + a_{n-1} D + a_n)\,y = F,$$

where D is the differentiation operator:

$$Dy = \frac{dy}{dx}, \quad D^k y = \frac{d^k y}{dx^k}.$$

If the coefficients a_i are constant, the expression $P_n(D)$ is a polynomial in D of degree n (with numbers as coefficients).

Solution of a homogeneous equation $P_n(D)\,y = 0$. To find the general solution we find the roots $r_1, r_2, \ldots, r_n$ of the algebraic equation $P_n(r) = 0$ called the *characteristic equation* (see pp. 164–167). Each root r_i $(i = 1, 2, \ldots, n)$ determines a solution $e^{r_i x}$ of the equation $P_n(D)\,y = 0$. If r_i is a k-fold root, then the functions $xe^{r_i x}, x^2 e^{r_i x}, \ldots, x^{k-1} e^{r_i x}$ are also solutions of this equation. The linear combination of solutions corresponding to all roots r_i taken with their multiplicities, i.e., the function

$$y = (C_1^{(1)} + C_2^{(1)} x + \ldots + C_{k_1}^{(1)} x^{k_1 - 1})\,e^{r_1 x} + \ldots + (C_1^{(i)} + C_2^{(i)} x + \ldots + C_{k_i}^{(i)} x^{k_i - 1})\,e^{r_i x} + \ldots,$$

where k_i is the multiplicity of the root r_i $(k_1 + k_2 + \ldots + k_i + \ldots = n)$, constitutes the general solution of the homogeneous equation.

If some roots of the characteristic equation [1] are complex then they are pairwise conjugate (e.g., $r_1 = \alpha + i\beta$, $r_2 = \alpha - i\beta$) and then, at the corresponding terms of the general solution we put the functions $e^{\alpha x}\cos\beta x$ and $e^{\alpha x}\sin\beta x$ instead of $e^{r_1 x}$ and $e^{r_2 x}$. We obtain expressions of the form $C_1\cos\beta x + C_2\sin\beta x$ which can also be written in the form $A\cos(\beta x + \varphi)$, where A and φ are arbitrary constants.

Example. $y^{(6)} + y^{(4)} - y'' - y = 0$.

The characteristic equation $r^6 + r^4 - r^2 - 1 = 0$ has the roots $r_1 = 1$, $r_2 = -1$, $r_3 = r_4 = i$, $r_5 = r_6 = -i$. The general solution has the form

$$y = C_1 e^x + C_2 e^{-x} + (C_3 + C_4 x)\cos x + (C_5 + C_6 x)\sin x$$

---

[1] The coefficients a_k are assumed to be real.

or

$$y = C_1 e^x + C_2 e^{-x} + A_1 \cos(x + \varphi_1) + x A_2 \cos(x + \varphi_2).$$

In vibration theory and in other applications it is important to know whether any solution of the given homogeneous linear equation with constant coefficient tends to zero, when $n \to \infty$. This is the case, if the real parts of all roots of the characteristic equation are negative.

Hurwitz's theorem. All roots of the equation

$$a_0 + a_1 x + a_2 x^2 + \ldots + a_n x^n = 0, \quad \text{where} \quad a_0 > 0,$$

have negative real parts if, and only if, the determinants

$$D_1 = a_1, \quad D_2 = \begin{vmatrix} a_1 & a_0 \\ a_3 & a_2 \end{vmatrix}, \quad \ldots, \quad D_n = \begin{vmatrix} a_1 & a_0 & 0 & \ldots & 0 \\ a_3 & a_2 & a_1 & \ldots & 0 \\ \ldots & \ldots & \ldots & \ldots & \ldots \\ a_{2n-1} & a_{2n-2} & a_{2n-3} & \ldots & a_n \end{vmatrix},$$

where $a_m = 0$ for $m > n$, are positive.

The solution of a non-homogeneous equation with constant coefficients can be found by use the method of variation of constants or by Cauchy's method (see p. 533). Another one is the operational method (see p. 542). The easiest to solve are equations whose right members have a special form.

Special form of the right member of a non-homogeneous equation. In some cases a particular solution of a non-homogeneous equation $P_n(D)y = F(x)$ can easily be found in an algebraic way.

If $F(x) = Ae^{kx}$ and $P_n(k) \neq 0$, then $y = Ae^{kx}/P_n(k)$ is a particular solution. If k is an m-fold root of the characteristic equation, i.e., if $P_n(k) = P_n'(k) = \ldots = P_n^{(m-1)}(k) = 0$, then $y = Ax^m e^{kx}/P_n^{(m)}(k)$ is a particular solution.

Using the decomposition theorem (p. 533), these formulas can also be used in the cases when $F(x) = Ae^{kx} \cos \omega x$ or $Ae^{kx} \sin \omega x$. The corresponding particular solution can be obtained as the real and the imaginary part of the solution of the same equation, when the right member is

$$F(x) = Ae^{kx}(\cos \omega x + i \sin \omega x) = Ae^{(k+i\omega)x}.$$

Examples. (1) The equation $y'' - 6y' + 8y = e^{2x}$ has the particular solution $y = -\frac{1}{2} x e^{2x}$, since $P(D) = D^2 - 6D + 8$, $P(2) = 0$, $P'(D) = 2D - 6$, $P'(2) = 2 \cdot 2 - 6 = -2$.

(2) For the equation $y'' + y' + y = e^x \sin x$, a particular solution

$y_1 = \frac{1}{13} e^x (2 \sin x - 3 \cos x)$ can be obtained as the imaginary part of the solution

$$y = \frac{e^{(1+i)x}}{(1+i)^2 + (1+i) + 1} = \frac{e^x(\cos x + i \sin x)}{2 + 3i}$$

of the equation $(D^2 + D + 1)y = e^{(1+i)x}$.

If the right member $F(x)$ has the form $Q_p(x)\, e^{kx}$, where $Q_p(x)$ is a polynomial of degree p, then a particular solution of the same form, $y = R(x)e^{kx}$, can always be found. If k is an m-fold root of the characteristic equation, then $R(x)$ is a polynomial of degree p multiplied by x^m. If we write this solution with undetermined coefficient and require it to satisfy the given equation, we obtain a linear algebraic equation for the undetermined coefficients which can easily be solved [1].

Example. $y^{(4)} + 2y''' + y'' = 6x + 2x \sin x$.

The characteristic equation has the roots $k_1 = k_2 = 0$, $k_3 = k_4 = -1$. By the superposition theorem (see p. 533), we can seek for the solutions corresponding to either summand of the right member of the equation. Putting $y_1 = x^2(ax + b)$ and substituting to the given equation, we obtain $12a + 2b + 6ax = 6x$, hence $a = 1$, $b = -6$. Similarly, for the second summand, we put $y_2 = (cx + d) \times \sin x + (fx + g) \cos x$, which gives $(2g + 2f - 6c + 2fx) \sin x - (2c + 2d + 6f + 2cx) \cos x = 2x \sin x$. Hence $c = 0$, $d = -3$, $f = 1$. $g = -1$. Finally, the general solution is

$$y = c_1 + c_2 x - 6x^2 + x^3 + (c_3 x + c_4) e^{-x} - 3 \sin x + (x - 1) \cos x.$$

Euler's equation of the form

$$\sum_{k=0}^{n} a_k (cx + d)^k y^{(k)} = F(x)$$

reduces to a linear equation with constant coefficients by the substitution $cx + d = e^t$.

Example. The equation $x^2 y'' - 5xy' + 8y = x^2$ can be reduced to the equation $\frac{d^2y}{dt^2} - 6\frac{dy}{dt} + 8y = e^{2t}$ considered on p. 535 by the substitution $x = e^t$. The general solution is

$$y = C_1 e^{2t} + C_2 e^{4t} - \tfrac{1}{2} t e^{2t} = C_1 x^2 + C_2 x^4 - \tfrac{1}{2} x^2 \ln x.$$

---

[1] This method can also be applied in the case, when $F(x) = Q_p(x)$, i.e., when $k = 0$, and also in the cases, when $F(x) = Q_p(x) e^{rx} \cos \omega x$ or $F(x) = Q_p(x)\, e^{rx} \sin \omega x$ which corresponds to the value $k = r \pm i\omega$. In the latter case we should seek for a solution in the form

$$y = x^m e^{rx} [M_p(x) \cos \omega x + N_p(x) \sin \omega x].$$

5. Systems of linear differential equations with constant coefficients

Systems of linear normal equations of the first order. The simplest case of a system of linear equations is a so-called *normal system*:

$$\text{(N)}\qquad \begin{aligned} y_1' &= a_{11}y_1 + a_{12}y_2 + \ldots + a_{1n}y_n, \\ y_2' &= a_{21}y_1 + a_{22}y_2 + \ldots + a_{2n}y_n, \\ &\cdots\cdots\cdots\cdots\cdots\cdots \\ y_n' &= a_{n1}y_1 + a_{n2}y_2 + \ldots + a_{nn}y_n. \end{aligned}$$

To find a general solution of such a system of equations, we should first solve the following algebraic equation called the *characteristic equation*:

$$\begin{vmatrix} a_{11}-r & a_{12} & \ldots & a_{1n} \\ a_{21} & a_{22}-r & \ldots & a_{2n} \\ \ldots & \ldots & \ldots & \ldots \\ a_{n1} & a_{n2} & \ldots & a_{nn}-r \end{vmatrix} = 0.$$

To each simple root r_i of the characteristic equation there corresponds a system of particular solutions

$$(*)\qquad y_1 = A_1 e^{r_i x}, \quad y_2 = A_2 e^{r_i x}, \quad \ldots, \quad y_n = A_n e^{r_i x},$$

where the coefficients A_k $(k = 1, 2, \ldots, n)$ are determined by the following system of linear homogeneous equations

$$\begin{aligned} (a_{11} - r_i) A_1 + a_{12}A_2 + \ldots + a_{1n}A_n &= 0, \\ \cdots\cdots\cdots\cdots\cdots\cdots \\ a_{n1}A_1 + a_{n2}A_2 + \ldots + (a_{nn} - r_i) A_n &= 0. \end{aligned}$$

Since this system of equations determines only a ratio of the coefficients A_k (see p. 179), the system of particular solutions obtained in this way will involve, for each r_i, one arbitrary constant.

If all roots of the characteristic equation are different, then the sum of such particular solutions will involve n independent arbitrary constants and will constitute the general solution of the system of equations.

If r_i is an m-fold root of the characteristic equation, then it determines a system of particular solutions of the form

$$y_1 = A_1(x)\, e^{r_i x}, \quad y_2 = A_2(x)\, e^{r_i x}, \quad \ldots, \quad y_n = A_n(x)\, e^{r_i x},$$

where $A_1(x)$, $A_2(x)$, ..., $A_n(x)$ are polynomials of degree not greater than $m-1$. By substituting these expressions with undetermined coefficients to the given system of equations and equating coefficients of the corresponding terms, we obtain a set of equations which enable us to express the undetermined coefficients by any m of the coefficients remaining arbitrary constants. In some cases the degree of the polynomial can be less than $m-1$. In particular, if the system (N) of equations is symmetric (i.e., $a_{ik} = a_{ki}$), we can take $A_i(x) = \text{const}$.

Example. For the system of equations

$$y_1' = 2y_1 + 2y_2 - y_3, \quad y_2' = -2y_1 + 4y_2 + y_3, \quad y_3' = -3y_1 + 8y_2 + 2y_3$$

the characteristic equation is

$$\begin{vmatrix} 2-r & 2 & -1 \\ -2 & 4-r & 1 \\ -3 & 8 & 2-r \end{vmatrix} = -(r-6)(r-1)^2 = 0.$$

For the simple root $r_1 = 6$, we obtain

$$-4A_1 + 2A_2 - A_3 = 0, \quad -2A_1 - 2A_2 + A_3 = 0,$$
$$-3A_1 + 8A_2 - 4A_3 = 0,$$

hence $A_1 = 0$, $A_2 = \frac{1}{2}$, $A_3 = C_1$ and $y_1 = 0$, $y_2 = C_1e^{6x}$, $y_3 = 2C_1e^{6x}$.

For the double root $r_2 = 1$, we obtain

$$y_1 = (P_1x + Q_1)e^x, \quad y_2 = (P_2x + Q_2)e^x, \quad y_3 = (P_3x + Q_3)e^x.$$

Substituting this function in the equation and dividing by e^x, we obtain

$$P_1x + (P_1 + Q_1) = (2P_1 + 2P_2 - P_3)x + (2Q_1 + 2Q_2 - Q_3),$$
$$P_2x + (P_2 + Q_2) = (-2P_1 + 4P_2 + P_3)x + (-2Q_1 + 4Q_2 + Q_3),$$
$$P_3x + (P_3 + Q_3) = (-3P_1 + 8P_2 + 2P_3)x + (-3Q_1 + 8Q_2 + 2Q_3),$$

hence

$$P_1 = 5C_2, \quad P_2 = C_2, \quad P_3 = 7C_2, \quad Q_1 = 5C_3 - 6C_2, \quad Q_2 = C_3,$$
$$Q_3 = 7C_3 - 11C_2.$$

The general solution of the system is

$$y_1 = (5C_2x + 5C_3 - 6C_2)e^x,$$
$$y_2 = C_1e^{6x} + (C_2x + C_3)e^x,$$
$$y_3 = 2C_1e^{6x} + (7C_2x + 7C_3 - 11C_2)e^x.$$

Systems of linear homogeneous equations of the first order. The general form of a *system of linear homogeneous equations of* *the first order with constant coefficients is*

$$\sum_{k=1}^{n} a_{ik}y'_k + \sum_{k=1}^{n} b_{ik}y_k = 0, \quad \text{where } i = 1, 2, \ldots, n.$$

If the determinant $|a_{ik}|$ (1) is not equal to zero, then this system can be reduced to a normal system. However, the solution can be obtained directly in the same way as for a normal system. The characteristic equation will have the form $|a_{ik}r + b_{ik}| = 0$ and the coefficients A_i in the solution (*) corresponding to a simple root r_j can be determined, in this case, from the equations

$$\sum_{k=1}^{n} (a_{ik}r_j + b_{ik}) A_k = 0, \text{ where } i = 1, 2, \ldots, n.$$

Besides the method is the same as in the case of a normal system.

The case $|a_{ik}| = 0$ requires additional consideration.

Example. $5y'_1 + 4y_1 - 2y'_2 - y_2 = 0$, $y'_1 + 8y_1 - 3y_2 = 0$.

The characteristic equation

$$\begin{vmatrix} 5r+4 & -2r-1 \\ r+8 & -3 \end{vmatrix} = 2r^2 + 2r - 4 = 0$$

has the roots $r_1 = 1$ and $r_2 = -2$. We find A_1 and A_2 for $r_1 = 1$: $9A_1 - 3A_2 = 0$, hence $A_2 = 3A_1 = 3C_1$. Similarly for $r_2 = -2$ we obtain $A_4 = 2A_3 = 2C_2$. The general solution

$$y_1 = C_1e^x + C_2e^{-2x}, \quad y_2 = 3C_1e^x + 2C_2e^{-2x}.$$

Systems of non-homogeneous equations of the first order. The general form of a *system of linear non-homogeneous equations of the first order with constant coefficients is*

$$\sum_{k=1}^{n} a_{ik}y'_k + \sum_{k=1}^{n} b_{ik}y_k = F_i(x), \quad \text{where} \quad i = 1,2, \ldots, n.$$

Superposition theorem. If $y_j^{(1)}$ and $y_j^{(2)}$, where $j = 1, 2, \ldots, n$, are solutions of two systems of non-linear equations whose left members are the same and whose right members are, respectively, $F_i^{(1)}$ and $F_i^{(2)}$, then the function $y_j = y_j^{(1)} + y_j^{(2)}$ is a solution of the same system of equations, but with the right member $F_i(x) = F_i^{(1)}(x) + F_i^{(2)}(x)$.

It follows that in order to obtain the general solution of a system of non-homogeneous equations it is sufficient to add its particular solution to the general solution of the corresponding system of homogeneous equations.

(1) This is an abbreviation for a determinant with elements a_{ik}.

A particular solution of a system of non-homogeneous equations can be found by use of the *method of variation of constants*:

We substitute the general solution of the system of homogeneous equations to the non-homogeneous system replacing the arbitrary constants $C_1, C_2, \ldots, C_n$ with unknown functions $C_1(x)$, $C_2(x), \ldots, C_n(x)$; in this process, terms involving the derivatives of the new unknown functions $C_k(x)$ will appear in the expressions for y_k'. After substituting to the given system of equations, only these additional terms will remain in the left members of the equations while the other ones will be cancelled, since $y_1, y_2, \ldots, y_n$ constitutes a solution of the system of homogeneous equations. Thus we shall obtain a non-homogeneous system of linear algebraic equations for $C_k'(x)$. Solving this system and integrating n times, we obtain the functions $C_1(x), C_2(x), \ldots, C_n(x)$. Finally, replacing the constants $C_1, C_2, \ldots, C_n$ with the functions $C_1(x), C_2(x), \ldots, C_n(x)$ in the solution of the system of homogeneous equations, we obtain the desired particular solution of the system of non-homogeneous equations.

Example. $5y_1' + 4y_1 - 2y_2' - y_2 = e^{-x}$, $y_1' + 8y_1 - 3y_2 = 5e^{-x}$.

The general solution of the homogeneous system (see p. 538) is $y_1 = C_1e^x + C_2e^{-2x}$, $y_2 = 3C_1e^x + 2C_2e^{-2x}$. Substituting this solution to the given system of non-homogeneous equations and regarding C_1 and C_2 as functions of x, we obtain

$$5C_1'e^x + 5C_2'e^{-2x} - 6C_1'e^x - 4C_2'e^{-2x} = e^{-x}, \qquad C_1'e^x + C_2'e^{-2x} = 5e^{-x},$$

or

$$C_2'e^{-2x} - C_1'e^x = e^{-x}, \qquad C_1'e^x + C_2'e^{-2x} = 5e^{-x}.$$

Hence $2C_1'e^x = 4e^{-x}$, $C_1 = -e^{-2x} + c_1$ and $2C_2'e^{-2x} = 6e^{-x}$, $C_2 = 3e^x + c_2$, where c_1 and c_2 are arbitrary constants. Taking $c_1 = 0$ and $c_2 = 0$ (for we want a particular solution), we obtain $y_1 = 2e^{-x}$, $y_2 = 3e^{-x}$. The general solution is

$$y_1 = 2e^{-x} + C_1e^x + C_2e^{-2x}, \qquad y_2 = 3e^{-x} + 3C_1e^x + 2C_2e^{-2x}.$$

In the case when the right members have a special form $Q_p(x)e^{kx}$, the method of undetermined coefficients can also be applied, as for one equation of order n (see p. 535).

Systems of equations of the second order. The methods described above can be applied to linear equations of higher order. In particular, for the system of equations

$$\sum_{k=1}^{n} a_{ik}y_k'' + \sum_{k=1}^{n} b_{ik}y_k' + \sum_{k=1}^{n} c_{ik}y_k = 0, \quad \text{where} \quad i = 1, 2, \ldots, n,$$

we can also seek for solutions in the form $y_i = A_i e^{r_i x}$, where r_i are the roots of the characteristic equation $a_{ik}r^2 + b_{ik}r + c_{ik} = 0$ and the coefficients A_i are determined by a system of algebraic linear homogeneous equations.

6. Operational method of solution of differential equations

Transform of a function. If $\varphi(t)$ is a function such that $\varphi(t) = 0$, for $t < 0$, and $|\varphi(t)| < Me^{at}$, where M and a are positive constants, for $t > 0$, then the *Carson-Heaviside transform* of $\varphi(t)$ is defined as a function of a complex variable p by the formula

$$(*) \qquad f(p) = p\int_0^\infty e^{-pt}\varphi(t)\,dt;$$

the function $\varphi(t)$ is called the *original function*.

Using the theory of complex functions, we can find a *formula for the inverse transform* which determines uniquely the original function $\varphi(t)$, when its transform $f(p)$ is known:

$$\varphi(t) = \frac{1}{2\pi i}\lim_{r\to\infty}\int_{s-ir}^{s+ir} e^{pt}\frac{f(p)}{p}\,dp,$$

s is chosen so that all singular points of the integrand lie on the left side of the straight line re $p = s$ (for integration of functions of a complex variable see p. 605).

We shall write the relation $(*)$ in the form $f(p) \div \varphi(t)$ [1].

Some simple functions and their transforms are given in the table on pp. 545, 546.

Fundamental properties of transforms. The following formulas can be obtained from formula $(*)$:

$$\frac{d\varphi}{dt} \div pf(p) - \varphi(0)p, \qquad \frac{d^2\varphi}{dt^2} \div p^2f(p) - p^2\varphi(0) - p\varphi'(0), \quad \dots$$

$$\dots, \quad \frac{d^n\varphi}{dt^n} \div p^nf(p) - p^n\varphi(0) - p^{n-1}\varphi'(0) - \dots - p\varphi^{(n-1)}(0) \text{ [2]},$$

$$\int_0^t \varphi(t)\,dt \div \frac{1}{p}f(p), \qquad \varphi(at) \div f(p/a) \qquad (a = \text{const} > 0).$$

---

[1] The function $f(p)/p$ is called the *Laplace transform* of $\varphi(t)$.

[2] $\frac{d^n\varphi}{dt^n}$ is assumed to satisfy the conditions introduced above under which the transform is defined.

If $\varphi_1(t) \div f_1(p)$, $\varphi_2(t) \div f_2(p)$, then

$$a_1\varphi_1(t) + a_2\varphi_2(t) \div a_1 f_1(p) + a_2 f_2(p).$$

Shifting theorem. If $\varphi(t) \div f(p)$, then

$$e^{-at}\varphi(t) \div \frac{p}{p+a} f(p+a).$$

Delaying theorem. If $\varphi(t) \div f(p)$ and $\lambda > 0$, then

$$e^{-\lambda p} f(p) \div \begin{cases} \varphi(t-\lambda) & \text{for} \quad t > \lambda, \\ 0 & \text{for} \quad t < \lambda. \end{cases}$$

Borel's theorem. If $\varphi_1(t) \div f_1(p)$, $\varphi_2(t) \div f_2(p)$, then

$$\int_0^t \varphi_1(t-\tau)\,\varphi_2(\tau)\,d\tau \div \frac{1}{p} f_1(p)\,f_2(p).$$

Impulse function. The original function of the transform $f(p) = p$ is the so-called *Dirac delta function* $\delta(t)$

$$\delta(t) = \begin{cases} 0 & \text{for} \quad t \neq 0, \\ +\infty & \text{for} \quad t = 0, \end{cases}$$

and such that

$$\int_{-\infty}^{+\infty} \delta(t)\,dt = 1.$$

This function can also be defined otherwise, e.g.:

$$\delta(t) = \lim_{h \to 0} f(t, h), \quad \text{where} \quad f(t, h) = \begin{cases} 1/h & \text{for } 0 < t < h, \\ 0 & \text{for } t < 0 \text{ and } t > h. \end{cases}$$

The delta function is used to represent mechanical or electric impulses acting for a very short time (see example (3) on p. 544).

Operational method. In the *operational method* of solving ordinary differential equations we pass from the equation for the unknown function to an equation for its transform (so-called *auxiliary equation*). This is no longer a differential equation but an algebraic one. Having found the transform, we use it to find the desired function. The main difficulty of the operational method lies not in solving the equation, but in passing from the function to its transform and conversely.

Linear equation with constant coefficients

$$L_n(D)y \equiv (D^n + a_1 D^{n-1} + a_2 D^{n-2} + \ldots + a_{n-1} D + a_n)y = F(t).$$

where D is the differentiation operator with respect to the independent variable t.

Let $y(t) \risingdotseq \bar{y}(p)$, $F(t) \risingdotseq \bar{F}(p)$. By the formulas of p. 542 we obtain the following auxiliary equation

$$(**) \qquad L_n(p)\,\bar{y} = \bar{F}(p) + \left(p^n y_0 + p^{n-1} y_0' + \ldots + p y_0^{(n-1)}\right) + \\ + a_1\left(p^{n-1} y_0 + p^{n-2} y_0' + \ldots + p y_0^{(n-2)}\right) + \\ + \ldots + a_{n-2}(p^2 y_0 + p y_0') + a_{n-1} p y_0 \equiv \bar{F}(p) + M(p),$$

where $y_0, y_0', \ldots, y_0^{(n-1)}$ are initial values of the function y and its derivatives for $t = 0$. In the simplest case, when $y_0 = y_0' = y_0^{(n-1)} = 0$, we have $M(p) \equiv 0$; a solution corresponding to these initial conditions is called a *normal solution*. If follows from (**) that

$$\bar{y} = \frac{\bar{F}(p) + M(p)}{L_n(p)}.$$

In many cases, y can be found from a resolution into partial fractions (see pp. 151, 152) and from formulas 2-9 in the table on pp. 545, 546; since these formulas contain the factor p in the numerator, hence we usually resolve the fractions with denominators $pL_n(p)$ and multiply the result by p. In the simplest case, when all the roots p_k of the denominator $L_n(p)$ are different and the numerator is a polynomial $P_m(p)$ of degree not greater than n, this process leads to *Heaviside's formula*

$$\frac{P_m(p)}{L_n(p)} \fallingdotseq \frac{P_m(0)}{L_n(0)} + \sum_{k=1}^{n} \left[\frac{P_m(p)}{pL_n'(p)}\right]_{p=p_k} e^{p_k t}.$$

If $\bar{F}(p)/L_n(p)$ is not a rational function, then we resolve the function $1/pL_k(p)$ into partial fractions and use the formulas

$$\frac{\bar{F}(p)}{p-a} \fallingdotseq e^{at} \int_0^t F(x)\, e^{-ax}\, dx$$

or

$$\frac{\bar{F}(p)}{(p-a)^m} \fallingdotseq \frac{e^{at}}{(m-1)!} \int_0^t F(x)\, e^{-ax}\, (t-x)^{m-1}\, dx.$$

If the equation $L_n(p) = 0$ has complex roots, the latter formulas can lead to complex quantities in intermediate calculations, but the final result can always be written in real form.

Examples. (1) Find the normal solution of the equation $y''' - y'' - y' + y = t$. We have

$$L(p) = (p+1)(p-1)^2, \quad M(p) = 0, \quad \bar{F}(p) = \frac{1}{p};$$

$$\bar{y} = \frac{1}{p(p+1)(p-1)^2} = \frac{p}{p^2(p+1)(p-1)^2}$$

$$= \frac{1}{p} + 1 + \frac{1}{4}\cdot\frac{p}{p+1} - \frac{5}{4}\cdot\frac{p}{p-1} + \frac{1}{2}\cdot\frac{p}{(p-1)^2}.$$

Hence, by formulas 1, 2 and 9 on pp. 545, 546 we obtain $y = t + 1 + \frac{1}{4}e^{-t} + (\frac{1}{2}t - \frac{5}{4})e^t$.

(2) Find the general solution of the equation $y'' + m^2y = a \sin mt$. We have

$$L(p) = p^2 + m^2, \quad \bar{F}(p) = \frac{amp}{m^2 + p^2}, \quad M(p) = p^2y_0 + py_0',$$

$$y = \frac{amp}{(p^2+m^2)^2} + \frac{p^2y_0 + py_0'}{p^2 + m^2}.$$

In order to use the formulas of the table on pp. 545, 546, we transform the first summand to the form

$$A\frac{p(p^2-m^2)}{(p^2+m^2)^2} + B\frac{pm}{p^2+m^2}.$$

Having found A and B by the method of undetermined coefficients, we obtain the solution by formulas 3, 4 and 8 on pp. 545, 546

$$y = \left(y_0 - \frac{a}{2m}t\right)\cos mt + \frac{a + 2my_0'}{2m^2}\sin mt.$$

(3) Find the law of motion of a particle m under an impulsive force A applied at the moment $t = 0$. The initial coordinate $x_0 = 0$, the initial velocity $x_0' = 0$.

Equation of the motion $m\dfrac{d^2x}{dt^2} = A\delta(t)$. Auxiliary equation $mp^2\bar{x} = Ap$. Hence $\bar{x} = A/mp$ or $x = At/m$.

Systems of linear equations with constant coefficients. After replacing each equation of the system by the corresponding auxiliary equation, as has been shown for one equation, we solve a system of algebraic linear equations with respect to the transform of the unknown functions, provided that the determinant of the system is different from zero [1]. To find original functions

---

[1] The case when the determinant is zero requires additional consideration.

from their transform, we usually use a resolution into partial fractions, as in the case of one equation.

Example. $(5D+4)y_1-(2D+1)y_2=e^{-x}$, $(D+8)y_1-3y_2=5e^{-x}$ with the initial condition $y_1=y_{10}$, $y_2=y_{20}$, for $x=0$.

Auxiliary equations

$$(5p+4)\bar{y}_1-(2p+1)\bar{y}_2=\frac{p}{p+1}+5py_{10}-2py_{20},$$

$$(p+8)\bar{y}_1-3\bar{y}_2=\frac{5p}{p+1}+py_{10}.$$

Solving these equations for $\bar{y}_1$ and $\bar{y}_2$, we obtain, after resolving into partial fractions,

$$\bar{y}_1=\frac{2p}{p+1}+(3y_{10}-y_{20}-3)\frac{p}{p+2}+(-2y_{10}+y_{20}+1)\frac{p}{p-1},$$

$$\bar{y}_2=\frac{3p}{p+1}+(6y_{10}-2y_{20}-6)\frac{p}{p+2}+(-6y_{10}+3y_{20}+3)\frac{p}{p-1},$$

hence

$$y_1=2e^{-x}+(3y_{10}-y_{20}-3)e^{-2x}+(1-2y_{10}+y_{20})e^{x},$$

$$y_2=3e^{-x}+(6y_{10}-2y_{20}-6)e^{-2x}+(3-6y_{10}+3y_{20})e^{x}.$$

For an application of geometric methods to solution of partial differential equations see p. 580.

Table of Carson-Heaviside's transforms of function

Number	$\varphi(t)$ $(t>0)$(¹)	$f(p)=p\int_0^\infty e^{-pt}\varphi(t)\,dt$
1	$\frac{t^n}{\Gamma(n+1)}$	$\frac{1}{p^n}$
2	e^{-at}	$\frac{p}{p+a}$
3	$\sin kt$	$\frac{pk}{p^2+k^2}$
4	$\cos kt$	$\frac{p^2}{p^2+k^2}$
5	$e^{-at}\sin kt$	$\frac{pk}{(p+a)^2+k^2}$
6	$e^{-at}\cos kt$	$\frac{p(p+a)}{(p+a)^2+k^2}$

(¹) We always assume $\varphi(t)=0$ for $t<0$.

Table of Carson-Heaviside's transforms of function

Number	$\varphi(t)$ $(t > 0)$	$f(p) = p \int_0^\infty e^{-pt} \varphi(t)\, dt$
7	$t \sin kt$	$\frac{2kp^2}{(p^2 + k^2)^2}$
8	$t \cos kt$	$\frac{p(p^2 - k^2)}{(p^2 + k^2)^2}$
9	$e^{-at} \frac{t^n}{n!}$	$\frac{p}{(p + a)^{n+1}}$
10	$\frac{(2t)^n}{1 \cdot 3 \cdot 5 \ldots (2n - 1) \sqrt{\pi t}}$	$\frac{\sqrt{p}}{p^n}$ (n natural)
11	$\frac{1}{\sqrt{\pi t}} e^{-a^2/4t}$	$\sqrt{p}\, e^{-a\sqrt{p}}$ $(a > 0)$
12	$\frac{a}{2\sqrt{\pi t^3}} e^{-a^2/4t}$	$p e^{-a\sqrt{p}}$
13	$1 - \Phi\left(\frac{a}{\sqrt{2t}}\right)$ $(a > 0)$ (¹)	$e^{-a\sqrt{p}}$
14	$J_0(t)$ (²)	$\frac{p}{\sqrt{1 + p^2}}$
15	$I_0(t)$ (²)	$\frac{p}{\sqrt{p^2 - 1}}$
16	$\int_t^\infty \frac{e^{-x}}{x} dx$	$\ln (1 + p)$

(¹) For a definition of the function $\Phi(x)$ see p. 744.
(²) For Bessel's function $J(x)$ and $I(x)$ see p. 549.

7. Linear equations of the second order

General methods.

The equation $y'' + p(x)y' + q(x)y = F(x)$. The general solution of this equation in the case, when $F(x) = 0$ (homogeneous equation) is of the form $y = C_1 y_1 + C_2 y_2$, where y_1 and y_2 are two linearly independent particular solutions (see p. 532). If one particular solution y_1 is known, then another particular solution can be obtained from the formula

$$y_2 = A y_1 \int \frac{e^{-\int p\,dx}}{y_1^2}\, dx, \qquad (*)$$

where A is an arbitrary constant; this formula is a corollary from

Liouville's formula (see p. 532). A particular solution of the non-homogeneous equation can be obtained, in this case, from the formula

$$y = \frac{1}{A} \int_{x_0}^{x} F(\xi) \, e^{\int p(\xi) d\xi} \left(y_2(x) \, y_1(\xi) - y_1(x) \, y_2(\xi) \right) d\xi,$$

where y_1 and y_2 are the particular solutions of the homogeneous equation mentioned above.

A particular solution of the non-homogeneous equation can also be found by use of the method of variation of constants (see p. 533).

If the functions $s(x)$, $p(x)$, $q(x)$, and $F(x)$ in the equation $s(x)y'' + p(x)y' + q(x)y = F(x)$ are polynomials or can be expanded into power series with respect to $x - x_0$ convergent in an interval, and $s(x_0) \neq 0$, then the solution of this equation can also be written as series of powers of $x - x_0$ convergent in the same interval. These solutions can be found by the method of undetermined coefficients. We write the desired solution as a series

$$y = a_0 + a_1(x - x_0) + a_2(x - x_0)^2 + \ldots$$

substitute it to the equation and, equating coefficients of equal powers of $x - x_0$, we obtain equations for the coefficients a_0, a_1, a_2, ...

Example. $y'' + xy = 0$.
Substituting

$$y = a_0 + a_1 x + a_2 x^2 + \ldots, \qquad y' = a_1 + 2a_2 x + 3a_3 x^2 + \ldots,$$

$$y'' = 2a_2 + 6a_3 x + \ldots,$$

we obtain $2a_2 = 0$, $6a_3 + a_0 = 0$, ..., $n(n-1)a_n + a_{n-3} = 0$, ... Solving successively these equations, we obtain $a_2 = 0$, $a_3 = -\frac{a_0}{2 \cdot 3}$, $a_4 = -\frac{a_1}{3 \cdot 4}$, $a_5 = 0$, ..., hence

$$y = a_0\left(1 - \frac{x^3}{2 \cdot 3} + \frac{x^6}{2 \cdot 3 \cdot 5 \cdot 6} - \ldots\right) + a_1\left(x - \frac{x^4}{3 \cdot 4} + \frac{x^7}{3 \cdot 4 \cdot 6 \cdot 7} - \ldots\right),$$

where a_0 and a_1 are arbitrary constants.

The equation $x^2y'' + xp(x)y' + q(x)y = 0$. If the functions $p(x)$ and $q(x)$ can be expanded into power series in x convergent in an interval, then the method of undetermined coefficients can be used to find a solution of the form

$$y = x^r(a_0 + a_1x + a_2x^2 + \dots);$$

we determine the exponent r from the following *defining equation*

$$r(r-1) + p(0)r + q(0) = 0.$$

If the roots of this equation are different and their difference is not an integral number, then we obtain two independent solutions of our equation. Otherwise, the method of undetermined coefficients gives only one solution.

The formula (*) given on p. 546 can be used directly to find another solution or else to determine the form in which this solution could be found by the method of undetermined coefficients.

Example. The method of undetermined coefficients enables us to find only one solution of Bessel's equation (see below) for an integral n; this solution has the form

$$y_1 = \sum_{k=0}^{\infty} a_k x^{n+2k}, \quad \text{where} \quad a_0 \neq 0,$$

and coincides with the function $J_n(x)$ up to a constant factor. Since, in this case, $e^{-\int p\,dx} = 1/x$, the second solution according to formula (*) has the form

$$y_2 = Ay_1 \int \frac{dx}{x \cdot x^{2n}\left(\sum a_k x^{2k}\right)^2}$$

$$= Ay_1 \int \frac{\sum c_k x^{2k}}{x^{2n+1}}\,dx = By_1 \ln x + x^{-n} \sum_{k=0}^{\infty} d_k x^{2k}.$$

Evaluation of successive coefficients c_k and d_k from a_k is troublesome, but the last expression can be used to find the solution by the method of undetermined coefficients (it is evident that the expansion of $Y_n(x)$ in a series has such a form, see below).

Bessel's equation $x^2y'' + xy' + (x^2 - n^2)y = 0$. Defining equation $r(r-1) + r - n^2 \equiv r^2 - n^2 = 0$, hence $r = \pm n$. Substituting $y = x^n(a_0 + a_1x + \dots)$ to the equation and equating to zero the coefficient of x^{n+k}, we have $k(2n+k)a_k + a_{k-2} = 0$. For $k = 1$, we have $(2n+1)a_1 = 0$. For k equal to 2, 3, ... we obtain $a_{2m+1} = 0$ $(m = 1, 2, \dots)$, hence

$$a_2 = -\frac{a_0}{2(2n+2)}, \quad a_4 = \frac{a_0}{2 \cdot 4(2n+2)(2n+4)}, \quad \dots,$$

where a_0 is an arbitrary constant.

Bessel's functions. The series obtained above for $a_0 = \dfrac{1}{2^n \Gamma(n+1)}$ [1] is called the *Bessel's function* (or *cylindrical function*) *of the first kind of index* n:

$$J_n(x) = \frac{x^n}{2^n \Gamma(n+1)}\left(1 - \frac{x^2}{2(2n+2)} + \frac{x^4}{2\cdot 4(2n+2)(2n+4)} - \dots\right)$$

$$= \sum_{k=0}^{\infty} \frac{(-1)^k (\frac{1}{2}x)^{n+2k}}{k!\,\Gamma(n+k+1)}.$$

The graphs of the functions J_0 and J_1 are shown in Fig. 357.

The general solution of Bessel's equation, in the case when n is not an integral number, has the form $y = C_1 J_n(x) + C_2 J_{-n}(x)$, where $J_{-n}(x)$ is defined by the series obtained from the series defining $J_n(x)$ by replacing n with $-n$. If n is an integer, then

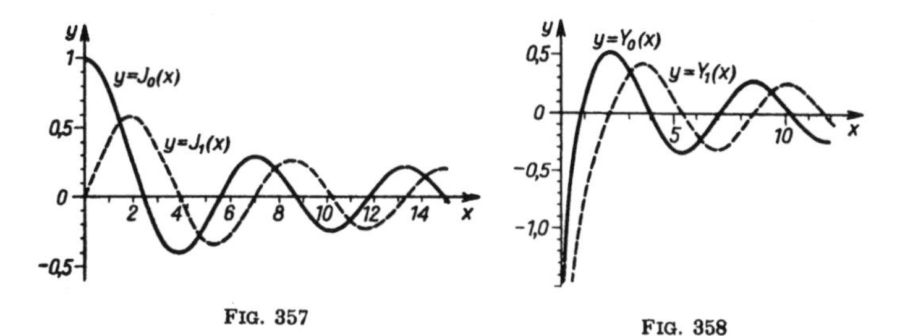

FIG. 357

FIG. 358

$J_{-n}(x) = (-1)^n J_n(x)$. In this case, the term $J_{-n}(x)$ in the general solution should be replaced with the *Bessel's function of the second kind* (the *Weber's function*) $Y_n(x)$ defined by the formula

$$Y_n(x) = \lim_{m\to n} \frac{J_m(x)\cos m\pi - J_{-m}(x)}{\sin m\pi} \quad [2].$$

The graphs of Y_0 and Y_1 are shown in Fig. 358.

Sometimes we use in applications the Bessel's functions of an imaginary argument and consider the products $i^{-n}J_n(ix)$ denoted usually by $I_n(x)$:

$$I_n(x) = i^{-n} J_n(ix) = \frac{(\frac{1}{2}x)^n}{\Gamma(n+1)} + \frac{(\frac{1}{2}x)^{n+2}}{1!\,\Gamma(n+2)} + \frac{(\frac{1}{2}x)^{n+4}}{2!\,\Gamma(n+3)} + \dots$$

---

[1] For the Γ function, see p. 192.

[2] This function is sometimes denoted by $N_n(x)$.

These functions are solutions of the differential equation

$$x^2y'' + xy' - (x^2 + n^2)y = 0.$$

As its second solution, the following *Macdonald's function* is usually taken

$$K_n(x) = \frac{\pi}{2} \cdot \frac{I_{-n}(x) - I_n(x)}{\sin n\pi}.$$

This expression tends to a definite limit, when n approaches an integer.

The graphs of I_0 and I_1 are shown in Fig. 359, and the graphs of K_0 and K_1—in Fig. 360. For tables of $J_0(x)$, $J_1(x)$, $Y_0(x)$, $Y_1(x)$, $I_0(x)$, $I_1(x)$, $K_0(x)$, $K_1(x)$ see pp. 88–90.

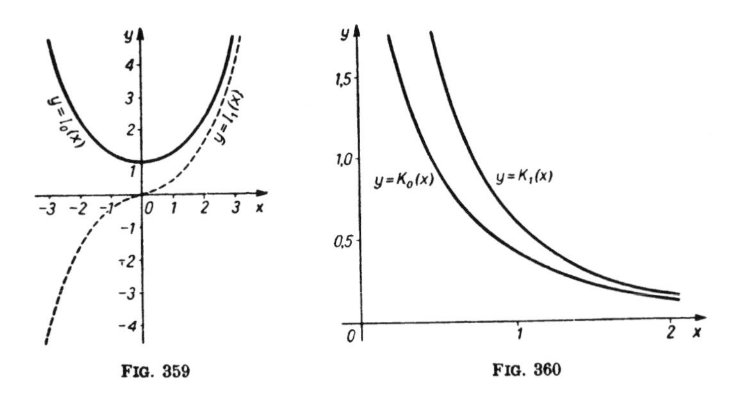

FIG. 359 FIG. 360

Fundamental formulas for Bessel's functions:

$$J_{n-1}(x) + J_{n+1}(x) = \frac{2n}{x} J_n(x), \quad \frac{dJ_n(x)}{dx} = -\frac{n}{x} J_n(x) + J_{n-1}(x)$$

these two formulas are also true for the function $Y_n(x)$);

$$I_{n-1}(x) - I_{n+1}(x) = \frac{2n\, I_n(x)}{x}, \quad \frac{dI_n(x)}{dx} = I_{n-1}(x) - \frac{n}{x} I_n(x);$$

$$K_{n+1}(x) - K_{n-1}(x) = \frac{2nK_n(x)}{x}, \quad \frac{dK_n(x)}{dx} = -K_{n-1}(x) - \frac{n}{x} K_n(x).$$

For an integral n:

$$J_{2n}(x) = \frac{2}{\pi} \int_0^{\pi/2} \cos (x \sin \varphi) \cos 2n\varphi \, d\varphi,$$

$$J_{2n+1}(x) = \frac{2}{\pi} \int_0^{\pi/2} \sin(x \sin \varphi) \sin(2n+1)\varphi \, d\varphi,$$

or, in the complex form,

$$J_n(x) = \frac{(-i)^n}{\pi} \int_0^{\pi} e^{ix \cos \varphi} \cos n\varphi \, d\varphi.$$

The functions $J_{n+1/2}(x)$ can be expressed in terms of elementary functions, in particular

$$J_{1/2}(x) = \sqrt{\frac{2}{\pi x}} \sin x, \qquad J_{-1/2}(x) = \sqrt{\frac{2}{\pi x}} \cos x.$$

Hence the expressions for $J_{n+1/2}(x)$, for an arbitrary integral n, can be obtained by successively applying the above recurrence formulas.

For large values of x, the following asymptotical formulas hold:

$$J_n(x) = \sqrt{\frac{2}{\pi x}} \left[\cos\left(x - \frac{n\pi}{2} - \frac{\pi}{4}\right) + O\left(\frac{1}{x}\right) \right],$$

$$I_n(x) = \frac{e^x}{\sqrt{2\pi x}} \left[1 + O\left(\frac{1}{x}\right) \right],$$

$$Y_n(x) = \sqrt{\frac{2}{\pi x}} \left[\sin\left(x - \frac{n\pi}{2} - \frac{\pi}{4}\right) + O\left(\frac{1}{x}\right) \right],$$

$$K_n(x) = \sqrt{\frac{\pi}{2x}} e^{-x} \left[1 + O\left(\frac{1}{x}\right) \right],$$

where $O(1/x)$ denotes an infinitesimal of the same order as $1/x$ (see p. 333).

Legendre's equation $(1 - x^2) y'' - 2xy' + n(n+1) y = 0$. If n is an integer, the solutions of this equation are so-called *Legendre's polynomials* (or *spherical functions*):

$$P_n(x) = \frac{1}{2^n n!} \cdot \frac{d^n\left((x^2 - 1)^n\right)}{dx^n}.$$

In particular, for $n = 0, 1, \ldots, 7$, we obtain the polynomials

$$P_0(x) = 1,$$

$$P_1(x) = x,$$

$$P_2(x) = \tfrac{1}{2}(3x^2 - 1),$$

$$P_3(x) = \tfrac{1}{2}(5x^3 - 3x),$$

$$P_4(x) = \tfrac{1}{8}(35x^4 - 30x^2 + 3),$$

$$P_5(x) = \tfrac{1}{8}(63x^5 - 70x^3 + 15x),$$

$$P_6(x) = \tfrac{1}{16}(231x^6 - 315x^4 + 105x^2 - 5),$$

$$P_7(x) = \tfrac{1}{16}(429x^7 - 693x^5 + 315x^3 - 35x);$$

the graphs of these polynomials are shown in Fig. 361 and their tables are given on pp. 90, 91.

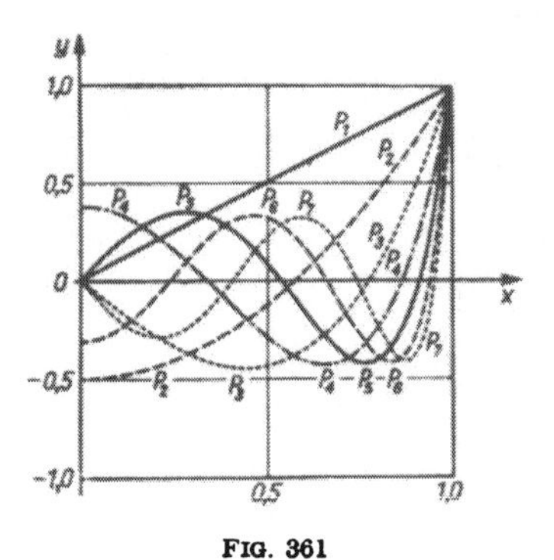

FIG. 361

Fundamental properties of Legendre's polynomials:

$$P_n(x) = \frac{1}{\pi}\int_0^{\pi}\left(x \pm \cos\varphi\sqrt{x^2-1}\right)^n d\varphi = \frac{1}{\pi}\int_0^{\pi}\frac{d\varphi}{\left(x \pm \cos\varphi\sqrt{x^2-1}\right)^{n+1}},$$

$$(n+1)\,P_{n+1}(x) = (2n+1)\,xP_n(x) - nP_{n-1}(x),$$

$$(x^2-1)\frac{dP_n(x)}{dx} = n\left(xP_n(x) - P_{n-1}(x)\right),$$

$$\int_{-1}^{+1} P_m(x)\,P_n(x)\,dx = 0 \quad \text{for} \quad m \neq n, \qquad \int_{-1}^{+1}\left(P_m(x)\right)^2 dx = \frac{2}{2m+1}.$$

Legendre's polynomials can also be obtained as coefficients in the expansion in power series of the function

$$(1 - 2xz + z^2)^{-1/2} = P_0(x) + P_1(x)\,z + P_2(x)\,z^2 + \dots \quad (|z| < 1).$$

Hypergeometric equation

$$x(1-x)\frac{d^2y}{dx^2}+\big(\gamma-(\alpha+\beta+1)x\big)\frac{dy}{dx}-\alpha\beta y=0,$$

where α, β and γ are parameters, comprises many particular cases. For example, if $\alpha=n+1$, $\beta=-n$, $\gamma=1$ and $x=\frac{1}{2}(1-z)$, then this equation reduces to Legendre's equation.

If γ is different from zero and from a negative integer, then the following *hypergeometric series* is a particular solution of the hypergeometric equation

$$(1)\quad F(\alpha,\beta,\gamma,x)=1+\frac{\alpha\cdot\beta}{1\cdot\gamma}x+\frac{\alpha(\alpha+1)\beta(\beta+1)}{1\cdot 2\cdot\gamma(\gamma+1)}x^2+\ldots+$$

$$+\frac{\alpha(\alpha+1)\ldots(\alpha+n)\beta(\beta+1)\ldots(\beta+n)}{1\cdot 2\ldots(n+1)\gamma(\gamma+1)\ldots(\gamma+n)}x^{n+1}+\ldots$$

It is absolutely convergent for $|x|<1$ [1]. If $2-\gamma$ is different from zero and from a negative integer, then

$$y=x^{1-\gamma}F(\alpha+1-\gamma,\ \beta+1-\gamma,\ 2-\gamma,\ x)$$

is a particular solution of the hypergeometric equation.

In some cases, the hypergeometric series can be reduced to elementary functions, for example,

$$F(1,\beta,\beta,x)=F(\alpha,1,\alpha,x)=\frac{1}{1-x},\qquad F(-n,\beta,\beta,-x)=(1+x)^n,$$

$$F(1,1,2,-x)=\frac{\ln(1+x)}{x},\qquad F(\tfrac{1}{2},\tfrac{1}{2},\tfrac{3}{2},x^2)=\frac{\arcsin x}{x}.$$

8. Boundary-value problems

Formulation of the problem. In many cases, in particular, in questions concerning equations of the mathematical physics, we encounter problems which, in contrast to the problems with initial conditions considered above, are called *boundary-value problems*, i.e., the desired solution of a differential is required to satisfy certain conditions at the ends of the considered interval of the independent variable. We confine ourselves to consider here the following boundary-value problem:

---

[1] The convergence of the hypergeometric series (1), for $x=1$ or $x=-1$, depends on the number $\delta=\gamma-\alpha-\beta$. For $x=1$, series (1) is absolutely convergent, if $\delta>0$, and divergent, if $\delta\leqslant 0$; for $x=-1$, the series is absolutely convergent, if $\delta>0$, conditionally convergent, if $-1<\delta\leqslant 0$, and divergent, if $\delta\leqslant -1$.

Find the solution $y(x)$ of the *self-conjugate differential equation*

$$(*) \qquad (py')' - qy + \lambda \varrho y = f,$$

satisfying the homogeneous conditions

$$A_0 y(a) + B_0 y'(a) = 0, \qquad A_1 y(b) + B_1 y'(b) = 0,$$

where the functions $p(x)$, $p'(x)$, $q(x)$, $\varrho(x)$, $f(x)$ are continuous in the interval $a < x < b$ [1] and $p(x) > p_0 > 0$, $\varrho(x) > \varrho_0 > 0$; λ is a constant (the parameter of the equation). Taking $f = 0$, we obtain a *homogeneous boundary-value problem* corresponding to the given *non-homogeneous problem*.

An equation of the second order $Ay'' + By' + Cy + \lambda Ry = F$ can be reduced to the form $(*)$ by multiplying by p/A, where $p = e^{\int (B/A)\,dx}$ provided that $A \neq 0$ in the interval in question. Then we put $q = -pC/A$, $\varrho = pR/A$.

The problem of finding a solution satisfying non-homogeneous conditions $A_0 y(a) + B_0 y'(a) = C_0$, $A_1 y(b) + B_1 y'(b) = C_1$ can be reduced to a problem with homogeneous conditions, but with a different right member $f(x)$; this can be achieved by the substitution $y = z + u$, where u is an arbitrary twice differentiable function satisfying non-homogeneous boundary conditions and z is a new unknown function satisfying the corresponding homogeneous conditions.

Sturm-Liouville's problem. For a fixed value of λ, the following possibilities can occur: either the non-homogeneous problem has a solution for any $f(x)$, and then this solution is unique, and the corresponding homogeneous problem has only a trivial (identically equal to zero) solution, or the corresponding homogeneous problem has non-trivial (different from zero) solutions, and then the non-homogeneous problem is solvable not for all right members $f(x)$, and, in the case, when a solution exists, it is not unique. The values of λ for which the second case occurs (the problem has non-trivial solutions) are called *eigenvalues* (or *characteristic values*) of the given boundary-value problem and the corresponding non-trivial solutions are called *eigenfunctions* (or *characteristic functions*) of the given eigenvalues. The problem of finding the eigenvalues and eigenfunctions of the equation $(*)$ is called *Sturm-Liouville's problem*.

Fundamental properties of eigenvalues and eigenfunctions.

(1) Eigenfunctions of a boundary problem constitute a sequence $\lambda_0 < \lambda_1 < \lambda_2 < \ldots < \lambda_n < \ldots$ of natural numbers tending to

[1] We shall assume, that the interval (a, b) is finite. In the case of an infinite interval, the results change essentially.

infinity. The eigenfunction corresponding to λ_n has precisely n zeros in the interval $a < x < b$.

(2) If $y(x)$ and $z(x)$ are two eigenfunctions corresponding to the eigenvalue λ, then $y(x) = cz(x)$, where c is a constant.

(3) If the eigenfunctions $y_1(x)$ and $y_2(x)$ correspond to the eigenvalues λ_1 and λ_2, then

$$\int_a^b y_1(x)\, y_2(x)\, \varrho(x)\, dx = 0.$$

This property says that $y_1(x)$ and $y_2(x)$ are orthogonal with the weight $\varrho(x)$.

(4) If we replace the coefficients $p(x)$ and $q(x)$ in the equation (*) by $\tilde{p}(x) > p(x)$ and $\tilde{q}(x) > q(x)$, then the eigenvalues will not be less, i.e., $\tilde{\lambda}_n > \lambda_n$, where $\tilde{\lambda}_n$ and λ_n are the eigenvalues of the corresponding equations. If we replace the coefficient $\varrho(x)$ by $\tilde{\varrho}(x) > \varrho(x)$, then the eigenvalues will not increase, i.e., $\tilde{\lambda}_n < \lambda_n$; moreover, the n-th eigenvalue λ_n depends continuously on the coefficients of the equation, i.e., to sufficiently small variations of the coefficients, there correspond arbitrary small variations of λ_n.

(5) When the interval $a < x < b$ decreases, the eigenvalues do not decrease.

Expansion in eigenfunctions. For every λ_n, we choose an eigenfunction $\varphi_n(x)$ such that

$$\int_a^b \left(\varphi_n(x)\right)^2 \varrho(x)\, dx = 1;$$

we call such an eigenfunction a *normed function*.

To every function $g(x)$ defined in the interval $a < x < b$, a Fourier series in the eigenfunctions of the given boundary-value problem can be related

$$g(x) \sim \sum_{n=0}^{\infty} c_n \varphi_n(x), \qquad c_n = \int_a^b g(x)\, \varphi_n(x)\, \varrho(x)\, dx,$$

provided that these integrals are not meaningless.

Expansion theorem. If the function $g(x)$ has a continuous derivative and satisfies the boundary conditions of the given problem, then the Fourier series of $g(x)$ in the eigenfunctions of this boundary-value problem is absolutely and uniformly convergent to $g(x)$.

For examples of eigenfunctions and expansions in the eigenfunctions see pp. 571–576.

Parseval's equation states that

$$\int_a^b [g(x)]^2 \varrho(x)\, dx = \sum_{n=0}^{\infty} c_n^2,$$

provided that the integral written on the left side has a sense. In this case, the Fourier series of $g(x)$ in the eigenfunctions of the boundary-value problem is convergent to $g(x)$, that is

$$\lim_{N\to\infty} \int_a^b \Big(g(x) - \sum_{n=0}^{N} c_n \varphi_n(x)\Big)^2 \varrho(x)\, dx = 0.$$

Singular cases. In applications of the Fourier method to problems of mathematical physics, we sometimes encounter boundary-value problems with singularities at the end points of the interval $a < x < b$, as, for example, when the function $p(x)$ vanishes. At such singular points we impose certain restrictions up on the behaviour of the solution, for example, we require the solution to be continuous or finite or to assume an infinite value of an order not exceeding a prescribed order. Such conditions play a rolè of a homogeneous boundary condition (see p. 569). Moreover, in some boundary-value problems we have to consider homogeneous boundary conditions connecting the values of the function and of its derivative at different end points of the interval. The most important of such conditions are $y(a) = y(b)$, $p(a)y'(a) = p(b)y'(b)$ which, in the case $p(a) = p(b)$, can be regarded as conditions of periodicity. For a boundary-value problem with such conditions, all the above remarks are true, except for statement (2) (see p. 555).

B. PARTIAL DIFFERENTIAL EQUATIONS

9. Equations of the first order

Linear equation. A *linear partial differential equation of the first order* is the equation of the form

$$X_1 \frac{\partial z}{\partial x_1} + X_2 \frac{\partial z}{\partial x_2} + \dots + X_n \frac{\partial z}{\partial x_n} = Y, \tag{1}$$

where z is the unknown function of the independent variables $x_1, x_2, \dots, x_n$, and $X_1, X_2, \dots, X_n$ and Y are given functions of the variables $x_1, x_2, \dots, x_n$. If the functions $X_1, X_2, \dots, X_n$ and Y in the equation (1) depend also on z, then the equation is called *quasi linear*. If $Y \equiv 0$, then the equation

$$X_1 \frac{\partial z}{\partial x_1} + X_2 \frac{\partial z}{\partial x_2} + \ldots + X_n \frac{\partial z}{\partial x_n} = 0 \tag{1a}$$

is called *homogeneous*.

The problem of integration of a linear homogeneous equation is equivalent to the problem of integration of so-called *characteristic system of equations*

$$\frac{dx_1}{X_1} = \frac{dx_2}{X_2} = \ldots = \frac{dx_n}{X_n} \text{ [1]}. \tag{2}$$

Each first integral of system (2) turns out to be a solution of the homogeneous linear equation (1a), and conversely: each solution of equation (1a) is a first integral of system (2) (see p. 531). Moreover, if the $n-1$ first integrals

$$\varphi_i(x_1, x_2, \ldots, x_n) = 0, \quad \text{where} \quad i = 1, 2, \ldots, n-1$$

are independent (see p. 532), then $z = \Phi(\varphi_1, \varphi_2, \ldots, \varphi_{n-1})$, where Φ is an arbitrary function of $n-1$ arguments, is the general solution of the homogeneous linear equation (1a).

In solving a non-homogeneous linear or quasi linear equation (1), we seek for a solution z in implicit form $V(x_1, x_2, \ldots, x_n, z) = C$. The function V is then a solution of a homogeneous linear equation with $n+1$ independent variables

$$X_1 \frac{\partial V}{\partial x_1} + X_2 \frac{\partial V}{\partial x_2} + \ldots + X_n \frac{\partial V}{\partial x_n} + Y \frac{\partial V}{\partial z} = 0;$$

the characteristic system of this equation

$$\frac{dx_1}{X_1} = \frac{dx_2}{X_2} = \ldots = \frac{dx_n}{X_n} = \frac{dz}{Y} \tag{2'}$$

is called the *characteristic system* of the original equation (1).

Geometric interpretation. In the case of an equation

$$P(x, y, z) \frac{\partial z}{\partial x} + Q(x, y, z) \frac{\partial z}{\partial y} = R(x, y, z), \tag{1_1}$$

with two independent variables $x_1 = x$ and $x_2 = y$, a solution

---

[1] In solving a system in this form, any of the variables x_k for which $X_k \neq 0$ can be taken as the independent variable. Then the system assumes the form $\frac{dx_j}{dx_k} = \frac{X_j}{X_k}$, $j = 1, 2, \ldots, n$. It is more convenient, however, to preserve the symmetry and introduce the parameter t as a new variable, letting $\frac{dx_j}{X_j} = dt$ or $\frac{dx_j}{dt} = X_j$.

$z = f(x, y)$ is geometrically represented in the space by a surface called the *integral surface* of this equation. Equation (1_1) means that the normal vector $\left\{\frac{\partial z}{\partial x}, \frac{\partial z}{\partial y}, -1\right\}$ to the surface $z = f(x, y)$ is orthogonal at every point of the surface to the vector $\{P, Q, R\}$ prescribed for that point. The system (2′) assumes the form

$$(2_1) \qquad \frac{dx}{P(x, y, z)} = \frac{dy}{Q(x, y, z)} = \frac{dz}{R(x, y, z)}.$$

It follows (see p. 632) that the integral curves of this system of equations, so-called *characteristics*, are tangent to the vector $\{P, Q, R\}$. Therefore a characteristic having a point in common with the integral surface $z = f(x, y)$ lies wholly on this surface. Through each point of the space there passes an integral curve of the characteristic system of equations (provided that the assumptions of the existence theorem are fulfilled, see p. 529) and the integral surfaces consist of characteristics.

Cauchy's problem. Let n functions of $n-1$ independent variables $t_1, t_2, \ldots, t_{n-1}$

$$(*) \qquad \begin{aligned} x_1 &= x_1(t_1, t_2, \ldots, t_{n-1}), \\ x_2 &= x_2(t_1, t_2, \ldots, t_{n-1}), \\ &\cdots\cdots\cdots\cdots \\ x_n &= x_n(t_1, t_2, \ldots, t_{n-1}), \end{aligned}$$

be given. *Cauchy's problem* for the equation (1) is the problem of finding a solution $z = \varphi(x_1, x_2, \ldots, x_n)$ such that the substitution $(*)$ changes it into a prescribed function $\psi(t_1, t_2, \ldots, t_{n-1})$:

$$\varphi\big(x_1(t_1, t_2, \ldots, t_{n-1}), x_2(t_1, t_2, \ldots, t_{n-1}), \ldots, x_n(t_1, t_2, \ldots, t_{n-1})\big) = \psi(t_1, t_2, \ldots, t_{n-1}).$$

In the case of two independent variables, this problem reduces to finding an integral surface passing through the given curve. If this curve has a tangent depending continuously on a point and is not tangent to the characteristic at any point, then Cauchy's problem has a unique solution in a certain neighbourhood of the curve.

The integral surface is then composed of all characteristics intersecting the given curve.

Examples. (1) $(mz - ny)\frac{\partial z}{\partial x} + (nx - lz)\frac{\partial z}{\partial y} = ly - mx$, where l, m and n are constants.

Equation of the characteristics

$$\frac{dx}{mz-ny}=\frac{dy}{nx-lz}=\frac{dz}{ly-mx}.$$

Integrals of this system are $lx+my+nz=C_1, x^2+y^2+z^2=C_2$. The characteristics are circumferences with centres lying on the straight line passing through the coordinate origin and with the direction cosines proportional to l, m, n. The integral surfaces are surfaces of revolution whose axis is that straight line.

(2) Find the integral surface of the equation $\frac{\partial z}{\partial x}+\frac{\partial z}{\partial y}=z$ passing through the curve $x=0,\ z=\varphi(y)$.

Equations of the characteristics $\frac{dx}{1}=\frac{dy}{1}=\frac{dz}{z}$. The characteristics passing through the point (x_0, y_0, z_0) have the equations $y=x-x_0+y_0, z=z_0e^{x-x_0}$. Taking $x_0=0, z_0=\varphi(y_0)$, we find $y=x+$ $+y_0, z=e^x\varphi(y_0)$; this is a parametric representation of the desired integral surface. Eliminating y_0, we obtain $z=e^x\varphi(y-x)$.

Non-linear equations. The general form of a partial equation of the first order is

$$F\left(x_1, x_2, \ldots, x_n, z, \frac{\partial z}{\partial x_1}, \frac{\partial z}{\partial x_2}, \ldots, \frac{\partial z}{\partial x_n}\right)=0. \tag{3}$$

A solution $z=\varphi(x_1, x_2, \ldots, x_n; a_1, a_2, \ldots, a_n)$ of equation (3) depending on n parameters $a_1, a_2, \ldots, a_n$ whose Jacobian $\frac{\partial(\varphi'_{x_1}, \varphi'_{x_2}, \ldots, \varphi'_{x_n})}{\partial(a_1, a_2, \ldots, a_n)}$ is not equal to zero for the considered values of $x_1, x_2, \ldots, x_n$ is called a *complete integral* of (3).

Integration of equation (3) reduces to the integration of the characteristic system of differential equations

$$\frac{dx_1}{P_1}=\ldots=\frac{dx_n}{P_n}=\frac{dz}{p_1P_1+\ldots+p_nP_n}=\frac{-dp_1}{X_1+p_1Z}=\ldots=\frac{-dp_n}{X_n+p_nZ}, \tag{4}$$

where

$$Z=\frac{\partial F}{\partial z},\quad X_i=\frac{\partial F}{\partial x_i},\quad p_i=\frac{\partial z}{\partial x_i},\quad P_i=\frac{\partial F}{\partial p_i}\quad (i=1, 2, \ldots, n).$$

Solutions of the characteristic system (4), satisfying the additional condition $F(x_1, x_2, \ldots, x_n, z, p_1, p_2, \ldots, p_n)=0$, are called *characteristic strips.*

Canonical systems. Sometimes it is more convenient to consider an equation not involving explicitly the unknown function z. Such an equation can be obtained by introducing an additional independent variable $x_{n+1} = z$ and an unknown function $V(x_1, x_2, \ldots, x_n, x_{n+1})$ such that the equation $V(x_1, x_2, \ldots, x_n, z) = C$ defines z as an implicit function of $x_1, x_2, \ldots, x_n$; at the same time, we substitute $-\frac{\partial V}{\partial x_i} \Big/ \frac{\partial V}{\partial x_{n+1}}$ instead of $\frac{\partial z}{\partial x_i}$ in equation (3), for $i = 1, 2, \ldots, n$. If, moreover, we solve the differential equation for an arbitrary partial derivative of the function V, denote this derivative by x and rearrange the remaining variables, then equation (3) will have the form

$$p + H(x_1, x_2, \ldots, x_n, x, p_1, p_2, \ldots, p_n) = 0, \tag{3'}$$

where

$$p = \frac{\partial V}{\partial x}, \quad p_i = \frac{\partial V}{\partial x_i} \quad (i = 1, 2, \ldots, n).$$

The system of characteristic differential equations is then transformed into the system

$$\frac{dx_i}{dx} = \frac{\partial H}{\partial p_i}, \quad \frac{dp_i}{dx} = -\frac{\partial H}{\partial x_i} \quad (i = 1, 2, \ldots, n) \tag{5}$$

and

$$\frac{\partial V}{dx} = p_1 \frac{\partial H}{\partial p_1} + \ldots + p_n \frac{\partial H}{\partial p_n} - H, \quad \frac{dp}{dx} = -\frac{\partial H}{\partial x}. \tag{6}$$

The equations (5) constitute a definite system of $2n$ ordinary differential equations. Such a system corresponding to an arbitrary function $H(x_1, x_2, \ldots, x_n, x, p_1, p_2, \ldots, p_n)$ with $2n + 1$ variables is called a *canonical system* of differential equations. Many problems of mechanics and theoretical physics lead to equations of this form. Knowledge of a complete integral

$$V = \varphi(x_1, x_2, \ldots, x_n, x, a_1, a_2, \ldots, a_n) + a$$

of equation (3′) enables us to find the general solution of the canonical system (5), since the equations

$$\frac{\partial \varphi}{\partial a_i} = b_i, \quad \frac{\partial \varphi}{\partial x_i} = p_i, \quad i = 1, 2, \ldots, n$$

with $2n$ parameters a_i and b_i determine a $2n$-parameter solution of the canonical system (5).

Clairaut's equation. The problem of finding a complete integral is particularly simple, when the equation has the form

$$z = x_1 p_1 + x_2 p_2 + \ldots + x_n p_n + f(p_1, p_2, \ldots, p_n),$$

where

$$p_i = \frac{\partial z}{\partial x_i} \quad (i = 1, 2, \ldots, n);$$

(*Clairaut's equation*). A complete integral of this equation is

$$z = a_1 x_1 + a_2 x_2 + \ldots + a_n x_n + f(a_1, a_2, \ldots, a_n),$$

where $a_1, a_2, \ldots, a_n$ are arbitrary parameters.

Example. (*The problem of two bodies*). The motion of two particles mutually attracting according to Newton's law takes place in one plane. Therefore, choosing the initial position of one of the points at the origin, we can write the equation of the motion in the form

$$\frac{d^2x}{dt^2} = \frac{\partial V}{\partial x}, \quad \frac{d^2y}{dt^2} = \frac{\partial V}{dy}, \quad V = \frac{k^2}{\sqrt{x^2 + y^2}}.$$

By introducing the Hamilton's function

$$H = \frac{1}{2}(p^2 + q^2) - \frac{k^2}{\sqrt{x^2 + y^2}}$$

we transform that system into the system of canonical differential equations

$$(*) \qquad \frac{dx}{dt} = \frac{\partial H}{\partial p}, \quad \frac{dy}{dt} = \frac{\partial H}{\partial q}, \quad \frac{dp}{dt} = -\frac{\partial H}{\partial x}, \quad \frac{dq}{dt} = -\frac{\partial H}{\partial y}$$

with respect to $x, y, p = dx/dt, q = dy/dt$. The corresponding partial equation is

$$\frac{\partial z}{\partial t} + \frac{1}{2}\left[\left(\frac{\partial z}{\partial x}\right)^2 + \left(\frac{\partial z}{\partial y}\right)^2\right] - \frac{k^2}{\sqrt{x_2 + y_2}} = 0.$$

After passing to polar coordinates ϱ, φ we can easily find a complete integral of this equation

$$z = -at - b\varphi + c - \int_{\varrho_0}^{\varrho} \sqrt{2a + \frac{2k^2}{r} - \frac{b^2}{r^2}}\, dr,$$

depending on parameters a, b, c. Therefore we find the general solution of the system $(*)$ from the equations $\frac{\partial z}{\partial a} = -t_0$, $\frac{\partial z}{\partial b} = -\varphi_0$.

Case of two independent variables $(x_1 = x, x_2 = y, p_1 = p, p_2 = q)$. In this case, the characteristic strip can be interpreted geometrically as a curve at every point (x, y, z) of which a plane $p(\xi - x) +$

$+q(\eta-y)=\zeta-z$ tangent to the curve is prescribed. Finding an integral surface of the equation

$$F\left(x, y, z, \frac{\partial z}{\partial x}, \frac{\partial z}{\partial y}\right)=0$$

passing through the given curve (Cauchy's problem) reduces to finding the characteristic strips passing through the points of the initial curve such that the corresponding tangent plane to each strip is tangent to that curve. The values of p and q are then defined by the relations $F(x, y, z, p, q)=0$ and $p\,dx+q\,dy=dz$ which, in the case of a non-linear equation have, in general, several solutions. Therefore, to obtain a definite solution, we should assume, in the formulation of Cauchy's problem, a pair of continuous functions p and q satisfying the above relations along the initial curve.

Example. For the equation $pq=1$ and the initial curve $y=x^3$, $z=2x^2$, we can assume $p=x$ and $q=1/x$ along the curve. The characteristic system has the form

$$\frac{dx}{dt}=q, \quad \frac{dy}{dt}=p, \quad \frac{dz}{dt}=2pq, \quad \frac{dp}{dt}=0, \quad \frac{dq}{dt}=0.$$

The characteristic strip with the initial conditions x_0, y_0, z_0, p_0, q_0, for $t=0$, is $x=x_0+q_0t$, $y=y_0+p_0t$, $z=2p_0q_0t+z_0$, $p=p_0$, $q=q_0$. In the case, when $p_0=x_0$, $q_0=1/x_0$, the curve belonging to the characteristic strip passing through the point (x_0, y_0, z_0) of the initial curve has the form

$$x=x_0+\frac{t}{x_0}, \quad y=x_0^3+tx_0, \quad z=2t+2x_0^2.$$

Eliminating the parameters x_0, t we obtain $z^2=4xy$.

If we assume other admissible values p and q along the curve, for example $p=3x$, $q=1/3x$, then we obtain another solution.

An enveloping surface of a one-parameter family of integral surfaces is also an integral surface. This fact can be used to solve Cauchy's problem by means of a complete integral by excluding the one-parameter family of solutions tangent to the planes at points of the initial curve and finding the enveloping surface of this family of surfaces.

Example. Given the equation $z-px-qy+pq=0$ (Clairaut's equation), find the integral surface passing through the curve $y=x$, $z=x^2$.

The equation has a complete integral $z=ax+by-ab$. Since we have to assume $p=q=x$ along the initial curve, hence the condition $a = b$ determines the desired one-parameter family of surfaces. Finding the enveloping surface of this family, we obtain $z = \frac{1}{4}(x + y)^2$.

Equations in total differentials. An *equation in total differentials* has the form

$$dz = f_1 dx_1 + f_2 dx_2 + \ldots + f_n dx_n, \tag{7}$$

where $f_1, f_2, \ldots, f_n$ are given functions of the variables $x_1, x_2, \ldots, x_n, z$. This equation is said to be *exact* or *completely integrable,* if there exists a unique relation between $x_1, x_2, \ldots, x_n, z$ with one arbitrary constant which implies equation (7). In this case, there exists a unique solution $z = z(x_1, x_2, \ldots, x_n)$ of equation (7) assuming the value z^0 for the initial values $x_1^0, x_2^0, \ldots, x_n^0$ of independent variables. If $n = 2$, this means that through each point of the space there passes a unique integral surface.

Equation (7) is completely integrable if, and only if, the following $\frac{1}{2}n(n-1)$ relations hold identically with respect to all variables $x_1, x_2, \ldots, x_n, z$:

$$\frac{\partial f_i}{\partial x_k} + f_k \frac{\partial f_i}{\partial z} = \frac{\partial f_k}{\partial x_i} + f_i \frac{\partial f_k}{\partial z} \quad \text{for} \quad i, k = 1, 2, \ldots, n.$$

If equation (7) is given in a symmetric form

$$f_1 dx_1 + f_2 dx_2 + \ldots + f_n dx_n = 0,$$

then a condition for complete integrability is the identities

$$f_i\left(\frac{\partial f_k}{\partial x_j} - \frac{\partial f_j}{\partial x_k}\right) + f_j\left(\frac{\partial f_i}{\partial x_k} - \frac{\partial f_k}{\partial x_i}\right) + f_k\left(\frac{\partial f_j}{\partial x_i} - \frac{\partial f_i}{\partial x_j}\right) = 0$$

for all systems of indices i, j, k.

If equation (7) is completely integrable, then its solution reduces to the integration of one ordinary equation with $n - 1$ parameters.

10. Linear equations of the second order

The general form of a linear equation of the second order in the case of two independent variables x, y and the unknown functions u is

$$A\frac{\partial^2 u}{\partial x^2} + 2B\frac{\partial^2 u}{\partial x \partial y} + C\frac{\partial^2 u}{\partial y^2} + a\frac{\partial u}{\partial x} + b\frac{\partial u}{\partial y} + cu = f, \tag{1}$$

where the coefficients A, B, C, a, b, c and the right member f are given functions of x and y.

Classification of equations. The type of a solution of such equation is greatly dependent on the sign of the discriminant $\delta = AC - B^2$. Equation (1) is called *hyperbolic* in a domain, if $\delta < 0$ in this domain; *parabolic*, if δ is identically equal to 0 in this domain; finally, it is called *elliptic*, if $\delta > 0$ in the domain in question.

If δ changes its sign in the domain, then equation (1) is said to be of a *mixed type*. The sign of the discriminant remains unchanged under an arbitrary transformation of independent variables (introducing a new coordinate system in the xy plane). Hence the sign of the discriminant is independent of the particular choice of independent variables.

The integral curves of the differential equation

$$A\,dy^2 - 2B\,dx\,dy + C\,dx^2 = 0 \quad \text{or} \quad \frac{dy}{dx} = \frac{B \pm \sqrt{-\delta}}{A}$$

are called *characteristics* of equation (1). A hyperbolic equation has two families of real characteristics, a parabolic equation has one family of real characteristics and an elliptic equation has no real characteristics. Any equation obtained from (1) by change of independent variables has the same characteristics as equation (1). If the family of characteristics coincides with that of coordinate lines, then equation (1) does not contain a term with the second derivative of the unknown function with respect to the corresponding independent variable; if moreover, the equation is of the parabolic type, then the term with the mixed derivative does not occur in the equation either.

Canonical form of the equation. By a change of independent variables $\xi = \varphi(x, y)$, $\eta = \psi(x, y)$, equation (1) can be reduced to one of the following three *canonical forms*:

$$\text{(a)} \qquad \frac{\partial^2 u}{\partial \xi^2} - \frac{\partial^2 u}{\partial \eta^2} + \ldots = 0 \quad (\delta < 0, \text{ a hyperbolic equation}),$$

$$\text{(b)} \qquad \frac{\partial^2 u}{\partial \eta^2} + \ldots = 0 \quad (\delta = 0, \text{ a parabolic equation}),$$

$$\text{(c)} \qquad \frac{\partial^2 u}{\partial \xi^2} + \frac{\partial^2 u}{\partial \eta^2} + \ldots = 0 \quad (\delta > 0, \text{ an elliptic equation}),$$

where the dots denote the terms not involving partial derivatives of the second order of the unknown function.

If, in the case of a hyperbolic equation, we choose the two families of characteristics as the families of coordinate lines, that is to say, if we let $\xi_1 = \varphi(x, y)$, $\eta_1 = \psi(x, y)$, where $\varphi(x, y) = \text{const}$ and $\psi(x, y) = \text{const}$ are equations of the families of characteristics, then the equation will have the form

$$\frac{\partial^2 u}{\partial \xi_1 \, \partial \eta_1} + \ldots = 0.$$

This form is also called the *canonical form of the hyperbolic equation*. By the transformation $\xi = \xi_1 + \eta_1$, $\eta = \xi_1 - \eta_1$, this form can be changed into the canonical form (a). In order to reduce a parabolic equation to the canonical form (b), we choose the single, in this case, family of characteristics as the family $\xi = \text{const}$ and take for η an arbitrary function of x and y independent of ξ. If the coefficients $A(x, y)$, $B(x, y)$ and $C(x, y)$ are analytic functions (see p. 598), then the characteristic equation of an elliptic equation determines two families of complex conjugate curves $\varphi(x, y) = \text{const}$, $\psi(x, y) = \text{const}$. The equation can be reduced to the canonical form (c) by the transformation $\xi = \varphi + \psi$, $\eta = i(\varphi - \psi)$.

All the above remarks about classification of equations and reducing to the canonical form can also be applied to equations of a slightly more general form

$$A(x, y)\frac{\partial^2 u}{\partial x^2} + 2B(x, y)\frac{\partial^2 u}{\partial x \, \partial y} + C(x, y)\frac{\partial^2 u}{\partial y^2} + F\left(x, y, u, \frac{\partial u}{\partial x}, \frac{\partial u}{\partial y}\right) = 0.$$

Case when the number of independent variables is greater than two. A *linear partial differential equation of the second order* with more than two independent variables has the form

$$\sum_{i,k} a_{ik} \frac{\partial^2 u}{\partial x_i \, \partial x_k} + \ldots = 0, \tag{2}$$

where a_{ik} are known functions of the independent variables and the dots denote the terms which do not involve the derivatives of the second order of the unknown function.

Equation (2) cannot, in general, be reduced to a simple, normal form by means of a change of variables. However, a classification important in applications and similar to that given above can also be produced for equations of the form (2).

Equations with constant coefficients. If all the coefficients of equation (2) are constant, then, by a linear homogeneous change of variables, equation (2) can be reduced to the following normal form

$$\sum_{i} \varkappa_i \frac{\partial^2 u}{\partial x_i^2} + \ldots = 0, \tag{2'}$$

where the coefficients $\varkappa_i$ are equal ± 1 or 0.

If all the coefficients $\varkappa_i$ are different from zero and have the same sign, then the equation is called *elliptic*. If all the coefficients $\varkappa_i$ are different from zero and one of them has a sign different from the remaining ones, then the equation is called *hyperbolic* [1]. If one of the coefficients $\varkappa_i$ is equal to zero and the remaining ones are different from zero and have the same sign, then the equation is called *parabolic*.

If a linear equation has constant coefficients not only of the higher derivatives, but also at the first derivatives of the unknown function, then we can free the equation from the terms containing derivatives of the first order with respect to these variables for which $\varkappa_i \neq 0$; this can be achieved by a change of the unknown function:

$$u = v \exp\left(-\frac{1}{2}\sum \frac{b_k}{\varkappa_k} x_k\right),$$

where b_k is the coefficient at $\frac{\partial u}{\partial x}$ in formula (2′) and the sum is extended to all $\varkappa_k \neq 0$. Thus any equation with constant coefficients can be reduced to the form $\Delta v + kv = g$, in the elliptic case, or to the form $\frac{\partial^2 v}{\partial t^2} - \Delta v + kv = g$, in the hyperbolic case, where Δ is Laplace's operator:

$$\Delta v = \frac{\partial^2 v}{\partial x_1^2} + \frac{\partial^2 v}{\partial x_2^2} + \ldots + \frac{\partial^2 v}{\partial x_n^2}.$$

Formulation of the problem. Investigation of various physical problems (mechanical, electric or thermal processes) in continuous media leads to partial differential equations called *equations of mathematical physics*. The most important of them and the most frequently encountered are linear equations of the second order. The solution of a physical problem satisfying differential equations is usually also required to satisfy certain additional *boundary* and *initial conditions*. The system of these conditions should determine the solution uniquely. Moreover, and this is by no means less important, the solution should be stable under slight modifications of the initial and boundary conditions, i.e., sufficiently small modifications of the initial and boundary conditions should induce small changes in the solution. In this case, we say that the problem is *correctly formulated*. Only if this condition is satisfied, is a mathematical solution of a differential equation suitable to describe real phenomena. Moreover, in problems leading

[1] If at least two positive and at least two negative coefficients occur in the equation, then the equation is called *ultrahyperbolic*.

to the hyperbolic equations, such as, in particular, problems concerning vibrations in continuous media, the correct one is "Cauchy's problem", prescribing the values of the unknown function and its derivative in a non-tangential direction (in particular, in the normal direction) of the initial *manifold* (a curve or surface). In problems leading to the elliptic equations, such as problems concerning stationary processes and equilibrium in continuous media the correct one is the boundary-value problem: prescribing the values of the unknown function (or its normal derivative) on the boundary of the considered region of independent variables; if this region is infinite, we usually impose certain condition upon the behaviour of the function at infinity, i.e., for infinitely increasing arguments.

Non-homogeneous condition and non-homogeneous equations. The solution of a linear (homogeneous or non-homogeneous) equation with non-homogeneous boundary or initial conditions (see p. 553), can be reduced to the solution of an equation differing from the original equation only by the member free from the unknown function, but with homogeneous conditions. This can be achieved by replacing the unknown function with the difference between this function and an arbitrary function having continuous derivatives and satisfying the given conditions.

The solution of a linear non-homogeneous equation with given boundary or initial conditions is the sum of the solution of the same equation with zero conditions and the solution of the corresponding homogeneous equation with the given conditions.

Finally, the solution of the non-homogeneous linear equation

$$\frac{\partial^2 u}{\partial t^2} - L[u] = g(x, t) \text{ [1]}$$

with homogeneous initial conditions $(u)_{t=0} = 0$, $\left(\frac{\partial u}{\partial t}\right)_{t=0} = 0$ can be reduced to the solution of a Cauchy problem of the corresponding homogeneous equation as follows:

$$u = \int_0^t \varphi(x, t; \tau)\, d\tau,$$

---

[1] In this formula, the symbol x is used to denote the system of n variables $x_1, x_2, \ldots, x_n$ regarded as a point of the n-dimensional space and $L[u]$ is a linear differential expression which may involve the derivative $\frac{\partial u}{\partial t}$ but which does not involve derivatives of higher order with respect to t.

where $\varphi(x, t; \tau)$ is a solution of the equation

$$\frac{\partial^2 u}{\partial t^2} - L[u] = 0$$

with the conditions $(u)_{t=\tau} = 0$, $\left(\frac{\partial u}{\partial t}\right)_{t=\tau} = g(x, \tau)$.

The most frequently encountered equations.

Wave equation or equation of dispersing vibrations in a homogeneous medium:

$$\frac{\partial^2 u}{\partial t^2} - a^2 \Delta u = Q(x, t),$$

where x denotes symbolically the variables $x_1, x_2, \ldots, x_n$ in the n-dimensional space and the right member $Q(x, t)$ is equal to zero, when no external force is applied.

For a homogeneous equation ($Q(x, t) = 0$), the solution satisfying the initial conditions $(u)_{t=0} = \varphi(x)$, $\left(\frac{\partial u}{\partial t}\right)_{t=0} = \psi(x)$ is given by the formulas:

for $n = 3$,

$$u(x_1, x_2, x_3, t) = \frac{1}{4\pi a^2}\left(\iint\limits_{s_{at}} \frac{\psi(\alpha_1, \alpha_2, \alpha_3)}{t}\, d\sigma + \frac{\partial}{\partial t}\iint\limits_{s_{at}} \frac{\varphi(\alpha_1, \alpha_2, \alpha_3)}{t}\, d\sigma\right),$$

where the integration is extended over the surface of the sphere with the equation $(\alpha_1 - x_1)^2 + (\alpha_2 - x_2)^2 + (\alpha_3 - x_3)^2 = a^2 t^2$ (*Kirchoff's formula*);

for $n = 2$,

$$u(x_1, x_2, t) = \frac{1}{2\pi a}\left(\iint\limits_{k_{at}} \frac{\psi(\alpha_1, \alpha_2)\, d\alpha_1 d\alpha_2}{\sqrt{a^2 t^2 - (\alpha_1 - x_1)^2 - (\alpha_2 - x_2)^2}} + \right.$$

$$\left. + \frac{\partial}{\partial t}\iint\limits_{k_{at}} \frac{\varphi(\alpha_1, \alpha_2)\, d\alpha_1 d\alpha_2}{\sqrt{a^2 t^2 - (\alpha_1 - x_1)^2 - (\alpha_2 - x_2)^2}}\right),$$

where the integration is extended over the circle $(\alpha_1 - x_1)^2 + (\alpha_2 - x_2)^2 \leqslant a^2 t^2$ (*Poisson's formula*);

for $n = 1$,

$$u(x_1, t) = \frac{\varphi(x_1 + at) + \varphi(x_1 - at)}{2} + \frac{1}{2a}\int_{x_1 - at}^{x_1 + at} \psi(\alpha)\, d\alpha$$

(*d'Alembert's formula*).

When the original equation is non-homogeneous, the following correction terms should be added to the right members of the above formulas:

for $n = 3$ (so-called *retarding potential*):

$$\frac{1}{4\pi a^2} \iiint\limits_{r \leqslant at} \frac{Q(\xi_1, \xi_2, \xi_3, t - r/a)}{r} \, d\xi_1 d\xi_2 d\xi_3,$$

where

$$r = \sqrt{(\xi_1 - x_1)^2 + (\xi_2 - x_2)^2 + (\xi_3 - x_3)^2};$$

for $n = 2$,

$$\frac{1}{2\pi a} \iiint\limits_{K} \frac{Q(\xi_1, \xi_2, \tau) \, d\xi_1 d\xi_2 d\tau}{\sqrt{a^2(t-\tau)^2 - (\xi_1 - x_1)^2 - (\xi_2 - x_2)^2}},$$

where K is the region of the $\xi_1 \, \xi_2 \, \tau$ space defined by the inequalities

$$0 < \tau < t, \qquad (\xi_1 - x_1)^2 + (\xi_2 - x_2)^2 < a^2(t-\tau)^2;$$

for $n = 1$,

$$\frac{1}{2a} \iint\limits_{T} Q(\xi, \tau) \, d\xi d\tau,$$

where T is the triangle $0 < \tau < t$, $|\xi - x_1| < a\,|t - \tau|$.

It follows from the above formulas that a is the velocity of spreading the disturbance.

Heat conduction equation (in a homogeneous medium):

$$\frac{\partial u}{\partial t} - a^2 \Delta u = Q(x, t) \text{ }(^1).$$

The right member $Q(x, t)$ is equal to zero, if no heat source or absorption occurs. The Cauchy problem for this equation is usually posed in the following form: find a solution bounded for $t > 0$ with the condition $(u)_{t=0} = f(x)$. The condition of boundedness provides the uniqueness of the solution.

For a homogeneous equation we have [2]

---

[1] Δ is the Laplace's operator with respect to n variables $x_1, x_2, \ldots, x_n$ (see p. 566).

[2] We use the notation $\exp x = e^x$.

$$u(x_1, \ldots, x_n, t) = \frac{1}{(2a\sqrt{\pi t})^n} \times$$

$$\times \int_{-\infty}^{+\infty} \ldots \int_{-\infty}^{+\infty} f(\alpha_1, \ldots, \alpha_n) \exp\left(-\frac{(x_1-\alpha_1)^2+\ldots+(x_n-\alpha_n)^2}{4a^2t}\right) d\alpha_1 \ldots d\alpha_n .$$

If $Q(x, t) \neq 0$, the following term should be added to the right member of the above formula

$$\int_0^t \left[\int_{-\infty}^{+\infty} \ldots \int_{-\infty}^{+\infty} \frac{Q(\alpha_1, \ldots, \alpha_n)}{[2a\sqrt{\pi(t-\tau)}]^n} \times \right.$$

$$\left. \times \exp\left(-\frac{(x_1-\alpha_1)^2+\ldots+(x_n-\alpha_n)^2}{4a^2t}\right) d\alpha_1 \ldots d\alpha_n \right] d\tau.$$

The problem of finding $u(x, t)$ for $t < 0$, when the values of $u(x, 0)$ are given appears to be incorrectly posed.

Potential theory equation

$$\Delta u = -4\pi\varrho \text{ (}^1\text{)},$$

where ϱ is a given function of a point (*Poisson's equation*). When $\varrho \equiv 0$, we obtain Laplace's equation $\Delta u = 0$ (see p. 648).

Methods of integration.

Separation of variables. For many equations of mathematical physics, special substitution enable us to obtain, if not the whole set of solutions, then at least a family of solutions dependent on arbitrary parameters. In linear differential equations, in particular in equations of the second order, we can often apply the substitution

$$u(x_1, x_2, \ldots, x_n) = \varphi_1(x_1)\, \varphi_2(x_2) \ldots \varphi_n(x_n).$$

To determine each of the functions $\varphi_k(x_k)$, we separate the variables after substituting this product in the original equation (see examples); we obtain an ordinary linear differential equation. In order that the solution of the original equation satisfy the required homogeneous boundary conditions, it may appear to be sufficient that a part of the functions $\varphi_1(x_1), \varphi_2(x_2), \ldots, \varphi_n(x_n)$ satisfy certain boundary conditions.

By means of summation, differentiation and integration, new

(1) Δ is the Laplace operator with respect to n variables $x_1, x_2, \ldots, x_n$ (see p. 566).

solutions can be acquired from the obtained ones; the parameters should be chosen so that the remaining boundary and initial conditions are satisfied (see examples). It should be remembered that a solution in the form of a series or an improper integral obtained by this method is only a formal solution; such a solution should necessarily be verified as to whether it has a sense (i.e., whether the series or the integral is convergent) and whether it satisfies the given equation and the boundary conditions (i.e., whether its termwise differentiation or passing to a limit is admissible and so on).

In all the following examples, the series and improper integrals are convergent, provided that certain restrictions are imposed upon the functions expressing the initial conditions (as, for example, the continuity of the second derivative in examples 1 and 2).

Examples. (1) Find a solution of the equation

$$\frac{\partial^2 u}{\partial t^2} = a^2 \frac{\partial^2 u}{\partial x^2}$$

satisfying the initial conditions $(u)_{t=0} = f(x)$, $\left(\frac{\partial u}{\partial t}\right)_{t=0} = \varphi(x)$ and the boundary conditions $(u)_{x=0} = 0$, $(u)_{x=l} = 0$ (the *vibrating string*).

We seek a solution in the form $u = X(x)T(t)$. Substituting in the equation we obtain $\frac{T''}{a^2 T} \equiv \frac{X''}{X}$ (the variables are separated). Since the left member is independent of x and the right one of t, hence each of them is a constant quantity. Denoting it by $-\lambda$ [1], we obtain $X'' + \lambda^2 X = 0$, $T'' + a^2\lambda^2 T = 0$. Moreover, from the boundary conditions, we have $X(0) = X(l) = 0$. Hence $X(x)$ is an eigenfunction of the Sturm-Liouville's boundary-value problem and λ^2 is an eigenvalue of this problem (see p. 554). From the equation for X and from the boundary conditions, we find $X(x) = C \sin \lambda x$ with $\sin \lambda l = 0$, i.e., $\lambda = n\frac{\pi}{l}$ $(n = 1, 2, \ldots)$. Now, integrating the equation $T'' + \lambda^2 a^2 T = 0$, we obtain a particular solution of the original equation in the form

$$u_n = \left(a_n \cos \frac{na\pi}{l} t + b_n \sin \frac{na\pi}{l} t\right) \sin \frac{n\pi}{l} x.$$

Requiring that the sum $u = \sum_{n=1}^{\infty} u_n$ is equal to $f(x)$ for $t = 0$ and

---

[1] It follows from the following discussion that the boundary conditions cannot be satisfied, when this constant is positive.

that the derivative $\frac{\partial}{\partial t}\sum_{n=1}^{\infty} u_n$ is equal to $\varphi(x)$, we obtain, using the formula for the expansion into Fourier series in sines (see p. 727),

$$a_n = \frac{2}{l}\int_0^l f(x)\sin\frac{n\pi x}{l}\,dx, \quad b_n = \frac{2}{na\pi}\int_0^l \varphi(x)\sin\frac{n\pi x}{l}\,dx.$$

(2) Considering *longitudinal vibrations of a bar* with one end free and the other end subjected to a constant force p leads to a differential equation discussed in example 1

$$\frac{\partial^2 u}{\partial t^2} = a^2\frac{\partial^2 u}{\partial x^2}$$

with the initial conditions $(u)_{t=0} = f(x)$, $\left(\frac{\partial u}{\partial t}\right)_{t=0} = \varphi(x)$ but with non-homogeneous boundary conditions $\left(\frac{\partial u}{\partial x}\right)_{x=0} = 0$ (for the free end), and $\left(\frac{\partial u}{\partial x}\right)_{x=l} = kp$. These conditions can be replaced by homogeneous ones $\left(\frac{\partial z}{\partial x}\right)_{x=0} = \left(\frac{\partial z}{\partial x}\right)_{x=l} = 0$ introducing a new unknown function $z = u - kpx^2/2l$, but then the equation becomes non-homogeneous:

$$\frac{\partial^2 z}{\partial t^2} = a^2\frac{\partial^2 z}{\partial x^2} + \frac{a^2kp}{l}.$$

We shall seek a solution in the form $z = v + w$, where v satisfies the homogeneous differential equation and the boundary and initial conditions for z, i.e., $(z)_{t=0} = f(x) - \frac{1}{2}kpx^2$, $\left(\frac{\partial z}{\partial t}\right)_{t=0} = \varphi(x)$, and w satisfies the non-homogeneous differential equation and the zero initial and boundary conditions. It is easy to observe that $w = ka^2pt^2/2l$.

Assuming $v = X(x)T(t)$ and substituting to the equation, we obtain, as above, $\frac{X''}{X} = \frac{T''}{a^2T} = -\lambda^2$. Integrating the equation for X with the boundary conditions $X'(0) = X'(l) = 0$, we find the eigenfunctions of the given problems $X_n = \cos(n\pi x/l)$ and the corresponding eigenvalues $\lambda_n^2 = n^2\pi^2/l^2$ $(n = 1, 2, \ldots)$. Proceeding as in the previous example, we obtain finally

$$u = \frac{ka^2pt^2}{2l} + \frac{kpx^2}{2l} + a_0 + \frac{a\pi}{l} b_0 t + \\ + \sum_{n=1}^{\infty} \left(a_n \cos \frac{an\pi t}{l} + \frac{b_n}{n} \sin \frac{an\pi t}{l} \right) \cos \frac{n\pi x}{l},$$

where a_n and b_n $(n = 0, 1, 2, \ldots)$ are the coefficients of the expansion into Fourier series in cosines of the functions $f(x) - \frac{kpx^2}{2}$ and $\frac{l}{a\pi} \varphi(x)$ in the interval $(0, l)$ (see p. 727).

(3) *Vibrations of a round membrane* fixed along the boundary lead to the equation

$$\frac{\partial^2 u}{\partial x^2} + \frac{\partial^2 u}{\partial y^2} = \frac{1}{a^2} \cdot \frac{\partial^2 u}{\partial t^2},$$

or in polar coordinates (see p. 376):

$$\frac{\partial^2 u}{\partial r^2} + \frac{1}{r} \cdot \frac{\partial u}{\partial r} + \frac{1}{r^2} \cdot \frac{\partial^2 u}{\partial \varphi^2} = \frac{1}{a^2} \cdot \frac{\partial^2 u}{\partial t^2},$$

with the conditions $(u)_{t=0} = f(r, \varphi)$, $\left(\frac{\partial u}{\partial t}\right)_{t=0} = F(r, \varphi)$, $(u)_{r=R} = 0$.

Let $u = U(r)\, \Phi(\varphi)\, T(t)$. Substituting in the equation, we obtain

$$\frac{U''}{U} + \frac{U'}{rU} + \frac{\Phi''}{r^2 \Phi} = \frac{1}{a^2} \cdot \frac{T''}{T}.$$

Hence, as above, $T'' + a^2 \lambda^2 T = 0$ and

$$\frac{r^2 U'' + rU'}{U} + \lambda^2 r^2 = -\frac{\Phi''}{\Phi} = \nu^2,$$

or $\Phi'' + \nu^2 \Phi = 0$.

From the conditions $\Phi(0) = \Phi(2\pi)$, $\Phi'(0) = \Phi'(2\pi)$, we find

$$\Phi(\varphi) = a_n \cos n\varphi + b_n \sin n\varphi, \qquad \nu^2 = n^2,$$

where $n = 0, 1, 2, \ldots$

To determine U and λ we have $[rU']' - \frac{n^2}{r} U = -\lambda^2 rU$ with the condition $U(R) = 0$. Adding the natural condition of boundedness of $U(r)$ for $r = 0$ and substituting $\lambda r = z$, we find

$$z^2 U'' + zU' + (z^2 - n^2) U = 0, \qquad \text{i.e.,} \qquad U(r) = J_n(z) = J_n\left(\mu \frac{r}{R}\right),$$

where J_n is the Bessel's function (see p. 549) and $\lambda = \mu/R$, $J_n(\mu) = 0$.

Let μ_{nk} be the k-th positive root of the function $J_n(z)$. The system of functions $U_{nk}(r) = J_n\left(\mu_{nk}\frac{r}{R}\right)$, where $k = 1, 2, \ldots$, is a complete system of eigenfunctions of the self-conjugate problem of Sturm-Liouville which are orthogonal with the weight r (see pp. 554, 555).

The solution of our problem has the form of a double series:

$$U = \sum_{n=0}^{\infty}\sum_{k=1}^{\infty}\Bigg((a_{nk}\cos n\varphi + b_{nk}\sin n\varphi)\cos\frac{a\mu_{nk}t}{R} +$$

$$+ (c_{nk}\cos n\varphi + d_{nk}\sin n\varphi)\sin\frac{a\mu_{nk}t}{R}\Bigg)J_n\left(\mu_{nk}\frac{r}{R}\right).$$

From the initial conditions at $t = 0$ we obtain

$$f(r, \varphi) = \sum_{n=0}^{\infty}\sum_{k=1}^{\infty}(a_{nk}\cos n\varphi + b_{nk}\sin n\varphi)\, J_n\left(\mu_{nk}\frac{r}{R}\right),$$

$$F(r, \varphi) = \sum_{n=0}^{\infty}\sum_{k=1}^{\infty}\frac{a\mu_{nk}}{R}(c_{nk}\cos n\varphi + d_{nk}\sin n\varphi)\, J_n\left(\mu_{nk}\frac{r}{R}\right),$$

whence

$$\begin{matrix} a_{nk} \\ b_{nk} \end{matrix} = \frac{2}{\pi R^2 J_{n-1}^2(\mu_{nk})}\int_0^{2\pi} d\varphi \int_0^R f(r, \varphi)\begin{matrix} \cos n\varphi \\ \sin n\varphi \end{matrix}\, J_n\left(\mu_{nk}\frac{r}{R}\right) r\, dr.$$

For $n = 0$, the numerator 2 should be changed into 1. The formulas for the coefficients c_{nk} and d_{nk} can be obtained from those for a_{nk} and b_{nk} by replacing $f(r, \varphi)$ with $F(r, \varphi)$ and multiplying by $R/a\mu_{nk}$.

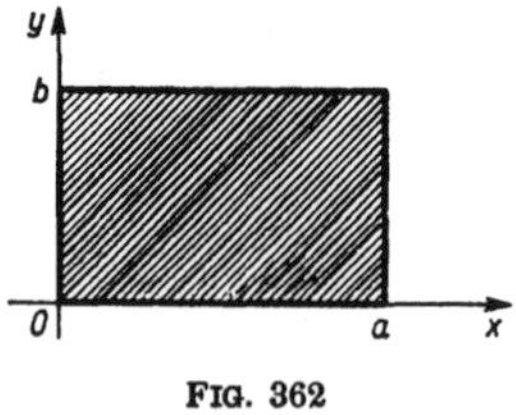

Fig. 362

(4) *Dirichlet's problem* (see p. 648) for the rectangle $0 < x < a$, $0 < y < b$ (Fig. 362). Find a function $u(x, y)$ satisfying Laplace's equation $\Delta u = 0$ and the conditions $u(0, y) = \varphi_1(y)$, $u(a, y) = \varphi_2(y)$, $u(x, 0) = \psi_1(x)$, $u(x, b) = \psi_2(x)$.

Let us first solve the problem for the case $\varphi_1(y) = \varphi_2(y) = 0$. Substituting $u = X(x)\,Y(y)$ to the equation, we obtain $\frac{X''}{X} = -\frac{Y''}{Y} = -\lambda^2$. Since $X(0) = X(a) = 0$, hence $X = C\sin\lambda x$, $\lambda = n\pi/a$, where $n = 1, 2, \ldots$ Writing the general solution of the equation $Y'' - \frac{n^2\pi^2}{a^2}Y = 0$ in the form

$$Y = a_n \sinh \frac{n\pi}{a}(b-y) + b_n \sinh \frac{n\pi}{a} y$$

we obtain a particular solution of the equation $\Delta u = 0$ satisfying the condition $u(0, y) = u(a, y) = 0$ in the form

$$u_n = \left(a_n \sinh \frac{n\pi}{a}(b-y) + b_n \sinh \frac{n\pi}{a} y\right) \sinh \frac{n\pi}{a} x.$$

Now, substituting $u = \sum u_n$, we find from the conditions at $y = 0$ and $y = b$ that

$$u = \sum_{n=1}^{\infty} \left(a_n \sinh \frac{n\pi}{a}(b-y) + b_n \sinh \frac{n\pi}{a} y\right) \sinh \frac{n\pi}{a} x,$$

where

$$a_n = \frac{2}{a \sinh(n\pi b/a)} \int_0^a \varphi_1(x) \sin \frac{n\pi}{a} x \, dx,$$

$$b_n = \frac{2}{a \sinh(n\pi b/a)} \int_0^a \varphi_2(x) \sin \frac{n\pi}{a} x \, dx.$$

Solving the analogous problem for $\varphi_1(x) = \varphi_2(x) = 0$, we find the solution of the general problem taking the sum of these two solutions.

(5) *Heat conduction in a homogeneous bar* with one end at infinity and the other kept at a constant temperature. We wish a bounded solution of the equation

$$\frac{\partial u}{\partial t} = a^2 \frac{\partial^2 u}{\partial x^2},$$

where $0 < x < +\infty$, $t > 0$ with the conditions $(u)_{t=0} = f(x)$, $(u)_{x=0} = 0$ (the constant temperature at one end is assumed to be zero). Substituting $u = X(x)T(t)$ in the equation, we obtain $\frac{T'}{a^2 T} = \frac{X''}{X} = -\lambda^2$. Hence $T(t) = C\lambda e^{-\lambda^2 a^2 t}$. From the condition of boundedness it follows that $\lambda^2 > 0$. Since $X(0) = 0$, hence $X(x) = C \sin \lambda x$ and therefore $u_\lambda = C_\lambda e^{-\lambda^2 a^2 t} \sin \lambda x$. In this formula, λ is an arbitrary real number and therefore we can consider a solution of the form

$$u(x, t) = \int_0^\infty C(\lambda) e^{-\lambda^2 a^2 t} \sin \lambda x \, d\lambda.$$

From the condition $(u)_{t=0} = f(x)$, we find $f(x) = \int_0^\infty C(\lambda) \sin \lambda x \, d\lambda$.

The latter equation will be satisfied, if we assume $C(\lambda) = \frac{2}{\pi}\int_0^\infty f(s)\sin\lambda s\,ds$ (see p. 727). Substituting $C(\lambda)$ and $u(x, t)$, we obtain

$$u(x, t) = \frac{2}{\pi}\int_0^\infty f(s)\left(\int_0^\infty e^{-\lambda^2 a^2 t}\sin\lambda s\sin\lambda x\,d\lambda\right)ds,$$

or, replacing the product of sines with one half of the difference of cosines (see p. 218) and using formula (5) on p. 481,

$$u(x, t) = \int_0^\infty f(s)\frac{1}{2a\sqrt{\pi t}}\left[\exp\left(-\frac{(x-s)^2}{4a^2t}\right) - \exp\left(-\frac{(x+s)^2}{4a^2t}\right)\right]ds.$$

Riemann's method of solving Cauchy's problem for the hyperbolic equation

$$\frac{\partial^2 u}{\partial x\partial y} + a\frac{\partial u}{\partial x} + b\frac{\partial u}{\partial y} + cu = F.$$

We find *Riemann's function* $v(x, y; \xi, \eta)$, where ξ and η are regarded as parameters, satisfying the homogeneous conjugate equation [1]

$$\frac{\partial^2 v}{\partial x\partial y} - \frac{\partial (av)}{\partial x} - \frac{\partial (bv)}{\partial y} + cv = 0$$

and the conditions

$$v(x, \eta; \xi, \eta) = \exp\left(\int_\xi^x b(s, \eta)\,ds\right), \qquad v(\xi, y; \xi, \eta) = \exp\left(\int_\eta^y a(\xi, s)\,ds\right).$$

The function $u(\xi, \eta)$ which satisfies the original equation and, along a given curve Γ (Fig. 363; the regular curve Γ should not have tangents parallel to the coordinate axes, that is to say, the curve should not be tangent to the characteristics), together with its derivative in the direction of the normal to the curve

---

[1] The equation *conjugate* to the linear equation

$$\sum_{i,k} a_{ik}\frac{\partial^2 u}{\partial x_i\partial x_k} + \sum_i b_i\frac{\partial u}{\partial x_i} + cu = f$$

is the equation

$$\sum_{i,k}\frac{\partial^2(a_{ik}v)}{\partial x_i\partial x_k} - \sum_i\frac{\partial(b_i v)}{\partial x_i} + cv = 0.$$

(see p. 279) assumes the preassigned values can be found from the formula

$$u(\xi, \eta) = \tfrac{1}{2}(uv)_P + \tfrac{1}{2}(uv)_Q -$$

$$-\int_{QP}\left[buv + \frac{1}{2}\left(v\frac{\partial u}{\partial x} - u\frac{\partial v}{\partial x}\right)\right]dx - \left[auv + \frac{1}{2}\left(v\frac{\partial u}{\partial y} - u\frac{\partial v}{\partial y}\right)\right]dy +$$

$$+\iint_{PMQ} Fv\,dxdy.$$

The line integral in this formula can be evaluated, for the values of both partial derivatives can be found from the values of the function and its derivative (in a non-tangential direction) along the arc. In the formulation of Cauchy's problem, we sometimes prescribe the values of one of the partial derivatives of the unknown function along the curve instead of the derivative in the normal direction. In this case, another form of Riemann's formula is convenient:

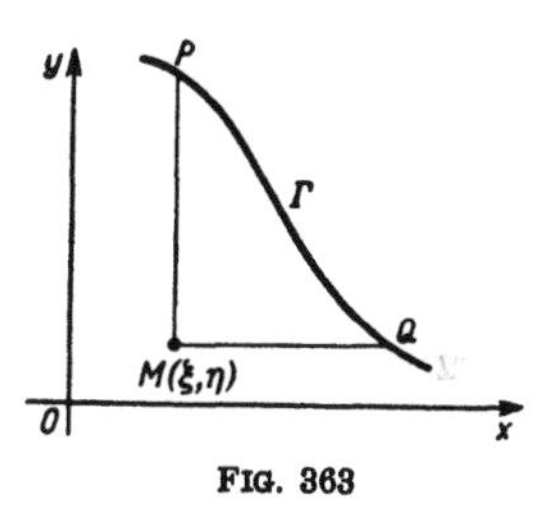

FIG. 363

$$u(\xi, \eta) = (uv)_P - \int_{QP}\left(buv - u\frac{\partial v}{\partial x}\right)dx - \left(auv + v\frac{\partial u}{\partial y}\right)dy + \iint_{PMQ} Fv\,dxdy,$$

if the values of $\dfrac{\partial u}{\partial y}$ are prescribed along the curve Γ.

Example. The *electric circuit equation* (*telegraphic equation*) has the form

$$a\frac{\partial^2 u}{\partial t^2} + 2b\frac{\partial u}{\partial t} + cu = \frac{\partial^2 u}{\partial x^2},$$

where $a > 0$ and b, c are constants.

By introducing a new unknown function $u = ze^{-(b/a)t}$, we reduce the equation to the form

$$\frac{\partial^2 z}{\partial t^2} = m^2\frac{\partial^2 z}{\partial x^2} + n^2 z,$$

where $m^2 = 1/a$, $n^2 = (b^2 - ac)/a^2$, and by the change of the independent variables $\xi = \dfrac{n}{m}(mt + x)$, $\eta = \dfrac{n}{m}(mt - x)$, we reduce it to the form

$$\frac{\partial^2 z}{\partial\xi\partial\eta} - \frac{z}{4} = 0.$$

Riemann's function $v(\xi, \eta; \xi_0, \eta_0)$ should satisfy this equation and assume the value one for $\xi = \xi_0$ and for $\eta = \eta_0$. If we seek for a solution v in the form $v = f(w)$, where $w = (\xi - \xi_0)(\eta - \eta_0)$, then $f(w)$ is a solution of the equation

$$w \frac{d^2 f}{dw^2} + \frac{df}{dw} - \frac{1}{4} f = 0$$

with the initial condition $f(0) = 1$. By means of the substitution $w = \alpha^2$, this equation can be reduced to Bessel's equation of order zero

$$\frac{d^2 f}{d\alpha^2} + \frac{1}{\alpha} \cdot \frac{df}{d\alpha} - f = 0$$

(see p. 548), hence $v = I_0\left(\sqrt{(\xi - \xi_0)(\eta - \eta_0)}\right)$.

If a solution of the original equation satisfying the initial conditions $(z)_{t=0} = f(x)$, $\left(\frac{\partial z}{\partial t}\right)_{t=0} = g(x)$ is required, then, substituting the found value of v and returning to the original variables, we obtain

$$z(x, t) = \tfrac{1}{2}[f(x - mt) + f(x + mt)] +$$

$$+ \frac{1}{2} \int_{x-mt}^{x+mt} \left(g(s) \frac{I_0\left[(n/m)\sqrt{m^2 t^2 - (s - x)^2}\right]}{m} - \right.$$

$$\left. - f(s) \frac{nt I_1\left[(n/m)\sqrt{m^2 t^2 - (s - x)^2}\right]}{\sqrt{m^2 t^2 - (s - x)^2}} \right) ds.$$

Green's method of solving the boundary-value problem for elliptic equations is very similar to Riemann's method of solving Cauchy's problem for hyperbolic equations. If a function $u(x, y)$ satisfying the equation

$$\frac{\partial^2 u}{\partial x^2} + \frac{\partial^2 u}{\partial y^2} + a \frac{\partial u}{\partial x} + b \frac{\partial u}{\partial y} + cu = f$$

in a certain domain and assuming the prescribed values on its boundary is desired, then we find first the *Green's function* $G(x, y; \xi, \eta)$ satisfying the following conditions (ξ, η are regarded as parameters):

(1) $G(x, y; \xi, \eta)$ satisfies the homogeneous conjugate equation [1]

$$\frac{\partial^2 G}{\partial x^2} + \frac{\partial^2 G}{\partial y^2} - \frac{\partial (aG)}{\partial x} - \frac{\partial (bG)}{\partial y} + cG = 0,$$

everywhere except for the point $x = \xi$, $y = \eta$;

---

[1] See footnote on p. 576.

(2) the function $G(x, y; \xi, \eta)$ has the form $U \ln (1/r) + V$, where U and V are functions continuous in the whole domain together with their derivatives of the second order and U assumes the value one at the point $x = \xi$, $y = \eta$, and $r = \sqrt{(x-\xi)^2 + (y-\eta)^2}$;

(3) the function $G(x, y; \xi, \eta)$ is equal to zero on the boundary of the domain in question.

The solution of the boundary-value problem is given, in terms of Green's function, by the formula

$$u(\xi, \eta) = \frac{1}{2\pi} \int_S u(x, y) \frac{\partial}{\partial n} G(x, y; \xi, \eta)\, ds - \\ - \frac{1}{2\pi} \iint_D f(x, y)\, G(x, y; \xi, \eta)\, dx dy,$$

where D is the domain in question, S is its boundary on which the function is assumed to be prescribed and $\frac{\partial}{\partial n}$ denotes the normal derivative directed to the interior of D.

Condition (3) depends on the formulation of the problem. For example, if the values of the derivative of the unknown function in the direction normal to the boundary of the domain are given instead of the values of the function itself, then we should require in condition (3) that the expression

$$\frac{\partial G}{\partial n} - (a \cos \alpha + b \cos \beta)\, G$$

is equal to zero on the boundary of the domain, where α and β denote the angles between the interior normal to the boundary of the domain and the coordinate axes. In this case, the solution is given by the formula

$$u(\xi, \eta) = -\frac{1}{2\pi} \int_S \frac{\partial u}{\partial n} G\, ds - \frac{1}{2\pi} \iint_D f G\, dx dy.$$

Green's method can also be applied to linear equations with three independent variables of the form

$$\Delta u + a \frac{\partial u}{\partial x} + b \frac{\partial u}{\partial y} + c \frac{\partial u}{\partial z} + eu = f.$$

To find a solution of this equation assuming the given values on the boundary of the domain, we find, as above, the Green's function (it now depends on three parameters ξ, η, ζ) which now satisfies the conjugate equation in the form

$$\Delta G - \frac{\partial (aG)}{\partial x} - \frac{\partial (bG)}{\partial y} - \frac{\partial (cG)}{\partial z} + eG = 0$$

and we require in condition (2) that G is of the form $U(1/r) + V$, where $r = \sqrt{(x-\xi)^2 + (y-\eta)^2 + (z-\zeta)^2}$. The solution of the problem is then given by the formula

$$u(\xi, \eta, \zeta) = \frac{1}{4\pi} \iint\limits_S u \frac{\partial G}{\partial n} ds - \frac{1}{4\pi} \iiint\limits_D fG \, dx dy dz.$$

In Riemann's method as well as in Green's, we have first to find a certain special solution of the differential equation which can be then used to obtain a solution with arbitrary boundary conditions. An essential difference between the Green's function and the Riemann's function is that the latter one depends only on the form of the left member of the differential equation, while the Green's function depends also on the domain in question. Finding the Green's function is, in practice, an extremely difficult problem, even if it is known to exist; therefore Green's method is used mostly in theoretical research.

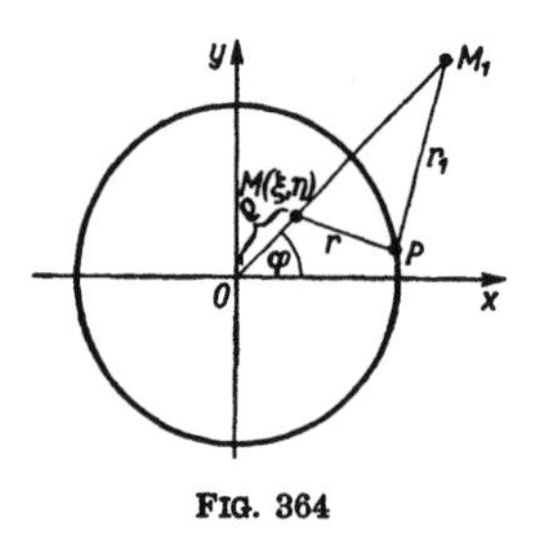

FIG. 364

Examples. (1) Green's function can be easily constructed for the Laplace's equation $\Delta u = 0$, when the given domain is a circle. If R is the radius of the circle and M_1 is the point symmetric to $M(\xi, \eta)$ with respect to the circumference, i.e., the points M and M_1 lie on one ray issuing from the centre O of the circle and $OM \cdot OM_1 = R^2$, then the Green's function is given by the formula

$$G(x, y; \xi, \eta) = \ln \frac{1}{r} + \ln \frac{\varrho r_1}{R},$$

where $r = MP$, $\varrho = OM$, $r_1 = M_1P$ (Fig. 364), and $P(x, y)$ is an arbitrary point of the circumference.

The formula for a solution of Direchlet's problem given above, after substituting the normal derivative of Green's function and after certain alterations, yields in this case the so-called *Poisson integral*

$$u(\xi, \eta) = \frac{1}{2\pi} \int\limits_0^{2\pi} \frac{R^2 - \varrho^2}{R^2 + \varrho^2 - 2R\varrho \cos(\psi - \varphi)} u(\varphi) \, d\varphi$$

(notation is the same as previously: $\xi = \varrho \cos \psi$, $\eta = \varrho \sin \psi$, $u(\varphi)$ is the given function defining the boundary values of the unknown function u).

(2) Green's function for the Dirichlet problem of Laplace's equation in space can be analogously constructed, when the domain in question is a sphere of radius R. Then the Green's function has the form

$$G(x, y, z; \xi, \eta, \zeta) = \frac{1}{r} - \frac{R}{r_1 \varrho},$$

where $\varrho = \sqrt{\xi^2 + \eta^2 + \zeta^2}$ is the distance from the point (ξ, η, ζ) to the centre of the sphere, r is the distance from the point (x, y, z) to the point (ξ, η, ζ), r_1 is the distance from the point (x, y, z) to the point $\left(\frac{R\xi}{\varrho}, \frac{R\eta}{\varrho}, \frac{R\zeta}{\varrho}\right)$ symmetric to (ξ, η, ζ) with respect to the surface of the sphere. Poisson integral has in this case (in the same notation) the form

$$u(\xi, \eta, \zeta) = \frac{1}{4\pi} \iint\limits_S \frac{R^2 - \varrho^2}{Rr^3} u \, ds.$$

Operational method. As for ordinary differential equations, we can apply the operational method, based on transition from the unknown function to its transform (see p. 541), to solution of partial differential equations. In this process, we regard the unknown function as a function of one variable and the remaining variables as parameters. We obtain then, for the unknown function, a differential equation (so-called *auxiliary equation*) containing one variable less, than the original equation. In particular, if the original equation involves two independent variables, then we obtain an ordinary differential equation for the transform. If we can find the transform of the unknown function from the obtained equation, then we determine the original function either from the table of transforms or from the formula for the inverse function.

Examples. (1) Consider the conduction of heat in a solid body bounded on one side ($x > 0$); the temperature on the boundary ($x = 0$) of the body varies according to the formula $u = k \cos \omega t$ for $t > 0$, and the temperature at the moment $t = 0$ is equal to zero.

The problem reduces to the solution of the equation

$$\frac{\partial u}{\partial t} = a^2 \frac{\partial^2 u}{\partial x^2},$$

in the domain $x > 0$, $t > 0$, with the conditions $(u)_{t=0, x>0} = 0$, $(u)_{x=0, t>0} = k \cos \omega t$. The auxiliary equation has the form

$$a^2 \frac{d^2 \bar{u}}{dx^2} - p\bar{u} = 0, \qquad x > 0$$

with the condition $\bar{u} = \frac{kp^2}{p^2 + \omega^2}$ for $x = 0$. The solution of the auxiliary equation, bounded for $x \to \infty$, is

$$\bar{u} = \frac{kp^2}{p^2 + \omega^2} \exp\left(-\frac{x}{a}\sqrt{p}\right).$$

Using formula 12 on p. 546 and Borel's theorem (see p. 542) for transition from the transform to the function, we obtain

$$u(x, t) = \frac{x}{2a\sqrt{\pi}} \int_0^t \cos \omega\tau \frac{\exp\left(-\frac{x^2}{4a^2(t-\tau)}\right)}{(t-\tau)^{3/2}} d\tau.$$

(2) A bar of length l is in a state of rest and its end $x = 0$ is fixed. At the moment $t = 0$, a force S (per unit of surface) is applied to the free end of the bar. The problem of examining the vibrations of the bar reduces to the solution of the equation

$$\frac{\partial^2 u}{\partial t^2} = a^2 \frac{\partial^2 u}{\partial x^2}$$

in the domain $0 < x < l$, $t > 0$, with the initial conditions $(u)_{t=0} = 0$, $\left(\frac{\partial u}{\partial t}\right)_{t=0} = 0$, where $0 < x < l$, and with the boundary conditions $(u)_{x=0} = 0$, $\left(\frac{\partial u}{\partial x}\right)_{x=l} = \frac{S}{E}$, where E is Young's modulus. The auxiliary equation has the form

$$\frac{d^2\bar{u}}{dx^2} - \frac{p^2}{a^2}\bar{u} = 0$$

with the conditions $(\bar{u})_{x=0} = 0$, $\left(\frac{\partial \bar{u}}{\partial x}\right)_{x=l} = \frac{S}{E}$. The solution is the function

$$\bar{u} = \frac{Sa}{Ep} \cdot \frac{\sinh (px/a)}{\cosh (pl/a)}.$$

Resolving the transform $\bar{u}$ into partial fractions or applying the inverse formula, we now obtain

$$u(x, t) = \frac{Sx}{E} - \frac{8Sl}{\pi^2 E} \sum_{n=0}^{\infty} \frac{(-1)^n}{2n+1} \sin \frac{(2n+1)\pi x}{2l} \cos \frac{(2n+1)\pi a t}{2l}.$$

Approximate methods. In solving particular problems related to the integration of partial differential equations, various approximate methods are widely applicable; the analytical approximate methods give an approximate analytical expression for the unknown function and the numerical methods enables us to obtain directly approximate values of the unknown function for certain definite values of the arguments.

The *numerical methods* are based on replacing the derivatives by quotients of finite increments; in this way, the differential equation is transformed into a system of algebraic equations which is a linear system, if the original equation is linear. Moreover, the analogue method is widely spread. It is based on the fact that one differential equation describes various physical phenomena. To solve a given equation, we construct a model in which one of the processes described by the equation takes place and we obtain the values of the unknown function directly from this model. The model usually involves elements which can be changed to a certain extent and, therefore, one model enables us sometimes to solve many differential equations.

PART FIVE

SUPPLEMENTARY CHAPTERS ON ANALYSIS

I. COMPLEX NUMBERS AND FUNCTIONS OF A COMPLEX VARIABLE

1. Fundamental concepts

Imaginary unit. The *imaginary unit* i [1] is formally defined as a number whose square is equal to -1. Introducing the imaginary unit leads to a generalization of the concept of a number, namely, to the *complex numbers* which play an important role in algebra and analysis and also have real interpretations in certain geometric and physical problems.

Complex numbers. The *general form* of a complex number is $a = \alpha + \beta i$, where α and β can be arbitrary real numbers. The number α is called the *real part* and the number β—the *imaginary part* of the complex number a. Notation

$$\alpha = \text{re}\, a, \qquad \beta = \text{im}\, a.$$

If $\beta = 0$, then $a = \alpha$ (a real number is a particular case of a complex number); if $\alpha = 0$, then $a = \beta i$ (*purely imaginary number*).

Geometric interpretation. Just as the real number can be represented by points of the number line, the complex numbers can be represented by points of the plane: the number $a = \alpha + \beta i$

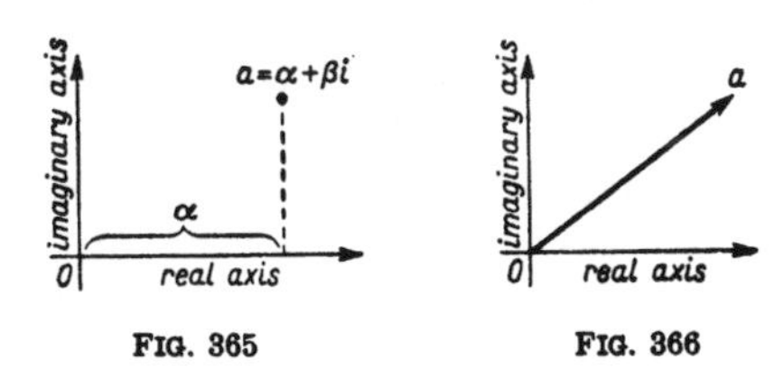

FIG. 365 FIG. 366

is represented by the point with abscissa α and ordinate β (Fig. 365). The points of the axis of abscissae (the *real axis*) represent the real

(1) In electricity, the letter j is used instead of i for the imaginary unit in order to avoid confusion with the notation i for current.

number and those of the axis of ordinates (the *imaginary axis*) represent the purely imaginary numbers. Since each point of the plane is completely determined by its radius vector (see p. 614), it follows that to each complex number, there corresponds a definite vector on the plane directed from the pole to the point representing that complex number (Fig. 366). Therefore, complex numbers can be represented either by points or by vectors.

Equality of complex numbers. Equality between complex numbers is defined as follows: two complex numbers are said to be *equal* if their real parts are equal and their imaginary parts are equal. In the geometric interpretation, two complex numbers are equal, if their corresponding points have equal abscissae and equal ordinates. Otherwise the numbers are not equal; the concepts such as "a greater number" or "a less number" do not exist for the complex numbers.

Trigonometric form of a complex number. The expression $a = \alpha + \beta i$ is called the *algebraic form* of a complex number; if we introduce polar coordinates of the point representing a complex number instead of Cartesian coordinates, then we obtain this complex number written in the *trigonometric form* (Fig. 367):

$$a = \varrho (\cos \varphi + i \sin \varphi),$$

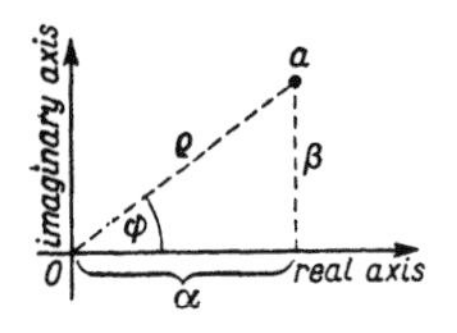

FIG. 367

where ϱ, the length of the radius vector, is called the *modulus* or *absolute value* of the complex number a and is denoted by $|a|$, and φ, the angle between the polar axis and the radius vector, expressed in radian measure is called the *argument* of the complex number a and denoted by arg a:

$$\varrho = |a|, \qquad \varphi = \arg a.$$

The relation between α, β and ϱ, φ is the same as between Cartesian and polar coordinates (see p. 237):

$$\alpha = \varrho \cos \varphi, \qquad \beta = \varrho \sin \varphi;$$

$$\varrho = \sqrt{\alpha^2 + \beta^2}, \qquad \cos \varphi = \frac{\alpha}{\sqrt{\alpha^2 + \beta^2}}, \qquad \sin \varphi = \frac{\beta}{\sqrt{\alpha^2 + \beta^2}},$$

where $0 < \varrho < \infty$, and φ can have an arbitrary value: $-\infty < \varphi < +\infty$; the argument of a given complex number has infinitely many values differing one from another by $2k\pi$ (k is an integer). The *principal value* of the argument is that contained in the interval $-\pi < \varphi \leq +\pi$. The number zero $(0 + 0i)$ has the modulus equal to zero; its argument is not defined.

Exponential form. Sometimes a complex number a with the modulus ϱ and the argument φ is written in the following form

$$a = \varrho\, e^{\varphi i} \qquad \text{(the } exponential\ form\text{)(1).}$$

Thus, for example, the number $1+\sqrt{3}i$ can be written as

$1+\sqrt{3}i$ (algebraic form),

$= 2\,(\cos\frac{1}{3}\pi + i\sin\frac{1}{3}\pi)$ (trigonometric form),

$= 2e^{\pi i/3}$ (exponential form).

Also, if we do not confine ourselves to the principal value of the argument, then

$$1+i\sqrt{3} = 2\,[\cos(\tfrac{1}{3}\pi+2k\pi)+i\sin(\tfrac{1}{3}\pi+2k\pi)] = 2e^{(\frac{1}{3}\pi+2k\pi)i},$$

$k = 0, 1, 2, \ldots$

Conjugate complex numbers. Two complex numbers are called *conjugate* (we write a and $\bar{a}$), if their real parts are equal and their imaginary parts differ only in sign:

$$\operatorname{re}\bar{a} = \operatorname{re}a, \qquad \operatorname{im}\bar{a} = -\operatorname{im}a.$$

In the geometric interpretation, the points representing two complex conjugate numbers are symmetric with respect to the abscissa axis. The moduli of two complex conjugate numbers are equal and the sum of their arguments is a multiple of 2π (in particular it may be zero):

$$a=\alpha+\beta i=\varrho(\cos\varphi+i\sin\varphi)=\varrho e^{\varphi i}, \qquad \bar{a}=\alpha-\beta i=\varrho(\cos\varphi-i\sin\varphi)=\varrho e^{-\varphi i}.$$

2. Algebraic operations with complex numbers

Addition and subtraction. *Addition* and *subtraction of two complex numbers* is defined by the formulas

$$(\alpha_1+\beta_1 i)+(\alpha_2+\beta_2 i) = (\alpha_1+\alpha_2)+(\beta_1+\beta_2)i,$$

$$(\alpha_1+\beta_1 i)-(\alpha_2+\beta_2 i) = (\alpha_1-\alpha_2)+(\beta_1-\beta_2)i.$$

In the geometric interpretation, to obtain the vector representing the sum or difference of complex numbers, we add or subtract the vectors representing the given numbers (Fig. 368, see also p. 614).

(1) For further details, see p. 593.

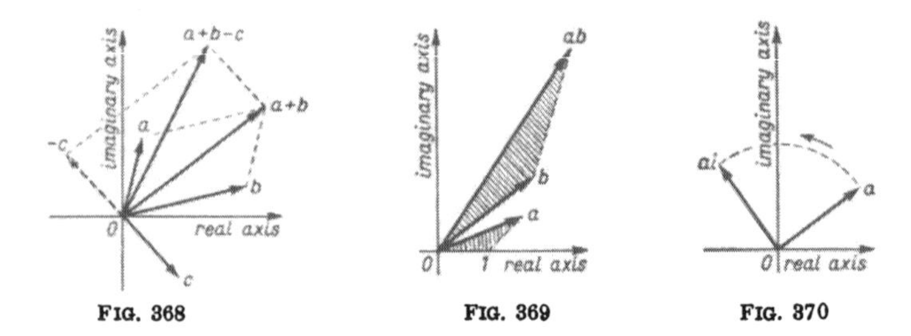

FIG. 368 FIG. 369 FIG. 370

Multiplication. The *multiplication of two complex numbers* is defined by the formula

$$(\alpha_1 + \beta_1 i)(\alpha_2 + \beta_2 i) = (\alpha_1\alpha_2 - \beta_1\beta_2) + (\alpha_1\beta_2 + \beta_1\alpha_2)i.$$

If the numbers are given in trigonometric form, then

$$[\varrho_1(\cos\varphi_1 + i\sin\varphi_1)][\varrho_2(\cos\varphi_2 + i\sin\varphi_2)] = \varrho_1\varrho_2[\cos(\varphi_1+\varphi_2) + i\sin(\varphi_1+\varphi_2)],$$

i.e., the modulus of the product is equal to the product of the moduli and the argument of the product is equal to the sum of the arguments of the factors. Geometrically, to obtain the vector representing ab, we rotate it counterclockwise through the angle arg b, and multiply its length by $|b|$. The product ab can also be obtained by aid of construction of the similar triangle (Fig. 369). In particular, in multiplication of the number a by i, the vector representing a rotates counterclockwise through the angle $\frac{1}{2}\pi$ and its length remains unchanged (Fig. 370).

Division. The *division of complex numbers* is defined as the operation inverse to the multiplication. In algebraic form

$$\frac{\alpha_1 + \beta_1 i}{\alpha_2 + \beta_2 i} = \frac{\alpha_1\alpha_2 + \beta_1\beta_2}{\alpha_2^2 + \beta_2^2} + \frac{\alpha_2\beta_1 - \alpha_1\beta_2}{\alpha_2^2 + \beta_2^2} i.$$

In trigonometric form

$$\frac{\varrho_1(\cos\varphi_1 + i\sin\varphi_1)}{\varrho_2(\cos\varphi_2 + i\sin\varphi_2)} = \frac{\varrho_1}{\varrho_2}[\cos(\varphi_1 - \varphi_2) + i\sin(\varphi_1 - \varphi_2)],$$

i.e., the modulus of the quotient of two complex numbers is equal to the quotient of the moduli of the dividend and divisor and the argument of the quotient is the difference of the arguments of the dividend and divisor. Division by zero is not allowed. Geometrically, to obtain the vector representing the quotient a/b, we rotate the vector representing a clockwise through the angle arg b, and divide its length by $|b|$.

General rule of performing four arithmetical operations. Formally, we calculate with complex numbers $\alpha + \beta i$ as with ordinary expressions assuming that $i^2 = -1$. In dividing one complex number by another, we eliminate the imaginary part of the denominator (as in rationalizing the denominator of a fraction) by multiplying both the numerator and denominator by the conjugate of the denominator using the equation $(\alpha + \beta i)(\alpha - \beta i) = \alpha^2 + \beta^2$ (a real number).

Example.

$$\frac{(3-4i)(-1+5i)^2}{1+3i} + \frac{10+7i}{5i}$$
$$= \frac{(3-4i)(1-10i-25)}{1+3i} + \frac{(10+7i)i}{5i \cdot i}$$
$$= \frac{-2(3-4i)(12+5i)}{1+3i} + \frac{7-10i}{5}$$
$$= \frac{-2(56-33i)(1-3i)}{(1+3i)(1-3i)} + \frac{7-10i}{5}$$
$$= \frac{-2(-43-201i)}{10} + \frac{7-10i}{5}$$
$$= \frac{1}{5}(50+191i) = 10 + 38.2i.$$

Raising to a power. *Raising a complex number to a power n* is performed according to *de Moivre's formula*

$$[\varrho(\cos\varphi + i\sin\varphi)]^n = \varrho^n(\cos n\varphi + i\sin n\varphi),$$

i.e., we raise the modulus to the power n and multiply the argument by n. De Moivre's formula can be used for any integral, fractional, positive or negative n. It should be pointed out that the result is not unique, when n is a fraction (see below).

In particular we have: $i^2 = -1$, $i^3 = -i$, $i^4 = 1$; in general,

$$i^{4n+k} = i^k.$$

Extracting a root. *Extracting a root*, as the operation inverse to raising to a power is performed by means of de Moivre's formula for a fractional exponent. If $a = \varrho(\cos\varphi + i\sin\varphi)$, then

$$(*) \qquad \sqrt[n]{a} = \sqrt[n]{\varrho}\left(\cos\frac{\varphi + 2k\pi}{n} + i\sin\frac{\varphi + 2k\pi}{n}\right).$$

Remark. Addition, subtraction, multiplication, division and raising to an integral exponent are unique operations, while extracting the n-th root always gives n different values. For, if we substitute $k = 0, 1, 2, \ldots, n-1$ in formula $(*)$, then $\arg \sqrt[n]{a}$

will assume the values

$$\frac{\varphi}{n}, \quad \frac{\varphi+2\pi}{n}, \quad \frac{\varphi+4\pi}{n}, \quad \ldots, \quad \frac{\varphi+2(n-1)\pi}{n}$$

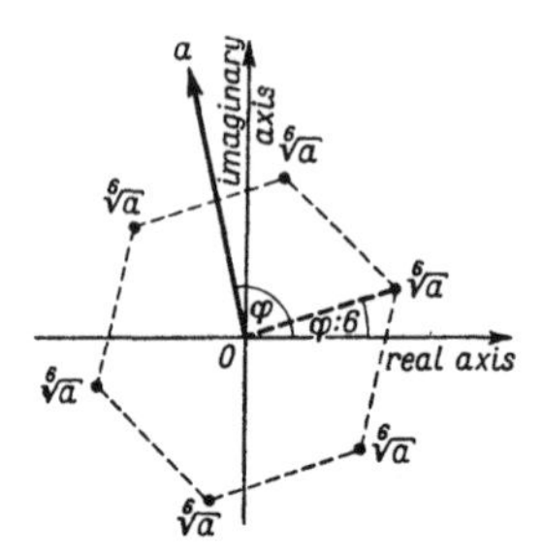

FIG. 371

differing one from another by $2\pi/n$; for further values of k, the values of arg $\sqrt[n]{a}$ repeat periodically. In geometric interpretation, the points representing $\sqrt[n]{a}$ are vertices of a regular polygon with n sides whose centre is at the pole (Fig. 371 shows 6 values of $\sqrt[6]{a}$).

Algebraic functions of a complex variable. If z is a complex variable (i.e., a quantity assuming arbitrary complex values), then the result of algebraic operations performed with z (and, possibly, with certain constants) is an *algebraic function* $w = f(z)$ of the variable z [1].

Examples.

$$w = az + b, \quad w = \frac{1}{z}, \quad w = z^2, \quad w = \frac{z+i}{z-i}, \quad w = \sqrt{z^2 - a^2}.$$

3. Elementary transcendental functions

Series with complex terms. An *infinite sequence* $z_1, z_2, \ldots, z_n, \ldots$ of complex numbers has the *limit* z ($z = \lim\limits_{n\to\infty} z_n$), if, for an arbitrary positive number ε, however small, $|z - z_n| < \varepsilon$ from a certain n onwards, i. e., the points representing the numbers $z_n, z_{n+1}, \ldots$, from n onwards, lie in the circle with centre z and radius ε.

---

[1] More generally, an *algebraic function* can be defined implicitly by an equation $a_1 z^{m_1} w^{n_1} + a_2 z^{m_2} w^{n_2} + \ldots + a_k z^{m_k} w^{n_k} = 0$, which may not be solvable explicitly (in terms of radicals) for w.

Example. $\lim \sqrt[n]{a} = 1$, for any a ($\sqrt[n]{a}$ denotes here the value of the root whose argument is the least, see Fig. 372).

The *infinite series*

$$a_1 + a_2 + \ldots + a_n + \ldots$$

with complex terms is *convergent* to the number s (the *sum* of the series), if $s = \lim_{n\to\infty} (a_1 + a_2 + \ldots + a_n)$. If the series is convergent, the end point of the broken line joining the points which represent $s_n = a_1 + a_2 + \ldots + a_n$ approaches infinitely the point s.

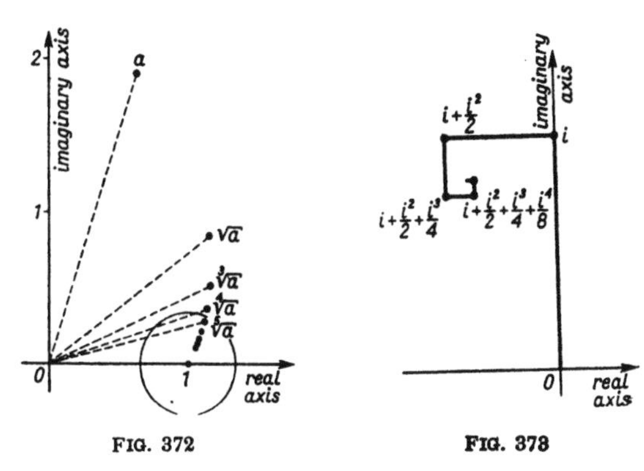

FIG. 372 FIG. 373

Examples. (1) $i + \frac{i^2}{2} + \frac{i^3}{3} + \frac{i^4}{4} + \ldots$,

(2) $i + \frac{i^2}{2} + \frac{i^3}{2^2} + \ldots$ (Fig. 373).

A series is *absolutely convergent*, if the series of moduli

$$|a_1| + |a_2| + |a_3| + \ldots$$

is convergent; it is *conditionally convergent*, if the series of moduli is divergent. In the example, series (1) converges conditionally, and series (2) converges absolutely.

A *series with variable terms*

$$f_1(z) + f_2(z) + \ldots + f_n(z) + \ldots$$

defines a certain function of z for the values of z for which it is convergent.

A *power series*

$$a_0 + a_1 z + a_2 z^2 + \ldots,$$

where a_i are complex constants, is absolutely convergent either for all values of z (on the whole plane), or for values lying inside a certain *circle of convergence* with centre at the origin; outside this circle the series is divergent. The radius of the circle of convergence is called the *radius of convergence* of the series[1]. For example, the radius of convergence of the series $1 + z + z^2 + \ldots$ is $R = 1$.

Simple exponential function. By definition,

$$e^z = 1 + \frac{z}{1!} + \frac{z^2}{2!} + \frac{z^3}{3!} + \cdots$$

This series converges absolutely in the whole plane. For a pure imaginary argument yi, $e^{yi} = \cos y + i \sin y$ (*Euler's formula*); e.g., $e^{\pi i} = -1$.

In the general case

$$e^z = e^{x+yi} = e^x(\cos y + i \sin y),$$

i.e.,

$$\operatorname{re} e^z = e^x \cos y, \qquad \operatorname{im} e^z = e^x \sin y, \qquad |e^z| = e^x, \qquad \arg e^z = y.$$

It follows from that the *exponential form* of a complex number

$$a + bi = \varrho e^{\varphi i + 2k\pi i}.$$

The function e^z is periodic with the period $2\pi i$. Thus

$$e^z = e^{z+2k\pi i}.$$

Example. $e^{2k\pi i} = e^0 = 1$, $e^{(2k+1)\pi i} = e^{\pi i} = -1$.

Euler's formulas for complex numbers:

$$e^{zi} = \cos z + i \sin z, \qquad e^{-zi} = \cos z - i \sin z$$

(for trigonometric functions of a complex argument see below).

Natural logarithm. By definition,

$$w = \operatorname{Ln} z, \quad \text{if} \quad z = e^w.$$

If $z = \varrho e^{\varphi i}$, then $\operatorname{Ln} z = \ln \varrho + (\varphi + 2k\pi) i$, i.e., $\operatorname{re}(\operatorname{Ln} z) = \ln \varrho$, $\operatorname{im}(\operatorname{Ln} z) = \varphi + 2k\pi$ ($k = 0, \pm 1, \pm 2, \ldots$). $\operatorname{Ln} z$ is a multiple-valued function. Confining ourselves to the principal value of φ (p. 586), we obtain the *principal value* of the logarithm (denoted by $\ln z$):

---

[1] The convergence of a series at the points lying on the circumference of the circle of convergence requires additional examining in each particular case.

$$\ln z = \ln \varrho + \varphi i, \quad \text{where} \quad -\pi < \varphi < +\pi.$$

Ln z exists for all complex numbers z except for zero.

General exponential function. By definition, $a^z = e^{z \operatorname{Ln} a}$ $(a \neq 0)$. a^z is a multiple-valued function. Its principal value is $e^{z \ln a}$.

Trigonometric and hyperbolic functions. By definition,

$$\sin z = z - \frac{z^3}{3!} + \frac{z^5}{5!} - \ldots = \frac{e^{zi} - e^{-zi}}{2i},$$

$$\cos z = 1 - \frac{z^2}{2!} + \frac{z^4}{4!} - \ldots = \frac{e^{zi} + e^{-zi}}{2},$$

$$\sinh z = z + \frac{z^3}{3!} + \frac{z^5}{5!} + \ldots = \frac{e^z - e^{-z}}{2},$$

$$\cosh z = 1 + \frac{z^2}{2!} + \frac{z^4}{4!} + \ldots = \frac{e^z + e^{-z}}{2}.$$

These series are convergent in the whole plane.

The functions $\sin z$ and $\cos z$ are periodic with the period 2π, the functions $\sinh z$ and $\cosh z$ are periodic with the period $2\pi i$.

The formulas for trigonometric and hyperbolic functions of a real number (pp. 216–218 and 229–232) remain true for functions of a complex variable. In particular, $\sin z$, $\cos z$, $\sinh z$, $\cosh z$, for $z = x + yi$, can be calculated by means of the formulas for $\sin (a+b)$, $\cos (a+b)$, $\sinh (a+b)$, $\cosh (a+b)$. For example,

$$\begin{aligned} \cos (x + yi) &= \cos x \cos yi - \sin x \sin yi \\ &= \cos x \cosh y - i \sin x \sinh y \end{aligned}$$

and, consequently,

$$\operatorname{re} (\cos z) = \cos (\operatorname{re} z) \cosh (\operatorname{im} z),$$

$$\operatorname{im} (\cos z) = - \sin (\operatorname{re} z) \sinh (\operatorname{im} z).$$

The functions $\tan z$, $\cot z$, $\tanh z$, $\coth z$ are defined by the formulas

$$\tan z = \frac{\sin z}{\cos z}, \qquad \cot z = \frac{\cos z}{\sin z},$$

$$\tanh z = \frac{\sinh z}{\cosh z}, \qquad \coth z = \frac{\cosh z}{\sinh z}.$$

Inverse trigonometric and hyperbolic functions. Arc sin z, Arc cos z, Arc tan z, Arc cot z, and Ar sinh z, Ar cosh z, Ar tanh z,

Ar coth z, are defined just as for a real variable (see pp. 223 and 232). For example, $w = \operatorname{Arc} \sin z$, if $z = \sin w$.

These functions have infinitely many values and can be expressed in terms of logarithms by the formulas

$$\operatorname{Arc}\sin z = -i \operatorname{Ln}\left(iz + \sqrt{1 - z^2}\right), \qquad \operatorname{Ar}\sinh z = \operatorname{Ln}\left(z + \sqrt{z^2 + 1}\right),$$

$$\operatorname{Arc}\cos z = -i \operatorname{Ln}\left(z + \sqrt{z^2 - 1}\right), \qquad \operatorname{Ar}\cosh z = \operatorname{Ln}\left(z + \sqrt{z^2 - 1}\right),$$

$$\operatorname{Arc}\tan z = \frac{1}{2i} \operatorname{Ln}\frac{1 + iz}{1 - iz}, \qquad \operatorname{Ar}\tanh z = \frac{1}{2} \operatorname{Ln}\frac{1 + z}{1 - z},$$

$$\operatorname{Arc}\cot z = -\frac{1}{2i} \operatorname{Ln}\frac{iz + 1}{iz - 1}, \qquad \operatorname{Ar}\coth z = \frac{1}{2} \operatorname{Ln}\frac{z + 1}{z - 1}.$$

The principal values of the inverse trigonometric and hyperbolic functions are expressed by the same formulas in terms of ln, i.e., the principal value of Ln:

$$\operatorname{arc}\sin z = -i \ln\left(iz + \sqrt{1 - z^2}\right), \qquad \operatorname{ar}\sinh z = \ln\left(z + \sqrt{z^2 + 1}\right),$$

$$\operatorname{arc}\cos z = -i \ln\left(z + \sqrt{z^2 - 1}\right), \qquad \operatorname{ar}\cosh z = \ln\left(z + \sqrt{z^2 - 1}\right),$$

$$\operatorname{arc}\tan z = \frac{1}{2i} \ln\frac{1 + iz}{1 - iz}, \qquad \operatorname{ar}\tanh z = \frac{1}{2} \ln\frac{1 + z}{1 - z},$$

$$\operatorname{arc}\cot z = -\frac{1}{2i} \ln\frac{iz + 1}{iz - 1}, \qquad \operatorname{ar}\coth z = \frac{1}{2} \ln\frac{z + 1}{z - 1}.$$

The following table gives the real and imaginary parts, moduli and arguments of the trigonometric and hyperbolic functions of the complex variable $z = x \pm iy$.

Expressions for re w and im w

Function $w = f(x \pm iy)$	Real part re w	Imaginary part im w
$\sin(x \pm iy)$	$\sin x \cosh y$	$\pm \cos x \sinh y$
$\cos(x \pm iy)$	$\cos x \cosh y$	$\mp \sin x \sinh y$
$\tan(x \pm iy)$	$\dfrac{\sin 2x}{\cos 2x + \cosh 2y}$	$\pm \dfrac{\sinh 2y}{\cos 2x + \cosh 2y}$
$\sinh(x \pm iy)$	$\sinh x \cos y$	$\pm \cosh x \sin y$
$\cosh(x \pm iy)$	$\cosh x \cos y$	$\pm \sinh x \sin y$
$\tanh(x \pm iy)$	$\dfrac{\sinh 2x}{\cosh 2x + \cos 2y}$	$\pm \dfrac{\sin 2y}{\cosh 2x + \cos 2y}$

Expressions for $|w|$ and $\arg w$

Function $w = f(x \pm iy)$	Modulus $\lvert w \rvert$	Argument $\arg w$
$\sin(x \pm iy)$	$\sqrt{\sin^2 x + \sinh^2 y}$	$\pm \arctan(\cot x \tanh y)$
$\cos(x \pm iy)$	$\sqrt{\cos^2 x + \sinh^2 y}$	$\mp \arctan(\tan x \tanh y)$
$\sinh(x \pm iy)$	$\sqrt{\sinh^2 x + \sin^2 y}$	$\pm \arctan(\coth x \tan y)$
$\cosh(x \pm iy)$	$\sqrt{\sinh^2 x + \cos^2 y}$	$\pm \arctan(\tanh x \tan y)$

4. Equations of curves in complex form

Complex function of a real variable. The points z representing a function $z = f(t)$, where $z = x + yi$ and t is a real variable, form a certain curve. Its *parametric equations* are $x = x(t)$, $y = y(t)$; equation in the *complex form*: $z = f(t)$.

Examples of curves in complex form.

(1) *Straight line*

(a) passing through the point z_1 and forming the angle φ with the x axis; equation: $z = z_1 + te^{\varphi i}$ (Fig. 374a);

(b) passing through two points z_1 and z_2; equation: $z = z_1 + t(z_2 - z_1)$ (Fig. 374b).

(2) *Circle*

(c) with radius r and centre at the origin; equation: $z = re^{ti}$ (Fig. 374c);

(d) with radius r and centre at the point z_1; equation: $z = z_1 + re^{ti}$ (Fig. 374d).

(3) *Hyperbola*

(e) in the canonical form $\frac{x^2}{a^2} - \frac{y^2}{b^2} = 1$; equation: $z = a \cosh t + ib \sinh t$ or $z = ce^t + \bar{c}e^{-t}$, where c and $\bar{c}$ are conjugate complex numbers, i.e., $c = \frac{a + bi}{2}$, $\bar{c} = \frac{a - bi}{2}$ (Fig. 374e).

(4) *Ellipse*

(f) in the canonical form $\frac{x^2}{a^2} + \frac{y^2}{b^2} = 1$; equation: $z = a \cos t + ib \sin t$ or $z = ce^{ti} + de^{-ti}$, where $c = \frac{a + b}{2}$, $d = \frac{a - b}{2}$ are arbitrary positive real numbers (Fig. 374f);

(g) in the general form, with the centre at the point z_1 and with the axes rotated through a certain angle; equation: $z = z_1 + ce^{ti} + de^{-ti}$, where c and d are arbitrary complex numbers which determine the lengths of the axes and the angle of rotation (Fig. 374g).

(5) *Logarithmic spiral*: equation: $z = ae^{bt}$, where a and b are arbitrary complex numbers (Fig. 374h).

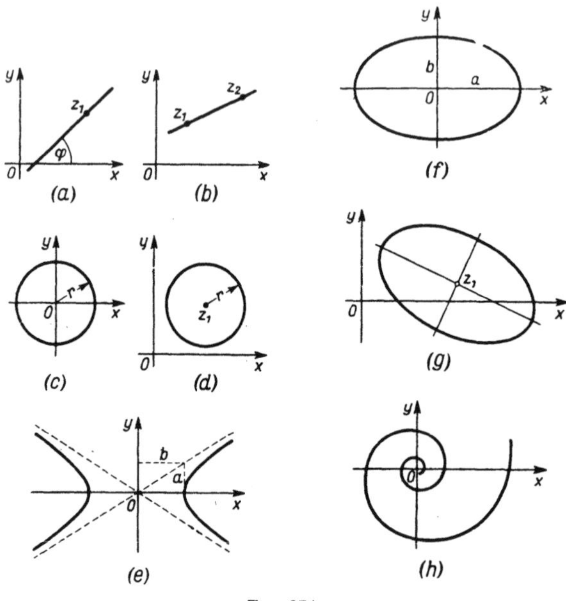

FIG. 374

5. Functions of a complex variable

Mapping of the plane. A function $w = f(z)$, where $z = x + yi$ and $w = u + vi$, is *defined*, if two functions of two real variables

$$u = u(x, y), \qquad v = v(x, y)$$

are known.

A function of a complex variable represents a mapping of the z plane into the plane w (¹); each point z_1 is sent to a corresponding point w_1, geometric objects (curves, regions) of the z plane are mapped into objects on the w plane. The curve $x = x(t)$, $y = y(t)$

(¹) If a function $w = f(z)$ is multiple-valued (as, for example, $\sqrt[n]{z}$, Ln z, Arc sin z, Ar tanh z), then the domain of values of w is a set of several or infinitely many surfaces placed one upon another, and to each value of the function there corresponds a point on one of the surfaces. These planes joined together along certain lines form the so-called *many-sheet Riemann surface*.

is mapped into the curve $u = u(x(t), y(t))$, $v = v(x(t), y(t))$ (t is a parameter).

The coordinate lines $y = c$ are mapped into $u = u(x, c)$, $v = v(x, c)$, where x is a parameter; the coordinate lines $x = c_1$ are mapped into $u = u(c_1, y)$, $v = v(c_1, y)$, where c_1 is a parameter.

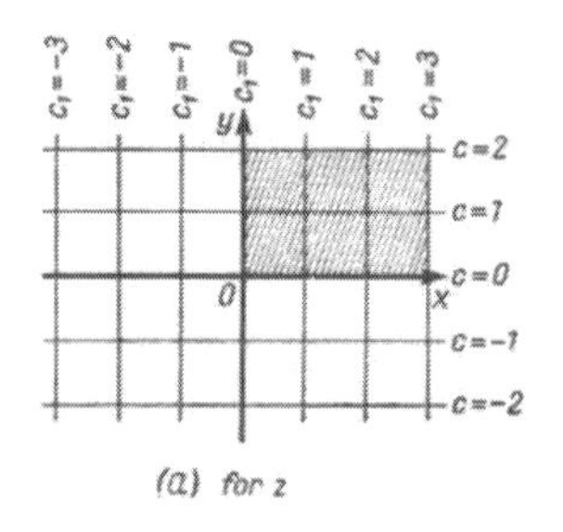

(a) for z

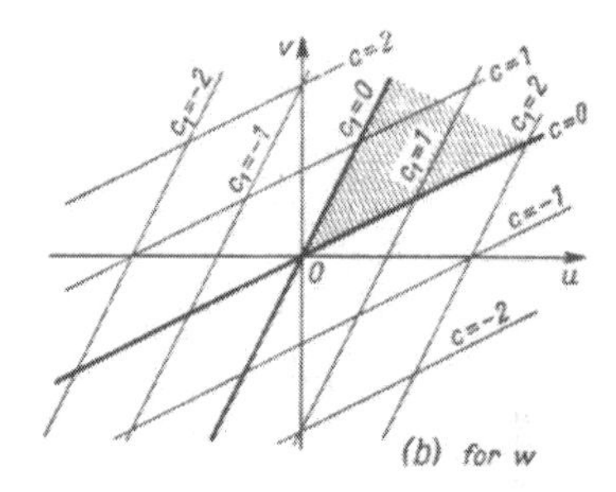

(b) for w

FIG. 375

Example of a mapping: $u = 2x + y$, $v = x + 2y$ (Fig. 375). The lines $y = c$ are mapped into $u = 2x + c$, $v = x + 2c$, i.e., into the straight lines $v = \frac{1}{2}u + \frac{3}{2}c$; the lines $x = c_1$ are mapped into the straight lines $v = 2u - 3c_1$; the region lined on the z plane is mapped onto the lined region on the w plane.

Limit, continuity, derivative. The concepts of a limit, continuity and derivative of a function $w = f(z)$ of a complex variable are defined formally in the same way as for a function of a real variable (see pp. 327, 334, 360).

The complex number A is said to be a *limit* of a function $f(z)$ when z tends to a:

$$A = \lim_{z \to a} f(z), \tag{*}$$

if, for an arbitrarily small positive number ε, we can find a positive number η such that, for any complex number z satisfying the condition $|a - z| < \eta$ (except, possibly, for $z = a$), the condition $|A - f(z)| < \varepsilon$ is satisfied.

Geometric significance (Fig. 376): each point z (except, possibly, the point a) lying within a circle with radius η is sent by the function $w = f(z)$ into the circle with centre A and radius ε.

If the function $w = f(z)$ has a limit for $x \to a$ and if

$$\lim_{z \to a} f(z) = f(a) \tag{**}$$

(i.e., the limit of the function is equal to its value at the limit point of the independent variable), then the function w is said to be *continuous at the point a*. An equivalent definition of the continui-

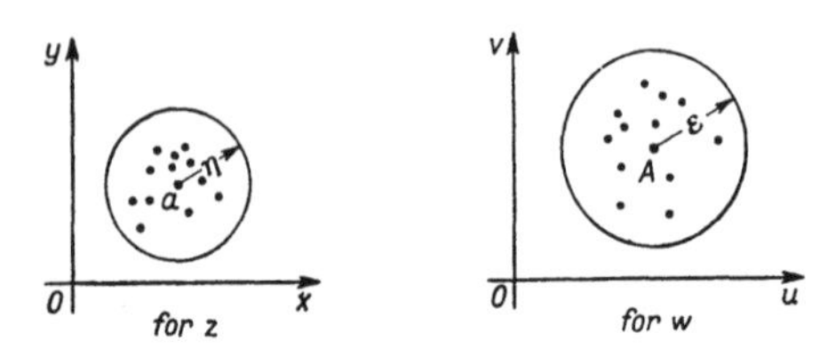

FIG. 376

ty: the function $w = f(z)$ is *continuous at the point z*, if the condition $|\Delta z| \to 0$ implies that

$$|\Delta w| = |f(z+\Delta z) - f(z)| \to 0 \quad \text{or} \quad \lim_{\Delta z \to 0} \Delta w = 0$$

(i.e., to infinitesimal increments of the independent variable there correspond infinitesimal increments of the function).

The *derivative* $w' = f'(z)$ of the function $w = f(z)$ is defined, for a given value z, by the equation

$$(\ast\ast) \qquad w' = f'(z) = \lim_{\Delta z \to 0} \frac{f(z+\Delta z) - f(z)}{\Delta z}.$$

If the limit (**) exists at a given point, then the function $f(z)$ is said to be *differentiable at this point*. (For a geometric significance of the modulus and argument of the derivative of a function of a complex variable see below, p. 601.)

Analytic functions. A function $w = f(z)$ differentiable at each point of a certain neighbourhood of a point z_0 (i.e., at each point of a circle with centre z_0 and an arbitrary small radius) is said to be *analytic at the point* z_0. A function is called *analytic in a connected domain* (see p. 341), if it is analytic at each point of this domain.

A necessary and sufficient [1] condition that the function $u(x, y) + iv(x, y) = f(x + yi)$ is analytic is that it satisfy the so-called *Cauchy-Riemann conditions*

$$\frac{\partial u}{\partial x} = \frac{\partial v}{\partial y}, \quad \frac{\partial u}{\partial y} = -\frac{\partial v}{\partial x}.$$

[1] In order that the condition shall be sufficient, we should require, in addition, that the partial derivatives involved in Cauchy-Riemann condition are continuous in the given domain.

Example. The function $w = z^2$ ($u = x^2 - y^2$, $v = 2xy$) is analytic in the whole plane; the function $w = u + vi$, where $u = 2x + y$, $v = x + 2y$, is nowhere analytic.

If the function $w = u + vi$ is analytic, then the functions u and v are *harmonic functions* of the real variables x and y, i.e., they satisfy Laplace's equation (see p. 648). If we know a harmonic function u, then we can determine up to a constant factor its conjugate harmonic function v from Cauchy-Riemann conditions:

$$v = \int \frac{\partial u}{\partial x} dy + \varphi(x) \quad \text{where} \quad \frac{d\varphi}{dx} = -\left(\frac{\partial u}{\partial y} + \frac{\partial}{\partial x}\int \frac{\partial u}{\partial x} dy\right).$$

Similarly, we can determine u from v.

The points at which the function is analytic are called *regular*. If the function is analytic in a certain domain except for some of its points, then such points are called *singular*. For examples and classification of singular points see p. 601. Elementary functions (algebraic and transcendental ones, see pp. 323, 324) are analytic in the whole plane except for certain isolated singular points.

Analytic functions have the derivative of an arbitrary order at each regular point. The derivatives of elementary functions of a complex variable can be calculated by the same rules as for a real variable.

Modulus of an analytic function. The *absolute value* (*modulus*)

$$|w| = |f(z)| = \sqrt{(u(x, y))^2 + (v(x, y))^2} = \varphi(x, y)$$

of a function plays an essential role in the theory and application of the functions of a complex variable. The surface $|w| = \varphi(x, y)$, where $|w|$ is the third coordinate, is called the *relief* of the function $w = f(z)$. For example, for the function $\sin z = \sin x \cosh y + i \cos x \sinh y$,

$$|\sin z| = \sqrt{\sin^2 x + \sinh^2 x}.$$

Figure 377a illustrates the relief of $\sin z$ and Fig. 377b—that of the function $w = e^{1/z}$.

Since the modulus of a function is non-negative, its relief lies always above the z plane except for the points at which $|f(z)| = 0$, and hence, also $f(z) = 0$. Such values of z, or the roots of the equation $f(z) = 0$, are called *zeros* of the function $f(z)$.

The function $f(z)$ is called *bounded* in a given domain, if there exists a positive number N with $|f(z)| < N$, for any z of the domain; if such a number N does not exist, the function is called *unbounded*.

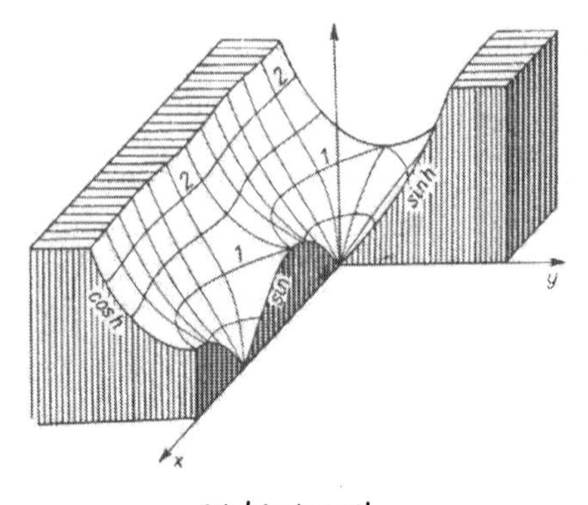

(a) $|\sin(x+yi)|$

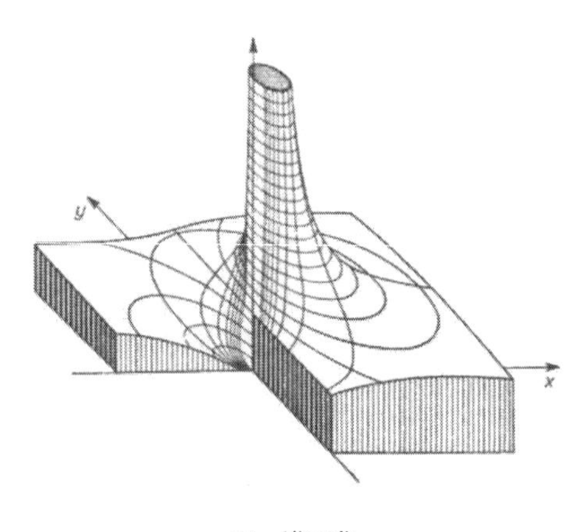

(b) $e^{1/(x+yi)}$

FIG. 377

Fundamental theorems on the modulus of an analytic function.

(1) If the function $w = f(z)$ is analytic in a closed domain, then its modulus reaches its maximum on the boundary of the domain.

(2) If the function $w = f(z)$ is analytic and bounded in the whole plane, then it is constant: $f(z) = \text{const}$ (*Liouville's theorem*).

Singular points. If the function $w = f(z)$ is analytic in a neighbourhood of a point $z = a$ [1] and bounded in this neighbourhood then the following two cases can occur:

(1) $f(a) = \lim_{z \to a} f(z)$. In this case the function is analytic also at the point a.

(2) $f(a)$ has a different value or the function is not defined at the point a. Then a is a singular point; such a singularity is called *removable*, for the function $f(z)$ can be made to be analytic also at the point a by replacing $f(a)$ with $\lim_{z \to a} f(z)$ [2].

If the function $w = f(z)$ is analytic in a neighbourhood of the point $z = a$, but is unbounded in this neighbourhood, then a is a singular point and two cases can then occur:

(1) $f(z) \to \infty$, when z approaches a in an arbitrary way. Such a point is called a *pole*. In this case we introduce the notation $f(a) = \infty$. For the order of a pole see p. 610.

(2) $|f(z)|$ does not tend to any number, when z approaches a: the sequences $f(z_1), f(z_2), \ldots, f(z_n), \ldots$ have various limits depending on the particular choice of the points z_n tending to the point a. In this case, the point a is called an *essentially singular point* [3].

Example. The point a is a pole of the function $w = \dfrac{1}{z - a}$; the point O is an essentially singular point of the function $w = e^{1/z}$ (see Fig. 377b).

Conformal mappings. A mapping of the plane induced by an analytic function has the following important properties in a neighbourhood of a point z at which $w' \neq 0$:

Infinitely small vectors attached at this point in all directions

(1) have their lengths multiplied by the same number $|w'|$ (with the accuracy up to infinitesimals of a higher order) and

(2) all rotate through the same angle equal to $\arg w'$.

Thus the figures lying in sufficiently small regions are mapped into similar figures and preserve their shape (Fig. 378). Such a mapping is called *conformal*. Figures of a finite size may be subjected to deformations, but the angle between two curves remains preserved (Fig. 379). In particular, the coordinate lines $x = \text{const}$

---

[1] I.e., within an arbitrarily small circle with centre a except, possibly, the point a.

[2] This case is analogous to that of a removable singularity of a function of a real variable (see p. 336).

[3] In this case, we can find a sequence $\{z_n\}$ tending to a such that the sequence $\{f(z_n)\}$ will tend to an arbitrary preassigned complex number (except, at most, for one complex number).

and $y = \text{const}$ are mapped into two families of mutually orthogonal curves.

Thus a variety of orthogonal curvilinear coordinate systems can be obtained by means of analytic functions.

Conversely, for each conformal mapping, there exists a net of orthogonal curves which is mapped into the rectangular Cartesian

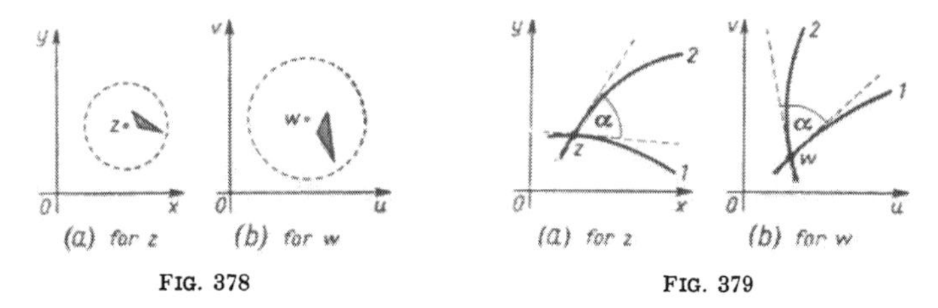

FIG. 378

FIG. 379

coordinate net. In the example $u = 2x + y$, $v = x + 2y$ (p. 597), the orthogonality was not preserved. In the example $w = z^2$, it is preserved: the coordinate lines are mapped into two families

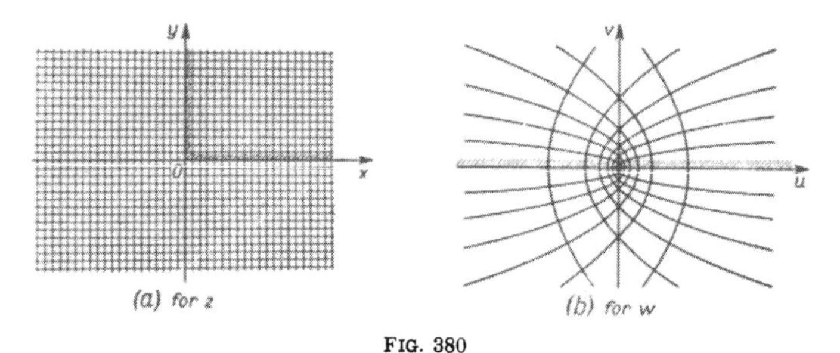

FIG. 380

of confocal parabolas (Fig. 380). At the point $z = 0$, we have $w' = 0$ and the mapping fails to be conformal. The first quadrant is mapped into the upper half-plane.

Conformal mappings have important applications in electricity, hydrodynamics, aerodynamics and in other branches of applied mathematics.

6. Simplest conformal mappings

We consider here the most frequently applied conformal mappings. In each case we give a graph of the orthogonal net of curves which is mapped into the Cartesian rectangular net.

The boundary of domains mapped into the upper half-plane are lined. The domains mapped onto the square with vertices (0,0), (0,1), (1,0), (1,1) (Fig. 381) are marked black.

(a) **Linear function** (Fig. 382a)

$$w = az + b \qquad (a = \varrho e^{\varphi i}).$$

The mapping can be decomposed into three mappings:

$t = e^{\varphi i}$ (rotation of the plane through the angle φ),

$s = \varrho t$ (similarity in the ratio ϱ),

$w = s + b$ (parallel translation by the vector b).

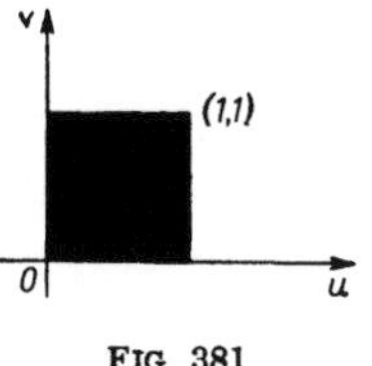

FIG. 381

The figures in the z plane are mapped into similar figures which are, moreover, rotated and translated. The points $z_1 = \dfrac{b}{1-a}$ and $z_2 = \infty$ are mapped into itself (fixed points).

(b) **Inversion** (Fig. 382b)

$$w = \frac{1}{z}.$$

The point z with the radius vector ϱ and polar angle φ is mapped into the point with the radius vector $1/\varrho$ and the polar angle $-\varphi$. The mapping is composed of the inversion with respect to the unit circle (¹) and the reflection in the x axis. The circles are mapped into circles (straight lines being regarded as circles with an infinite radius). The point O is sent to ∞, $+1$ and -1 are fixed points. The mapping fails to be conformal at $z=0$.

(c) **Homographic (or fractional-linear) function** (Fig. 382c)

$$w = \frac{az+b}{cz+d}, \qquad \text{where} \quad c \neq 0 \text{ and } ad - bc \neq 0.$$

The mapping can be decomposed into three mappings

$t = cz + d$ (linear function),

$s = \dfrac{1}{t}$ (inversion),

$w = \dfrac{a}{c} + \dfrac{bc - ad}{c} s$ (linear function).

(¹) The *inversion* with respect to a circle of the radius R is a mapping of the plane by which a point M_1 lying at the distance d_1 from the centre of the circle is sent into the point M_2 lying on the same ray OM_1 at the distance $OM_2 = d_2 = R^2/d_1$; thus the point M_2 is sent into M_1. The points lying inside the circle are mapped outside and conversely.

A homographic function maps a circle into a circle (straight lines being regarded as a particular case of circles); there are two fixed points satisfying the equation $z = \dfrac{az + b}{cz + d}$.

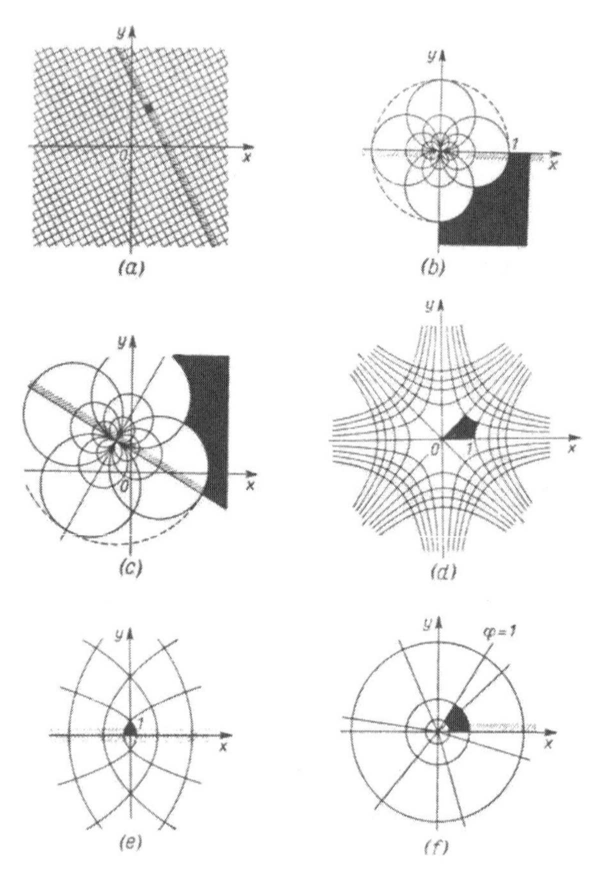

FIG. 382

(d) **Quadratic function** (Fig. 382d)

$$w = z^2.$$

The whole z plane is mapped in the twofold covered w plane. The isotermic net of the z plane consists of two families of hyperbolas $u = x^2 - y^2$, $v = 2xy$. The mapping fails to be conformal at $z = 0$. 0 and 1 are fixed points.

(e) **Square root** (Fig. 382e)

$$w = \sqrt{z}.$$

The function is two-valued. The whole plane is mapped (1) onto the upper half-plane, (2) onto the lower half-plane. The isotermic net on the z plane consists of two families of confocal parabolas whose focus is at the origin and whose axes are directed in the positive and negative direction of the x axis. The mapping fail to be conformal at $z = 0$. 0 and $+1$ are fixed points.

(f) **Logarithm** (Fig. 382f)

$$w = \operatorname{Ln} z, \qquad u = \ln \varrho, \qquad v = \varphi + 2k\pi.$$

The isotermic net consists of the circles $\ln \varrho = \text{const}$ and of the half-lines $\varphi = \text{const}$, i.e., is the net of polar coordinates. The function is infinitely many valued; the principal value of the logarithm maps the whole plane onto the strip bounded by the lines $v = -\pi$ and $v = \pi$, the latter line being excluded.

7. Integrals in the domain of complex numbers

Definition. The *integral* of a function $w = f(z)$ of a complex variable taken along an arc $\breve{AB}$ of a curve (*path of integration*) lying in the z plane is a complex number obtained as follows (Fig. 383):

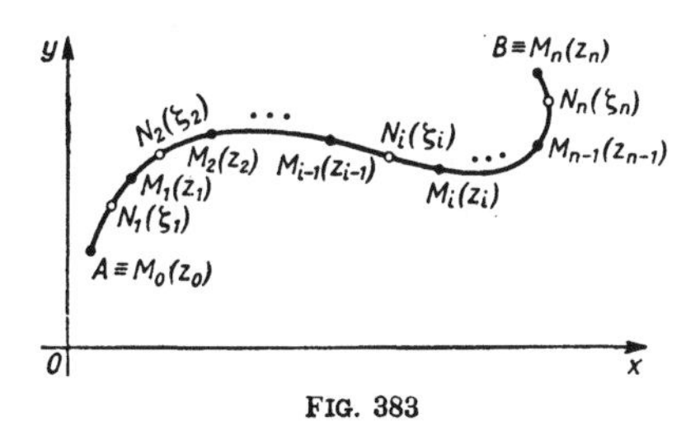

FIG. 383

(1) we divide the arc $\breve{AB}$ by means of arbitrary points

$$M_1(z_1),\ M_2(z_2),\ \ldots,\ M_{n-1}(z_{n-1})\ (^1)$$

and we put $A = M_0(z_0)$, $B = M_n(z_n)$;

[1] The complex number in brackets following the name of a point is the value of the complex variable represented by this point.

(2) we select an arbitrary point $N_i(\zeta_i)$ inside or on the boundary of each arc $\overset{\frown}{M_{i-1}M_i}$;

(3) we multiply the values of the function $f(z)$ at the points ζ_i by the corresponding differences $z_i - z_{i-1}$ (the increments of the independent variable);

(4) we add the obtained n products $f(\zeta_i)(z_i - z_{i-1})$;

(5) we determine the limit of the sum

$$\sum_{i=1}^{n} f(\zeta_i)(z_i - z_{i-1}),$$

when $n \to \infty$ and all the increments of the independent variable tend to zero.

If this limit exists and is independent of the particular choice of the points M_i and N_i, then it is called the *integral of the function* $f(z)$ *along the arc* $\breve{AB}$ and is denoted by

$$(*) \qquad \int_{\breve{AB}} f(z)\,dz.$$

Properties of the integral. Integral $(*)$ has the same properties as the line integral of the second type (see p. 491): when the direction of the path of integration is reversed, the integral changes its sign; if the path of integration is divided into several parts, then the integral along the whole path is equal to the sum of the integrals along the successive parts of the path.

Estimation of the integral. If the length of the path $\breve{AB}$ is s and if the modulus $|f(z)|$ does not exceed a positive number M ($|f(z)| < M$), for any value z of the path, then

$$\left|\int_{\breve{AB}} f(z)\,dz\right| < Ms.$$

Evaluation of the integral. If the integrand $f(z)$ has the form $u(x, y) + iv(x, y)$, and the path $\breve{AB}$ is defined by parametric equations $x = x(t)$, $y = y(t)$, where the values of t at the initial and end points of the path of integration are, respectively, t_A and t_B, then integral $(*)$ can be expressed by two line integrals of functions of a real variable:

$$\int_{\breve{AB}} f(z)\,dz = \int_{\breve{AB}} u(x, y)\,dx - v(x\ y)\,dy + i \int_{\breve{AB}} v(x, y)\,dx + u(x, y)\,dy,$$

which can be evaluated by the rules given on p. 490.

Independence of the path of integration. In order that the integral $(*)$ of a function of a complex variable defined in a simply connected domain [1] is independent of the particular choice of the path joining two fixed points $A(z_A)$ and $B(z_B)$ it is necessary and sufficient that the function $f(z)$ is analytic in this domain, i.e., satisfies the Cauchy-Riemann conditions (see p. 598). If, under these conditions, we fix the initial point $A_0(z_0)$ of the path of integration and let its end point $M(z)$ vary, then

$$\int_{\overset{\smile}{A_0M}} f(z)\,dz = F(z),$$

with $F'(z) = f(z)$; the function $F(z)$ is called a *primitive function* of the analytic function $f(z)$. A primitive function depends on the choice of the initial point A_0; the general form of all possible primitive functions of $f(z)$ is

$$F(z) + C = \int f(z)\,dz$$

(*indefinite integral*). Indefinite integrals of elementary functions of a complex variable are evaluated by the same formulas as in the case of functions of a real variable.

Fundamental theorem of the integral calculus. The integral (*) of an analytic function $f(z)$ is equal to the increment of a primitive function between the initial and the end point of the path of integration:

$$\int_{\overset{\smile}{AB}} f(z)\,dz = F(z_B) - F(z_A).$$

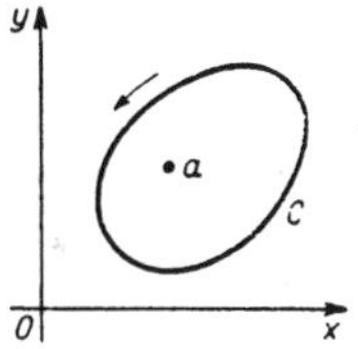

Integral around a closed curve. If the function $f(z)$ is analytic in a whole simply connected domain bounded by a closed curve C, then the integral of $f(z)$ along C is zero (*Cauchy's theorem*); if, however, the domain contains singular points, then, to evaluate the integral, we use the theorem on residues (see p. 611).

FIG. 384

In particular, for the function $f(z) = \dfrac{1}{z-a}$ with a single singular point $z = a$, the integral around a closed curve C described counterclockwise and including the point a inside (Fig. 384) is equal to

$$\oint_{\overset{\leftarrow}{C}} \frac{dz}{z-a} = 2\pi i.$$

[1] For a simply connected domain see p. 342. In the case of a multi-connected domain, the condition may not be sufficient.

Cauchy's formulas. If the function $f(z)$ is analytic in a certain simply connected domain bounded by a closed curve C, then its value at an arbitrary point z of the domain and also the values of its derivatives of an arbitrary order can be expressed by means of its values on the curve C (Fig. 385) by means of *Cauchy's formulas*:

$$f(z)=\frac{1}{2\pi i}\int\limits_{\overleftarrow{C}}\frac{f(\zeta)}{\zeta-z}\,d\zeta,$$

$$f'(z)=\frac{1}{2\pi i}\int\limits_{\overleftarrow{C}}\frac{f(\zeta)}{(\zeta-z)^2}\,d\zeta,$$

$$(**)\qquad f''(z)=\frac{2}{2\pi i}\int\limits_{\overleftarrow{C}}\frac{f(\zeta)}{(\zeta-z)^3}\,d\zeta,$$

$$\cdots\cdots\cdots\cdots\cdots\cdots\cdots\cdots$$

$$f^{(n)}(z)=\frac{n!}{2\pi i}\int\limits_{\overleftarrow{C}}\frac{f(\zeta)}{(\zeta-z)^{n+1}}\,d\zeta,$$

where ζ is a variable of integration and the integrals are taken along the curve C counterclockwise.

If, however, the function $f(z)$ is analytic in the whole domain of the plane lying outside the curve C, then the values of $f(z)$ and its derivatives at an arbitrary point of this domain (Fig. 386) are expressed by the same formulas (**), but the integrals are taken clockwise $\vec{C}$.

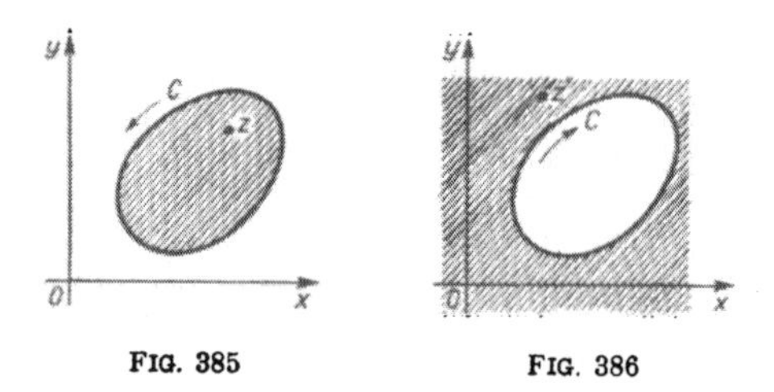

FIG. 385 FIG. 386

Cauchy's formulas enable us to evaluate certain definite integrals.

Example. Putting $f(z)=e^z$ (this is a function analytic in the whole plane) and integrating it along the circle C with the centre at z and radius r (its equation is $\zeta=z+re^{i\varphi}$, see p. 595), we obtain from the last formula of (**)

$$e^z = \frac{n!}{2\pi i}\int\limits_{\bar{C}} \frac{e^\zeta}{(\zeta - z)^{n+1}} d\zeta = \frac{n!}{2\pi i}\int\limits_0^{2\pi} \frac{e^{z+re^{i\varphi}}}{r^{n+1}e^{i\varphi(n+1)}} ire^{i\varphi}\,d\varphi$$

$$= \frac{n!}{2\pi r^n}\int\limits_0^{2\pi} e^{z+r\cos\varphi+ir\sin\varphi-in\varphi}\,d\varphi,$$

hence

$$\frac{2\pi r^n}{n!} = \int\limits_0^{2\pi} e^{r\cos\varphi+i(r\sin\varphi-n\varphi)}\,d\varphi$$

$$= \int\limits_0^{2\pi} e^{r\cos\varphi}\cos(r\sin\varphi - n\varphi)\,d\varphi + i\int\limits_0^{2\pi} e^{r\cos\varphi}\sin(r\sin\varphi - n\varphi)\,d\varphi.$$

Since the imaginary part is equal to zero, the value of the integral is

$$\int\limits_0^{2\pi} e^{r\cos\varphi}\cos(r\sin\varphi - n\varphi)\,d\varphi = \frac{2\pi r^n}{n!}.$$

8. Expansion of analytic functions into power series

Taylor's series. Any function $f(z)$ analytic within a circle with centre a can be written uniquely at each point of this circle in the form of a power series

$$f(z) = \sum_{n=0}^{\infty} c_n(z-a)^n,$$

where the coefficients c_n of the expansion are complex numbers given by the formula

$$c_n = \frac{f^{(n)}(a)}{n!}.$$

Thus we obtain the *Taylor's series*

$$f(z) = f(a) + \frac{f'(a)}{1!}(z-a) + \frac{f''(a)}{2!}(z-a)^2 + \ldots + \frac{f^{(n)}(a)}{n!}(z-a)^n + \ldots$$

For expansion of the functions e^z, $\sin z$, $\cos z$, $\sinh z$, $\cosh z$ into power series see pp. 592, 593.

Laurent's series. Any function $f(z)$ analytic within an annulus between two circles with a common centre a [1] can be written uniquely in the form of the so-called *Laurent power series*:

$$(*)\quad f(z) = \sum_{n=-\infty}^{\infty} c_n(z-a)^n$$

$$= c_0 + c_1(z-a) + c_2(z-a)^2 + \ldots + c_n(z-a)^n + \ldots +$$

$$+ c_{-1}(z-a)^{-1} + c_{-2}(z-a)^{-2} + \ldots + c_{-n}(z-a)^{-n} + \ldots,$$

where the coefficients c_n of the expansion are complex numbers given by the formula

$$c_n = \frac{1}{2\pi i}\int_{\overleftarrow{C}} (\zeta - a)^{-n-1} f(\zeta)\, d\zeta,$$

for $n = 0, \pm 1, \pm 2, \ldots$ and $\overleftarrow{C}$ is a closed curve lying inside the annulus and described counterclockwise around the point a.

Singular points. If the function $f(z)$ is analytic in the neighbourhood of the point a [2], then the type of the point a can be determined from the form of the Laurent expansion $(*)$ of $f(z)$ in a neighbourhood of this point as follows:

(1) If the series $(*)$ does not contain powers of $z - a$ with negative indices ($c_n = 0$ for $n < 0$), then the Laurent's series becomes a Taylor series [3] and the function $f(z)$ is analytic also at the point a, if $f(a) = c_0$, or else a is a removable singular point.

(2) If the series $(*)$ contains a finite number of powers of $z - a$ with negative indices, (there exists an m such that $m < 0$, $c_m \neq 0$ and all $c_n = 0$ for $n < m < 0$), then a is called a *pole* (see p. 601) *of the m-th order* or an *m-fold pole.*

(3) If the series $(*)$ contains an infinite number of powers of $z - a$ with negative indices, then a is an *essential singular point* (see p. 601).

Residues. In cases (2) and (3), the coefficient c_{-1} at $(z-a)^{-1}$ in Laurent's series is called the *residue* of the function $f(z)$ at the

---

[1] The radius of the inner circle can be equal to zero and then the annulus becomes a circle with the centre removed.

[2] See footnote on p. 601.

[3] In this case, the coefficients of the series are, by Cauchy's formulas of p. 608, equal to

$$c_n = \frac{1}{2\pi i}\int_{\overleftarrow{C}} (\zeta - a)^{-n-1} f(\zeta)\, d\zeta = \frac{f^{(n)}(a)}{n!}.$$

point $z = a$:

(1) $$\operatorname{res} f(z)_{z=a} = \frac{1}{2\pi i} \int_{\overleftarrow{C}} f(\zeta)\, d\zeta.$$

Definition (1) implies the following theorem which enables us to evaluate integrals around closed curves containing singular points inside (p. 607):

Theorem of residues. The integral

$$\int_{\overleftarrow{C}} f(z)\, dz$$

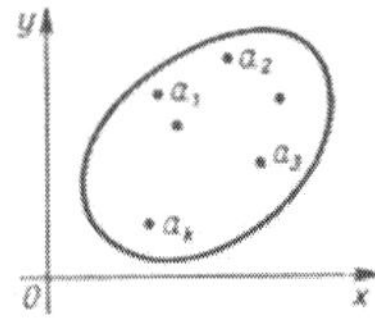

FIG. 387

of function $f(z)$ analytic in a simply connected domain bounded by a closed curve $\overleftarrow{C}$ except at a finite number of points $a_1, a_2, \dots, a_k$ (Fig. 387) taken around the curve C counterclockwise is equal to the sum of the residues of $f(z)$ at these points multiplied by $2\pi i$:

$$\int_{\overleftarrow{C}} f(z)\, dz = 2\pi i\, [\operatorname{res} f(z)_{z=a_1} + \operatorname{res} f(z)_{z=a_2} + \dots + \operatorname{res} f(z)_{z=a_k}].$$

The residue of $f(z)$ at a pole of the m-th order can be found from the formula

(2) $$\operatorname{res} f(z)_{z=a} = \frac{1}{(m-1)!} \cdot \frac{d^{m-1}}{dz^{m-1}} [f(z)(z-a)^m] \Big|_{z=a}.$$

If $f(z) = \dfrac{\varphi(z)}{\psi(z)}$, where $\varphi(z)$ and $\psi(z)$ are functions analytic at the point $z = a$ and a is a simple root (see p. 164) of the equation $\psi(z) = 0$, but not a root of the equation $\varphi(z) = 0$ (i.e., $\psi(a) = 0$, but $\psi'(a) \neq 0$ and $\varphi(a) \neq 0$), then the point $z = a$ is a pole of order 1 of $f(z)$ and formula (2) gives

$$\operatorname{res} \left[\frac{\varphi(z)}{\psi(z)}\right]_{z=a} = \frac{\varphi(a)}{\psi'(a)}.$$

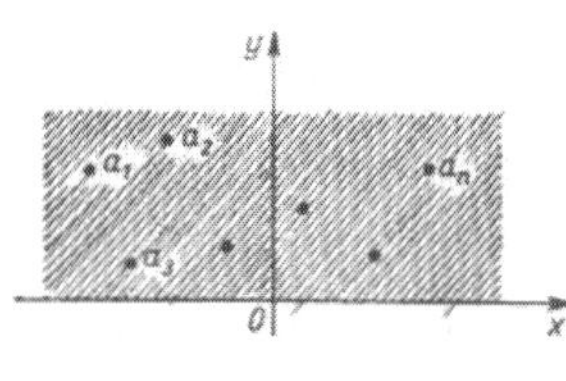

FIG. 388

If a is an m-fold root of the equation $\psi(z) = 0$ but is not a root of $\varphi(z) = 0$ (i.e., $\psi(a) = \psi'(a) = \dots = \psi^{(m-1)}(a) = 0$, but $\psi^{(m)}(a) \neq 0$ and $\varphi(a) \neq 0$), then the point $z = a$ is a pole of the m-th order of $f(z)$.

Application to evaluation of definite integrals. The theory of residues enables us to evaluate certain definite integrals of functions of a real variable:

If the function $f(z)$ is analytic in the open upper half-plane with the real axis excluded, except at a finite number of singular points $a_1, a_2, \ldots, a_n$ lying above the real axis (Fig. 388), and the number zero is an m-fold root of the equation $f(\frac{1}{z}) = 0$ with $m > 2$ [1], then

$$\text{(3)} \qquad \int_{-\infty}^{+\infty} f(x)dx = 2\pi i \sum_{i=1}^{n} \operatorname{res} f(z)_{z=a_i}.$$

Example. Evaluate the definite integral

$$\int_{-\infty}^{+\infty} \frac{dx}{(1+x^2)^3}.$$

The equation

$$f\left(\frac{1}{x}\right) = \frac{1}{(1+1/x^2)^3} = \frac{x^6}{(x^2+1)^3} = 0$$

has a 6-fold root $x = 0$. In the upper half-plane the function $w = \dfrac{1}{(1+z^2)^3}$ has a single singular point $z = i$ which is a pole of order 3 [2]. According to formula (2),

$$\operatorname{res} \frac{1}{(1+z^2)^3}_{z=i} = \frac{1}{2!} \cdot \frac{d^2}{dz^2}\left[\frac{(z-i)^3}{(1+z^2)^3}\right]_{z=i}.$$

Evaluating $\dfrac{d^2}{dz^2}\left(\dfrac{z-i}{1+z^2}\right)^3 = \dfrac{d^2}{dz^2}(z+i)^{-3} = 12(z+i)^{-5}$, we obtain

$$\operatorname{res} \frac{1}{(1+z^2)^3}_{z=i} = 6(z+i)^{-5}_{z=i} = \frac{6}{(2i)^5} = -\frac{3}{16}i$$

and, by formula (3),

$$\int_{-\infty}^{+\infty} f(x)\,dx = 2\pi i\left(-\tfrac{3}{16}i\right) = \tfrac{3}{8}\pi.$$

---

(1) See p. 164.

(2) The equation $(1+z^2)^3 = 0$ has two 3-fold roots: i and $-i$, but only the first of them lies in the upper half-plane.

II. VECTOR CALCULUS

A. VECTOR ALGEBRA AND VECTOR FUNCTIONS OF A SCALAR

1. Fundamental concepts

Scalar and vector quantities. Quantities whose values can be expressed by means of positive or negative numbers (*scalars*) are called *scalar quantities* (e.g., mass, temperature, work and so on). Quantities whose specification requires both the *magnitude* and the *direction in the space* (e.g., force, velocity, acceleration, magnetic field intensity and so on) are called *vector quantities* and can be expressed by means of vectors.

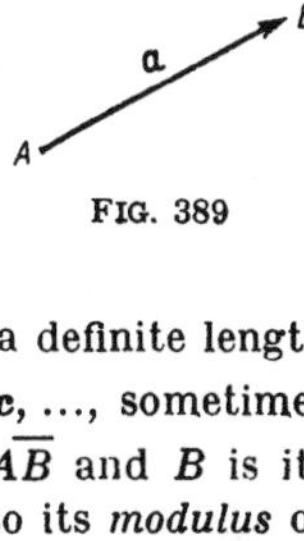

FIG. 389

Vectors. A *vector* (Fig. 389) is a segment with a definite length and a definite direction (notation: $\overline{AB}$, $\boldsymbol{a}$, $\boldsymbol{b}$, $\boldsymbol{c}$, ..., sometimes $\bar{a}$, $\bar{b}$, $\bar{c}$, ...). A is the *initial point* of the vector $\overline{AB}$ and B is its *final point*. The length of the vector $\boldsymbol{a}$ (called also its *modulus* or *absolute value*) is denoted by $|\boldsymbol{a}|$ or a. The *zero vector* ($\boldsymbol{0}$) is a vector whose initial and final points coincide; its length is zero and its direction is not defined. Two vectors $\boldsymbol{a}$ and $\boldsymbol{b}$ are said to be *equal*, if their lengths are equal and their directions coincide (i.e., the vectors $\boldsymbol{a}$ and $\boldsymbol{b}$ are parallel and oriented on one side) (1).

Vectors parallel to a given straight line are called *colinear*, and vectors parallel to one plane are called *conplanar*. The *negative of a vector* is a vector of the same length but opposite direction: $\overline{AB} = \boldsymbol{a}$ and $\overline{BA} = -\boldsymbol{a}$. A vector of length 1 is called a *unit vector*. The unit vector whose direction coincides with that of the vector

---

(1) According to this definition, a vector remains unchanged by a parallel displacement of its initial point to an arbitrary point of the space; such vectors are called *free*. In some problems of mechanics we consider vectors whose initial point is fixed in the space (so-called *bound vectors*) or vectors which can be displaced only along a straight line (*sliding vectors*).

$\boldsymbol{a}$ is called the *unit vector of the vector* $\boldsymbol{a}$ and denoted $\boldsymbol{a}^0$. Any vector $\boldsymbol{a}$ can be written in the form $\boldsymbol{a} = a\boldsymbol{a}^0$, where a is the length of $\boldsymbol{a}$. The unit vectors in the positive directions of the coordinate axes [1] are denoted by $\boldsymbol{i}, \boldsymbol{j}, \boldsymbol{k}$ (Fig. 390).

Radius vector. A point M in the space can be determined by the vector $\overline{OM}$ (Fig. 390) with the initial point at the origin and with the final point M. The vector $\overline{OM}$ is called the *radius vector* of the point M and denoted by $\boldsymbol{r}$. In this case the origin O is called the *pole*.

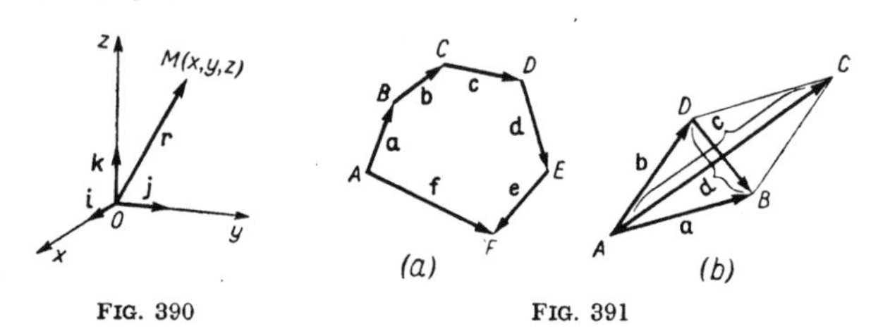

FIG. 390 FIG. 391

Linear combinations of vectors. The *sum* of several vectors $\boldsymbol{a}, \boldsymbol{b}, \boldsymbol{c}, \ldots, \boldsymbol{e}$ is a vector $\boldsymbol{f} = \overline{AF}$ such that the broken line $ABC\ldots EF$ formed by the successive summands is closed (Fig. 391a). The *sum* of two vectors $\overline{AB} = \boldsymbol{a}$ and $\overline{AD} = \boldsymbol{b}$ is the vector $\overline{AC} = \boldsymbol{c}$ represented by the diagonal of the parallelogram $ABCD$. Main properties of the sum of vectors:

$$\boldsymbol{a} + \boldsymbol{b} = \boldsymbol{b} + \boldsymbol{a}, \quad (\boldsymbol{a} + \boldsymbol{b}) + \boldsymbol{c} = \boldsymbol{a} + (\boldsymbol{b} + \boldsymbol{c}), \quad |\boldsymbol{a}| + |\boldsymbol{b}| \geqslant |\boldsymbol{a} + \boldsymbol{b}|.$$

The *difference* $\boldsymbol{a} - \boldsymbol{b}$ of two vectors $\boldsymbol{a}$ and $\boldsymbol{b}$ is the sum of $\boldsymbol{a}$ and $(-\boldsymbol{b})$ (the diagonal DB in Fig. 391b): $\boldsymbol{a} - \boldsymbol{b} = \boldsymbol{a} + (-\boldsymbol{b}) = \boldsymbol{d}$. Properties of the difference: $\boldsymbol{a} - \boldsymbol{a} = \boldsymbol{0}$ (the zero vector), $|\boldsymbol{a} - \boldsymbol{b}| > |\boldsymbol{a}| - |\boldsymbol{b}|$.

The *product of a vector* $\boldsymbol{a}$ *and a scalar* α (denoted $\alpha\boldsymbol{a}$ or $\boldsymbol{a}\alpha$) is a vector colinear with $\boldsymbol{a}$; its length is $|\alpha| \cdot |\boldsymbol{a}|$ and its direction coincides with that of $\boldsymbol{a}$, if $\alpha > 0$, and is opposite, if $\alpha < 0$. Properties of the product of a vector by a scalar:

$$\boldsymbol{a}\alpha = \alpha\boldsymbol{a}, \quad \alpha(\beta\boldsymbol{a}) = \alpha\beta\boldsymbol{a}, \quad (\alpha + \beta)\,\boldsymbol{a} = \alpha\boldsymbol{a} + \beta\boldsymbol{a},$$
$$\alpha(\boldsymbol{a} + \boldsymbol{b}) = \alpha\boldsymbol{a} + \alpha\boldsymbol{b}.$$

The *linear combination of the vectors* $\boldsymbol{a}, \boldsymbol{b}, \ldots, \boldsymbol{d}$ with the coefficients $\alpha, \beta, \ldots, \delta$ is the vector

$$(*) \qquad \boldsymbol{l} = \alpha\boldsymbol{a} + \beta\boldsymbol{b} + \ldots + \delta\boldsymbol{d}.$$

---

[1] In this chapter the coordinate system is assumed to be right-handed (see p. 256).

Given three non-coplanar vectors $\boldsymbol{u}, \boldsymbol{v}, \boldsymbol{w}$, any vector $\bar{\boldsymbol{a}}$ can be represented as a sum

$$(**) \qquad \boldsymbol{a} = \alpha\boldsymbol{u} + \beta\boldsymbol{v} + \gamma\boldsymbol{w};$$

(Fig. 392a); the summands $\alpha\boldsymbol{u}, \beta\boldsymbol{v}, \gamma\boldsymbol{w}$ are called the *components* of $\boldsymbol{a}$ with respect to $\boldsymbol{u}, \boldsymbol{v}, \boldsymbol{w}$, and the coefficients α, β, γ are called the

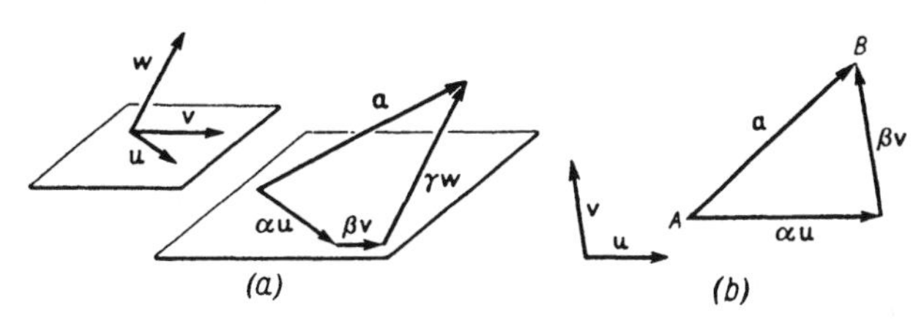

FIG. 392

coordinates of the vector $\boldsymbol{a}$ with respect to $\boldsymbol{u}, \boldsymbol{v}, \boldsymbol{w}$. Vectors parallel to one plane can be written in the form $\alpha\boldsymbol{u} + \beta\boldsymbol{v}$, where $\boldsymbol{u}$ and $\boldsymbol{v}$ are two non-colinear vectors (Fig. 392b).

Coordinates of a vector.

Cartesian rectangular coordinates. According to formula (*) any vector $\overline{AB} = \boldsymbol{a}$ in the space can be resolved uniquely into a sum of three vectors parallel to the unit vectors $\boldsymbol{i}$, $\boldsymbol{j}$, $\boldsymbol{k}$ (see p. 614):

$$(1) \qquad \boldsymbol{a} = a_x\boldsymbol{i} + a_y\boldsymbol{j} + a_z\boldsymbol{k}.$$

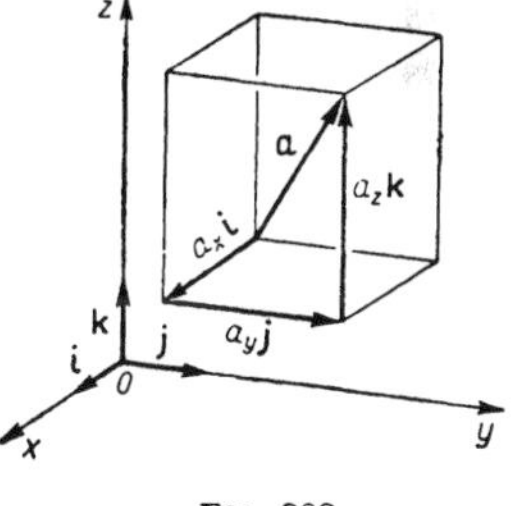

FIG. 393

The scalars a_x, a_y, a_z are called *Cartesian rectangular coordinates* of $\boldsymbol{a}$ in the system $\boldsymbol{i}, \boldsymbol{j}, \boldsymbol{k}$; we also write

$$(2) \qquad \boldsymbol{a}\{a_x, a_y, a_z\};$$

formulas (1) and (2) are equivalent. The rectangular coordinates of a vector are its projections onto the x, y and z axes (Fig. 393). Coordinates of a vector remain unchanged under a parallel displacement of the vector.

The coordinates of a linear combination of several vectors are equal to the corresponding linear combinations of the coordinates of these vectors: the vector equation (*) implies three scalar equalities

$$k_x = \alpha a_x + \beta b_x + \ldots + \delta d_x,$$
(3) $$k_y = \alpha a_y + \beta b_y + \ldots + \delta d_y,$$
$$k_z = \alpha a_z + \beta b_z + \ldots + \delta d_z.$$

In particular, for a sum or difference of two vectors, we have

(4) $$c_x = a_x \pm b_x, \quad c_y = a_y \pm b_y, \quad c_z = a_z \pm b_z.$$

Cartesian coordinates of the radius vector $\boldsymbol{r}$ of a point $M(x, y, z)$ are equal to the corresponding coordinates of M:

$$r_x = x, \quad r_y = y, \quad r_z = z; \quad \boldsymbol{r} = x\boldsymbol{i} + y\boldsymbol{j} + z\boldsymbol{k}.$$

Affine coordinates. A generalization of Cartesian rectangular coordinates of a vector are its *affine coordinates* in a system of three non-coplanar vectors $\boldsymbol{e}_1, \boldsymbol{e}_2, \boldsymbol{e}_3$, i.e., the coefficients a^1, a^2, a^3 of the resolution of $\boldsymbol{a}$ into the vectors $\boldsymbol{e}_1, \boldsymbol{e}_2, \boldsymbol{e}_3$:

(1′) $$\boldsymbol{a} = a^1 \boldsymbol{e}_1 + a^2 \boldsymbol{e}_2 + a^3 \boldsymbol{e}_3,$$

or, in an equivalent notation,

(2′) $$\boldsymbol{a}\{a^1, a^2, a^3\}\ (^1).$$

The vectors $\boldsymbol{e}_1, \boldsymbol{e}_2, \boldsymbol{e}_3$ are called *base vectors*. Formulas (1) and (2) are particular cases of (1′) and (2′), when $\boldsymbol{e}_1 = \boldsymbol{i}$, $\boldsymbol{e}_2 = \boldsymbol{j}$, $\boldsymbol{e}_3 = \boldsymbol{k}$.

For coordinates of a linear combination (*) and for a sum and difference (4) of vectors, similar formulas hold:

$$k^1 = \alpha a^1 + \beta b^1 + \ldots + \delta d^1,$$
(3′) $$k^2 = \alpha a^2 + \beta b^2 + \ldots + \delta d^2,$$
$$k^3 = \alpha a^3 + \beta b^3 + \ldots + \delta d^3,$$

(4′) $$c^1 = a^1 \pm b^1, \quad c^2 = a^2 \pm b^2, \quad c^3 = a^3 \pm b^3.$$

2. Multiplication of vectors

Scalar product of vectors. The *scalar product of two vectors* $\boldsymbol{a}$ and $\boldsymbol{b}$ is a scalar $\boldsymbol{ab}$ defined by the equation $\boldsymbol{ab} = ab \cos \varphi$, where φ is the angle between the vectors $\boldsymbol{a}$ and $\boldsymbol{b}$ reduced to a common initial point (Fig. 394).

Vector product of vectors. The *vector product of two vectors* $\boldsymbol{a}$ and $\boldsymbol{b}$ is a vector $\boldsymbol{c}$ (denoted by $\boldsymbol{a} \times \boldsymbol{b}$ or $[\boldsymbol{ab}]$) such that the length

---

(1) The upper indices should not be confused with the exponents of powers. Such a notation is convenient, for the scalars a^1, a^2, a^3 are contravariant coordinates of the vector $\boldsymbol{a}$ (see p. 621).

of $\boldsymbol{c}$ is equal to $ab \sin \varphi$ (i.e., is equal to the area of the parallelogram spanned on the vectors $\boldsymbol{a}$ and $\boldsymbol{b}$ as its sides), the vector $\boldsymbol{c}$ is perpendicular to $\boldsymbol{a}$ and $\boldsymbol{b}$ and is directed so that three vectors $\boldsymbol{a}, \boldsymbol{b}, \boldsymbol{c}$ constitute a right-handed system (that is, if the vectors $\boldsymbol{a}$,

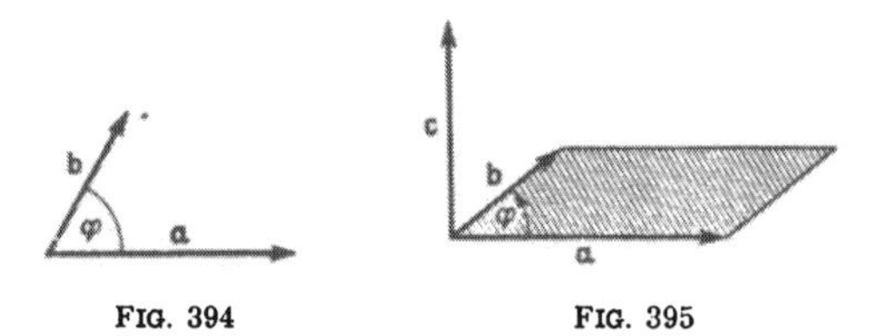

FIG. 394 FIG. 395

$\boldsymbol{b}, \boldsymbol{c}$ are displaced to a common origin, and, if we look at the vectors $\boldsymbol{a}$ and $\boldsymbol{b}$ from the final point of the vector $\boldsymbol{c}$, we see that the shortest rotation from the direction of the vector $\boldsymbol{a}$ to that of $\boldsymbol{b}$ is a counterclockwise rotation, see Fig. 395).

Properties of the products of vectors [1].

$\boldsymbol{ab} = \boldsymbol{ba}$ (commutative law), but $\boldsymbol{a} \times \boldsymbol{b} = -\boldsymbol{b} \times \boldsymbol{a}$ (anticommutative law for the vector product; the direction of the vector product is reversed, if we interchange the order of the factors);

$\alpha(\boldsymbol{ab}) = (\alpha\boldsymbol{a})\boldsymbol{b}$ and $\alpha(\boldsymbol{a} \times \boldsymbol{b}) = (\alpha\boldsymbol{a}) \times \boldsymbol{b}$ (associative law in multiplication of the scalar or vector product by a scalar α);

$\boldsymbol{a}(\boldsymbol{bc}) \neq (\boldsymbol{ab})\boldsymbol{c}$ and $\boldsymbol{a} \times (\boldsymbol{b} \times \boldsymbol{c}) \neq (\boldsymbol{a} \times \boldsymbol{b}) \times \boldsymbol{c}$ (in this case, the associative law fails);

$\boldsymbol{a}(\boldsymbol{b} + \boldsymbol{c}) = \boldsymbol{ab} + \boldsymbol{ac}$ and $\boldsymbol{a} \times (\boldsymbol{b} + \boldsymbol{c}) = \boldsymbol{a} \times \boldsymbol{b} + \boldsymbol{a} \times \boldsymbol{c}$ (distributive law);

$\boldsymbol{ab} = 0$, if $\boldsymbol{a} \perp \boldsymbol{b}$ (condition for perpendicularity of vectors);

$\boldsymbol{a} \times \boldsymbol{b} = \boldsymbol{0}$, if $\boldsymbol{a} \| \boldsymbol{b}$ (condition for parallelness or colinearity of veetors);

$\boldsymbol{aa} = \boldsymbol{a}^2 = a^2$, but $\boldsymbol{a} \times \boldsymbol{a} = \boldsymbol{0}$.

Linear combinations of vectors can be multiplied like scalar polynomials, but, if the order of the factors of a vector product is interchanged (as, for example, in grouping similar terms), then its sign should be reversed.

Examples.

(1) $(3\boldsymbol{a} + 5\boldsymbol{b} - 2\boldsymbol{c})(\boldsymbol{a} - 2\boldsymbol{b} - 4\boldsymbol{c})$

$= 3\boldsymbol{a}^2 + 5\boldsymbol{ba} - 2\boldsymbol{ca} - 6\boldsymbol{ab} - 10\boldsymbol{b}^2 + 4\boldsymbol{cb} - 12\boldsymbol{ac} - 20\boldsymbol{bc} + 8\boldsymbol{c}^2$

$= 3\boldsymbol{a}^2 - 10\boldsymbol{b}^2 + 8\boldsymbol{c}^2 - \boldsymbol{ab} - 14\boldsymbol{ac} - 16\boldsymbol{bc}.$

---

[1] We consider here only non-zero vectors.

(2) $(3\boldsymbol{a}+5\boldsymbol{b}-2\boldsymbol{c})\times(\boldsymbol{a}-2\boldsymbol{b}-4\boldsymbol{c})$

$$= 3\boldsymbol{a}\times\boldsymbol{a}+5\boldsymbol{b}\times\boldsymbol{a}-2\boldsymbol{c}\times\boldsymbol{a}-6\boldsymbol{a}\times\boldsymbol{b}-10\boldsymbol{b}\times\boldsymbol{b}+4\boldsymbol{c}\times\boldsymbol{b}-12\boldsymbol{a}\times\boldsymbol{c}-$$
$$-20\boldsymbol{b}\times\boldsymbol{c}+8\boldsymbol{c}\times\boldsymbol{c}$$
$$=\boldsymbol{0}-5\boldsymbol{a}\times\boldsymbol{b}+2\boldsymbol{a}\times\boldsymbol{c}-6\boldsymbol{a}\times\boldsymbol{b}+\boldsymbol{0}-4\boldsymbol{b}\times\boldsymbol{c}-12\boldsymbol{a}\times\boldsymbol{c}-20\boldsymbol{b}\times\boldsymbol{c}+\boldsymbol{0}$$
$$=-11\boldsymbol{a}\times\boldsymbol{b}-10\boldsymbol{a}\times\boldsymbol{c}-24\boldsymbol{b}\times\boldsymbol{c}=11\boldsymbol{b}\times\boldsymbol{a}+10\boldsymbol{c}\times\boldsymbol{a}+24\boldsymbol{c}\times\boldsymbol{b}.$$

Successive multiplication of vectors. The *double vector product* $\boldsymbol{a}\times(\boldsymbol{b}\times\boldsymbol{c})$ is a vector coplanar with $\boldsymbol{b}$ and $\boldsymbol{c}$ and can be found from the formula

$$\boldsymbol{a}\times(\boldsymbol{b}\times\boldsymbol{c})=\boldsymbol{b}(\boldsymbol{ac})-\boldsymbol{c}(\boldsymbol{ab}).$$

The *mixed triple product* (*box product*) $(\boldsymbol{a}\times\boldsymbol{b})\,\boldsymbol{c}$ is equal to the volume of the parallelepiped spanned by the vectors $\boldsymbol{a}$, $\boldsymbol{b}$, and $\boldsymbol{c}$ taken with the sign "+", if $\boldsymbol{a}$, $\boldsymbol{b}$, $\boldsymbol{c}$ form a right-handed system, and with the sign "−", if $\boldsymbol{a}$, $\boldsymbol{b}$, $\boldsymbol{c}$, form a left-handed system (see above). The brackets and the sign of multiplication are often omitted in the triple product: $(\boldsymbol{a}\times\boldsymbol{b})\,\boldsymbol{c}=\boldsymbol{abc}$.

$\boldsymbol{abc}=\boldsymbol{bca}=\boldsymbol{cab}=-\boldsymbol{acb}=-\boldsymbol{bac}=-\boldsymbol{cba}$ (interchanging two factors reverses the sign of the triple product; a cyclic transposition of three factors does not change the product).

Formulas for composite products

$$(\boldsymbol{a}\times\boldsymbol{b})\,(\boldsymbol{c}\times\boldsymbol{d})=(\boldsymbol{ac})\,(\boldsymbol{bd})-(\boldsymbol{bc})\,(\boldsymbol{ad})\qquad (\textit{Lagrange's identity}),$$

$$\boldsymbol{abc}\cdot\boldsymbol{efg}=\begin{vmatrix}\boldsymbol{ae} & \boldsymbol{af} & \boldsymbol{ag}\\ \boldsymbol{be} & \boldsymbol{bf} & \boldsymbol{bg}\\ \boldsymbol{ce} & \boldsymbol{cf} & \boldsymbol{cg}\end{vmatrix}.$$

Expression of the products in Cartesian rectangular coordinates. If the vectors $\boldsymbol{a}$, $\boldsymbol{b}$, $\boldsymbol{c}$ are given by their Cartesian rectangular coordinates

$$\boldsymbol{a}\{a_x, a_y, a_z\},\qquad \boldsymbol{b}\{b_x, b_y, b_z\},\qquad \boldsymbol{c}\{c_x, c_y, c_z\},$$

then their products can be found from the formulas:

Scalar product:

(1) $$\boldsymbol{ab}=a_xb_x+a_yb_y+a_zb_z.$$

Vector product:

(2) $$\boldsymbol{a}\times\boldsymbol{b}=(a_yb_z-a_zb_y)\,\boldsymbol{i}+(a_zb_x-a_xb_z)\,\boldsymbol{j}+(a_xb_y-a_yb_x)\boldsymbol{k}=\begin{vmatrix}\boldsymbol{i} & \boldsymbol{j} & \boldsymbol{k}\\ a_x & a_y & a_z\\ b_x & b_y & b_z\end{vmatrix}.$$

Triple mixed product:

$$abc = \begin{vmatrix} a_x & a_y & a_z \\ b_x & b_y & b_z \\ c_x & c_y & c_z \end{vmatrix}.$$

Expression of the products in affine coordinates.

Metric coefficients and reciprocal systems of vectors. If affine coordinates of two vectors $\boldsymbol{a}$ and $\boldsymbol{b}$ in a system $\mathbf{e}_1, \mathbf{e}_2, \mathbf{e}_3$ of base vectors are known:

$$\boldsymbol{a} = a^1\mathbf{e}_1 + a^2\mathbf{e}_2 + a^3\mathbf{e}_3, \qquad \boldsymbol{b} = b^1\mathbf{e}_1 + b^2\mathbf{e}_2 + b^3\mathbf{e}_3,$$

then, in order to compute the scalar product

$$\text{(A)} \quad \boldsymbol{ab} = a^1b^1\mathbf{e}_1\mathbf{e}_1 + a^2b^2\mathbf{e}_2\mathbf{e}_2 + a^3b^3\mathbf{e}_3\mathbf{e}_3 + \\ + (a^1b^2 + a^2b^1)\,\mathbf{e}_1\mathbf{e}_2 + (a^2b^3 + a^3b^2)\,\mathbf{e}_2\mathbf{e}_3 + (a^3b^1 + a^1b^3)\,\mathbf{e}_3\mathbf{e}_1$$

or the vector product

$$\text{(B)} \quad \boldsymbol{a}\times\boldsymbol{b} = (a^2b^3 - a^3b^2)\,\mathbf{e}_2\times\mathbf{e}_3 + (a^3b^1 - a^1b^3)\,\mathbf{e}_3\times\mathbf{e}_1 + \\ + (a^1b^2 - a^2b^1)\,\mathbf{e}_1\times\mathbf{e}_2$$

(since $\mathbf{e}_1\times\mathbf{e}_1 = \mathbf{e}_2\times\mathbf{e}_2 = \mathbf{e}_3\times\mathbf{e}_3 = \mathbf{0}$), it is necessary to know the pairwise products of base vectors. Thus, for the scalar product, six numbers, so-called *metric coefficients*,

$$g_{11} = \mathbf{e}_1\mathbf{e}_1, \qquad g_{22} = \mathbf{e}_2\mathbf{e}_2, \qquad g_{33} = \mathbf{e}_3\mathbf{e}_3,$$
$$g_{12} = \mathbf{e}_1\mathbf{e}_2 = \mathbf{e}_2\mathbf{e}_1, \qquad g_{23} = \mathbf{e}_2\mathbf{e}_3 = \mathbf{e}_3\mathbf{e}_2, \qquad g_{31} = \mathbf{e}_3\mathbf{e}_1 = \mathbf{e}_1\mathbf{e}_3,$$

are necessary, and, for the vector product, a system of three vectors

$$\mathbf{e}^1 = \Omega(\mathbf{e}_2\times\mathbf{e}_3), \qquad \mathbf{e}^2 = \Omega(\mathbf{e}_3\times\mathbf{e}_1), \qquad \mathbf{e}^3 = \Omega(\mathbf{e}_1\times\mathbf{e}_2),$$

called *reciprocal* to the system $\mathbf{e}_1, \mathbf{e}_2, \mathbf{e}_3$, is necessary; the coefficient Ω equal to the reciprocal to the mixed product of base vectors:

$$\Omega = \frac{1}{\mathbf{e}_1\,\mathbf{e}_2\,\mathbf{e}_3}$$

has been introduced to simplify further formulas.

Multiplication table of base vectors

Scalar product

	$\mathbf{e}_1$	$\mathbf{e}_2$	$\mathbf{e}_3$
$\mathbf{e}_1$	g_{11}	g_{12}	g_{13}
$\mathbf{e}_2$	g_{21}	g_{22}	g_{23}
$\mathbf{e}_3$	g_{31}	g_{32}	g_{33}

$(g_{ki} = g_{ik})$

Vector product

Multipliers

Multiplicands	$\mathbf{e}_1$	$\mathbf{e}_2$	$\mathbf{e}_3$
$\mathbf{e}_1$	$\mathbf{0}$	$\mathbf{e}^3/\Omega$	$-\mathbf{e}^2/\Omega$
$\mathbf{e}_2$	$-\mathbf{e}^3/\Omega$	$\mathbf{0}$	$\mathbf{e}^1/\Omega$
$\mathbf{e}_3$	$\mathbf{e}^2/\Omega$	$-\mathbf{e}^1/\Omega$	$\mathbf{0}$

In Cartesian rectangular coordinates ($\boldsymbol{e}_1 = \boldsymbol{i}$, $\boldsymbol{e}_2 = \boldsymbol{j}$, $\boldsymbol{e}_3 = \boldsymbol{k}$) the metric coefficients are

$$g_{11} = g_{22} = g_{33} = 1, \quad g_{12} = g_{23} = g_{31} = 0, \quad \Omega = \frac{1}{\boldsymbol{ijk}} = 1,$$

the reciprocal system of vectors $\boldsymbol{e}^1 = \boldsymbol{i}$, $\boldsymbol{e}^2 = \boldsymbol{j}$, $\boldsymbol{e}^3 = \boldsymbol{k}$ coincides with the base vectors and the multiplication table has the form

Scalar product

	$\boldsymbol{i}$	$\boldsymbol{j}$	$\boldsymbol{k}$
$\boldsymbol{i}$	1	0	0
$\boldsymbol{j}$	0	1	0
$\boldsymbol{k}$	0	0	1

Vector products

Multipliers

Multiplicands	$\boldsymbol{i}$	$\boldsymbol{j}$	$\boldsymbol{k}$
$\boldsymbol{i}$	$\boldsymbol{0}$	$\boldsymbol{k}$	$-\boldsymbol{j}$
$\boldsymbol{j}$	$-\boldsymbol{k}$	$\boldsymbol{0}$	$\boldsymbol{i}$
$\boldsymbol{k}$	$\boldsymbol{j}$	$-\boldsymbol{i}$	$\boldsymbol{0}$

Expression of products in coordinates. According to formula (A) for the scalar product, we have

$$\text{(1'')} \qquad \boldsymbol{ab} = \sum_{m=1}^{3}\sum_{n=1}^{3} g_{mn}\, a^m a^n = g_{\alpha\beta}\, a^\alpha a^\beta \ (^1).$$

In Cartesian rectangular coordinates formula (1″) reduces to (1) on p. 618.

According to formula (B), we have for the vector product

$$\text{(2'')} \qquad \boldsymbol{a} \times \boldsymbol{b} = \frac{1}{\boldsymbol{e}_1\,\boldsymbol{e}_2\,\boldsymbol{e}_3}\begin{vmatrix} \boldsymbol{e}^1 & \boldsymbol{e}^2 & \boldsymbol{e}^3 \\ a^1 & a^2 & a^3 \\ b^1 & b^2 & b^3 \end{vmatrix}$$

$$= \frac{1}{\boldsymbol{e}_1\,\boldsymbol{e}_2\,\boldsymbol{e}_3}\left[(a^2b^3 - a^3b^2)\,\boldsymbol{e}^1 + (a^3b^1 - a^1b^3)\,\boldsymbol{e}^2 + (a^1b^2 - a^2b^1)\,\boldsymbol{e}^3\right].$$

In Cartesian rectangular coordinates formula (2″) reduces to (2) on p. 618.

---

(1) The last member of equation (1″) is written as a summation convention used in the tensor calculus: instead of the whole sum, only its general term is written. The same index denoted by a Greek letter (the summation index) which appears in the general term as a subscript and a superscript assumes the values 1, 2, 3, so that this term stands for the sum of the terms obtained by giving the *summation index* each of the values 1, 2, 3. Thus

$$g_{\alpha\beta}a^\alpha a^\beta = g_{11}a^1b^1 + g_{12}a^1b^2 + g_{13}a^1b^3 + g_{21}a^2b^1 + g_{22}a^2b^2 + g_{23}a^2b^3 +$$
$$+ g_{31}a^3b^1 + g_{32}a^3b^2 + g_{33}a^3b^3.$$

Triple mixed product:

$$\boldsymbol{abc} = \begin{vmatrix} a^1 & a^2 & a^3 \\ b^1 & b^2 & b^3 \\ c^1 & c^2 & c^3 \end{vmatrix}.$$

Vector equations. Notation: $\boldsymbol{x}$ the unknown vector; $\boldsymbol{a}, \boldsymbol{b}, \boldsymbol{c}, \boldsymbol{d}$ known vectors; x, y, z unknown scalars; α, β, γ known scalars.

(1) $\boldsymbol{x}+\boldsymbol{a}=\boldsymbol{b}$; solution $\boldsymbol{x}=\boldsymbol{b}-\boldsymbol{a}$.

(2) $\boldsymbol{x}\alpha=\boldsymbol{a}$; $\alpha\neq 0$; solution $\boldsymbol{x}=\dfrac{\boldsymbol{a}}{\alpha}$.

(3) $\boldsymbol{xa}=\alpha$. The equation is indefinite; if we displace all the vectors $\boldsymbol{x}$ satisfying the equation to a common initial point, then their final points will lie in a plane perpendicular to the vector $\boldsymbol{a}$. Equation (3) is called a *vector equation of* this *plane*.

(4) $\boldsymbol{x}\times\boldsymbol{a}=\boldsymbol{b}$, where $\boldsymbol{b}\perp\boldsymbol{a}$. The equation is indefinite; if we displace all the vectors satisfying the equation to a common initial point, then their final points will lie on a straight line parallel to the vector $\boldsymbol{a}$. Equation (4) is called a *vector equation of this line*.

(5) $\boldsymbol{xa}=\alpha$, $\boldsymbol{x}\times\boldsymbol{a}=\boldsymbol{b}$, where $\boldsymbol{b}\perp\boldsymbol{a}$; solution $\boldsymbol{x}=\dfrac{\boldsymbol{a}\alpha+\boldsymbol{a}\times\boldsymbol{b}}{\boldsymbol{a}^2}$.

(6) $\boldsymbol{xa}=\alpha$, $\boldsymbol{xb}=\beta$, $\boldsymbol{xc}=\gamma$; solution

$$\boldsymbol{x}=\frac{\alpha(\boldsymbol{b}\times\boldsymbol{c})+\beta(\boldsymbol{c}\times\boldsymbol{a})+\gamma(\boldsymbol{a}\times\boldsymbol{b})}{\boldsymbol{abc}}=\alpha\tilde{\boldsymbol{a}}+\beta\tilde{\boldsymbol{b}}+\gamma\tilde{\boldsymbol{c}},$$

where $\tilde{\boldsymbol{a}}, \tilde{\boldsymbol{b}}, \tilde{\boldsymbol{c}}$ is the system of vectors reciprocal to $\boldsymbol{a}, \boldsymbol{b}, \boldsymbol{c}$ (see p. 619).

(7) $\boldsymbol{d}=x\boldsymbol{a}+y\boldsymbol{b}+z\boldsymbol{c}$; solution $x=\dfrac{\boldsymbol{dbc}}{\boldsymbol{abc}}$, $y=\dfrac{\boldsymbol{adc}}{\boldsymbol{abc}}$, $z=\dfrac{\boldsymbol{abd}}{\boldsymbol{abc}}$.

(8) $\boldsymbol{d}=x(\boldsymbol{b}\times\boldsymbol{c})+y(\boldsymbol{c}\times\boldsymbol{a})+z(\boldsymbol{a}\times\boldsymbol{b})$; solution $x=\dfrac{\boldsymbol{da}}{\boldsymbol{abc}}$, $y=\dfrac{\boldsymbol{db}}{\boldsymbol{abc}}$, $z=\dfrac{\boldsymbol{dc}}{\boldsymbol{abc}}$.

3. Covariant and contravariant coordinates of a vector

Definitions. Affine coordinates a^1, a^2, a^3 of a vector $\boldsymbol{a}$ defined by the formula

$$\boldsymbol{a}=a^1\boldsymbol{e}_1+a^2\boldsymbol{e}_2+a^3\boldsymbol{e}_3=a^\alpha\boldsymbol{e}_\alpha\ [1]$$

are also called its *contravariant coordinates* in contrast to the *covariant coordinates* of the vector $\boldsymbol{a}$ defined as the coefficients

[1] See footnote on p. 620.

of the resolution of the vector $\boldsymbol{a}$ into the system $\boldsymbol{e}^1$, $\boldsymbol{e}^2$, $\boldsymbol{e}^3$ of vectors reciprocal to $\boldsymbol{e}_1$, $\boldsymbol{e}_2$, $\boldsymbol{e}_3$ (see p. 619). We denote the covariant coordinates of the vector $\boldsymbol{a}$ by a_1, a_2, a_3:

$$\boldsymbol{a} = a_1\boldsymbol{e}^1 + a_2\boldsymbol{e}^2 + a_3\boldsymbol{e}^3 = a_\alpha\boldsymbol{e}^\alpha.$$

In Cartesian rectangular coordinate system the covariant and contravariant coordinates coincide.

Expression of coordinates by scalar products. A covariant coordinate of a vector $\boldsymbol{a}$ is equal to the scalar product of $\boldsymbol{a}$ by the corresponding base vector:

$$\text{(a)} \qquad a_1 = \boldsymbol{a}\boldsymbol{e}_1, \qquad a_2 = \boldsymbol{a}\boldsymbol{e}_2, \qquad a_3 = \boldsymbol{a}\boldsymbol{e}_3.$$

A contravariant coordinate of a vector $\boldsymbol{a}$ is equal to the scalar product of $\boldsymbol{a}$ by the corresponding reciprocal vector:

$$\text{(b)} \qquad a^1 = \boldsymbol{a}\boldsymbol{e}^1, \qquad a^2 = \boldsymbol{a}\boldsymbol{e}^2, \qquad a^3 = \boldsymbol{a}\boldsymbol{e}^3.$$

In Cartesian rectangular coordinate system formulas (a) and (b) are identical:

$$a_x = \boldsymbol{a}\boldsymbol{i}, \qquad a_y = \boldsymbol{a}\boldsymbol{j}, \qquad a_z = \boldsymbol{a}\boldsymbol{k}.$$

Expression of the scalar product by coordinates. Formula (1″) on p. 620 expresses the scalar product of two vectors in terms of their contravariant coordinates. The corresponding formula in covariant coordinates has the form

$$\boldsymbol{a}\boldsymbol{b} = g^{\alpha\beta}\, a_\alpha\, a_\beta,$$

where $g^{mn} = \boldsymbol{e}^m\boldsymbol{e}^n$ are the metric coefficients in the reciprocal system of vectors; these coefficients are related to the coefficients g_{mn} by the formula

$$g^{mn} = \frac{(-1)^{m+n} A^{mn}}{\begin{vmatrix} g_{11} & g_{12} & g_{13} \\ g_{21} & g_{22} & g_{23} \\ g_{31} & g_{32} & g_{33} \end{vmatrix}},$$

where A^{mn} is the minor of the determinant standing in the denominator obtained by removing the row and column intersecting at the element g_{mn}.

If the vector $\boldsymbol{a}$ is given by its contravariant coordinates and the vector $\boldsymbol{b}$ by its covariant coordinates, then their scalar product is equal to

$$\boldsymbol{a}\boldsymbol{b} = a^1b_1 + a^2b_2 + a^3b_3 = a^\alpha b_\alpha,$$

and, similarly,

$$\boldsymbol{ab} = a_\alpha b^\alpha.$$

4. Geometric applications of vector algebra

Name	Symbol	Vector formula	Formula of analytic geometry (in Cartesian rectangular coordinates)
Length of the vector $\boldsymbol{a}$	a	$\sqrt{\boldsymbol{a}^2}$	$\sqrt{a_x^2 + a_y^2 + a_z^2}$
Area of the parallelogram spanned by the vectors $\boldsymbol{a}$ and $\boldsymbol{b}$	S	$\lvert \boldsymbol{a} \times \boldsymbol{b} \rvert$	$\sqrt{\begin{vmatrix} a_y & a_z \\ b_y & b_z \end{vmatrix}^2 + \begin{vmatrix} a_z & a_x \\ b_z & b_x \end{vmatrix}^2 + \begin{vmatrix} a_x & a_y \\ b_x & b_y \end{vmatrix}^2}$
Volume of the parallelepiped spanned by the vectors $\boldsymbol{a}$, $\boldsymbol{b}$, $\boldsymbol{c}$	V	$\lvert \boldsymbol{abc} \rvert$	$V = \begin{vmatrix} a_x & a_y & a_z \\ b_x & b_y & b_z \\ c_x & c_y & c_z \end{vmatrix}$
Angle between the vectors $\boldsymbol{a}$ and $\boldsymbol{b}$	$\cos\varphi$	$\dfrac{\boldsymbol{ab}}{\sqrt{\boldsymbol{a}^2\boldsymbol{b}^2}}$	$\dfrac{a_x b_x + a_y b_y + a_z b_z}{\sqrt{a_x^2 + a_y^2 + a_z^2}\,\sqrt{b_x^2 + b_y^2 + b_z^2}}$

For applications of vector algebra to analytic geometry (vector equations of a plane and of a line) see p. 621 and pp. 263–269.

5. Vector function of a scalar variable

Definition. A variable vector $\boldsymbol{a}$ is called a *vector function of a scalar variable* t, if to each value of t there corresponds a definite vector $\boldsymbol{a}$. Notation:

$$\boldsymbol{a} = \boldsymbol{f}(t).$$

Coordinate definition of a vector function

$$\boldsymbol{a} = a_x\boldsymbol{i} + a_y\boldsymbol{j} + a_z\boldsymbol{k}$$

consists of three scalar functions of a scalar variable t:

$$a_x = f_x(t), \qquad a_y = f_y(t), \qquad a_z = f_z(t).$$

If we represent the variable vector $\boldsymbol{a}$ as a radius vector $\overline{OM} = \boldsymbol{r} = \boldsymbol{r}(t)$ of a point M, then, when t varies, the point M describes a space curve (Fig. 396) called the *hodograph* of the given vector function; its definition in coordinates consists of three functions

$$x = x(t), \qquad y = y(t), \qquad z = z(t).$$

Thus $\boldsymbol{r} = x\boldsymbol{i} + y\boldsymbol{j} + z\boldsymbol{k}$.

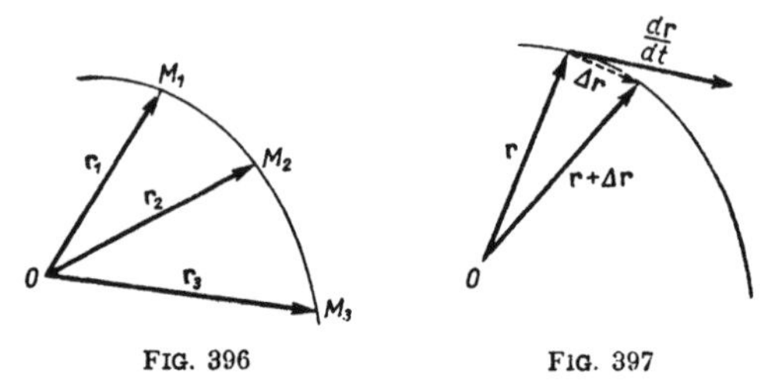

FIG. 396 FIG. 397

The derivative of the vector function $\boldsymbol{a} = \boldsymbol{f}(t)$:

$$\frac{d\boldsymbol{a}}{dt} = \lim_{\Delta t \to 0} \frac{\boldsymbol{f}(t + \Delta t) - \boldsymbol{f}(t)}{\Delta t}$$

is a new vector function of t. Geometric interpretation of the derivative of the radius vector: $\frac{d\boldsymbol{r}}{dt}$ is a vector tangent to the hodograph at the corresponding point (Fig. 397). The length of $\frac{d\boldsymbol{r}}{dt}$ depends on the choice of the parameter t. If t denotes the time, then the function $\boldsymbol{r}(t)$ describes the motion of a point in the space and the vector $\frac{d\boldsymbol{r}}{dt}$, in its length and direction, is equal to the velocity of the motion. If t denotes the length of the hodograph from a certain fixed point M_0 to the point M, then $\left|\frac{d\boldsymbol{r}}{dt}\right| = 1$.

Rules for differentiation of vectors:

$$\frac{d}{dt}(\boldsymbol{a} + \boldsymbol{b} + \boldsymbol{c}) = \frac{d\boldsymbol{a}}{dt} + \frac{d\boldsymbol{b}}{dt} + \frac{d\boldsymbol{c}}{dt}.$$

$\frac{d}{dt}(\varphi\boldsymbol{a}) = \frac{d\varphi}{dt}\boldsymbol{a} + \varphi\frac{d\boldsymbol{a}}{dt}$, where φ is a scalar function of t.

$$\frac{d}{dt}(\boldsymbol{a}\boldsymbol{b}) = \frac{d\boldsymbol{a}}{dt}\boldsymbol{b} + \boldsymbol{a}\frac{d\boldsymbol{b}}{dt}.$$

$\frac{d}{dt}(\boldsymbol{a} \times \boldsymbol{b}) = \frac{d\boldsymbol{a}}{dt} \times \boldsymbol{b} + \boldsymbol{a} \times \frac{d\boldsymbol{b}}{dt}$ (interchanging of factors is not allowed, see p. 617).

$$\frac{d}{dt}\boldsymbol{a}[\varphi(t)] = \frac{d\boldsymbol{a}}{d\varphi}\cdot\frac{d\varphi}{dt}.$$

If the vector $\boldsymbol{r}$ has a constant length, then $\boldsymbol{r}\dfrac{d\boldsymbol{r}}{dt} = 0$ (the hodograph is a spherical curve and the tangent is perpendicular to the radius vector).

Taylor series of a vector function:

$$\boldsymbol{a}(t+h) = \boldsymbol{a}(t) + h\frac{d\boldsymbol{a}}{dt} + \frac{h^2}{2}\cdot\frac{d^2\boldsymbol{a}}{dt^2} + \ldots + \frac{h^n}{n!}\cdot\frac{d^n\boldsymbol{a}}{dt^n} + \ldots$$

The convergence of this series (and of an arbitrary series whose terms are vectors) is defined in the same way as that of a series with complex terms (see p. 591). We can speak of an expansion of a vector function into a Taylor series only when this series is convergent.

The *differential* of the vector function $\boldsymbol{a}(t)$ is defined by the formula

$$d\boldsymbol{a} = \frac{d\boldsymbol{a}}{dt}\Delta t,$$

where Δt is an increment of t.

B. FIELD THEORY

6. Scalar field

Functions of a point. A scalar value U which assumes, at each point M of the space, a definite value $U = U(M)$ is called a *scalar function* of M or a *scalar field* (as, for example, the field of temperature, of potential, of density in a non-homogeneous medium and so on). A field can be regarded as a scalar function of a vector variable, namely, of the radius vector $\boldsymbol{r}$ of the point M with a given pole O (see p. 614):

$$U = U(\boldsymbol{r}) \tag{1}$$

A field defined only for points of a plane is called a *plane field* [1].

---

[1] Sometimes the term plane field is used to mean a field defined for points of the space such that it is constant at all points of a straight line parallel to a given direction. Such a field should rather be called a *plane-parallel field*; its investigation reduces to the investigation of the field in a plane perpendicular to the given fixed direction.

Central field and axial field. If a function assumes equal values at all points lying at the same distance from a fixed point $C(\mathbf{r}_1)$ called the *centre*, then the field of this function is called a *central* or *spherical field*; the value U depends only on the distance $CM = r$. For example, $U = r$ (the distance of a point from the centre), $U = \frac{c}{r^2}$ (the field of brightness with a point-like source of light at the pole). In general,

$$U = f(r). \tag{2}$$

If the function U has the same value at all points lying at an equal distance from a certain straight line (axis of the field), then the field is called *axial* or *cylindrical*.

Coordinate definition of a field. Expressing the point M by its coordinates (Cartesian x, y, z, cylindrical ϱ, φ, z, or spherical r, θ, φ coordinates [1]), we represent a scalar field (1) as a function of three variables:

$$U = \Phi(x, y, z), \qquad U = \Psi(\varrho, \varphi, z) \quad \text{or} \quad U = \Theta(r, \theta, \varphi). \tag{1a}$$

Similarly, a plane field can be represented as a function of two variables (Cartesian or polar coordinates):

$$U = \Phi(x, y) \quad \text{or} \quad U = \Psi(\varrho, \varphi). \tag{1b}$$

We assume that the functions U in formulas (1a) and (1b) are single-valued and continuous everywhere except, possibly, at certain separate points, lines or surfaces of discontinuity.

Expression of a central field in coordinates:

$$U = U\left(\sqrt{x^2 + y^2 + z^2}\right) = U\left(\sqrt{\varrho^2 + z^2}\right) = U(r), \tag{2a}$$

expression of an axial field in coordinates:

$$U = U\left(\sqrt{x^2 + y^2}\right) = U(\varrho) = U(r \sin \theta). \tag{3}$$

For a central field the most convenient are spherical coordinates, and for an axial field cylindrical coordinates.

Level surfaces and level lines of a field. The points at which the function (1) has the same value

$$U = \text{const} \tag{4}$$

form a surface in the space called a *level* or *equipotential surface* of the field. Its equations in coordinates are

---

[1] See p. 257.

$$U = \Phi(x, y, z) = c, \qquad U = \Psi(\varrho, \varphi, z) = c, \text{ or} \qquad \text{(4a)}$$
$$U = \Theta(r, \theta, \varphi) = c.$$

For different values of c we obtain various surfaces. Through each point of the space (except for points where the function U is not uniquely defined), there passes one level surface.

Examples. (1) The level surfaces of the field $U = \mathbf{cr} = c_x x + c_y y + c_z z$ are parallel planes.

(2) The level surfaces of the field $U = x^2 + 2y^2 + 4z^2$ are similar ellipsoids in a similar position with respect to the pole O.

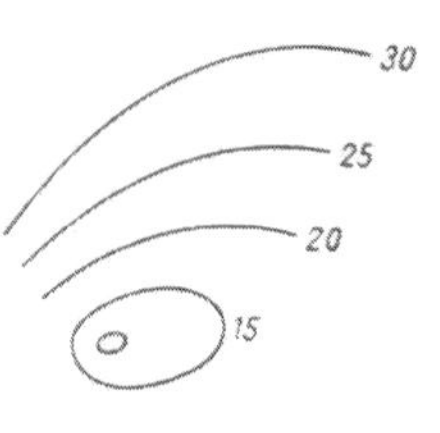

FIG. 398

The level surfaces of a central field are concentric spheres and the level surfaces of an axial field are coaxial cylinders.

For a plane field, the equation $U =$ const represents a line called a *level line* or *equipotential line* (also called a *niveau line*). In coordinates, a level line has the equation

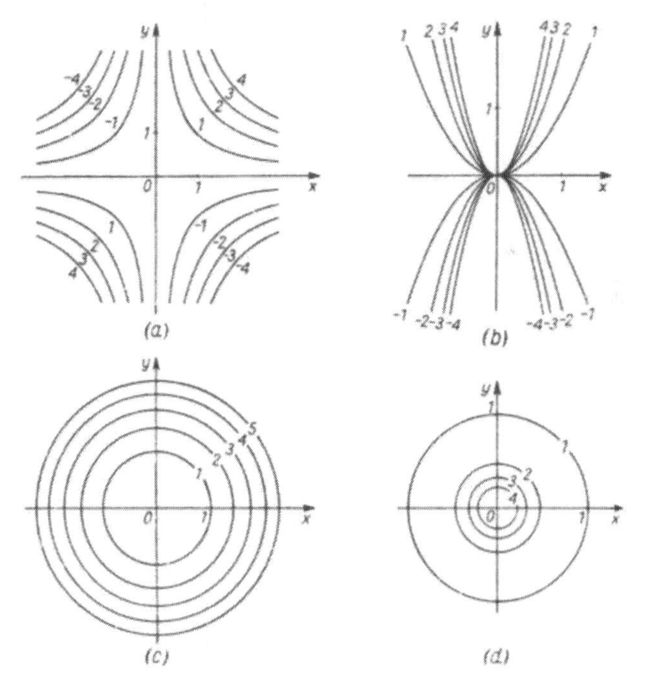

FIG. 399

(4b) $$U(x, y) = c \quad \text{or} \quad U(\varrho, \varphi) = c.$$

The level lines on graphs are usually drawn so as to correspond to equal intervals of U and each of them is marked by the corresponding value of U (Fig. 398). For example, the isobaric lines on a synoptic map or contour lines on topographic maps are level lines. In particular cases, level lines can degenerate to separate points and level surfaces—to lines or points.

Examples. (1) $U = xy$ (Fig. 399 a). (2) $U = \frac{y}{x^2}$, (Fig. 399 b). (3) $U = r^2$ (Fig. 399 c). (4) $U = \frac{1}{r}$ (Fig. 399 d).

7. Vector field

Vector function of a point. A variable vector quantity $\boldsymbol{V}$ which assumes a definite value at each point M of the space is called a *vector function* $\boldsymbol{V} = \boldsymbol{V}(M)$ of a point M or a *vector field.*

Examples: the velocity field of a fluid in motion, a field of force, a magnetic or electric intensity field.

A vector field can be regarded as a vector function

(1) $$\boldsymbol{V} = \boldsymbol{V}(\boldsymbol{r})$$

of a vector variable $\boldsymbol{r}$.

If all values of $\boldsymbol{r}$ as well as $\boldsymbol{V}$ lie in one plane, the field is called a *plane vector field* (¹).

Frequently occurring types of vector fields.

A *central vector field* is a field in which all the vectors $\boldsymbol{V}$ lie on straight lines passing through a fixed point called the *centre* (Fig. 400a). If we locate the pole at the centre, then the field is defined by the formula

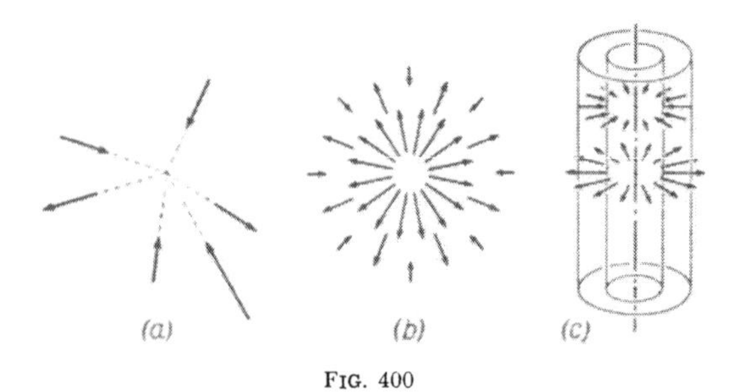

FIG. 400

(¹) See footnote on p. 625. An analogous remark is true for a vector field.

$$V = f(\boldsymbol{r})\boldsymbol{r},$$

where $f(\boldsymbol{r})$ is a scalar and each vector $\boldsymbol{V}$ has the direction of a radius vector. It is convenient to express such a field by the formula

$$V = \varphi(\boldsymbol{r})\frac{\boldsymbol{r}}{r},$$

where $\varphi(\boldsymbol{r})$ is the length and $\boldsymbol{r}/r$ is the unit vector of the vector of the field.

An important case of a central field is a *spherical vector field* (Fig. 400b):

$$V = \varphi(r)\frac{\boldsymbol{r}}{r},$$

when the length $\varphi(r)$ of $\boldsymbol{V}$ depends only on the length of the vector $\boldsymbol{r}$ but not on its direction as, for example, in the case of Newton-Coulomb gravitational field:

$$V = \frac{c}{r^3}\boldsymbol{r} = \frac{c}{r^2}\cdot\frac{\boldsymbol{r}}{r}.$$

For a plane field, that case is called a *circular field.*

A *cylindrical vector field* (Fig. 400c) is a field in which all the vectors $\boldsymbol{V}$ lie on straight lines intersecting a certain line called the *axis*, are perpendicular to the axis and, moreover, the vectors at the points lying at a constant distance from the axis have equal lengths and are all directed either toward the axis or away from it. If we locate the pole on the axis of the field, then the field can be expressed by the formula

$$V = \varphi(\varrho)\frac{\boldsymbol{r}'}{\varrho},$$

where $\boldsymbol{r}'$ is the projection of the vector $\boldsymbol{r}$ on a plane perpendicular to the axis; if $\boldsymbol{c}$ is the unit vector of the axis, then $\boldsymbol{r}' = \boldsymbol{c}\times(\boldsymbol{r}\times\boldsymbol{c})$. By intersecting a cylindrical vector field with planes perpendicular to its axis, we obtain equal circular fields.

Coordinate definition of a field. A vector field (1) can be defined by means of three scalar fields $V^1(\boldsymbol{r})$, $V^2(\boldsymbol{r})$, $V^3(\boldsymbol{r})$ which are the coefficients of the resolution of $\boldsymbol{V}$ into any three non-complenar base vectors $\boldsymbol{e}_1, \boldsymbol{e}_2, \boldsymbol{e}_3$:

$$\boldsymbol{V} = V^1\boldsymbol{e}_1 + V^2\boldsymbol{e}_2 + V^3\boldsymbol{e}_3. \tag{2}$$

If we take the coordinate unit vectors $\boldsymbol{i}, \boldsymbol{j}, \boldsymbol{k}$ as the base vectors and express the coefficients V^1, V^2, V^3 in terms of Cartesian coordinates x, y, z then we obtain

(2a) $$V = V_x(x, y, z)\,i + V_y(x, y, z)\,j + V_z(x, y, z)\,k.$$

Hence a vector field is defined by means of three scalar functions of three variables (definition of a field in Cartesian coordinates).

In cylindrical and spherical coordinates, the coordinate unit

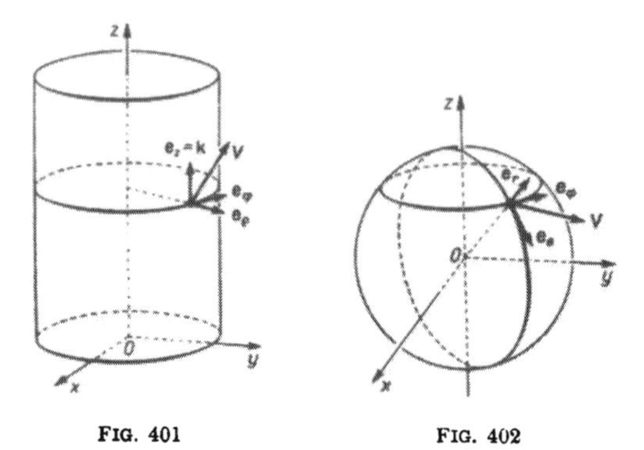

FIG. 401 FIG. 402

vectors e_ϱ, e_φ, $e_z(=k)$ (Fig. 401) and $e_r(=r/r)$, e_φ, e_θ (Fig. 402) are tangent to the coordinate lines at each point and, expressing the coefficients by the corresponding coordinates, we have

(2b) $$V = V_\varrho(\varrho, \varphi, z)\,e_\varrho + V_\varphi(\varrho, \varphi, z)\,e_\varphi + V_z(\varrho, \varphi, z)\,e_z.$$

(2c) $$V = V_r(r, \varphi, \theta)\,e_r + V_\varphi(r, \varphi, \theta)\,e_\varphi + V_\theta(r, \varphi, \theta)\,e_\theta.$$

In both cases the unit vectors have different directions at various points, but remain mutually perpendicular.

Transformation formulas from one coordinate system to another.

The expression of Cartesian coordinates by cylindrical coordinates:

$$V_x = V_\varrho \cos\varphi - V_\varphi \sin\varphi, \qquad V_y = V_\varrho \sin\varphi + V_\varphi \cos\varphi, \qquad V_z = V_z.$$

The expression of cylindrical coordinates by Cartesian coordinates:

$$V_\varrho = V_x \cos\varphi + V_y \sin\varphi, \qquad V_\varphi = -V_x \sin\varphi + V_y \cos\varphi, \qquad V_z = V_z.$$

The expression of Cartesian coordinates by spherical coordinates:

$$V_x = V_r \sin\theta\cos\varphi - V_\varphi \sin\varphi + V_\theta \cos\varphi\cos\theta,$$
$$V_y = V_r \sin\theta\sin\varphi + V_\varphi \cos\varphi + V_\theta \sin\varphi\cos\theta,$$
$$V_z = V_r \cos\theta - V_\theta \sin\theta.$$

The expression of spherical coordinates by Cartesian coordinates:

$$V_r = V_x \sin\theta\cos\varphi + V_y \sin\theta\sin\varphi + V_z\cos\theta,$$
$$V_\varphi = -V_x\sin\varphi + V_y\cos\varphi,$$
$$V_\theta = V_x\cos\theta\cos\varphi + V_y\cos\theta\sin\varphi - V_z\sin\theta.$$

The expression of a spherical vector field in Cartesian coordinates

$$\boldsymbol{V} = \varphi\left(\sqrt{x^2+y^2+z^2}\right)(x\boldsymbol{i} + y\boldsymbol{j} + z\boldsymbol{k}).$$

The expression of a cylindrical vector field in Cartesian coordinates:

$$\boldsymbol{V} = \varphi\left(\sqrt{x^2+y^2}\right)(x\boldsymbol{i} + y\boldsymbol{j}).$$

In the case of a spherical field, the spherical coordinates ($\boldsymbol{V} = V(r)\boldsymbol{e}_r$) are most convenient, and, for a cylindrical field, the cylindrical coordinates ($\boldsymbol{V} = V(\varrho)\boldsymbol{e}_\varrho$) are most convenient. In the case of a plane field (Fig. 403), we have

$$\boldsymbol{V} = V_x(x, y)\,\boldsymbol{i} + V_y(x, y)\,\boldsymbol{j} = V_\varrho(x, y)\,\boldsymbol{e}_\varrho + V_\varphi(x, y)\,\boldsymbol{e}_\varphi,$$

and for a circular field

$$\boldsymbol{V} = \varphi\left(\sqrt{x^2+y^2}\right)(x\boldsymbol{i} + y\boldsymbol{j}) = \varphi(\varrho)\,\boldsymbol{e}_\varrho.$$

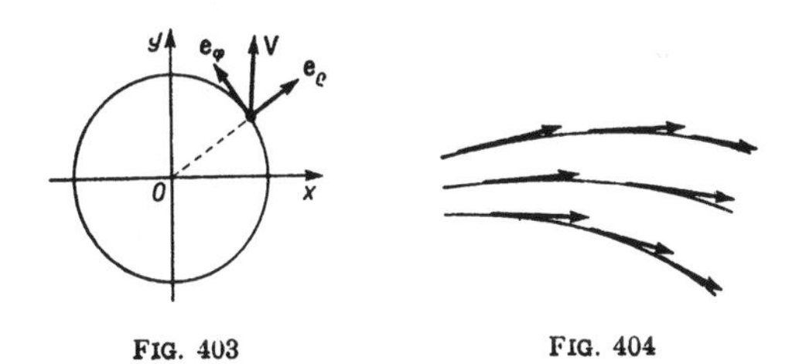

FIG. 403 FIG. 404

Lines of force. A curve Γ at every point $M(\boldsymbol{r})$ of which the vector $\boldsymbol{V}(\boldsymbol{r})$ is tangent to Γ is called a *line of force* (Fig. 404). The lines of force are mutually disjoint (except at the points where the function $\boldsymbol{V}$ is not defined or $\boldsymbol{V} = \boldsymbol{0}$).

Examples. A line of force of a central field is a ray issuing from the centre O and passing through M. Lines of force of the field $\boldsymbol{V} = \boldsymbol{c}\times\boldsymbol{r}$ are circles lying in planes perpendicular to the vector $\boldsymbol{c}$ and having their centres on the axis parallel to $\boldsymbol{c}$.

Differential equations of lines of force of a field expressed in Cartesian coordinates are [1]:

$$\frac{dx}{V_x} = \frac{dy}{V_y} = \frac{dz}{V_z},$$

and, for a plane field,

$$\frac{dx}{V_x} = \frac{dy}{V_y}.$$

8. Gradient

Derivative of a scalar field. The *derivative of the scalar field* $U = U(\boldsymbol{r})$ at a given point $\boldsymbol{r}$ with respect to the vector $\boldsymbol{c}$ (or in the direction of $\boldsymbol{c}$) (Fig. 405) is the limit

$$\frac{\partial U}{\partial \boldsymbol{c}} = \lim_{\varepsilon \to 0} \frac{U(\boldsymbol{r} + \varepsilon \boldsymbol{c}) - U(\boldsymbol{r})}{\varepsilon}.$$

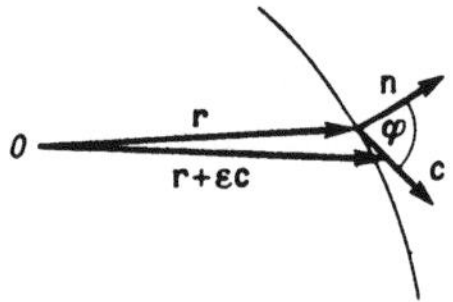

FIG. 405

The derivative of the field U with respect to the unit vector $\boldsymbol{c}^0$ is $\frac{\partial U}{\partial \boldsymbol{c}^0}$. The derivatives with respect to the vector $\boldsymbol{c}$ and to its unit vector $\boldsymbol{c}^0$ are related by the formula

$$\frac{\partial U}{\partial \boldsymbol{c}} = |\boldsymbol{c}| \frac{\partial U}{\partial \boldsymbol{c}^0}.$$

The derivative $\frac{\partial U}{\partial \boldsymbol{c}^0}$ represents the speed of increasing of the function U in the direction of the vector $\boldsymbol{c}^0$ at each point. The greatest of all derivatives with respect to all unit vectors at a given point is the derivative $\partial U/\partial \boldsymbol{n}$, where $\boldsymbol{n}$ is the normal unit vector to the level surface of the field pointing in the direction of increasing U. Any other derivative is given by the formula

$$\frac{\partial U}{\partial \boldsymbol{c}^0} = \frac{\partial U}{\partial \boldsymbol{n}} \cos(\boldsymbol{c}^0, \boldsymbol{n}) = \frac{\partial U}{\partial \boldsymbol{n}} \cos \varphi.$$

Gradient of a field. The *gradient of the scalar field* $U(r)$ is a vector denoted by grad U or ∇U [2] defined at each point of the field; it is normal to the level surface, pointing in the direction of increasing U and with the length equal to $\frac{\partial U}{\partial \boldsymbol{n}}$. Thus

---

(1) For solution of such differential equations see pp. 517, 529.

(2) For the operator ∇ (*nabla*) see p. 643.

$$\operatorname{grad} U = \boldsymbol{n}\frac{\partial U}{\partial \boldsymbol{n}}.$$

The derivative $\frac{\partial U}{\partial \mathbf{c}^0}$ is equal to the projection of grad U onto the direction of $\mathbf{c}^0$:

$$\frac{\partial U}{\partial \mathbf{c}^0} = \mathbf{c}^0 \operatorname{grad} U.$$

Coordinates of the gradient:
in Cartesian coordinate system

$$\operatorname{grad} U = \frac{\partial U}{\partial x}\boldsymbol{i} + \frac{\partial U}{\partial y}\boldsymbol{j} + \frac{\partial U}{\partial z}\boldsymbol{k},$$

in cylindrical coordinate system

$$\operatorname{grad} U = \frac{\partial U}{\partial \varrho}\boldsymbol{e}_\varrho + \frac{1}{\varrho}\cdot\frac{\partial U}{\partial \varphi}\boldsymbol{e}_\varphi + \frac{\partial U}{\partial z}\boldsymbol{e}_z,$$

in spherical coordinate system

$$\operatorname{grad} U = \frac{\partial U}{\partial r}\boldsymbol{e}_r + \frac{1}{r\sin\theta}\cdot\frac{\partial U}{\partial \varphi}\boldsymbol{e}_\varphi + \frac{1}{r}\cdot\frac{\partial U}{\partial \theta}\boldsymbol{e}_\theta.$$

At the points of the field where the level lines drawn according to the condition on p. 626 are more dense, the absolute value of the gradient is greater. At points of maximum or minimum of the field, where the level surface and lines degenerate to a point, grad $U = 0$.

Differential of a scalar field. The *differential of the scalar field U* is the total differential of the function U (see p. 363):

$$dU = \operatorname{grad} U\, d\boldsymbol{r} = \frac{\partial U}{\partial x}dx + \frac{\partial U}{\partial y}dy + \frac{\partial U}{\partial z}dz.$$

Rules for computing the gradient [1]

$$\operatorname{grad} c = 0, \qquad \operatorname{grad}(U_1 + U_2) = \operatorname{grad} U_1 + \operatorname{grad} U_2,$$

$$\operatorname{grad}(cU) = c \operatorname{grad} U,$$

$$\operatorname{grad}(U_1 U_2) = U_1 \operatorname{grad} U_2 + U_2 \operatorname{grad} U_1, \qquad \operatorname{grad}\varphi(U) = \frac{d\varphi}{dU}\operatorname{grad} U,$$

$\operatorname{grad}(\boldsymbol{V}_1\boldsymbol{V}_2) = (\boldsymbol{V}_1 \operatorname{grad})\boldsymbol{V}_2 + (\boldsymbol{V}_2 \operatorname{grad})\boldsymbol{V}_1 + \boldsymbol{V}_1 \times \operatorname{rot}\boldsymbol{V}_2 + \boldsymbol{V}_2 \times \operatorname{rot}\boldsymbol{V}_1$ [2];

in particular, $\operatorname{grad}(\boldsymbol{r}\mathbf{c}) = \mathbf{c}$.

---

[1] We assume here and in the following that c and $\mathbf{c}$ are constants.
[2] For the expressions $(\boldsymbol{V}\operatorname{grad})\boldsymbol{W}$ and rot $\boldsymbol{V}$ see pp. 644, 642.

The gradient of a central field: grad $U(r) = U'(r)\,\frac{\boldsymbol{r}}{r}$ (a spherical field); in particular, grad $r = \frac{\boldsymbol{r}}{r}$ (the field of unit vectors).

Gradient as the space derivative. The *space derivative* of a scalar field (see p. 640) is a vector equal to the gradient of the field. This property can be taken as a definition of the gradient:

$$\operatorname{grad} U = \lim_{v\to 0} \frac{\oint\limits_{\Sigma} U\,d\boldsymbol{S}}{v}.$$

9. Line integral and potential in a vector field [1]

Definition. The *line integral*

$$\int\limits_{\breve{AB}} \boldsymbol{V}(\boldsymbol{r})\,d\boldsymbol{r}$$

of a vector function $\boldsymbol{V}(\boldsymbol{r})$ taken along the arc $\breve{AB}$ is a scalar P obtained as follows:

(1) We divide the path $\breve{AB}$ of integration (Fig. 406) into n elementary arcs by means of points

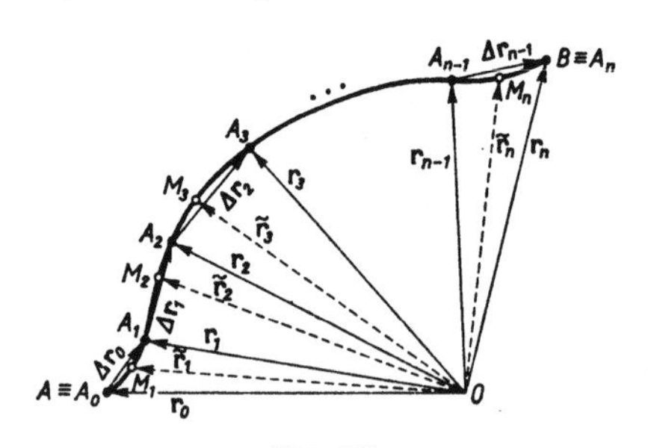

FIG. 406

$$A = A_0,\ A_1(\boldsymbol{r}_1),\ A_2(\boldsymbol{r}_2),\ \dots,\ A_{n-1}(\boldsymbol{r}_{n-1}),\ A_n = B.$$

The elementary arcs are approximately represented by the vectors $\boldsymbol{r}_i - \boldsymbol{r}_{i-1} = \Delta\boldsymbol{r}_{i-1}$.

(2) We select an arbitrary point M_i with the radius vector $\tilde{\boldsymbol{r}}_i$ inside or on the boundary of each elementary arc $\overset{\frown}{A_{i-1}A_i}$.

---

[1] In this section an exposition of the line integral of the second type in the general form is given.

(3) We form the scalar products of the value $\boldsymbol{V}(\tilde{\boldsymbol{r}}_i)$ of the vector function at M_i times the vector $\Delta \boldsymbol{r}_{i-1}$.

(4) We add the obtained n scalar products.

(5) We determine the limit of the obtained sum

$$\sum_{i=1}^{n} \boldsymbol{V}(\tilde{\boldsymbol{r}}_i)\, \Delta \boldsymbol{r}_{i-1},$$

when the length of each elementary vector $\Delta \boldsymbol{r}_{i-1}$ tends to zero (hence $n \to \infty$).

If this limit exists and is independent of the particular choice of the points $A_1, A_2, \ldots, A_{n-1}$ and $M_0, M_1, \ldots, M_{n-1}$, then we call it the *line integral*

$$\int_{\breve{AB}} \boldsymbol{V}(\boldsymbol{r})\, d\boldsymbol{r} = \lim_{\substack{\Delta \boldsymbol{r} \to 0 \\ n \to \infty}} \sum_{i=1}^{n} \boldsymbol{V}(\tilde{\boldsymbol{r}}_i)\, \Delta \boldsymbol{r}_{i-1}.$$

Existence theorem. If the function $\boldsymbol{V}(\boldsymbol{r})$ is continuous [1] and the arc $\breve{AB}$ is continuous and has a tangent varying continuously, then the line integral $\int_{\breve{AB}} \boldsymbol{V}(\boldsymbol{r})\, d\boldsymbol{r}$ exists.

Mechanical interpretation of a line integral. If $\boldsymbol{V}$ is a field of force, then the line integral $\int_{\breve{AB}} \boldsymbol{V}(\boldsymbol{r})\, d\boldsymbol{r}$ represents the work done by the force $\boldsymbol{V}$ in displacement along the path $\breve{AB}$.

Properties of the line integral:

$$\int_{\breve{ABC}} \boldsymbol{V}(\boldsymbol{r})\, d\boldsymbol{r} = \int_{\breve{AB}} \boldsymbol{V}(\boldsymbol{r})\, d\boldsymbol{r} + \int_{\breve{BC}} \boldsymbol{V}(\boldsymbol{r})\, d\boldsymbol{r} \quad \text{(Fig. 407)},$$

$$\int_{\breve{AB}} \boldsymbol{V}(\boldsymbol{r})\, d\boldsymbol{r} = -\int_{\breve{BA}} \boldsymbol{V}(\boldsymbol{r})\, d\boldsymbol{r},$$

$$\int_{\breve{AB}} [\boldsymbol{V}(\boldsymbol{r}) + \boldsymbol{W}(\boldsymbol{r})]\, d\boldsymbol{r} = \int_{\breve{AB}} \boldsymbol{V}(\boldsymbol{r})\, d\boldsymbol{r} + \int_{\breve{AB}} \boldsymbol{W}(\boldsymbol{r})\, d\boldsymbol{r},$$

$$\int_{\breve{AB}} c\boldsymbol{V}(\boldsymbol{r})\, d\boldsymbol{r} = c\int_{\breve{AB}} \boldsymbol{V}(\boldsymbol{r})\, d\boldsymbol{r}.$$

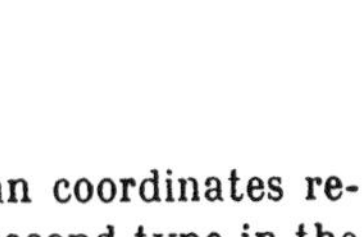

FIG. 407

The evaluation of a line integral in Cartesian coordinates reduces to the evaluation of a line integral of the second type in the general form (see pp. 490, 491):

$$\int_{\breve{AB}} \boldsymbol{V}(\boldsymbol{r})\, d\boldsymbol{r} = \int_{\breve{AB}} (V_x\, dx + V_y\, dy + V_z\, dz).$$

---

[1] The vector function $\boldsymbol{V}$ is continuous if all the coefficients of its resolution into three base vectors $\boldsymbol{e}_1, \boldsymbol{e}_2, \boldsymbol{e}_3$ are continuous.

Circulation. The *circuit integral* of the vector field $\boldsymbol{V}(\boldsymbol{r})$ about a closed curve C is its line integral along C. It is denoted by $\oint\limits_C \boldsymbol{V}\,d\boldsymbol{r}$ and called the *circulation* of the field $\boldsymbol{V}$.

Potential or conservative field. The vector field $\boldsymbol{V}$ is called a *potential* or *conservative field* if the line integral $\int\limits_{\breve{AB}} \boldsymbol{V}(\boldsymbol{r})\,d\boldsymbol{r}$ depends solely on the end points A and B but not on the particular choice of the path integration $\breve{AB}$ joining them. A circuit integral in a potential field is always equal to zero; a potential field is irrotational:

$$\text{(1)} \qquad \operatorname{rot} \boldsymbol{V} = 0$$

(see p. 647). Equation (1) together with the continuity of the corresponding partial derivatives of the coordinates of the field constitutes a necessary and sufficient condition for a field to be a potential field.

Condition (1) in Cartesian coordinates has the form

$$\text{(1a)} \qquad \frac{\partial V_x}{\partial y} = \frac{\partial V_y}{\partial x}, \quad \frac{\partial V_y}{\partial z} = \frac{\partial V_z}{\partial y}, \quad \frac{\partial V_z}{\partial x} = \frac{\partial V_x}{\partial z} \ (^1).$$

For a plane field we take only two first of equalities (1a).

Potential of a conservative field. If we fix the initial point $A(\boldsymbol{r}_0)$ in a potential field and let the final point $B(\boldsymbol{r})$ vary, then the integral $\int\limits_{\breve{AB}} \boldsymbol{V}(\boldsymbol{r})\,d\boldsymbol{r}$ (which can be written as $\int\limits_{\boldsymbol{r}_0}^{\boldsymbol{r}} \boldsymbol{V}(\boldsymbol{r})\,d\boldsymbol{r}$) is a scalar function

$$\int\limits_{\boldsymbol{r}_0}^{\boldsymbol{r}} \boldsymbol{V}(\boldsymbol{r})\,d\boldsymbol{r} = \varphi(\boldsymbol{r})$$

of the vector $\boldsymbol{r}$ and the scalar function $\varphi(\boldsymbol{r})$ is called a *potential function* or a *potential* of the field $\boldsymbol{V}(\boldsymbol{r})$ ([2]).

The potential of a field is determined up to an additive constant which depends on the lower limit $\boldsymbol{r}_0$ of integration. *Difference of potentials*:

$$\varphi(\boldsymbol{r}_2) - \varphi(\boldsymbol{r}_1) = \int\limits_{\boldsymbol{r}_1}^{\boldsymbol{r}_2} \boldsymbol{V}(\boldsymbol{r})\,d\boldsymbol{r}.$$

([1]) This is a condition for integrability (see p. 492).

([2]) This is a primitive function (see p. 493). In physics, the integral with the opposite sign is sometimes called a *potential* at the point $\boldsymbol{r}$: $= \int\limits_{\boldsymbol{r}_0}^{\boldsymbol{r}} \boldsymbol{V}(\boldsymbol{r})\,d\boldsymbol{r}$.

Relation between the gradient, line integral and potential of a field. If $\boldsymbol{V}(\boldsymbol{r}) = \text{grad } U(\boldsymbol{r})$, then $U(\boldsymbol{r})$ is a potential of $\boldsymbol{V}(\boldsymbol{r})$ [1] and conversely.

The evaluation of the potential U of a conservative field $\boldsymbol{V} = V_x\boldsymbol{i} + V_y\boldsymbol{j} + V_z\boldsymbol{k}$ in Cartesian coordinates is equivalent to the evaluation of the function U from its total differential $dU = V_x dx + V_y dy + V_z dz$ (V_x, V_y, V_z should satisfy conditions (1a)). The function U can be found from the system of equations

$$\frac{\partial U}{\partial x} = V_x, \quad \frac{\partial U}{\partial y} = V_y, \quad \frac{\partial U}{\partial z} = V_z.$$

In practice, we find the potential by integration along a broken line (Fig. 408) with sides parallel to the coordinate axes (see evaluation of a primitive function on p. 493):

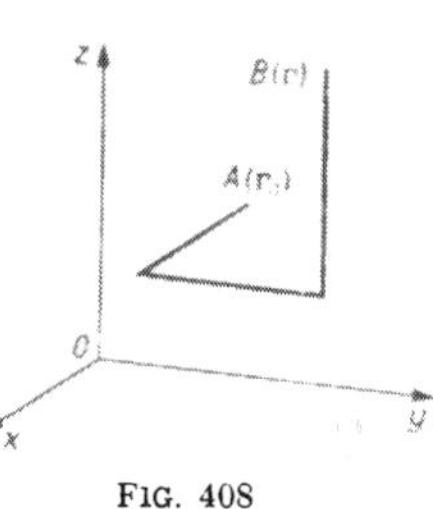

FIG. 408

$$U = \int_{r_0}^{r} \boldsymbol{V} d\boldsymbol{r} = U(x_0, y_0, z_0) + \int_{x_0}^{x} V_x(x, y_0, z_0)\, dx + \int_{y_0}^{y} V_y(x, y, z_0)\, dy + \int_{z_0}^{z} V_z(x, y, z)\, dz.$$

10. Surface integral [2]

Normal vector. The *normal vector* to an oriented plane sheet of surface Σ bounded by a closed curve C (Fig. 409a) is a vector $\boldsymbol{S}$ whose length is equal to the area S of Σ, whose direction is per-

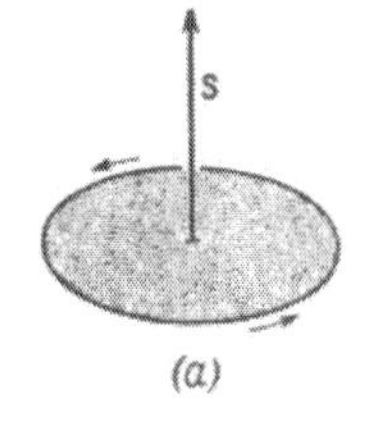

(a)

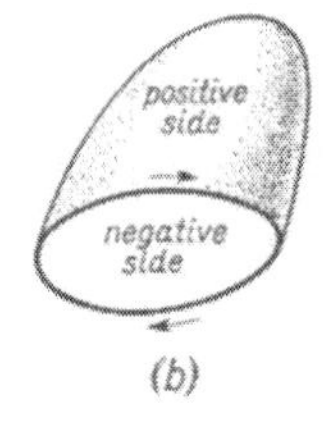

(b)

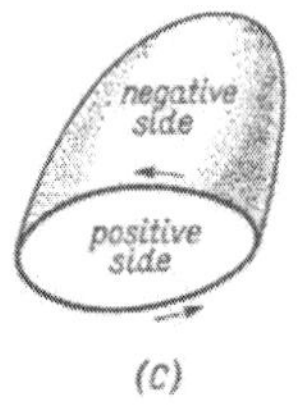

(c)

FIG. 409

[1] Or else a potential with the opposite sign, see the previous footnote.

[2] In this section we give a vectorial treatment of the theory of the surface integral of the second type in the general form.

pendicular to the plane of Σ and which is pointing so that if we attach the initial point of the vector $\boldsymbol{S}$ to the surface Σ and look at Σ from the final point of the vector $\boldsymbol{S}$, we see the positive side of Σ (i.e., the side on which the closed curve C is directed counter-clockwise). This definition can be extended to an arbitrary oriented sheet of surface bounded by a closed curve (Fig. 409b and c).

Three forms of surface integrals (over an oriented sheet of surface Σ bounded by a closed curve or over a closed surface).

Definition. *Surface integrals* in a scalar of vector field are quantities obtained as follows:

(1) We divide the surface Σ on which a positive side is chosen (in the case of a closed surface, we take the exterior side as positive) into n elementary regions dS_i in an arbitrary way (Fig. 410); we regard each of the elementary regions approximately as a plane region and to each of them we attach its normal vector $d\boldsymbol{S}_i$;

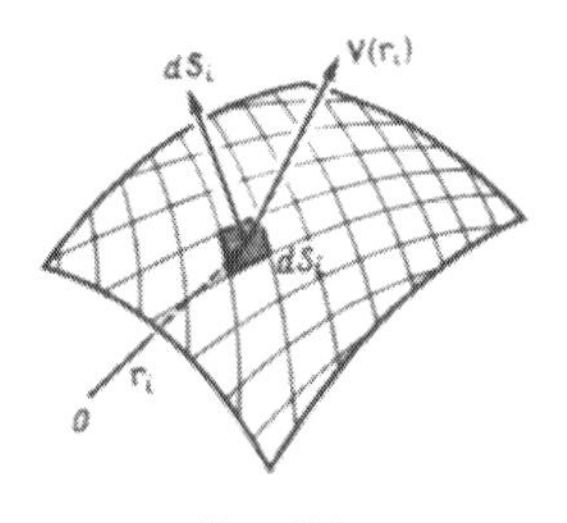

Fig. 410

(2) we select an arbitrary point $\boldsymbol{r}_i$ inside or on the boundary of each elementary region dS_i;

(3) in the case of a scalar field we form the product $U(\boldsymbol{r}_i)d\boldsymbol{S}_i$, and, in the case of a vector field, we form the scalar product $\boldsymbol{V}(\boldsymbol{r}_i)d\boldsymbol{S}_i$ or the vector product $\boldsymbol{V}(\boldsymbol{r}_i)\times d\boldsymbol{S}_i$;

(4) we add the products obtained for each particular elementary region;

(5) we determine the limit of the obtained sum, when dS_i or $d\boldsymbol{S}_i$ tends to zero (hence $n\to\infty$), provided that this limit exists, and is independent on the decomposition of the surface of integration into the elementary regions or of the choice of the points $\boldsymbol{r}_i$ (1).

A. *Flow of a scalar field*:

$$\boldsymbol{P} = \lim_{dS_i\to 0} \sum U(\boldsymbol{r}_i)\, d\boldsymbol{S}_i = \int_{\Sigma} U(\boldsymbol{r})\, d\boldsymbol{S}$$

B. *Scalar flow of a vector field*:

$$Q = \lim_{dS_i\to 0} \sum \boldsymbol{V}(\boldsymbol{r}_i)\, d\boldsymbol{S}_i = \int_{\Sigma} \boldsymbol{V}(\boldsymbol{r})\, d\boldsymbol{S}.$$

C. *Vector flow of a vector field*:

(1) Each elementary region dS_i is contracted to a point in the sense of the footnote on p. 496.

$$\boldsymbol{R} = \lim_{dS_i \to 0} \sum \boldsymbol{V}(\boldsymbol{r}_i) \times d\boldsymbol{S}_i = \int_{\Sigma} \boldsymbol{V}(\boldsymbol{r}) \times d\boldsymbol{S}\ (^1).$$

The evaluation of a surface integral in Cartesian coordinates reduces to the evaluation of surface integrals of the second type (see p. 508) and can be performed according to the formulas:

A. $$\int_{\Sigma} U d\boldsymbol{S} = \iint_{\Sigma_{yz}} U dydz\, \boldsymbol{i} + \iint_{\Sigma_{zx}} U dzdx\, \boldsymbol{j} + \iint_{\Sigma_{xy}} U dxdy\, \boldsymbol{k}.$$

B. $$\int_{\Sigma} \boldsymbol{V} d\boldsymbol{S} = \iint_{\Sigma_{yz}} V_x dydz + \iint_{\Sigma_{zx}} V_y dzdx + \iint_{\Sigma_{xy}} V_z dxdy.$$

C. $$\int_{\Sigma} \boldsymbol{V} \times d\boldsymbol{S} = \iint_{\Sigma_{yz}} (V_z \boldsymbol{j} - V_y \boldsymbol{k})\, dydz + \iint_{\Sigma_{zx}} (V_x \boldsymbol{k} - V_z \boldsymbol{i})\, dzdx + \\ + \iint_{\Sigma_{xy}} (V_y \boldsymbol{i} - V_x \boldsymbol{j})\, dxdy.$$

In the above formulas each of the integrals is taken over projection of $\boldsymbol{\Sigma}$ on the corresponding coordinate plane [1] (Fig. 411); in each expression under the integral sign one of the variables x, y, z should be expressed respectively by two remaining ones using the equation of the surface $\boldsymbol{\Sigma}$.

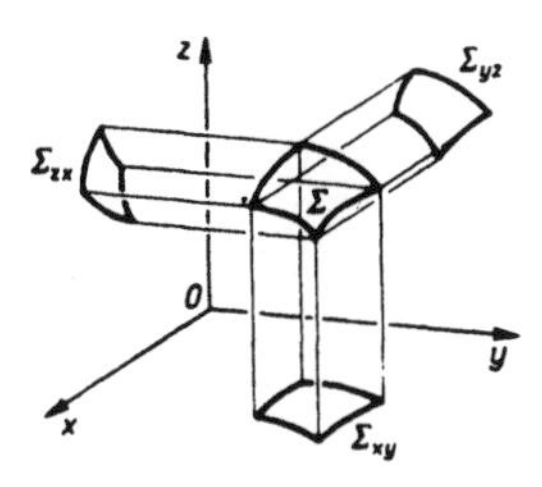

FIG. 411

Examples. A. $\boldsymbol{P} = \int_{\Sigma} xyz\, d\boldsymbol{S}$, where $\boldsymbol{\Sigma}$ is the part of the plane $x + y + z = 1$ contained between three coordinate planes (the upper side of $\boldsymbol{\Sigma}$ is positive). We have

$$\boldsymbol{P} = \iint_{yz} (1 - y - z)\, yz\, dydz\, \boldsymbol{i} + \iint_{zx} (1 - x - z)\, xz\, dzdx\, \boldsymbol{j} + \\ + \iint_{xy} (1 - x - y)\, xy\, dxdy\, \boldsymbol{k};$$

$$\iint_{yz} (1 - y - z)\, yz\, dydz = \int_0^1 \int_0^{1-z} (1 - y - z)\, yz\, dydz = \tfrac{1}{120};$$

(1) Furnished with the signs "+" or "−" (see p. 508).

two remaining integrals are analogous. The result:

$$\boldsymbol{P} = \tfrac{1}{120}(\boldsymbol{i} + \boldsymbol{j} + \boldsymbol{k}).$$

B. $Q = \int_{\Sigma} \boldsymbol{r}\, d\boldsymbol{S} = \iint_{yz} x\, dydz + \iint_{zx} y\, dzdx + \iint_{xy} z\, dxdy$, where the surface Σ is the same as in example A. We have

$$\iint_{xy} z\, dxdy = \int_0^1 \int_0^{1-x} (1 - x - y)\, dydx = \tfrac{1}{6};$$

two remaining integrals are analogous. The result $Q = \frac{1}{6} + \frac{1}{6} + \frac{1}{6} = \frac{1}{2}$.

C. $\boldsymbol{R} = \int_{\Sigma} \boldsymbol{r} \times d\boldsymbol{S} = \int_{\Sigma} (x\boldsymbol{i} + y\boldsymbol{j} + z\boldsymbol{k}) \times (dydz\, \boldsymbol{i} + dzdx\, \boldsymbol{j} + dxdy\, \boldsymbol{k})$ over the same surface Σ. Analogous computation yields $\boldsymbol{R} = 0$.

Integrals over a closed surface are denoted by the symbols

$$\oint_{\Sigma} U\, d\boldsymbol{S}, \quad \oint_{\Sigma} \boldsymbol{V}\, d\boldsymbol{S}, \quad \oint_{\Sigma} \boldsymbol{V} \times d\boldsymbol{S}.$$

11. Space differentiation

Definition. *Space derivatives* of a scalar or vector field at a point $\boldsymbol{r}$ are quantities of three forms obtained as follows:

(1) We surround the point $\boldsymbol{r}$ of the scalar field $U(\boldsymbol{r})$ or vector field $\boldsymbol{V}(\boldsymbol{r})$ with a closed surface Σ;

(2) we evaluate the circuit integrals

$$\oint_{\Sigma} U(\boldsymbol{r})\, d\boldsymbol{S}, \quad \oint_{\Sigma} \boldsymbol{V}(\boldsymbol{r})\, d\boldsymbol{S} \quad \text{or} \quad \oint_{\Sigma} \boldsymbol{V}(\boldsymbol{r}) \times d\boldsymbol{S};$$

around Σ;

(3) we determine the limit of the ratio of the integral to the volume interior to Σ, when the region interior to Σ is contracted to a point (in the sense of the footnote on p. 496).

The space derivative of a scalar field is equal to its gradient (see p. 634), and the scalar derivative and vector derivative of a vector field lead to the concepts of the divergence and rotation of the field.

12. Divergence of a vector field

Definition. The *divergence* of the vector field $\boldsymbol{V}$ is a scalar (denoted div $\boldsymbol{V}$ or $\nabla \boldsymbol{V}$ [1]) defined at each point of the field and

[1] For the symbol ∇ (nabla), see p. 643.

equal to the space scalar derivative of the field:

$$\operatorname{div} \boldsymbol{V} = \lim_{v \to 0} \frac{\oint_{\Sigma} \boldsymbol{V}\, d\boldsymbol{S}}{v}.$$

Formulas and rules for evaluation of the divergence:
In Cartesian coordinates

$$\operatorname{div} \boldsymbol{V} = \frac{\partial V_x}{\partial x} + \frac{\partial V_y}{\partial y} + \frac{\partial V_z}{\partial z}.$$

In cylindrical coordinates

$$\operatorname{div} \boldsymbol{V} = \frac{1}{\varrho} \cdot \frac{\partial}{\partial \varrho} (\varrho V_\varrho) + \frac{1}{\varrho} \cdot \frac{\partial V_\varphi}{\partial \varphi} + \frac{\partial V_z}{\partial z}.$$

In spherical coordinates

$$\operatorname{div} \boldsymbol{V} = \frac{1}{r^2} \left(\frac{\partial}{\partial r} (r^2 V_r) \right) + \frac{1}{r \sin \theta} \cdot \frac{\partial V_\varphi}{\partial \varphi} + \frac{1}{r \sin \theta} \left(\frac{\partial}{\partial \theta} (\sin \theta\, V_\theta) \right).$$

Moreover we have the relations

$$\operatorname{div} \boldsymbol{c} = 0, \qquad \operatorname{div} (\boldsymbol{V}_1 + \boldsymbol{V}_2) = \operatorname{div} \boldsymbol{V}_1 + \operatorname{div} \boldsymbol{V}_2, \qquad \operatorname{div} (c\boldsymbol{V}) = c \operatorname{div} \boldsymbol{V},$$

$$\operatorname{div} (U\boldsymbol{V}) = U \operatorname{div} \boldsymbol{V} + \boldsymbol{V} \operatorname{grad} U, \text{ in particular, } \operatorname{div} \boldsymbol{r}\boldsymbol{c} = \frac{\boldsymbol{r}\boldsymbol{c}}{r},$$

$$\operatorname{div} (\boldsymbol{V}_1 \times \boldsymbol{V}_2) = \boldsymbol{V}_2 \operatorname{rot} \boldsymbol{V}_1 - \boldsymbol{V}_1 \operatorname{rot} \boldsymbol{V}_2.$$

Divergence of a central field:

$$\operatorname{div} \boldsymbol{r} = 3, \qquad \operatorname{div} \varphi(r) \boldsymbol{r} = 3\varphi(r) + r\varphi'(\boldsymbol{r}).$$

13. Rotation of a vector field

Definition. The *rotation* (or *curl*) of the vector field $\boldsymbol{V}$ is a vector (denoted rot $\boldsymbol{V}$, curl $\boldsymbol{V}$ or $\nabla \times \boldsymbol{V}$ (¹)) defined at each point of the field and equal to the space vector derivative of the field taken with the opposite sign:

$$\operatorname{rot} \boldsymbol{V} = - \lim_{v \to 0} \left(\frac{1}{v} \int_{\Sigma} \boldsymbol{V} \times d\boldsymbol{S} \right) (^2).$$

(¹) For the symbol ∇ (nabla) see p. 643.
(²) We can remove the sign "–" by interchanging the factors under the integral sign: $\int d\boldsymbol{S} \times \boldsymbol{V}$ (see p. 617).

Another definition: the *rotation* of the vector field $\boldsymbol{V}$ is a vector obtained as follows:

(1) We put a small elementary surface sheet S through the point $\boldsymbol{r}$ (Fig. 412);

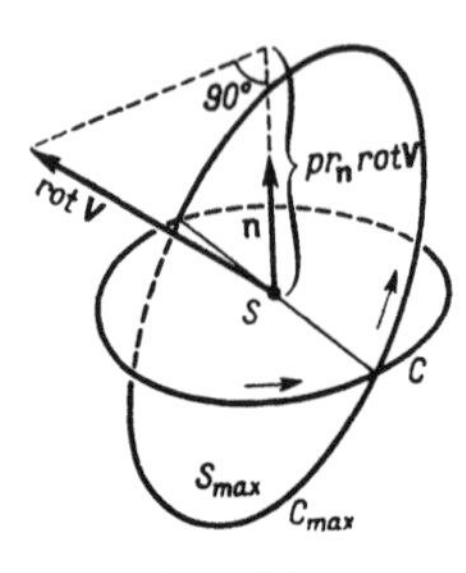

FIG. 412

(2) evaluate the circuit integral $\oint\limits_C \boldsymbol{V} d\boldsymbol{r}$ about the boundary C of the sheet (see p. 636);

(3) determine the limit of the ratio of this integral to the area of the elementary sheet S, when S is contracted to a point and the position of the plane of S remains unchanged;

(4) by changing the position of the plane of S, we find its position $S_{\max}$ such that the obtained limit assumes a maximum;

(5) at the point $\boldsymbol{r}$ we attach a vector rot $\boldsymbol{r}$ whose length is equal to the obtained maximum and whose direction is that of the vector normal to $S_{\max}$:

$$|\operatorname{rot} \boldsymbol{V}| = \lim_{S \to 0} \frac{\int\limits_{C_{\max}} \boldsymbol{V} d\boldsymbol{r}}{S_{\max}};$$

the projection of the vector rot $\boldsymbol{V}$ on the normal to S is equal to

$$\lim_{S \to 0} \frac{\int\limits_{C} \boldsymbol{V} d\boldsymbol{r}}{S}.$$

The rotation of a potential field is equal to zero (a corollary from Stokes' theorem, see p. 646).

Formulas for coordinates of the rotation:

In Cartesian coordinates

$$\operatorname{rot} \boldsymbol{V} = \left(\frac{\partial V_z}{\partial y} - \frac{\partial V_y}{\partial z}\right)\boldsymbol{i} + \left(\frac{\partial V_x}{\partial z} - \frac{\partial V_z}{\partial x}\right)\boldsymbol{j} + \left(\frac{\partial V_y}{\partial x} - \frac{\partial V_x}{\partial y}\right)\boldsymbol{k}.$$

In cylindrical coordinates

$$\operatorname{rot} \boldsymbol{V} = \left(\frac{1}{\varrho} \cdot \frac{\partial V_z}{\partial \varphi} - \frac{\partial V_\varphi}{\partial z}\right)\boldsymbol{e}_\varrho + \left(\frac{\partial V_\varrho}{\partial z} - \frac{\partial V_z}{\partial \varrho}\right)\boldsymbol{e}_\varphi + \\ + \left(\frac{1}{\varrho} \cdot \frac{\partial(\varrho V_\varphi)}{\partial \varrho} - \frac{1}{\varrho} \cdot \frac{\partial V_\varrho}{\partial \varphi}\right)\boldsymbol{e}_z.$$

In spherical coordinates

$$\operatorname{rot} \boldsymbol{V} = \left[\frac{1}{r \sin\theta}\left(\frac{\partial}{\partial\theta}(\sin\theta\, V_\varphi) - \frac{\partial V_\theta}{\partial\varphi}\right)\right]\boldsymbol{e}_r +$$
$$+\left[\frac{1}{r}\cdot\frac{\partial(rV_\theta)}{\partial r} - \frac{1}{r}\cdot\frac{\partial V_r}{\partial\theta}\right]\boldsymbol{e}_\varphi + \left[\frac{1}{\sin\theta}\cdot\frac{\partial V_r}{\partial\varphi} - \frac{1}{r}\left(\frac{\partial}{\partial r}(rV_\varphi)\right)\right]\boldsymbol{e}_\theta.$$

Moreover we have the relations

$$\operatorname{rot}(\boldsymbol{V}_1 + \boldsymbol{V}_2) = \operatorname{rot}\boldsymbol{V}_1 + \operatorname{rot}\boldsymbol{V}_2, \qquad \operatorname{rot}(c\boldsymbol{V}) = c \operatorname{rot}\boldsymbol{V},$$
$$\operatorname{rot}(U\boldsymbol{V}) = U \operatorname{rot}\boldsymbol{V} + \operatorname{grad} U \times \boldsymbol{V},$$
$$\operatorname{rot}(\boldsymbol{V}_1 \times \boldsymbol{V}_2) = (\boldsymbol{V}_2 \operatorname{grad})\boldsymbol{V}_1 - (\boldsymbol{V}_1 \operatorname{grad})\boldsymbol{V}_2 + \boldsymbol{V}_1 \operatorname{div}\boldsymbol{V}_2 - \boldsymbol{V}_2 \operatorname{div}\boldsymbol{V}_1 \text{ [1]}.$$

Lines of force of the field rot $\boldsymbol{V}$ (see p. 631) are called the *curl lines* of the field $\boldsymbol{V}$.

14. The operators ∇ (Hamilton's operator), $(a\nabla)$ and Δ (Laplace's operator)

Hamilton's operator. *Hamilton's operator* ∇ (*nabla*) is a symbolical vector used to replace the symbols grad, div and rot:

$$\nabla U = \operatorname{grad} U, \qquad \nabla\boldsymbol{V} = \operatorname{div}\boldsymbol{V}, \qquad \nabla \times \boldsymbol{V} = \operatorname{rot}\boldsymbol{V}.$$

Introduction of this symbol simplifies computations in the vector analysis. Expression for Hamilton's operator in Cartesian coordinates:

$$\nabla = \frac{\partial}{\partial x}\boldsymbol{i} + \frac{\partial}{\partial y}\boldsymbol{j} + \frac{\partial}{\partial z}\boldsymbol{k}.$$

Multiplying this vector formally by the scalar U or forming the scalar or vector product of ∇ and $\boldsymbol{V}$, we obtain formulas for the gradient (p. 632), divergence (p. 640) and rotation (p. 641) in Cartesian coordinates.

Rules of operations with ∇.

(1) If ∇ is applied to a linear combination $\sum a_i X_i$, where a_i are constant and X_i are scalar functions or vector functions of a point, then

$$\nabla\left(\sum a_i X_i\right) = \sum a_i \nabla X_i.$$

(2) If ∇ is applied to a product of scalar functions or vector functions X, Y, Z of a point, then we apply it successively to each of the factors (writing above the corresponding function the sign $\downarrow$) and add the results:

---

[1] For expression $\boldsymbol{V}$ grad, see p. 644.

$$\nabla(XYZ) = \nabla(\overset{\downarrow}{X}YZ) + \nabla(X\overset{\downarrow}{Y}Z) + \nabla(XY\overset{\downarrow}{Z}),$$

and then we transform the obtained products according to vector algebra so as the operator ∇ to be applied to only one factor with the sign $\downarrow$; having performed the computation we can omit that sign.

Examples.

$$\text{(1)}\quad \operatorname{div}(U\boldsymbol{V}) = \nabla(U\boldsymbol{V}) = \nabla(\overset{\downarrow}{U}\boldsymbol{V}) + \nabla(U\overset{\downarrow}{\boldsymbol{V}}) = \boldsymbol{V}\cdot\nabla U + U\cdot\nabla\boldsymbol{V}$$
$$= \boldsymbol{V}\operatorname{grad} U + U\operatorname{div}\boldsymbol{V}.$$

$$\text{(2)}\quad \operatorname{div}(\boldsymbol{V}_1\times\boldsymbol{V}_2) = \nabla(\boldsymbol{V}_1\times\boldsymbol{V}_2) = \nabla(\overset{\downarrow}{\boldsymbol{V}_1}\times\boldsymbol{V}_2) + \nabla(\boldsymbol{V}_1\times\overset{\downarrow}{\boldsymbol{V}_2})$$
$$= \nabla\overset{\downarrow}{\boldsymbol{V}_1}\boldsymbol{V}_2 + \nabla\boldsymbol{V}_1\overset{\downarrow}{\boldsymbol{V}_2} = \boldsymbol{V}_2\nabla\boldsymbol{V}_1 - \boldsymbol{V}_1\nabla\boldsymbol{V}_2$$
$$= \boldsymbol{V}_2(\nabla\times\boldsymbol{V}_1) - \boldsymbol{V}_1(\nabla\times\boldsymbol{V}_2) = \boldsymbol{V}_2\operatorname{rot}\boldsymbol{V}_1 - \boldsymbol{V}_1\operatorname{rot}\boldsymbol{V}_2.$$

Operator $(\boldsymbol{a}\nabla)$. In computations, the following operational expression may arise:

$$(\boldsymbol{a}\nabla) = a_x\frac{\partial}{\partial x}\boldsymbol{i} + a_y\frac{\partial}{\partial y}\boldsymbol{j} + a_z\frac{\partial}{\partial z}\boldsymbol{k}.$$

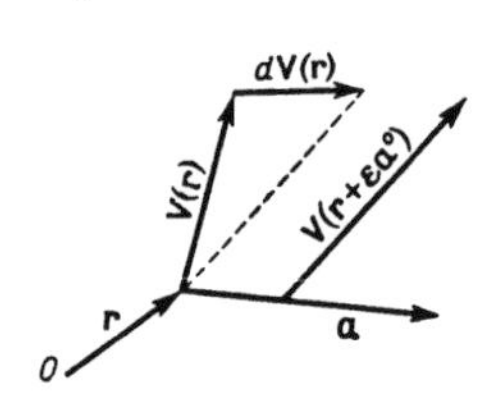

FIG. 413

The vector $(\boldsymbol{a}\nabla)\boldsymbol{V} = (\boldsymbol{a}\operatorname{grad})\boldsymbol{V}$ is called the *gradient* of the vector field $\boldsymbol{V}$ with respect to the vector $\boldsymbol{a}$; it is equal to the derivative of $\boldsymbol{V}$ with respect to $\boldsymbol{a}$ (Fig. 413):

$$(\boldsymbol{a}\nabla)\boldsymbol{V} = (\boldsymbol{a}\operatorname{grad})\boldsymbol{V}$$
$$= |\boldsymbol{a}|\cdot\lim_{\varepsilon\to 0}\frac{\boldsymbol{V}(\boldsymbol{r}+\varepsilon\boldsymbol{a}^0) - \boldsymbol{V}(\boldsymbol{r})}{\varepsilon}.$$

Example. $\operatorname{grad}(\boldsymbol{V}_1\boldsymbol{V}_2) = \nabla(\boldsymbol{V}_1\boldsymbol{V}_2) = \nabla(\overset{\downarrow}{\boldsymbol{V}_1}\boldsymbol{V}_2 + \nabla(\boldsymbol{V}_1\overset{\downarrow}{\boldsymbol{V}_2})$.

By the formula $\boldsymbol{b}(\boldsymbol{ac}) = (\boldsymbol{ab})\boldsymbol{c} + \boldsymbol{a}\times(\boldsymbol{b}\times\boldsymbol{c})$ (see p. 617), we obtain

$$\operatorname{grad}(\boldsymbol{V}_1\boldsymbol{V}_2) = (\boldsymbol{V}_2\nabla)\boldsymbol{V}_1 + \boldsymbol{V}_2\times(\nabla\times\boldsymbol{V}_1) + (\boldsymbol{V}_1\nabla)\boldsymbol{V}_2 + \boldsymbol{V}_1\times(\nabla\times\boldsymbol{V}_2)$$
$$= (\boldsymbol{V}_2\operatorname{grad})\boldsymbol{V}_1 + \boldsymbol{V}_2\times\operatorname{rot}\boldsymbol{V}_1 + (\boldsymbol{V}_1\operatorname{grad})\boldsymbol{V}_2 + \boldsymbol{V}_1\times\operatorname{rot}\boldsymbol{V}_2.$$

The expression $(\boldsymbol{a}\nabla)\boldsymbol{V}$ can be transformed according to

$$2(\boldsymbol{a}\nabla)\boldsymbol{V} = \operatorname{rot}(\boldsymbol{V}\times\boldsymbol{a}) + \operatorname{grad}(\boldsymbol{a}\boldsymbol{V}) + \boldsymbol{a}\operatorname{div}\boldsymbol{V} - \boldsymbol{V}\operatorname{div}\boldsymbol{a} - \boldsymbol{a}\times\operatorname{rot}\boldsymbol{V} -$$
$$-\boldsymbol{V}\times\operatorname{rot}\boldsymbol{a}.$$

Operator ∇ twice applied; operator Δ. For any field we have

(1) $\nabla(\nabla\times\boldsymbol{V}) = \operatorname{div}\operatorname{rot}\boldsymbol{V} = 0,$

(2) $\nabla\times(\nabla U) = \operatorname{rot}\operatorname{grad} U = 0,$

(3) $\nabla(\nabla U) = \operatorname{div}\operatorname{grad} U = \Delta U,$
where Δ (also $\nabla\nabla$ or Δ^2) is the *Laplace's operator.*

Formulas for the Laplace's operator:
In Cartesian coordinates

$$\Delta U = \frac{\partial^2 U}{\partial x^2} + \frac{\partial^2 U}{\partial y^2} + \frac{\partial^2 U}{\partial z^2}.$$

In cylindrical coordinates

$$\Delta U = \frac{1}{\varrho}\cdot\frac{\partial}{\partial\varrho}\left(\varrho\frac{\partial U}{\partial\varrho}\right) + \frac{1}{\varrho^2}\cdot\frac{\partial^2 U}{\partial\varphi^2} + \frac{\partial^2 U}{\partial z^2}.$$

In spherical coordinates

$$\Delta U = \frac{\partial^2 U}{\partial r^2} + \frac{2}{r}\cdot\frac{\partial U}{\partial r} + \frac{1}{r^2\sin^2\theta}\cdot\frac{\partial^2 U}{\partial\varphi^2} + \frac{1}{r^2}\cdot\frac{\partial^2 U}{\partial\theta^2} + \frac{1}{r^2}\cot\theta\frac{\partial U}{\partial\theta}.$$

(4) $\nabla(\nabla \boldsymbol{V}) = \operatorname{grad}\operatorname{div}\boldsymbol{V},$

(5) $\nabla\times(\nabla\times\boldsymbol{V}) = \operatorname{rot}\operatorname{rot}\boldsymbol{V}.$

The expressions (4) and (5) are related by the formula

$$\nabla(\nabla\boldsymbol{V}) - \nabla\times(\nabla\times\boldsymbol{V}) = \Delta\boldsymbol{V},$$

where $\Delta\boldsymbol{V} = (\nabla\nabla)\boldsymbol{V}$ is the Laplace's operator applied to the vector $\boldsymbol{V}$:

$$\begin{aligned}\Delta\boldsymbol{V} &= \Delta V_x\boldsymbol{i} + \Delta V_y\boldsymbol{j} + \Delta V_z\boldsymbol{k}\\ &= \left(\frac{\partial^2 V_x}{\partial x^2} + \frac{\partial^2 V_x}{\partial y^2} + \frac{\partial^2 V_x}{\partial z^2}\right)\boldsymbol{i} + \left(\frac{\partial^2 V_y}{\partial x^2} + \frac{\partial^2 V_y}{\partial y^2} + \frac{\partial^2 V_y}{\partial z^2}\right)\boldsymbol{j} +\\ &\quad + \left(\frac{\partial^2 V_z}{\partial x^2} + \frac{\partial^2 V_z}{\partial y^2} + \frac{\partial^2 V_z}{\partial z^2}\right)\boldsymbol{k}.\end{aligned}$$

15. Integral theorems [1]

Gauss-Ostrogradsky theorem:

$$\oint_{\Sigma} \boldsymbol{V}\,d\boldsymbol{S} = \int_{v} \operatorname{div}\boldsymbol{V}\,dv.$$

This states that the scalar flow of the field $\boldsymbol{V}$ through a closed surface $\boldsymbol{\Sigma}$ is equal to the integral of the divergence of $\boldsymbol{V}$ over the region v bounded by $\boldsymbol{\Sigma}$.

In Cartesian coordinates

---

[1] See pp. 512 and 513.

$$\iint_{\Sigma} (V_x dydz + V_y dzdx + V_z dxdy) = \iiint_{v} \left(\frac{\partial V_x}{\partial x} + \frac{\partial V_y}{\partial y} + \frac{\partial V_z}{\partial z} \right) dxdydz,$$

where V_x, V_y, V_z are functions of three variables x, y, z.

Stokes' theorem:

$$\oint_C \boldsymbol{V} d\boldsymbol{r} = \int_{\Sigma} \operatorname{rot} \boldsymbol{V} d\boldsymbol{S}.$$

This means that the circulation of the field about a closed curve C is equal to the flow of the rotation of the field through an arbitrary surface Σ bounded by C [1].

In Cartesian coordinates

$$\int_C (V_x dx + V_y dy + V_z dz)$$

$$= \iint_{\Sigma} \left(\frac{\partial V_z}{\partial y} - \frac{\partial V_y}{\partial z} \right) dydz + \left(\frac{\partial V_x}{\partial z} - \frac{\partial V_z}{\partial x} \right) dzdx + \left(\frac{\partial V_y}{\partial x} - \frac{\partial V_x}{\partial y} \right) dxdy.$$

For a plane closed curve (*Green's formula*):

$$\int_C (V_x dx + V_y dy) = \iint_{\Sigma} \left(\frac{\partial V_y}{\partial x} - \frac{\partial V_x}{\partial y} \right) dxdy,$$

where V_x and V_y are functions of two variables x and y.

Green's theorem:

$$(1) \quad \int_{\Sigma} U_1 \operatorname{grad} U_2 \, d\boldsymbol{S} = \int_{v} (U_1 \Delta U_2 + \operatorname{grad} U_1 \operatorname{grad} U_2) dv,$$

$$(2) \quad \int_{\Sigma} (U_1 \operatorname{grad} U_2 - U_2 \operatorname{grad} U_1) d\boldsymbol{S} = \int_{v} (U_1 \Delta U_2 - U_2 \Delta U_1) dv,$$

where U_1 and U_2 are scalar fields and v is a space region bounded by the surface Σ. In particular, for $U_1 = 1$, we have

$$(3) \quad \int_{\Sigma} \operatorname{grad} U d\boldsymbol{S} = \int_{v} \Delta U dv.$$

In Cartesian coordinates theorem (3) has the form

$$\oint_{\Sigma} \frac{\partial U}{\partial x} dydz + \frac{\partial U}{\partial y} dzdx + \frac{\partial U}{\partial z} dxdy = \int_{v} \left(\frac{\partial^2 U}{\partial x^2} + \frac{\partial^2 U}{\partial y^2} + \frac{\partial^2 U}{\partial z^2} \right) dv.$$

---

[1] For a more accurate formulation, see p. 512.

16. Irrotational and solenoidal vector fields

Definitions. The field $\boldsymbol{V}$ is called *irrotational*, if its rotation is everywhere equal to zero. If rot $\boldsymbol{V}=0$, then $\boldsymbol{V}=\operatorname{grad} U$; the function U, or the potential of the field $\boldsymbol{V}$ [1], can be expressed at any point M by the formula

$$U=-\frac{1}{4\pi}\int\frac{\operatorname{div}\boldsymbol{V}\,dv}{r}, \tag{1}$$

where r is the distance from the elementary region dv to the point M and the integral is extended over the whole space [2].

A vector field is called *solenoidal* if its divergence is everywhere equal to zero. If div $\boldsymbol{V}=0$, then there exists a solenoidal field $\boldsymbol{W}$ (the vector potential of the field $\boldsymbol{V}$) such that $\boldsymbol{V}=\operatorname{rot}\boldsymbol{W}$ and, at any point M, the vector potential $\boldsymbol{W}$ can be expressed by the formula

$$\boldsymbol{W}=\frac{1}{4\pi}\int\frac{\operatorname{rot}\boldsymbol{V}\,dv}{r}, \tag{2}$$

where r has the same meaning as in formula (1) and the integral is extended over the whole space [3].

Any vector field $\boldsymbol{V}$ which decreases sufficiently rapidly, when r (the distance from a fixed point) increases infinitely, can be resolved uniquely into a sum $\boldsymbol{V}=\boldsymbol{V}_1+\boldsymbol{V}_2$, where $\boldsymbol{V}_1$ is an irrotational field and $\boldsymbol{V}_2$ is a solenoidal field, according to the formula

$$\boldsymbol{V}_1=-\frac{1}{4\pi}\operatorname{grad}\int\frac{\operatorname{div}\boldsymbol{V}\,dv}{r},\qquad \boldsymbol{V}_2=\frac{1}{4\pi}\operatorname{rot}\int\frac{\operatorname{rot}\boldsymbol{V}\,dv}{r}.$$

The field with point sources. The *Newton-Coulomb field*

$$\boldsymbol{E}=\frac{e}{r^3}\boldsymbol{r}$$

is irrotational and solenoidal everywhere except at the pole O (a *source of the field*). Its potential is equal to $U=-e/r$ [4]. The scalar flow $\oint_S \boldsymbol{E}\,d\boldsymbol{S}$ is zero, if the source is outside the surface S, and is equal to $4\pi e$, if the source is inside the surface S; the quantity e is called the *intensity of the source.*

---

[1] Or the negative of the potential, see footnote [2] on p. 636.

[2] Formula (1) is true provided that the divergence of $\boldsymbol{V}$ is a differentiable function and sufficiently rapidly decreases to zero, when r increases infinitely.

[3] Formula (2) is true provided that the rotation of the field $\boldsymbol{V}$ is a differentiable function and sufficiently rapidly decreases to zero, if r increases infinitely.

[4] Or $+e/r$, see footnote [2] on p. 636.

A Newton field with the source at the point $\boldsymbol{r}_1$:

$$\boldsymbol{E} = \frac{e_1}{|\boldsymbol{r} - \boldsymbol{r}_1|^3}(\boldsymbol{r} - \boldsymbol{r}_1).$$

A Newton field with several sources $\boldsymbol{r}_1, \boldsymbol{r}_2, \boldsymbol{r}_3, \ldots$ whose intensities are, respectively, $e_1, e_2, e_3, \ldots$:

$$\boldsymbol{E} = \sum \frac{e_i}{|\boldsymbol{r} - \boldsymbol{r}_i|^3}(\boldsymbol{r} - \boldsymbol{r}_i).$$

The scalar flow is equal to zero if the interior of the surface S is free of sources, and it is equal to $4\pi\sum' e_i$, if some sources lie inside S, the sum being extended over all the sources inside S.

17. Laplace's and Poisson's equations

Laplace's equation. Determining a scalar field U for which $\Delta U = 0$ (div grad $U = 0$) leads to a partial differential equation called *Laplace's equation*:

$$\frac{\partial^2 U}{\partial x^2} + \frac{\partial^2 U}{\partial y^2} + \frac{\partial^2 U}{\partial z^2} = 0,$$

or, in the plane,

$$\frac{\partial^2 U}{\partial x^2} + \frac{\partial^2 U}{\partial y^2} = 0.$$

Functions (continuous and with continuous derivatives of the first and second order) satisfying these equations are called *Laplace's functions* or *harmonic functions*. If the values of a harmonic function at points of a closed surface Σ are known, then its values at all points within the surface are uniquely determined. The problem of finding these values is *Dirichlet's problem*.

Let Σ be a closed surface. If the values of a harmonic function U and of its derivative $\frac{\partial U}{\partial \boldsymbol{n}}$ in the direction of the exterior normal vector to the surface are known, then its value U_M at the point M can be found from the formula

$$U_M = \frac{1}{4\pi}\oint_{\Sigma} \frac{1}{r} \cdot \frac{\partial U}{\partial \boldsymbol{n}}\, dS - \frac{1}{4\pi}\oint_{\Sigma} \frac{\partial(1/r)}{\partial \boldsymbol{n}}\, U\, dS,$$

where r is the distance from the elementary region dS of the surface to the point M.

Poisson's equation. Given the divergence $\varrho(x, y, z)$ of the gradient of a scalar field U, determining the field U leads to the *Poisson's equation*

$$\Delta U = \varrho(x, y, z) \quad \text{or} \quad \frac{\partial^2 U}{\partial x^2} + \frac{\partial^2 U}{\partial y^2} + \frac{\partial^2 U}{\partial z^2} = \varrho(x, y, z).$$

If ϱ is a continuous function and if it is known that the function U tends to zero sufficiently rapidly, when $r \to \infty$ (i.e., when the distance of a point from a fixed point increases infinitely), then the Newton potential of the function ϱ is a solution of Poisson's equation. It is defined by the formula

$$U_M = -\frac{1}{4\pi} \int \frac{\varrho \, dv}{r},$$

where r is the distance from the elementary space region dv to the point M and the integral is extended over the whole space.

III. THE CALCULUS OF VARIATIONS

1. Fundamental principles

The *calculus of variations* concerns itself with the problem of determining from some previously given class of functions one or more functions which, in a given single or multiple integral, according to the type of functions, yields an extremum; i.e., assumes a maximal or minimal value. Many problems of theoretical physics, engineering, and geometry lead to such problems.

Examples. 1. Through two distinct points P_1 and P_2 with different altitudes we wish to determine a curve lying in such a manner that under the influence of gravity a point mass descends along this curve in the shortest possible amount of time (the friction is considered negligible). We establish our coordinate system so that the abscissa passes through P_1 and the positive half of the ordinate axis goes downward from P_1. When the point mass has fallen a distance y, its velocity due to the acceleration of the earth's gravity g, is $\sqrt{2gy}$. Since the velocity equals the derivative of the distance with respect to time, we get

$$\frac{ds}{dt} = \sqrt{2gy}.$$

Since $ds = \sqrt{1 + y'^2}\,dx$ it follows that (cf. p. 271)

$$dt = \frac{ds}{\sqrt{2gy}} = \frac{\sqrt{1 + y'^2}}{\sqrt{2gy}}\,dx$$

or, after integration

$$t = \frac{1}{\sqrt{2g}} \int_{x_1}^{x_2} \frac{\sqrt{1 + y'^2}}{\sqrt{y}}\,dx.$$

Thus our problem amounts to finding the curve (i.e. the function $y = y(x)$) which will yield a minimum in the integral

$$\int_{x_1}^{x_2} \frac{\sqrt{1+y'^2}}{\sqrt{y}}\, dx.$$

We call the integrand $\frac{\sqrt{1+y'^2}}{\sqrt{y}}$ the *base function.*

2. In xyz-space let an orthogonal coordinate system and a surface determined by the equation $G(x, y, z) = 0$ be given. On it lie the points P_1 and P_2 with coordinates (x_1, y_1, z_1) and (x_2, y_2, z_2), $x_1 < x_2$. The two points P_1 and P_2 are to be joined by the shortest differentiable space curve $y = y(x)$, $z = z(x)$ which lies on the given surface $G(x, y, z) = 0$ (the *problem of the geodesic line*). From differential geometry we know that the solution will yield a minimum of the integral

$$J = \int_{x_1}^{x_2} \sqrt{1 + y'^2 + z'^2}\, dx$$

subject to the side condition

$$G\big(x, y(x), z(x)\big) = 0$$

for all $x_1 < x < x_2$. Here two functions are sought and a *side condition* is to be satisfied.

Such side conditions may be of very different natures. In the previous example it is an ordinary algebraic equation; in other problems, in its place, may appear a differential equation or any other functional equation.

3. The coordinate origin and a fixed point $P(l,0)$ on the x-axis are to be joined by a continuous differentiable curve of fixed length L in such a manner that the curve and the x-axis enclose the greatest possible area. Thus the solution will yield a maximum value for

$$J = \int_0^l y\, dx$$

while simultaneously fulfilling the side condition

$$\int_0^l \sqrt{1 + y'^2}\, dx = L.$$

Here the side condition is in the form of a given fixed integral; we call such a problem an *isoperimetric* variation problem.

2. The simple variation problem with one unknown function

(Necessary conditions for an extremum)

The simple variation problem is: For a given base function $F(x, y, y')$ to determine a curve $y = y(x)$ such that for x, $y = y(x)$, and $y' = y'(x)$ the integral

$$J = \int_{x_1}^{x_2} F(x, y, y')\, dx$$

will assume an extreme value; i.e., the greatest or smallest value in comparison with the value which any other of the continuous differentiable curves under consideration yield in this integral. The base function will be determined by the concrete physical, technical or geometric problem; in example 1 it was $F = \sqrt{1 + y'^2}/\sqrt{y}$.

Assume that the curve $y = y(x)$ is an extreme value of the integral J. In passing from the curve $y = y(x)$ to another curve $\bar{y} = \bar{y}(x)$ we may deduce that the increment of the integral $\Delta J = (J\bar{y}(x)) - J(y(x))$ will have the same sign for any curve $\bar{y} = \bar{y}(x)$. Thus for a minimum $\Delta J > 0$ and for a maximum $\Delta J < 0$. We call the difference $\bar{y}(x) - y(x)$ the *variation* of the function $y(x)$ and denote it by δy. Clearly this difference considered at the end points of the curve is equal to zero:

$$\delta y = 0 \quad \text{for} \quad x = x_1 \quad \text{and} \quad x = x_2.$$

In the variation problem the variation of the function plays a role similar to that of the increment of the independent variable in the extreme value problem of differential calculus. It is moreover a function of x and indeed $\frac{d}{dx}\delta y = \delta(y')$ holds. From the curve $y = y(x)$ we get the considered curve $\bar{y} = \bar{y}(x)$, by adding to $y(x)$ its variation δy:

$$\bar{y}(x) = y(x) + \delta y.$$

Ordinarily the integral will be examined at a so-called relative extremum; i.e., at an extremum which represents a minimum or maximum value in the case of neighbouring comparable curves. Here the notion of "neighbourhood" may be interpreted differently. If we assume that in the equations of the values of the functions $\bar{y} = y(x) + \delta y$ the largest value of $|\delta y|$ is small, while $|\delta y'|$ may take any value, then if the curve $y = y(x)$ is an extremum, we call this a *strong extremum*. If we obtain an extreme value from the curve $y = y(x)$ only in the case when we compare it with the restricted class of curve for which both $|\delta y|$ and $|\delta y'|$ are small, i.e., in the case of curves, which not only with respect to their position but also with respect to the directions of their tangents differ only a little from $y(x)$, then we speak of a *weak extremum*.

A strong extremum is at the same time a weak extremum but the reverse situation does not always hold true.

To derive a *necessary condition* for the yielding of an extremum we calculate the increment of the integral J for the passage from $y(x)$ to a curve $\bar{y}(x) = y(x) + \delta y$ and we get

$$\Delta J = \int_{x_1}^{x_2} F(x, \bar{y}, \bar{y}')\, dx - \int_{x_1}^{x_2} F(x, y, y')\, dx$$

and using the mean value theorem

$$\Delta J = \int_{x_1}^{x_2} \left[\left(\frac{\partial \hat{F}}{\partial y}\right)\delta y + \left(\frac{\partial \hat{F}}{\partial y'}\right)\delta y'\right] dx.$$

The symbol $\hat{}$ denotes that the partial derivative is evaluated for some intermediate value of the second and third variables. We assume that the partial derivatives $\partial F/\partial y$ and $\partial F/\partial y'$ are continuous, thus ΔJ may be rewritten as

$$\Delta J = \int_{x_1}^{x_2} \left[\frac{\partial F}{\partial y}\delta y + \frac{\partial F}{\partial y'}\delta y'\right] dx + R(y, y')$$

where $R(y, y')$ is in comparison with δy and $\delta y'$ an infinitesimal of higher order.

As an analoque to the notion of differentials we consider the main part of the variation ΔJ of our integral which is linear in δy and $\delta y'$ and denote this by δJ:

$$\delta J = \int_{x_1}^{x_2} \left[\frac{\partial F}{\partial y}\delta y + \frac{\partial F}{\partial y'}\delta y'\right] dx.$$

If $\delta J \neq 0$ then for sufficiently small δy and $\delta y'$ the sign of ΔJ is dominated by the sign of δJ. As δy changes its sign, the sign of δJ also changes and thus the sign of ΔJ cannot remain constant for arbitrary δy. Thus the *condition that the variation vanishes, $\delta J = 0$ is a necessary condition for the existence of an extreme value of our integral*. Thus at an extremum we get

$$\int_{x_1}^{x_2} \left[\frac{\partial F}{\partial y}\delta y + \frac{\partial F}{\partial y'}\delta y'\right] dx = 0.$$

It follows from partial integration of the second summand and from an examination of the boundary conditions $y = 0$ for $x = x_1$, $x = x_2$ that

$$\int_{x_1}^{x_2} \left[\frac{\partial F}{\partial y} - \frac{d}{dx}\left(\frac{\partial F}{\partial y'}\right)\right] \delta y \, dx = 0.$$

This integral must vanish for any choice of δy. This is possible only when the first factor is identically zero; i.e., when

$$\frac{\partial F}{\partial y} - \frac{d}{dx}\left(\frac{\partial F}{\partial y'}\right) = 0.$$

The function $y(x)$ for which our integral assumes an extreme value must therefore satisfy this so-called *Euler differential equation* of the variational calculus. More explicitly, this equation takes the form

$$\frac{\partial F}{\partial y} - \frac{\partial^2 F}{\partial y'^2} y'' - \frac{\partial^2 F}{\partial y' \partial y} y' - \frac{\partial^2 F}{\partial y' \partial x} = 0.$$

The curves which are solutions to this equation are called *extremal* in the sense of the calculus of variations. ***This condition is indeed necessary but not sufficient.*** In general the solution $y = y(x)$ of a second order ordinary differential equation has two undetermined constants of integration. To determine these constants two boundary conditions are required. In many problems the boundary conditions are of the form $y_1 = y(x_1)$ and $y_2 = y(x_2)$; i.e., the equation $y = y(x)$ is to determine a curve which passes through the two points $P_1(x_1, y_1)$ and $P_2(x_2, y_2)$. In practice, when the Euler differential equation may be easily integrated, this leads immediately to the solution of the variation problem in question. We would like, therefore, before we derive a sufficient condition, to give an example of the use of the Euler differential equation.

Example.

$$\int_0^1 (\tfrac{1}{2}y'^2 - y^2 - 2xy)dx = \text{extremum}.$$

Boundary conditions: $y(0) = 0$, $y(1) = 0$.

For this particular base function the Euler differential equation becomes

$$-2y - 2x - 2y'' = 0 \quad \text{or} \quad y'' + y + x = 0.$$

The general solution is $y = C_1 \cos x + C_2 \sin x - x$ (cf. the section

"Ordinary differential equations"). The boundary conditions yield for the constants of integration

$$C_1 = 0, \quad C_2 = \frac{1}{\sin 1}.$$

Accordingly the function we seek is $y = \sin x/\sin 1 - x$.

Special cases.

(a) The variable x does not occur in the base function:

$$\int_{x_1}^{x_2} F(y, y')\, dx = \text{extremum}.$$

In this case the Euler differential equation may be written in the form

$$\left[\frac{\partial F}{\partial y} - \frac{d}{dx}\left(\frac{\partial F}{\partial y'}\right)\right] y' = 0$$

or

$$\frac{d}{dx}\left[F - \frac{\partial F}{\partial y'} y'\right] = 0.$$

The intermediate integral is

$$F - \frac{\partial F}{\partial y'} y' = C_1.$$

From this equation the desired solution of the Euler differential equation may be obtained by another integration.

Example.

$$\int_{x_1}^{x_2} \frac{\sqrt{1 + y'^2}}{\sqrt{y}}\, dx = \text{extremum} \quad (\text{cf. p.650}).$$

Boundary conditions: $f(x_1) = y_1$; $f(x_2) = y_2$.

The intermediate integral is

$$\text{(A)} \qquad \frac{\sqrt{1 + y'^2}}{\sqrt{y}} - \frac{y'^2}{\sqrt{y}\sqrt{1 + y'^2}} = C_1.$$

From this we deduce that

$$y' = \frac{\sqrt{1 - C_1^2 y}}{C_1 \sqrt{y}}$$

or, after a separation of variables

$$dx = \frac{dy\, C_1 \sqrt{y}}{\sqrt{1 - C_1^2 y}}.$$

After integration we obtain

$$x = C_2 - \frac{1}{C_1}\sqrt{y - C_1^2 y^2} - \frac{1}{C_1^2}\arcsin\sqrt{1 - C_1^2 y}.$$

This form of the solution is not clear: By a suitable transformation it may be changed to parametric form. From the equation (A) we obtain

$$\sqrt{y(1 + y'^2)} = \frac{1}{C_1}$$

and writing for simplification

$$\frac{1}{C_1} = \sqrt{2K_1},$$

it follows that

$$y(1 + y'^2) = 2K_1.$$

We next introduce the parameter t through the substitution

$$\text{(B)} \qquad y' = -\tan\tfrac{1}{2}t = -\frac{\sin t}{1 + \cos t}$$

and we find that

$$\text{(C)} \qquad y(1 + \tan^2\tfrac{1}{2}t) = \frac{y}{\cos^2\frac{1}{2}t} = 2K_1,$$

$$y = 2K_1\cos^2\tfrac{1}{2}t = K_1(1 + \cos t).$$

Further from (B) we get,

$$\frac{dx}{dt} = \frac{dx}{dy}\cdot\frac{dy}{dt} = -\frac{1 + \cos t}{\sin t}(-K_1 \sin t)$$

or, after simplification

$$\frac{dx}{dt} = K_1(1 + \cos t).$$

It follows by integration that

$$x = K_1(t + \sin t) + K_2.$$

Joining this with equation (C) we deduce finally that

$$x = K_1(t + \sin t) + K_2, \qquad y = K_1(1 + \cos t).$$

But this is the known parametric form of a cycloid.

(b) The variable y does not occur in the base function

$$\int_{x_1}^{x_2} F(x, y')\, dx = \text{extremum}.$$

The Euler differential equation is simple in this case, and it follows that

$$\frac{\partial^2 F}{\partial y'^2} y'' + \frac{\partial^2 F}{\partial y' \partial x} = 0$$

or

$$\frac{d}{dx}\left(\frac{\partial F}{\partial y'}\right) = 0.$$

We obtain for the intermediate integral

$$\frac{\partial F}{\partial y'} = C_1.$$

Example. The Euler differential equation for the problem

$$\int_{x_1}^{x_2} x\sqrt{1 + y'^2}\, dx = \text{extremum}$$

is clearly

$$\frac{d}{dx}\left(\frac{xy'}{\sqrt{1 + y'^2}}\right) = 0.$$

The intermediate integral

$$\frac{xy'}{\sqrt{1 + y'^2}} = C_1$$

yields the general solution $y = C_1 \ln |x + \sqrt{x^2 - C_1^2}| + C_2$.

(c) The base function contains neither x nor y

$$\int_{x_1}^{x_2} F(y')dx = \text{extremum}.$$

In this case the Euler differential equation simplifies to

$$\frac{\partial^2 F}{\partial y'^2} y'' = 0,$$

thus, if $\partial^2 F/\partial y'^2$ is not identically zero,

$$y'' = 0,$$

from which we obtain as the solution, the line $y = C_1 x + C_2$.

Example. The two fixed points $P_1(x_1, y_1)$ and $P_2(x_2, y_2)$ are to be joined by the shortest possible continuously differentiable curve.

Solution: The base function $\sqrt{1 + y'^2}$ in the integral

$$\int_{x_1}^{x_2} \sqrt{1 + y'^2}\, dx$$

depends only on y'.

Since $\partial^2 F/\partial y'^2 \not\equiv 0$ the curve which is the solution of the Euler differential equation is the line passing through the two points P_1 and P_2:

$$y - y_1 = \frac{y_2 - y_1}{x_2 - x_1}(x - x_1).$$

(d) $\partial^2 F/\partial y'^2 \equiv 0$.

This is precisely the case when the base function is linear in y'. It can then be written in the form

$$F(x, y, y') = \varphi(x, y) + \psi(x, y)\, y'.$$

The Euler differential equation becomes

$$\frac{\partial \psi}{\partial x} - \frac{\partial \varphi}{\partial y} = 0.$$

There are now two cases: If $\partial\psi/\partial x$ is not identical with $\partial\varphi/\partial y$ then it is an ordinary equation of a curve with no undetermined constants. The boundary conditions will in general not be fulfilled. If $\partial\psi/\partial x$ is identical with $\partial\varphi/\partial y$ it follows that $[\varphi(x, y) + \psi(x, y)y']dx$ is an exact differential (cf. p. 364). The value of the integral

$$\int_{x_1, y_1}^{x_2, y_2} F(x, y, y')dx$$

thus depends only on the coordinates x_1, y_1 and x_2, y_2 but not on the nature of the curve joining the two points. This is not an extremal problem in the sense of the calculus of variations.

Examples.

1. $\int_{x_1}^{x_2} [xy + (xy^2 + y^3)y']dx = \text{extremum}.\quad \partial\varphi/\partial y = x, \quad \partial\psi/\partial x = y^2.$

Thus the solution is the parabola $y^2 - x = 0$.

2. $\int_{x_1}^{x_2} [3x^2 + 2xy + y^2 + (x^2 + 2xy + 3y^2)\, y'] = \text{extremum}$

Boundary conditions: $f(0) = 0$; $f(1) = 1$.

Thus

$$\frac{\partial \varphi}{\partial y} = 2x + 2y = \frac{\partial \psi}{\partial x},$$

$$\int_{x=0,\, y=0}^{x=1,\, y=1} [(3x^2 + 2xy + y^2)\, dx + (x^2 + 2xy + 3y^2)\, dy]$$

$$= \left[x^3 + x^2 y + xy^2 + y^3\right]_{x=0,\; y=0}^{x=1,\; y=1} = 4.$$

This value is not dependent on the path one takes from the point (0,0) to the point (1,1).

3. Sufficient conditions for the assumption of an extremum

It follows from the Taylor formula that the increment of the integral for the simple variation problem with one unknown function may be written in the following form:

$$\Delta J = \int_{x_1}^{x_2} F(x, y + \delta y, y' + \delta y')\, dx - \int_{x_1}^{x_2} F(x, y, y')\, dx$$

$$= \int_{x_1}^{x_2} \left(\frac{\partial F(x, y, y')}{\partial y} \delta y + \frac{\partial F(x, y, y')}{\partial y'} \delta y' \right) dx +$$

$$+ \frac{1}{2} \int_{x_1}^{x_2} \left(\frac{\partial^2 F}{\partial y^2} \delta y^2 + 2 \frac{\partial^2 F}{\partial y \partial y'} \delta y \delta y' + \frac{\partial^2 F}{\partial y'^2} \delta y'^2 \right) dx + R,$$

where R is an infinitesimal depending on δy and $\delta y'$ of order higher than two. With respect to the second differential we call the term

$$\int_{x_1}^{x_2} \left(\frac{\partial^2 F}{\partial y^2} \delta y^2 + 2 \frac{\partial^2 F}{\partial y \partial y'} \delta y \delta y' + \frac{\partial^2 F}{\partial y'^2} \delta y'^2 \right) dx$$

the *second variation* and denote it by $\delta^2 J$. If $\delta J = 0$ and $\delta^2 J \neq 0$ then for all sufficiently small δy and $\delta y'$ the sign of ΔJ agrees with the sign of the second variation. From this we can derive (in conjunction with the main condition $\delta J = 0$) not only a condition which is necessary for an extremum in the general case, but also a condition which is sufficient for a weak extremum. At an extremum, a minimum or maximum, of our integral it follows that it is necessary that the *condition of Legendre* $\partial^2 F/\partial y'^2 > 0$ or $\partial^2 F/\partial y'^2 < 0$, respectively, be fulfilled. Let us further suppose that the extreme curve is so embedded in a one parameter extremal set that for some neighbourhood, each curve of the set does not cross any other (*the condition of Jacobi*). This forms a so-called extremal field. Thus the condition $\partial^2 F/\partial y'^2 > 0$ or $\partial^2 F/\partial y'^2 < 0$ for a minimum or maximum, respectively, is sufficient for a weak extremum.

Example. We seek an extremum for the integral

$$J = \int_0^a (y'^2 - y^2)\, dx$$

with boundary conditions $y(0) = 0$ and $y(a) = 0$, where $a > 0$ and $a \neq n\pi$ (n an integer). The Euler differential equation has, in this case, the form $y'' + y = 0$. The general solution is $y = C_1 \cos x + C_2 \sin x$ or $y = \overline{C}_1 \sin (x + \overline{C}_2)$. The boundary conditions imply that for the constants of integration either $C_1 = C_2 = 0$ or $\overline{C}_1 = \overline{C}_2 = 0$, respectively. If $a > \pi$ every curve of the family $y = C_1 \cos x + C_2 \sin x$ crosses our extremal curve so it is not able to be imbedded in this family of curves. The condition of Jacobi is thus unsatisfied in this case. If $a < \pi$ it follows that the line $y = 0$ is in the family of curves $y(x) = \overline{C}_1 \sin (x + \varepsilon)$ with $0 < \varepsilon < \pi - a$, and in the interval $(0, a)$ no curves of this family cross any other curve of this family. Further $\partial^2 F/\partial y'^2 = 2 > 0$; it follows that for $a < \pi$ both the conditions for the existence of a minimum of our integral are fulfilled.

Sufficient conditions for a strong extremum have been given by Weierstrass and rely on another method of argument. It is assumed that the desired extremal curve may belong to an extreme field $y = y(x, \alpha)$; let $p(x, y)$ be the so-called gradient of the field; i.e., the directional coefficient of the tangent to extremum of the field at the point (x, y). The increment of the integral may then be written in the form

$$\Delta J = \int_{(L)} \left[F(x, y, y') - F(x, y, p) - \frac{\partial F}{\partial p} (y' - p) \right] dx$$

where the integral is to be taken over an arbitrary class of curves $y = \overline{y}(x)$, denoted by (L). The integrand of the last equation is called the *Weierstrass function* and it is denoted by $E(x, y, p, y')$. Since we have

$$\Delta J = \int_{(L)} E(x, y, p, y')\, dx.$$

It suffices for a weak minimum that the function $E(x, y, p, y')$ is not negative for all points lying sufficiently near the extremum and for all values of y' sufficiently near p.

For a strong minimum the condition given by the equation will be, however, that the function $E(x, y, p, y')$ is positive for all values of y' not only for those values which differ little from p.

The sufficient conditions for a strong and weak maximum may be formulated similarly.

The conditions under consideration may be simplified. We expand the base function in the Taylor series

$$F(x, y, y') = F(x, y, p) + \frac{\partial F}{\partial p}(y' - p) + \frac{(y' - p)^2}{2} \cdot \frac{\partial^2 F(x, y, \bar{p})}{\partial p^2},$$

where $\bar{p}$ denotes a value lying between p and y'. Thus

$$E(x, y, p, y') = \frac{(y' - p)^2}{2} \cdot \frac{\partial^2 F(x, y, \bar{p})}{\partial p^2}.$$

The sign of E is determined by that of $\partial^2 F(x, y, \bar{p})/\partial p^2$ and we further infer for an extremum which is a weak minimum the condition of Legendre, $\partial^2 F/\partial y'^2 > 0$. Because of the assumed continuity of the derivative it follows that for some neighbourhood of values of the variables $\partial^2 F/\partial y'^2 > 0$. For a strong minimum it suffices then that this inequality holds for all points (x, y) which lie in a neighbourhood of the extremum and for any value of y'. In the first case, as well as in the second, it is necessary that the condition of Jacobi is fulfilled.

Even so, as in the differential calculus we frequently reject considerations of the sufficient conditions and show the existence of an extreme value in an indirect manner. Thus we will calculate the extremum from the practical problem, which leads to the variation problem with the help of the necessary conditions, and frequently the Euler differential equation, and will demonstrate the existence of an extremum through the practical nature of the physical, technical, or geometrical problem. This occurs, for example, in the problem of the geodesic line on a curved surface; it is obvious that between two points of the surface there must exist at least one curve which joins the points and which has arc-length less than or equal to that of any other such curve. However, one must always proceed with great caution in such intuitive arguments.

4. The variation problem in polar coordinates

There are two possible cases, (a) and (b).

(a) $$\int_{r_1}^{r_2} F(r, \varphi, \varphi')\, dr = \text{extremum}.$$

The Euler differential equation then becomes:

$$\frac{\partial F}{\partial \varphi} - \frac{d}{dr}\left(\frac{\partial F}{\partial \varphi'}\right) = 0.$$

Example.

$$\int_{r_1}^{r_2} \frac{\sqrt{1 + r^2 \varphi'^2}}{r}\, dr = \text{extremum}.$$

The Euler differential equation

$$0 - \frac{d}{dr}\left(\frac{r^2 \varphi'}{r\sqrt{1 + r^2 \varphi'^2}}\right) = 0$$

has the solution

$$r = K_2 e^{K_1 \varphi}.$$

(b) $$\int_{\varphi_1}^{\varphi_2} F(\varphi, r, r')\, d\varphi = \text{extremum}.$$

In this case the Euler differential equation becomes

$$\frac{\partial F}{\partial r} - \frac{d}{d\varphi}\left(\frac{\partial F}{\partial r'}\right) = 0.$$

Example.

$$\int_{\varphi_1}^{\varphi_2} (r^2 + r'^2 + 2r \sin \varphi)\, d\varphi = \text{extremum}.$$

For the Euler differential equation

$$2r + 2 \sin \varphi - 2r'' = 0$$

we find that the solution is $r = C_1 e^{\varphi} + C_2 e^{-\varphi} - \frac{1}{2} \sin \varphi$.

5. The inverse problem of the variational calculus

The simple variation problem with one unknown function leads to an ordinary differential equation of the second order.

Conversely, every second order ordinary differential equation yields a variation problem of the form

$$J = \int_{x_1}^{x_2} F(x, y, y')\, dx = \text{extremum}.$$

In accordance with the differential equation

$$y'' = \varphi(x, y, y'),$$

we seek the general solution $z(x, y, y')$ of the following first order partial differential equation:

$$\frac{\partial z}{\partial x} + y' \frac{\partial z}{\partial y} + \varphi(x, y, y') \frac{\partial z}{\partial y'} + \frac{\partial \varphi(x, y, y')}{\partial y'} z = 0.$$

The base function $F(x, y, y')$ of the variation problem corresponding to the ordinary differential equation $y'' = \varphi(x, y, y')$ is thus given as the solution to the following second order partial differential equation:

$$\frac{\partial^2 F}{\partial y'^2} = z(x, y, y').$$

In the solution appear two undetermined functions of x and y. These are to be so determined that

$$\varphi(x, y, y') \frac{\partial^2 F}{\partial y'^2} + \frac{\partial^2 F}{\partial y' \partial x} + \frac{\partial^2 F}{\partial y' \partial y} y' - \frac{\partial F}{\partial y} = 0$$

holds.

Example. We are to find all the base functions for which the extrema are straight lines. The second order differential equation of a linear function is $y'' = 0$, thus $\varphi(x, y, y') \equiv 0$. The first order partial differential equation

$$\frac{\partial z}{\partial x} + y' \frac{\partial z}{\partial y} = 0$$

has the general solution

$$z = \Phi(y', xy' - y).$$

The second order partial differential equation

$$\frac{\partial^2 F}{\partial y'^2} = \Phi(y', xy' - y)$$

has the solution

$$F = \int_0^{y'} (y' - t)\,\Phi(t, xt - y)\,dt + \varphi(x, y)\,y' + \omega(x, y).$$

The undetermined functions $\varphi(x, y)$ and $\omega(x, y)$ we determine in accordance with the condition

$$\frac{\partial \varphi}{\partial x} - \frac{\partial \omega}{\partial y} = 0,$$

thus

$$\varphi = \frac{\partial \Omega}{\partial y}, \qquad \omega = \frac{\partial \Omega}{\partial x},$$

where Ω is any function of x and y.

There is also a special case in which one can find the variation problem which corresponds to a second order ordinary differential equation without solving the partial differential equation. Thus to the differential equation

$$y''\alpha(x) + y'\alpha'(x) - y\beta(x) - \gamma(x) = 0$$

with boundary conditions $f(0) = y_0$ and $f(x_1) = y_1$ corresponds the variation problem

$$J = \int_0^{x_1} [\alpha(x)y'^2 + \beta(x)y^2 + 2\gamma(x)y]\,dx = \text{extremum}.$$

Example. The differential equation

$$y'' + y + x = 0 \qquad \text{with} \qquad f(0) = 0,\ f(1) = 0$$

corresponds to the variation problem

$$J = \int_0^1 [\tfrac{1}{2}y'^2 - y^2 - 2xy]\,dx = \text{extremum}.$$

6. The variation problem in parametric form

In many problems of the variational calculus it is more expedient to consider the desired function to be in parametric form

$$x = x(t), \quad y = y(t), \quad x^{\cdot} = \frac{dx}{dt}, \quad y^{\cdot} = \frac{dy}{dt},$$

where

$$y' = \frac{dy}{dx} = \frac{y^{\cdot}}{x^{\cdot}}.$$

The variation problem

$$\int_{t_1}^{t_2} F(x(t),\, y(t),\, x^{\cdot},\, y^{\cdot})\, dt = \text{extremum}$$

will be solved through the Euler equations

$$\frac{\partial F}{\partial x} - \frac{d}{dt}\left(\frac{\partial F}{\partial x^{\cdot}}\right) = 0, \quad \frac{\partial F}{\partial y} - \frac{d}{dt}\left(\frac{\partial F}{\partial y^{\cdot}}\right) = 0.$$

These two equations are, however, not independent of one another (1). We choose as the parameter the angle τ formed by the tangent at the point $P(x, y)$ and the abscissa-axis, thus $y^{\cdot} = x^{\cdot} \tan \tau$. In this manner the two Euler equations can be replaced by one equation (2).

There is, however, one matter which must be carefully attended to in the treatment of the problem in parametric form: under certain circumstances the value of the integral

$$\int_{t_1}^{t_2} F(x, y, x^{\cdot}, y^{\cdot})\, dt$$

is dependent on the choice of the parameter t. In this case the question concerning the extremum of the integral is without meaning (3). The following theorem will help decide on the above matter:

The value of an extremum to be obtained from an integral is independent of the choice of parametric form precisely in the case when the base function $F(x, y, x^{\cdot}, y^{\cdot})$ is a positive-homogeneous function of the first order in $x^{\cdot}$ and $y^{\cdot}$: i.e., when the identity $F(x, y, kx^{\cdot}, ky^{\cdot}) = kF(x, y, x^{\cdot}, y^{\cdot})$ holds for $k > 0$ (4).

---

(1) This is only true in the case of the geometric variation problem where this system of equations is equivalent to one single Euler differential equation, namely, the Euler differential equation in Weierstrass form.

(2) This is only the case when, as in the following example, either $x(t)$ or $y(t)$ does not appear in the base function.

(3) To assert this is only permissible in the geometric variation problem.

(4) The concept of a positive-homogeneous function is weaker than the general concept of homogeneous function. For a positive-homogeneous function the identity $f(x, y, kx^{\cdot}, ky^{\cdot}) = kf(x, y, x^{\cdot}, y^{\cdot})$ is fulfilled only for positive values of k, while for the general notion of homogeneity this identity must hold not only for positive but also for negative values of k. Thus, for example, the two functions

$$\sqrt{x'^2 + y'^2}, \quad xy' - xy' + \sqrt{x'^2 + y'^2}$$

are positive-homogeneous, but not homogeneous in the ordinary sense.

Example.

$$\int_{x_1}^{x_2} \frac{\sqrt{1+y'^2}}{\sqrt{y}}\, dx = \text{extremum}.$$

We write the integral in the form:

$$\int_{t_1}^{t_2} \frac{\sqrt{\dot{x}^2+\dot{y}^2}}{\sqrt{y}}\, dt.$$

From the equation

$$\frac{\sqrt{(k\dot{x})^2+(k\dot{y})^2}}{\sqrt{y}} = k\,\frac{\sqrt{\dot{x}^2+\dot{y}^2}}{\sqrt{y}} \qquad (\text{for } k>0)$$

it follows that the value of the integral is independent of the method of parametrization. The Euler differential equation

$$\frac{\partial F}{\partial x} - \frac{d}{dt}\left(\frac{\partial F}{\partial \dot{x}}\right) = 0$$

has the simplified form

$$\frac{d}{dt}\left(\frac{\dot{x}}{\sqrt{y}\,\sqrt{\dot{x}^2+\dot{y}^2}}\right) = 0.$$

Thus

$$\frac{\dot{x}}{\sqrt{y}\,\sqrt{\dot{x}^2+\dot{y}^2}} = c_1.$$

The solution of this differential equation will be facilitated through the use of the tangential angle τ as the parameter. It then follows that

$$\frac{\dot{x}}{\sqrt{\dot{x}^2+\dot{y}^2}} = \frac{dx}{ds} = \cos\tau \qquad \text{and} \qquad \dot{y} = \dot{x}\tan\tau.$$

We deduce that

$$\frac{\cos\tau}{\sqrt{y}} = c_1 \qquad \text{or} \qquad y = \frac{\cos^2\tau}{c_1^2} = \frac{1+\cos 2\tau}{2c_1^2}.$$

Further

$$\dot{y} = -\frac{\sin 2\tau}{c_1^2}$$

and therefore

$$x^{\cdot} = y^{\cdot} \cot \tau = -\frac{2 \cos^2 \tau}{c_1^2}$$

which, after integration, yields

$$x = -\frac{2}{c_1^2}\left(\tfrac{1}{2}\tau + \tfrac{1}{2} \sin \tau \cos \tau\right) + c_2.$$

The two equations

$$x = -\frac{1}{2c_1^2}(2\tau + \sin 2\tau) + c_2$$

and

$$y = \frac{1 + \cos 2\tau}{2c_1^2}$$

represent the ordinary cycloid given in parametric form.

7. Base functions involving derivatives of higher orders

It is assumed that the integral

$$\int_{x_1}^{x_2} F(x, y, y', y'', \ldots, y^{(n)})\, dx$$

has an extreme value. The Euler differential equation becomes

$$\frac{\partial F}{\partial y} - \frac{d}{dx}\left(\frac{\partial F}{\partial y'}\right) + \frac{d^2}{dx^2}\left(\frac{\partial F}{\partial y''}\right) - \frac{d^3}{dx^3}\left(\frac{\partial F}{\partial y'''}\right) + \ldots + (-1)^n \frac{d^n}{dx^n}\left(\frac{\partial F}{\partial y^{(n)}}\right) = 0.$$

Examples.

1. $\int_{x=0}^{x=1} (y'')^2\, dx = \text{extremum}.$

Boundary conditions: $y(0) = y'(0) = 0,\ y(1) = 2,\ y'(1) = 5.$

From the Euler differential equation

$$\frac{d^2}{dx^2}(2y'') = 2\frac{d^4y}{dx^4} = 0$$

it follows that the solution is

$$y = C_1 + C_2x + C_3x^2 + C_4x^3.$$

The boundary conditions determine the constants

$$C_1 = C_2 = 0, \qquad C_3 = C_4 = 1,$$

thus

$$y = x^2 + x^3.$$

2. $\int\limits_0^{\frac{1}{2}\pi} (y''^2 - 2y'^2 + y^2)\, dx = \text{extremum}.$

Boundary conditions: $y(0) = y'(0) = 0$; $y(\frac{1}{2}\pi) = 1$, $y'(\frac{1}{2}\pi) = \frac{1}{2}\pi$.

The Euler differential equation has the form:

$$2y + 4\frac{d^2y}{dx^2} + 2\frac{d^4y}{dx^4} = 0.$$

The general solution of this differential equation of fourth order is:

$$y = (C_1 x + C_2) \cos x + (C_3 x + C_4) \sin x.$$

Thus

$$C_2 = 0, \qquad C_1 + C_4 = 0,$$

$$C_3 \cdot \tfrac{1}{2}\pi + C_4 = 1, \qquad -(C_1 \cdot \tfrac{1}{2}\pi + C_2) + C_3 = \tfrac{1}{2}\pi.$$

From these we find that

$$C_1 = -1, \quad C_2 = 0, \quad C_3 = 0, \quad C_4 = 1,$$

from which it follows that the solution is

$$y = -x \cos x + \sin x.$$

8. The Euler differential equations for the variation problem with n unknown functions

Suppose that in the base function there appear n unknown functions

$$y_1 = f_1(x), \quad y_2 = f_2(x), \quad \ldots, \quad y_n = f_n(x)$$

of the independent variable x together with their first derivatives

$$y_1' = f_1'(x), \quad y_2' = f_2'(x), \quad \ldots, \quad y_n' = f_n'(x);$$

thus in order to determine the n unknown functions which yield an extremum in the integral

$$\int\limits_{x_1}^{x_2} F(x, y_1, y_2, \ldots, y_n, y_1', y_2', \ldots, y_n')\, dx$$

there are the n differential equations

$$\frac{\partial F}{\partial y_1} - \frac{d}{dx}\left(\frac{\partial F}{\partial y_1'}\right) = 0,$$

$$\frac{\partial F}{\partial y_2} - \frac{d}{dx}\left(\frac{\partial F}{\partial y_2'}\right) = 0,$$

$$\cdots\cdots\cdots\cdots\cdots$$

$$\frac{\partial F}{\partial y_n} - \frac{d}{dx}\left(\frac{\partial F}{\partial y_n'}\right) = 0$$

to solve. The number of boundary conditions, in general, is equal to $2n$.

For the variation problem with two unknown functions

$$\int_{x_1}^{x_2} F(x, y, z, y', z')\, dx = \text{extremum}$$

we deduce the Euler differential equations

$$\frac{\partial F}{\partial y} - \frac{d}{dx}\left(\frac{\partial F}{\partial y'}\right) = 0, \quad \frac{\partial F}{\partial z} - \frac{d}{dx}\left(\frac{\partial F}{\partial z'}\right) = 0\,.$$

In general four boundary conditions are necessary to determine the constants in this variation problem.

Examples.

1. $$\int_{x_1}^{x_2} (y'^2 + z'^2 + 2yz' + 2zy')\, dx = \text{extremum}.$$

The two unknown functions are $y = f(x)$ and $z = g(x)$. The boundary conditions are given by

$$y_1 = f(x_1), \quad z_1 = g(x_1),$$

$$y_2 = f(x_2), \quad z_2 = g(x_2).$$

Solution:

$$2z' - (2y'' + 2z') = 0,$$

$$2y' - (2z'' + 2y') = 0$$

or

$$y'' = 0, \quad z'' = 0,$$

$$y = C_1 x + C_2, \quad z = C_3 x + C_4\,.$$

On the basis of the boundary conditions we deduce that

$$\frac{x-x_1}{x_2-x_1}=\frac{y-y_1}{y_2-y_1}=\frac{z-z_1}{z_2-z_1}.$$

2. $$\int_{x_1}^{x_2}(y'^2+z'^2-2yz+2y+2z)\,dx=\text{extremum}.$$

The Euler differential equations yield:

$$\begin{aligned}&-2z+2-2y''=0,\\&-2y+2-2z''=0\end{aligned}\qquad\text{or}\qquad\begin{aligned}y''&=1-z,\\z''&=1-y.\end{aligned}$$

The general solution of this simultaneous system is

$$\begin{aligned}y&=C_1e^x+C_2e^{-x}+C_3\cos x+C_4\sin x+1,\\z&=-C_1e^x-C_2e^{-x}+C_3\cos x+C_4\sin x+1.\end{aligned}$$

9. The extremum of a multiple integral

The unknown function $z=f(x,y)$ is to be so determined that the double integral

$$\iint_B F\left(x,y,z,\frac{\partial z}{\partial x},\frac{\partial z}{\partial y}\right)dxdy$$

over a closed domain B of the xy-plane assumes an extreme value. We introduce the notation $\partial z/\partial x=z_x$, $\partial z/\partial y=z_y$. Then it follows that the Euler differential equation is:

$$\frac{\partial F}{\partial z}-\frac{\partial}{\partial x}\left(\frac{\partial F}{\partial z_x}\right)-\frac{\partial}{\partial y}\left(\frac{\partial F}{\partial z_y}\right)=0.$$

In general this is a partial differential equation of the second order.

Examples.

1. $$\iint_B\left[\left(\frac{\partial z}{\partial x}\right)^2+\left(\frac{\partial z}{\partial y}\right)^2+2zf(x,y)\right]dxdy=\text{extremum}.$$

The Euler differential equation yields

$$\frac{\partial^2 z}{\partial x^2}+\frac{\partial^2 z}{\partial y^2}=f(x,y).$$

2. $$\iint_B\left[\left(\frac{\partial z}{\partial x}\right)^2-\left(\frac{\partial z}{\partial y}\right)^2\right]dxdy=\text{extremum}.$$

The Euler differential equation

$$\frac{\partial^2 z}{\partial x^2} - \frac{\partial^2 z}{\partial y^2} = 0$$

has the general solution

$$z = \varphi(x + y) + \psi(x - y).$$

For the n-fold integral

$$\iiint_B \cdots \int F\left(x_1, x_2, \ldots, x_n, z, \frac{\partial z}{\partial x_1}, \frac{\partial z}{\partial x_2}, \ldots, \frac{\partial z}{\partial x_n}\right) dx_1 dx_2 \ldots dx_n$$

the following is the Euler differential equation:

$$\frac{\partial F}{\partial z} - \frac{\partial}{\partial x_1}\left(\frac{\partial F}{\partial z_{x_1}}\right) - \frac{\partial}{\partial x_2}\left(\frac{\partial F}{\partial z_{x_2}}\right) - \cdots - \frac{\partial}{\partial x_n}\left(\frac{\partial F}{\partial z_{xn}}\right) = 0.$$

Example.

$$\iiint_B \left[\left(\frac{\partial u}{\partial x}\right)^2 + \left(\frac{\partial u}{\partial y}\right)^2 + \left(\frac{\partial u}{\partial z}\right)^2\right] dx dy dz = \text{extremum}.$$

As the Euler differential equation we find:

$$\frac{\partial^2 u}{\partial x^2} + \frac{\partial^2 u}{\partial y^2} + \frac{\partial^2 u}{\partial z^2} = 0.$$

Multiple integrals with higher partial derivatives. We restrict our attention to base functions in which, besides the variables x, y and z, the partial derivatives of first and second order

$$\frac{\partial z}{\partial x} = z_x, \quad \frac{\partial z}{\partial y} = z_y, \quad \frac{\partial^2 z}{\partial x^2} = z_{xx}, \quad \frac{\partial^2 z}{\partial x \partial y} = z_{xy}, \quad \frac{\partial^2 z}{\partial y^2} = z_{yy}$$

may also occur. It is also assumed that the double integral

$$\iint_B F\left(x, y, z, \frac{\partial z}{\partial x}, \frac{\partial z}{\partial y}, \frac{\partial^2 z}{\partial x^2}, \frac{\partial^2 z}{\partial x \partial y}, \frac{\partial^2 z}{\partial y^2}\right) dx dy$$

has an extremum.

The Euler differential equation for this problem is

$$\frac{\partial F}{\partial z} - \frac{\partial}{\partial x}\left(\frac{\partial F}{\partial z_x}\right) - \frac{\partial}{\partial y}\left(\frac{\partial F}{\partial z_y}\right) + \frac{\partial^2}{\partial x^2}\left(\frac{\partial F}{\partial z_{xx}}\right) + \frac{\partial^2}{\partial x \partial y}\left(\frac{\partial F}{\partial z_{xy}}\right) + \frac{\partial^2}{\partial y^2}\left(\frac{\partial F}{\partial z_{yy}}\right) = 0.$$

Examples.

$$1. \iint\limits_B \left[\left(\frac{\partial^2 z}{\partial x^2}\right)^2 + \left(\frac{\partial^2 z}{\partial y^2}\right)^2 + 2\left(\frac{\partial^2 z}{\partial x \partial y}\right)^2 - 2zf(x, y)\right] dxdy = \text{extremum}.$$

We find for the Euler differential equation

$$\frac{\partial^4 z}{\partial x^4} + 2\frac{\partial^4 z}{\partial x^2 \partial y^2} + \frac{\partial^4 z}{\partial y^4} = f(x, y).$$

2. In the derivation of the equations of motion of the vibrations of an elastic rod with the help of the Hamiltonian principle it becomes necessary to solve the following variation problem:

$$\int\limits_{t_1}^{t_2}\int\limits_0^1 \frac{1}{2}\left\{\varrho q(x)\left(\frac{\partial y}{\partial t}\right)^2 - EJ(x)\left(\frac{\partial^2 y}{\partial x^2}\right)^2\right\} dxdt = \min.$$

We derive as the Euler differential equation

$$\varrho q(x)\frac{\partial^2 y}{\partial t^2} + E\frac{\partial^2}{\partial x^2}\left(J(x)\frac{\partial^2 y}{\partial x^2}\right) = 0.$$

10. The variation problem with side conditions

It is assumed that the desired function $y = f(x)$ in addition to the condition of extremity must also satisfy some further condition. The methods employed in solving such problems are in essence the same methods which one employs in differential calculus in order to determine extreme values subject to given side conditions. The most important method for dealing with such problems is the *method of Lagrange multipliers* (cf. p. 384).

Let us consider a variation problem with two unknown functions and a side condition, i.e., an integral

$$J = \int\limits_{x_1}^{x_2} F(x, y, z, y', z')\, dx$$

which is to be made extreme and at the same time a side condition $G(x, y, z) = 0$ which is to be fulfilled. We first construct the base function $H(x, y, z, y', z') = F(x, y, z, y', z') + \lambda(x)G(x, y, z)$ in which, as already indicated by the notation $\lambda = \lambda(x)$ may be a function of x.

The two desired functions $y = f(x)$ and $z = g(x)$ as well as the parameter function $\lambda = \lambda(x)$ satisfy the Euler equations

$$\frac{\partial H}{\partial y} - \frac{d}{dx}\left(\frac{\partial H}{\partial y'}\right) = 0, \quad \frac{\partial H}{\partial z} - \frac{d}{dx}\left(\frac{\partial H}{\partial z'}\right) = 0$$

because of the side condition $G(x, y, z) = 0$. The Euler differential equations can be written out as

$$\frac{\partial F}{\partial y} + \lambda(x)\frac{\partial G}{\partial y} - \frac{d}{dx}\left(\frac{\partial F}{\partial y'}\right) = 0,$$

$$\frac{\partial F}{\partial z} + \lambda(x)\frac{\partial G}{\partial z} - \frac{d}{dx}\left(\frac{\partial F}{\partial z'}\right) = 0.$$

In the variation problem with side conditions it is often expedient to introduce a parameter t. In this case the treatment of the problem follows the techniques of the calculus of parametric forms in the following manner.

We again use the notation

$$\frac{dx}{dt} = x^{\cdot}, \quad \frac{dy}{dt} = y^{\cdot}, \quad \frac{dz}{dt} = z^{\cdot}.$$

It is assumed that

$$J = \int_{t_1}^{t_2} F\big(x(t), y(t), z(t), x^{\cdot}, y^{\cdot}, z^{\cdot}\big)\, dt$$

has an extremum subject to the side condition $G(x, y, z) = 0$. The Euler differential equations have the form:

$$\frac{\partial F}{\partial x} + \lambda(t)\frac{\partial G}{\partial x} - \frac{d}{dt}\left(\frac{\partial F}{\partial x^{\cdot}}\right) = 0,$$

$$\frac{\partial F}{\partial y} + \lambda(t)\frac{\partial G}{\partial y} - \frac{d}{dt}\left(\frac{\partial F}{\partial y^{\cdot}}\right) = 0,$$

$$\frac{\partial F}{\partial z} + \lambda(t)\frac{\partial G}{\partial z} - \frac{d}{dt}\left(\frac{\partial F}{\partial z^{\cdot}}\right) = 0.$$

Examples. 1. Given a surface $G(x, y, z) = 0$ and two points $P_1(x_1, y_1, z_1)$ and $P_2(x_2, y_2, z_2)$ lying on it we are to determine the shortest curve lying on the surface which joins the two points (cf. p. 650). It is assumed that

$$\int_{t_1}^{t_2} \sqrt{x^{\cdot 2} + y^{\cdot 2} + z^{\cdot 2}}\, dt$$

has a minimum subject to the side condition $G(x, y, z) = 0$ (the problem of the geodesic line).

The Euler equations yield:

$$\lambda(t)\frac{\partial G}{\partial x} - \frac{d}{dt}\left(\frac{x^{\cdot}}{\sqrt{x^{\cdot 2} + y^{\cdot 2} + z^{\cdot 2}}}\right) = 0,$$

$$\lambda(t)\frac{\partial G}{\partial y} - \frac{d}{dt}\left(\frac{y^{\cdot}}{\sqrt{x^{\cdot 2} + y^{\cdot 2} + z^{\cdot 2}}}\right) = 0,$$

$$\lambda(t)\frac{\partial G}{\partial z} - \frac{d}{dt}\left(\frac{z^{\cdot}}{\sqrt{x^{\cdot 2} + y^{\cdot 2} + z^{\cdot 2}}}\right) = 0$$

or, when we introduce the arc length s and use the notation $\lambda = \mu(\partial s/\partial t)$

$$\mu\frac{\partial G}{\partial x} = \frac{d^2x}{ds^2}, \qquad \mu\frac{\partial G}{\partial y} = \frac{d^2y}{ds^2}, \qquad \mu\frac{\partial G}{\partial z} = \frac{d^2z}{ds^2}.$$

These three equations express the fact that at each point of the geodesic curve the normal to the given surface and the principal normal of the curve coincide.

2. This example is a determination of the geodesic line on the circular cylinder $x^2 + y^2 = R^2$. The solution of this problem is greatly simplified when we write the circular cylinder in parametric form as

$$x = R\cos t, \qquad y = R\sin t.$$

The desired curve must lie in the cylinder and also satisfy the system

$$x = R\cos t, \qquad y = R\sin t.$$

Thus it remains only to determine the z-function. It follows from this that in the condition

$$\int_{t_1}^{t_2} \sqrt{x^{\cdot 2} + y^{\cdot 2} + z^{\cdot 2}}\, dt = \text{min.},$$

because $x^{\cdot} = -R\sin t$ and $y^{\cdot} = R\cos t$ the base function reduces to $F = \sqrt{R^2 + z^{\cdot 2}}$.

We deduce from the Euler equation that

$$-\frac{d}{dt}\left(\frac{z^{\cdot}}{\sqrt{R^2 + z^{\cdot 2}}}\right) = 0.$$

Thus we obtain the solution

$$z = \frac{C_1 R t}{\sqrt{1 - C_1^2}} + C_2.$$

This equation in conjunction with $x = R\cos t$, $y = R\sin t$ is the parametric representation of a cylindrical helix.

11. The isoperimetric problem of the calculus of variations

Assume that the integral

$$J = \int_{x_1}^{x_2} F(x, y, y')\, dx$$

has an extremum subject to the side condition

$$\int_{x_1}^{x_2} G(x, y, y')\, dx = k.$$

We construct the base function

$$H(x, y, y') = F(x, y, y') + \lambda G(x, y, y')$$

in which λ is a parametric constant. The Euler differential equation then has the form

$$\frac{\partial H}{\partial y} - \frac{d}{dx}\left(\frac{\partial H}{\partial y'}\right) = 0.$$

Example.

$$\int_{x_1=0}^{x_2=l} y\, dx = \text{extremum},$$

$$\int_{x_1=0}^{x_2=l} \sqrt{1 + y'^2}\, dx = L \qquad \text{(cf. p. 651).}$$

Here $H(x, y, y') = y + \lambda\sqrt{1 + y'^2}$.

Since the base function $y + \lambda\sqrt{1 + y'^2}$ has no occurrences of the variable x, it follows from the intermediate integral of the Euler differential equation that

$$y + \lambda\sqrt{1 + y'^2} - \frac{\lambda y'^2}{\sqrt{1 + y'^2}} = C_1.$$

Thus we obtain $y' = \sqrt{\lambda^2 - (C_1 - y)^2}/(C_1 - y)$. The integral of this first order differential equation

$$(x - C_2)^2 + (y - C_1)^2 = \lambda^2$$

yields a circle. The values of C_1, C_2, and λ are determined by the conditions that the curve must pass through the two points $O(0, 0)$ and $P(l, 0)$ and that the curve to the point P must have length L.

This yields a transcendental equation for λ which can be solved by approximation methods.

In many cases, by means of a suitable parametrization, one can replace a problem which requires the use of the method of Lagrange multipliers by an isoperimetric problem.

Example. In the following example the point P is not completely fixed, but is free to move on the x-axis. Thus the problem becomes

$$\int_0^x y\,dx = \text{extremum}$$

subject to the side condition that

$$\int_0^x \sqrt{1+y'^2}\,dx = L.$$

We use the formula

$$dx^2 + dy^2 = ds^2$$

or

$$dx = \sqrt{1-(dy/ds)^2}\,ds.$$

Thus the problem reduces to making an extremum in the integral

$$J = \int_{s=0}^{s=L} y\sqrt{1-(dy/ds)^2}\,ds.$$

Since the base function

$$F = y\sqrt{1-(dy/ds)^2}$$

the independent variable s does not appear, it follows that the intermediate integral of the Euler differential equation is

$$y\sqrt{1-(dy/ds)^2} + \frac{dy}{ds}\cdot\frac{y\,dy/ds}{\sqrt{1-(dy/ds)^2}} = c_1$$

or

$$\frac{dy}{ds} = \frac{1}{c_1}\sqrt{c_1^2 - y^2}.$$

Through another integration we deduce

$$\text{(I)} \qquad y = c_1 \sin\left(\frac{s}{c_1} + c_2\right).$$

Further

$$dx = \sqrt{1-(dy/ds)^2}\,ds = \frac{y}{c_1}\,ds,$$

thus

$$\text{(II)} \qquad x = -c_1 \cos\left(\frac{s}{c_1} + c_2\right) + c_3.$$

The equations (I) and (II) are a parametric representation of a circle. Elimination of the parameter s yields

$$\text{(III)} \qquad (x - c_3)^2 + y^2 = c_1^2.$$

The boundary conditions $x = 0$, $y = 0$, $s = 0$ and $y = 0$, $s = L$ determine the constants

$$c_1 = c_3 = L/\pi \quad \text{and} \quad c_2 = 0.$$

The final solution is thus

$$(x - L/\pi)^2 + y^2 = (L/\pi)^2.$$

12. Two geometric variation problems with two independent variables

1. Problem: To determine the surface with the smallest area which can be enclosed by a given space curve.

Solution: It is assumed that the integral

$$J = \iint_B \sqrt{1 + \left(\frac{\partial z}{\partial x}\right)^2 + \left(\frac{\partial z}{\partial y}\right)^2}\, dx dy$$

possesses a minimum. After the introduction of the customary notation

$$\frac{\partial z}{\partial x} = z_x, \quad \frac{\partial z}{\partial y} = z_y,$$

etc., we obtain the Euler differential equation

$$\frac{\partial}{\partial x}\left(\frac{z_x}{\sqrt{1 + z_x^2 + z_y^2}}\right) + \frac{\partial}{\partial y}\left(\frac{z_y}{\sqrt{1 + z_x^2 + z_y^2}}\right) = 0$$

or

$$\frac{z_{xx}(1 + z_y^2) - 2 z_x z_y z_{xy} + z_{yy}(1 + z_x^2)}{(1 + z_x^2 + z_y^2)^{3/2}} = 0.$$

On the left side of this equation, as is well known from differential geometry, stands the expression for the standard curvature of the surface, $1/\varrho_1 + 1/\varrho_2$ (cf. p. 311). This says that for the desired surface the standard curvature is equal to zero, or, in other words, that the principal curvatures $1/\varrho_1$ and $1/\varrho_2$ are equal to the negatives of one another. Such surfaces are called *minimal surfaces*.

2. Problem: Which surface, with a given surface area, encloses the greatest volume (an isoperimetric problem)?

Solution: $\iint_B z\,dxdy = \max.$

Side condition $\iint_B \sqrt{1 + z_x^2 + z_y^2}\,dxdy = k.$

We construct the base function

$$H(x, y, z, z_x, z_y) = z + \lambda\sqrt{1 + z_x^2 + z_y^2}.$$

Using the Euler differential equation we get

$$\frac{\partial H}{\partial z} - \frac{\partial}{\partial x}\left(\frac{\partial H}{\partial z_x}\right) - \frac{\partial}{\partial y}\left(\frac{\partial H}{\partial z_y}\right) = 0,$$

thus

$$1 - \lambda\frac{z_{xx}z_y^2 - z_xz_yz_{xy} + z_{xx} + z_{yy}z_x^2 - z_xz_yz_{xy} + z_{yy}}{(1 + z_x^2 + z_y^2)^{3/2}} = 0,$$

or

$$\frac{z_{xx}(1 + z_y^2) - 2z_xz_yz_{xy} + z_{yy}(1 + z_x^2)}{(1 + z_x^2 + z_y^2)^{3/2}} = \frac{1}{\lambda},$$

i.e.,

$$\frac{1}{\varrho_1} + \frac{1}{\varrho_2} = \frac{1}{\lambda}.$$

Thus the solution is given by a surface with constant standard curvature.

13. Ritz's method of solution of variation problems

In many problems of the calculus of variation the exact solution of the Euler differential equations is either very difficult or impossible. It is for that reason that approximation methods for the solution of variation problems are developed by means of which the Euler differential equations may be avoided. The commonest approximation method of the calculus of variations is the method of W. Ritz (W. Ritz, *Über eine neue Methode zur Lösung gewisser Variationsprobleme der mathematischen Physik*, Journal f. d. Reine und Angewandte Math. 135 (1909)).

In order to solve the problem

$$\text{(I)} \qquad J = \int_{x_1}^{x_2} F(x, y, y')\, dx = \text{extremum}$$

by this approximation method, we write for the desired function $y = f(x)$ the form

$$\text{(II)} \qquad y = c_1\varphi_1(x) + c_2\varphi_2(x) + \ldots + c_n\varphi_n(x).$$

The functions $\varphi_1(x), \varphi_2(x), \ldots, \varphi_n(x)$ are formed such that the boundary conditions are satisfied. Our problem then consists only of the determination of the constants $c_1, c_2, \ldots, c_n$. To this end we put the function (II) into (I) and obtain $J(c_1, c_2, \ldots, c_n) = \text{extremum}$. The desired values of the c's are given by the following n equations:

$$\frac{\partial J}{\partial c_1} = 0, \quad \frac{\partial J}{\partial c_2} = 0, \quad \ldots, \quad \frac{\partial J}{\partial c_n} = 0.$$

Examples.

1. $\int_0^1 (y'^2 - y^2 - 2xy)\, dx = \text{extremum}.$

Boundary conditions

$$f(0) = f(1) = 0.$$

Form:

$$y = c_1 x(1 - x) + c_2 x^2(1 - x),$$

thus

$$y' = c_1(1 - 2x) + c_2(2x - 3x^2).$$

Placing these expressions in the integral we obtain

$$J(c_1, c_2) = \int_0^1 [c_1^2(1 - 2x)^2 + 2c_1c_2(1 - 2x)(2x - 3x^2) + c_2^2(2x - 3x^2)^2 - \\ - c_1^2 x^2(1 - x)^2 - 2c_1c_2x^3(1 - x)^2 - c_2^2 x^4(1 - x)^2 - \\ - 2c_1x^2(1 - x) - 2c_2x^3(1 - x)]\, dx = \text{extremum}$$

or

$$J(c_1, c_2) = c_1^2 \cdot \tfrac{3}{10} + 2c_1c_2 \cdot \tfrac{3}{20} + c_2{}^2 \cdot \tfrac{13}{105} - 2c_1 \cdot \tfrac{1}{12} - 2c_2 \cdot \tfrac{1}{20} = \text{extremum}.$$

Thus it must hold that

$$\frac{\partial J}{\partial c_1} = 2c_1 \cdot \tfrac{3}{10} + 2c_2 \cdot \tfrac{3}{20} - 2 \cdot \tfrac{1}{12} = 0,$$

$$\frac{\partial J}{\partial c_2} = 2c_1 \cdot \tfrac{3}{20} + 2c_2 \cdot \tfrac{13}{105} - 2 \cdot \tfrac{1}{20} = 0.$$

From these two linear equations we obtain

$$c_1 = \tfrac{71}{369}, \qquad c_2 = \tfrac{7}{41}.$$

Solution:

$$y = \tfrac{71}{369}x - \tfrac{8}{369}x^2 - \tfrac{7}{41}x^3.$$

We already have (p. 655) an exact solution for this problem, namely,

$$y = \frac{\sin x}{\sin 1} - x.$$

The following table gives various points on the curves of both the exact and the approximate solutions:

x	y (exact)	y (approx.)
0	0	0
0.2	0.0361	0.0362
0.4	0.0627	0.0626
0.6	0.0710	0.0708
0.8	0.0525	0.0526
1.0	0	0

The difference between the exact and the approximate solutions has the order of magnitude of 10^{-4}.

2. $\int_0^1 (x^2y'^2 + xy)\,dx = \text{extremum}.$

Boundary conditions:

$$f(0) = 0, \qquad f(1) = 0.$$

Form:

$$\varphi_1 = x(x-1), \qquad \varphi_2 = x^2(x-1);$$

$$y = c_1x(x-1) + c_2x^2(x-1), \qquad y' = c_1(2x-1) + c_2(3x^2 - 2x).$$

$$J(c_1, c_2) = \int_0^1 [c_1^2(4x^4 - 4x^3 + x^2) + 2c_1c_2(6x^5 - 7x^4 + 2x^3) +$$

$$+ c_2^2(9x^6 - 12x^5 + 4x^4) + c_1(x^3 - x^2) + c_2(x^4 - x^3)]\,dx$$

$$= \tfrac{2}{15}c_1^2 + \tfrac{1}{5}c_1c_2 + \tfrac{3}{35}c_2^2 - \tfrac{1}{12}c_1 - \tfrac{1}{20}c_2,$$

$$\frac{\partial J}{\partial c_1} = \tfrac{4}{15}c_1 + \tfrac{1}{5}c_2 - \tfrac{1}{12} = 0,$$

$$\frac{\partial J}{\partial c_2} = \tfrac{1}{5}c_1 + \tfrac{6}{35}c_2 - \tfrac{1}{20} = 0,$$

$$c_1 = \tfrac{3}{4}, \qquad c_2 = -\tfrac{7}{12}.$$

Solution:

$$y = \tfrac{3}{4}(x^2 - x) - \tfrac{7}{12}(x^3 - x^2) = -\tfrac{7}{12}x^3 + \tfrac{4}{3}x^2 - \tfrac{3}{4}x.$$

IV. INTEGRAL EQUATIONS

1. General notions

By an *integral equation* is meant an equation to determine an unknown function $\varphi(x)$ ($\varphi(x)$ is defined in $a < x < b$) such that in the equation appears an integral in which the integrand is dependent on the desired function $\varphi(x)$. Naturally in such an equation there can occur other terms—not necessarily in the form of an integral—which depend directly on $\varphi(x)$.

Examples. 1. $\int_a^b K(x, y)\varphi(y)\,dy + f(x) = 0$,

2. $\int_a^b K(x, y)\varphi(y)\,dy + f(x) = \varphi(x)$,

3. $\int_a^x K(x, y)\varphi(y)\,dy + f(x) = 0$,

4. $\int_a^x K(x, y)\varphi(y)\,dy + f(x) = \varphi(x)$.

In these four examples $f(x)$ and $K(x, y)$ are given functions. The known function which appears under the integral will be called the *kernel* of the integral equation. $K(x, y)$ must be defined in the rectangle $a < x < b$, $a < y < b$; while $f(x)$ must be defined in the interval $a < x < b$.

Integral equations in which the unknown function appears linearly are called *linear integral equations*. The above four examples are linear integral equations. In the sequel we will deal mainly with linear integral equations. We assume in advance, unless stated otherwise, that the functions $K(x, y)$ and $f(x)$ are continuous in their domains of definition (cf. p. 347 and p. 335; the domains of definition of the functions $f(x)$ and $K(x, y)$ are closed).

Furthermore, from the beginning we consider only those solutions $\varphi(x)$ which are continuous in their domains of definition $a < x < b$. This is a consequence of the usual (Riemann) conception of the integral (cf. the existence theorem for the integral of a continuous function on p. 455).

We call integral equations in which both limits of integration are constants *Fredholm integral equations* (examples 1 and 2). If only one of the limits of integration is a constant, we speak of a *Volterra integral equation* (examples 3 and 4). By an *integral equation of the first kind* we denote an integral equation in which the unknown function occurs only under the integral (examples 1 and 3). If the unknown function occurs not only under the integral sign but also in some other manner then the equation is called an *integral equation of the second kind* (examples 2 and 4). Equations in which every term contains the unknown function are called *homogeneous integral equations.* An integral equation which contains one term in which the unknown function does not appear is called *inhomogeneous.* The term in which the unknown function does not occur, denoted by $f(x)$ in the previous examples, is called the *perturbation function.*

2. Simple integral equations which can be reduced to ordinary differential equations by differentiation

(1) John Bernoulli-(16671748) considered in his *Ersten Integralrechnung* (Ostwalds Klassiker der exakten Wissenschaften, Nr. 194, p. 35) the following problem: It is to determine the nature of the curve OB (i.e., the equation $y = \varphi(x)$) which is of such a shape that the area OAB is one third of the area of the circumscribed rectangle $OABC$ (Fig. 414).

Solution: $\int\limits_a^x \varphi(x)\,dx = \frac{1}{3}x\varphi(x)$.

This Volterra integral equation with kernel $K(x, y) \equiv 1$ can be reduced by differentiation to

$$\varphi(x) = \frac{1}{3}[x\varphi'(x) + \varphi(x)]$$

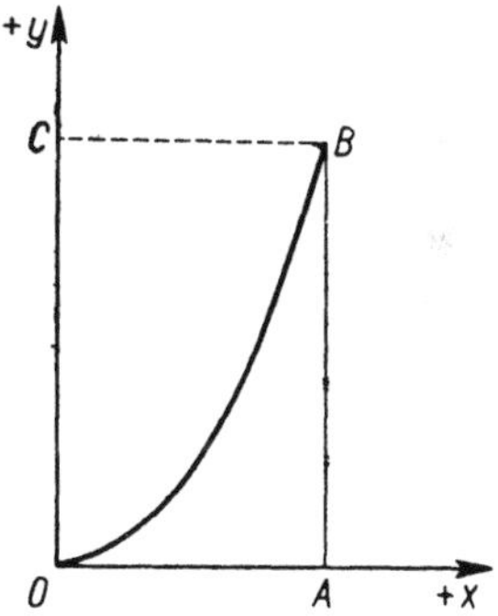

FIG. 414

or

$$\frac{d\varphi(x)}{\varphi(x)} = 2\frac{dx}{x}.$$

As the solutions of this homogeneous linear differential equation we obtain the set of second order parabolas $y = \varphi(x) = Cx^2$ (for $0 < x < +\infty$).

(2) The solid of rotation schematically denoted in the accompanying diagram (Fig. 415) is of density γ and is to have its upper end fastened. Further, the pull from the weight Q and its own

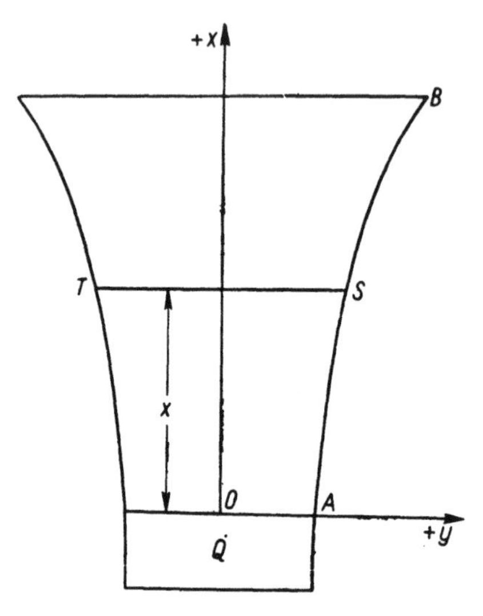

FIG. 415

weight is to be such that in each cross-section parallel to the base of the surface there is a constant tensile stress of magnitude σ. It is desired to determine the shape of this body, written by means of the equation $y = \varphi(x)$ of the curve AB.

Solution: The total force acting on the cross-section ST will be

$$Q + \gamma\pi \int_0^x [\varphi(x)]^2 \, dx.$$

The area of the cross-section is $\pi[\varphi(x)]^2$. Therefore we must have

$$Q + \gamma\pi \int_0^x [\varphi(x)]^2 \, dx = \sigma\pi[\varphi(x)]^2.$$

This Volterra integral equation may be solved by differentiation:

$$\gamma\pi[\varphi(x)]^2 = \sigma\pi 2\varphi(x) \frac{d\varphi(x)}{dx}.$$

The integral of this differential equation is

$$y = \varphi(x) = Ce^{\gamma x/2\sigma}.$$

By substitution in the above integral equation we find:

$$Q + \gamma\pi \int_0^x C^2 e^{\gamma x/\sigma} dx = \sigma\pi C^2 e^{\gamma x/\sigma},$$

i.e.,

$$Q + \gamma\pi C^2 [e^{\gamma x/\sigma} - 1] = \sigma\pi C^2 e^{\gamma x/\sigma}.$$

From this it follows that

$$C = \sqrt{Q/\sigma\pi}.$$

The desired equation is thus

$$y = \varphi(x) = \sqrt{Q/\sigma\pi}\, e^{\gamma x/2\sigma}.$$

3. Integral equations which can be solved by differentiation

Under this category fall, first of all, Volterra integral equations of the first kind. The differentiation with respect to x of the integral

$$\int_{y=a}^{y=x} K(x, y)\varphi(y)\, dy$$

is performed in (2′), p. 479.

We consider x as the parameter and obtain (the partial derivative with respect to x of the kernel function $K(x, y)$ is assumed to exist and be continuous)

$$\frac{d}{dx}\int_{y=a}^{y=x} K(x, y)\, \varphi(y)\, dy = \int_{y=a}^{y=x} \frac{\partial K(x, y)}{\partial x}\varphi(y)\, dy + K(x, x)\, \varphi(x).$$

Example. $\int_0^x e^{-x}\varphi(y)\, dy = e^{-x} + x - 1.$

Differentiation by x yields:

$$-\int_0^x e^{-x}\varphi(y)\, dy + e^{-x}\varphi(x) = -e^{-x} + 1$$

or

$$-e^{-x} - x + 1 + e^{-x}\varphi(x) = -e^{-x} + 1.$$

It thus follows that

$$\varphi(x) = xe^x.$$

The method of differentiation may be regularly employed when the kernel of a Volterra integral equation of the first kind is a polynomial.

Example.

$$\int_0^x [(x-y)^3 - 2]\varphi(y)\,dy = -4x. \tag{1}$$

Threefold differentiation yields

$$2\int_0^x [x-y]\varphi(y)\,dy - 2\varphi(x) = -4, \tag{2}$$

$$2\int_0^x \varphi(y)\,dy - 2\varphi'(x) = 0, \tag{3}$$

$$\varphi(x) - \varphi''(x) = 0 \tag{4}$$

from which it follows that

$$\varphi(x) = Ae^x + Be^{-x}. \tag{5}$$

In order to determine A and B we substitute the value 0 for x in (2) and (3):

$$\varphi(0) = 2, \quad \varphi'(0) = 0.$$

It follows that $A = B = 1$, and the desired solution of (1) is

$$\varphi(x) = e^x + e^{-x}.$$

4. The Abel integral equation

By *Abel integral equation* one denotes the Volterra integral equation of the first kind

$$\int_0^x \frac{\varphi(y)}{\sqrt{x-y}}\,dy = f(x)$$

in which the kernel $1/\sqrt{x-y}$ is infinite for $y = x$.

We multiply both sides of the equation by $1/\sqrt{\eta - x}$ and integrate from 0 to η with respect to x:

$$\int_0^\eta \frac{1}{\sqrt{\eta - x}}\left(\int_0^x \frac{\varphi(y)}{\sqrt{x-y}}\,dy\right)dx = \int_0^\eta \frac{f(x)}{\sqrt{\eta - x}}\,dx.$$

The double integration on the left side of this equation is so written that first it will be integrated in the y-direction from 0 to x. The region of integration therefore is the triangle lying above the diagonal $x = y$ in the yx-plane (cf. Fig. 416). One changes the order of integration so that one first integrates in the x-direction from $x = y$ to $x = \eta$ and afterward in the y-direction from $y = 0$ to $y = \eta$. One thus obtains:

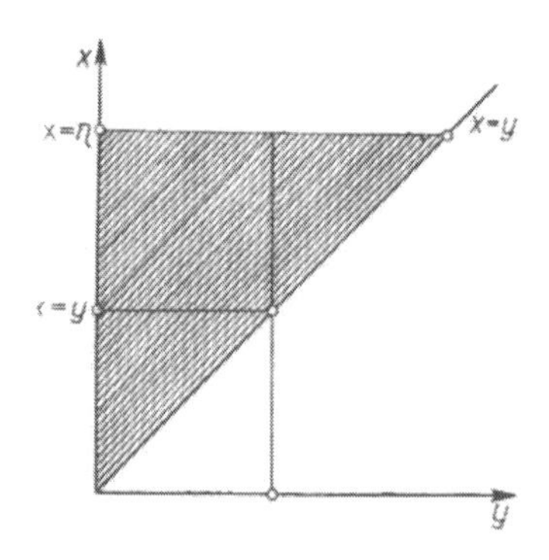

FIG. 416

$$\int_0^\eta \varphi(y)\left(\int_y^\eta \frac{1}{\sqrt{\eta - x}} \cdot \frac{1}{\sqrt{x - y}}\, dx\right) dy = \int_0^\eta \frac{f(x)}{\sqrt{\eta - x}}\, dx.$$

Since

$$\int_y^\eta \frac{dx}{\sqrt{\eta - x}\sqrt{x - y}} = \pi \ (^1)$$

it follows when we replace η by y that

$$\int_0^y \varphi(y)\, dy = \frac{1}{\pi} \int_0^y \frac{f(x)}{\sqrt{y - x}}\, dx.$$

---

(1) This integral can be evaluated through the substitution $x = y + (\eta - y)u$. It follows that

$$\int_0^1 \frac{du}{\sqrt{u - u^2}} = \left[\arcsin (2u - 1)\right]_0^1 = \pi \quad \text{(cf. p. 432, formula (264))}.$$

The desired function $\varphi(y)$ is then obtained through differentiation:

$$\varphi(y) = \frac{1}{\pi} \cdot \frac{d}{dy}\left[\int_0^y \frac{f(x)}{\sqrt{y-x}}\, dx\right].$$

Despite the singularity of the integrand this integral convergers. One can see this for example from the second criterion for convergence of an integral on p. 478. The solution function $\varphi(x)$ was from the first assumed to be continuous in order to insure the existence of the integral (in the sense of Riemann).

Example. $\int_0^x \frac{\varphi(y)}{\sqrt{x-y}}\, dy = x.$

$$\varphi(y) = \frac{1}{\pi} \cdot \frac{d}{dy}\left[\int_0^y \frac{x}{\sqrt{y-x}}\, dx\right] \quad \text{(cf. p. 421, formula (125))}$$

$$= \frac{1}{\pi} \cdot \frac{d}{dy}\left[-\tfrac{2}{3}(x+2y)\sqrt{y-x}\right]_0^y$$

$$= \frac{1}{\pi} \cdot \frac{d}{dy}\left[\tfrac{4}{3} y^{3/2}\right]$$

$$= \frac{1}{\pi} 2\sqrt{y}.$$

Occasionally a Volterra integral equation of the second kind may also be solved by differentiation.

Example.

$$\int_0^x (x-y)\,\varphi(y)\, dy + f(x) = \varphi(x). \tag{1}$$

Two differentiations with respect to x yield:

$$\int_0^x \varphi(y)\, dy + f'(x) = \varphi'(x), \tag{2}$$

$$\varphi(x) + f''(x) = \varphi''(x). \tag{3}$$

The function $f(x)$ naturally must be assumed to have a continuous second derivative. One must also assume, from the beginning, that the desired function $\varphi(x)$ is doubly differentiable. The solution to this second order ordinary differential equation with constant coefficients and perturbation function $f''(x)$ is (cf. p. 546)

$$(4) \qquad \varphi(x) = \tfrac{1}{2} e^x \left[C_1 + \int_0^x f''(t)\, e^{-t}\, dt\right] - \tfrac{1}{2} e^{-x} \left[C_2 + \int_0^x f''(t)\, e^t\, dt\right],$$

or after partial integration (cf. p. 459)

$$(4a) \qquad \varphi(x) = f(x) + \tfrac{1}{2} e^x \left[C_1 - f'(0) - f(0) + \int_0^x f(t)\, e^{-t}\, dt\right] -$$

$$- \tfrac{1}{2} e^{-x} \left[C_2 - f'(0) + f(0) + \int_0^x f(t)\, e^t\, dt\right].$$

In order to determine C_1 and C_2 we differentiate equation (4) with respect to x:

$$(5) \qquad \varphi'(x) = \tfrac{1}{2} e^x \left[C_1 + \int_0^x f''(t)\, e^{-t}\, dt\right] + \tfrac{1}{2} e^{-x} \left[C_2 + \int_0^x f''(t)\, e^t\, dt\right].$$

Upon substitution of the value 0 for x in (1), (2), (4) and (5) it follows that

$$(6) \qquad \varphi(0) = f(0),$$

$$(7) \qquad \varphi'(0) = f'(0),$$

$$(8) \qquad \varphi(0) = \tfrac{1}{2}[C_1 - C_2],$$

$$(9) \qquad \varphi'(0) = \tfrac{1}{2}[C_1 + C_2].$$

From (6), (7), (8) and (9) we deduce that

$$C_1 = f'(0) + f(0), \qquad C_2 = f'(0) - f(0).$$

Placing these values in (4a) yields the solution of (1):

$$\varphi(x) = f(x) + \tfrac{1}{2} e^x \int_0^x f(t)\, e^{-t}\, dt - \tfrac{1}{2} e^{-x} \int_0^x f(t)\, e^t\, dt.$$

5. Integral equations with product kernels

If the kernel $K(x, y)$ of an integral equation appears as a sum of products of a function in x alone times a function in y alone then we speak of a *product kernel* or *degenerate kernel*. An inhomogeneous Fredholm integral equation of the second kind with a product kernel has the form:

$$(1) \qquad \lambda \int_a^b [\alpha_1(x)\, \beta_1(y) + \alpha_2(x)\, \beta_2(y) + \ldots + \alpha_n(x)\, \beta_n(y)]\, \varphi(y)\, dy +$$

$$+ f(x) = \varphi(x).$$

λ is in this case a (in general, complex) numerical factor. The solution procedure is essentially dependent on the choice of such numerical factors λ. It will be called the *parameter of the integral equation*. The functions $\alpha_1(x), \ldots, \alpha_n(x)$ and $\beta_1(x), \ldots, \beta_n(x)$ respectively must be defined and continuous in the interval $a < x < b$. Furthermore, we assume from the beginning that the functions $\alpha_1(x), \ldots, \alpha_n(x)$ on the one hand and the functions $\beta_1(x), \ldots, \beta_n(x)$ on the other hand are linearly independent. One defines linear independence of a system of functions as follows: n functions $g_1(x), \ldots, g_n(x)$ all of which are defined in the interval $a < x < b$ are said to be *linearly independent* if a relation of the type

$$\sum_{i=1}^{n} c_i\, g_i(x) \equiv 0$$

with constant coefficients implies that all the coefficients which appear in it must be zero:

$$c_1 = c_2 = \ldots = c_n = 0.$$

If the functions $g_1(x), \ldots, g_n(x)$ are linearly dependent then there exists a relation $c_1 g_1(x) + \ldots + c_n g_n(x) \equiv 0$ in which not all the coefficients are zero. If, let us say, $c_1 \neq 0$ then we can write the function $g_1(x)$ in terms of $g_2(x), \ldots, g_n(x)$:

$$g_1(x) = \frac{1}{c_1}\left(c_2\, g_2(x) + \ldots + c_n\, g_n(x)\right).$$

We can thus reduce every linear combination of $g_1(x), \ldots, g_n(x)$ to a linear combination of $g_2(x), \ldots, g_n(x)$. Since we can apply this process to the $\alpha_i(x)$ or $\beta_i(x)$ respectively we can attain a situation in which the $\alpha_i(x)$ on the one hand and the $\beta_i(x)$ on the other hand are linearly independent systems. For this reason it is assumed that in the sequel the functions are given in such a form. We set

$$\int_a^b \beta_1(y)\,\varphi(y)\,dy = B_1, \quad \int_a^b \beta_2(y)\,\varphi(y)\,dy = B_2, \quad \ldots,$$

$$\int_a^b \beta_n(y)\,\varphi(y)\,dy = B_n,$$

and deduce that

$$\varphi(x) = \lambda B_1 \alpha_1(x) + \lambda B_2 \alpha_2(x) + \ldots + \lambda B_n \alpha_n(x) + f(x), \tag{2}$$

and thus also that

(3) $$\varphi(y) = \lambda B_1\alpha_1(y) + \lambda B_2\alpha_2(y) + \ldots + \lambda B_n\alpha_n(y) + f(y).$$

Equation (1) can be written in the following form:

(1a) $$\varphi(x) = \lambda\alpha_1(x) \int_a^b \beta_1(y)\varphi(y)\,dy + \lambda\alpha_2(x) \int_a^b \beta_2(y)\varphi(y)\,dy + \ldots$$
$$\ldots + \lambda\alpha_n(x) \int_a^b \beta_n(y)\varphi(y)\,dy + f(x).$$

To determine the unknowns $B_1, B_2, \ldots, B_n$ we place the expression for $\varphi(y)$ given by (3) into (1a) and equate this to (2):

$$\lambda B_1\alpha_1(x) + \lambda B_2\alpha_2(x) + \ldots + \lambda B_n\alpha_n(x)$$
$$= \lambda\alpha_1(x) \int_a^b \beta_1(y)[\lambda B_1\alpha_1(y) + \lambda B_2\alpha_2(y) + \ldots + \lambda B_n\alpha_n(y) + f(y)]\,dy +$$
$$+ \lambda\alpha_2(x) \int_a^b \beta_2(y)[\lambda B_1\alpha_1(y) + \lambda B_2\alpha_2(y) + \ldots + \lambda B_n\alpha_n(y) + f(y)]\,dy +$$
$$\cdots\cdots\cdots\cdots\cdots\cdots\cdots\cdots\cdots$$
$$+ \lambda\alpha_n(x) \int_a^b \beta_n(y)[\lambda B_1\alpha_1(y) + \lambda B_2\alpha_2(y) + \ldots + \lambda B_n\alpha_n(y) + f(y)]\,dy.$$

From the linear independence of the functions $\alpha_1(x), \ldots, \alpha_n(x)$ it follows that the coefficients of $\alpha_1(x), \ldots, \alpha_n(x)$ must be the same on both sides of the equality. Using the simplifying notation

$$\int_a^b \beta_\mu(y)\alpha_\nu(y)dy = a_{\mu\nu} \quad \text{and} \quad \int_a^b \beta_\mu(y)f(y)dy = b_\mu$$

we obtain the equations:

(4) $$\begin{array}{llll} B_1(1-\lambda a_{11}) & - B_2\lambda a_{12} & - \ldots - B_n\lambda a_{1n} & = b_1, \\ -B_1\lambda a_{21} & + B_2(1-\lambda a_{22}) & - \ldots - B_n\lambda a_{2n} & = b_2, \\ \cdots & \cdots & \cdots & \cdots \\ -B_1\lambda a_{n1} & - B_2\lambda a_{n2} & - \ldots + B_n(1-\lambda a_{nn}) & = b_n. \end{array}$$

If the functions $\beta_1(x), \ldots, \beta_n(x)$ were linearly dependent, then we would obtain at least one system of coefficients $c_1, \ldots, c_n$ (not all of which were zero) such that

$$\sum_{i=1}^{n} c_i\beta_i(x) \equiv 0$$

in the whole interval $a < x < b$.

From the defining equations for $a_{\mu\nu}$ and b_μ it would however follow that

$$\sum_{i=1}^{n} c_i b_i = 0$$

and

$$\sum_{i=1}^{n} c_i a_{i\nu} = 0 \qquad \text{for all } \nu.$$

The system of equations (4) would in this case thus be reducible

The decision on the solvability of the system (4) of linear equations will be determined with the help of determinant theory in the following manner: (cf. pp. 174-182).

For all values of λ for which the value of the determinant of the coefficients

$$\text{(5)} \qquad \begin{vmatrix} (1-\lambda a_{11}) & -\lambda a_{12} & \dots & -\lambda a_{1n} \\ -\lambda a_{21} & (1-\lambda a_{22}) & \dots & -\lambda a_{2n} \\ \dots & \dots & \dots & \dots \\ -\lambda a_{n1} & -\lambda a_{n2} & \dots & (1-\lambda a_{nn}) \end{vmatrix} = D(\lambda)$$

is different from zero, the system (4) and thus also the *inhomogeneous integral equation* (1) has one and only one solution. This solution of the integral equation is then given by (2).

For all λ-values which are roots of the equation $D(\lambda) = 0$ the homogeneous integral equation

$$\text{(6)} \quad \lambda \int_a^b [\alpha_1(x)\beta_1(y) + \alpha_2(x)\beta_2(y) + \dots + \alpha_n(x)\beta_n(y)]\varphi(y)\,dy = \varphi(x)$$

is solvable. The system of equations (4) is in this case homogeneous. It has in the homogeneous case (also when $D(\lambda) \neq 0$) the trivial solution $B_1 = B_2 = \dots = B_n = 0$. In the case $D(\lambda) = 0$ the equation system (4) (homogeneous case) also admits non-trivial solutions $B_1, \dots, B_n$ (for which thus not all the B_i are zero). The number of linearly independent solutions of the homogeneous equations (4) depends on the rank of the determinant of coefficients of the system of equations (4) (cf. p. 179). The solution of (6) has the form

$$\varphi(x) = \lambda[B_1\alpha_1(x) + B_2\alpha_2(x) + \dots + B_n\alpha_n(x)]$$

or

$$\text{(7)} \qquad \varphi(x) = C[B_1\alpha_1(x) + B_2\alpha_2(x) + \dots + B_n\alpha_n(x)]$$

respectively.

The function $B_1\alpha_1(x)+\ldots+B_n\alpha_n(x)$ cannot vanish identically for a non-trivial solution $B_1, \ldots, B_n$ of the homogeneous system of equations for if it were zero the functions $\alpha_1(x), \ldots, \alpha_n(x)$ would be linearly dependent. Thus the function

$$\varphi(x) = C[B_1\alpha_1(x)+\ldots+B_n\alpha_n(x)]$$

cannot vanish identically for $C \neq 0$. Thus one can determine the constant C such that the function $\varphi(x)$ satisfies the condition

$$\int_a^b [\varphi(x)]^2\,dx = 1. \tag{8}$$

One calls this the *normalized characteristic function.*

Assume finally that an integral equation of form (1) for which the value of λ is a root of the equation $D(\lambda)=0$ is solvable. Then the perturbation function $f(x)$ must satisfy certain conditions.

The integral equation

$$\lambda\int_a^b [\alpha_1(y)\,\beta_1(x)+\ldots+\alpha_n(y)\,\beta_n(x)]\,\psi(y)\,dy + f(x) = \psi(x) \tag{9}$$

is called the *transposed integral equation* of the integral equation (1). (If the kernel of the integral equation (1) is $K(x, y)$ then the kernel of the transposed integral equation is $K(y, x)$). The condition on $f(x)$ takes on the following form:

The inhomogeneous integral equation (1) is solvable for a value of λ for which $D(\lambda)=0$ if and only if the perturbation function $f(x)$ satisfies the orthogonality condition

$$\int_a^b f(x)\,\psi(x)\,dx = 0$$

for all functions $\psi(x)$ which satisfy the transposed homogeneous integral equation for the same value of λ; namely,

$$\lambda\int_a^b [\alpha_1(y)\,\beta_1(x)+\ldots+\alpha_n(y)\,\beta_n(x)]\,\psi(y)\,dy = \psi(x).$$

We can strengthen this theorem as follows: A value of λ is a characteristic value of the transposed integral equation (9) if and only if it is a characteristic value of the integral equation (1). Moreover for every value of λ which is a characteristic value the two equations (1) and (9) always possess the same finite number of linearly independent characteristic functions. The theorem however, is not limited only to integral equations with product kernels, but extends to integral equations with more general kernel types (cf. Section 9).

Examples. 1. $\int_0^1 [xy + \sqrt{xy}]\,\varphi(y)\,dy + x = \varphi(x)$,

$$\alpha_1(x) = x, \quad \beta_1(y) = y, \quad \alpha_2(x) = \sqrt{x}, \quad \beta_2(y) = \sqrt{y}, \quad f(x) \equiv x,$$

$$a_{11} = \int_0^1 y^2\,dy = \tfrac{1}{3}, \qquad a_{12} = \int_0^1 y^{3/2}\,dy = \tfrac{2}{5},$$

$$a_{21} = \int_0^1 y^{3/2}\,dy = \tfrac{2}{5}, \qquad a_{22} = \int_0^1 y\,dy = \tfrac{1}{2},$$

$$b_1 = \int_0^1 y^2\,dy = \tfrac{1}{3}, \qquad b_2 = \int_0^1 y^{3/2}\,dy = \tfrac{2}{5}.$$

From

$$(1 - \tfrac{1}{3})\,B_1 - \tfrac{2}{5}\,B_2 = \tfrac{1}{3}, \qquad -\tfrac{2}{5}\,B_1 + (1 - \tfrac{1}{2})\,B_2 = \tfrac{2}{5}$$

we find that

$$B_1 = \tfrac{49}{26} \quad \text{and} \quad B_2 = \tfrac{30}{13}.$$

The solution of the above integral equation is

$$\varphi(x) = \tfrac{49}{26}x + \tfrac{30}{13}\sqrt{x} + x$$

or

$$\varphi(x) = \tfrac{75}{26}x + \tfrac{30}{13}\sqrt{x}.$$

2. $\lambda \int_1^2 [xy + 1/xy]\,\varphi(y)\,dy = \varphi(x)$,

$$\alpha_1(x) = x, \quad \beta_1(y) = y, \quad \alpha_2(x) = 1/x, \quad \beta_2(y) = 1/y,$$

$$a_{11} = \int_1^2 y^2\,dy = \tfrac{7}{3}, \quad a_{12} = \int_1^2 dy = 1 = a_{21}, \quad a_{22} = \int_1^2 \frac{dy}{y^2} = \tfrac{1}{2}.$$

We obtain the system of homogeneous linear equations

$$(1 - \tfrac{7}{3}\lambda)\,B_1 - \lambda B_2 = 0, \qquad -\lambda B_1 + (1 - \tfrac{1}{2}\lambda)\,B_2 = 0.$$

This system is solvable only when the determinant

$$\begin{vmatrix} (1 - \tfrac{7}{3}\lambda) & -\lambda \\ -\lambda & (1 - \tfrac{1}{2}\lambda) \end{vmatrix} = \Delta(\lambda)$$

vanishes. From $1-\frac{17}{6}\lambda+\frac{1}{6}\lambda^2=0$ we obtain the characteristic values

$$\lambda_1=\frac{17+\sqrt{265}}{2}\approx 16.6394,\qquad \lambda_2=\frac{17-\sqrt{265}}{2}\approx 0.3606.$$

These equations give $B_2\approx-2.2732B_1$ for λ_1, and $B'_2\approx+0.4399B'_1$ for λ_2. The characteristic functions are, accordingly

$$\varphi_1(x)\approx 16.6394B_1\left[x-2.2732\frac{1}{x}\right]$$

and

$$\varphi_2(x)\approx 0.3606B'_1\left[x+0.4399\frac{1}{x}\right].$$

Since B_1 and B'_1 are undetermined constants we may write

$$\varphi_1(x)\approx C_1\left[x-2.2732\frac{1}{x}\right],$$

$$\varphi_2(x)\approx C_2\left[x+0.4399\frac{1}{x}\right].$$

3. The integral equation

$$\lambda\int_{-1}^{+1}[xy+x^2y^2]\,\varphi(y)\,dy+f(x)=\varphi(x)\tag{1}$$

may be solved as an equation with a product kernel (cf. p. 689) by setting

$$\alpha_1(x)=x,\qquad \beta_1(y)=y,\qquad \alpha_2(x)=x^2,\qquad \beta_2(y)=y^2,$$

$$a_{11}=\tfrac{2}{3},\qquad a_{12}=a_{21}=0,\qquad a_{22}=\tfrac{2}{5},$$

$$b_1=\int_{-1}^{+1}yf(y)\,dy,\qquad b_2=\int_{-1}^{+1}y^2f(y)\,dy.$$

From the system of equations

$$B_1(1-\tfrac{2}{3}\lambda)=\int_{-1}^{+1}yf(y)\,dy,\qquad B_2(1-\tfrac{2}{5}\lambda)=\int_{-1}^{+1}y^2f(y)\,dy$$

it follows that the solution of (1) is

$$\varphi(x)=\frac{x\lambda\int_{-1}^{+1}yf(y)\,dy}{1-\frac{2}{3}\lambda}+\frac{x^2\lambda\int_{-1}^{+1}y^2f(y)\,dy}{1-\frac{2}{5}\lambda}+f(x).$$

The homogeneous integral equation

$$\lambda \int_{-1}^{+1} [xy + x^2 y^2] \varphi(y) \, dy = \varphi(x) \tag{1a}$$

is solvable only for the characteristic values $\lambda_1 = \frac{3}{2}$ and $\lambda_2 = \frac{5}{2}$ obtained from the equation

$$\begin{vmatrix} (1 - \frac{2}{3}\lambda) & 0 \\ 0 & (1 - \frac{2}{5}\lambda) \end{vmatrix} = 0.$$

The corresponding characteristic functions are $\varphi_1(x) = C_1 x$ and $\varphi_2(x) = C_2 x^2$. From $\int_{-1}^{+1} C_1 x C_2 x^2 \, dx = 0$ it is clear that the two characteristic functions are orthogonal to one another. Equation (1) is solvable only for the characteristic values $\lambda_1 = \frac{3}{2}$ and $\lambda_2 = \frac{5}{2}$ when $f(x)$ is orthogonal to the characteristic functions $\varphi_1(x) = C_1 x$ or $\varphi_2(x) = C_2 x^2$ respectively. These conditions are fulfilled for example by:

for $\lambda_1 = \frac{3}{2}$; functions of the form $f(x) = k_0 + k_2 x^2 + k_4 x^4 + \ldots$

for $\lambda_2 = \frac{5}{2}$; functions of the form $f(x) = k_0 + k_1 x + k_3 x^3 + k_5 x^5 + \ldots$

The normalized characteristic functions will be obtained by determining C_1 and C_2 so that

$$\int_{-1}^{+1} C_1^2 x^2 dx = 1 \quad \text{or} \quad \int_{-1}^{+1} C_2^2 x^4 dx = 1$$

respectively. We find that

$$C_1 = \tfrac{1}{2}\sqrt{6} \quad \text{and} \quad C_2 = \tfrac{1}{2}\sqrt{10},$$

thus

$$\varphi_1(x) = \tfrac{1}{2}\sqrt{6}\, x \quad \text{and} \quad \varphi_2(x) = \tfrac{1}{2}\sqrt{10}\, x^2.$$

6. The Neumann series (successive approximation)

Ordinary first order differential equations can be solved by Picard's method of *successive approximation* (cf. p. 526). There is also an iteration method based on the same principle for the solving of linear integral equations. Consider the following Fredholm integral equation:

$$\lambda \int_a^b K(x, y) \varphi(y) \, dy + f(x) = \varphi(x). \tag{1}$$

As an (zeroth) approximation to the desired solution $\varphi(x)$ the function $\varphi_0(x) \equiv 0$ is taken. One substitutes this for the unknown function which occurs under the integral sign in (1), and thus obtains the first approximation:

$$\varphi_1(x) \equiv f(x). \tag{2}$$

One substitutes (2) into (1) and obtains as the second approximation

$$\varphi_2(x) = \lambda \int_a^b K(x, \eta) f(\eta)\, d\eta + f(x)\ (^1). \tag{3}$$

This function when substituted into (1) yields the third approximation:

$$\varphi_3(x) = \lambda \int_a^b K(x, y) f(y)\, dy + \lambda^2 \int_a^b \int_a^b K(x, y) K(y, \eta) f(\eta)\, d\eta dy + f(x). \tag{4}$$

Introducing the notation

$$K^{(1)}(x, y) = K(x, y),$$

$$K^{(2)}(x, y) = \int_a^b K(x, t) K(t, y)\, dt$$

one may rewrite equation (4), when one changes the order of integration of the double integral which occurs in (4), in the following form:

$$\varphi_3(x) = \lambda \int_a^b K^{(1)}(x, y) f(y)\, dy + \lambda^2 \int_a^b K^{(2)}(x, \eta) f(\eta)\, d\eta + f(x). \tag{4a}$$

After substitution of (4a) into (1) and replacing

$$\int_a^b K(x, \eta) K^{(2)}(\eta, y)\, d\eta$$

by the symbol $K^{(3)}(x, y)$ we obtain:

$$\varphi_4(x) = \lambda \int_a^b K^{(1)}(x, y) f(y)\, dy + \lambda^2 \int_a^b K^{(2)}(x, y) f(y)\, dy + \lambda^3 \int_a^b K^{(3)}(x, y) f(y)\, dy + f(x). \tag{5}$$

We continue this process, and after the introduction of the notation

---

(1) In order to avoid confusion we replace y by η.

(6) $$K^{(n)}(x, y) = \int_a^b K^{(n-1)}(x, y)\, K(\eta, y)\, d\eta$$

we get as the n-th approximate solution of the integral equation

(7) $$\varphi_n(x) = \lambda \int_a^b K^{(1)}(x, y) f(y)\, dy + \lambda^2 \int_a^b K^{(2)}(x, y) f(y)\, dy + \ldots + \lambda^n \int_a^b K^{(n)}(x, y) f(y)\, dy + f(x).$$

We call the expression $K^{(n)}(x, y)$ the $(n-1)$-th iteration kernel, where $K^{(1)}(x, y) = K(x, y)$. Passing to the limit as $n \to \infty$ one obtains the so-called Neumann series:

(8) $$\varphi(x) = \lim_{n\to\infty} \varphi_n(x) = \sum_{n=1}^{\infty} \lambda^n \int_a^b K^{(n)}(x, y) f(y)\, dy + f(x).$$

This can be rewritten as

(8a)
$$\varphi(x) = \lambda \int_a^b [K^{(1)}(x, y) + \lambda K^{(2)}(x, y) + \lambda^2 K^{(3)}(x, y) + \ldots] f(y)\, dy + f(x).$$

The expression appearing in the brackets,

(9) $$K^{(1)}(x, y) + \lambda K^{(2)}(x, y) + \lambda^2 K^{(3)}(x, y) + \ldots = \Gamma(x, y, \lambda),$$

is called the *solving kernel* or *resolvent*. The solution of the integral equation (1) appears then in the form:

(10) $$\varphi(x) = \lambda \int_a^b \Gamma(x, y, \lambda) f(y)\, dy + f(x).$$

The Neumann series converges uniformly and absolutely in the square $a < x < b$, $a < y < b$ when

(11) $$|\lambda| < \frac{1}{\sqrt{\int_a^b \int_a^b |K(x, y)|^2\, dx dy}}.$$

If M is the maximum of $|K(x, y)|$ in the square $a < x < b$, $a < y < b$ [1] then one deduces from the estimate (cf. p. 457)

[1] For the function $K(x, y)$ defined in $a \leqslant x \leqslant b$, $a \leqslant y \leqslant b$ there exists such a number G that

(a) in the entire domain of definition $|K(x, y)| \leqslant G$,

(b) for every $\varepsilon > 0$ there exists a point $(x_\varepsilon, y_\varepsilon)$ in the domain of definition for which $|K(x_\varepsilon, y_\varepsilon)| > G - \varepsilon$ holds.

$$\int_a^b \int_a^b |K(x, y)|^2 \, dx dy \leqslant M^2 (b-a)^2$$

that

$$\frac{1}{M(b-a)} < \frac{1}{\sqrt{\int_a^b \int_a^b |K(x, y)|^2 \, dx dy}}.$$

The Neumann series certainly converges for all λ such that

$$|\lambda| < \frac{1}{M(b-a)}. \tag{12}$$

The radius of convergence given by (12) is however in comparison with (11) an unnecessarily small limit.

With the help of the Neumann series one can also solve Volterra integral equations. Consider the equation

$$\lambda \int_a^x K(x, y) \varphi(y) \, dy + f(x) = \varphi(x). \tag{13}$$

In order to be able to integrate over the closed interval $a \leqslant x \leqslant b$, one sets

$$K(x, y) \equiv 0 \quad \text{for} \quad y > x.$$

The iteration kernels ($y \leqslant x$) will be given by the equations

$$\begin{aligned} K^{(1)}(x, y) &= K(x, y), \\ K^{(2)}(x, y) &= \int_a^b K(x, t) K(t, y) \, dt \\ &= \int_y^x K(x, t) K(t, y) \, dt, \\ K^{(3)}(x, y) &= \int_y^x K(x, t) K^{(2)}(t, y) \, dt, \\ &\cdots\cdots\cdots\cdots \\ K^{(n)}(x, y) &= \int_y^x K(x, t) K^{(n-1)}(t, y) \, dt, \end{aligned} \tag{14}$$

where $K(x, t) \equiv 0$ for $t > x$ and $K(t, y) \equiv 0$ for $t < y$.

G is uniquely determined and called the *upper bound* of the function $|K(x, y)|$. Since $|K(x, y)|$ is continuous and the domain of definition is closed and bounded there is a point (x_1, y_1) at which $|K(x, y)|$ assumes the value of the upper bound: $|K(x_1, y_1)| = G$. In this case we speak of the maximum $M (= G)$ of the function $|K(x, y)|$ instead of the upper bound.

For $y > x$ the iteration kernels based on this function also vanish identically. The Neumann series then has the form:

$$\varphi(x) = \lambda \sum_{n=1}^{\infty} \lambda^{n-1} \int_a^x K^{(n)}(x, y)\, f(y)\, dy + f(x) \tag{15}$$

or

$$\varphi(x) = \lambda \int_a^{\pi} \sum_{n=1}^{\infty} \lambda^{n-1} K^{(n)}(x, y)\, f(y)\, dy + f(x). \tag{15a}$$

A special examination into the convergence of this Neumann series is not necessary since for Volterra integral equations of form (13) the corresponding Neumann series ((15) and (15a)) converges for all finite values of λ.

Example.

$$\lambda \int_0^{\pi} \sin(x+y)\, \varphi(y)\, dy + 1 = \varphi(x),$$

$$K^{(1)}(x, y) = \sin(x+y), \quad f(x) = 1,$$

$$K^{(2)}(x, y) = \int_0^{\pi} \sin(x+\eta) \sin(\eta+y)\, d\eta$$

$$= \tfrac{1}{2}\pi [\sin x \sin y + \cos x \cos y],$$

$$K^{(3)}(x, y) = \tfrac{1}{2}\pi \int_0^{\pi} [\sin x \sin \eta + \cos x \cos \eta] [\sin y \cos \eta + \cos y \sin \eta]\, d\eta$$

$$= (\tfrac{1}{2}\pi)^2 [\sin x \cos y + \cos x \sin y],$$

$$K^{(4)}(x, y) = (\tfrac{1}{2}\pi)^3 [\sin x \sin y + \cos x \cos y],$$

$$K^{(5)}(x, y) = (\tfrac{1}{2}\pi)^4 [\sin x \cos y + \cos x \sin y],$$

$$K^{(6)}(x, y) = (\tfrac{1}{2}\pi)^5 [\sin x \sin y + \cos x \cos y], \qquad \text{etc.}$$

Since

$$\int_0^{\pi} [\sin x \cos y + \cos x \sin y]\, dy = 2 \cos x$$

and

$$\int_0^{\pi} [\sin x \sin y + \cos x \cos y]\, dy = 2 \sin x,$$

it follows that the solution of the integral equation under consideration is

$$\varphi(x) = 2\lambda \cos x[1 + \lambda^2(\tfrac{1}{2}\pi)^2 + \lambda^4(\tfrac{1}{2}\pi)^4 + \ldots] +$$
$$+ \lambda^2\pi \sin x[1 + \lambda^2(\tfrac{1}{2}\pi)^2 + \lambda^4(\tfrac{1}{2}\pi)^4 + \ldots] + 1$$

or

$$\varphi(x) = \frac{2\lambda \cos x + \lambda^2 \pi \sin x}{1 - \lambda^2 (\frac{1}{2}\pi)^2} + 1.$$

Since $M = 1$, the interval of convergence given by (12),

$$|\lambda| > \frac{1}{1 \cdot |\pi - 0|}$$

will be the interval from $-1/\pi$ to $+1/\pi$.

The inequality (11), since $\int\limits_0^\pi \int\limits_0^\pi \sin^2(x+y)\,dx dy = \frac{1}{2}\pi^2$ yields the somewhat larger interval from $-\sqrt{2}/\pi$ to $+\sqrt{2}/\pi$.

But in this case the Neumann series appears as two geometric series with quotient $\lambda^2(\frac{1}{2}\pi)^2$; thus the radius of uniform convergence can be computed in its entirety. It is $|\lambda| < 2/\pi$.

The characteristic values, given by $1 - \lambda^2(\frac{1}{2}\pi)^2 = 0$ are $\lambda = \pm 2/\pi$. For these values the denominator of $\varphi(x)$ vanishes.

An example of the solution of a Volterra integral equation of the second kind by means of the Neumann series:

$$\int\limits_0^x xy\,\varphi(y)\,dy + 1 = \varphi(x),$$

$$K^{(1)} = xy,$$

$$K^{(2)}(x, y) = \frac{x^4 y - xy^4}{3},$$

$$K^{(3)}(x, y) = \frac{x^7 y - 2x^4 y^4 + xy^7}{18},$$

$$K^{(4)}(x, y) = \frac{x^{10} y - 3x^7 y^4 + 3x^4 y^7 - xy^{10}}{162},$$

$$\varphi(x) = \frac{x^3}{2} + \frac{x^6}{2 \cdot 5} + \frac{x^9}{2 \cdot 5 \cdot 8} + \frac{x^{12}}{2 \cdot 5 \cdot 8 \cdot 11} + \ldots + 1.$$

A further example of a Volterra integral equation of the second kind:

$$\lambda \int\limits_0^x (x - y)\,\varphi(y)\,dy + (x + 1) = \varphi(x),$$

$$K^{(1)} = x - y, \qquad K^{(2)} = \frac{(x-y)^3}{3!}, \qquad K^{(3)} = \frac{(x-y)^5}{5!}, \qquad \ldots$$

$$\varphi(x) = \lambda\left(\frac{x^2}{2!} + \frac{x^3}{3!}\right) + \lambda^2\left(\frac{x^4}{4!} + \frac{x^5}{5!}\right) + \ldots + x + 1.$$

For $\lambda = 1$ it follows that $\varphi(x) = e^x$. From the method on p. 689 it follows that

$$\varphi(x) = x + 1 + \tfrac{1}{2} e^x \int_0^x (t+1)\, e^{-t}\, dt - \tfrac{1}{2} e^{-x} \int_0^x (t+1)\, e^t\, dt$$
$$= x + 1 - \tfrac{1}{2} x - 1 + e^x - \tfrac{1}{2} x = e^x.$$

7. The method of solution of Fredholm

The Fredholm integral equation of the second kind

$$(1) \qquad \lambda \int_a^b K(x, y)\, \varphi(y)\, dy + f(x) = \varphi(x)$$

in which the functions $K(x, y)$ and $f(x)$ are continuous throughout the domain $a < x < b$, $a < y < b$, can be looked upon as the limiting case of systems of linear equations. We divide the limits of integration into n equal parts. Let

$$\Delta = \frac{b-a}{n}$$

and write:

$$x_1 = y_1 = a, \quad x_2 = y_2 = a + \Delta, \quad \ldots, \quad x_n = y_n = a + (n-1)\Delta.$$

We then get the approximate equation

$$(2) \quad \int_a^b K(x, y)\, \varphi(y)\, dy \approx [K(x, y_1)\, \varphi(y_1) + K(x, y_2)\, \varphi(y_2) + \ldots$$
$$\ldots + K(x, y_n)\, \varphi(y_n)]\, \Delta.$$

Equation (1) then takes the form

$$(3) \quad \lambda [K(x, y_1)\, \varphi(y_1) + K(x, y_2)\, \varphi(y_2) + \ldots$$
$$\ldots + K(x, y_n)\, \varphi(y_n)]\, \Delta + f(x) \approx \varphi(x).$$

This equation must hold for all values of x in the interval $a < x < b$.

Upon introduction of the notation

$$f(x_1) = f_1, \quad f(x_2) = f_2, \quad \ldots, \quad f(x_n) = f_n,$$
$$\varphi(x_1) = \varphi_1, \quad \varphi(x_2) = \varphi_2, \quad \ldots, \quad \varphi(x_n) = \varphi_n,$$
$$K(x_1, y_1) = k_{11}, \quad K(x_1, y_2) = k_{12}, \quad \ldots,$$
$$K(x_i, y_j) = k_{ij}, \quad \ldots, \quad K(x_n, y_n) = k_{nn},$$

we obtain an approximation for the integral equation (1) in terms of the system of linear equations

$$\begin{array}{llll} \varphi_1[1-\lambda\Delta k_{11}] - \varphi_2\lambda\Delta k_{12} - \ldots - & & \varphi_n\lambda\Delta k_{1n} & = f_1, \\ -\varphi_1\lambda\Delta k_{21} & + \varphi_2[1-\lambda\Delta k_{22}] - \ldots - & \varphi_n\lambda\Delta k_{2n} & = f_2, \\ \ldots & & & \\ -\varphi_1\lambda\Delta k_{n1} & -\varphi_2\lambda\Delta k_{n2} - \ldots + & \varphi_n[1-\lambda\Delta k_{nn}] & = f_n. \end{array}$$

The unknown φ-values are approximate values for the integral equation (1) satisfying the function values for $x_1, x_2, \ldots, x_n$. Regarding the solvability of this system of equations, see the method on p. 175. With the aid of this system we can also determine approximations for the solution and characteristic values of a Fredholm integral equation of the second kind. The practical applications of this method are limited, since in most cases in order to determine the values of φ the number n must be relatively large.

We consider now the limit as $n \to \infty$. Then for the integral equation

$$\lambda \int_a^b K(x, y)\,\varphi(y)\,dy + f(x) = \varphi(x)$$

the solution is given by

$$\varphi(x) = \lambda \int_a^b \Gamma(x, y, \lambda)\,f(y)\,dy + f(x).$$

The "solving kernel" $\Gamma(x, y, \lambda)$ will be given by

$$\Gamma(x, y, \lambda) = \frac{\sum_{n=0}^{\infty} (-1)^n K_n(x, y)\,\lambda^n}{\sum_{n=0}^{\infty} (-1)^n \delta_n \lambda^n},$$

where $\delta_0 = 1$, $K_0(x, y) = K(x, y)$, and the further terms of these two series are given by the recursion formulas

$$\delta_n = \frac{1}{n} \int_a^b K_{n-1}(x, x)\,dx$$

and

$$K_n(x, y) = K(x, y)\,\delta_n - \int_a^b K(x, t)\,K_{n-1}(t, y)\,dt.$$

Examples of the determination of the characteristic values by the Fredholm approximation method:

$$(1)\quad \lambda \int_0^\pi \sin(x+y)\,\varphi(y)\,dy = \varphi(x).$$

We divide the interval from 0 to π into $n = 3$ parts [1], thus:

$$x_1 = y_1 = 0, \qquad x_2 = y_2 = \tfrac{1}{3}\pi, \qquad x_3 = y_3 = \tfrac{2}{3}\pi.$$

We obtain:

$$\begin{aligned}
k_{11} &= \sin 0 = 0; & k_{12} &= \sin \tfrac{1}{3}\pi = 0.866, & k_{13} &= \sin \tfrac{2}{3}\pi = 0.866,\\
k_{21} &= \sin \tfrac{1}{3}\pi = 0.866, & k_{22} &= \sin \tfrac{2}{3}\pi = 0.866, & k_{23} &= \sin \pi = 0,\\
k_{31} &= \sin \tfrac{2}{3}\pi = 0.866, & k_{32} &= \sin \pi = 0, & k_{33} &= \sin \tfrac{4}{3}\pi = -0.866.
\end{aligned}$$

From the determinant

$$\begin{vmatrix} 1 & -0.907\lambda & -0.907\lambda \\ -0.907\lambda & (1-0.907\lambda) & 0 \\ -0.907\lambda & 0 & (1+0.907\lambda) \end{vmatrix} = 0$$

or $1 - 3 \cdot 0.907^2 \lambda^2 = 0$ it follows that

$$\lambda = \pm \frac{1}{0.907 \cdot \sqrt{3}} = \pm 0.6365.$$

The exact values are (cf. page 699)

$$\lambda = \pm 2/ = \pm\pi 0.6366.$$

$$(2)\quad \lambda \int_0^1 \left[xy + \sqrt{xy}\right] \varphi(y)\,dy + x = \varphi(x).$$

Thus:

$$\delta_0 = 1, \qquad K_0(x, y) = xy + \sqrt{xy};$$

$$\delta_1 = \int_0^1 [x^2 + x]\,dx = \tfrac{5}{6}.$$

$$\begin{aligned}
K_1(x, y) &= \left[xy + \sqrt{xy}\right]\tfrac{5}{6} - \int_0^1 \left[xt + \sqrt{xt}\right]\left[ty + \sqrt{ty}\right] dt\\
&= \tfrac{1}{2}xy + \tfrac{1}{3}\sqrt{xy} - \tfrac{2}{5}\left[x\sqrt{y} + y\sqrt{x}\right];
\end{aligned}$$

$$\delta_2 = \tfrac{1}{2}\int_0^1 \left[\tfrac{1}{2}x^2 + \tfrac{1}{3}x - \tfrac{4}{5}x^{3/2}\right] dx = \tfrac{1}{150}, \qquad K_2(x, y) = 0.$$

---

[1] In this example the exceptionally small number $n = 3$ suffices.

Since $K_2(x, y)$ vanishes identically it follows immediately that δ_3 vanishes, from which it follows that $K_3(x, y)$ also vanishes, and hence that all the further δ_i as well as the $K_i(x, y)$ vanish identically. The expression for the solving kernel will thus be a quotient of two broken off power series in λ:

$$\Gamma(x, y, \lambda) = \frac{\left[xy + \sqrt{xy}\right] - \left[\frac{1}{2}xy + \frac{1}{3}\sqrt{xy} - \frac{2}{5}\left(x\sqrt{y} + y\sqrt{x}\right)\right]\lambda}{1 - \frac{5}{6}\lambda + \frac{1}{150}\lambda^2}.$$

Consequently, the solution is

$$\varphi(x) = \lambda \int_0^1 \Gamma(x, y, \lambda) y \, dy + x.$$

thus

$$\varphi(x) = \frac{150x + \lambda\left[60\sqrt{x} - 75x\right]}{\lambda^2 - 125\lambda + 150}.$$

For $\lambda = 1$ we obtain $\varphi(x) = \frac{75}{26}x + \frac{30}{13}\sqrt{x}$ (cf. p. 693). From $\lambda^2 - 125\lambda + 150 = 0$ it follows that the characteristic values are $\lambda = \frac{5}{2}\left[25 \pm \sqrt{601}\right]$, for which the above inhomogeneous integral equation has no solution ([1]). Accordingly the homogeneous integral equation

$$\lambda \int_0^1 \left[xy + \sqrt{xy}\right]\varphi(y) \, dy = \varphi(x)$$

can be solved only for these characteristic values.

Further example:

$$\lambda \int_0^\pi \sin(x + y)\,\varphi(y) \, dy + 1 = \varphi(x).$$

([1]) From the process on p. 693 it follows that for $\lambda = \frac{5}{2}[25 \pm \sqrt{601}]$ the integral equation has the following two linearly independent characteristic solutions:

$$\varphi_1 = C_1\left[6x - 119\sqrt{x} - 5\sqrt{601}\sqrt{x}\right],$$

$$\varphi_2 = C_2\left[6x - 119\sqrt{x} + 5\sqrt{601}\sqrt{x}\right].$$

Since an inhomogeneous integral equation in the case of its characteristic values is solvable if and only if the perturbation function $f(x)$ is orthogonal to all the characteristic solutions of the transposed integral equation, the conditions of solvability are, when one notices, that the integral equation coincides with its transpose:

$$\int_0^1 f(x)\varphi_1(x) \, dx = 0, \qquad \int_0^1 f(x)\varphi_2(x) \, dx = 0.$$

One sees easily that in the case $f(x) = x$ this condition will not be fulfilled.

$$\delta_0 = 1, \quad K_0(x, y) = \sin(x + y), \quad \delta_1 = \int_0^\pi \sin 2x\, dx = 0,$$

$$K_1(x, y) = 0 - \int_0^\pi \sin(x + t) \sin(t + y)\, dt = -\tfrac{1}{2}\pi \cos(x - y),$$

$$\delta_2 = -\tfrac{1}{2} \int_0^\pi \tfrac{1}{2}\pi \cos 0\, dx = -\tfrac{1}{4}\pi^2,$$

$$K_2(x, y) = -\tfrac{1}{4}\pi^2 \sin(x + y) + \int_0 \sin(x + t)\, \tfrac{1}{2}\pi \cos(t - y)\, dt = 0,$$

$$\Gamma(x, y, \lambda) = \frac{\sin(x + y) + \frac{1}{2}\pi \cos(x - y)\lambda}{1 - \frac{1}{4}\pi^2\lambda^2}.$$

$$\varphi(x) = \frac{\lambda}{1 - \frac{1}{4}\pi^2\lambda^2} \int_0^\pi [\sin(x + y) + \tfrac{1}{2}\pi \cos(x - y)\lambda]\cdot 1 dy + 1$$

or

$$\varphi(x) = \frac{2\lambda \cos x + \lambda^2 \pi \sin x}{1 - \frac{1}{4}\pi^2\lambda^2} + 1 \qquad \text{(cf. p. 699).}$$

8. The Nyström method of approximation for the solution of Fredholm integral equations of the second kind

The Fredholm method of approximation is of small practical significance since, in most cases, for small values of n the approximation is unsatisfactory. The *Nyström method of approximation*, founded upon the quadrature formula of Gauss, is more useful.

Following Gauss one can, in order to approximately determine the value of a given integral $\int_b^a f(x)\, dx$, proceed in the following manner:

One first considers the Legendre polynomial (cf. pp. 551, 552 and pp. 90, 91)

$$P_n(x) = \frac{1}{2^n n!} \cdot \frac{d^n[(x^2 - 1)^n]}{dx^n}.$$

The Legendre polynomials are thus the n-th derivatives of the $2n$-th degree polynomials $(x^2 - 1)^n = [(x - 1)(x + 1)]^n$ for which the points $x = -1$ and $x = +1$ are both zeros of order n [1]. The first derivative

[1] A zero x_0 of a function is said to be of order n if it is also a zero of the first $n - 1$ derivatives but not of the n-th. Accordingly, a simple zero is a zero for which the value of the first derivative is different from zero.

$\frac{d}{dx}[(x-1)(x+1)]^n$ has zeros of order $n-1$ at both the points $x=-1$ and $x=+1$; further there is a simple zero lying between the points -1 and $+1$ (by Rolle's theorem, cf. p. 378). One further derives that the second derivative $\frac{d^2}{dx^2}[(x-1)(x+1)]^n$ has zeros of order $n-2$ at both the points $x=-1$ and $x=+1$, and also that there are two simple zeros between -1 and $+1$. By a further continuation of this process one reaches the conclusion that the Legendre polynomial $P_n(x)$ has altogether n simple zeros, t_ν $(\nu = 1, 2, \ldots, n)$ lying between -1 and $+1$. The points -1 and $+1$ are not zeros of $P_n(x)$. Naturally the t_ν depend on n.

The interval $a < x < b$ will now be transformed by the relationship

$$t = \frac{1}{a-b}[a+b-2x]$$

into the interval $-1 < t < +1$. Conversely each t-value of this interval thus corresponds to the value

$$x = \tfrac{1}{2}(a+b) + \tfrac{1}{2}(b-a)t$$

of the interval $a < x < b$. In particular, corresponding to the n zeros t_ν of the n-th Legendre polynomial are the n values

$$x_\nu = \tfrac{1}{2}(a+b) + \tfrac{1}{2}(b-a)t_\nu.$$

Further

$$dx = \frac{b-a}{2}dt.$$

Next one can write the given integral in the form

$$(*) \qquad \int_a^b f(x)\,dx = \tfrac{1}{2}(b-a)\int_{-1}^{+1} f[\tfrac{1}{2}(a+b) + \tfrac{1}{2}(b-a)t]\,dt.$$

The integrand of the right-hand integral takes on the value $f(x_\nu)$ for $t = t_\nu$.

Now one constructs the n auxiliary functions

$$F_\nu(t) = \frac{(t-t_1)(t-t_2)\ldots(t-t_{\nu-1})(t-t_{\nu+1})\ldots(t-t_n)}{(t_\nu-t_1)(t_\nu-t_2)\ldots(t_\nu-t_{\nu-1})(t_\nu-t_{\nu+1})\ldots(t_\nu-t_n)}.$$

One easily sees that $F_\nu(t_\nu)=1$ and $F_\nu(t_\mu)=0$ for $\mu \neq \nu$ $(1 < \nu, \mu < n)$.

Thus the function

$$\sum_1^n F_\nu(t)f(x_\nu)$$

takes on the function value $f(x_\nu)$ at the point t_ν. Thus one may approximately calculate the right-hand integral of (*) when one replaces the integrand by the polynomial $\sum F_\nu(t) f(x_\nu)$. One now writes for brevity

$$A_\nu = \tfrac{1}{2}\int_{-1}^{+1} F_\nu(t)\,dt$$

$$= \tfrac{1}{2}\int_{-1}^{+1} \frac{(t-t_1)(t-t_2)\ldots(t-t_{\nu-1})(t-t_{\nu+1})\ldots(t-t_n)}{(t_\nu-t_1)(t_\nu-t_2)\ldots(t_\nu-t_{\nu-1})(t_\nu-t_{\nu+1})\ldots(t_\nu-t_n)}\,dt$$

(the A_ν can thus be considered as mean values of the functions $F_\nu(t)$ in the interval $-1 < t < +1$), thus one obtains as an approximate expression for the above integral (*):

$$\int_a^b f(x)\,dx \approx [b-a]\,[A_1 f(x_1) + A_2 f(x_2) + \ldots + A_n f(x_n)].$$

It is assumed, as should be noticed, that the function $f(x)$ may be approximated directly at the zeros of the Legendre polynomial. One can show that this approximation is especially good.

For $n = 1, 2, \ldots, 6$ we give the values of t and A in the following table:

n	t	A	n	t	A
1	$t_1=0.5$	$A_1=1$	5	$t_1=0.0469$	$A_1=0.1185$
				$t_2=0.2308$	$A_2=0.2393$
2	$t_1=0.2113$	$A_1=0.5$		$t_3=0.5$	$A_3=0.2844$
	$t_2=0.7887$	$A_2=0.5$		$t_4=0.7692$	$A_4=0.2393$
				$t_5=0.9531$	$A_5=0.1185$
3	$t_1=0.1127$	$A_1=0.2778$			
	$t_2=0.5$	$A_2=0.4444$	6	$t_1=0.0338$	$A_1=0.0857$
	$t_3=0.8873$	$A_3=0.2778$		$t_2=0.1694$	$A_2=0.1804$
				$t_3=0.3807$	$A_3=0.2340$
4	$t_1=0.0694$	$A_1=0.1739$		$t_4=0.6193$	$A_4=0.2340$
	$t_2=0.3300$	$A_2=0.3261$		$t_5=0.8306$	$A_5=0.1804$
	$t_3=0.6700$	$A_3=0.3261$		$t_6=0.9662$	$A_6=0.0857$
	$t_4=0.9306$	$A_4=0.1739$			

The integral equation

$$\lambda \int_a^b K(x, y)\,\varphi(y)\,dy + f(x) = \varphi(x)$$

will then be approximately given by the following system of linear equations:

$$\begin{aligned}
\varphi_1[1-\lambda A_1k_{11}]-\varphi_2\lambda A_2k_{12}-\ldots-\quad &\varphi_n\lambda A_nk_{1n}=f_1,\\
-\varphi_1\lambda A_1k_{21}\quad +\varphi_2[1-\lambda A_2k_{22}]-\ldots-&\varphi_n\lambda A_nk_{2n}=f_2,\\
\cdots\cdots\cdots\cdots\cdots\cdots\cdots&\cdots\\
-\varphi_1\lambda A_1k_{n1}\quad -\varphi_2\lambda A_2k_{n2}-\ldots+&\varphi_n[1-\lambda A_nk_{nn}]=f_n.
\end{aligned}$$

The φ_i, f_i and k_{ij} are the values of the corresponding functions at the point x_i or x_i, x_j respectively.

Example. $\int_0^1 [xy+\sqrt{xy}]\varphi(y)\,dy + x = \varphi(x)$. Here $a=0$, $b=1$ and $x_\nu = t_\nu$.

If $n=2$:

$$t_1 = 0.2113, \quad A_1 = 0.5,$$
$$x_1 = y_1 = 0.2113,$$
$$t_2 = 0.7887, \quad A_2 = 0.5,$$
$$x_2 = y_2 = 0.7887,$$
$$k_{11} = 0.2559, \quad k_{12} = k_{21} = 0.5750, \quad k_{22} = 1.4107,$$
$$\varphi_1 \cdot 0.8720 - \varphi_2 \cdot 0.2875 = 0.2113,$$
$$-\varphi_1 \cdot 0.2875 + \varphi_2 \cdot 0.2946 = 0.7887.$$
$$\varphi_1 = 1.659, \quad \varphi_2 = 4.296.$$

We compare this solution with the exact solution (cf. p. 692).

$$\varphi_1 = \tfrac{75}{26} \cdot 0.2113 + \tfrac{20}{13}\sqrt{0.2113} = 1.670,$$
$$\varphi_2 = \tfrac{75}{26} \cdot 0.7887 + \tfrac{20}{13}\sqrt{0.7887} = 4.325.$$

The above approximation, in view of the low value of $n=2$ is thoroughly satisfactory.

9. The Fredholm alternative theorem for Fredholm integral equations of the second kind with symmetric kernel

Suppose that a Fredholm integral equation of the second kind

$$(*1) \qquad \lambda \int_a^b K(x, y)\,\varphi(y)\,dy + f(x) = \varphi(x)$$

with a continuous kernel $K(x, y)$ is given. The corresponding homogeneous integral equation is then

$$\lambda \int_a^b K(x, y)\,\varphi(y)\,dy = \varphi(x). \tag{*2}$$

Every λ for which the integral equation (*2) has a continuous solution $\varphi(x)$ which is not identically zero is called a *characteristic value* of the homogeneous integral equation (*2) or of the kernel $K(x, y)$. (One uses this expression also in the degenerate case as in Section 5.) The corresponding solution is called the *characteristic solution.*

Let λ be a characteristic value of the integral equation (*2). We can then prove that there can be at most finitely many linearly independent characteristic functions for this characteristic value (for an explanation of the concept of a linearly independent system of functions one should see p. 345). It thus follows that there are r linearly independent functions $\varphi_1(x), \ldots, \varphi_r(x)$, such that every characteristic function for this characteristic value can be written in the form $c_1\varphi_1(x) + \ldots + c_r\varphi_r(x)$ where the c_i are constants. Naturally r can be different for different characteristic values.

We now consider the transposes of the integral equations (*1) and (*2)

$$\lambda \int_a^b K(y, x)\,\psi(y)\,dy + f(x) = \psi(x) \tag{**1}$$

and

$$\lambda \int_a^b K(y, x)\,\psi(y)\,dy = \psi(x). \tag{**2}$$

The solution procedure for a Fredholm integral equation of the second kind can be characterized by the following theorem [1]:

Fredholm's alternative theorem.

Either λ is not a characteristic value of the integral equation (*2), i.e., the equation (*2) yields no non-trivial solutions. In this case the inhomogeneous integral equations (*1) and (**1) are solvable for each specific continuous $f(x)$.

Or λ is a characteristic value of the integral equation (*2). In this case it is also a characteristic value of the transposed integral equation (**2) and the number r of linearly independent characteristic solutions for this characteristic value are the same for both of the equations (*2) and (**2). The inhomogeneous integral equation (*1) is under this hypothesis solvable precisely

---

[1] The reader should compare this with the section on degenerate kernels.

when the perturbation function $f(x)$ is orthogonal to all the characteristic functions of the transposed integral equation (**2) (cf. pp. 555 and 755), i.e., the equations

$$\int_a^b f(x)\,\psi_1(x)\,dx = 0, \qquad \dots, \qquad \int_a^b f(x)\,\psi_r(x)\,dx = 0$$

must hold, where $\psi_1(x), \dots, \psi_r(x)$ is a linearly independent system of characteristic function of the integral equation (**2) for the characteristic value λ; since every characteristic function can be written as a linear combination of the functions in this system it follows that $f(x)$ is orthogonal to every characteristic function.

In the sequel it will be assumed that the kernel is *symmetric*, i.e.,

$$K(x, y) = K(y, x).$$

For symmetric kernels the following theorems hold:

(a) Every real symmetric kernel has at least one characteristic value.

(b) The characteristic values of a real symmetric kernel are real.

(c) The characteristic values can have no finite points of accumulation.

(d) For two different characteristic values λ and λ^* the corresponding characteristic functions $\varphi(x)$ and $\varphi^*(x)$ are orthogonal; i.e., they satisfy the condition

$$\int_a^b \varphi(x)\,\varphi^*(x)\,dx = 0.$$

The Fredholm alternative theorem is simplified for symmetric kernels in as much as the integral equation and its transpose coincide for such kernels.

Example. For the integral equation

$$\lambda \int_1^2 \left[xy + \frac{1}{xy}\right] \varphi(y)\,dy = \varphi(x) \qquad \text{(cf. p. 694)}$$

the characteristic values, obtained from $\lambda^2 - 17\lambda + 6 = 0$ are $\lambda_1 = \frac{1}{2}(17 + \sqrt{265})$ and $\lambda_2 = \frac{1}{2}(17 - \sqrt{265})$, i.e. $\lambda_1 + \lambda_2 = 17$ and $\lambda_1\lambda_2 = 6$. The characteristic solutions are

$$\varphi_1(x) = K_1\left[x + \frac{\lambda_1}{(1 - \frac{1}{2}\lambda_1)\,x}\right] \quad \text{and} \quad \varphi_2(x) = K_2\left[x + \frac{\lambda_2}{(1 - \frac{1}{2}\lambda_2)\,x}\right].$$

We consider the integral

$$\int_1^2 \varphi_1(x)\,\varphi_2(x)\,dx$$

which is equal to

$$K_1 K_2 \int_1^2 \left[x + \frac{\lambda_1}{(1-\frac{1}{2}\lambda_1)\,x}\right]\left[x + \frac{\lambda_2}{(1-\frac{1}{2}\lambda_2)\,x}\right] dx$$

$$= K_1 K_2 \int_1^2 \left[x^2 + \frac{\lambda_1+\lambda_2-\lambda_1\lambda_2}{1-\frac{1}{2}(\lambda_1+\lambda_2)+\frac{1}{4}\lambda_1\lambda_2} + \frac{\lambda_1\lambda_2}{[1-\frac{1}{2}(\lambda_1+\lambda_2)+\frac{1}{4}\lambda_1\lambda_2]x^2}\right] dx$$

$$= K_1 K_2 \int_1^2 \left[x^2 - \frac{11}{6} - \frac{1}{x^2}\right] dx = 0;$$

i.e., $\varphi_1(x)$ and $\varphi_2(x)$ are orthogonal to one another, as follows from theorem (d) for symmetric kernels.

10. The operator method in the theory of integral equations

In the sequel it will be shown how we can treat a Fredholm integral equation of the second kind

$$\lambda \int_a^b K(x, y)\,\varphi(y)\,dy + f(x) = \varphi(x) \tag{1}$$

with a continuous and symmetric kernel from the standpoint of modern analysis.

(a) *The concept of a function space.* The functions $f(x)$ and $K(x, y)$ in (1) are continuous in the interval $a < x < b$ and the square $a < x < b$, $a < y < b$, respectively. Consequently, the only functions $\varphi(x)$ which are considered as possible solutions are those which are continuous in the interval $a < x < b$ [1]. Accordingly, it is convenient in the study of the integral equation (1) to examine the totality of functions $g(x)$ which are continuous

[1] If the function $\varphi(x)$ were only integrable (cf. the footnote on p. 454) then from the continuity of the function $K(x, y)$ considered as a function of x it would follow that the integral $\int_a^b K(x, y)\varphi(y)dy$ is also a continuous function of x. On the left side of (1) would therefore stand the sum of two continuous functions of x, and therefore the function $\varphi(x)$ would have to be continuous also.

in the interval $a < x < b$. This totality will be denoted by R for conciseness. Since R is a totality of functions R will be called a *function space*. The fact that a function $g(x)$ belongs to the collection R is symbolically denoted by the expressions

$$g(x) \varepsilon R \quad \text{or} \quad g \varepsilon R.$$

$g \varepsilon R$ thus signifies: The function $g(x)$ is defined and continuous in the entire interval $a < x < b$. If c is a real constant and $g(x)$ is continuous in the interval $a < x < b$, then $cg(x)$ is also continuous in $a < x < b$. Using the notation introduced above this can be rewritten as: From $g \varepsilon R$ it follows that $cg \varepsilon R$. Since the sum of two continuous functions is also continuous, one further obtains:

From $g_1 \varepsilon R$ and $g_2 \varepsilon R$ it follows that $(g_1 + g_2) \varepsilon R$.

Further if $g(x)$ is a function in R, i.e., $g \varepsilon R$, then

$$\int_a^b K(x, y)\, g(y)\, dy \tag{2}$$

is also a continuous function of x in the entire interval $a < x < b$. The integral thus assigns to each function in R a new function in R. This correspondence will be symbolically denoted by

$$\int_a^b K(x, y)\, g(y)\, dy \equiv \mathfrak{K}[g] \varepsilon R. \tag{3}$$

$\mathfrak{K}$ is called an *integral operator*; to each function of R it assigns a new function of R.

From the defining equation (3) we further deduce that the integral operator $\mathfrak{K}$ satisfies the following two characteristic conditions:

$$\mathfrak{K}[cg] = c\mathfrak{K}[g] \quad \text{and} \quad \mathfrak{K}[g_1 + g_2] = \mathfrak{K}[g_1] + \mathfrak{K}[g_2]. \tag{4}$$

Using the symbol $\mathfrak{K}$ one can write the integral equation (1) in the form

$$\lambda \mathfrak{K}[\varphi] + f = \varphi. \tag{5}$$

(b) Definition of a scalar product in the function space R. There are certain analogies between the totality of all vectors in three dimensional Euclidean space, the so-called *vector space* (cf. p. 613) and the totality of all continuous functions in the interval $a < x < b$ (thus the function space R). In particular, analogously to the scalar product of two vectors (cf. pp. 616 and 617) we can also define a *scalar product* for two functions in R

by means of the following equation

$$(6) \qquad (g_1, g_2) = \int_a^b g_1(x)\, g_2(x)\, dx.$$

This scalar product has the following characteristic properties:

$$(7) \qquad \begin{aligned} (g_1, g_2) &= (g_2, g_1), \\ (g_1 + g_2, g_3) &= (g_1, g_3) + (g_2, g_3), \\ (cg_1, g_2) &= c(g_1, g_2). \end{aligned}$$

Further $(g_1, g_1) = \int_a^b [g_1(x)]^2\, dx$ is always non-negative and vanishes only for the function $g_1(x) \equiv 0$.

In analogy to the concept of length of a vector, we take the positive square root of (g_1, g_1) as the norm of the function g_1. Symbolically:

$$(8) \qquad \|g_1\| = \sqrt{(g_1, g_1)}.$$

We can then estimate the scalar product of two functions $g_1, g_2 \in R$ by

$$(9) \qquad |(g_1, g_2)| \leq \|g_1\| \cdot \|g_2\|$$

(Schwarz's inequality). The equality is possible only in the case when the function g_2 has the form $g_2 = cg_1$. In analogy to the fact that two vectors, the scalar product of which vanishes (cf. p. 617), are perpendicular (orthogonal) to one another, we define two functions $g_1, g_2 \in R$ to be orthogonal if their scalar product (g_1, g_2) vanishes (cf. p. 555).

Finally, we can by the substitution of (3) into (6) obtain for an integral operator $\mathfrak{K}$ defined by a symmetric kernel function $K(x, y)$ $(K(x, y) = K(y, x))$ the important statement that

$$(10) \qquad (\mathfrak{K}[g_1], g_2) = (g_1, \mathfrak{K}[g_2]).$$

From (10) we can further derive, for example, that the characteristic values of a real symmetric kernel are real (cf. Section 9).

(c) The Schmidt orthonormalization process. Let $g_1(x), \ldots, g_n(x)$ be n linearly independent functions of R (cf. Section 5, p. 689) i.e., it is assumed that no relation

$$\sum_1^n c_i g_i(x) \equiv 0$$

with constant coefficients c_i not all zero can hold. We can then form from the functions $g_i(x)$ n new functions $\tilde{g}_i(x)$ as follows:

1. *The* $\tilde{g}_i(x)$ *are orthonormal, i.e.,*

$$(\tilde{g}_i, \tilde{g}_j) = \delta_{ij} \text{ (}^1\text{)}.$$

2. *Each function* $g(x)$ *which is a linear combination of the* $g_i(x)$:

$$g(x) = c_1 g_1(x) + \ldots + c_n g_n(x)$$

is also a linear combination of the $\tilde{g}_i$:

$$g(x) = \tilde{c}_1 \tilde{g}_1(x) + \ldots + \tilde{c}_n \tilde{g}_n(x).$$

The existence of such functions $\tilde{g}_i$ will be shown by their construction through the Schmidt process.

For this purpose we set

$$\tilde{g}_1(x) = \frac{1}{\|g_1\|^2} g_1(x).$$

This formula is possible since $\|g_1\| \neq 0$. Otherwise we must have $g_1(x) \equiv 0$ and the n functions $g_1, \ldots, g_n$ would not be linearly independent since $1 \cdot g_1(x) + 0 \cdot g_2(x) + \ldots + 0 \cdot g_n(x)$ would be an identically vanishing linear combination in which all the coefficients are not zero.

Further we put

$$\tilde{g}_2(x) = \frac{1}{\|g_2 - (g_2, g_1) g_1\|^2} \big(g_2(x) - (g_2, g_1) g_1(x)\big),$$

$$\tilde{g}_3(x) = \frac{1}{\|g_3 - (g_3, g_1) g_1 - (g_3, g_2) g_2\|^2} \big(g_3(x) - (g_3, g_1) g_1(x) - \\ - (g_3, g_2) g_2(x)\big),$$

. . . .

$$\tilde{g}_n(x) = \frac{1}{\|g_n - \sum_1^{n-1} (g_n, g_i) g_i(x)\|^2} \Big(g_n(x) - \sum_1^{n-1} (g_n, g_i) g_i(x)\Big).$$

None of the functions $g_k(x) - \sum_1^{k-1} (g_k, g_i) g_i(x)$ can vanish identically for indeed if they did the functions $g_1, \ldots, g_n$ would be linearly dependent (in this relation the coefficient of g_k is 1). Thus in the defining equations for the $\tilde{g}_i$ none of the denominators can be equal to zero.

---

(1) δ_{ij} is the Kronecker symbol. That is, it equals 1 for $i = j$ and zero for $i \neq j$.

As we can verify by direct calculation the $\tilde{g}_i$ are orthonormal, i.e., they satisfy the condition of 1.

Conversely from the defining equations of the $\tilde{g}_i$ we can obtain all the g_i as linear combinations of the $\tilde{g}_i$. Thus condition 2 is also fulfilled.

(d) The Hilbert method for the construction of characteristic values and characteristic functions. For this construction the following theorem of Arzela is essential:

Let $g_n(x)$ be a sequence of functions of R whose norms are 1: $\|g_n\| = 1$. Then there exists at least one subsequence $g_{n_i}(x)$ of this sequence so that the functions

$$f_{n_i}(x) = \mathfrak{K}[g_{n_i}]$$

converge uniformly to a function $f(x) \varepsilon R$.

By uniform convergence we here mean the following:

The functions $f_{n_i}(x)$ converge uniformly to $f(x)$ (symbolically $f_{n_i} \rightrightarrows f$) if for previously given $\varepsilon > 0$ there always exists an integer $N(\varepsilon)$ such that for all $n_i > N(\varepsilon)$ the inequality

$$|f_{n_i}(x) - f(x)| < \varepsilon$$

is fulfilled for all x in the interval $a \leq x \leq b$.

The Hilbert construction process runs as follows:

(1) We consider the set of all

$$(\mathfrak{K}[g], g) \quad \text{for} \quad g \varepsilon R, \ \|g\| = 1. \tag{11}$$

M_1 and m_1 are respectively the upper and lower limits of this set. Thus we have

$$m_1 \leq (\mathfrak{K}[g], g) \leq M_1$$

and further for every $\varepsilon > 0$ there is a $g_1 \varepsilon R$ ($\|g_1\| = 1$) such that $(\mathfrak{K}[g_1], g_1) < m_1 + \varepsilon$. And also there is a $g_2 \varepsilon R$, $\|g_2\| = 1$ such that $M_1 - \varepsilon < (\mathfrak{K}[g_2], g_2)$.

Now set $k_1 = M_1$ or $k_1 = m_1$ accordingly as $|M_1| \geq |m_1|$ or $|M_1| < m_1$.

If $k_1 \neq 0$ then at least one of the numbers m_1 and M_1 is not zero. This is true for every continuous symmetric kernel which is not identically zero. Since k_1 is equal to one of the two limits of the set of all $(\mathfrak{K}[g], g)$ for $g \varepsilon R$, $\|g\| = 1$, it follows that there is a sequence of $g_n \varepsilon R$, $\|g_n\| = 1$, such that

$$\lim_{n \to \infty} (\mathfrak{K}[g_n], g_n) = k_1.$$

From the theorem of Arzela we obtain a subsequence g_{n_i} such that

$$\mathfrak{K}[g_{n_i}] \rightrightarrows \tilde{\varphi}_1(x) \in R.$$

It then follows that $\lambda_1 = 1/k_1$ is a characteristic value and that $\varphi_1(x) = \tilde{\varphi}_1(x)/\|\tilde{\varphi}_1\|$ is a corresponding normalized characteristic function.

(2) To construct the next characteristic function we consider the set of all

$$(12) \qquad (\mathfrak{K}[g], g) \quad \text{for} \quad g \in R, \ \|g\| = 1, \ (g, \varphi_1) = 0,$$

where φ_1 is the characteristic function constructed in (1). Let the upper and lower limits of this set be M_2 and m_2:

$$m_2 < (\mathfrak{K}[g], g) < M_2.$$

Similarly to (1) we set $k_2 = M_2$ or $k_2 = m_2$, respectively, according to whether $|M_2| > |m_2|$ or $|m_2| > |M_2|$.

There are two possible cases:

Case (a). It is $k_2 = 0$. Then λ_1 is the sole characteristic value of the integral equation and every characteristic function has the form $c\varphi_1(x)$. The Hilbert process has in this case already come to its conclusion.

Case (b). It is $k_2 \neq 0$. In this case we construct a second sequence $g_n \in R$, $\|g_n\| = 1$, $(g_n, \varphi_1) = 0$ such that

$$\lim_{n\to\infty} (\mathfrak{K}[g_n], g_n) = k_2.$$

Again from the theorem of Arzela we obtain a subsequence g_{n_i} for which

$$\mathfrak{K}[g_{n_i}] \rightrightarrows \varphi_2(x) \in R.$$

It is then true that $\lambda_2 = 1/k_2$ is a characteristic value and that $\varphi_2(x) = \tilde{\varphi}_2(x)/\|\tilde{\varphi}_2\|$ is a corresponding normalized characteristic function. Thus $(\varphi_2, \varphi_1) = 0$. From formula (10) it follows that $(\mathfrak{K}[g_{n_i}], \varphi_1) = (g_{n_i}, \mathfrak{K}[\varphi_1])$. Now, however, since φ_1 is a characteristic function for the characteristic value λ_1 it follows that $\mathfrak{K}[\varphi_1] = \varphi_1/\lambda_1$ and $(\mathfrak{K}[g_{n_i}], \varphi_1) = (g_{n_i}, \varphi_1)/\lambda_1 = 0$. Since the $\mathfrak{K}[g_{n_i}]$ also converge uniformly to $\tilde{\varphi}_2$ it follows by passing to the limit that $(\tilde{\varphi}_2, \varphi_1) = 0$ and thus that $(\varphi_2, \varphi_1) = 0$.

(3) This process will be continued in the following manner: When we have already constructed the orthogonal characteristic functions

$$\varphi_1(x), \ \ldots, \ \varphi_\mu(x)$$

we consider, just as in (11) and (12), in this case the set of all $(\mathfrak{K}[g], g)$ for which $g \in R$, $\|g\| = 1$, and g is orthogonal to all the previously constructed characteristic functions. That is, we consider the set of all

$$(13) \quad (\mathfrak{K}[g], g) \quad \text{for} \quad g \in R,\ \|g\| = 1,\ (g, \varphi_1) = (g, \varphi_2) = \dots \\ \dots = (g, \varphi_\mu) = 0.$$

The upper and lower limits of the set (13) will, similarly, be denoted by $M_{\mu+1}$ and $m_{\mu+1}$ respectively:

$$m_{\mu+1} \leq (\mathfrak{K}[g], g) \leq M_{\mu+1}.$$

$k_{\mu+1}$ is similarly equal to $M_{\mu+1}$ or $m_{\mu+1}$ respectively according to whether $|M_{\mu+1}| \geq |m_{\mu+1}|$ or $|m_{\mu+1}| > |M_{\mu+1}|$.

Either $k_{\mu+1} = 0$, in which case there are no further characteristic values and all the characteristic functions can be written as a linear combination of the previously constructed functions $\varphi_1, \dots, \varphi_\mu$ or $k_{\mu+1} \neq 0$, in which case there is a sequence g_n $(g_n \in R,\ \|g_n\| = 1,$ $(g_n, \varphi_1) = \dots = (g_n, \varphi_\mu) = 0)$ such that

$$\lim_{n \to \infty} (\mathfrak{K}[g_n], g_n) = k_{\mu+1} \neq 0.$$

Once more from the theorem of Arzela it follows that there is a subsequence g_{n_i} such that $\mathfrak{K}[g_{n_i}]$ converges uniformly to a function $\tilde{\varphi}_{\mu+1} \in R$.

Again we find that $\lambda_{\mu+1} = 1/k_{\mu+1}$ is a characteristic value and that $\varphi_{\mu+1}(x) = \tilde{\varphi}_{\mu+1}(x)/\|\tilde{\varphi}_{\mu+1}\|$ is a corresponding normalized characteristic function.

The function $\varphi_{\mu+1}$ is orthogonal to all the previously constructed functions $\varphi_1, \dots, \varphi_\mu$. This may be seen from the following argument: From formula (10) we obtain

$$(\mathfrak{K}[g_{n_i}], \varphi_j) = (g_{n_i}, \mathfrak{K}[\varphi_j]) = \left(g_{n_i}, \frac{1}{\lambda_j} \varphi_j\right) = \frac{1}{\lambda_j} (g_{n_i}, \varphi_j) = 0$$

for all j between 1 and μ. Since the $\mathfrak{K}[g_{n_i}]$ converge uniformly to $\tilde{\varphi}_{\mu+1}$ we find upon passage to the limit that

$$(\tilde{\varphi}_{\mu+1}, \varphi_j) = 0 \quad \text{thus} \quad (\varphi_{\mu+1}, \varphi_j) = 0 \quad \text{for all } 1 \leq j \leq \mu.$$

(e) The significance of the Hilbert method. From equation (13) we can show that

$$m_\mu \leq m_{\mu+1} \quad \text{and} \quad M_{\mu+1} \leq M_\mu$$

Thus $|k_{\mu+1}|<|k_\mu|$, i.e. for the characteristic values

(14) $$|\lambda_{\mu+1}|>|\lambda_\mu|.$$

It may happen that $k_{\mu+1}=k_\mu$ or even that

$$k_\mu=k_{\mu+1}=\ldots=k_{\mu+r}.$$

But after finitely many steps this situation must come to an end, so that

$$k_{\mu+r+1}\neq k_{\mu+r}.$$

Proof. To each characteristic value may correspond only finitely many linearly independent characteristic functions (cf. Section 9, p. 709).

We construct from these functions an orthonormal system according to the Schmidt orthonormalization process. We then know that all the characteristic functions appear as linear combinations of these finitely many orthonormal characteristic functions (cf. Section 10(c)). Since in the above process we obtain another orthonormal characteristic function, we must obtain after finitely many steps a $k_{\mu+r+1}\neq k_{\mu+r}$.

Since the characteristic values of an integral equation with a continuous symmetric kernel can have no finite accumulation point (cf. Section 9, p. 709) we find by this process all the characteristic values $\lambda_1, \lambda_2, \ldots$, and indeed in an order such that $|\lambda_1|<|\lambda_2|<\ldots$ (cf. formula (14)). The corresponding characteristic functions are normalized $\|\varphi_i\|=1$ and orthogonal to one another: $(\varphi_i, \varphi_j)=0$ for $i\neq j$.

(f) Examples. 1. Let the integral equation $\lambda\int_0^1 xy\varphi(y)\,dy=\varphi(x)$ be given. For the integral operator $\mathfrak{K}$ we obtain from this (cf. formula (3))

$$\mathfrak{K}[g]=\int_0^1 xyg(y)\,dy=x\int_0^1 yg(y)\,dy=(x, g)\,x\ (^1),$$

where it further holds that

(15) $$(\mathfrak{K}[g], g)=(x, g)^2.$$

We apply the Schwartz inequality (9) to the scalar product (x, g) and get $|(x, g)|<\|x\|\cdot\|g\|$.

---

[1] Where (x, g) denotes the scalar product of the function x with the function $g(x)$; $(x, g)=\int_0^1 xg(x)dx=\int_0^1 yg(y)dy$.

Considering the fact that

$$\|x\|^2 = \int_0^1 x^2\, dx = \tfrac{1}{3}$$

we obtain, when one restricts himself to those functions $g \in R$ for which $\|g\| = 1$,

$$(x, g)^2 < \tfrac{1}{3}.$$

Since equality can hold in the Schwartz inequality only when $g = cx$ and since the function $g_1 = \sqrt{3}x$ has norm $\|g_1\| = 1$, one sees that $k_1 = \frac{1}{3}$. Moreover it is clear that

$$g_1, g_1, g_1, \ldots$$

is a sequence of functions for which $(\mathfrak{K}[g_1], g_1) = k_1$. Since this sequence is composed of only one function every subsequence is identical with the whole sequence; it follows as a consequence of Arzela's theorem that

$$\tilde{\varphi}_1(x) = \mathfrak{K}[g_1] = x \int_0^1 x\sqrt{3}\, x\, dx = \frac{1}{\sqrt{3}} x$$

is the characteristic function which corresponds to the characteristic value $\lambda_1 = 1/k_1 = 3$. In order to find the further linearly independent characteristic functions one constructs the set (12). Since $\tilde{\varphi}_1(x) = x/\sqrt{3}$ and $\varphi_1 = \tilde{\varphi}_1/\|\tilde{\varphi}_1\|$ the condition $(g, \varphi_1) = 0$ takes on the form $(x, g) = 0$. Thus all the $(\mathfrak{K}[g], g)$ for $\|g\| = 1$, $(\varphi_1, g) = 0$ are zero. Hence in this case $k_2 = 0$, from which we see that $\lambda = 3$ is the sole characteristic value and that all the characteristic functions have the form $\varphi(x) = cx$.

2. We consider the integral equation

$$\lambda \int_{-\pi}^{+\pi} (\sin x \sin y - \tfrac{1}{2} \cos x \cos y)\, \varphi(y)\, dy = \varphi(x).$$

The integral operator (3) then has the form

$$\mathfrak{K}[g] = (g, \sin x) \sin x - \tfrac{1}{2} (g, \cos x) \cos x \text{ (}^1\text{)}$$

---

(1) Where $(g, \sin x)$ again is the scalar product

$$(g, \sin x) = \int_{-\pi}^{+\pi} g(x) \sin x\, dx = \int_{-\pi}^{+\pi} g(y) \sin y\, dy$$

and analogously

$$(g, \cos x) = \int_{-\pi}^{+\pi} g(x) \cos x\, dx = \int_{-\pi}^{+\pi} g(y) \cos y\, dy.$$

from which follows:

$$(\mathfrak{K}[g], g) = (g, \sin x)^2 - \tfrac{1}{2}(g, \cos x)^2. \tag{16}$$

For $g \in R$ with $\|g\| = 1$ it follows from the Schwartz inequality (9) that

$$(g, \sin x)^2 < \|\sin x\|^2 = \pi, \qquad \tfrac{1}{2}(g, \cos x)^2 < \tfrac{1}{2}\|\cos x\|^2 = \tfrac{1}{2}\pi,$$

where the equality sign holds only for $g_1 = \dfrac{1}{\sqrt{\pi}} \sin x$ or $g_2 = \dfrac{1}{\sqrt{\pi}} \cos x$ ($\|g_1\| = \|g_2\| = 1$), respectively. Since $(g_1, \cos x) = 0$, it follows that $M_1 = \pi$. Similarly it follows from $(g_2, \sin x) = 0$ that $m_1 = \tfrac{1}{2}\pi$. Clearly $k_1 = \pi$.

The sequence

$$g_1, g_1, g_1, \cdots$$

has, moreover, the property that $(\mathfrak{K}[g_1], g_1) = k_1 = \pi$. From the theorem of Arzela it is clear that

$$\tilde{\varphi}_1(x) = \mathfrak{K}[g_1] = \sqrt{\pi} \sin x$$

is a characteristic function for the characteristic value $\lambda_1 = 1/\pi$.

In order to obtain the further characteristic functions we must consider (16) under the conditions that $\|g\| = 1$ and $(g, \tilde{\varphi}_1) = \sqrt{\pi}(g, \sin x) = 0$ [1]. In this case (16) becames

$$(\mathfrak{K}[g], g) = -\tfrac{1}{2}(g, \cos x)^2,$$

from which we find that

$$m_2 = -\tfrac{1}{2}\pi, \qquad M_2 = 0, \qquad \text{thus} \qquad k_2 = -\tfrac{1}{2}\pi.$$

Since for the above function $g_2 = \dfrac{1}{\sqrt{\pi}} \cos x$

$$(\mathfrak{K}[g_2], g_2) = k_2$$

it follows that the function $\tilde{\varphi}_2(x) = \mathfrak{K}[g_2] = -\tfrac{1}{2}\sqrt{\pi} \cos x$ is a characteristic function for the characteristic value $\lambda_2 = -2/\pi$.

In order to continue this process we must consider (16) under the assumptions that $\|g\| = 1$, $(g, \tilde{\varphi}_1) = 0$, and $(g, \tilde{\varphi}_2) = 0$. By substituting the particular functions for φ_1 and φ_2 we see that $(g, \sin x) = 0$ and $(g, \cos x) = 0$. Thus for all the functions g which come under

---

[1] Since $\varphi_1 = \tilde{\varphi}_1/\|\tilde{\varphi}\|$ the conditions $(g, \tilde{\varphi}_1) = 0$ and $(g, \varphi_1) = 0$ are equivalent.

consideration we have

$$(\mathfrak{K}[g], g) = 0.$$

In other words $m_3 = M_3 = k_3 = 0$. Thus there are no characteristic functions other than those previously constructed.

11. The Schmidt series

In the handling of a problem with boundary conditions the *development of the characteristic function* (cf. p. 554) plays an important role. By a similar method we also make use of the development of normalized characteristic functions for the solution of Fredholm integral equations of the second kind with symmetric kernels.

The Hilbert method for construction of the characteristic values and characteristic functions (cf. Section 10(d) and (e)) yields all the characteristic values $\lambda_1, \lambda_2, \ldots$ of a symmetric integral equation as well as corresponding characteristic functions $\varphi_1, \varphi_2, \ldots$ Here we have

$$|\lambda_1| < |\lambda_2| < \cdots$$

The functions φ_i are normalized, $\|\varphi_i\| = 1$, and orthogonal to one another, $(\varphi_i, \varphi_j) = 0$ for $i \neq j$. Such a sequence of functions in which all functions are normalized and pairwise orthogonal is called an *orthonormal system*. Every characteristic function of the integral equation appears as a linear combination of the known characteristic functions φ_i.

In the following procedure therefore the expansions in terms of the characteristic functions φ_i will be presented.

We say that an orthogonal system is complete if for every continuous function $f(x)$ one can give such a system $c_1, c_2, \ldots$ of real numbers that for the functions defined by

$$\chi_N = f(x) - \sum_{i=1}^{N} c_i \varphi_i(x)$$

the norm $\|\chi_N\|$ tends to zero as N tends to infinity. We can show that it is necessary for the c_i to have the form

$$c_i = (f, \varphi_i).$$

The series $\sum_{i=1}^{\infty} c_i \varphi_i(x)$ is called the *Fourier series* of the function $f(x)$. The c_i are called the *Fourier coefficients*. A complete orthonormal system is not derivable from the characteristic functions of all integral equations with symmetric kernels (see above). Indeed we have the following theorem:

A system of characteristic functions of a symmetric kernel is complete if and only if the kernel is closed.

A kernel is called *closed* if there exists no continuous function $h(y)$ $(h(y) \not\equiv 0)$ for which

$$\int_a^b K(x, y)\, h(y)\, dy = 0$$

for all x in the interval (a, b).

Thus we now take as a further assumption the fact that the kernel is closed, so that we can develop every continuous function in a Fourier series of characteristic functions of the kernel. In general, this Fourier series will not always converge pointwise to the function $f(x)$. It only follows that the series converges "in mean"; i.e., the above functions χ_N converge in the norm to zero.

However the function $f(x)$ is produceable by integration from the kernel $K(x, y)$; i.e., there is a continuous function $g(x)$ such that

$$f(x) = \int_a^b K(x, y)\, g(y)\, dy$$

so that under the assumption of the continuity of the kernel we can show that the Fourier series of the function $f(x)$ even converges uniformly to the function $f(x)$.

We now consider once more the integral equation

$$\lambda \int_a^b K(x, y)\, \varphi(y)\, dy + f(x) = \varphi(x). \tag{1}$$

The function (in x)

$$\lambda \int_a^b K(x, y)\, \varphi(y)\, dy$$

is integrably produceable. Thus one can write the equation (1) in the form

$$\varphi(x) - f(x) = \sum_{\nu=1}^{\infty} c_\nu \varphi_\nu(x), \tag{1a}$$

where the series is uniformly convergent. Thus,

$$\varphi(x) = \sum_{\nu=1}^{\infty} c_\nu \varphi_\nu(x) + f(x). \tag{2}$$

By substituting from (2) into (1) we obtain

$$\sum_{\nu=1}^{\infty} c_\nu \varphi_\nu(x) + f(x) = \lambda \sum_{\nu=1}^{\infty} c_\nu \int_a^b K(x, y)\, \varphi_\nu(y)\, dy + \tag{3}$$

$$+ \lambda \int K(x, y) f(y)\, dy + f(x)$$

$$= \lambda \sum_{\nu=1}^{\infty} \frac{c_\nu \varphi_\nu(x)}{\lambda_\nu} + \lambda \sum_{\nu=1}^{\infty} \frac{f_\nu \varphi_\nu(x)}{\lambda_\nu} + f(x).$$

By equating the coefficients of the functions $\varphi_\nu(x)$ in (3) we get:

$$c_\nu = \frac{\lambda c_\nu}{\lambda_\nu} + \lambda \frac{f_\nu}{\lambda_\nu}, \tag{4}$$

i.e.,

$$c_\nu = \frac{\lambda f_\nu}{\lambda_\nu - \lambda}, \tag{5}$$

where

$$f_\nu = \int_a^b f(y)\, \varphi_\nu(y)\, dy.$$

The solution to the integral equation (1) thus becomes

$$\varphi(x) = \lambda \sum_{\nu=1}^{\infty} \frac{f_\nu \varphi_\nu(x)}{\lambda_\nu - \lambda} + f(x).$$

If in the equation (1) λ is equal to a characteristic value of the kernel; i.e., $\lambda = \lambda_K$ [1] then according to the Fredholm alternative theorem on symmetric kernels (cf. p. 710) (1) is solvable only for those values of λ_K for which the perturbation function $f(x)$ is orthogonal to the corresponding characteristic functions $\varphi_K(x)$. In this case the terms $\frac{f_K}{\lambda_K - \lambda} \varphi_K(x)$ disappear. It follows that the solution $\varphi(x)$ may be expanded by a sum of the (finitely many) summands $C\, \varphi_K(x)$.

Examples of the solution of an integral equation by means of the Schmidt series:

$$\int_{-1}^{+1} [xy + x^2 y^2]\, \varphi(y)\, dy + (x + 1)^2 = \varphi(x). \tag{a}$$

[1] There can be only finitely many K such that $\lambda_K = \lambda$. Thus for each characteristic value there can exist only finitely many linearly independent characteristic functions.

Here $\lambda = 1$. The characteristic values are $\lambda_1 = \frac{3}{2}$ and $\lambda_2 = \frac{5}{2}$. The corresponding normalized characteristic functions become $\varphi_1(x) = \frac{1}{2}\sqrt{6}\,x$ and $\varphi_2(x) = \frac{1}{2}\sqrt{10}\,x^2$ (cf. p. 710). We find

$$f_1 = \int_{-1}^{+1} (y^2 + 2y + 1)\tfrac{1}{2}\sqrt{6}\,y\,dy = \tfrac{2}{3}\sqrt{6}$$

and

$$f_2 = \int_{-1}^{+1} (y^2 + 2y + 1)\tfrac{1}{2}\sqrt{10}\,y^2\,dy = \tfrac{8}{15}\sqrt{10}.$$

Thus

$$\varphi(x) = \frac{\frac{2}{3}\sqrt{6}}{\frac{3}{2} - 1}\cdot\tfrac{1}{2}\sqrt{6}\,x + \frac{\frac{8}{15}\sqrt{10}}{\frac{5}{2} - 1}\cdot\tfrac{1}{2}\sqrt{10}\,x^2 + (x+1)^2$$

or

$$\varphi(x) = \tfrac{25}{9}x^2 + 6x + 1.$$

(b)
$$\tfrac{3}{2}\int_{-1}^{+1} [xy + x^2y^2]\varphi(y)\,dy + x^2 + 1 = \varphi(x).$$

For the characteristic value of the kernel $\lambda = \frac{3}{2}$; the perturbation function $x^2 + 1$ is orthogonal to the corresponding characteristic function $\varphi_1(x) = Cx$, further $f_1 = 0$ and

$$f_2 = \int_{-1}^{+1} (y^2 + 1)\tfrac{1}{2}\sqrt{10}\,y^2\,dy = \tfrac{8}{15}\sqrt{10}.$$

The solution thus is

$$\varphi(x) = \tfrac{3}{2}\cdot\frac{\frac{8}{15}\sqrt{10}}{\frac{5}{2} - \frac{3}{2}}\cdot\tfrac{1}{2}\sqrt{10}\,x^2 + x^2 + 1,$$

i.e.,

$$\varphi(x) = 5x^2 + 1.$$

However $\varphi(x) = 5x^2 + 1 + Cx$ is also a solution.

Further example:

$$\lambda\int_a^b k(x)\,k(y)\,\varphi(y)\,dy + f(x) = \varphi(x).$$

It follows that the characteristic value is

$$\lambda_1 = \frac{1}{\int_a^b [k(y)]^2\,dy}.$$

The corresponding normalized characteristic function is

$$\varphi_1(x) = \frac{k(x)}{\sqrt{\int_a^b [k(x)]^2 \, dx}},$$

$$f_1 = \frac{1}{\sqrt{\int_a^b [k(x)]^2 \, dx}} \int_a^b f(y) k(y) \, dy.$$

For

$$\lambda \neq \frac{1}{\int_a^b [k(y)]^2 \, dy}$$

the solution is

$$\varphi(x) = \frac{\lambda f_1}{\lambda_1 - \lambda} \varphi_1(x) + f(x);$$

thus

$$\varphi(x) = \frac{\lambda k(x) \int_a^b f(y) \, k(y) \, dy}{1 - \lambda \int_a^b [k(y)]^2 \, dy} + f(x).$$

If

$$\lambda = \frac{1}{\int_a^b [k(y)]^2 \, dy}$$

then $f(x)$ must satisfy the condition $\int_a^b f(x) k(x) dx = 0$. The solution is then

$$\varphi(x) = f(x) + C k(x).$$

V. FOURIER SERIES

1. General information

Fundamental concepts. In many problems (differential equations, vibration theory) it is sometimes necessary to replace, accurately or approximately, a given periodic function $f(x)$ with period T by a trigonometric sum

$$s_n(x) = \tfrac{1}{2} a_0 + a_1 \cos \omega x + a_2 \cos 2\omega x + \ldots + a_n \cos n\omega x +$$
$$+ b_1 \sin \omega x + b_2 \sin 2\omega x + \ldots + b_n \sin n\omega x,$$

where $\omega = \frac{2\pi}{T}$ (if $T = 2\pi$, then $\omega = 1$). The sum $s_n(x)$ is the best approximation of the function $f(x)$ (in the sense of p. 728), if we take for a_k and b_k the so-called *Fourier coefficients* defined by the following formulas due to Euler (*Euler formulas*):

$$a_k = \frac{2}{T} \int_0^T f(x) \cos k\omega x \, dx = \frac{2}{T} \int_{x_0}^{x_0+T} f(x) \cos k\omega x \, dx$$

$$= \frac{2}{T} \int_0^{T/2} [f(x) + f(-x)] \cos k\omega x \, dx, \quad k = 0, 1, 2, \ldots, n,$$

$$b_k = \frac{2}{T} \int_0^T f(x) \sin k\omega x \, dx = \frac{2}{T} \int_{x_0}^{x_0+T} f(x) \sin k\omega x \, dx$$

$$= \frac{2}{T} \int_0^{T/2} [f(x) - f(-x)] \sin k\omega x \, dx, \quad k = 1, 2, \ldots, n.$$

If, for a certain set of values of x, the sum $s_n(x)$ tends to a definite limit, when $n \to \infty$, then, for these values of x, we obtain a convergent *Fourier series* of the given function $f(x)$:

$$s(x) = \tfrac{1}{2}a_0 + a_1 \cos \omega x + a_2 \cos 2\omega x + \ldots + a_n \cos n\omega x + \ldots + \\ + b_1 \sin \omega x + b_2 \sin 2\omega x + \ldots + b_n \sin n\omega x + \ldots$$

A Fourier series can also be written in the form

$$s(x) = \tfrac{1}{2}a_0 + A_1 \sin(\omega x + \varphi_1) + A_2 \sin(2\omega x + \varphi_2) + \ldots + \\ + A_n \sin(n\omega x + \varphi_n) + \ldots,$$

where $A_k = \sqrt{a_k^2 + b_k^2}$ and $\tan \varphi_k = a_k/b_k$. In the complex form a Fourier series can be written as follows

$$s(x) = \sum_{n=-\infty}^{+\infty} c_n e^{in\omega x},$$

where

$$c_n = \frac{1}{T}\int_0^T f(x)e^{-in\omega x}\,dx = \begin{cases} \tfrac{1}{2}(a_n - ib_n) & \text{for } n > 0, \\ \tfrac{1}{2}(a_{-n} + ib_{-n}) & \text{for } n < 0. \end{cases}$$

Determining the Fourier series of a given function $f(x)$ is a problem of *harmonic analysis*.

Fundamental properties of Fourier series.

(1) In replacing the function $f(x)$ with an approximate trigonometric sum

$$s_n(x) = \tfrac{1}{2}\alpha_0 + \sum_{k=1}^{n} \alpha_k \cos k\omega x + \sum_{k=1}^{n} \beta_k \sin k\omega x$$

the mean quadratic error (see p. 755)

$$\delta^2 = \frac{1}{T}\int_0^T [f(x) - s_n(x)]^2\,dx$$

has a least value, if the coefficients α_k and β_k are the Fourier coefficients of $f(x)$.

(2) If the function $f(x)$ is bounded and piecewise continuous in the interval $0 < x < T$ (see p. 335), then its Fourier series *converges in the mean* to $f(x)$:

$$\int_0^T [f(x) - s_n(x)]^2\,dx \to 0 \quad \text{for} \quad n \to \infty.$$

This implies *Parseval relation*

$$\frac{2}{T}\int_0^T [f(x)]^2\,dx = \frac{a_0^2}{2} + \sum_{k=1}^{\infty} (a_k^2 + b_k^2).$$

(3) If the function $f(x)$ satisfies *Dirichlet conditions*, i.e.,

(a) the interval in which the function is defined can be decomposed into a finite number of subintervals such that in each of which the function $f(x)$ is continuous and monotonic;

(b) at each point of discontinuity of $f(x)$ there exists a right-hand limit $f(x+0)$ and a left-hand limit $f(x-0)$ (see pp. 335, 329), then the Fourier series of $f(x)$ is convergent and its sum is equal to $f(x)$ at the points of continuity, and is equal to $\frac{f(x-0)+f(x+0)}{2}$ at the points of discontinuity of $f(x)$.

(4) If the periodic function $f(x)$ is continuous and has continuous derivatives up to the order k, then $a_n n^{k+1} \to 0$ and $b_n n^{k+1} \to 0$ for $n \to \infty$.

Symmetry. If $f(x)$ is an *even* function, i.e., if $f(-x) = f(x)$ (*symmetry of the first kind*, Fig. 417), then

$$a_k = \frac{4}{T}\int_0^{T/2} f(x)\cos k\frac{2\pi x}{T}\,dx, \qquad b_k = 0,$$

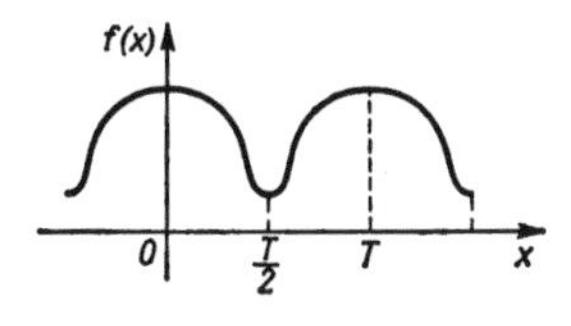

FIG. 417

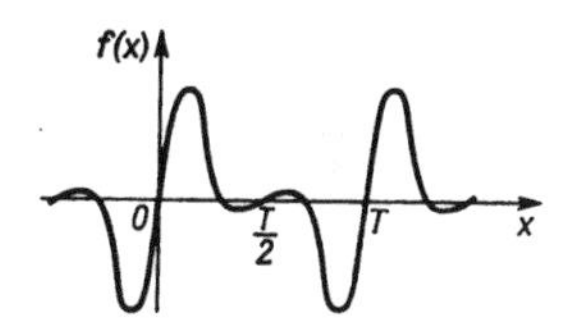

FIG. 418

for $k = 0, 1, 2, \ldots$ If $f(x)$ is an *odd* function, i.e., if $f(-x) = -f(x)$ (*symmetry of the second kind*, Fig. 418), then

$$a_k = 0, \qquad b_k = \frac{4}{T}\int_0^{T/2} f(x)\sin k\frac{2\pi x}{T}\,dx,$$

for $k = 0, 1, 2, \ldots$ If $f(x + \frac{1}{2}T) = -f(x)$ (*symmetry of the third kind*, Fig. 419), then

$$a_{2k+1} = \frac{4}{T}\int_0^{T/2} f(x)\cos(2k+1)\frac{2\pi x}{T}\,dx, \qquad a_{2k} = 0,$$

$$b_{2k+1} = \frac{4}{T}\int_0^{T/2} f(x)\sin(2k+1)\frac{2\pi x}{T}\,dx, \qquad b_{2k} = 0,$$

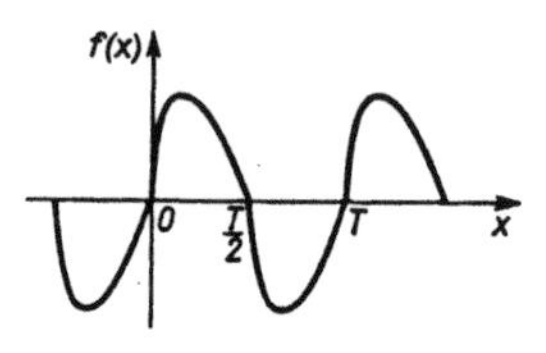

FIG. 419

for $k = 0, 1, 2, \ldots$ If the function $f(x)$ is odd and, moreover, has a symmetry of the third kind (Fig. 420a), then

$$a_k = b_{2k} = 0, \quad b_{2k+1} = \frac{8}{T}\int_0^{T/4} f(x)\sin(2k+1)\frac{2\pi x}{T}\,dx,$$

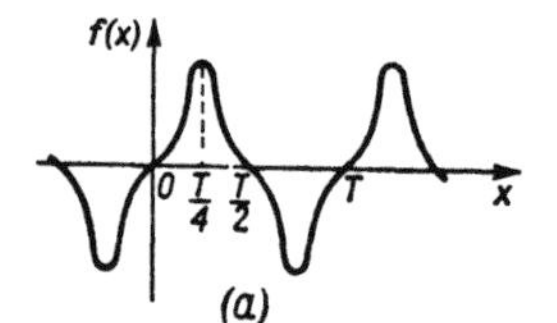

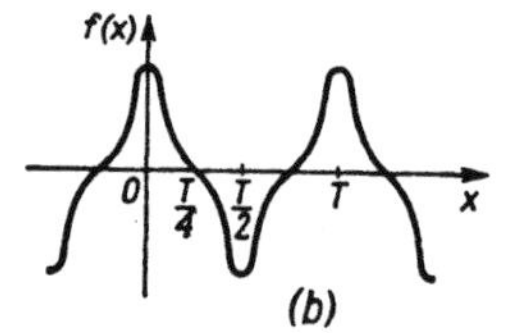

FIG. 420

for $k = 0, 1, 2, \ldots$ If the function $f(x)$ is even and, moreover, has a symmetry of the third kind (Fig. 420b), then

$$b_k = a_{2k} = 0, \quad a_{2k+1} = \frac{8}{T}\int_0^{T/4} f(x)\cos(2k+1)\frac{2\pi x}{T}\,dx,$$

for $k = 0, 1, 2, \ldots$

Expansion of a non-periodic function into Fourier series Any function $f(x)$ satisfying Dirichlet's conditions in the interval

$0 < x < l$ (see p. 729) can be expanded in this interval into a convergent series of one of the following forms

$$(1)\quad f_1(x) = \frac{a_0}{2} + a_1 \cos\frac{2\pi x}{l} + a_2 \cos 2\frac{2\pi x}{l} + \ldots + a_n \cos n\frac{2\pi x}{l} + \ldots +$$

$$+ b_1 \sin\frac{2\pi x}{l} + b_2 \sin 2\frac{2\pi x}{l} + \ldots + b_n \sin n\frac{2\pi x}{l} + \ldots,$$

$$(2)\quad f_2(x) = \frac{a_0}{2} + a_1 \cos\frac{\pi x}{l} + a_2 \cos 2\frac{\pi x}{l} + \ldots + a_n \cos n\frac{\pi x}{l} + \ldots,$$

$$(3)\quad f_3(x) = b_1 \sin\frac{\pi x}{l} + b_2 \sin 2\frac{\pi x}{l} + \ldots + b_n \sin n\frac{\pi x}{l} + \ldots$$

The function $f_1(x)$ is periodic with the period $T = l$ and coincides with the function $f(x)$ in the interval $0 < x < l$ [1] (Fig. 421); we

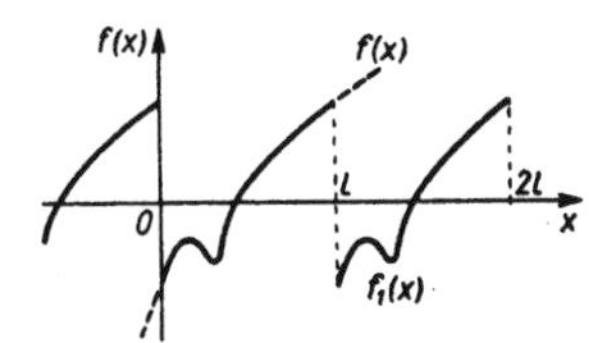

FIG. 421

determine the coefficients of the expansion from Euler's formulas (see p. 727) putting $\omega = 2\pi/l$. The function $f_2(x)$ is periodic with period $T = 2l$, has a symmetry of the first kind and coincides with the function $f(x)$ in the interval $0 < x < l$ (Fig. 422); we determine

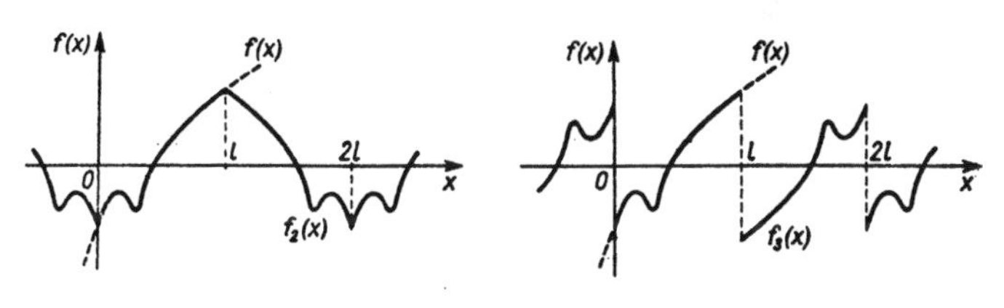

FIG. 422 FIG. 423

the coefficients of the expansion of $f_2(x)$ from the corresponding formulas for the symmetry of the first kind putting $T = 2l$. The function $f_3(x)$ is periodic with period $T = 2l$, has a symmetry of

[1] At points of discontinuity we assume $f(x) = \frac{f(x-0) + f(x+0)}{2}$.

the second kind and coincides with the function $f(x)$ in the interval $0 < x < l$ (Fig. 423); we determine the coefficients of the expansion of $f_2(x)$ from the corresponding formulas for the symmetry of the second kind putting $T = 2l$.

Fourier integral. If the function $f(x)$ satisfies Dirichlet's conditions (see p. 729) in any finite interval and if, in addition, the integral $\int_{-\infty}^{+\infty} |f(x)| dx$ is convergent (see pp. 471–472), then the following formula (*Fourier integral*) holds

$$f(x) = \frac{1}{2\pi} \int_{-\infty}^{+\infty} e^{iux}\, du \int_{-\infty}^{+\infty} f(t)\, e^{-iut}\, dt = \frac{1}{\pi} \int_{0}^{\infty} du \int_{-\infty}^{+\infty} f(t) \cos u(t - x) dt.$$

This formula can be regarded as a limiting case, when $l \to \infty$, of the formula for the expansion of a non-periodic function $f(x)$ into a trigonometric series in the interval $(-l, l)$. For, while a Fourier series represents a periodic function with period T as a sum of harmonic vibrations with frequences $u_n = n2\pi/T$ and amplitudes A_n $(n = 1, 2, \ldots)$, the Fourier integral represents $f(x)$ as in the form of a sum of an infinite number of harmonic vibrations with a frequency u which varies continuously. We say that the Fourier integral gives an expansion of a function into a *continuous spectrum*. Corresponding to the frequency u we have the *density of the spectrum*

$$g(u) = \frac{1}{2\pi} \int_{-\infty}^{+\infty} f(t)\, e^{-iut}\, dt.$$

The Fourier integral has a simple form if the function $f(x)$ is even:

$$f(x) = \frac{2}{\pi} \int_{0}^{\infty} \cos ux\, du \int_{0}^{\infty} f(t) \cos ut\, dt,$$

or if it is odd:

$$f(x) = \frac{2}{\pi} \int_{0}^{\infty} \sin ux\, du \int_{0}^{\infty} f(t) \sin ut\, dt.$$

Example. The density of the spectrum of the even function $f(x) = e^{-|x|}$ is equal to

$$g(u) = \frac{2}{\pi}\int_0^\infty e^{-t}\cos ut\,dt = \frac{2}{\pi}\cdot\frac{1}{u^2+1}, \text{ that is, } e^{-|x|} = \frac{2}{\pi}\int_0^\infty \frac{\cos ux}{u^2+1}\,du.$$

2. Table of certain Fourier expansions

Expansion into trigonometric series of some simple functions defined in a certain interval and then periodically extended are given below. Many other simple periodic functions can be reduced to one of the forms by this table; this can be done by a change of the unit of measure on the x axis and y axis and also by a translation of the coordinate axes.

Example. The periodic function y with period T (Fig. 424) defined by the equations

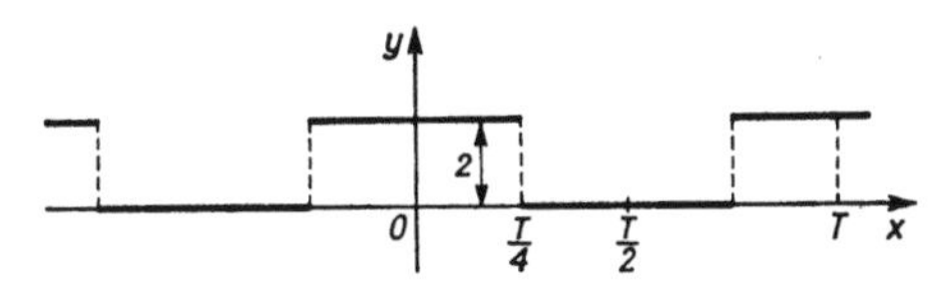

FIG. 424

$$y = \begin{cases} 2 & \text{for} \quad 0 < x < \frac{1}{4}T, \\ 0 & \text{for} \quad \frac{1}{4}T < x < \frac{1}{2}T, \end{cases}$$

can be reduced to the form 5 (for $a = 1$) by introducing the variables

$$Y = y - 1, \quad X = \frac{2\pi x}{T} + \frac{\pi}{2}.$$

Since

$$\sin(2n+1)\left(\frac{2\pi x}{T} + \frac{\pi}{2}\right) = (-1)^n \cos(2n+1)\frac{2\pi x}{T},$$

hence, substituting the new variables to the series 5, we obtain for our function

$$y = 1 + \frac{4}{\pi}\left(\cos\frac{2\pi x}{T} - \frac{1}{3}\cos 3\frac{2\pi x}{T} + \frac{1}{5}\cos 5\frac{2\pi x}{T} - \dots\right).$$

Table.

1. $y = x$ for $0 < x < 2\pi$ (Fig. 425):

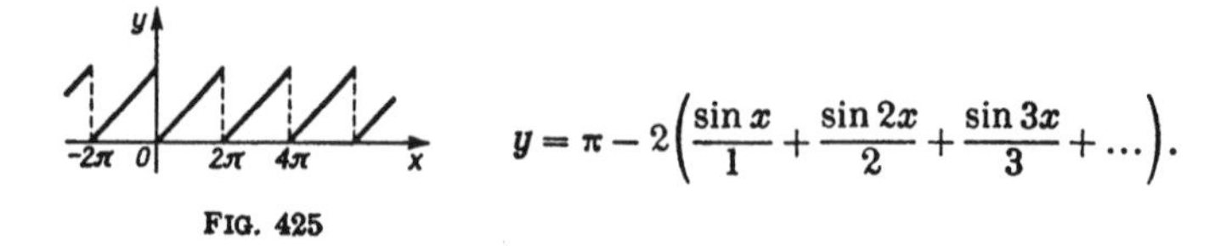

$$y = \pi - 2\left(\frac{\sin x}{1} + \frac{\sin 2x}{2} + \frac{\sin 3x}{3} + \dots\right).$$

FIG. 425

2. $y = x$ for $0 < x < \pi$ (Fig. 426):

$$y = \frac{\pi}{2} - \frac{4}{\pi}\left(\cos x + \frac{\cos 3x}{3^2} + \frac{\cos 5x}{5^2} + \dots\right).$$

FIG. 426

3. $y = x$ for $-\pi < x < \pi$ (Fig. 427):

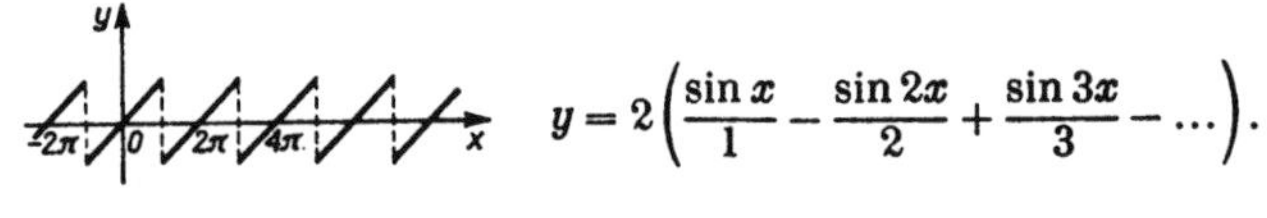

$$y = 2\left(\frac{\sin x}{1} - \frac{\sin 2x}{2} + \frac{\sin 3x}{3} - \dots\right).$$

FIG. 427

4. $y = x$ for $-\frac{1}{2}\pi < x < \frac{1}{2}\pi$ (Fig. 428):

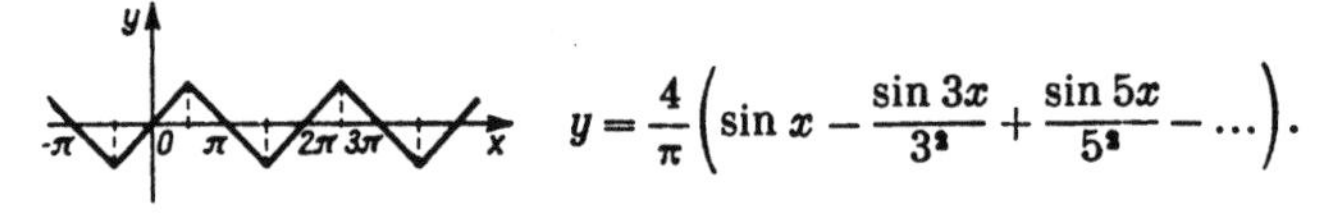

$$y = \frac{4}{\pi}\left(\sin x - \frac{\sin 3x}{3^2} + \frac{\sin 5x}{5^2} - \dots\right).$$

FIG. 428

5. $y = a$ for $0 < x < \pi$ (Fig. 429):

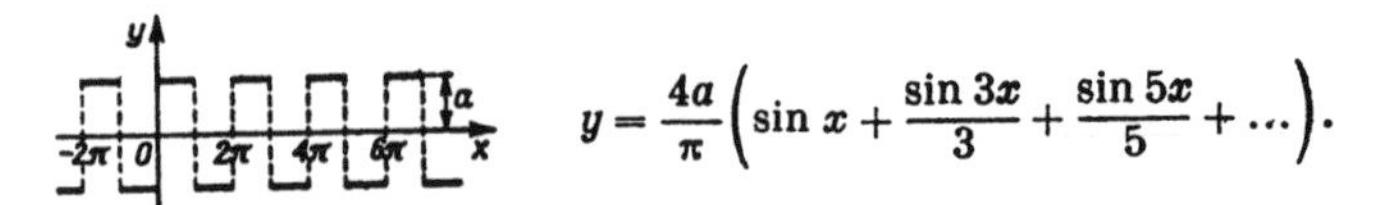

$$y = \frac{4a}{\pi}\left(\sin x + \frac{\sin 3x}{3} + \frac{\sin 5x}{5} + \dots\right).$$

FIG. 429

6. $y = 0$ for $0 < x < \alpha$ and $\pi - \alpha < x < \pi$, $y = a$ for $\alpha < x < \pi - \alpha$ (Fig. 430):

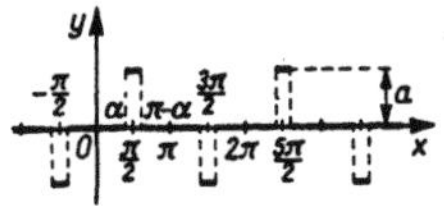

$$y = \frac{4a}{\pi}\left(\cos\alpha \sin x + \frac{1}{3}\cos 3\alpha \sin 3x + \frac{1}{5}\cos 5\alpha \sin 5x + \ldots\right).$$

FIG. 430

7. $y = ax/\alpha$ for $0 < x < \alpha$, $y = a$ for $\alpha < x < \pi - \alpha$, $y = a(\pi - x)/\alpha$ for $\pi - \alpha < x < \pi$ (Fig. 431):

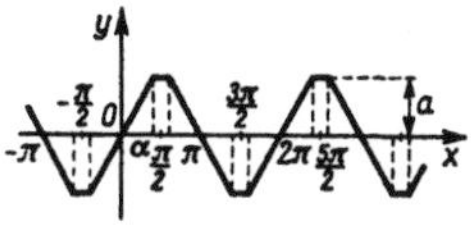

$$y = \frac{4}{\pi}\cdot\frac{a}{\alpha}\left(\sin\alpha \sin x + \frac{1}{3^2}\sin 3\alpha \sin 3x + \frac{1}{5^2}\sin 5\alpha \sin 5x + \ldots\right).$$

FIG. 431

In particular, for $\alpha = \frac{1}{3}\pi$, we have

$$y = \frac{6\sqrt{3}\,a}{\pi^2}\left(\sin x - \frac{1}{5^2}\sin 5x + \frac{1}{7^2}\sin 7x - \frac{1}{11^2}\sin 11x + \ldots\right).$$

8. $y = x^2$ for $-\pi < x < \pi$ (Fig. 432):

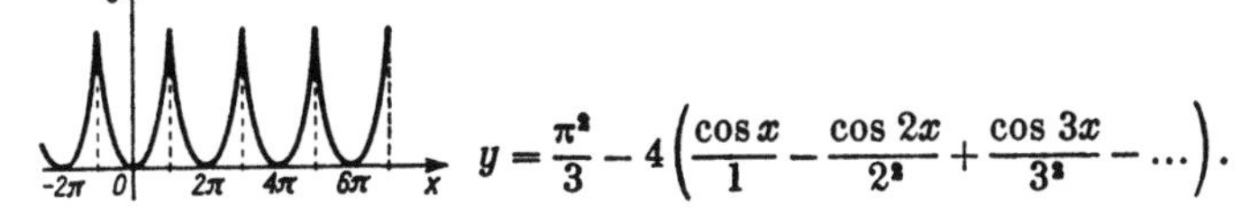

$$y = \frac{\pi^2}{3} - 4\left(\frac{\cos x}{1} - \frac{\cos 2x}{2^2} + \frac{\cos 3x}{3^2} - \ldots\right).$$

FIG. 432

9. $y = x(\pi - x)$ for $0 < x < \pi$ (Fig. 433):

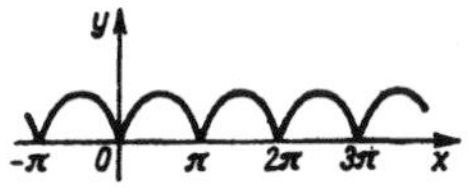

$$y = \frac{\pi^2}{6} - \left(\frac{\cos 2x}{1^2} + \frac{\cos 4x}{2^2} + \frac{\cos 6x}{3^2} + \ldots\right).$$

FIG. 433

10. $y = x(\pi - x)$ for $0 < x < 2\pi$ (Fig. 434):

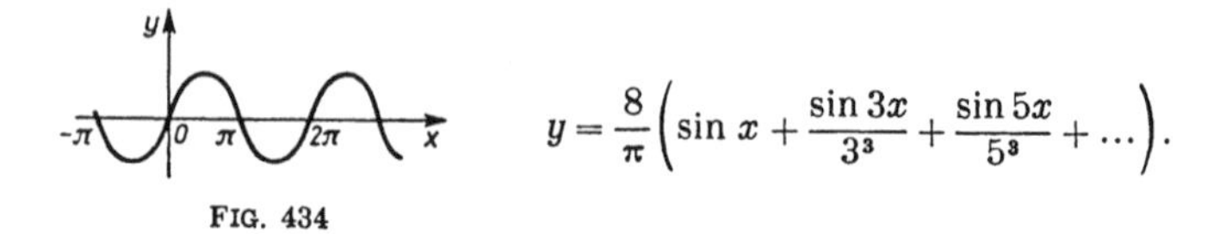

$$y = \frac{8}{\pi}\left(\sin x + \frac{\sin 3x}{3^3} + \frac{\sin 5x}{5^3} + \dots\right).$$

FIG. 434

11. $y = \sin x$ for $0 < x < \pi$ (Fig. 435):

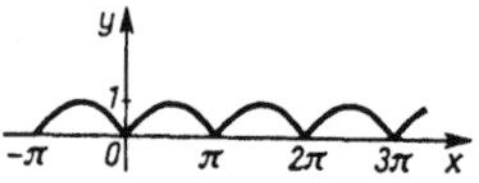

$$y = \frac{2}{\pi} - \frac{4}{\pi}\left(\frac{\cos 2x}{1\cdot 3} + \frac{\cos 4x}{3\cdot 5} + \frac{\cos 6x}{5\cdot 7} + \dots\right).$$

FIG. 435

12. $y = \cos x$ for $0 < x < \pi$ (Fig. 436):

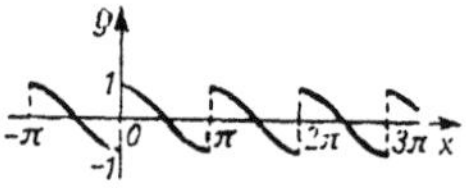

$$y = \frac{4}{\pi}\left(\frac{2\sin 2x}{1\cdot 3} + \frac{4\sin 4x}{3\cdot 5} + \frac{6\sin 6x}{5\cdot 7} + \dots\right).$$

FIG. 436

13. $y = \sin x$ for $0 < x < \pi$, $y = 0$ for $\pi < x < 2\pi$ (Fig. 437):

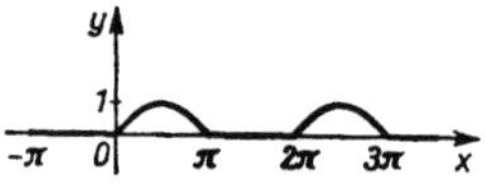

$$y = \frac{1}{\pi} + \frac{1}{2}\sin x - \frac{2}{\pi}\left(\frac{\cos 2x}{1\cdot 3} + \frac{\cos 4x}{3\cdot 5} + \frac{\cos 6x}{5\cdot 7} + \dots\right).$$

FIG. 437

14. $y = \cos ux$ for $-\pi < x < \pi$:

$$y = \frac{2u\sin u\pi}{\pi}\left[\frac{1}{2u^2} - \frac{\cos x}{u^2 - 1} + \frac{\cos 2x}{u^2 - 4} + \frac{\cos 3x}{u^2 - 9} + \dots\right],$$

where u is an arbitrary non-integral number.

15. $y = \sin ux$ for $-\pi < x < \pi$:

$$y = \frac{2\sin u\pi}{\pi}\left(\frac{\sin x}{1 - u^2} - \frac{2\sin 2x}{4 - u^2} + \frac{3\sin 3x}{9 - u^2} - \dots\right),$$

where u is an arbitrary non-integral number.

16. $y = x\cos x$ for $-\pi < x < \pi$:

$$y = -\frac{1}{2}\sin x + \frac{4\sin 2x}{1\cdot 3} - \frac{6\sin 3x}{3\cdot 5} + \frac{8\sin 4x}{5\cdot 7} - \dots$$

17. $y = -\ln(2\sin\frac{1}{2}x)$ for $0 < x < \pi$:

$$y = \cos x + \tfrac{1}{2}\cos 2x + \tfrac{1}{3}\cos 3x + \dots$$

18. $y = \ln(2\cos\frac{1}{2}x)$ for $0 < x < \pi$:

$$y = \cos x - \tfrac{1}{2}\cos 2x + \tfrac{1}{3}\cos 3x - \dots$$

19. $y = \frac{1}{2}\ln\cot\frac{1}{2}x$ for $0 < x < \pi$:

$$y = \cos x + \tfrac{1}{3}\cos 3x + \tfrac{1}{5}\cos 5x + \dots$$

A great number of formulas for expansion of functions into trigonometric series can be obtained from power series of functions of a complex variable.

Example. From the expansion

$$\frac{1}{1-z} = 1 + z + z^2 + \dots, \qquad (|z| < 1),$$

substituting $z = ae^{i\varphi}$ and separating the real and imaginary parts, we obtain

$$1 + a\cos\varphi + a^2\cos 2\varphi + \dots + a^n\cos n\varphi + \dots = \frac{1 - a\cos\varphi}{1 - 2a\cos\varphi + a^2},$$

$$a\sin\varphi + a^2\sin 2\varphi + \dots + a^n\sin n\varphi + \dots = \frac{a\sin\varphi}{1 - 2a\cos\varphi + a^2},$$

for $|a| < 1$.

3. Approximate harmonic analysis

Bessel's formulas. Approximate evaluation of the coefficients of Fourier series is based on replacing the integrals in Euler's formulas (see p. 727) with finite sums by means of one of the formulas for approximate integration. The most convenient in this case is the trapezoid formula (p. 462). It leads to the following *Bessel formulas* of approximate harmonic analysis:

We divide the period T into $2n$ equal parts (Fig. 438). Let $x_k = kT/2n$ be the abscissae of the points of division and $y_k = f(x_k)$ be the corresponding ordinates, $k = 0, 1, \dots, 2n$. Then we have approximately

$$na_0 = \sum_{k=0}^{2n-1} y_k, \quad na_m = \sum_{k=0}^{2n-1} y_k \cos\frac{km\pi}{n}, \quad nb_m = \sum_{k=0}^{2n-1} y_k \sin\frac{km\pi}{n},$$

for $m = 1, 2, \ldots, n$ and always $b_n = 0$.

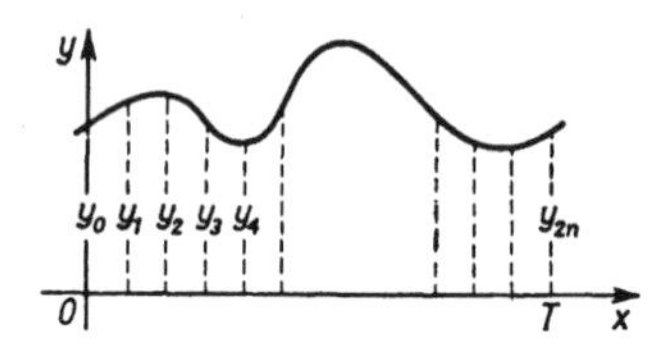

FIG. 438

If we form the trigonometric sum

$$s_r(x) = \frac{a_0}{2} + \sum_{k=1}^{r} a_k \cos\frac{2k\pi x}{T} + \sum_{k=1}^{r} b_k \sin\frac{2k\pi x}{T},$$

for $r < n$, then this sum will be the best approximation of the function defined by the ordinates y_k $(k = 1, 2, \ldots, 2n)$ in the sense of the least squares (see p. 757), if its coefficients are computed from the Bessel's formulas. In the case $r = n$, the trigonometric sum

$$s_n(x) = \frac{a_0}{2} + a_1 \cos\frac{2\pi x}{T} + a_2 \cos 2\frac{2\pi x}{T} + \ldots + \frac{a_n}{2} \cos n\frac{2\pi x}{T} +$$

$$+ b_1 \sin\frac{2\pi x}{T} + b_2 \sin 2\frac{2\pi x}{T} + \ldots + b_{n-1} \sin (n-1)\frac{2\pi x}{T},$$

with coefficients computed from the Bessel's formulas assumes the values y_k at the points $x = x_k$ and hence solves the problem of *trigometric interpolation* for a periodic function (see p. 757).

Computing patterns and instruments. Special patterns and computing instruments are in use in computation according to Bessel's formulas. Some computing patterns of harmonic analysis corresponding to a division of the period into 12 and 24 parts are given below. If the function $f(x)$ is defined by a graph, then, besides Bessel's formulas, special instruments called *harmonic analysers* can be used. After tracing the graph of the given function by a pin of the analyser, special computers give approximate values of the coefficients of the Fourier series.

Schemes for approximate harmonic analysis.

Scheme I. We divide the period T into 12 equal parts. Let $y_0, y_1, y_2, \ldots, y_{11}$ be the ordinates of the points of division. We evaluate the sums and differences according to the scheme:

$\pm$	y_0	y_1	y_2	y_3	y_4	y_5	y_6	s_0	s_1	s_2	s_3	d_1	d_2	d_3
		y_{11}	y_{10}	y_9	y_8	y_7		s_6	s_5	s_4		d_5	d_4	
Sums	s_0	s_1	s_2	s_3	s_4	s_5	s_6	σ_0	σ_1	σ_2	σ_3	δ_1	δ_2	δ_3
Differences		d_1	d_2	d_3	d_4	d_5		τ_0	τ_1	τ_2		γ_1	γ_2	

Further computations are performed according to the following scheme:

	Terms involving cosines								Terms involving sines					
1	σ_0 / σ_2	σ_1 / σ_3	τ_0		σ_0	$-\sigma_3$	τ_0	τ_2	δ_3				δ_1	δ_3
1 − 0.134 (= 0.866)				τ_1						δ_2	γ_1	γ_2		
0.5			τ_2		$-\sigma_2$	σ_1			δ_1					
Sums	I	II	I	II	I	II	I	II	I	II	I	II	I	II
Sums I + II	$6a_0$ (¹)		$6a_1$		$6a_2$		—		$6b_1$		$6b_2$		—	
Differences I − II	$6a_6$ (¹)		$6a_5$		$6a_4$		$6a_3$		$6b_5$		$6b_4$		$6b_3$	

(¹) We should remember that the trigonometric interpolation polynomial (see p. 738) involves $\frac{1}{2}a_0$ and $\frac{1}{2}a_n$ but not a_0 and a_n.

In computation according to this scheme we have to put, instead of σ, τ, δ and γ, the corresponding values multiplied by the factors standing in the same row on the left side (1 − 0.134 has been put into the table instead of 0.866, for the multiplication by 0.134 by aid of a slide rule can be done more accurately than by 0.866).

Scheme II. We divide the period T into 24 equal parts, and write the ordinates $y_0, y_1, y_2, \ldots, y_{23}$ at the points of division as follows:

y_0	y_2	y_4	y_6	y_8	y_{10}	y_{12}
	y_{22}	y_{20}	y_{18}	y_{16}	y_{14}	
y_3	y_5	y_7	y_9	y_{11}	y_{13}	y_{15}
	y_1	y_{23}	y_{21}	y_{19}	y_{17}	

For each group of ordinates separately we perform the computation according to the above scheme for 12 ordinates. Let A_k and B_k the coefficients obtained from the first group of ordinates, and A'_k and B'_k those obtained from the second group. We compute $\overline{A}_k$ and $\overline{B}_k$ from the formulas

$$\overline{A}_0 = A'_0, \quad \overline{A}_1 = \frac{1}{\sqrt{2}}(A'_1 - B'_1), \quad \overline{A}_2 = -B'_2, \quad A_3 = -\frac{1}{\sqrt{2}}(A'_3 + B'_3),$$

$$\overline{A}_4 = -A'_4, \quad \overline{A}_5 = -\frac{1}{\sqrt{2}}(A'_5 - B'_5);$$

$$\overline{B}_1 = \frac{1}{\sqrt{2}}(A'_1 + B'_1), \quad \overline{B}_2 = A'_2, \quad \overline{B}_3 = \frac{1}{\sqrt{2}}(A'_3 - B'_3),$$

$$\overline{B}_4 = -B'_4, \quad \overline{B}_5 = -\frac{1}{\sqrt{2}}(A'_5 + B'_5), \quad \overline{B}_6 = -A'_6$$

(in fact we have to find not $\overline{A}_k$ and $\overline{B}_k$, but their values multiplied by 6, see below). Then we find the sums and differences which yield the desired coefficients:

	$6A_0$	$6A_1$	$6A_2$	$6A_3$	$6A_4$	$6A_5$	$6A_6$
	$6\overline{A}_0$	$6\overline{A}_1$	$6\overline{A}_2$	$6\overline{A}_3$	$6\overline{A}_4$	$6\overline{A}_5$	
Sums	$12a_0$(¹)	$12a_1$	$12a_2$	$12a_3$	$12a_4$	$12a_5$	$12a_6$
Differences	$12a_{12}$(¹)	$12a_{11}$	$12a_{10}$	$12a_9$	$12a_8$	$12a_7$	
	$6\overline{B}_1$	$6\overline{B}_2$	$6\overline{B}_3$	$6\overline{B}_4$	$6\overline{B}_5$	$6\overline{B}_6$	
	$6B_1$	$6B_2$	$6B_3$	$6B_4$	$6B_5$		
Sums	$12b_1$	$12b_2$	$12b_3$	$12b_4$	$12b_5$	$12b_6$	
Differences	$12b_{11}$	$12b_{10}$	$12b_9$	$12b_8$	$12b_7$		

(¹) See footnote on the preceding page.

Synthesis. By a *synthesis* we understand the evaluation of the values of a periodic function $f(x)$ defined by its Fourier series. If the admissible accuracy allows us to confine ourselves to the first six harmonic terms (i.e., to put $a_k = b_k = 0$ for $k > 6$), then the evaluation of $y_k = f(x_k)$ at the points $x_0, x_1, x_2, \ldots, x_{11}$ dividing the period into 12 equal parts can be performed by aid of the above scheme I (see p. 739). In order to do that, we replace $s_0, s_1, \ldots, s_6$ by the given coefficients $a_0, a_1, \ldots, a_6$, and $d_1, d_2, \ldots, d_5$ by $b_1, b_2, \ldots,$

b_5 (1) and accomplish the computation. We denote the numbers obtained in the last two rows of the table on p. 739 (instead of $6a_0, 6a_1, \ldots, 6a_6, 6b_1, 6b_2, \ldots, 6b_5$) by $\alpha_0, \alpha_1, \ldots, \alpha_6, \beta_1, \beta_2, \ldots, \beta_5$, and then, to obtain the desired values of the function, it remains only to perform the addition and subtractions according to the scheme:

	α_0	α_1	α_2	α_3	α_4	α_5	α_6
		β_1	β_2	β_3	β_4	β_5	
Sums	y_0	y_1	y_2	y_3	y_4	y_5	y_6
Differences		y_{11}	y_{10}	y_9	y_8	y_7	

(1) The coefficient b_6 can be discarded, for, as can easily be observed, the orresponding term of the series is immaterial for the values of the function at the onsidered points.

PART SIX

INTERPRETATION OF EXPERIMENTAL RESULTS

I. FOUNDATIONS OF THE THEORY OF PROBABILITY AND THE THEORY OF ERRORS

1. Theory of probability

Random event. If a certain event, in given circumstances, may occur or may not occur, then it is called a *random event*. The *probability* of a random event is a quantitative estimation of the probability of its occurrence.

Definition of probability. If, in certain circumstances, one of n mutually exclusive random events must occur, but none of them is expected to have a greater chance, then these events have an *equal probability* $p = 1/n$.

If a random event A appears as a result of the occurrence of any m events from among n mutually exclusive and equally likely random events, then the *probability* of the event A is the number $p = m/n$. Probability 0 corresponds to an impossible event, and probability 1 to a certain event. The probability of any event lies between 0 and 1.

Addition and multiplication of probabilities. The probability of the occurrence of any one of several mutually exclusive random events is equal to the sum of the probabilities of each event. The probability of the simultaneous occurrence of several events is equal to the product of the probabilities of these events; if, however, the events follow successively one another, then, estimating the probability of each of them, we should take into account a possible influence of the preceding events.

Example. A box contain 5 black, 3 white and 2 red balls. The probability of drawing a white ball is 0.3; the probability of drawing a red ball is 0.2; the probability of drawing a red or a white ball is $0.3 + 0.2 = 0.5$. The probability of drawing successively a white ball and then a red one is $0.3 \cdot 0.2 = 0.06$, provided that the first drawn ball is replaced to the box; if the first ball is not replaced, then the probability is equal to $0.3 \cdot \frac{2}{9} = 0.067$.

Repeated trials. If n independent trials are being done and in each of them the probability of the event A is p, then the probability that the event A will occur m times is equal to

(1) $$p_{m,n} = \binom{n}{m} p^m (1-p)^{n-m}.$$

This probability will have a greatest value, if $np + p - 1 < m < np + p$. For large m and n, an approximate value of $p_{m,n}$ can be obtained by means of Stirling's formula (see p. 190)

(2) $$p_{m,n} \approx \frac{1}{\sigma\sqrt{2\pi}} e^{-x^2/2},$$

where

$$\sigma = \sqrt{np(1-p)}, \quad x = \frac{m-np}{\sigma}.$$

For small values of p, the following *Poisson formula* is more accurate [1]

$$p_{m,n} \approx \frac{y^m}{m!} e^{-y} \quad \text{where} \quad y = np.$$

If the number n of trials increases, the most probable frequency m/n of the event A approaches its probability p. Moreover, the probability that the frequency of A is contained in the interval

$$p - \frac{a\sigma}{n} \quad \text{and} \quad p + \frac{a\sigma}{n}$$

approaches the limit

(*) $$\Phi(a) = \frac{2}{\sqrt{2\pi}} \int_0^a e^{-x^2/2}\, dx$$

(*Laplace's theorem*). The function (*) is called the *Gauss probability integral* [2]. (For tables of $\Phi(x)$ see pp. 93–95).

Examples. (1) What is the probability that in 400 tosses of a coin the frequency of obtaining a head will differ from $p = \frac{1}{2}$ less than by $\frac{1}{25}$, i.e., that the number of obtaining a head will be contained between 184 and 216?

---

[1] If the value of p is near to 1, Poisson's formula can also be applied by considering the event "non-A" (the complementary event); its probability is $q = 1 - p$, hence is small.

[2] Sometimes the function

$$\operatorname{Erf} x = \frac{2}{\sqrt{\pi}} \int_0^x e^{-t^2}\, dt = \Phi(x\sqrt{2}).$$

is called the *probability integral*.

Since

$$\sigma = \sqrt{400 \cdot \tfrac{1}{2} \cdot \tfrac{1}{2}} = 10 \text{ and } \frac{a\sigma}{n} = \frac{1}{25}, \quad \text{hence } a = \frac{400}{10 \cdot 25} = 1.6.$$

By Laplace's theorem, the desired probability is equal approximately to $\Phi(1.6) = 0.8904$.

(2) Let the probability of obtaining a defective article be equal to 0.01. What is the probability that in a lot composed of 100 pieces at most 3 defective pieces occur?

The desired probability is equal to $p = p_{0,100} + p_{1,100} + p_{2,100} + p_{3,100}$. By Laplace's formula, we obtain

$$p = \frac{1}{e}\left(1 + 1 + \frac{1}{2} + \frac{1}{6}\right) = 0.9810.$$

Gauss integral yields in this case a result which is too inaccurate ($p = 0.928$). By formula (2) we obtain a slightly better result ($p = 0.938$). The exact value is $p = 0.9816$.

Law of large numbers. A corollary from Laplace's theorem: We can expect with the probability arbitrarily near to 1 that, by selecting the number of trials large enough, the frequency of the event A can be made as near to its probability as desired (*theorem of Bernoulli*).

Random variables. A *random variable* is a variable whose values depend on chance. Examples of random variables: the number of hits to a target for a given number of shots, the number obtained in rolling a die, the velocity of particles of a gas.

In order to characterize a random variable, the set of all its possible values and also the probabilities of occurrence of particular values must be known. These data form the *distribution of a random variable*. If a random variable X can assume an arbitrary value in an interval (a, b), then it is called *continuous*; the probability that the continuous random variable X will assume a definite value x is equal to 0, for the number of all mutually exclusive cases is infinite. If we assume that, for each sufficiently small interval $(x, x + dx)$ contained in the interval (a, b) of admissible values of X, the probability that the value of X falls in the interval $(x, x + dx)$ is proportional to its length dx, then the random variable X can be characterized by the probability $\varphi(x)dx$ of the occurrence of the inequality $x < X < x + dx$. The function $\varphi(x)$ is called the *density of the distribution of the probability* of the random variable X (or *probability density*). It follows from the theorem on addition of probabilities that the probability that X falls in the interval from x_0 to x is equal to $\int_{x_0}^{x_1} \varphi(x)\,dx$. For the random variable always assumes a certain value, hence $\int_{-\infty}^{+\infty} \varphi(x)\,dx = 1$.

Mean value of a random variable. If the random variable X can assume the values $x_1, x_2, \ldots, x_n$ with the corresponding probabilities $p_1, p_2, \ldots, p_n$, then

$$\mathrm{E}X = x = \sum_{i=1}^{n} p_i x_i$$

is called the *mean value* or *mathematical expectation* of X. The mathematical expectation of a continuous random value X with the probability density $\varphi(x)$ is equal to

$$\mathrm{E}X = \bar{x} = \int_{-\infty}^{+\infty} x\varphi(x)\,dx.$$

Examples. (1) In a lottery for 1000 tickets, one ticket wins \$ 1000, 10 tickets win \$ 100 each and 100 tickets win \$ 20 each. The distribution of the random variable X—the winnings for one ticket—is given in the table:

a_i	1000	100	20	0
p_i	0.001	0.01	0.1	0.889

The mean value of the random variable X is $\mathrm{E}X = 1000 \times \times 0.001 + 100 \cdot 0.01 + 20 \cdot 0.1 = \$\, 4$.

(2) The Maxwell distribution law (in the kinetic theory of gas):

$$\varphi(v) = 4\sqrt{\frac{k^3}{\pi}}\, v^2 e^{-kv^2} \quad (k \text{ is a positive constant}).$$

The probability that the velocity of a molecule of a homogeneous gas being in a thermic equilibrium is contained between v and $v + dv$ is equal to $\varphi(v)dv$. The mean value is

$$\mathrm{E}v = \bar{v} = \int_0^{\infty} v\varphi(v)\,dv = \frac{2}{\sqrt{k\pi}}$$

(see p. 480).

Variance. The *variance* is a measure of dispersion of a random variable X. The variance D^2X or σ^2 of the random variable X is the mean value of the square of the deviation of X from the mean value of X.

Example. The variance of the Maxwell distribution law (see above) is equal to

$$D^2 v = \sigma^2 = \overline{(v - \bar{v})^2} = \int_0^\infty (v - \bar{v})^2 \varphi(v)\, dv = \int_0^\infty v^2 \varphi(v)\, dv - (\bar{v})^2$$

$$= \frac{3}{2k} - \frac{4}{\pi k} = \frac{0.227}{k}.$$

The equality $\overline{(v - \bar{v})^2} = \overline{v^2} - (\bar{v})^2$ obtained in this example is identically satisfied and is used in computing the variance.

2. Theory of errors

Random errors. Experimental results always involve errors which may come from various causes. Among them, we should distinguish the *systematic errors* and the *random errors*. The systematic errors are due to definite causes acting according to definite laws and can either be removed or taken into account with sufficient accuracy (for example, errors caused by an irregularity of the scale of an instrument or by external conditions of an experiment). The random errors are usually due to a large number of causes acting in a different way in particular measurements. These errors cannot be completely excluded; only mean corrections can be introduced, but a knowledge of laws governing the random errors is then necessary.

The measured quantity will be denoted by A, and the random error in a measurement by x. The random error x can assume arbitrary values, it is therefore a continuous random variable characterized by its probability distribution (see p. 745). Experiments show that the probability density $\varphi(x)$ of a random error should have, in most cases, the following properties:

(1) The density $\varphi(x)$ is an even function: $\varphi(-x) = \varphi(x)$, i.e., the errors with different signs are equally likely.

(2) The density $\varphi(x)$ is a decreasing function for $x > 0$, i.e., the greater errors are less likely.

(3) The mean value of the absolute value of the error

$$2\int_0^\infty x\varphi(x)\, dx$$

is finite.

Certain definite distributions of errors can be obtained by imposing additional conditions.

Normal distribution. The simplest and, in most cases, the best approximation to reality, is the so-called *normal distribution* of a random error:

$$\varphi(x) = \frac{1}{\sigma\sqrt{2\pi}} \exp\left(-\frac{x^2}{2\sigma^2}\right).$$

This distribution can be obtained as a result of certain theoretical assumptions, in particular from the requirement that the most likely value of a unknown quantity measured immediately by a sequence of measurements with the same accuracy is the arithmetic mean of the results of all measurements. The quantity σ^2 is a parameter of the normal distribution; it can assume arbitrary values. Since

$$\bar{x} = \int_{-\infty}^{+\infty} x\varphi(x)\,dx = 0 \quad \text{and} \quad \overline{x^2} = \int_{-\infty}^{+\infty} x^2\varphi(x)\,dx = \sigma^2,$$

hence $\sigma^2 = \overline{x^2} - (\bar{x})^2$ is the variance of the random error x. When σ^2 increases, the maximum of $\varphi(x)$ decreases; it corresponds to $x = 0$ and is equal to $\frac{1}{\sigma\sqrt{2\pi}}$. For the area between the graph of $\varphi(x)$ and the x axis remains unchanged (Fig. 439) and equal to

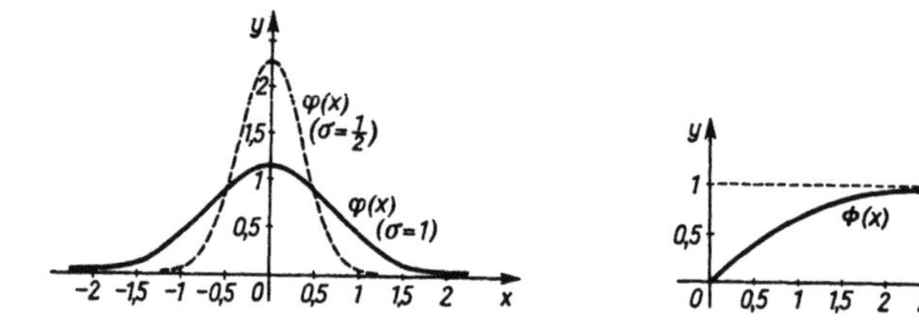

FIG. 439

1 (see p. 745), then, to an increasing variance, there corresponds an increasing probability of greater errors. The probability that the error x in the absolute does not exceed a is, in the case of a normal distribution, equal to

$$\Phi\left(\frac{a}{\sigma}\right) = \frac{2}{\sqrt{2\pi}} \int_0^{a/\sigma} \exp\left(-\frac{t^2}{2}\right) dt.$$

The random variable $x = \lambda_1 x_1 + \lambda_2 x_2 + \ldots + \lambda_n x_n$ constituting a linear combination of random variables $x_1, x_2, \ldots, x_n$ with normal distributions and variances $\sigma_1^2, \sigma_2^2, \ldots, \sigma_n^2$ has also a normal distribution with the variance σ^2:

$$\sigma^2 = \lambda_1^2\sigma_1^2 + \lambda_2^2\sigma_2^2 + \ldots + \lambda_n^2\sigma_n^2.$$

Besides the variance σ^2, the following quantities are used in characterization a normal distribution:

(1) *Simple mean error* η constituting the mathematical expectation of the absolute value of the error:

$$\eta = \overline{|x|} = 2\int_0^\infty |x|\varphi(x)\,dx.$$

(2) *Mean square error* or *standard error* σ equal to the square root from the variance σ^2.

(3) *Probable error* r; this is a quantity such that the probability of an error less than r in the absolute value is equal to $\frac{1}{2}$:

$$\int_{-r}^{+r} \varphi(x)\,dx = \Phi\left(\frac{r}{\sigma}\right) = \frac{1}{2}.$$

(4) *Measure of accuracy* $h = 1/\sqrt{2}\,\sigma$.

All these quantities are related by the formulas [1]:

$$\eta = \frac{1}{\sqrt{\pi}\,h} = \sqrt{\frac{2}{\pi}}\,\sigma = \frac{r}{\varrho\sqrt{\pi}}, \qquad \sigma = \frac{1}{\sqrt{2}\,h} = \sqrt{\frac{\pi}{2}}\,\eta = \frac{r}{\sqrt{2}\,\varrho},$$

$$r = \frac{\varrho}{h} = \varrho\sqrt{2}\,\sigma = \varrho\sqrt{\pi}\,\eta, \qquad h = \frac{1}{\sqrt{\pi}\,\eta} = \frac{1}{\sqrt{2}\,\sigma} = \frac{\varrho}{r},$$

where

$$\frac{1}{\sqrt{2}} = 0.7071 = \frac{1}{1.4142}, \quad \frac{1}{\sqrt{\pi}} = 0.5642, \quad \sqrt{\frac{2}{\pi}} = 0.7979 = \frac{1}{1.2533},$$

$$\varrho = 0.4769, \quad \varrho\sqrt{2} = 0.6745 = \frac{1}{1.4826}, \quad \varrho\sqrt{\pi} = 0.8454 = \frac{1}{1.1829}.$$

Determining the variance from experimental data. If n values a_i ($i = 1, 2, \ldots, n$) have been obtained for a certain value A by means of immediate measurements with the same accuracy and if the errors of A have a normal distribution, then the arithmetic mean

$$a = \frac{1}{n}\sum_{i=1}^{n} a_i$$

is the most likely value of A.

[1] The number ϱ is determined by the equation $\Phi(\varrho\sqrt{2}) = \frac{1}{2}$.

Let ε_i be the deviation of the result a_i of a measurement of A from the arithmetic mean a: $\varepsilon_i = a_i - a$, $i = 1, 2, \ldots, n$.

In determining the variance of the distribution of errors, we use in this case the formula

$$\sigma^2 = \frac{1}{n-1} \sum_{i=1}^{n} \varepsilon_i^2 \ (^1)$$

or determine σ from the mean error which can be found by the formula

$$\eta = \frac{1}{\sqrt{n(n-1)}} \sum_{i=1}^{n} |\varepsilon_i| = \frac{1}{n-\frac{1}{2}} \sum_{i=1}^{n} |\varepsilon_i| .$$

If the values of σ obtained in two ways differ considerably, then we infer that the normal distribution is not applicable to this case.

If the particular values a_i $(i = 1, 2, \ldots, n)$ of A have been obtained with various accuracy characterized by the standard error σ_i, then the *weighted mean*

$$a = \frac{w_1 a_1 + w_2 a_2 + \ldots + w_n a_n}{w_1 + w_2 + \ldots + w_n},$$

where the weights w_i are certain numbers inversely proportional to the corresponding standard errors, is the most likely value of A. The standard error of each particular value a_i with the weight w_i is equal to

$$\sigma_i = \sqrt{\frac{1}{(n-1)\, w_i} \sum_{i=1}^{n} w_i \varepsilon_i^2},$$

where ε_i is the deviation of a_i from the weighted mean. According to the formula for the variance of a linear combination, the standard errors of the arithmetic and weighted mean are determined by the formulas

$$\sigma = \sqrt{\frac{1}{n(n-1)} \sum_{i=1}^{n} \varepsilon_i^2} \quad \text{and} \quad \sigma = \sqrt{\frac{1}{(n-1)\, w} \sum_{i=1}^{n} w_i \varepsilon_i^2},$$

where $w = w_1 + w_2 + \ldots + w_n$.

---

(1) In the calculus of probability and mathematical statistics the following notation due to Gauss is in use: $[\varepsilon\varepsilon]$ for $\sum \varepsilon_i^2$ and $[ab]$ for $\sum a_i b_i$ and so on.

Example. The interior diameter d and exterior diameter D of a hollowed cylindrical vessel have been determined by five measurements. The results of measurements are given in the table:

Number i of the measurement	d	D	ε_{d_i}	$\varepsilon^2_{d_i}$	ε_{D_i}	$\varepsilon^2_{D_i}$
1	17.3	22.7	0.06	0.0036	−0.08	0.0064
2	17.0	22.8	−0.24	0.0576	0.02	0.0004
3	17.3	23.0	0.06	0.0036	0.22	0.0484
4	17.4	22.8	0.16	0.0256	0.02	0.0004
5	17.2	22.6	−0.04	0.0016	−0.18	0.0324
$\sum_i$	86.2	113.9	0.56[1]	0.0920	0.52[1]	0.0880

[1] In this column the sum of the absolute values is computed.

Having found the arithmetic means $d = 17.24$ and $D = 22.78$ we compute the deviations ε_{d_i} and ε_{D_i}. According to the above formulas we find for each measurement d separately:

$$\sigma = \sqrt{\frac{0.0920}{4}} = 0.152 \quad \text{or} \quad \eta = \frac{0.56}{\sqrt{20}} = 0.125$$

which gives $\sigma = 0.157$, and, for each measurement D separately:

$$\sigma = \sqrt{\frac{0.0880}{4}} = 0.148 \quad \text{or} \quad \eta = \frac{0.52}{\sqrt{20}} = 0.116$$

which gives $\sigma = 0.146$. Conformity of the values of σ obtained in two ways is quite satisfactory. For the arithmetic means:

$$\sigma_d = \frac{0.152}{\sqrt{5}} = 0.068, \quad \sigma_D = \frac{0.148}{\sqrt{5}} = 0.066.$$

The standard error for the thickness $m = \frac{1}{2}(D - d) = 2.77$ of the lateral faces of the vessel is

$$\sigma_m = \sqrt{\tfrac{1}{4}\sigma_D^2 + \tfrac{1}{4}\sigma_d^2} = 0.047.$$

Method of least squares. If the values f_i of certain m function

$$\varphi_i(x_1, x_2, \ldots, x_n) \qquad (i = 1, 2, \ldots, m)$$

of unknown quantities $x_1, x_2, \ldots, x_n$ are determined from an experiment, then, in order to determine the unknowns $x_1, x_2, \ldots, x_n$, the system of conditional equations

$$\varphi_i(x_1, x_2, \dots, x_n) - f_i = 0 \qquad (i = 1, 2, \dots, m)$$

has to be solved. Such a system is, in general, inconsistent for $m > n$ and we seek the most likely values of $x_1, x_2, \dots, x_n$. If the errors of $f_1, f_2, \dots, f_n$ have a normal distribution (which is usually assumed), then, for the most likely system of values of the unknowns, the sum of squares of the deviations $\varepsilon_i = \varphi_i - f_i$ will have a least value.

If the conditional equations are linear:

$$\begin{aligned} a_1x_1 + b_1x_2 + \dots + l_1x_n &= f_1, \\ a_2x_1 + b_2x_2 + \dots + l_2x_n &= f_2, \\ &\dots\dots\dots \\ a_mx_1 + b_mx_2 + \dots + l_mx_n &= f_m, \end{aligned}$$

then the requirement of a minimum (see p. 384) leads to the system of *normal linear equations* (¹):

$$\begin{aligned} [aa]x_1 + [ab]x_2 + \dots + [al]x_n &= [af], \\ [ba]x_1 + [bb]x_2 + \dots + [bl]x_n &= [bf], \\ &\dots\dots\dots \\ [la]x_1 + [lb]x_2 + \dots + [ll]x_n &= [lf]. \end{aligned}$$

To obtain the k-th normal equation, we multiply each conditional equation by the coefficient of x_k and then add all equations.

In the case of non-linear relations, we find roughly approximate values $x_1^0, x_2^0, \dots, x_n^0$ of the desired quantities $x_1, x_2, \dots, x_n$ and expand the functions $\varphi_i(x_1, x_2, \dots, x_n)$ into series of powers of the differences $\xi_1 = x_1 - x_1^0$, $\xi_2 = x_2 - x_2^0$, ..., $\xi_n = x_n - x_n^0$. Disregarding the terms of degree greater than one, we obtain linear conditional equations by means of which we determine the most likely values of the corrections ξ_i.

The above method is most appropriate in the case when all values are equally accurate. Otherwise each conditional equation should be first multiplied by its weight inversely proportional to the mean square error of the corresponding value of f_i.

Example. The results of measurements of the electric resistance R of a copper bar at various temperatures (t°C) are listed in the following table (in the first and second column):

(¹) For Gauss' notation see footnote on p. 750.

t	R	t^2	tR	R (computed)
19.1	76.30	364.8	1457.3	76.26
25.0	77.80	625.0	1945.0	77.96
30.1	79.75	906.0	2400.5	79.43
36.0	80.80	1296.0	2908.8	81.13
40.0	82.35	1600.0	3294.0	82.28
45.1	83.90	2034.0	3783.9	83.76
50.0	85.10	2500.0	4255.0	85.16
Σ 245.3	566.00	9325.8	20044.5	

If we seek for a dependence of R on t in the form $R = a + bt$, then we obtain seven conditional equations of the form $R_i = a + bt_i$, where t_i and R_i are the corresponding values of t and R, for the constants a and b.

The normal equations are

$$7a+[t]b = [R], \qquad [t]a+[t^2]b = [tR],$$

or

$$7a + 245.3\,b = 566.0, \qquad 245.3\,a + 9325.8\,b = 20044.5.$$

Solving these equations we obtain $a = 70.76$ and $b = 0.288$. The values of R computed from the formula $R = 70.76 + 0.288\,t$ are given in the last column of the table.

II. EMPIRICAL FORMULAS AND INTERPOLATION

1. Approximate representation of a functional dependence

Formulation of the problem. We sometimes encounter a need to choose an analytic expression which would represent approximately a function given only by a table or a graph. A similar problem can arise for a function given by a formula, when this formula is too complicated or not appropriate to a given purpose as, for example, when the function has to be integrated while the integral cannot be expressed in terms of elementary functions. Formulas representing a functional dependence obtained from an experiment in the form of a table or a graph are called *empirical formulas*. To represent approximately a given function $f(x)$, we usually choose an *approximating function* $\varphi(x)$ from among functions of a definite form; for example, we seek for a function $\varphi(x)$ in the form of a polynomial

$$\varphi(x) = a_0 + a_1 x + \ldots + a_n x^n$$

or in the form

$$\varphi(x) = Ae^{rx} + Be^{sx} + \ldots,$$

requiring that the function $\varphi(x)$ approaches the function $f(x)$ in a certain interval $a < x < b$ as closely as possible. According to the manner of estimating the approximation of the function $f(x)$ with $\varphi(x)$, we obtain various systems of parameters of the function $\varphi(x)$ yielding the best approximation of $f(x)$.

Uniform approximation. It is theoretically suitable to require that the best approximation $\varphi(x)$ of $f(x)$ satisfies the condition that the maximum of the function $|f(x) - \varphi(x)|$ in the interval $a < x < b$ a as small as possible in comparison with another choice of the ppproximating function $\varphi(x)$. However, effective methods of obtaining such uniform approximation do not exist, except for several isarticular cases. If, for example, the function $f(x)$ has the second derivative of a constant sign in the interval $a < x < b$, then a linear function giving the best uniform approximation of $f(x)$ in this interval can be found as follows (Fig. 440). We seek for a point P on the graph of the function $y = f(x)$ such that the tangent to the curve is parallel to the chord MN, where M and N are the points of the curve with the abscissae a and b. The straight line passing through the mid-points of the chords MP and PN is the graph of the desired linear function in the interval $a < x < b$.

The uniform approximation is used in theoretical consideration. The theory of uniform approximation was developed in a great part by Tschebyscheff.

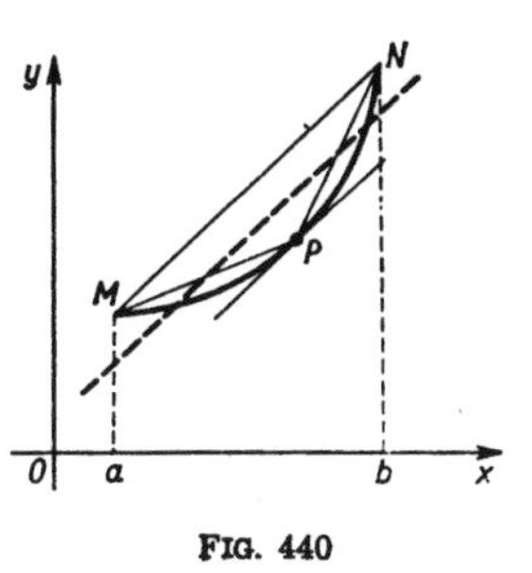

FIG. 440

Approximation by the least squares method. The most frequently used approximation $\varphi(x)$ of the function $f(x)$ is such that the integral

$$M = \int_a^b \big(f(x) - (x)\big)^2 dx$$

has a least value. Requiring that the partial derivatives of the integral M with respect to the parameters determining the function $\varphi(x)$ (see p. 385) are equal to zero, we obtain a system of equations which enables us to find the best values (in the sense mentioned above) of the parameters. The number $\delta = \sqrt{M/(b-a)}$ is called in this case the *mean square error*.

If we seek for the function $\varphi(x)$ in the form of a linear combination of certain definite functions [1]:

$$\varphi(x) = a_0\varphi_0(x) + a_1\varphi_1(x) + \ldots + a_n\varphi_n(x),$$

then, for the coefficients $a_0, a_1, \ldots, a_n$, we obtain the system of linear equations

$$\frac{1}{2}\cdot\frac{\partial M}{\partial a_k} = \sum_{i=0}^{n} a_i \int_a^b \varphi_i(x)\,\varphi_k(x)\,dx - \int_a^b f(x)\varphi_k(x)\,dx = 0 \quad (k = 0, 1, \ldots, n).$$

This system of equations has a particulary simple form, if the functions $\varphi_i(x)$ are *orthogonal* [2] in the interval (a, b), that

[1] For example, $\varphi(x)$ is a polynomial, if $\varphi_0 = 1$, $\varphi_1 = x, \ldots, \varphi_n = x^n$, or is a trigonometric polynomial, if $\varphi_0 = 1$, $\varphi_1 = \cos x$, $\varphi_2 = \sin x, \ldots, \varphi_{2n-1} = \cos nx$, $\varphi_{2n} = \sin nx$.

[2] Two examples of orthogonal systems of functions:

(1) $1, \cos x, \cos 2x, \ldots, \cos nx; \sin x, \sin 2x, \ldots, \sin nx$, in the interval $(0, 2\pi)$.

(2) Legendre's polynomials $P_i(x)$ in the interval $(-1, +1)$ (see pp. 551, 552).

is, when

$$\int_a^b \varphi_i(x)\varphi_k(x)\,dx = 0 \quad \text{for} \quad i \neq k.$$

In this case

$$a_k \int_a^b [\varphi_k(x)]^2 dx = \int_a^b f(x)\varphi_k(x)\,dx \qquad (k = 0, 1, 2, \ldots, n)$$

(see Euler's formulas, p. 727).

In connection with this simplification it is more convenient, if an approximating polynomial $b_0 + b_1 x + \ldots + b_n x^n$ has to be found, to transform the given interval (a, b) into the interval $(-1, +1)$ by means of the substitution

$$x = \frac{a+b}{2} + \frac{b-a}{2} t$$

and seek for the polynomial in the form

$$\varphi(x) = a_0 P_0 + a_1 P_1 + \ldots + a_n P_n,$$

where $P_k(t)$ are Legendre's polynomials (see pp. 551, 552).

Example. Find the best approximation of the function $y = \sin x$ in the form of a quadratic trinomial in the interval $0 < x < \pi$.

Substituting $x = \frac{1}{2}\pi(t+1)$ we transform the interval $(0, \pi)$ into the interval $(-1, +1)$. We seek for the approximating function in the form

$$\varphi = a_0 + a_1 P_1(t) + a_2 P_2(t).$$

Therefore (see p. 433),

$$a_0 = \tfrac{1}{2} \int_{-1}^{1} \sin \tfrac{1}{2}\pi(t+1)\,dt = \frac{2}{\pi},$$

$$a_1 = \tfrac{3}{2} \int_{-1}^{1} t \sin \tfrac{1}{2}\pi(t+1)\,dt = 0,$$

$$a_2 = \tfrac{5}{2} \int_{-1}^{1} (\tfrac{3}{2}t^2 - \tfrac{1}{2}) \sin \tfrac{1}{2}\pi(t+1)\,dt = \frac{10}{\pi}\left(1 - \frac{12}{\pi^2}\right).$$

Hence

$$\sin x \approx \frac{2}{\pi} + \frac{10}{\pi}\left(1 - \frac{12}{\pi^2}\right)\left(\frac{3}{2}t^2 - \frac{1}{2}\right) \approx 0.980 - 0.418\left(x - \frac{\pi}{2}\right)^2.$$

Approximation at separate points. In many cases, in particular when the function $f(x)$ is defined by a table or a graph, in order to estimate the approximation, we do not consider the differences $f(x)-\varphi(x)$ for all points of the interval in which the function $f(x)$ is to be approximated, but only for separate prescribed points $x_0, x_1, \dots, x_n$. The function $\varphi(x)$ is regarded to be the best approximation of $f(x)$ in the *least squares method* if the sum

$$S=\sum_{i=0}^{n}\left(f(x_i)-\varphi(x_i)\right)^2$$

has the least value in comparison with other functions from which the desired approximation is chosen (¹).

If the function $\varphi(x)$ is completely determined by the parameters $k, l, m, \dots$, then the best values (in the above sense) of the parameters can be found from the system of equations

$$\frac{\partial S}{\partial k}=0, \quad \frac{\partial S}{\partial l}=0, \quad \frac{\partial S}{\partial m}=0, \quad \dots$$

If the number of parameters defining the function $\varphi(x)$ is equal to the number of $n+1$ selected points, then the function $\varphi(x)$ can, in general, be chosen so that the equalities $\varphi(x_i)=f(x_i), i=0, 1, \dots, n$, hold; this can be done by solving a system of $n+1$ equations with $n+1$ unknowns. In this case the function $\varphi(x)$ is called an *interpolating function* and the process of finding and computing the values of $\varphi(x)$ is called *interpolation.*

The most frequently used is *parabolic interpolation* in which the interpolating function is a polynomial $\varphi(x)=a_0+a_1x+\dots+a_nx^n$. For periodic functions, trigonometric interpolation (see p. 738) is used.

For approximation by the method of means see p. 764.

2. Parabolic interpolation

General case. For any function $f(x)$ and for any arbitrary choice of *nodes of interpolation* $x_0, x_1, \dots, x_n$, there exists always a unique polynomial $\varphi_n(x)$ of degree n which assumes at the points $x_0, x_1, \dots, x_n$ the same values as the given function $f(x)$: $\varphi(x_i)=f(x_i)$, for $i=0, 1, \dots, n$. To find the interpolating polynomial the following *interpolation formula* due to Lagrange can be used:

$$\varphi_n(x)=L_0(x)f_0+L_1(x)f_1+\dots+L_n(x)f_n,$$

(¹) The best approximation of $f(x)$ can be defined, as above (see the uniform approximation, p. 754), as a function $\varphi(x)$ such that the maximum of $|f(x_i)-\varphi(x_i)|$ has the least value. However, determining the approximation in this way is, practically, troublesome.

where

$$L_i(x) = \frac{(x - x_0) \dots (x - x_{i-1})(x - x_{i+1}) \dots (x - x_n)}{(x_i - x_0) \dots (x_i - x_{i-1})(x_i - x_{i+1}) \dots (x_i - x_n)}$$

and

$$f_i = f(x_i).$$

If the value of $\varphi_n(x)$ for a definite value of x has to be computed, then the following cross scheme can be applied; it is particularly suitable when computing machines are used:

$$\begin{array}{l|l}
x_0 - x & f_0 \\
x_1 - x & f_1 \ (f_0, f_1) \\
x_2 - x & f_2 \ (f_0, f_2) \ (f_0, f_1, f_2) \\
\dots\dots & \dots\dots\dots\dots\dots\dots \\
x_n - x & f_n \ (f_0, f_n) \ (f_0, f_1, f_n) \ \dots \ (f_0, f_1, \dots, f_n)
\end{array}$$

Each symbol $(f_0, f_1, \dots, f_k)$ denotes the value of the interpolating polynomial formed according to the nodes $x_0, x_1, \dots, x_k$ at the point x. The columns of these numbers are determined successively as follows: We compute the numbers of the (f_0, f_k) column from the formula

$$(f_0, f_k) = \frac{(x_0 - x) f_k - (x_k - x) f_0}{(x_0 - x) - (x_k - x)}.$$

Then we obtain each column from the previous one according to the same formula, for example

$$(f_0, f_1, f_k) = \frac{(x_1 - x)(f_0, f_k) - (x_k - x)(f_0, f_1)}{(x_1 - x) - (x_k - x)}$$

and so on. The location of the nodes can be chosen arbitrarily.

Example. Compute sin 50° using five figure values of the sine of 0°, 30°, 45°, 60°, 90°.

In this case the cross scheme is the following

−50	0.00000				
−20	0.50000	0.83333			
−5	0.70711	0.78568	0.7 6980		
10	0.86603	0.72169	0.7 5890	66 17	
40	1.00000	0.55556	0.7 4074	66 57	04

If the initial figures in any column are equal (in our example they are separated), then they need not be used in further computation. For example, in computing the last column, we use the final figures of the preceding result:

$$\frac{10 \cdot 57 - 40 \cdot 17}{10 - 40} = 04.$$

Finally $\sin 50° = 0.76604$.

Nodes located at equal distances. Tables of differences. We often encounter the case when the interpolating nodes are located at equal distances. In this case the constant interval $h = x_{i+1} - x_i$ is called the *step* of the table of the given function $f(x)$. We have then $x_k = x_0 + hk$ (we preserve this notation also for $k < 0$).

The *first differences* (or *differences of the first order*) of the function for the given step h are defined by the formulas:

$$\Delta f(x) = f(x+h) - f(x), \qquad \Delta f_i = f_{i+1} - f_i.$$

The differences of the first differences form the *second differences* (or *differences of the second order*):

$$\Delta^2 f(x) = \Delta f(x+h) - \Delta f(x), \qquad \Delta^2 f_i = \Delta f_{i+1} - \Delta f_i.$$

Differences of higher order are defined similarly.

The differences of the function can be expressed by the given values of the function:

$$\Delta^k f_0 = f_k - k f_{k-1} + \frac{k(k-1)}{2} f_{k-2} - \dots + (-1)^k f_0.$$

This can be written symbolically in the form

$$\Delta^k f_0 = (E-1)^k f_0,$$

where we use the notation $E^i f_0 = f_i$.

For the purpose of interpolation between given values of a function, we form the table of differences according to the scheme:

x	$f(x)$	$\Delta f(x)$	$\Delta^2 f(x)$	$\Delta^3 f(x)$	$\Delta^4 f(x)$	
...	...					
		Δf_{-3}				N_{II}
x_{-2}	f_{-2}		$\Delta^2 f_{-3}$		$\Delta^4 f_{-4}$	
		Δf_{-2}		$\Delta^3 f_{-3}$		
x_{-1}	f_{-1}		$\Delta^2 f_{-2}$		$\Delta^4 f_{-3}$	
		Δf_{-1}		$\Delta^3 f_{-2}$		
x_0	f_0		$\Delta^2 f_{-1}$		$\Delta^4 f_{-2}$	S
		Δf_0		$\Delta^3 f_{-1}$		B
x_1	f_1		$\Delta^2 f_0$		$\Delta^4 f_{-1}$	
		Δf_1		$\Delta^3 f_0$		
x_2	f_2		$\Delta^2 f_1$		$\Delta^4 f_0$	N_I
		Δf_2		$\Delta^3 f_1$		
x_3	f_3		$\Delta^2 f_2$			
...	...					

In this table each number (except those appearing in the two initial columns) is the difference of two numbers from the preceding column which are placed one half of the row above and below the number in question [1]. In setting up a table of differences we should bear in mind that an error not exceeding ε in the absolute value appearing in the first column can lead to errors not exceeding 2ε in the second column, to 4ε in the third column, and, in general, to errors not exceeding $2^{m-1}\varepsilon$ in the m-th column. Therefore even slight errors in the values of the function (for example those due to rounding off) can have a great influence on the differences of higher order. The calculation of differences should be stopped, if all differences in one of the columns turn out to be almost equal (i.e., if the difference is constant). The differences of order m are constant in the case of a polynomial of degree m. Therefore the fact that the m-th difference is constant indicates that the given function $f(x)$ can be approximated with a sufficient accuracy by a polynomial of degree m. (For the table on p. 761, $m = 3$; the fourth differences are superfluous.)

Difference interpolation formulas. By using the differences, an interpolating polynomial can be found according to one of the following formulas (the notation $u = (x - x_0)/h$ is introduced):

Newton's formulas

$$N_{\mathrm{I}}(x) = f_0 + u\Delta f_0 + \frac{u(u-1)}{2}\Delta^2 f_0 + \dots + \frac{u(u-1)\dots(u-n+1)}{n!}\Delta^n f_0,$$

$$N_{\mathrm{II}}(x) = f_0 + u\Delta f_{-1} + \frac{u(u+1)}{2}\Delta^2 f_{-2} + \dots + \frac{u(u+1)\dots(u+n-1)}{n!}\Delta^n f_{-n}.$$

Stirling's formula

$$S(x) = f_0 + u\frac{\Delta f_0 + \Delta f_{-1}}{2} + \frac{u^2}{2}\Delta^2 f_{-1} + \frac{u(u^2-1)}{3!}\cdot\frac{\Delta^3 f_{-2} + \Delta^3 f_{-1}}{2} + \frac{u^2(u^2-1)}{4!}\Delta^4 f_{-2} + \dots + \frac{u^2(u^2-1)\dots[u^2-(n-1)^2]}{(2n)!}\Delta^{2n} f_{-n}.$$

[1] For an example of such a scheme see p. 761.

Bessel's formula

$$B(x) = f_0 + u\Delta f_0 + \frac{u(u-1)}{2}\cdot\frac{\Delta^2 f_{-1} + \Delta^2 f_0}{2} +$$

$$+ \frac{u(u-1)\,(u-0.5)}{3!}\Delta^3 f_{-1} + \frac{u(u^2-1)\,(u-2)}{4!}\cdot\frac{\Delta^4 f_{-2} + \Delta^4 f_{-1}}{2} + \ldots +$$

$$+ \frac{(u-0.5)\,u(u^2-1)\ldots[u^2-(n-1)^2]\,(u-n)}{(2n+1)!}\Delta^{2n+1} f_{-1}.$$

Newton's formulas give an interpolation polynomial, if x_0 is either the first or the last one of the interpolation nodes, while in Bessel's and Stirling's formulas x_0 is the middle one or one of two middle interpolation nodes. The differences used in computation according to one of the formulas are shown in the preceding scheme (p. 759). The interpolation formulas are used mainly in computation an intermediate value of a function defined by a table. By a suitable choice of x_0, we can always make $|u| < 1$. When $|u| < 0.25$, the Stirling formula is more appropriate; when $0.25 < u < 0.75$, Bessel's formula should be recommended. Newton's formulas are used, when the application of the formulas $S(x)$ or $B(x)$ is impossible, i.e., when x lies near to the beginning or the end of the table.

Example. Compute the value of the function $f(x)$ for $x = 22$, when $f(x)$ is given by the following table[1]:

x	f	Δf	$\Delta^2 f$	$\Delta^3 f$	$\Delta^4 f$
0	0				
		487			
5	4.87		78		
		565		29	
10	10.52		107		2
		672		31	
15	17.24		138		3
		810		34	
20	25.34		172		−7
		982		27	
25	35.16		199		5
		1181		32	
30	46.97		231		1
		1412		33	
35	61.09		264		
		1676			
40	77.85				

[1] The decimal point is usually omitted in the table of differences and the difference is expressed in the units of the last significant figure.

As it has already been mentioned, we confine ourselves to three differences. If we put $x_0 = 20$, then $u = \frac{22-20}{5} = 0.4$.

According to Bessel's formula:

$$f(22) = 25.34 + 0.4 \cdot 9.82 - \frac{0.4 \cdot 0.6}{2} \cdot \frac{1.72 + 1.99}{2} + \\ + \frac{0.4 \cdot 0.6 \cdot 0.1}{6} 0.27 = 29.05.$$

According to Stirling's formula:

$$f(22) = 25.34 + 0.4 \frac{8.10 + 9.82}{2} + \frac{0.16}{2} 1.72 - \\ - \frac{0.4 \cdot 0.84}{6} \cdot \frac{0.34 + 0.27}{2} = 29.04.$$

According to Newton's formula:

$$f(22) = 25.34 + 0.4 \cdot 9.82 - \frac{0.4 \cdot 0.6}{2} 1.99 + \frac{0.4 \cdot 0.6 \cdot 1.6}{6} 0.32 = 29.05.$$

If we confine ourselves to the second differences, then we obtain 29.05 from Bessel's formula, 29.06 from Stirling's formula and 29.03 from Newton's formula.

Error of interpolation. If the function $f(x)$ is defined analytically and has a sufficient number of continuous derivatives in the considered interval, then the error appearing in replacing the function $f(x)$ by the interpolating polynomial is, by the mean value theorem, equal to

$$f(x) - \varphi_n(x) = \frac{f^{(n+1)}(\xi)}{(n+1)!} (x - x_0)(x - x_1) \dots (x - x_n),$$

where ξ is an intermediate value between the greatest and the least of the numbers $x_0, x_1, \dots, x_n$. For the difference formulas we have

$$f(x) - N_{\mathrm{I}}(x) = \frac{h^{n+1}}{(n+1)!} f^{(n+1)}(\xi) \cdot u(u-1) \dots (u-n),$$

$$f(x) - N_{\mathrm{II}}(x) = \frac{h^{n+1}}{(n+1)!} f^{(n+1)}(\xi) \cdot u(u+1) \dots (u+n),$$

$$f(x) - S(x) = \frac{h^{2n+1}}{(2n+1)!} f^{(2n+1)}(\xi) \cdot u(u^2-1) \dots (u^2-n^2),$$

$$f(x) - B(x) = \frac{h^{2n+2}}{(2n+2)!} f^{(2n+2)}(\xi) \cdot u(u^2-1) \dots (u^2-n^2)(u-n-1).$$

It is assumed here that the number of terms of the polynomials $N_{\mathrm{I}}(x)$, $N_{\mathrm{II}}(x)$, $S(x)$ and $B(x)$ is the same as on pp. 760, 761 and ξ is an intermediate value between the interpolation nodes; ξ is different in different formulas and depends on x.

Application of interpolation formulas. The interpolation formulas can be used to the approximate differentiation and integration: we replace a given function $f(x)$ by the interpolating polynomial $\varphi(x)$ and perform the corresponding operation with it. For example, the following formula for an approximate value of the derivative of $f(x)$ at $x = x_0$ can be obtained by applying the Stirling formula:

$$\left[\frac{df}{dx}\right]_{x=x_0} = \frac{1}{h}\left[\frac{\Delta f_0 + \Delta f_{-1}}{2} - \frac{1}{6}\cdot\frac{\Delta^3 f_{-1} + \Delta^3 f_{-2}}{2} + \frac{1}{30}\cdot\frac{\Delta^5 f_{-2} + \Delta^5 f_{-3}}{2} - \ldots\right].$$

The most convenient formulas for the approximate integration derived from the interpolation formulas are given on p. 463.

3. Selection of empirical formulas

Comparison of the graphs. The process of selecting or fitting an empirical formula to a functional relation $y = f(x)$ obtained from an experiment is composed of two steps: we select first the form of the formula and then determine the numerical values of the parameters so as to provide the best approximation of the given function. If the selection of the form of a formula is not determined by any theoretical reasons, we usually select the formula from the simplest ones comparing their graphs with the graph of the given function. Since the similarity of the graphs determined roughly, "at sight", can be deceptive, having selected a formula and before determining its parameters, we have to verify as whether or not it is appropriate. This can be done by the *rectification method*.

The rectification method can be described as follows. Assuming that the relationship between y and x has a definite form, we find certain quantities $X = \varphi(x, y)$ and $Y = \psi(x, y)$ which, under our assumption, are related linearly one to another (if, for example, $y = x/(a + bx)$, then we put $X = x$, $Y = x/y$ or $X = 1/x$, $Y = 1/y$). Computing the values of X and Y for the given x and y and plotting them on the graph we easily see whether the relation between X and Y is near to a linear one (i.e., whether the plotted points lie approximately along a straight line); thus we see whether or not the selected formula fits the given relationship.

For hints on the rectification of certain simple functions with reference to their graphs see pp. 765–769. An example is given on p. 769.

Determining the parameters. The most accurate method of determining the parameters is the *method of least squares* (see pp. 751, 757). In most cases, however, we can apply certain simpler methods, in particular, the *method of mean values*. If a formula obtained by this method turns out to be not accurate enough, then the least squares method can be used to make it more accurate; since the approximate values of the parameters are known, the computation will then be less troublesome (see p. 751). We first use the method of mean values to determine a linear relation between the "rectified" variables X and Y: $Y = aX + b$. In order to do that, we divide the conditional equations $Y_i = aX_i + b$ for the given values of X_i and Y_i into equal (or almost equal) groups arranged in the order of increasing variable X_i or Y_i. Adding the equations of each group, we obtain two equations from which we determine a and b. Expressing the variables X and Y by the original variables we obtain the desired relation connecting x and y. If not all parameters are then determined, we apply the same method again with other quantities X and Y rectified (see, for example, case 13 on p. 768); an example is on p. 769.

Most common empirical formulas. Some simple formulas with the corresponding graphs are given below. In each figure several graphs for various values of the parameters involved in the formula are shown (for the influence of parameters upon the form of the curve see the chapter "Graphs", pp. 96–131). When considering graphs we should always remember that, given an empirical formula, we use only a part of the curve corresponding to a certain interval of the independent variable. Therefore, for example, we should not think that the formula $y = x^2 + bx + c$ is suitable only in the case when the curve $y = f(x)$ has a maximum or minimum (see below).

(1) $y = ax^b$ (Fig. 441). The graphs are shown in Figs. 6, 12, 15 and 16 and explanations are on pp. 99, 103, 105. We rectify $X = \log x$ and $Y = \log y$:

$$Y = \log a + bX.$$

(2) $y = ae^{bx}$ (Fig. 442). The graphs are shown in Fig. 17 and explanation is on p. 106. We rectify $X = x$ and $Y = \log y$:

$$Y = \log a + b \log e \cdot x.$$

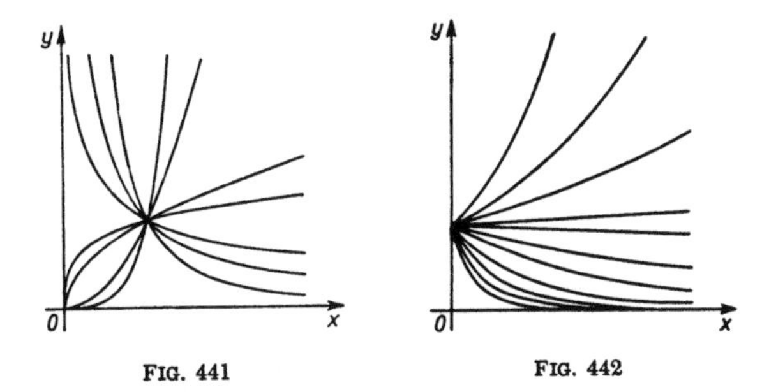

FIG. 441 FIG. 442

(3) $y = ax^b + c$ (Fig. 443). The graphs are the same as in case 1 but shifted in the direction of the y axis. If b is given, we rectify $X = x^b$ and $Y = y$:

$$y = aX + c.$$

If b is not known, we rectify $X = \log x$ and $Y = \log (y - c)$:

$$Y = \log a + bX,$$

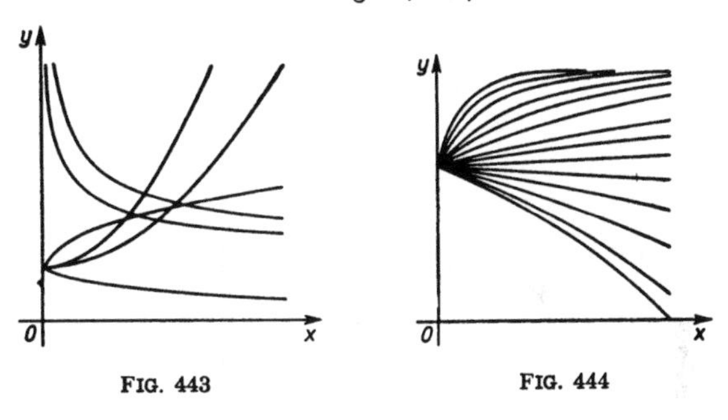

FIG. 443 FIG. 444

determining first c. To do this, we find three points with the abscissae x_1, x_2 and $x_3 = \sqrt{x_1 x_2}$ and the corresponding ordinates y_1, y_2, y_3 on the graphs of the function (x_1 and x_2 can be chosen arbitrarily) and assume

$$c = \frac{y_1 y_2 - y_3^2}{y_1 + y_2 - 2y_3} \quad (^1).$$

(4) $y = ae^{bx} + c$ (Fig. 444). The graph are the same as in case 2 but shifted in the direction of the y axis. We rectify $X = x$ and

---

(1) After determining a and b, we can choose c equal to the mean value of $y - ax^b$.

$Y = \log(y - c)$:

$$Y = \log a + b \log e \cdot x,$$

determining first c. To do this, we find three points with the abscissae x_1, x_2 and $x_3 = \frac{1}{2}(x_1 + x_2)$ and the corresponding ordinates y_1, y_2, y_3 on the graph of the function; x_1 and x_2 can be chosen arbitrarily and we assume

$$c = \frac{y_1 y_2 - y_3^2}{y_1 + y_2 - 2y_3} \text{ (1)}.$$

(5) $y = ax^2 + bx + c$ (Fig. 445). The graph is shown in Fig. 3 and explanation is on p. 96. Selecting an arbitrary point (x_1, y_1) on the graph of the function, we rectify $X = x$ and $Y = (y - y_1)/(x - x_1)$:

$$Y = (b + ax_1) + ax.$$

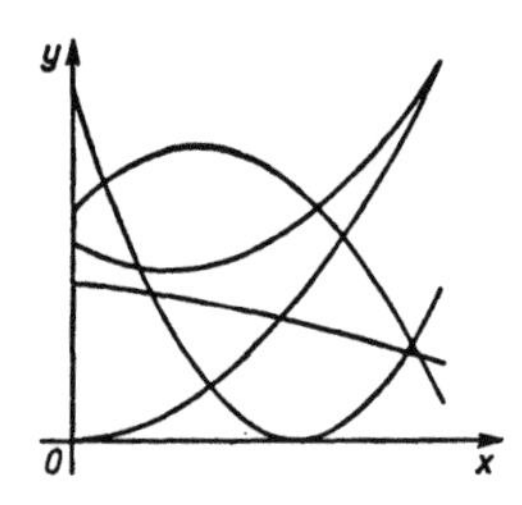

FIG. 445

If the given values form an arithmetic progression with the difference h, then we rectify $Y = \Delta y$ and $X = x$:

$$Y = (bh + ah^2) + 2ahx.$$

In either case, after determining a and b, we find c from the equation

$$\sum y = a \sum x^2 + b \sum x + nc,$$

where n is the number of given values of x to which the summation is extended.

(6) $y = \dfrac{ax + b}{cx + d}$. The graph is shown in Fig. 8 and explanation is on p. 100. We select any point (x_1, y_1) on the graph of the function and rectify $X = x$ and $Y = (x - x_1)/(y - y_1)$:

$$Y = A + Bx.$$

---

(1) After determining a and b, we can anew take c equal to the mean value of $y - ae^{bx}$.

We confine ourselves here to determining A and B and write the obtained formula in the form

$$y = y_1 + \frac{x - x_1}{A + Bx}.$$

Sometimes we can confine ourselves to a formula in the form

$$y = \frac{x}{cx + d} \quad \text{or} \quad y = \frac{1}{cx + d}.$$

In this case we rectify $X = 1/x$ and $Y = 1/y$ or $X = x$ and $Y = x/y$ for the first formula and $X = x$ and $Y = 1/y$ for the second formula.

(7) $y^2 = ax^2 + bx + c$ (Fig. 446). The graph is shown in Fig. 14 and explanation is on p. 105. If we introduce the new variable $\bar{y} = y^2$, then further computation can be performed as in case (5).

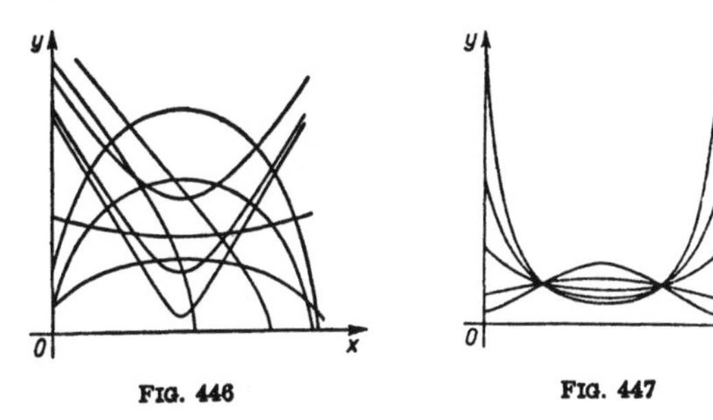

FIG. 446 FIG. 447

(8) $y = ae^{bx+cx^2}$ or $\log y = \log a + \log e \cdot bx + \log e \cdot cx^2$ (Fig. 447). The graph is shown in Fig. 21 and explanation is on p. 108.

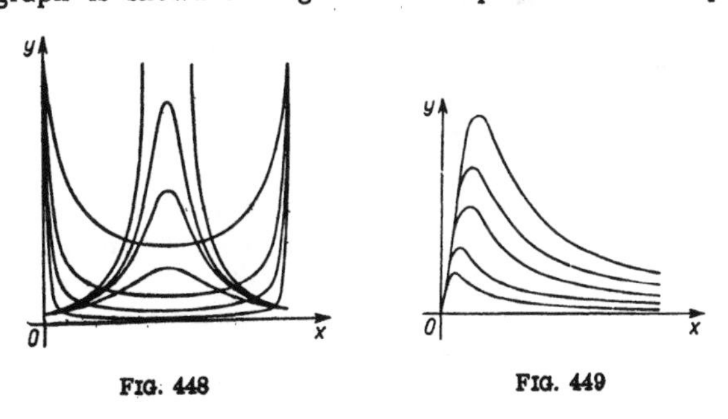

FIG. 448 FIG. 449

Introducing the new variable $\bar{y} = \log y$ we reduce this case to case (5).

(9) $y = \dfrac{1}{ax^2 + bx + b}$ (Fig. 448). The graph is shown in Fig. 10

and explanation is given on p. 101. Introducing the new variable $\bar{y} = 1/y$, we reduce this case to case (5).

(10) $y = \dfrac{x}{ax^2 + bx + c}$ (Fig. 449). The graph is given in Fig. 11 and explanation on p. 102. Introducing the new variable $\bar{y} = x/y$ we reduce this case to case (5).

(11) $y = a + \dfrac{b}{x} + \dfrac{c}{x^2}$ (Fig. 450). The graph is given in Fig. 9 and explanation on p. 101. Introducing the new variable $\bar{x} = 1/x$ we reduce this case to case (5).

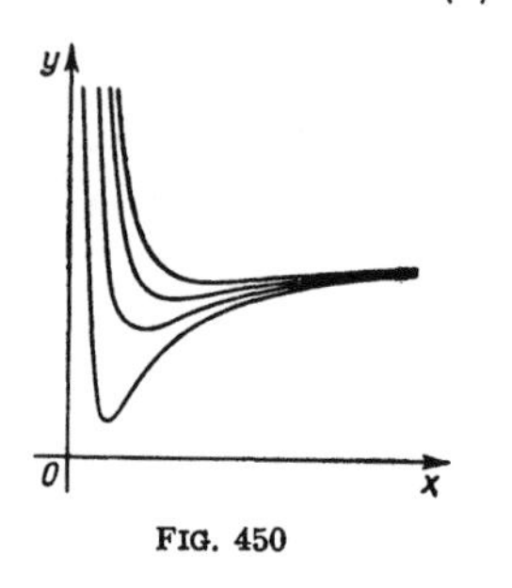

FIG. 450

FIG. 451

(12) $y = ax^b e^{cx}$ (Fig. 451). The graph is given in Fig. 22 and explanation on p. 109. If the given values of x form an arithmetical progression, we rectify $Y = \Delta \log y$ and $X = \Delta \log x$:

$$y = hc \log e + bX.$$

If the given values form a geometric progression with the ratio q, then we rectify $Y = \Delta_1 \log y$ and $X = x$:

$$Y = b \log q + c(q - 1) \log e \cdot x,$$

where Δ_1 is the difference of two successive values of y. After determining b and c, we take the logarithms of both sides of the equation and find $\log a$ in the same way as c was found in case (5).

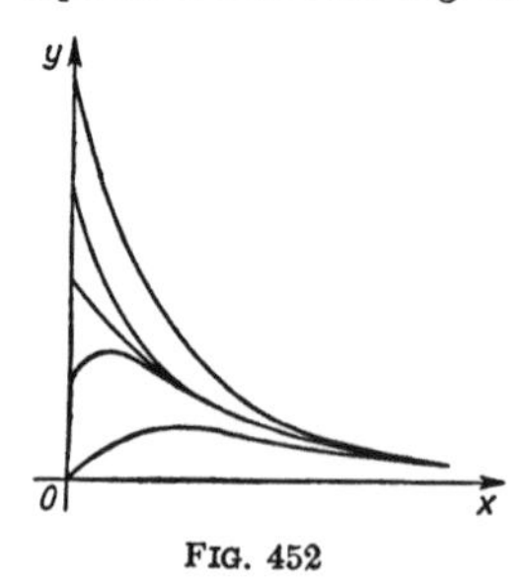

FIG. 452

(13) $y = ae^{bx} + ce^{dx}$ (Fig. 452). The graph is given in Fig. 20 and explanation on pp. 107, 108. If the given values form an arithmetical progression with difference h and y_1, y_2, y_3 are arbitrary successive values of the given function, then we rectify $Y = y_2/y$ and $X = y_1/y$:

$$Y = (e^{bh} + e^{dh})X - e^{bh} \cdot e^{dh}.$$

Having determined b and d from

this equation, we rectify $\overline{Y} = ye^{-dx}$ and $\overline{X} = e^{(b-d)x}$:

$$\overline{Y} = a\overline{X} + c.$$

Example. Find an empiric formula for the functional dependance y on x defined by the first and second column of the following table:

x	y	x/y	$\Delta(x/y)$	$\log x$	$\log y$	$\Delta \log x$	$\Delta \log y$	$\Delta_1 \log y$	y computed
0.1	1.78	0.056	0.007	−1.000	0.250	0.301	0.252	0.252	1.78
0.2	3.18	0.063	0.031	−0.699	0.502	0.176	+0.002	−0.097	3.15
0.3	3.19	0.094	0.063	−0.523	0.504	0.125	−0.099	−0.447	3.16
0.4	2.54	0.157	0.125	−0.398	0.405	0.097	−0.157	−0.803	2.52
0.5	1.77	0.282	0.244	−0.301	0.248	0.079	−0.191	−1.134	1.76
0.6	1.14	0.526	0.488	−0.222	0.057	0.067	−0.218	−1.455	1.14
0.7	0.69	1.014	0.986	−0.155	−0.161	0.058	−0.237	—	0.70
0.8	0.40	2.000	1.913	−0.097	−0.398	0.051	−0.240	—	0.41
0.9	0.23	3.913	3.78	−0.046	−0.638	0.046	−0.248	—	0.23
1.0	0.13	7.69	8.02	0.000	−0.886	0.041	−0.269	—	0.13
1.1	0.07	15.71	14.29	0.041	−1.155	0.038	−0.243	—	0.07
1.2	0.04	30.0	—	0.079	−1.398	—	—	—	0.04

Plotting the graph (Fig. 453) and comparing it with the graphs on pp. 765–768, we come to the conclusion that the cases (10) and (12) fit our case. In case (10) we have to rectify $\Delta(x/y)$ and x, but the computation indicates that the relationship between x and $\Delta(x/y)$ is far from a linear one. To verify whether formula (12) is suitable

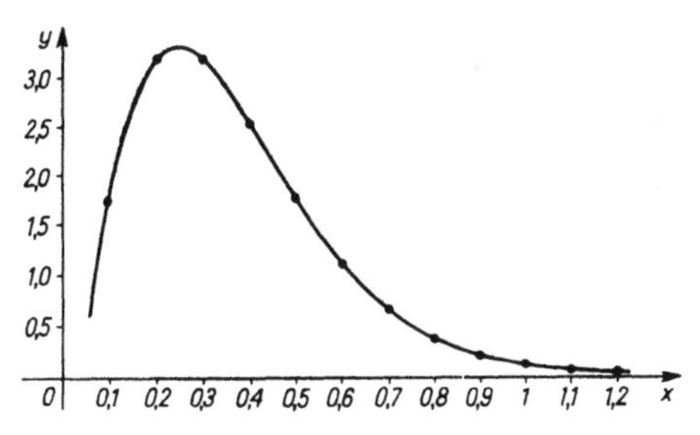

FIG. 453

we plot the graph of the relation between $\Delta \log x$ and $\Delta \log y$ (for $h = 0.1$; Fig. 454), and also between $\Delta_1 \log y$ and x (for $q = 2$, Fig. 455). In either case the distribution of the points along a straight line can be assumed to be good enough and the formula $y = ax^b e^{cx}$ can be used.

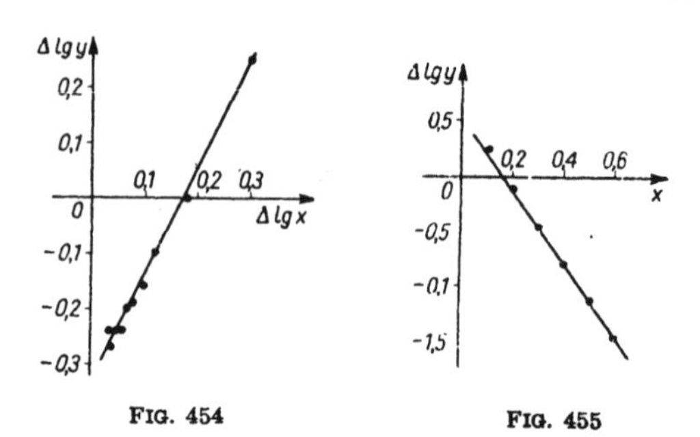

FIG. 454

FIG. 455

To determine the constants a, b and c, we seek a linear relation between x and $\Delta_1 \log y$ by the method of mean values. Adding the conditional equations

$$\Delta_1 \log y = b \log 2 + cx \log e$$

in groups of three equations each, we obtain

$$-0.292 = 0.903b + 0.2606c, \qquad -3.392 = 0.903b + 0.6514c,$$

hence $b = 1.966$ and $c = -7.932$. To determine a we add the equations of the form $\log y = \log a + b \log x + c \log e \cdot x$ which yields

$$-2.670 = 12 \log a - 6.529 - 26.87,$$

hence $\log a = 2.561$, $a = 364$. The values of y computed from the formula $y = 364x^{1.966}\, e^{-7.932x}$ are given in the last column of the above table.

INDEX

INDEX